Tenth Edition

COLLEGE ZOOLOGY

Richard A. Boolootian

Science Software Systems, Inc.
Director, Institute of Visual Medicine
Los Angeles, California
Formerly University of California, Los Angeles

Karl A. Stiles

Late Professor Emeritus of Zoology
Michigan State University

Macmillan Publishing Co., Inc.
New York

Collier Macmillan Publishers
London

The following photographic illustrations have been provided through the courtesy of Science Software Systems, Inc., West Los Angeles, Calif.: Figures 2-12, 2-15, 2-18, 2-22, 3-2, 3-3, 3-4, 3-5, 3-12, 5-2, 6-1, 6-5, 6-7, 6-9, 7-1, 7-7, 8-5, 9-1, 10-1, 10-3, 10-6, 11-2, 11-9, 11-12, 11-13, 11-14, 12-1, 13-1, 13-3, 13-5, 14-2, 14-7, 15-3, 16-4, 16-11, 16-13, 16-15, 16-17, 17-10, 17-11, 18-7, 20-3, 20-5, 21-11, 21-14, 21-18, 22-4, 22-5, 22-6, 22-7, 22-8, 22-9, 24-2, 26-2, 26-3, 26-6, 26-11, 26-12, 26-13, 27-18, 27-20, 27-21, 27-22, 28-4, 28-11, 28-13, 28-14, 28-15, 28-19, 28-21, 28-22, 28-23, 28-24, 28-25, 28-26, 28-27, 28-31, 29-17, 29-18, 29-20, 29-21, 29-22, 29-25, 30-4, 30-5, 30-10, 30-11, 30-12, 30-13, 30-15, 30-16, 30-19, 31-6, 31-8, 32-5, 35-6, 37-8, 38-18, 40-4, 40-5, 40-7, 40-8, 40-9, 40-10, 42-1, 42-2.

(Special Note: Graphic records appearing on pages 486 and 489 were also contributed by Science Software Systems, Inc.)

MACMILLAN PUBLISHING CO., INC.
866 THIRD AVENUE, NEW YORK, NEW YORK 10022

COLLIER MACMILLAN CANADA, LTD.

Library of Congress Cataloging in Publication Data

Boolootian, Richard A
   College zoology.

   Includes bibliographies and index.
   1. Zoology.   I. Stiles, Karl Amos, 1897–1968,
joint author.   II. Title.
QL45.2.H4 1981   591   80-22446
ISBN 0-02-311990-X

Printing: 1 2 3 4 5 6 7 8   Year: 0 1 2 3 4 5 6 7 8

to
Mary Jo

Preface

It is compelling yet comforting to note the evolution of *College Zoology*. First written in 1912, this text is in its tenth revision as the body of biological information has increased. Material has been added, expanded, and deleted over the years. But more than a simple update of the facts, each new edition of the text presents a more finely honed and developed knowledge of zoology.

As you study zoology and meet the myriad of organisms that once lived or with whom you presently cohabit the earth, keep in mind the unifying themes that connect all living things: respiration, reproduction, motility, and irritability. Examine all organisms as to how they eat, breathe, move and reproduce. Notice how these phenomena become increasingly diverse and complex as you move up the evolutionary tree.

Zoology is a dynamic discipline. Advances in research elucidate finer measures of biological similarity and diversity between animals. Zoologists experiment daily and postulate new hypothesis. Issues of concern for human health and welfare change as everyday life becomes more complex. Although this text retains the same basic approach as past editions, new facts and theories are presented, outdated sections have been deleted, and several chapters have been reorganized. New headings have been employed to better organize the chapters. For example, the organization of headings concerning anatomy and physiology of a phylum, class or representative species was made more consistent among all chapters. These changes improve the overall organization and avoid the neglecting of minor phyla and organ systems that were discussed in less detail.

The text is divided into four parts, containing forty-two chapters, followed by a classification appendix and a glossary.

Part I, Molecules, Cells and Tissues, introduces basic principles of living organisms, namely cell chemistry, morphology, histology and organology. New information and diagrams on nucleic acids, lipids, proteins, and membrane structure are included.

Diversity and adaptation are covered in Part II. Taxonomic and phylogenetic treatments have been updated. The order of chapters now reflects the increasing evolutionary complexity in each group discussed. In the phylum by phylum discussions (Chapters 5–23), consistent descriptions of representative species preceed the more comparative "Origins" and "Relationships" sections. New sections on Phylum Gnathostomulida and Phylum Priapulida are included.

In the class by class discussions of the chordates (Chapters 24–30), the increased familiarity of these organisms allows an evolutionary discussion to introduce each chapter. In addition, new information concerning chordate evolution, especially on the reptiles, mammals and birds, has been added. New views on the origins of birds and endothermy are included. Chapter 30 has a new section on Order Pinnipedia and Order Lagomorpha.

Part III, Morphology, Physiology, and Continuity of Animals (Chapters 31–39), presents updated information about animal behavior, a survey of the organ systems, and discussions of genetics, including a discussion of molecular genetics, heredity and embryology. New information has been added to the discussion on the molecular mechanisms of muscle contractions and nervous control. There is increased emphasis on the lymph system and an additional section on the immune system.

The treatment of zoogeography and ecology, evolution, and *Homo sapiens* comprise Part IV, Species Dynamics (Chapters 40–42).

I am personally indebted to Peter Dalby, Ernest C. Anderson, and Walter K. Taylor, who offered invaluable comments during the writing of this edition. I am especially grateful to Professor Harold L. Zimmack for his careful reviews of the galleys and page proofs. To Thomas Broderick and Charles Manske, I extend my appreciation for their superb work on manuscript research and preparation. To Kathleen Marce, my sincere thanks for her administrative talents which served to expedite the publication of *College Zoology*. And to my wife, my deepest gratitude for her continued support.

Because the author has been the final judge for all that is presented in this book, he alone is responsible for errors or misinterpretations of facts. Suggestions for improvement are not only welcome, but greatly appreciated.

R. A. B.

Contents

‖‖‖ Morphology, Physiology, and Continuity of Animals 475

‖V Species Dynamics 635

Molecules, Cells, and Tissues

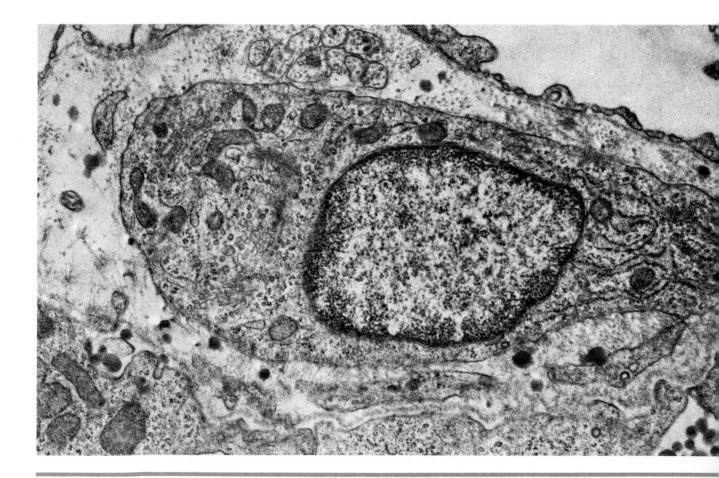

Introduction

Prehistoric cave paintings provide us with some of the first evidences of human self-awareness relative to the rest of the animal world. Hunting scenes depict bison, horses, reindeer, and other prey, even dolphins, though the interpretation of such figures is open to some dispute. Some say that they may have been purely representational paintings; others believe that such scenes served as a medium for casting magic spells. In any event these paintings indicate a differentiation of the human mind from the natural and physical environment. Since then, our dependence and awe of the animal world has never ceased, but our perspective has become increasingly particularized and specialized.

The beginnings of human involvement with animals was mainly based on various kinds of dependence. At first, animals provided primarily food, but soon they were also used for the construction of hide-covered shelters; for clothing in the form of furs, wool, and hides; and for transportation. The transitions of man from hunter to shepherd to farmer are milestones along the road toward civilization.

Animals have also been worshipped through the ages. The ancient Egyptians venerated Hathor, the goddess of love, who was represented variously as having the horns, ears, or the entire head of a cow. The Minoan culture in early Greece, about 200 B.C., worshipped the Minotaur, a diety that was half man, half bull. In India even today cattle are considered sacred by the Hindus.

In addition, animals have been important to man in other ways. Human affection for animals has existed since ancient times; the companionship of a pet is highly valued by many. Animals have also served as scientific inspiration; airplanes and jets are patterned after birds, submarines after fish. One important modern use of animals is in scientific research. Hundreds of thousands of studies using animals have contributed to health improvements for people and domesticated animals. Practically every medication presently available was tested first on animals to determine probable effects and toxicity. The recent interest in environment and wildlife protection and other nature-oriented activities reflects an appreciation of animals as part of a delicately balanced global ecosystem.

Modern zoology can trace its roots back to Aristotle, the natural philosopher, who lived in Athens in about 350 B.C. Aristotle based his theories of the living world upon extensive empirical observations. He collected animals and plants and compared their external body forms, internal anatomies, behaviors, and embryological developments. From such observational data he distinguished patterns that allowed him to classify some 540 animal species according to gradations in structural organization. When Aristotle began to study animals carefully, looking for similarities and differences among them, he generated a classification scheme (taxonomy). Today this scheme reflects evolutionary relationships that are intimately mingled with form–function considerations. More than a million types of animals have now been identified and described, and new ones are rapidly being added to the list. The taxonomy used today is of a different nature than that employed by Aristotle; Aristotle's scheme was essentially descriptive, whereas modern taxonomists look to traits and principles of anatomy and physiology that reflect evolutionary relationships.

It is important to remember that zoological studies

include not only those of animals but also those of humans, as humans are a significant link in the evolutionary chain. Some human characteristics, although not totally unique to us, are much more developed; these include our ability to construct and implement complex tools, our capacity to communicate and deal with abstractions, and our ability to manipulate our environments.

Scientific Method

The scientific method is an important part of the science of zoology, and one objective of any zoology course is to understand and experience this method of discovery. The scientific method is a system of inquiry composed of five steps: observation, problem, hypothesis, experiment, and theory. Its aim is to predict the principles and forces that underlie what we observe.

The scientific method begins with *observation.* A phenomenon is noticed either directly by the senses or indirectly by instruments that extend the range of our receptors. The mind of the scientist becomes activated; he or she wants to know *how, what,* and *why.*

From the observation, a *problem* is recognized. One asks a question about the observation. This question must be both relevant and testable.

The third step of the scientific method is the formation of a *hypothesis;* that is, one "guesses" what the answer to the problem might be. The hypothesis is based on the researcher's training and intuition. It is not critical to the scientist that the hypothesis be correct, for either a refutation or a confirmation will help to answer the problem.

Next the hypothesis is *tested,* either through pure observation or, more frequently, through *experiment.* An experiment is simply a group of observations that are made in a specified fashion. The scientist attempts to hold all conditions constant except the one that he or she is testing, so that any result observed can be attributed to that variable. In addition, these results almost invariably are compared with those of a *control.* A control is, in the case of a zoological study, an animal or group of animals similar to the experimental organism, but on which no manipulation occurs. The control is an attempt to produce a natural state against which the results of the experimental manipulation

may be compared. In some cases an animal may act as its own control if base rates (data on the animal before manipulation) are considered. All experiments must be repeated. Repetition by the original investigator and also by independent researchers ensures that the observed results are reliable and not due to chance (luck) or artifacts (errors).

When all the data are tabulated and analyzed, a *conclusion* based on these results is reached, and the hypothesis is either accepted or rejected. Note that the hypothesis may be refuted by one contrary case but that supportive evidence cannot prove a hypothesis; it can only give credence to it. Experimental evidence for a hypothesis is the basis for the final step in the scientific method: the formulation of a *theory.* Theories are statements about the probabilities of future observations of the tested phenomenon. All theories are broader than the specific conditions tested in the experiment. Theories are predictions of the principles and forces that underlie what we observe.

Functioning within the strict confines of the scientific method requires a particular set of attitudes on the part of the experimenter. These include intellectual honesty, open-mindedness, caution in reaching conclusions, and vigilance in searching for flaws in hypotheses, theories, evidence, and conclusions. Ancient philosopher-scientists constructed broad, sweeping generalizations and then tried to fit their observations of the natural world into this preconceived framework. Modern science attempts to do just the opposite; theories are constructed from observations whose underlying causes are tested to the greatest degree of confidence by experiments. Scientific inquiry is thus a dynamic process. The results of each tested observation suggest theories and future observations that provide an increasingly greater probability of our predicting the "how," "what," and "why" of the phenomena that we observe.

General Characteristics of Animal Life

The study of zoology is practically limitless in its complexity. To the diversity of contemporary animals may be added the fossil record of nearly 4 billion years of evolution. Nevertheless, some characteristics are shared by all animals.

All animals are composed of *cells,* small volumes of protoplasm enclosed within a membrane. Living things show a constancy of morphology (form), and so animals can usually be distinguished from each other because of their characteristic structures. Morphology also clearly distinguishes (in most cases) plants from animals; plant cells have thick cell walls, and many manufacture food through photosynthesis. Most zoologists agree that form and function are inseparable. In a sense, form is the evolutionary outcome of demands placed on the organism by the environment. This point will become more obvious as the student becomes familiar with the classification categories.

Animals and even their individual cells respond quickly to certain changes in their environments, such as fluctuations in temperature or blood solute concentration. This capacity to respond is referred to as *irritability.* Most animals can move at least part of their bodies. *Locomotion* of the entire body is a trait of the majority of animal species.

A major distinction between living systems and nonliving matter is *metabolism,* the process transforming environmental material into specifically organized and active substances, combined with the ability to break down such substances when necessary and derive usable energy from them. The building-up process, or synthesis, is called *anabolism;* the breaking-down process is called *catabolism.*

Living systems also have the capacity for *growth,* although the capacity has definite limits (see Chapter 3). Growth is the result of anabolic activity. Nonliving things may increase in size or mass, but, because the new material is merely added to the outside, the term growth is inappropriate for the inorganic realm. Other metabolic processes (catabolism) must exist to liberate the stored energy for a variety of life activities other than growth. The body must be fed and maintained, injuries must be healed, old or dead cells must be replaced, waste materials must be removed, and so on. An organism is by nature a dynamic *steady-state system;* that is, energy is always expended to maintain the complex ordered structure of an organism. Growth will occur only when internal synthesis exceeds the maintenance requirements of the organism, all of which occur at the expense of energy extracted from the environment via *metabolism.*

Living organisms are usually capable of *reproduc-tion.* This is the process whereby new individuals of the same species are formed, either sexually or asexually (Chapter 38).

Certain characteristics are shared by all animals: a cellular structure, consistent morphology, adaptation of form and function to environmental demands, irritability, locomotion, metabolism, growth, maintenance, and reproduction.

Maintenance of the Species

Just as individual animals do not live isolated from their physical and biological environments, so too does each belong to the historical continuity of life. Every animal belongs to a *species,* a population of potentially interbreeding individuals that resemble one another more closely than they do members of other species. The success of any species is measured in terms of its capacity to produce new individuals that will manifest its traits (characteristics). Assuming that each individual is equipotent in terms of reproductive ability, there still exist potential hindrances to perpetuation of the species.

There must be sufficient food, and, almost as important, there must be sufficient living space. Species require suitable *habitats* in which to live. Marine animals do not, for example, roam throughout the ocean; they are restricted to specific habitats by such factors as water temperature and water depth. The physical environment of this planet provides three major habitats for life: marine, freshwater, and terrestrial. Each presents a very different array of environmental demands to which animals must adapt. A fourth major habitat is provided by the bodies of living animals ans plants. Indeed, the body surfaces and internal organs of living organisms provide the greatest diversity of habitats on the planet. It is not surprising that these habitats have been invaded by a bewildering variety of *parasites* that gain food and shelter at the expense of their *host.* A lack of either food or space will cause reduction in population size or even extinction of the species.

Reproductive success is influenced by competition with other species within the same habitat. Competition for food and living space is often severe, but so too is *predation* by members of other species. Every species is equipped with a bewildering constellation of morphological and physiological adaptations that help to ensure its viability through its reproductive lifetime. Animals must also be able to contend with parasitic organisms and, in some cases, coexist with other organisms in symbiotic and other relationship forms in which both parties benefit. For example, microorganisms within the stomachs of a cow digest the cellulose in the grass that the cow ingests, and then the cow digests them.

The Realm of Zoology

Scientific observation has uncovered so many aspects of animal life that zoology has diversified into a myriad of allied sciences (Fig. 1-1). Each of these subdivisions has direct relevance to every member of the animal kingdom, and it is the zoologist's job to coordinate the data of these allied disciplines to formulate a universal view of the animal kingdom.

The function of this text is to introduce you, the beginning student, to the fascinating yet sometimes confusing world of animals by presenting current basic information in this realm of science, known as *zoology*, the study of animals.

The author believes that certain areas of zoology merit separate consideration, apart from the standard taxonomic considerations. These include animal behavior; body systems such as the skeletal, digestive, circulatory, reproductive, and nervous systems; and such topics as genetics, embryology, zoogeography, ecology, evolution, and human biology.

The Value of Zoology

As you undertake the study of zoology, its value to your everyday life will become evident.

Zoology has both direct and indirect application to all the sciences listed in Figure 1-1, but its relevance

Figure 1-1. The main subdivisions of zoology with descriptions.

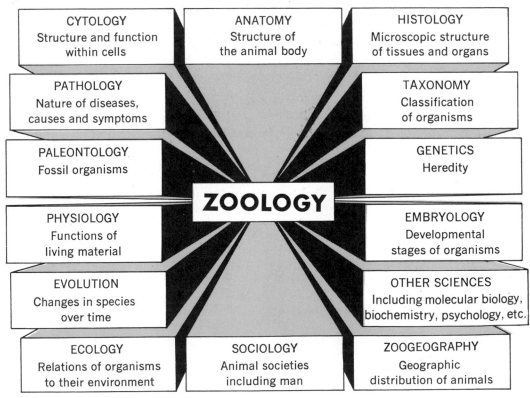

does not stop there. Medicine, dentistry, optometry, animal husbandry, agriculture, and many other disciplines partially rely on, and interact with, zoology. Zoology can offer much to the understanding of current world problems such as population growth, famine, war, space travel, pollution, energy sources, and, most important, the coexistence and survival of humans and other species on this earth.

General Information for the Student

Zoology is a vast science, and no single text can present an all-inclusive view of the animal kingdom. Many details are not provided in this textbook, for the following reasons:

1. Some zoological data are clearly beyond the scope of this introductory text.
2. Many aspects of zoology are still unknown—zoology is a dynamic field of scientific inquiry.
3. The scope of zoology—animals and their interrelationships with their environments—is in a constant state of flux; evolution is an ongoing process.

At various points in the text, in-depth information about particular subjects is presented. Usually the student is not required to "know it cold" but, rather, to understand the implications of what is presented. Such detailed treatments are included to point out trends, culminations of trends, or general zoological principles. For example, in Chapters 2 and 3, the student will learn about subcellular organelles called mitochondria. It will be noted that mitochondria function in energy metabolism and that, generally, more mitochondria are present in relatively active cells. The student will then be expected to predict where mitochondria concentrations would be high, and where they would be low. As an example, a spermatozoan (Chapter 38) has a tail that whips about, pushing the sperm forward; the middle piece of the sperm is composed of vast amounts of mitochondria that provide energy for the tail locomotion. Thus students should attempt to integrate such general principles with their developing awareness of the particulars of the animal kingdom.

The sets of *review questions* at the end of each chapter are important aids to such comprehensive understanding, for they frequently require an integration of concepts. Some knowledge of taxonomy (naming and classification) and nomenclature (vocabulary) will be necessary as well, but their true utility comes in providing the framework for the communication and synthesis of general zoological concepts, particularly when different animal groups are compared and contrasted.

One appendix and a glossary are included in the text. The *classification appendix* is more exhaustive than the taxonomy information supplied in the body of the text and can serve as a quick reference throughout your study of zoology. The concluding extensive glossary provides concise definitions of zoological terms. The *index* at the end of the book cross-references topics for easy location in the main body of the text.

The author's final suggestion is to look for trends throughout the text. That will make zoology easier to comprehend and more relevant to human concerns. Aim for intuitive understanding rather than for mechanical recitation of facts. The science of zoology will become more intriguing, and this deeper level of comprehension of its intricacies will prove useful to your understanding of life.

Cell Chemistry and Morphology

Ancient philosophers first erected an overall theory of the natural world and then tried to fit the particulars of nature into their preconceived theories. Modern scientists do just the opposite: the scientific method begins with careful observations and, from these, theories are formulated only after thorough testing has been undertaken. Observations provide the data of science, and it is not surprising that the history of science has been accompanied by progressively more careful observations. Zacharias Janssen discovered the principle of the compound microscope about 1590. Later Malpighi wrote a detailed microscopic anatomical account of the silkworm in 1669. A milestone in science occurred when Antony van Leeuwenhoek (1632–1723), a Dutch linen merchant and amateur lens grinder, reported his discoveries of freshwater algae, protozoans, micrometazoans, and bacteria to the Royal Society in London. He even observed the circulation of blood through capillaries in the tail of a tadpole. A previously unexplored world had become accessible to the human senses.

Another important discovery occurred in 1663 when the English scientist Robert Hooke focused his microscope upon thin sections of cork. Hooke observed that this tree bark was composed of numerous empty chambers, which reminded him of the *cells* in a monastery.

Almost 200 years later, in 1838, the German botanist Matthias Schleiden stated that the cell was the basic unit of all vegetable matter. This concept was extended during the following year by Theodore Schwann, a German zoologist, who held that both plants and animals were aggregates of cells. Modern **cell theory** recognizes cells to be the basic units of life and further recognizes that the cells of multicellular organisms are intimately interdependent.

Fundamental Chemical Overview

To understand the processes of life, one must first acquire a working knowledge of the chemical constituents of cells.

Because students enter this course with various backgrounds in chemical theory, this section is presented to fill in any gaps in basic chemical knowledge. Those students with sufficient chemical training may use this section for a quick review, but those with a more limited exposure to chemistry should benefit from a more careful reading.

The Atom and Its Forms

All matter is composed of **atoms.** Atoms in turn are composed of three types of particles: protons, neutrons, and electrons.

A **proton** is a positively charged particle with an atomic weight of approximately 1. **Neutrons** are es-

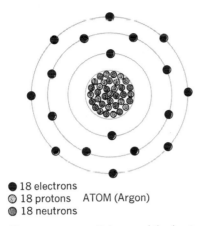

● 18 electrons
◉ 18 protons   ATOM (Argon)
◉ 18 neutrons

Figure 2-1. The atom, argon (Ar), one of the inert gases, has three complete shells and is thus highly unreactive. All nucleons (protons and neutrons) and electrons are shown diagrammatically as orbiting the nucleus; the actual localities of the electrons are within spherical or lobe-shaped areas surrounding the nucleus.

sentially uncharged protons. An **electron** is a negatively charged particle with an atomic weight equal to approximately 1/1800 that of a proton or neutron.

An atom is composed of a central **nucleus** with one or more orbiting electrons. A nucleus is composed of one proton, one proton and one neutron, or several protons and neutrons. Electrons revolve around the nucleus, confined to specific regions called **shells.** Modern atomic theory visualizes electron shells as probability clouds of electron density (i.e., the electron does not act as a single particle, but more like a diffuse charged cloud), but for our purposes atomic structure may be visualized by the model in Fig-

ure 2-1. Each concentric shell can hold a specific number of electrons. Inner shells that contain a full complement of electrons are chemically stable. The outermost shell is often incomplete and can potentially acquire more electrons. Atomic stability and chemical activity are determined in great measure by the completeness or incompleteness of electron shells. Chemical reactions usually involve the electrons in the outermost shell. Atoms tend to gain or lose electrons so as to attain a complete outermost shell—the most stable configuration. You might imagine that, if a shell has room for just one more electron, the atom may be quite reactive (unstable), "wanting" to fill the shell. Conversely, if the outermost shell contains only one electron, that atom will tend to lose it—this also represents a relatively unstable state. In contract, the **inert gases** (also known as the **noble gases**), such as helium, neon, argon, and others, are atoms with complete electron shells; such atoms are highly unreactive.

An atom is unique by virtue of the number of its protons. An **element** is the set of all atoms with the same number of protons. The numbers of neutrons and electrons may vary somewhat in the atoms of an element, as described in the next paragraph. There are 92 naturally occurring elements, and 13 more have been synthesized by nuclear physicists. The elements are identified by common names (e.g., carbon, hydrogen, oxygen) and by chemical symbols (e.g., C, H, O).

The number of neutrons in an atomic nucleus may vary somewhat. For example, most atoms of the element carbon (C) are composed of six protons, six neutrons, and six electrons. These carbon atoms are symbolized as C^{12}, with the superscript indicating the

Figure 2-2. Three isotopes of carbon. The extra neutrons change the atomic mass but not the charge of the basic C^{12} atom. With increased mass comes decreased atomic stability.

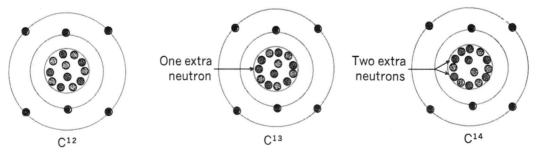

C^{12} One extra neutron → C^{13} Two extra neutrons → C^{14}

total **atomic weight** (the sum of the protons and neutrons). However, other atoms of carbon have been observed. These **isotopes** have different numbers of neutrons and hence different atomic weights, for example, C^{13} and C^{14} (Fig. 2-2). Many isotopes are radioactive because of instabilities in the nucleus as a result of these extra neutrons.

An atom may also have varying numbers of electrons. Normally, any atom has the same number of electrons as protons, and, because the positive and negative charges cancel, such an atom is electrically neutral. Frequently, however, one or more electrons are removed from or added to an atom, resulting in a charged atom, known as an ion. If electrons are removed, a net positive charge results, and the ion is referred to as a **cation.** Addition of electrons produces a net negative charge, creating an ion known as an **anion.**

Chemical Bonds

In nature it is rare to find a free atom. Atoms are usually combined with other atoms in definite proportions, forming **compounds. A molecule** is the smallest unit that retains the characteristic properties of the compound itself. Molecules are complexes of atoms that are held together by chemical bonds—attractions that result from the transfer or sharing of electrons between atoms (Fig. 2-3). **Ionic bonds** involve a complete transfer of one or more electrons from one atom to another. As you know, opposite charges attract each other.* If a chloride ion, Cl^- (where the superscript represents the net charge of the ion), is in the presence of a sodium cation (Na^+), the two may be drawn together into a salt, which in this case is common table salt, sodium chloride. This compound may be represented as Na^+Cl^- or, more simply, as $NaCl$. Ionic bonds may occur between more than two atoms in a molecule, and each bond involves an anion donat-

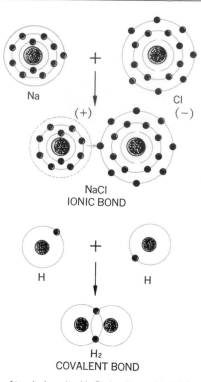

Figure 2-3. Atomic bonds. NaCl (sodium chloride), common table salt, is ionically bonded through transfer of one of sodium's electrons to chloride: The atoms become charged (Na^+, Cl^-), and the resulting mutual ionic attraction keeps them together. In its natural state, NaCl is not a single molecule but a lattice of evenly spaced Na^+ and Cl^- ions. H_2 (molecular hydrogen) is covalently bonded; the electrons are shared by both atoms.

* You may wonder why, if opposite charges attract, an atom does not collapse. Classical atomic theorists wondered the same thing. The answer lies in a realm of physics outside the scope of this book. Briefly, electrons seem to be confined to certain "quantum states" of very specific energy levels, and they cannot easily switch levels. The lowest level permitted is at a minimum but discrete distance from the nucleus.

ing one or more electrons to a cation. Ionic bonds tend to dissociate in aqueous solution, leaving the ions free in solution. Evaporation of the water causes the ions to recombine into molecules of salt.

A second type of chemical bond is the **covalent bond,** in which electrons are shared by the participating atoms. As in ionic bonding, multiple bonding is possible: more than one electron may be shared between two atoms. In covalent bonds the participating electrons become community property, whereas in ionic bonding electrons are transferred from the anion to the cation. About 95 percent of the molecules in cells are covalently bonded.

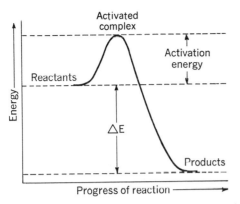

Figure 2-4. Activation energy diagram. As one proceeds toward the right along the abscissa (horizontal axis), the energy of the entire system changes. Energy is initially added to the reactants, and, when sufficient activation energy has been added, the reaction proceeds spontaneously. The decrease in the net energy of the system, ΔE, results in an increase in stability. This reaction is exothermic.

Both types of chemical bonds result from electrical interactions between adjacent atoms. The chemical bond is a form of potential energy that may be used by the cell to do work when the bond is broken. During a chemical reaction, bonds may be broken, formed, or both. For the bonds of a stable molecule to be broken, a certain amount of **activation energy** must be added to the system. Once sufficient activation energy has been added to the system, the reaction proceeds spontaneously, much as a combustible material will burst into flames once its kindling temperature has been reached. Figure 2-4 illustrates that, after the

proper amount of activation energy is put into the system, the rest of the reaction is downhill; that is, a decrease in the total energy of the system, E, leaves the system more chemically stable (less reactive). Cells are able to harness some of the energy that is released during biochemical reactions and use that energy to do metabolic work.

Generally, the higher the potential energy in a system, the less stable it is. The possibility of a reaction's occurrence depends upon the activation energy requirements; if too much energy is needed, the reaction will not occur. A reaction that absorbs energy is called an **endothermic** reaction. The formation of sugar during photosynthesis in green plants is an endothermic reaction in which solar energy is stored in chemical form. An **exothermic** reaction gives off heat, which is equivalent to saying that it releases energy. The dephosphorylation of adenosine triphosphate (ATP) to adenosine diphosphate (ADP), the principal means by which cells obtain energy for their metabolic activities, is an example of an exothermic reaction.

One feature of some molecules that has direct relevance to zoology is the existence of a **dipole.** In covalently bonded water molecules (H_2O), for example, the oxygen atom has a somewhat greater affinity for the shared electrons than do the hydrogen atoms; oxygen is said to be more **electronegative** than hydrogen. The oxygen atom tends to pull the shared electrons toward it, thereby attaining a slightly negative charge with respect to each hydrogen atom (Fig. 2-5). This separation of charge within a molecule is known as a dipole. The water molecule is said to be polar,

Figure 2-5. Molecular polarity. Left: H_2O molecule. Positive portion of dipole is on the hydrogen "end." Also shown are three hydrogen-bonded water molecules. Right: CO is polar, but the dipoles of CO_2 cancel, leaving the linear molecule without a dipole.

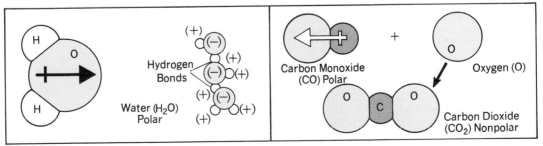

and this is the major reason that water is so universal a solvent; polar molecules tend to dissolve in polar solvents. Polarity plays a major role in the permeability of cell membranes and so, in part, determines which substances may enter or leave a cell. Nonpolar molecules have no dipole (Fig. 2-5); note that carbon dioxide (CO_2) is nonpolar because, although oxygen is more electronegative than carbon, the net effect of the electronegativity is zero because of the spatial arrangement of the atoms.

The dipoles of individual water molecules result in the attraction of the relatively negative oxygen to the relatively positive hydrogen of another water molecule, forming what is called a **hydrogen bond.** Hydrogen bonds also occur frequently between a hydrogen atom and an electronegative atom, such as an oxygen, atom, in organic molecules; they are important in holding together the structures of **proteins** and **nucleic acids.** It is exactly this bonding that allows such a small molecule as H_2O to be liquid at temperatures and pressures at which nonpolar molecules such as CO_2 would be gaseous. Hydrogen bonds are much weaker than ionic or covalent bonds, but, if vibrational energy is reduced enough (i.e., if the temperature lowered), water molecules form the hydrogen-bonded crystalline lattice, **ice.**

Acids, Bases, and Salts

An **acid** is any substance that in aqueous solution gives off hydrogen ions (H^+). For example, HCl, hydrochloric acid, dissociates into hydrogen ions (H^+) and chloride ions (Cl^-). The pH of a solution is a notation that designates the negative logarithm of the hydrogen ion concentration.

pH ranges from 1 to 14. A very low pH, for example, pH = 1, represents an extremely acidic solution—one with many free hydrogen ions. A **base** is a substance that gives off hydroxide ions (OH^-), for example, NaOH, sodium hydroxide. pH values of greater than 7 characterize basic (alkaline) solutions. Neutral is pH 7. At this point, the number of free hydrogen ions is balanced by an equal number of free hydroxide ions, and so the solution is neither acidic nor basic.

An acid is said to be strong if it dissociates completely or almost completely in aqueous solution. Analogously, a base is strong if it dissociates to a great extent. Note that nothing is implied about toxic or corrosive effects by saying that an acid or base is "strong."

A compound resulting from the chemical interaction of an acid and a base is called a **salt.** For example, if HCl and NaOH are dissolved in water, Cl^- and Na^+ ions will be formed from their respective dissociations; these ions will bond together into NaCl molecules (table salt) if water is then evaporated from the solution. *What happens to the hydrogen and hydroxide ions?*

Cellular Constituents

Chemical analyses of cells indicate that they are made up of the same elements that occur in nonliving matter but that some 20 elements (see Table 2-1) are essential components of cells. These elements are generally combined to form organic and inorganic compounds. **Organic compounds** contain carbon and usually hydrogen and oxygen; they occur naturally in living plants and animals. Organic compounds are for the most part proteins, lipids, nucleic acids, and carbohydrates. **Inorganic compounds** generally do not contain carbon atoms and include principally water and salts. The relative percentages of each in an average cell are as follows:

Compounds	Approximate PerCent of Cellular Volume
Water	80.0
Proteins	12.0
Fats	3.0
Nucleic acids	2.0
Carbohydrates	1.0
Inorganic salts	1.0
Steroids	0.5
Other substances	0.5

Table 2-1. Essential Elements for Cell Life (Percentages are slightly variable.)

Essential Elements	Symbols	Per Cent in Cytoplasm
Oxygen	O	63.00
Carbon	C	20.00
Hydrogen	H	10.00
Nitrogen	N	2.50
Calcium	Ca	2.50
Phosphorus	P	1.14
Potassium	K	0.11
Sulfur	S	0.14
Chlorine	Cl	0.10
Fluorine	F	0.10
Sodium	Na	0.10
Magnesium	Mg	0.07
Iron	Fe	0.01
Copper	Cu	Trace
Cobalt	Co	Trace
Zinc	Zn	Trace
Silicon	Si	Trace
Manganese	Mn	Trace
Iodine	I	Trace
Nickel	Ni	Trace

Small Molecules

Water molecules (H_2O) are composed of two atoms of hydrogen bonded to one atom of oxygen. Water is the most abundant compound in the cell and usually accounts for 80 to 95 percent of cellular volume. Water's importance lies in its versatility. Water acts as a very good solvent and serves as a medium for material transport both inside and outside the cell; water molecules also participate in several critical reactions, notably in cellular respiration.

Molecular oxygen (O_2) and carbon dioxide (CO_2) also play vital roles in cellular respiration. More detail on this will be presented in Figure 2-11. Another small molecule, ammonia (NH_3), is a necessary by-product of protein decomposition. It is a highly toxic substance; blood concentrations of only 0.005 percent NH_3 are fatal to rabbits. Consequently, ammonia must be excreted immediately or rendered nontoxic through the synthesis of larger molecules. The result of this reaction is usually the formation of **urea** or **uric acid**, a molecule of limited toxicity that is eventually excreted as a major component of urine.

Lipids

The first group of large organic molecules to be considered, the **lipids,** are a structurally heterogeneous group of substances that share solubility in a number of organic solvents and only slight (or no) solubility in water. Lipids are important structural components of cells, especially in membranes. **Fats** are employed in energy storage, as they contain many high-energy bonds. **Waxes** are involved in the formation and maintenance of skin and fur. Lipids in the form of four-ringed **steroids** act as chemical messengers (hormones) that coordinate the metabolism of an organism as a whole.

There are three major types of lipids: true fats, phospholipids, and steroids. **True,** or **neutral, fats,** important in long-term energy storage, are composed of three fatty acid molecules bonded to a molecule of glycerol. A fatty acid contains a single hydrocarbon chain that is nonpolar in nature, but it ends with the polar and acidic carboxyl group $\left(\begin{smallmatrix} O \\ \| \\ -C-OH \end{smallmatrix}\right)$. This polar end is lost when reacted with glycerol, resulting in the neutral, true fat.

Phospholipids have structures similar to fats, except that one of the three fatty acids is replaced by a nitrogen-containing base plus a phosphate group, giving the overall molecule a polar "head" and a nonpolar fatty acid "tail." The bringing together of these opposite chemical natures allows phospholipids to form membranes spontaneously.

Steroids are complex molecules in which the carbon atoms are arranged to form a particular four-ringed structure. Cholesterol is a common steroid; an important function of many steroids is to act as chemical messengers (**hormones**) in the regulation of metabolism. The basic lipid structures are presented in Figure 2-6.

Carbohydrates

Carbohydrates also play a major role in cellular chemistry. These organic molecules are generally

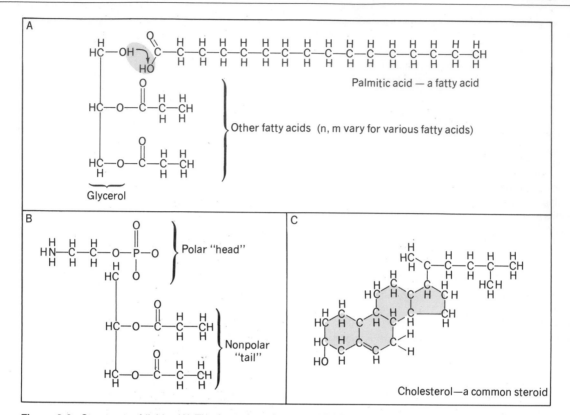

Figure 2-6. Structures of lipids. (A) The formation of a neutral fat from glycerol and three fatty acid molecules; shaded area shows one of the three water molecules given off during synthesis. The already attached fatty acids illustrate the shorthand notation for the long carbon chains. (B) A phospholipid, showing polar group. (C) Cholesterol, a representative steroid; shaded region shows the four-ringed structure common to all steroids.

composed of carbon, hydrogen, and oxygen atoms in a $1:2:1$ ratio, respectively. Carbohydrates occur as long chains, or **polymers,** of smaller units called **monomers,** which are **sugars** in this case. The simplest sugar is a chain of three carbons, whereas the largest has seven carbons. The most common sugar in animals is **glucose,** with the molecular formula $C_6H_{12}O_6$, indicating six atoms of carbon, twelve of hydrogen, and six of oxygen (Fig. 2-7).

Carbohydrates are used by animal cells mainly for energy storage. **Glycogen,** also known as animal starch, the carbohydrate used by animals for short-term energy storage, is composed of a chain of glucose molecules (Fig. 2-7). In vertebrates glycogen is found

especially in liver and muscle cells. Glycogen storage is rather limited, and, as a result, if the intake of carbohydrates is high, those carbohydrates will be converted into fats for long-term energy storage.

Nucleic Acids

The functions of **nucleic acids** are central to the existence of life. Even the simplest one-celled organism is an amazing complex of chemicals and reactions. The continued existence of any species is accomplished by **inheritance** (the process of **heredity**) of a much simpler representation of the complex orga-

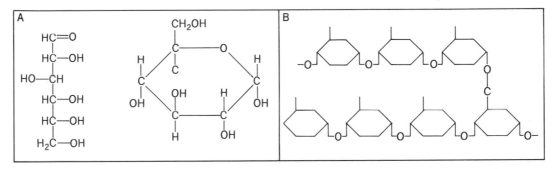

Figure 2-7. Structures of glucose and glycogen. (A) Two different forms of glucose: left, straight; right, cyclic. (B) Glycogen is a complex of repeating glucose units. Note example of branching that may occur. Hydroxyls and hydrogens are omitted for clarity.

nism—the information contained in the chromosomes of cells that is passed from generation to generation. This information is not only hereditary, but it is also needed for all cellular activities during the entire lifetime of an organism. The chromosomes contain this information by means of biochemical language of nucleic acids, the **genetic code** (Chapter 38).

Nucleic acids are another type of polymer, but, unlike carbohydrates, they are always linear; there may be different monomers within one polymeric chain, so the chain is called a **heteropolymer.** Each nucleic acid monomer, the **nucleotide,** is composed of a sugar, phosphate, and **nitrogenous base** (Fig. 2-8); the type of sugar and bases determines the two classes of nucleic acids: **deoxyribonucleic acid** (**DNA**) and **ribonucleic acid** (**RNA**). The variability of monomers allows these molecules to contain **coded information;** that is, the linear sequence of nucleotides contains messages, as do the series of letters in the sentences that you are reading. The messages in the DNA determine what kind of **proteins** will be made (see Chapter 38). All of an organism's morphology and physiology is determined by proteins, either directly as structural proteins or indirectly by the **enzymes** that are needed to build all the nonprotein parts of an organism.

One chromosome is a single, very long molecule of DNA. The size of the molecule is quite amazing. An average human chromosome includes a DNA polymer containing approximately *200 million* nucleotides, which is 7 cm long (and only 10^{-6} cm wide!). Obviously, this molecule must be carefully packed (by forming helices, and helices of helices, etc.) into the 10^{-4} cm (10μ)* long chromosome; this is further complicated by the fact that we have 46 chromosomes per cell. A region along this DNA of a few thousand nucleotides contains the message for one protein and is called a **gene.** There may be many thousands of genes (and thus proteins) per chromosome, which is an indication of the large amount of information needed to make and manage a living organism. Much more detail on nucleic acid structure and function is presented in Chapter 38.

Proteins

The **proteins** are a class of extremely important organic molecules. Not only are proteins the chief structural components of cells, but proteins also include enzymes and some hormones and are important components of chromosomes and membranes. Each species of animal has a characteristic array of proteins. Because this reflects the genetic makeup of an organism, it is useful in determining evolutionary similarity or divergence (discussed in Chapter 40). Proteins contain carbon, hydrogen, oxygen, and nitrogen atoms, with sometimes sulfur, phosphorus, and iodine.

* One micron (μ) = 10^{-6} m.

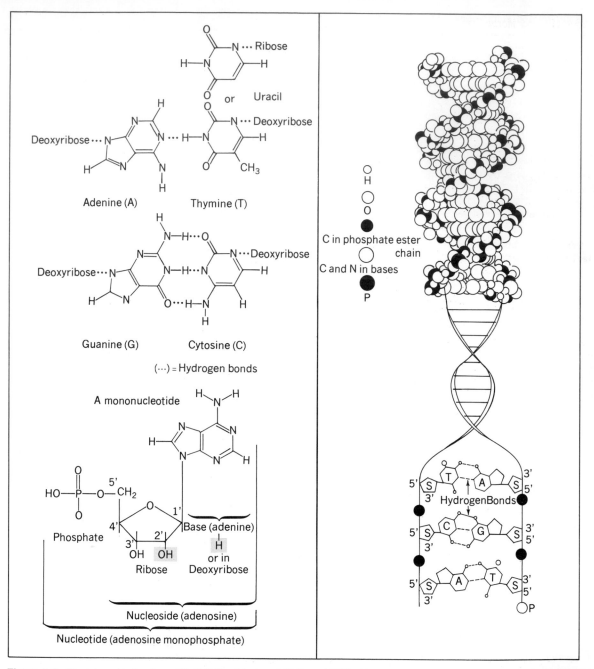

Figure 2-8. Nucleic acid structure. (A) Building blocks of DNA and RNA. Nitrogenous bases with possible base pairing are shown on top; the manner in which these bases are joined to the sugar and phosphate is shown on bottom. Note that RNA uses ribose and bases A, G, C, U; DNA contians deoxyribose and A, G, C, T. (B) DNA structure. Diagrammatically shown at bottom is the manner in which nucleotides join to form the base paired double helix; upper drawing shows a "space-filling" molecular model that better depicts the actual shape of DNA. Numbered carbons on the sugars in both (A) and (B) illustrate directionality of polymers.

As with carbohydrates, proteins are polymers—long chains of simpler subunits, in this case **amino acids,** but, as with nucleic acids, the monomers are variable. There are about 25 amino acids used by organisms, but only about 20 of these are important in animal cells. The single-asterisked ones in Table 2-2 are essential for human nutrition. **Essential amino acids** (which vary slightly for different organism) cannot be synthesized by the organism but must be ingested in the diet. When a protein is ingested, digestion breaks the polymer apart and frees the amino

Table 2-2. The Important Amino Acids for Man (See key and text for explanation.)

Name	Abbreviation	Notes
Alanine	Ala	
Arginine	Arg	
Asparagine	Asn	
Aspartic Acid	Asp	
Cysteine	Cys	
Cystine	—	**
Di-iodotryosine	—	**
Glutamine	Gln	
Glutamic Acid	Glu	
Glycine	Gly	
Histidine	His	
Hydroxyproline	—	**
Isoleucine	Ile, Iln	*
Leucine	Leu	*
Lysine	Lys	*
Methionine	Met	*
Phenylalanine	Phe	*
Proline	Pro	
Serine	Ser	
Threonine	Thr	
Thyroxine	—	***
Tri-iodothyronine	—	***
Tryptophan	Try	*
Tyrosine	Tyr	
Valine	Val	*

Key:
*Essential amino acids—the body must ingest these since it cannot manufacture them itself.
** Considered by some to be necessary for human life. These may be formed from existing or ingested precursors in the body.
*** These are thyroid gland hormones which incorporate iodine. They appear in the form of amino acids, however, and so are listed here. See also **. The "R" group possesses the iodine-containing loci.

acids for absorption and assimilation. The other **nonessential amino acids** can be synthesized by the organism from precursors in the diet including other amino acids.

Generally, an amino acid has the structure

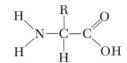

where R represents 1 of some 20 **functional groups** composed of carbon and hydrogen atoms. The

$$\begin{array}{c} H \\ \diagdown \\ N— \\ \diagup \\ H \end{array}$$ portion of the molecule is known as the

amino end, and the $—C\diagup^{O}_{\diagdown OH}$ part is called the

carboxyl end. Figure 2-9 illustrates the structures of a few selected amino acid molecules.

When a protein is synthesized, the amino acids are bonded to each other by **peptide bonds,** in which the amino group of one amino acid is linked to the adjacent carboxyl group of another, forming a polypeptide. For example, the protein hemoglobin, found in our red blood corpuscles, contains approximately 574 amino acids arranged in four polypeptide chains. A related protein, myoglobin, contains one polypeptide chain that is homologous to each hemoglobin polymer; it is indicated in Figure 2-9. Proteins may be identified by listing the amino acids in sequence, starting from the amino end. Every protein consists of a distinct sequence of amino acids. This sequence is genetically determined and totally accounts for each protein's unique structural and functional properties; that is, the chemical nature of the amino acids determines the shape that the protein will assume and exactly what chemicals may interact with it as part of its functions. For example, all enzymes are composed of protein, but many attract other chemical groups that become part of the enzyme structure and are necessary for its catalytic activity (see Enzymes).

Enzymes

A cell's existence and function rely completely on its ability to make use of accessible nutrients. A multi-

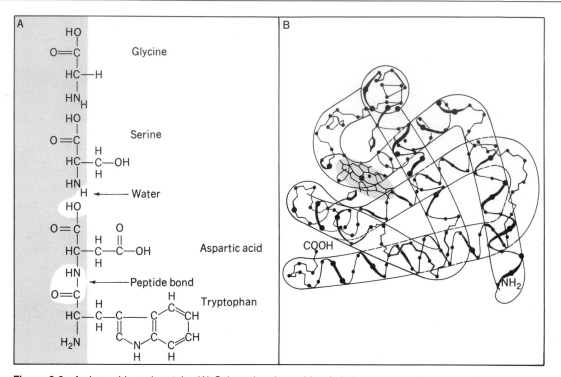

Figure 2-9. Amino acids and protein. (A) Selected amino acids; circled areas show the water released and the peptide bond. Note variety in size and chemistry of the side groups; the part of amino acids common to all is in shaded area. (B) Myoglobin, a protein of 151 amino acids that carries oxygen. Note how the long chain of amino acids (thick line) winds into helices (shown outlined) that fold into the overall globular shape. Shaded area is the four-ringed nonproteinaceous heme group that carries the O_2 or CO_2 molecules.

tude of chemical reactions occur within the cell, the sum of which constitutes **metabolism.** Metabolic processes may be either synthetic or degradative. All a cell's synthetic activities are termed **anabolism,** and such processes usually require an input of energy; that is, they are **endothermic.** Some of this energy is stored in the chemical bonds of the anabolic products, and this potential energy is later released by **catabolism.** Animal metabolism is primarily catabolic, requiring the breakdown of ingested organic compounds (food) for energy acquisition (**heterotrophy**). This is a major distinction between animals and plants, for most plants can, through **photosynthesis,** synthesize complex organic compounds from simple inorganic molecules using sunlight as an energy source (**autotrophy**).

Metabolic reactions are facilitated by **enzymes,** which are biological catalysts. An enzyme is usually named by adding the suffix -**ase** to the name of the **substrate** upon which it acts; for example, lipase acts on lipids. Effective in minute concentrations, enzymes perform their catalytic functions without being used up themselves. They are not incorporated into the reaction products and may be reused many times.

All enzymes are composed of one or more polypeptide chains that are folded into a globular structure, but many are complexes of a proteinaceous **apoenzyme** and a smaller organic molecule, a **coenzyme,** that is essential for catalytic action. **Vitamins** such as thiamine, nicotinic acid (niacin), riboflavin, or pyridoxine act as coenzymes. Most coenzymes are chemical derivatives of **nucleotides.** Metallic elements such

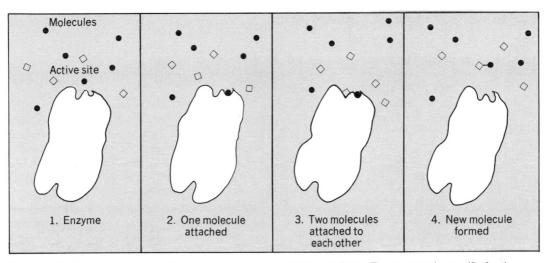

Figure 2-10. Enzymatic action. The enzyme may be reused several times. The enzyme is specific for the molecule formed or destroyed with its aid.

as iron, manganese, copper, magnesium, molybdenum, and zinc also enhance the activity of certain enzymes and are known as **cofactors.**

Enzymes promote chemical change by lowering the **activation energies** required to initiate specific chemical reactions. Frequently an atom or group of atoms must be transferred from one compound to another, and the enzyme physically bonds to the transient group, carrying it to the recipient. Another function is that of a **template** (or platform) for a reaction; the enzyme provides a surface upon which the substrates are brought into close proximity to one another, in the region of the enzyme called the **active site** (Fig. 2-10). The active site is extremely specific for its substrate(s); that is, the chemical structures of the amino acids in the active site allow only certain chemicals to bind to the enzyme, much like a key fits a particular lock. This means that each type of biochemical reaction needs a different enzyme. Thus, although a cell requires thousands of enzymes, the rates of all metabolic processes can be controlled in a highly coordinated fashion. An enzyme also provides the proper chemical environment for a particular reaction. For example, many biosynthetic reactions do not proceed well in water; the enzyme may have nonpolar amino acids in its active site that keep water away during the reaction. The enzyme's unique phys-

ical and electrical configuration permits the reactant molecules to come together more often than by chance alone. In this way enzymes may increase the rate of a chemical reaction several thousandfold (in either direction). After the reaction has occurred on the enzyme's surface, the product or products split off, leaving the enzyme free to act again. Enzymes are generally large and complex molecules that are very sensitive to temperature and pH; most are inactivated outside the optimal temperature range of the organism that contains them.

Types of Chemical Reactions

Chemical reactions share some common characteristics. Their rates depend on temperature, pressure, and the concentrations of reactants and products.

Some chemical reactions proceed in only one direction, but most reactions are reversible. This is indicated in chemical equations by a double arrow, as in the dissociation of water molecules:

$$H_2O \rightleftharpoons H^+ + OH^-$$

Note that this equation is in *balance:* the number of atoms of each element is the same in the reactants as in the products. The law of **conservation of mass** states that the number of atoms of each element on the reactant side of the equation must be balanced by an equal number of atoms on the product side. Each reversible reaction proceeds at a given temperature and pressure until a definite ratio of concentration of products to concentration of reactants results. That ratio, called the **equilibrium constant,** is represented by the symbol K:

$$K_{[H_2O]} = \frac{[H^+][OH^-]}{[H_2O]}$$

Chemical equilibrium is established when the ratio of concentrations of products to reactants has reached the characteristic equilibrium constant. At equilibrium, the number of reactant molecules being transformed into products is balanced by an equal number of product molecules being changed back into reactants, and so there is no *net* change.

Four types of chemical reactions are most prominent in cellular metabolism: transfer, synthesis, hydrolysis, and reduction–oxidation reactions.

Transfer reactions involve the transfer of an atom or group of atoms from one molecule to another. Coenzymes usually act as carriers in such transfers. Commonly carried substances are hydrogen ions (H^+), amino groups, and phosphate groups.

Synthesis reactions involve the formation of a larger molecule from smaller ones, usually accompanied by a simultaneous formation and release of a water molecule from the reactants (*e.g.,* see Figs. 2-6 and 2-9); the process is thus called **dehydration synthesis.** An example of a synthesis reaction is the bonding together of amino acids to form a protein molecule.

Hydrolysis is the opposite of synthesis. Such catabolic processes involve a splitting of the substance that is broken down by the addition of water at the site of cleavage. Examples of such **hydrolyses,** or "water-dissolving" reactions, include the digestion of large lipid, nucleic acid, carbohydrate, and protein molecules into their smaller subunits.

Redox reactions are coupled **reduction** and **oxidation** reactions. Such reactions involve an exchange of electrons and *always* occur together. When an atom or molecule loses an electron, it is said to be oxidized; as an atom or molecule gains an electron, it is said to

be reduced. Oxidations are energy-liberating reactions, whereas reductions consume energy.

Basic Energetics of Animal Metabolism

Let us now consider some implications of the biochemical basis of life. We will limit this discussion to adult organisms, those that have reached mature stages of physiological development.

Though an animal may have ceased its growth, it still maintains its tissues very actively. With the exceptions of most nervous and much bone tissue, each animal is involved in a continuous synthesis process as old cells die and must be replaced. Also, each animal continually expends energy (e.g., in metabolic activity or in locomotion), and that energy must be replaced if life is to continue.

Life involves a continuous renewal of raw materials and energy. Both must be taken into the organism from its environment. In unicellular animals such as protozoans, this intake occurs directly through the cell membrane, whereas in multicellular animals the intake frequently involves elaborate digestive and respiratory systems.

Once the nutrients are obtained, certain energetic principles become evident. These principles are best understood in terms of efficiency. Efficiency is defined as the ratio of output to input times 100, which yields a percentage:

$$\% \text{ Efficiency} = \frac{\text{Output}}{\text{Input}} \times 100$$

If there is any "waste" (and there always is), efficiency is necessarily less than 100 percent. The higher the efficiency, the less the waste.

The input that concerns the zoologist is the amount of potential energy in the food ingested by an animal. Animals are very inefficient: they consume much more energy than they can use constructively.

The greatest source of energy waste in animal metabolism is heat, which is given off especially as byproduct of muscular contraction. **Homeothermic** (warm-blooded) animals generate metabolic heat to maintain a constant body temperature, and this greatly increases their energy requirements.

If an animal takes in more food energy than it re-

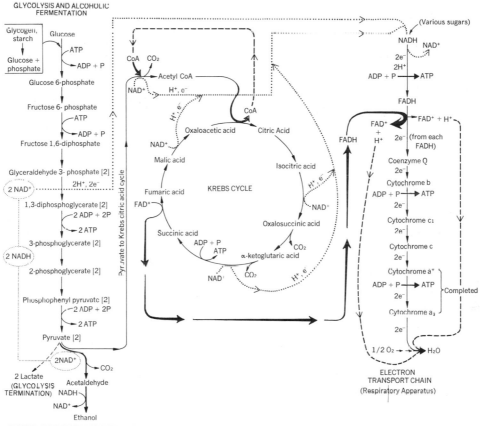

Figure 2-11. Central metabolism of animal cells. These biochemical processes are the main energy-deriving reactions and the starting points for the biosynthesis of most all the cells' biochemicals. During **glycolysis, glucose** is broken down into **pyruvate.** Energy in the chemical bonds is transferred from the glycolytic intermediates to **adenosine diphosphate (ADP)** and phosphate to form **adenosine triphosphate (ATP),** the energy currency of all cells. Almost all energy-requiring processes derive energy from the hydrolysis of ATP to ADP + P, thus releasing the stored energy. Glycolysis is an **anaerobic** (without oxygen) process; many anaerobic micoorganisms metabolize pyruvate to **ethanol** (e.g., in all alcoholic beverages), whereas muscle cells (during strenuous activity) will convert it to **lactate** if there is not enough oxygen to utilize the succeeding reactions (this is called **oxygen debt;** (see Chapter 32). Under **aerobic** conditions, pyruvate is broken down (oxidized) all the way to CO_2 in the **Krebs cycle.** As shown, the intermediate steps capture chemical energy in the form of electrons and protons carried by the coenzymes **FAD^+** and **NAD^+** (forming **FADH** and **NADH**); the electrons are transferred from protein to protein (the **cytochromes**) in the **electron transport chain.** The transported electrons and protons (in solution) combine with oxygen (O_2) to yield water; much more ATP is formed during this process, which is also called **cellular respiration.** The intermediate of glycolysis and the Krebs cycle are starting materials for many biosynthetic pathways, such as the synthesis of many amino acids. The diagram reveals many of the intricacies of biochemistry; the essential concept is the formation of ATP by oxidation of glucose, quickly, yet incompletely (glycolysis), under anaerobic conditions and by the complete aerobic oxidation to CO_2 and H_2O (Krebs cycle and electron transport). The latter produces 19 times more ATP than does glycolysis, as shown by the overall reactions:

Glycolysis: 1 Glucose $\longrightarrow$ 2 pyruvate (2 ATPs made)
Cellular respiration: 1 Glucose + $6O_2 \longrightarrow 6O_2 + 6H_2O$ (38 ATPs made)

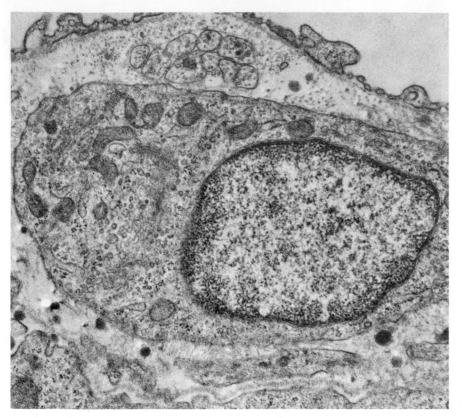

Figure 2-12. Top: An electron micrograph of a typical animal cell. Bottom: Diagram of a cell. Can you identify the structures shown in the diagram with those in the photograph?

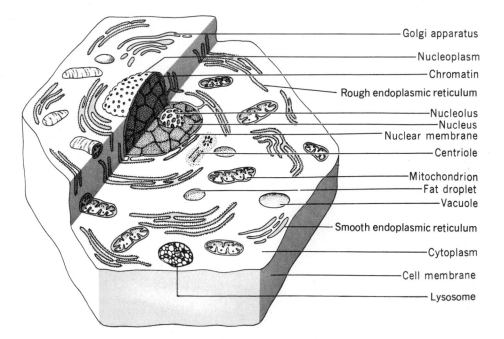

Golgi apparatus

Nucleoplasm

Chromatin

Rough endoplasmic reticulum

Nucleolus
Nucleus
Nuclear membrane

Centriole

Mitochondrion
Fat droplet
Vacuole

Smooth endoplasmic reticulum

Cytoplasm

Cell membrane

Lysosome

quires for steady-state existence, it may do one of two things: eliminate the excess or store it for future use. Most often excess energy is stored, especially if the diet contains a large amount of fats or lipids, both high-energy substances. Animals can store limited amounts of excess energy as glycogen, a carbohydrate polymer that is deposited in the liver, subcutaneously (under the skin), and in muscles. Proteins have a lower energy content than do fats or lipids (their bonds, when broken, liberate less energy), but proteins are more important structurally.

What happens when food intake is drastically reduced? The animal must still expend energy to continue its existence. But if the required energy (and raw materials) cannot be obtained from its external environment, the animal turns to its *internal* environment and begins to catabolize its energy reserves, including stored fats and glycogen. A normal human adult can subsist for approximately two weeks completely on its glycogen reserves (but only for a few days without water). Once those high-energy reserves have been used up, an animal must catabolize the fats and proteins in its own tissues to obtain energy. Eventually, starvation becomes complete and the animal dies.

An animal's capacity for energy storage permits a certain flexibility in the input–output ratio over short periods of time. Nevertheless, prolonged periods of minimal input have drastic effects, and those effects may be understood on a biochemical basis. A brief summary of the biochemical pathways that constitute the central core of metabolism appears in Figure 2-11.

Cell Morphology

Let us now consider the general substructure of animal cells. The term **protoplasm,** developed by the Czechoslovakian physiologist Johannes Purkinje in 1840, has become synonymous with "living substance." Protoplasm is the complex conglomeration of materials that make up a cell. Current usage tends to employ "protoplasm" as merely a descriptive term and focuses on the cell itself as the true unit of life.

Most cells range in size from $0.5~\mu$ to $20~\mu$, but some bacteria are as small as 250 millimicrons (mμ; $1000~m\mu = 1~\mu$). Lower limits for living cells vary

around a mean diameter of 225 mμ. Upper limits are represented by very large eggs and by some nerve cells, which in humans may attain lengths of 2 m.

Subcellular Organization

Microscopically, protoplasm has the appearance of a grayish jellylike substance in which one may see globules, droplets, granules, and crystals of various shapes and sizes. Cells may be stained with various dyes so that these normally colorless structures are rendered more distinct.

The most basic subdivision of a cell is that between cytoplasm and nucleoplasm (karyoplasm). **Cytoplasm** consists of the semifluid substance outside the nucleus, whereas **nucleoplasm** consists of the material within the nucleus. Suspended in both are several kinds of structures called **organelles** that allow a compartmentalization of form and function within the cell. The most typical organelles that make up the ultrastructure of animal cells are indicated in Figure 2-12.

The protoplasm, in which organelles and other inclusions are suspended, is in the form of a colloid, with the general consistency of a viscous syrup. Depending on conditions, the protoplasm may be either in a **sol** state, with the characteristics of a liquid, or in a **gel** state, tending toward the consistency of gelatin. The colloid may vary between the sol and gel states (Fig. 2-13).

In 1827, the Scottish botanist Robert Brown noticed that, when the cellular contents were in a sol state, the visible particles seemed to be moving around. This phenomenon, called **Brownian motion,** is due to invisible particles', mostly water molecules, striking against the larger granules. Brownian movement also occurs in water and other liquids as well as in gases, and so it is not exclusively a characteristic of life.

Membranes and Transport Phenomena

Both cells and subcellular organelles are surrounded by **membranes.** Passage of materials through these membranes may take place in a variety of ways.

Diffusion is the movement of molecules from an area of higher concentration (more molecules per unit volume) to an area of lower concentration (fewer mol-

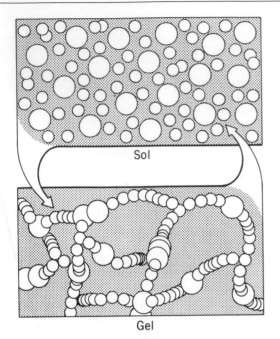

Figure 2-13. Colloidal states. Ultramicroscopic structure of a sol and a gel (diagrammatic). Top: Sol state. The colloidal particles are represented as circles of different diameters and the water particles (molecules) as dots. Such a solution has the physical properties of a liquid. Bottom: Gel state. The colloidal particles adhere together to form a continuous network. Such a substance has the physical properties of a semisolid substance (jellylike) that tends to be elastic. The arrows show that the sol and gel states are reversible under appropriate conditions.

ecules per unit volume). Diffusion is a natural, physical tendency that results from random molecular movements. This phenomenon is easily noticed when one opens a bottle of perfume. The volatile molecules diffuse from the surface of the liquid (where their concentration is greatest) into the air (where their concentration is initially zero) and eventually are detected by olfactory receptors in one's nasal passages. Note that air is **permeable** to the perfume molecules; that is, the perfume molecules can pass through the air relatively unobstructed. Analogously, a cellular

Figure 2-14. Diagram to illustrate ordinary diffusion and osmosis. (A) Ordinary diffusion. The battery jar is divided into two chambers, a and b, by a permeable membrane that offers practically no hindrance to the diffusion of both water and sugar molecules (particles). In ordinary diffusion, any kind of molecule tends to diffuse (move) from where it is more abundant, per volume of space, to where the molecule is less abundant. Diffusion through a permeable membrane continues until every component reaches equal concentration; therefore in A, the water and sugar molecules are of equal concentration on both sides of the permeable membrane. (B) Osmosis. The battery jar is divided into two chambers, a and b, by a semipermeable membrane—that is, one that is permeable to water but hinders the passage of sugar molecules. Under these conditions, water molecules will diffuse through the membrane more rapidly into chamber b than into chamber a. In accordance with the law of diffusion, the water molecules move in greater numbers from the place of higher water molecule concentration to the region of lower water molecule concentration.

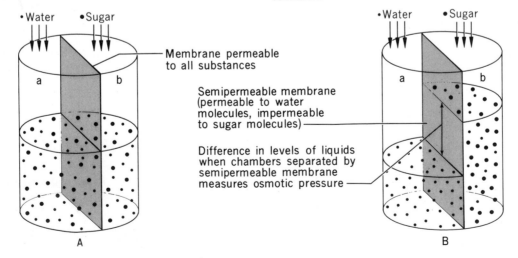

membrane may be permeable to certain substances. The membrane, in this case, does not affect the direction of molecular movement. Molecules move across the membrane until their concentrations reach a dynamic equilibrium on both sides of the membrane. At this point, the concentrations of a particular molecule on both sides of the membrane are equal. Molecules continue to move across the membrane, but, for each entering molecule (into any given side), another leaves, and so there is no net change (Fig. 2-14).

But cellular and subcellular membranes are *not* completely permeable; if they were, any substance could diffuse into the cell, and the cell could not maintain a chemical identity distinct from its environment. Indeed, cells have **semipermeable membranes,** which allow only selected substances free access in and out.

Osmosis is defined as the diffusion of water through a semipermeable membrane (Fig. 2-14). The implications for cells of osmosis are tremendous. If a cell is placed in an **isotonic** solution (in which the concentrations of solutes, the substances dissolved in solution, are equal on both sides of the cell membrane), the cell will remain undisturbed. But, if placed in a **hypotonic** solution (in which the concentration of solutes is greater inside the cell), the cell will swell as the water, but not the solutes, diffuses along its concentration gradient into the cell. Cells frequently burst in hypotonic solutions, hence the extreme danger of intravenous injection of pure water. Cells in hypotonic solution are depicted in Figure 2-15B. The osmotic movement of water into a cell when in a hypotonic solution is known as **deplasmolysis.** The reverse phenomenon, **plasmolysis,** occurs when a cell is placed in a **hypertonic** solution. In this case, the concentration of solutes outside the cell is higher than the concentration within the cell, and so water diffuses from the cell, leaving it shriveled (Fig. 2-15C). The maintenance of a constant internal environment is therefore of vital importance to all multicellular animals.

A membrane may be semipermeable by virtue of minute **pores,** submicroscopic holes that physically limit the size of molecules that may pass through them.* Lipids and lipid-soluble substances diffuse

*These pores are probably channels only for very small molecules such as ions or gases, as the pores constitute only 0.1 percent of the cell membrane.

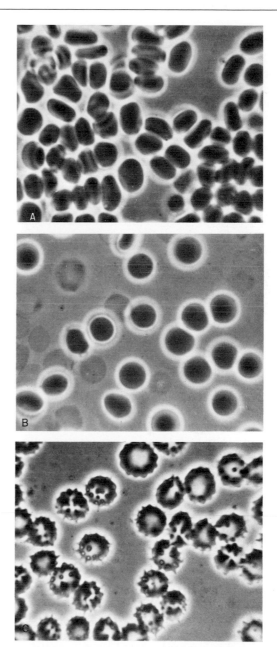

Figure 2-15. Plasmolysis and deplasmolysis. (A) Human erythrocytes in isotonic solution. (B) Erythrocytes in hypotonic solution; fluids move into cell to equalize solute concentration—pressure in cells increases. (C) Erythrocytes in hypertonic solution; fluids move out of cell (plasmolysis) to equalize solute concentration—cells shrink. 2000 ×.

through the membrane quite rapidly, whereas water-soluble substances are restricted; smaller molecules generally pass through cellular membranes more readily than do larger molecules. A further refinement of the semipermeability concept requires the realization that membranes are also **selectively permeable.** A molecule's electrical polarity and other physical characteristics may affect its ability to cross a membrane. This further limits a cell's ability to *passively* obtain nutrition and remove wastes.

A major method by which materials enter and leave cells is **facilitated transport.** As mentioned earlier, even if there is no adverse osmotic concentration gradient, a molecule may be prevented from penetrating a membrane because of its polarity. If a molecule is polar, it may interact with the charges on a polar membrane, and a **carrier** may be necessary. The carrier in this case need not necessarily provide energy for transport across the membrane, but it must neutralize or mask the polarity of the entering molecule. Carrier molecules are primarily proteins embedded in the membrane that migrate back and forth or simply rotate within cell membranes to facilitate the movement of the molecule across the membrane in either direction. The net flow is determined by the concentration gradient.

A cell frequently requires substances that cannot enter by diffusion because of a reverse concentration gradient—a high internal solute concentration prevents more solute molecules from entering. This problem may be circumvented by a process called **active transport,** which moves substances against a concentration gradient. Active transport is not a passive process but involves metabolic work. The energy for this work comes from adenosine triphosphate (ATP), which is the energy source for most cellular functions. As with facilitated transport, the transported molecule is first chemically bound to a carrier molecule, usually a protein specific for that substance, and the carrier then physically escorts the molecule through the membrane. The energy released by hydrolysis of an ATP molecule is converted into the mechanical energy manifested by the carrier's movements, allowing the molecule to be forced across the membrane against a concentration gradient.

Another method by which a cell may obtain nutrition is by **phagocytosis,** the engulfment of large particles. An amoeba, for instance, preys upon smaller unicellular organisms that are too large to diffuse through its cell membrane. When feeding, the amoeba invaginates a portion of its surface membrane next to the prey; the victim is then drawn into the crevice, and the cytoplasm closes over it. Such engulfment produces a **phagosome** (**food vacuole**). This sequence is depicted in Figure 6-7 for an amoeba feeding on the ciliate **Tetrahymena.** Small volumes of impermeable liquids may also taken in or ejected from a cell by a process closely analogous to phagocytosis, called **pinocytosis. Endocytosis** is a general term used to describe both phagocytosis and pinocytosis.

Membrane Structure

The basic function of membranes is the regulation of substances entering and leaving the cell or subcellular organelles that they enclose. The structure of cellular membranes was first elucidated by H. Davson and J. F. Danielli in 1935. Their hypothesis is now known as the **unit membrane model,** named by D. Robertson, who substantiated it in 1959 with the electron microscope. According to Davson and Danielli, a membrane is something like a sandwich, and, indeed, under the electron microscope most stained membranes appear as two parallel dark lines with a clear zone between 75 Å and 100 Å in diameter.* The dark-stained lines are believed to be made of proteins, which constitute about 50 percent of the membrane's bulk. The other 50 percent is made up of lipid molecules, which have both polar and nonpolar ends. According to the unit membrane model, the clear zone of a unit membrane is composed of nonpolar fatty acid side chains, whereas the polar ends of the lipids reside in the outer proteinaceous layers of the membrane (Fig. 2-16).

The unit membrane model has received extensive experimental support, but some studies have cast doubt on its simplicity. For example, when either is applied to certain membranes, the lipid portions are destroyed. Nevertheless, the membrane may still retain its shape, indicating that some nonlipid substances (possible proteins) may exist in the central region of the sandwich. In addition, the application of phospholipidase, an enzyme that hydrolyzes phos-

*One Angstrom (Å) = 10^{-10} m.

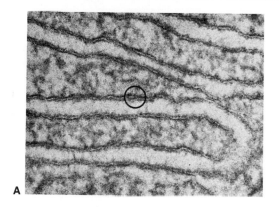

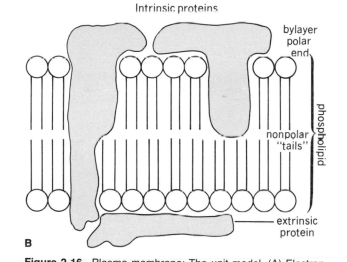

Figure 2-16. Plasma membrane: The unit model. (A) Electron micrograph of plasma membrane (circled). 400,000 ×. [Micrograph courtesy Dr. Sjostrand.] (B) Fluid mosaic unit membrane model. Note how some proteins penetrate by layer (intrinsic) whereas others are external (extrinsic).

pholipids but does not affect proteins, destroys certain cell membranes. This indicates that membranes are not 100 percent coated by a homogenous protein coat, as direct access to the lipid components of the membrane must have been available. To these points has been added the objection that the simple Davson–Danielli model cannot account for selective active transport. These inconsistencies have spurred molecular biologists to seek explanations more consistent with experimental findings.

One line of experiments has shown that a unit membranelike bilayer can be formed artificially and spontaneously by adding phospholipids to a water solution. The nonpolar "tails" are not soluble in water and are called **hydrophobic** (Gr. *hydro*, water + *phobia*, fear); they tend to associate with another to exclude water (the same manner by which oil droplets form in a water–oil emulsion). The polar ends form a boundary near the water, as shown in Figure 2-16. Recent studies on a variety of membranes show that it is the proteins that are highly variable in quality and quantity in different cells. In some, there is an extensive protein framework rather than a smooth outer layer that includes proteins spaning across the entire membrane, connecting to other proteins inside and outside. This explains both the ether and the phospholipidase experimental results. The so-called **fluid mosaic** model of membrane structure reflects this idea of proteins occurring outside, inside, and sometimes within the lipid bilayer matrix (Figure 2-16), the observed relative mobilities of lipid and protein molecules within the bilayer accounts for the "fluid" description of the membrane.

The **plasma membranes** of cells are covered by a complex of carbohydrates as well as by proteins or by **glycoproteins** that contain carbohydrate and protein portions. For example, cells lining the digestive tract are covered with a glycoprotein that protects cells from digestive juices. Cell surface glycoproteins differ among organisms and thus are important "cell signatures" that allow an organism to recognize its own cells but to invade foreign cells via the **immune response** (Chapter 35).

The cells of multicellular animals are usually separated from each other by a nonliving matrix that keeps them 200–300 Å apart. Numerous exceptions are known, however. In some tissues, intercellular spaces may be essentially nonexistent, and adjacent cells are in contact. This kind of a junction between cells is known as a **tight junction;** it is exemplified by the cells of the intestinal lining that control transport into cells by allowing close communication among adjacent cells. Another type of cell junction, called a **desmosome,** strongly binds two cells (e.g., skin cells) by a network of thick fibrils that reach into the cytoplasm of the adjoining cells. A third type of connection is the **gap junction,** which allows direct continuity of the cytoplasm of two cells through small tubular gaps in the bordering membranes. This type of junction occurs in cardiac muscle tissue and is responsible for the synchronous contraction of large numbers of cells.

The Nucleus and Nucleolus

The most conspicuous subcellular organelle is the **nucleus** (Fig. 2-17), which was named by Robert Brown in 1833. All cells of higher plants and animals contain nuclei except mature erythrocytes of mammals. Most animals cells have only a single nucleus, but some contain more than one. In most cases, nuclei are spherical or discoidal, their contents being segregated from the cytoplasm by a porous double unit membrane (Fig. 2-18). Suspended within the **nucleoplasm** is **chromatin** (threadlike strands of deoxyribonucleic acid, DNA, that condense into **chromo-**

somes during mitosis and meiosis; ribonucleic acid, RNA; and protein). The chromosomes contain the **genes,** as noted earlier. In addition, glycoproteins and various hydrolytic enzymes, such as ribonuclease, alkaline phosphatase, and dipeptidase are found within the nucleus.

A subnuclear organelle called the **nucleolus** usually appears as a dark-stained body in microscopic preparations. Nucleoli contain large quantities of RNA and protein. The nucleolus has been shown to be the synthesis site of cytoplasmic *ribosomes* (see The Endoplasmic Reticulum).

The Endoplasmic Reticulum

A system of membranes, called the **endoplasmic reticulum,** was found throughout the cytoplasm by

Figure 2-17. The nucleus. Electron micrograph. 30,000 ×. [Courtesy Dr. K. Porter.]

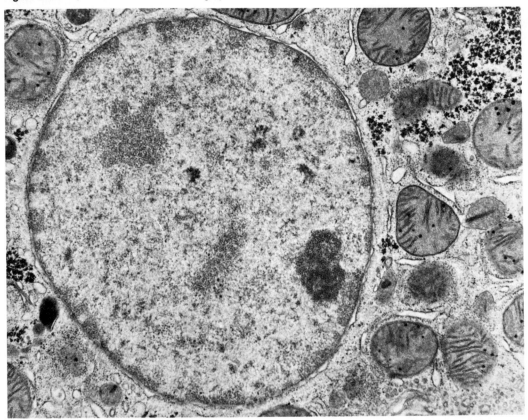

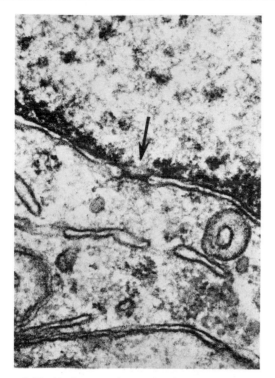

Figure 2-18. Nuclear membrane showing nuclear pore (arrow). The nucleus appears above the membrane. 75,000 ×.

Porter, Claude, and Fullman in 1945 using the electron microscope. The endoplasmic reticulum is essentially a complex system of interconnecting channels made up of unit membranes. The function of the endoplasmic reticulum is to provide a means for the orderly distribution of proteins and other molecules throughout the cell.

Two types of endoplasmic reticula have been identified. **Rough endoplasmic reticulum** appears as a membrane with granules attached along its length (Fig. 2-19). These granules are **ribosomes,** on which protein synthesis occurs (Chapter 38). Rough endoplasmic reticulum is especially prominent in cells where there are high rates of protein synthesis, such as in those of the pancreas that secrete digestive enzymes.

The second type is called **smooth endoplasmic reticulum** (Fig. 2-19). Ribosomes are not attached here. The function of the smooth endoplasmic reticulum appears to encompass the synthesis and transport of

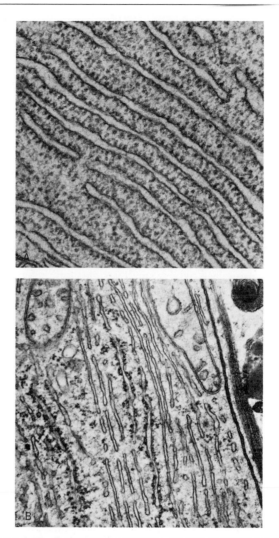

Figure 2-19. Endoplasmic reticulum. (A) Rough ER from bat pancreas. 30,000 ×. [Courtesy Dr. K. Porter.] (B) Smooth ER. 24,000 ×.

lipids and glycogen. Smooth endoplasmic reticulum is especially abundant in liver cells, gland cells, mature leukocytes, and spermatocytes.

Parts of the endoplasmic reticulum may be locally expanded into **vacuoles.** Such cavities, conveniently located along the cytoplasmic channels, may serve as waste dumps and nutrient storage depots within the cell. Interestingly, *Paramecium* and some other freshwater protozoans use vacuoles for osmotic regulation by collecting excess water in vacuoles and emptying them outside the cell. The majority of endoplasmic reticular structures are flattened, and these portions are known as **lamellae,** a general term for a flattened and layered morphological configuration. Endoplasmic reticula exhibit characteristic morphologies in various cell types.

The Golgi Apparatus

The **Golgi apparatus,** or **Golgi bodies,** was discovered in 1898 by the Italian scientist Camillo Golgi. Under the electron microscope Golgi bodies appear as flattened, stacked sacks, with the edges forming a tubular network (Fig. 2-20). Some of the tubes are blind, whereas others interconnect with other Golgi units. The Golgi apparatus consists of two basic areas: the forming face, where small vesicles join to form the flattened stacks, and the maturation face, where vesicles pinch of the stacks. Golgi bodies have been implicated in the synthesis and secretion of polysaccharides and complexes of carbohydrates and proteins. It is thought that substances are transported to the Golgi bodies via smooth endoplasmic reticula. Golgi vesicles have been observed to bud off the main body of the apparatus, move through the cytoplasm to the cell surface, and there release their contents.

Many cytologists believe that endoplasmic reticula maintain a continuity with Golgi bodies, but evidence does not support this theory. For one thing, Golgi bodies have no physical association with the ribosomes on rough endoplasmic reticula. Golgi body membranes are thicker than those of the endoplasmic reticulum, and, as opposed to endoplasmic reticula, there is considerable morphological similarity among species.

Lysosomes

Lysosomes (Fig. 2-21) are believed to be formed from the maturation face of the Golgi apparatus. A lysosome is essentially a bag of enzymes packaged in a Golgi body that can potentially be released into the cytoplasm. The major enzymes in lysosomes are known as lysosomal hydrolases; about 30 such enzymes catalyze the digestion of proteins, nucleic acids, some carbohydrates, and possibly some fats—most substances in the cell. Recall that a cell may engulf a solid food particle by phagocytosis (Figs. 6-6 and 6-7), and the resulting intracellular food packet is called a phagosome. Lysosomes fuse with phagosomes, forming **digestion vacuoles,** and the lysosomal enzymes break down the food material within the confines of the vacuole; nutrient by-products of this

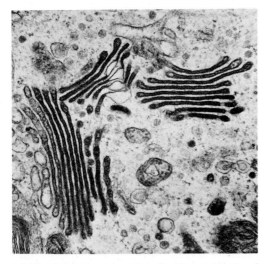

Figure 2-20. Golgi apparatus. Note forming and maturation faces. 11,500 ×. [Courtesy Dr. Alan Bell.]

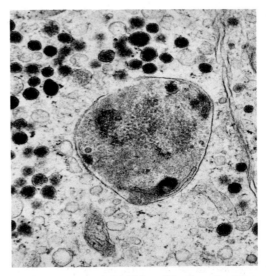

Figure 2-21. Lysosome. 120,000 ×. [Courtesy Dr. Sjostrand.]

digestion are absorbed into the cytoplasm. After digestion is completed, the digestion vacuole moves to the plasma membrane and fuses with it, and waste materials are expelled from the cell.

In the case of white blood cells, it is the lysosomes that actually digest (lyse) the foreign particles that they phagocytose. And during internal fertilization, a lysosome in the sperm head dissolves the membrane surrounding the oocyte (egg), thereby permitting sperm penetration. Certain cells that dissolve bone also utilize lysosomes.

If a cell is starving, **autophagy** may occur: lysosomes break down proteinaceous components of the cell itself, either from the general cytoplasm or sometimes by destroying organelles. Autophagy is a cell's last resort if no other energy-containing substances are available. When a cell dies, its lysosomes burst and release their digestive enzymes into the surrounding cytoplasm, which is the process occurring when we "age" meat: lysosomal breakdown tenderizes by enzymic predigestion of meat.

Lysosomes are involved in several **pathogenic** (disease-causing) and other abnormal conditions. In arthritis, joint inflammation may be aggravated by burst lysosomes. The drug cortisone stabilizes lysosomal membrane, thus making lysosomal enzyme release less likely and reducing inflammation. In the lung, fibers of silicone (in silicosis) and asbestos (in asbestosis) may pierce lysosomes, releasing their enzymes and causing local tissue destruction.

Centrioles, Basal Bodies, Cilia, and Flagella

Near the nucleus is usually found two organelles called **centrioles**, which together are commonly termed the **centrosome** (Fig. 2-22). Centrioles appear as tiny cylinders situated at 90° to each other, forming a "T" or cross. Centrioles usually have radiating fibers called **asters** that extend into the cytoplasm during cell division. Figure 2-22A shows an electron micrographic cross-section of a centriole. Note the nine triples of longitudinal fibers that appear as circles in cross-section. Such fibers are called microtubules and are composed of **tubulin** and other proteins. Centrioles function during animal cell division by anchoring the spindle fibers (also microtubules) that pull the chromosomes apart. (see Cell Division: Mitosis).

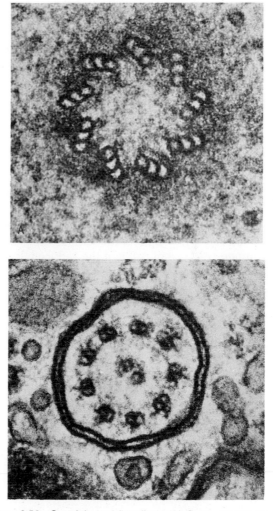

Figure 2-22. Centriole and flagellum. (A) Cross-section of centriole with nine groups of three fibers. 78,000 ×. [Courtesy Dr. Sjostrand.] (B) Cross-section of flagellum. Note 9 + 2 arrangement of fibers. Inner two are not double. 82,000 ×.

Cilia and **flagella** are motile organelles that extend from **basal bodies,** organelles identical to centrioles in structure and found beneath the surfaces of certain cells. The internal morphologies of cilia and flagella consist of nine double pairs of microtubules plus an additional set of two single tubules in the center. The internal morphologies of centrioles and basal bodies are similar, but lack the central microtubules. The major differences between cilia and flagella are that cilia are usually shorter than flagella and that flagella number one to a few per cell whereas many cilia usually adorn ciliated cells. Cilia vary in length from 5 μ to 10 μ, whereas flagella may reach lengths of 150 μ. Experimental studies have shown that cilia and flagella are identical both biochemically and metabolically.

Mitochondria

Mitochondria are the most numerous cellular organelles; a typical animal cell contains hundreds of mitochondria. In very active cells, such as secretory cells and heart muscle, 500 to 1000 mitochondria may be observed; sea urchin eggs contain about 1.5×10^5 mitochondria. They appear in various shapes, depending on the type of cell and the intracellular condition. Mitochondria range in size from 0.5 μ to 3.0 μ in diameter and up to 10 μ in length.

A mitochondrion consists of two membrane systems and is essentially a sac within a sac. The outer membrane is smooth, but the inner membrane is comprised of complex folds, called **cristae,** that are usually perpendicular to the long axis of the mito-

chondrion (Fig. 2-23). The outer membrane conforms to the unit bilayer model. The cristae are usually about 70 Å apart and are composed of subunit membranes with globular lipid–protein aggregates. Each crista membrane is covered with substructures that resemble balls on the ends of filaments (Fig. 2-24). The cristae enclose a matrix material that contains, among other things, ribosomes and DNA.

Mitochondria have been called the powerhouses of the cell because they are the sites of **cellular respiration,** also called **oxidative phosphorylation** because ATP is synthesized as a result of a chain of oxidation–reduction reactions with O_2 as the terminal oxidizer (Fig. 2-11). They account for almost all the molecular oxygen (O_2) consumed in a cell, as well as 90

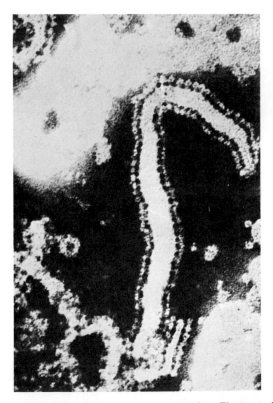

Figure 2-24. Mitochondrial crista cross-section. Electron micrograph exhibiting "ball" structures lining the membrane; these are the proteins on which ATP is made. 230,000 ×. [Courtesy Dr. D. Parsons, *Science* 142 (1963).]

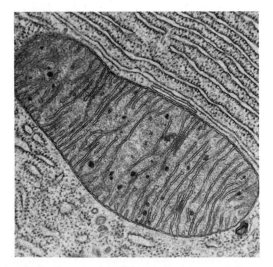

Figure 2-23. The mitochondrian. Note transverse cristae and ribosomes. From bat pancreas. 53,000 ×. [Courtesy Dr. K. Porter.]

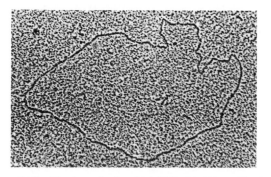

Figure 2-25. Circular mitochondrial DNA. 38,400 ×. The isolation and electron microscopy was done by M. M. K. Nass [*Proc. Nat. Acad. Sci.* 56 (1966):1215].

percent of the ATP formed. The respiratory enzymes are sequestered in an orderly array on the cristae.

Many investigators believe that mitochondria are remnants of once independent organisms, perhaps similar to bacteria, that successfully parasitized other cells and then evolved into a useful permanent organelle. This **endosymbiotic theory** is somewhat speculative, but it is supported by the fact that each mitochondrion has its own circular strand of DNA, whose nucleotide sequences for many genes resemble bacterial genes more than those of the nucleus within the same cell. Figure 2-25 is an electron micrograph of the circular DNA of a mitochondrion. **Endosymbiosis** is more strongly supported for the origin of chloroplasts from symbiotic blue-green algal-like cells.

Mitochondria are useful indicators of cellular trauma: If the cell is damaged, mitochondria may fragment, swell, or degenerate. Mitochondrial disintegration is a sure sign of impending cell death. Mitochondria also appear abnormal in the disease known as scurvy, which is prevented by adequate vitamin C in the diet.

Cell Division: Mitosis

With few exceptions (notably roundworms), animals produce new cells throughout their lives; as old cells die, they are replaced. **Somatic** (i.e., nonreproductive) cells generally grow until they reach a certain size and then divide. Somatic cell division is called **mitosis.** The two new **daughter cells** will proceed to grow and then divide, continuing the process generation after generation. Some cells, however, notably those formed during the development of **oocytes** (egg cells), grow very little during the period between successive divisions. Exactly why cells divide when they do is not known, but the relative quantities of cytoplasm and nucleoplasm are usually maintained in each kind of cell. It has been suggested that the cell divides when this critical ratio of cytoplasm to nucleoplasm is upset.

The process of mitosis can be somewhat arbitrarily subdivided into several stages. This is useful for understanding, but it should be kept in mind that mitosis is a continuous process.

Figure 2-26 illustrates the basic features of the mitotic cycle in animal cells.

Interphase

Interphase is commonly referred to as the "resting stage" between cell divisions, but during this period the cell is by no means at rest. Normal metabolic activities are carried on during interphase, and the cell grows to normal size. Chromosomal **replication** (duplication) also occurs: each chromosome divides into a pair of identical **chromatids** (see Chapter 38).

The chromosomes may be visible as thin, randomly coiled threads (**chromatin**) of DNA, or they may not be visible at all. Also visible are the nucleolus within the nucleus and the centrosome complex outside the nuclear membrane. Thin filaments (the asters) may be seen radiating from the centrioles.

Prophase

Prophase initiates the onset of cell division, sometimes rather abruptly. Prophase may take varying lengths of time, depending on the organism and type of cell. In certain cells of a chick, at 39°C, prophase lasts between 5 and 50 minutes, averaging about 30 minutes.

An animal cell often rounds out somewhat during prophase, and its cytoplasm becomes more viscous. The **chromosomes,** each now composed of a pair of

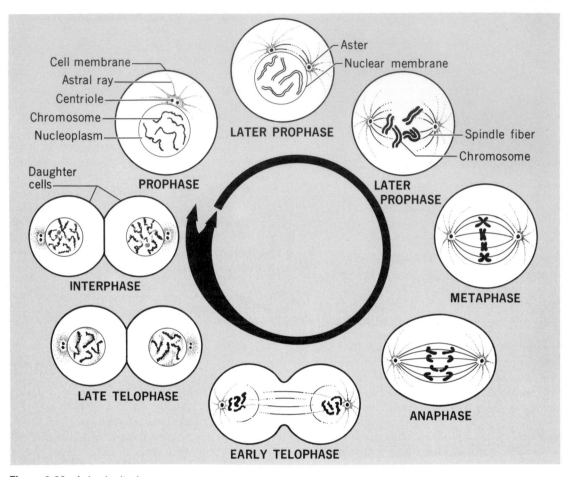

Cell membrane
Astral ray
Centriole
Chromosome
Nucleoplasm

Daughter cells

PROPHASE

LATER PROPHASE

Aster
Nuclear membrane

LATER PROPHASE

Spindle fiber
Chromosome

METAPHASE

INTERPHASE

LATE TELOPHASE

EARLY TELOPHASE

ANAPHASE

Figure 2-26. Animal mitosis.

chromatids with identical genes, are joined by, at a common site, the **centromere.** The chromatids now become visibly distinct as they coil and shorten, with resultant thickening. During early prophase the chromosomes migrate toward the nuclear membrane, which begins to disintegrate. The **centrioles** split apart and begin to migrate from their interphase position near the nucleus to opposite poles of the cell.

Eventually the nuclear membrane completely disintegrates and the nucleoli become less distinct. About this time, the middle of prophase, the centrioles begin spindle formation.

By the end of prophase, the two centrioles have reached opposite poles of the cell. Their spindle fibers appear to reach out toward the chromosomes, which by this time have almost completed their migration to the center of the cell.

Metaphase

Metaphase begins when the chromosomes have reached the equator of the cell, known as the **metaphase plate,** or **equatorial plane.** In chick mesenchyme this stage usually lasts from 2 to 10 minutes. As the chromosomes move into the metaphase orientation, they look as if they are being dragged along by their centromeres. Spindle fibers extend from the centrioles and attach to the centromeres.

Anaphase

This stage is rather short; in chick mesenchyme cells it lasts only 2 or 3 minutes. During **anaphase** the centromeres split, forming a separate chromosome from each chromatid. The spindle fibers appear to shorten and pull the now separated identical chromosomes toward opposite poles of the cell.

Telophase

In chick mesenchyme cells, **telophase** usually takes from 70 to 180 minutes. At this point the chromosomes have reached the poles of the cell. The spindle filaments begin to disintegrate, and **cytokinesis,** or division of the cytoplasm, begins as a cleavage furrow divides the cytoplasm. Nuclei form in each new daughter cell. The centrioles subsequently replicate to renew a pair. The chromosomes begin to uncoil and become less distinct. Nucleoli appear in the daughter nuclei. Cytokinesis continues to completion, and the result is two genetically identical daughter cells, each with about half the volume of the original mother cell. Interphase follows, and the cycle repeats.

Sex cell (**gamete**) division, called **meiosis,** will be discussed in Chapter 37. For now, simply note the following points about meiosis: The mother cells (**spermatogonia** in males and **oogonia** in females) contain the same number of chromosomes as the somatic cells of the organism. But, during meiosis, sperm and eggs are formed that have only *half* the original number of chromosomes. This halved chromosome number in the gametes is termed **haploid,** as opposed to the **diploid** number in somatic cells. When a haploid sperm fertilizes a haploid ovum, a diploid zygote, the one-celled embryo, results and grows into the multicellular embryo by mitosis.

Summary

Various combinations of protons, neutrons, and electrons form atoms. Molecules are made up of two or more atoms linked by chemical bonds, which are most commonly either ionic or covalent. Chemical reactions involve bond breaking (releasing energy) or bond formation. The much weaker hydrogen bonds link H_2O molecules in water and ice and also aid in the structures of macromolecules.

Acids involve hydrogen ion loss; bases involve hydroxide ion loss; salts are ionically bonded anions and cations.

Water, carbon dioxide, molecular oxygen, and ammonia are small molecules that are intimately involved in cellular metabolism. Large organic molecules include lipids, carbohydrates, nucleic acids, and proteins.

Enzymes are biological catalysts and are composed of proteins with vitamin coenzymes and metallic cofactors. Enzymes may be reused continuously.

Metabolic reactions include transfer, synthetic, hydrolytic, or redox reactions.

Cells, the basic units of life, contain a set of subcellular organelles, each performing a specific function that contributes to the maintenance of the cell. Typical organelles include membranes, nucleus, nucleoli, endoplasmic reticulum, Golgi apparatus, lysosomes, centrioles, basal bodies, cilia, flagella, and mitochondria.

Review Questions

1. How many microns are in a millimeter?
2. Define "osmosis."
3. Which subcellular organelle is most easily likened to a freeway?
4. Facilitated transport is necessary to overcome what type of barrier?
5. What is a phagosome?
6. What is the normal ratio of carbon to hydrogen to oxygen atoms in a carbohydrate molecule?
7. If a hemoglobin molecule is composed of 574 amino acid subunits, how many peptide linkages are there?
8. What is the animal equivalent of plant starch?
9. Is the formation of protein from amino acids an anabolic or catabolic process? Name the bonds made in protein synthesis.
10. Comparing cilia to flagella, which type would displace

more fluid (assuming, of course, that they were in a liquid medium)?

11. Adenosine triphosphate molecules contain large amounts of potential energy. What organelle produces most of the cellular ATP?

12. What subcellular organelle is described by the phrase "bag of enzymes?"

13. Predict the atomic weight of one possible isotope of an element with atomic weight of 35.

14. When chromosomes have reached the center line of the cell, what mitotic stage has begun? How much time is there between the end of this stage and the beginning of the next?

15. "Amphoteric" means capable of acting as either an acid or as a base. If an amphoteric compound were placed in a neutral solution and it acted as a base, would the pH of the solution become higher or lower than neutrality?

16. Membranes are composed mostly of which two substances?

17. What is the function of urea?

18. Define "activation energy" and predict what happens when there isn't enough of it.

19. Define and give an example of isotonic, hypotonic, and hypertonic solutions.

20. Explain the different ways in which a cell may obtain nutrition.

21. Knowing something about hydrogen bonds and the chemistry of water, explain why ice floats.

Selected References

Altman. J. *Organic Foundations of Animal Behavior*. New York: Holt, Rinehart and Winston, Inc., 1966.

Baer, A. W., W. E. Hazen, D. L. Jameson, and W. C. Sloan. *Central Concepts of Biology*. New York: Macmillan Co., 1971.

Balinsky, B. I. *An Introduction to Embryology*, 3rd ed. Philadelphia: W. B. Saunders Co., 1970.

Bloom, W., and D. W. Fawcett. *Textbook of Histology*. Philadelphia: W. B. Saunders Co., 1962.

Bonner, J. T. *Cells and Societies*. Princeton: Princeton University Press, 1955.

Bourne, G. H. *Cytology and Cell Physiology*, 3rd ed. New York: Oxford University Press, 1964.

Butler, J. A. V. *Life of the Cell*. New York: Basic Books, 1964.

Chance, B., and D. F. Parsons. "Cytochrome Function in Relation to Inner Membrane Structure of Mitochondria," *Science*, **142:** 1176, 1963.

CRM Books. *Biology Today*, 3rd ed. Del Mar, California: Communications Research Machines, 1980.

De Robertis, E. D. P., W. W. Nowinski, and F. A. Saez. *Cell Biology*, 4th ed. New York: Oxford University Press, 1965.

Gordon, M. S. *Animal Physiology*. New York: Macmillan Co., 1972.

Hokin, L. E., and M. R. Hokin. "The Chemistry of Cell Membranes," *Scientific American* 213:78, 1965.

Karlson, P. *Introduction to Modern Biochemistry*. New York: Academic Press, 1965.

Leake, L. D. Comparative Histology: an introduction to the microscopic structure of animals. New York, Academic Press, 1975.

Lehninger, A. L. *Biochemistry*. New York: Worth Publishers, 1970.

Loewy, A. G., and P. Siekevitz. *Cell Structure and Function*. New York: Holt, Rinehart and Winston, Inc., 1963.

Mazia, D., and A. Tyler (eds.). *The Physiology of Cell Specialization*. New York: McGraw-Hill Book Co., Inc., 1963.

Morrison, J. H. *Functional Organelles*. New York: Reinhold Publishing Corp, 1966.

Nason, A., and R. L. De Haan. *The Biological World*. New York: John Wiley & Sons, Inc., 1973.

Nelson, G. E., G. G. Robinson, and R. A. Boolootian. *Fundamental Concepts of Biology*, 2nd ed. New York: John Wiley & Sons, Inc., 1970.

Porter, K. R., and M. A. Bonneville. *An Introduction to the Fine Structure of Cells and Tissues*. Philadelphia: Lea & Febiger, 1963.

Ramsey, J. A. and V. B. Wigglesworth (eds.). *The Cell and the Organism*. Cambridge: Cambridge University Press, 1961.

Rhodin, J. A. G. *Histology: A Text and Atlas*. New York: Oxford University Press, 1974.

Swanson, C. P. *The Cell*, 2nd ed. Englewood Cliffs: Prentice-Hall, Inc., 1964.

Watson, J. D. *Molecular Biology of the Gene*, 3rd ed., Menlo Park, Calif.: W. A. Benjamin, 1976.

Welch, U., and V. Starch. *Comparative Animal Cytology and Histology*. Seattle: University of Washington Press, 1976.

Wilson, G. B., and J. H. Morrison. *Cytology*. New York: Reinhold Publishing Corp., 1966.

Histology and Organology

A major principle of evolutionary theory is that animals with complex morphologies and physiologies have evolved from less complex ancestors over hundreds of millions of years. The evolution of life is considered to have encompassed chemical, biochemical, and biological phases. Microfossils of unicellular organisms similar to modern bacteria have been discovered by microscopic examination of sedimentary rocks that are believed to be over 3 billion years old. The remains of multicellular animals do not become abundant in the fossil record until 600 million years ago. The evolution of multicellular organization required much time, but its advantages are manifested by the diversity of the present-day animal kingdom.

Introduction to Histology

Unicellular organization has some potential disadvantages. As explained in Chapter 2, a cell communicates with its external environment exclusively through its plasma membrane. There is a limit to the amount of nutrients or wastes that the membrane is able to handle. If it cannot take in sufficient nutrients, the cell will cease growing or die. If toxic wastes cannot be expelled from the cell fast enough, their accumulation may also lead to cell death. A given area of cell membrane must be capable of servicing a given volume of protoplasm. The greater the cell volume, the greater the amounts of the nutrients required and wastes produced and, therefore, the greater the required surface area of plasma membrane.

The surface area of a sphere (a good approximation of general cellular shape) increases as the *square* of its radius ($A = 4\pi r^2$). Its volume, however, increases as the *cube* of the radius ($V = 4\pi r^3/3$). That is, volume increases much more rapidly than does surface area (Fig. 3-1). Consequently, as a cell grows, its capacity for transporting materials across its membrane eventually reaches a limit at which its volume becomes too great for its membrane's surface area to adequately supply. The cell, by dividing, forms two cells with much more favorable ratios of surface area to volume.

Some protozoans are **colonial** in organization. Simple colonies are aggregates of virtually identical cells, with little interdependence or specialization of form and function. Colonies represent a transitional stage between unicellular and true multicellular organization. Colonial organization led to a new innovation—layers of cells. Cells began to differentiate, each performing a specific cooperative function in addition to general metabolism. Groups of cells performing a similar function are called **tissues.** A tissue may consist of either one type of cell or of more than one type. The Porifera (sponges) are the simplest animals that possess weakly defined tissue layers.

Though constantly undergoing renovation, the general morphological pattern of a particular tissue remains about the same. Old cells die and are replaced. Tissue physiology is dynamic; under hormonal, enzymatic, and vitamin control, a steady flux of substances is produced and passed through an organism's tissues.

Histology, the study of tissues, distinguishes between several kinds of tissues. This chapter presents an overview of the major types, including epithelial, connective, muscular, and nervous tissues.

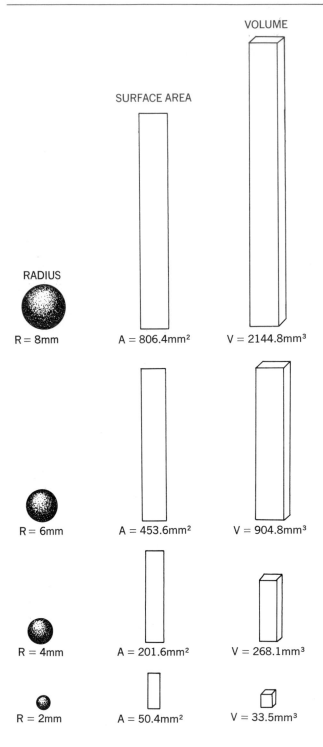

VOLUME

SURFACE AREA

RADIUS

R = 8mm   A = 806.4mm²   V = 2144.8mm³

R = 6mm   A = 453.6mm²   V = 904.8mm³

R = 4mm   A = 201.6mm²   V = 268.1mm³

R = 2mm   A = 50.4mm²   V = 33.5mm³

Figure 3-1. Interrelationships of radius, surface area, and volume of various spheres. Note that volume increases more rapidly than does surface area for a given increase in radius. Drawn to scale.

Epithelial Tissue

Tissues that cover internal or external surfaces of the body are known as **epithelia** (Figs. 3-2 and 3-7). Epithelial tissues may serve several functions, including protection, lubrication, absorption, secretion, and excretion. Most epithelia are packed closely together, with one surface free and the other lying on vascular connective tissue, such as the cells facing the **lumen** (cavity) of the intestine. Epithelial tissues are often attached to a supportive **basement membrane,** an extracellular secretion of the underlying connective tissue.

Several types of epithelium are classified on the basis of number of layers and cell morphology. **Simple epithelium** consists of only one layer of cells (unilaminar). Many invertebrate epithelia are of the simple variety; they appear as somewhat irregular tiles on a floor. In humans, **squamous epithelium** occurs in the endothelium (inner lining) of blood vessels and in the peritoneum (the membrane lining the body cavity or coelom).

Cuboidal epithelium, with cells approximating the shape of a cube, is usually found in glands and ducts. The kidney contains much cuboidal epithelium. **Columnar epithelium** is found in the stomach and intestines; columnar cells resemble pillars. The outer layer of terrestrial vertebrate skin is frequently flat and hard *(cornified)* and is derived from underlying columnar cells. *What purpose might such cornification serve?*

Epithelial tissues that are composed of more than one layer of cells are called **stratified epithelia.** Stratification commonly occurs in the skin, sweat glands, and urethra. **Pseudostratified** epithelium appears to

Figure 3-2. Epithelia. (A) Simple squamous, from inside of cornea. Arrow indicates the epithelial layer. 320 ×. (B) Simple cuboidal, from kidney. 800 ×. (C) Columnar, from villus in the intestine. 2000 ×. (D) Stratified squamous, from human skin. 250 ×. (E) Pseudostratified columnar, from trachea. 800 ×. (F) Stratified squamous and cuboidal, from outer surface of cornea. 500 ×.

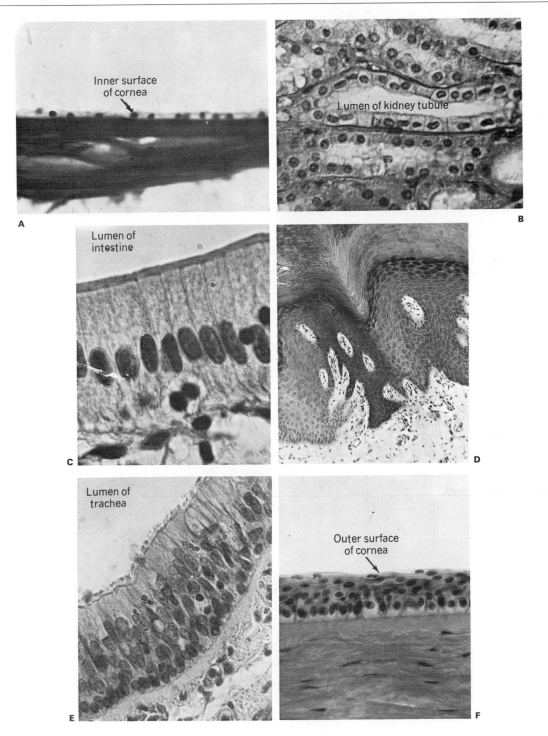

Inner surface of cornea

Lumen of kidney tubule

A

B

Lumen of intestine

C

D

Lumen of trachea

Outer surface of cornea

E

F

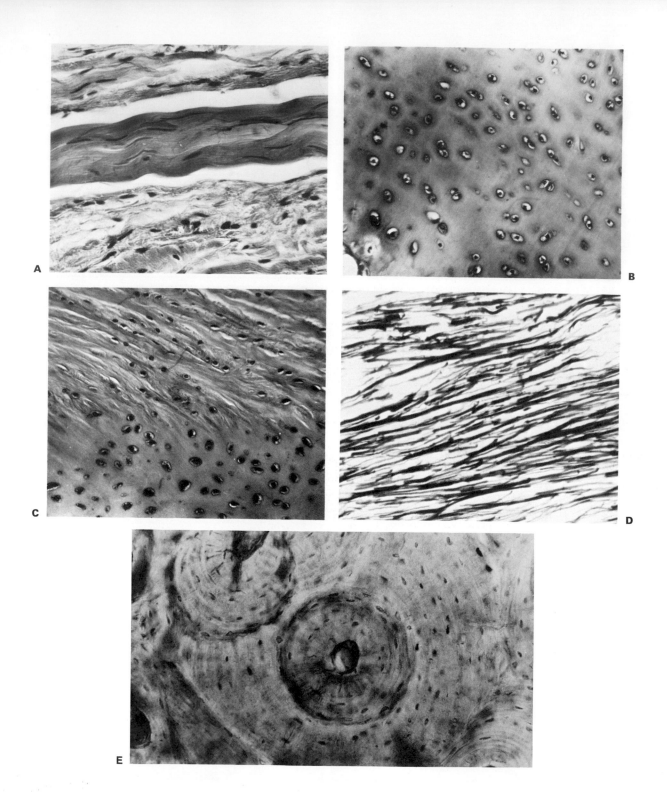

consist of several layer of overlapping cells with variable heights. This tissue lines the trachea and the ducts of some glands. **Stratified squamous epithelium** is found in human epideris and lines the nasal cavities and mouth.

The free surfaces of epithelial cells may be ciliated or flagellated. **Cuboidal ciliated epithelium** has been found in the sperm ducts of earthworms. Epithelial cilia beat in synchrony and can thus propel fluids past their surfaces. **Columnar ciliated epithelium** is found, among other places, in the air passages of terrestrial vertebrates. **Flagellated epithelial cells** are found in the digestive cavities of cnidarians (e.g., jellyfish).

Epithelial tissues may contain receptors of the sensory system; some secrete mucus and other complex materials.

Connective (Supportive) Tissue

Connective tissues (Fig. 3-3) are supporting and uniting structures that are found throughout an animal's body. Their main functions are to form rigid structures capable of resisting shocks and pressures of various kinds and to bind together various parts of the body.

Unlike epithelium, connective tissue consists predominantly of extracellular material with relatively few cells. Nonliving **fibers** are embedded in an extracellular matrix, both of which are manufactured by the scattered connective tissue cells. The matrix varies in consistency from loose to gelatin like; the density and type of fibers make some connective tissues soft and rubbery, whereas others are hard and tough. Connective tissue has more varied forms than any of the other general tissue types. Connective tissue can be classified in several ways, based on the qualities of the intercellular material.

Fibrous connective tissue may be of three types. **White (collagenous) fibers** are found lying parallel to each other in bundles. The bundles may be crossed or

Figure 3-3. Connective (supportive) tissues. (A) Tendon tissue. 500 ×. (B) Hyaline cartilage. 320 ×. (C) Fibrocartilage, from monkey heel. 320 ×. (D) Ligament, note elastin. 500 ×. (E) Cross-section of bone tissue showing Haversion canal system. 160 ×.

interlaced, but are not branched. **Tendons**, which connect muscles to bone, are composed of collagenous fibers.

Yellow (elastic) fibers, which may be either branched or bent, are found in the walls of some of the larger blood vessels; they also bind organs to one another, such as the skin to underlying muscles. Both elastic and collagenous fibers occur in the dermis and intestines of vertebrates. **Ligaments,** which link bone to bone, are similar to tendons except that elastic fibers are more abundant; ligaments are consequently more elastic than tendons. **Reticular (branched) fibers** form fine intercellular networks that hold many other types of tissues together and are particularly evident in the liver, spleen, and lymphatic tissues.

Other supporting connective tissues include the **cartilages. Simple cartilage** is made of a firm extracellular matrix with elastic properties known as **chondrin;** the living connective tissue cells lie within particular spaces termed **lacunae.** The chondrin is enveloped by a thin fibrous membrane, the **perichondrium,** which supplies the nutritional needs of the underlying cells.

The type of cartilage that covers joint surfaces and rib ends is known as **hyaline cartilage.** This is the skeletal cartilage of all vertebrate embryos and is also found in adult sharks and rays. It has considerable elasticity and appears bluish-white and somewhat translucent. This homogenous connective tissue also occurs in the human nose and trachea. **Elastic cartilage** is more flexible than hyaline. Yellowish in color, it contains many elastic fibers that branch in all directions. Elastic cartilage is found in the external ears and Eustachian tubes of mammals. The strongest type of cartilage is **fibrocartilage.** This contains thick, compact, parallel bundles of collagenous fibers, with very few cells and little matrix. Fibrocartilage appears at areas of stress; for example, joints that experience strain, such as the pubic symphysis of the pelvis and between the vertebrate.

Bone, or osseous tissue, is found only in the skeletons of bony fish and higher vertebrates (see Chapter 33). Bone is composed of about 45 percent calcium salts $[Ca_3(PO_4)_2 \cdot CaCO_3]$ and other inorganic salts, 30 percent collagen, and 25 percent water. Unlike cartilage, bone has a direct blood supply and lacunae connected by minute passageways called **canaliculi.** In

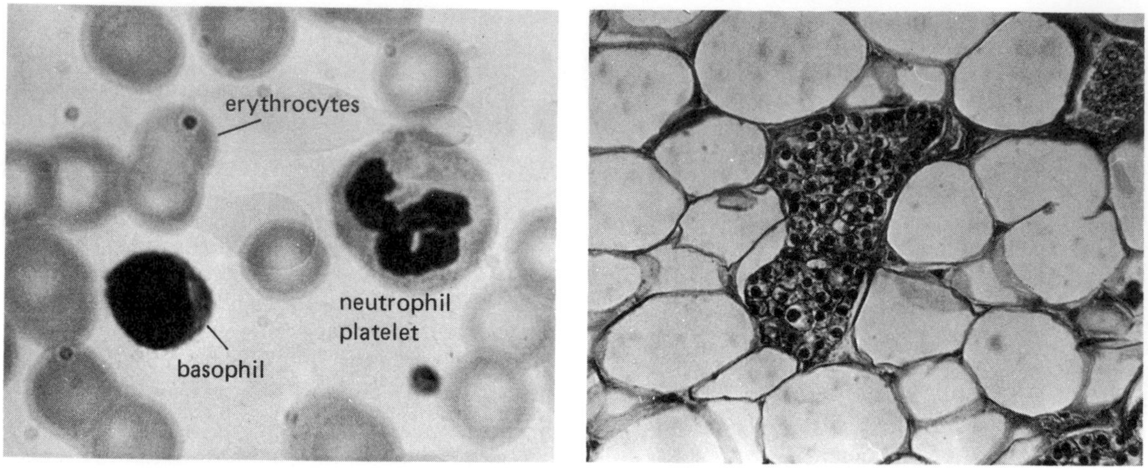

Figure 3-4. Vascular and adipose tissue. Left: Human blood vascular tissue. Note the variety of cell types. 2000 ×. Right: White adipose tissue (light-colored spheridial bodies); from parathyroid gland. 800 ×.

compact bone these are arranged in characteristic concentric patterns, called **Haversian systems** (Fig. 3-3). The outside of bone is covered by a thin, tough, fibrous membrane called the **periosteum.**

Vascular connective tissue, or **blood,** is a circulating fluid composed of white blood cells (**leukocytes**), red blood **corpuscles** (**erythrocytes**), blood **platelets** (**thrombocytes**), and liquid **plasma** (Fig. 3-4). Blood plasma transports the blood cells, nutrients, and wastes to and from the cells of the body; erythrocytes transport oxygen and carbon dioxide; and leukocytes engulf bacteria and other foreign particles that invade the body. The platelets are small particles important to the clotting mechanism. **Interstitial fluid** arises from blood by diffusion of plasma through the walls of capillaries into tissue spaces, where it bathes the body's cells. **Lymph** consists of interstitial fluid and leukocytes that have entered the lymphatic vessels for eventual return to the bloodstream (see Fig. 3-8).

Adipose connective tissue is mainly composed of fat cells. It is used for energy storage and insulation. The polygonal or rounded cells enclose fat droplets and a nonspherical nucleus. There are two main types of adipose tissue. **White adipose tissue** makes up most of the body fat of an animal; actually, "white" adipose tissue is rather yellowish. **Brown fat tissue** is much less abundant than white, being found only in restricted locations within most organisms, and it is most abundant in hibernating animals. Brown adipose tissue is extremely thermogenic; that is, its catabolism produces heat that helps arouse the hibernating animal.

Muscular (Contractile) Tissue

Muscle tissue (Fig. 3-5) is composed of cells specialized for **contraction** (i.e., shortening). Muscles are the most abundant tissues in most animals. Muscle tissue is composed of elongated cells, called **muscle fibers,** that are able to contract with great force when stimulated by nerve impulses (see Chapter 32). The fiber cells contain an unspecialized cytoplasm, or **sarcoplasm,** that surrounds contractile filaments called **myofibrils.** In some muscle fibers, the myofibrils line up parallel to one another, giving the muscle a banded, or striated, appearance.

Smooth muscle tissue has no such striations. Smooth muscle tissue is found in the walls of the viscera and arterial blood vessels. Smooth muscle cells are rather small and spindle shaped, and there is only one nucleus in each fiber. Smooth muscles are not under direct conscious control.

Two major types of **striated muscle tissue** are recog-

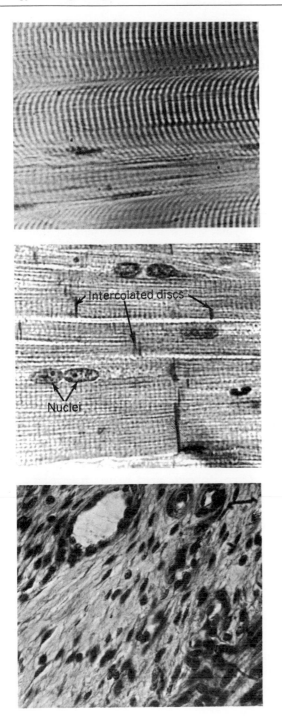

nized. **Cardiac muscle tissue** is found in the heart. These cells are characterized by a multinucleate condition (several nuclei per cell), with the nuclei embedded deep within the sarcoplasm. Cardiac fibers form a branching network. Dark-staining areas known as **intercalated discs** occur where the fibers abut one another. Cardiac muscle is also not under direct conscious control.

Skeletal (striated) muscles are under conscious, involuntary, and reflexive control—all exerted through the nervous system. Striated skeletal muscle is the most abundant muscle tissue in vertebrates, comprising about 40 percent of human body weight. As in the case of cardiac muscle, skeletal muscle fibers are multinucleate, but their nuclei are located near the surface and no intercalated discs are present.

Nervous Tissue

Nervous tissue, composed of cells called **neurons,** is specialized for conductivity and irritability. A typical neuron consists of a **cell body** (or **soma**), a single **axon** conducting impulses away from the cell body, and many **dendrites** conducting impulses toward the cell body (Fig. 3-6). In most animals, the soma are restricted to the brain and spinal cord, known as the **central nervous system** and peripheral **ganglia,** which are groups of cell bodies usually located outside the central nervous system, except in some invertebrates. The axons and dendrites extend to the receptors—muscles and glands—in the extremities. Neurons may also be linked together by means of their dendrites and axons at **synapses** (or junctions) where nerve impulses are chemically transmitted from neuron to neuron, or from neuron to muscle for axons of **motor neurons** (Fig. 3-6).

Three major types of neurons are recognized. **Sensory neurons** relay information from external and internal **receptors** to integration centers in the central nervous system. Motor neurons transmit instructions to **effectors** (muscles or glands), stimulating

Figure 3-5. Muscular (contractile) tissues. Top: Striated, voluntary. 384 ×. Middle: Cardiac. 320 ×. Bottom: Smooth, involuntary 240 ×.

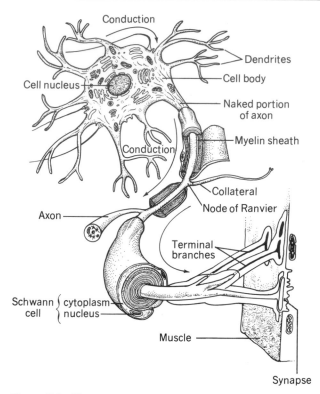

Figure 3-6. Motor neuron structure, showing the spiral structure of the myelin sheath that is formed from the membrane of a Schwann cell that wraps around the axon as it grows.

them to perform their functions. **Association neurons** or **interneurons** may occur between any two (or more) neurons.

Some neurons are extensively coated with an insulating, fatty layer, the **myelin sheath,** that is formed from membranes of a **Schwann cell.** Nonmyelinated neurons, which lack myelin, are covered by a thin **neurilemma,** a delicate, tubelike sheath of cells that envelop the nerve fiber. Myelinated neurons conduct nerve impulses faster than nonmyelinated fibers, and velocity is enhanced by periodic **nodes of Ranvier** (gaps in the myelin coating). The disease **multiple sclerosis** involves a progressive destruction of the myelin sheaths, resulting in loss of muscular control.

In **nerves,** which are essentially bundles of nerve fibers, the neurons are separated by supporting **neuroglia** and nutritive **glia** cells.

Introduction to Organology

Just as a tissue is an aggregate of cells that performs a specialized function, an **organ** is a group of tissues cooperating in the performance of a common and often intricate function. Most organs are composed of more than one type of tissue. For example, the intestine is covered and lined by epithelia, it has a framework of connective tissue, its activities are coordinated by nervous tissue, and vascular tissue serves all the intestinal tissues and provides for the transport of nutrients absorbed by the intestine (Fig. 3-7).

Generally, no matter how many types of tissue may compose an organ, it is the predominant tissue, the **parenchyma,** that is most important for that organ's function. The tissues of secondary importance, the **stroma,** are frequently of a supportive nature. For instance, in the pancreas, the glandular epithelia that secrete digestive juices are the parenchyma, whereas connective and other tissues form the stroma.

Several organs may work together to perform a major bodily function. Ten major **organ systems** are generally recognized. An overview follows, covering the circulatory and lymphatic, respiratory, digestive, endocrine, nervous, skeletal, reproductive, muscular, execcretory, and integumentary systems. Human biology is stressed here, but more detailed information on other animals will be presented in subsequent chapters.

The Circulatory and Lymphatic Systems

These organ systems transport nutrients and wastes throughout the animal body. The medium for the **circulatory system** is generally **blood,** which helps to maintain a stable internal fluid environment for all the body's cells functions. Blood carries oxygen from the **lungs** to the tissues, carbon dioxide from tissues to the lungs, food substances to the tissues, **hormones** and other internal secretions to various parts of the body, and metabolic wastes to the excretory organs; it also helps to maintain a constant temperature in "warm-blooded" (**homeothermic**) animals. Blood is generally pumped from a **heart** through **arteries** to thin-walled **capillaries** and then back through **veins** to the heart (Fig. 3-8).

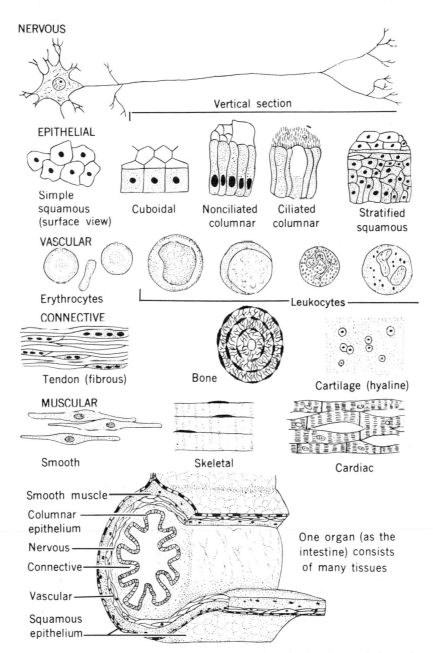

NERVOUS

Vertical section

EPITHELIAL

Simple squamous (surface view)

Cuboidal

Nonciliated columnar

Ciliated columnar

Stratified squamous

VASCULAR

Erythrocytes

Leukocytes

CONNECTIVE

Tendon (fibrous)

Bone

Cartilage (hyaline)

MUSCULAR

Smooth

Skeletal

Cardiac

Smooth muscle

Columnar epithelium

Nervous

Connective

Vascular

Squamous epithelium

One organ (as the intestine) consists of many tissues

Figure 3-7. Types of tissue. One organ (such as the intestine) consists of many tissues that complement to form the functional whole.

This most pervasive organ system ramifies throughout the organism. The body's cells are rarely more than three or four cells away from a life-sustaining capillary. The exchange of nutrients and wastes between the blood and interstitial fluid occurs by diffusion through the thin capillary walls.

Some of the interstitial fluid that diffuses out of the capillaries is not immediately replaced. This **lymph**

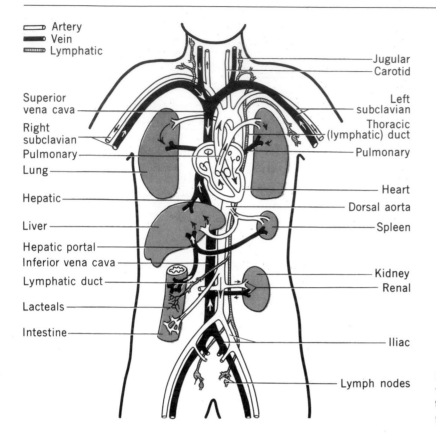

Artery
Vein
Lymphatic

Jugular
Carotid
Superior vena cava
Left subclavian
Right subclavian
Thoracic (lymphatic) duct
Pulmonary
Pulmonary
Lung
Hepatic
Heart
Dorsal aorta
Liver
Spleen
Hepatic portal
Inferior vena cava
Lymphatic duct
Kidney
Renal
Lacteals
Intestine
Iliac
Lymph nodes

Figure 3-8. The larger blood and lymphatic vessels together with some of the organs that they serve. The lymph flows in only one general direction, toward the heart.

is eventually returned to the blood after passing through a series of **lymphatic vessels** to enlarged junctions, the **lymph nodes,** where foreign bacteria are removed by **leukocytes** and **antibody**-producing leucocytes are activated.

The Respiratory System

Metabolic activities require oxygen, and the **respiratory system** functions to deliver oxygen to the body's cells; it also removes carbon dioxide, a by-product of cellular metabolism, from the body. In humans, air enters the respiratory system by way of the **nostrils** and **mouth** and then passes through the **larynx, trachea,** and **bronchial tubes** into the **lungs** (Fig. 3-9). Here *external respiration* occurs: the blood gains about 5 percent oxygen and loses about 4 percent carbon dioxide. The circulatory system then transports

oxygen throughout the body. *Internal respiration* involves the exchange of oxygen and carbon dioxide between the capillaries and tissues.

The respiratory system exhibits a multitude of forms, as do many organ systems, in different animals. Many terrestrial animals have lungs, whereas aquatic animals employ gills or simply utilize their general body surface for respiratory gas exchange. Insects acquire oxygen-rich air through small slits, called **spiracles,** that open into **tracheal tubes** that branch extensively throughout the internal tissues.

The Digestive System

The chief functions of the digestive system (Fig. 3-10) are the **ingestion, digestion,** and **absorption** of food and the **elimination** of solid wastes. Within the mouth, the **teeth,** assisted by the **tongue, masticate**

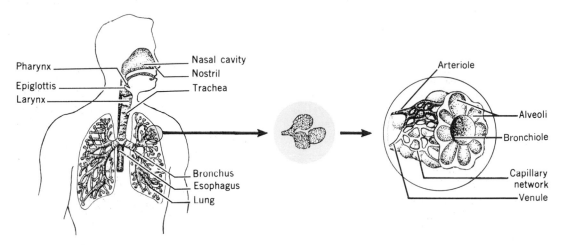

Figure 3-9. The lungs of man, each enclosed in a double-walled pleural sac. The right lung is cut open to show the internal details.

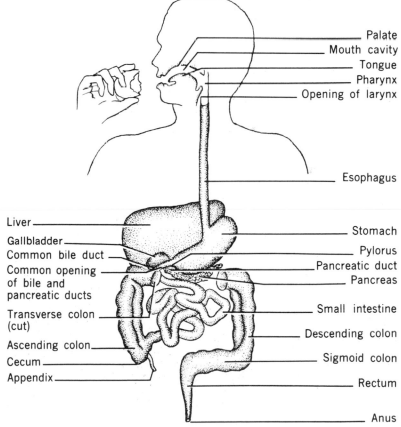

Figure 3-10. Diagrammatic human digestive system.

(chew) food. **Salivary glands** secrete **saliva** that lubricates and begins to digest the food. The masticated food passes through the **pharynx** and **esophagus** into the **stomach**, where **gastric juices** are added. Here the food undergoes further mechanical and chemical breakdown. Digestion continues in the **small intestine,** where digestive enzymes are released by the intestinal wall and from the **pancreas; bile** from the **liver** also empties into the small intestine. Absorption of nutrients occurs mainly through the wall of the small intestine, whereas both limited digestion and substantial water absorption continue in the **large intestine (colon).** Undigested wastes are eliminated from the end of the colon, or **rectum,** through the **anus.**

The Endocrine System

This system is composed of ductless glands that secrete **hormones** into the bloodstream. Hormones are chemical messengers that regulate the metabolic activities of specific **target organs** that may be in distant parts of the body.

The **endocrine system** is one of the most complex systems in the body. Different types of animals often have different sets of **endocrine glands.** The major

Figure 3-11. Left: A diagram to show the approximate location of some of the endocrine glands in man. Although the pineal body and thymus are included, they are not known definitely to be organs of internal secretion. Right: Section of a pituitary gland showing lobes and microscopic structure of the pancreas.

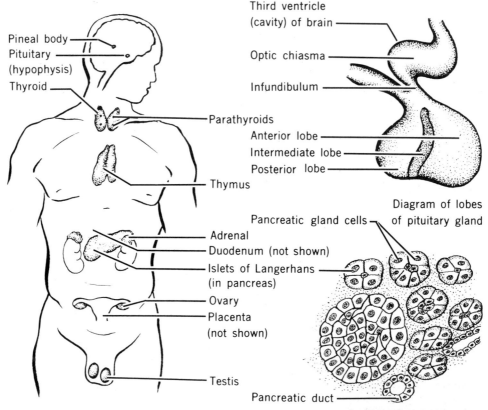

endocrine glands in humans, indicated in Figure 3-11, include the **pituitary, thyroid, parathyroid, adrenal,** parts of the **ovaries** and **testes,** and various **digestive glands.** The endocrine system responds to environmental fluctuations less rapidly and in a more general fashion than does the nervous system.

The Nervous System

The complex activities of an animal require a high degree of coordination, which is the function of the **nervous system.** It coordinates the different parts of the body, exerts control over the internal organs, and is responsible for human thought processes. The **central nervous system** consists of the **brain** and **spinal cord.** The **peripheral nervous system** is that part of the nervous system outside the central nervous system and includes receptors in **sense organs** and nerves connecting the central nervous system with the muscles and glands of the body. Nervous control and regulation is under both voluntary and involuntary control.

The Skeletal System

This system serves for both support and muscle attachment. It varies dramatically between different types of animals (Fig. 3-12). Vertebrates have an **endoskeleton** of bone and/or cartilage contained within the animal. Many invertebrate animals are surrounded by a hard **exoskeleton** that holds the organism together and provides protection. Sponges (Porifera) are unique in having skeletons composed of loose conglomerations of mineral **spicules** and/or elastic **spongin.**

The Reproductive System

This system concerns itself with the perpetuation of the species. The essential reproductive organs are the **ovaries,** in which **eggs,** (**ova**) develop, and the **testes,** in which **sperm** are formed. Collectively these organs are termed **gonads.** Accessory reproductive organs include those that supply yolk and various secretions, ducts that convey the sperm or eggs, and copulatory organs that ensure fertilization. Some animals are **hermaphroditic,** containing both sets of sex organs, but self-fertilization is rare.

The Muscular System

Motion and locomotion are the main functions of the **muscular system. Skeletal, smooth,** and **cardiac muscles** make up the muscular system. Many animals contain literally thousands of muscles. Skeletal muscles move the limbs and other body parts, cardiac muscles pump blood throughout the body, and smooth muscles are responsible for moving materials through the digestive tract, bladder, certain glands, and even shunting blood through the capillary beds, based on the needs of various tissues.

Figure 3-12. Skeletal systems. Left: Endoskeleton of *Felis domesticus* (common cat). Right: Exoskeleton of a representative of order Coleoptera (beetle). See also Figure 10-4 for skeletal spicules.

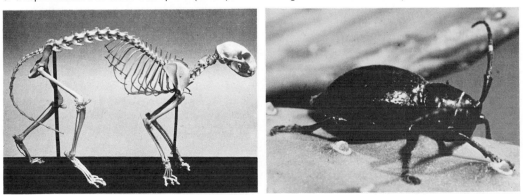

The Excretory System

This system concerns itself with eliminating the waste products of metabolism, especially nitrogenous wastes from protein metabolism. The **kidneys** extract these wastes from the blood. **Urine,** consisting largely of water and urea, passes from the kidneys through **ureters** into the **bladder,** where it is temporarily stored before being eliminated through the **urethra.** **Egestion** of undigested materials and excretion of metabolic wastes is also a secondary function of the digestive system (solid feces), lungs (carbon dioxide), and **sweat glands** (soluble wastes) in the skin. Invertebrate animals exhibit a wide variety of excretory systems.

The Integumentary System

This system consists of the **skin** and its associated structures, which include **hair, nails, claws, scales,** and **feathers.** The skin's function is primarily for protection, but skin also functions in excretion and temperature regulation (through sweat) and in reception of a number of external stimuli (touch, pressure, temperature, and pain).

General Considerations of Tissues and Organ Systems

The organ systems are interdependent. Distinguishing them as separate systems is merely a convenience in beginning to understand their interrelations. Perhaps the best way to understand these systems is to think of them as comprising one system: the **organism.**

A tissue depends on its constituent cells. An organ is made up of one or many tissues, all of which function cooperatively. In large multicellular organisms, no organ can function without nourishment via the circulatory system. The circulatory system, in turn, depends upon the muscular system, especially the heart and arterial muscles. Contraction of muscle cells is regulated by the nervous system. For all these tissues to function, energy must be available, and the animal must obtain that energy through the digestion of food. The respiratory system is indispensable because cellular metabolism requires oxygen and elimination of carbon dioxide. The excretory system removes the potentially toxic wastes of cellular metabolism.

The endocrine system controls growth and development and exerts continuous control over most major metabolic activities. The skeletal system is necessary not only for support, but also for the production of red blood cells (manufactured in the bone marrow). The integumentary system protects against injuries, dehydration, and bacterial invasion.

Perhaps the only system unnecessary for an individual's survival is the reproductive system, but its necessity lies in the continuation of the species.

Of course, much of what has been summarized here has no direct application to unicellular animals. Protozoans have no organs; rather they have organelles, as explained in Chapters 2 and 5. Nevertheless, the functions of subcellular organelles correspond roughly to those of the organ systems of a multicellular animal just considered: for example, digestive system—lysome, integumentary system—plasma membrane. *What other examples can you think of?*

Finally, it should be emphasized that extreme variability exists in the anatomies of the different animal groups. Even among individuals of the same species, there are variations in the exact arrangement of blood vessels and nerves, and even the number of bones in the skeleton.

Summary

Histology is the study of tissues. A group of cells cooperating in the performance of a particular function forms a tissue.

Epithelial tissues cover surfaces and appear in various forms (simple, cuboidal, columnar, stratified, pseudostratified, ciliated, flagellated, etc.).

Muscular tissues are specialized for contraction and may be smooth or striated, voluntary or involuntary.

Connective tissue includes the supportive and binding fibrous connective tissues, tendons, ligaments, cartilage, and bone. These consist mostly of various types and configurations of extracellular fibers. Vascular connective tissue is fluid: blood, interstitial fluid, and lymph transport nutrients, wastes, and hormones. Adipose (fat) tissue functions as an energy storage depot within the body.

Nervous tissue, composed of neurons, conducts electrical impulses throughout the animal. Sensory neurons, motor neurons, and interneurons make up the central and peripheral divisions of the nervous system.

Organs, consisting of various types of tissues, perform more complex functions. Only an overview of organs and organ systems has been presented in this chapter. More details will be discussed throughout the text.

Review Questions

1. Would a large or small ratio of surface area to volume be more advantageous to a cell?
2. Distinguish between an organ and a tissue.
3. Can epithelium be both pseudostratified and ciliated?
4. Every capillary has a precapillary sphincter; this circular muscle can contract, thereby closing the lumen (cavity) of the capillary. What type of muscle tissue would you expect to find in such a sphincter? What are the reasons for your decision?
5. Parenchyma is
   (a) a specialized organelle of nervous tissue.
   (b) the functionally most important tissue in an organ.
   (c) an endocrine gland.
   (d) a gelatinous material found in some connective tissues.
6. Do jellyfish have osseous tissue?
7. What type of connective tissue would you expect to find between your ulna (largest bone in forearm) and your humerus (bone between forearm and shoulder)?
8. A lymph node is considered part of the:
   (a) digestive system.
   (b) circulatory system.
   (c) endocrine system.
   (d) nervous system.
9. What is the insect equivalent of human nostrils?
10. Describe the function of the endocrine system.

Selected References

Bloom, W., and D. W. Fawcett. *A Textbook of Histology*, 9th ed. Philadelphia: W. B. Saunders Company, 1968.

Bonner, J. T. *Cells and Societies.* Princeton: Princeton University Press, 1955.

Bourne, G. H. *Cytology and Cell Physiology*, 3rd ed. New York: Oxford University Press, 1964.

Butler, J. A. V. *Life of the Cell.* New York: Basic Books, 1964.

DeRobertis, E. D. P., W. W. Nowinski, and F. A. Saez. *Cell Biology*, 4th ed. New York: Oxford University Press, 1965.

Galigher, A. E., and E. N. Kozloff. *Essentials of Practical Microtechnique.*, 2nd ed., Philadelphia: Lea & Febiger, 1971.

Gardner. E. *Fundamentals of Neurology*, 5th ed. Philadelphia: W. B. Saunders Company, 1968.

Gray, P. *The Use of the Microscope.* New York: McGraw-Hill Book Company, Inc., 1967.

Hickman, C. P. *Integrated Principles of Zoology*, 4th ed. St. Louis: C. V. Mosby Company, 1970.

Hokin, L. E., and M. R. Hokin. "The Chemistry of Cell Membranes," *Scientific American*, **213**:78, 1965.

Loewy, A. G., and P. Siekevitz. *Cell Structure and Function.* New York: Holt, Rinehart and Winston, Inc., 1963.

Mazia, D., and A. Tyler (eds.). *The Physiology of Cell Specialization.* New York: McGraw-Hill Book Company, Inc., 1963.

Porter, K. R., and M. A. Bonneville. *An Introduction to the Fine Structure of Cells and Tissues.* Philadelphia: Lea & Febiger, 1963.

Ramsey, J. A., and V. B. Wigglesworth (eds.). *The Cell and the Organism.* Cambridge: Cambridge University Press, 1961.

Slayter, E. M. *Optical Methods in Biology.* New York: Wiley—Interscience, 1970.

Storer, T. I., R. L. Usinger, and J. W. Nybakken. *Elements of Zoology.* New York: McGraw-Hill Book Company, 1968.

Swanson, C. P. *The Cell*, 2nd ed. Englewood Cliffs: Prentice-Hall, Inc., 1964.

Wyckoff, W. G. *The World of the Electron Microscope.* New Haven: Yale University Press, 1958.

Diversity and Adaptation

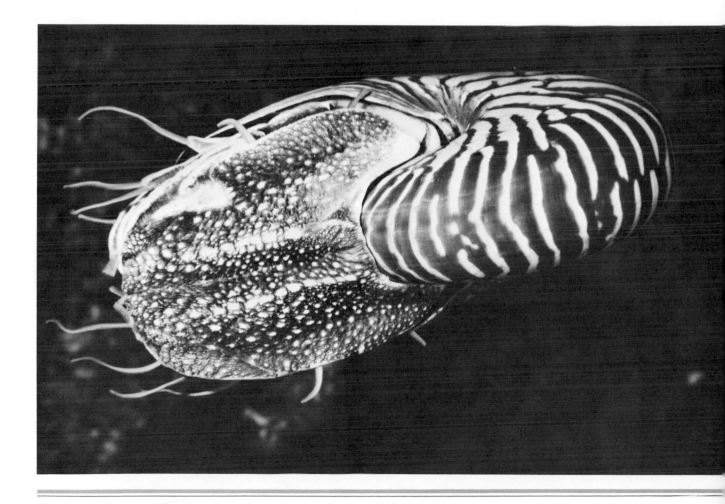

Taxonomy and Synopsis of Animal Phyla

Taxonomy is defined as any system of classification of organisms, usually by means of their resemblances and differences according to established rules. The practice of classification is common to most all cultures, its objective being to bring order to the seeming chaos of the natural world. For instance, the Papuan aborigines of New Guinea devised ways of distinguishing 137 of the 138 types of birds that share their mountainous habitat. Careful observation is the first step in designing any taxonomy. The Papuans are especially interested in the habits of birds because these animals are important parts of their diet.

Artificial taxonomies group animals according to arbitrary and superficial resemblances in structure, color, habitat, and the like; for example, "aquatic" animals live in water, "terrestrial" animals live on land. Animals may also be classified according to diet: carnivores eat flesh, herbivores eat plants, and omnivores consume both animals and plant tissues. Such artificial classifications are sometimes useful, but their categories are too broad for scientific use. Modern biology tries to construct a **natural taxonomy,** one that classifies living organisms in a way that reflects their evolutionary relationships; this evolutionary classification process is known as **systematics.**

History of Scientific Taxonomy

Perhaps, one of the first persons to collect and organize an animal classification system was the Greek philosopher Aristotle (384–322 B.C.), who characterized animals according to their way of living, their actions, their habits, and their bodily parts. Aristotle classified about 500 types of animals into 11 categories according to the complexity of their structural form and their degree of development at birth. Aristotle arranged the animal world in a **hierarchy,** a graded series in which he ranked different categories one above the other. One of Aristotle's pupils, the eminent botanist Theophrastus, arranged some 500 types of plants in a similar manner.

These remarkable advances in scientific thought were largely forgotten until the scientific renaissance of the sixteenth and seventeenth centuries. By then, the sheer numbers of animal and plant types that were being recognized (some 70,000 by 1800) dictated a systematic arrangement. Two naturalists contributed significantly to this task: an Englishman, John Ray (1627–1705), and a Swede, Carolus Linnaeus (1707–1778). Ray classified some 18,600 species of plants according to general differences in anatomical organization. He then did the same for animals, arranging them into six classes: quadrupeds, birds, amphibians, fish, insects, and worms. A century later, Linnaeus devised a method the main features of which are still in use today. During his lifetime, Linnaeus personally classified some 18,000 species of plants. He also classified animals, minerals, and diseases. His *Systema Naturae*, first published in 1737, was continually expanded during his lifetime. The tenth edition (1758) classifies 4,162 animal

species. Linnaeus grouped animals that most closely resembled one another into a species, species resembling one another into a **genus,** similar genera into an **order,** and like orders into a **class.** By the tenth edition, Linnaeus consistently assigned a double name —the name of the genus and the name of the species itself—to each animal, for example, *Felis domesticus* (the house cat), *Felis leo* (the lion), *Felis tigris* (the tiger). This **binomial nomenclature** is still in use today.

The taxonomies of Linnaeus and his immediate successors were mainly descriptions of local floras and faunas. Approximately 100 years later, the publication of Charles Darwin's *Origin of Species* (1859) crystallized the concept of organic evolution. Emphasis has since shifted from simply cataloging species to attempts at understanding their relationships. Another major refinement of taxonomic science was introduced in 1866 by Ernst Haekel, who represented *phylogeny* (the evolutionary history of a group of species or higher categories) by means of branching tree-like diagrams. Thus, from that time on, most taxonomists were also *systematists,* although it must be recognized that a working taxonomy is still very useful to biology, even if it does not reflect true evolutionary relationships. This is especially true for simpler and more ancient organisms, such as protozoans and many invertebrates; there exists a very useful taxonomy of these species, but a comprehensive phylogeny is still subject to controversy and uncertainty.

Purpose

More than one million species of animals have been described in the zoological literature, and every year about 10,000 more species are identified. Imagine the task of gathering information about these species if one had to sort through literally tons of unclassified references!

The purpose of taxonomy is to provide a useful, convenient, and universal system into which all observations about a given organism or group of organisms can be compiled. The system is hierarchical, and for it to function, several prerequisites must be met. A taxonomy must be able to discriminate among different types of organisms; it must provide criteria for such discriminations; and it must have the capacity for grouping more specific *taxa* (singular, *taxon*) into more inclusive taxa. A taxon is a hierarchical position (*e.g.*, species, genus, family, etc.) within a taxonomy.

The criteria for taxonomic discriminations constitute the first part of this chapter. Then an overview of the major types of animals will be presented.

The Taxonomic System

Biological taxonomy is based on the principle of evolution and represents the efforts of many scientists to express the genetic relationships among living organisms. The taxonomic system employs a hierarchy of taxa, the base of which is the

SPECIES. One or more similar species are grouped into a
GENUS. One or more similar genera form a
FAMILY. One or more similar families form an
ORDER. One or more similar orders form a
CLASS. One or more similar classes form a
PHYLUM. One of more similar phyla form a
KINGDOM. All kingdoms taken together comprise the living world.

A species represents the smallest taxon, although subgroups, such as races, are also used. Note that there are more species than there are genera, more genera than families, and so on. The hierarchy takes the form of a pyramid, with higher taxa containing progressively greater numbers of species.

The most meaningful definition of a *species* is *a group of potentially freely interbreeding organisms that are genetically distinct and reproductively isolated from other such groups.* The members of a species form a reproductively isolated group; their genes are able to recombine continually and be sexual reproductive within the group, but they are reproductively isolated from outsiders. As a consequence, their genetic compositions, their anatomies, their physiologies, and their behaviors are more similar to each other than to those of members of different species.

The next higher taxon is the *genus*. A genus *includes one or more species with an assumed common*

Table 4-1. Classification Divisions

Kingdom
Subkingdom
Branch
Grade
Division
Subdivision
Superphylum
Phylum
Subphylum
Superclass
Class
Subclass
Infraclass (Cohort) (Series)
Superorder
Order
Suborder
Section
Superfamily
Family *(-idae) (-oidea)*
Subfamily *(inae)*
Tribe *(-ini)*
Supergenus
Genus
Subgenus
Superspecies
Species
Subspecies
Variety
Form or Race

phylogenetic origin as expressed by many shared characteristics; this set of species is distinct from all other genera.

Genera are subsets of their respective **families.** Similar families compose an **order,** similar orders a **class,** and similar classes a **phylum.*** As we progress further in the taxonomic sequence, the categories become more and more inclusive and, therefore, more general. A phylum is *a group of animals (or plants) that exhibits a common body plan.*

The six major taxa may be subdivided to more finely distinguish among related groups (Table 4-1). Also, sometimes a species can form a genus, family,

---

* The following mnemonic may prove useful in remembering the sequence of taxons: **S**everal **G**ray **F**oxes **O**verload the **C**ourts **P**ending **K**ickoff.

order, and even class of its own. For example, in the protozoan phylum Sarcomastigophora, there are three superclasses, one of which (Opalinata) contains only one order (Opalinida).

A Word of Caution

Keep in mind that all taxonomies are simply conceptual aids to understand the relationships among living organisms. As such, they are dynamic and are constantly being updated and revised as new knowledge of living and fossil organisms reveals the finer branches of evolutionary history. Therefore a truly universal, totally acceptable, taxonomy does not exist. Biologists sometimes differ in their interpretations of the major and minor events of evolutionary history, and so several different taxonomic names for a particular taxon may be supported by different segments of the scientific community. The taxonomic scheme presented in this text probably receives the widest consensus among zoologists at this time (Table 4-2 and classification appendix).

Taxonomic inconsistencies are especially prevalent within the higher taxa. Some biologists favor simplified taxonomies and so tend to group several taxa together. Others insist that such consolidations conceal too much information. But extremely subdivided taxonomies lose much of their usefulness as indices. The different nomenclatures that result from this conflict of opinion between "lumpers" and "splitters" reflect the different ways in which taxonomies can be used as tools for understanding.

Inconsistencies within lower taxa generally result from other causes. For one thing, a species is occasionally misnamed when it is first identified. Another reason for taxonomic confusion is that animals may differ on very fine points. Similar characteristics may in fact be merely **analogous** (serving similar functions) and not **homologous** (reflecting evolutionary descent from a common ancestor). Such distinctions can become highly subjective in the absence of critical data.

Rules of Nomenclature

An International Commission on Zoological Nomenclature has served since 1901 to establish rules by which animal species are named. The adopted no-

Table 4-2. Classification Scheme

Kingdom Monera (procaryotes)

Phylum Schizophyta, or Schizomycetes (bacteria)
Phylum Cyanophyta, or Myxophyta (blue-green algae)

Kingdom Protista (protists)

Phylum Chrysophyta (Golden algae and diatoms)
Phylum Xanthophyta (Yellow-green algae)
Phylum Hypochytridiomycota (hyphochytrids)
Phylum Plasmodiophoromycota (plasmodiophores)
Phylum Sarcomastigophora (amoebas, flagellates)
Phylum Ciliophora (ciliates)
Phylum Sporozoa (sporozoans)
Phylum Cnidospora (cnidosporidians)

Kingdom Plantae (plants)

Phylum Rhodophyta (red algae)
Phylum Phaeophyta (brown algae)
Phylum Chlorophyta (green algae)
Phylum Charophyta (stoneworts)
Phylum Bryophyta (liverworts, hornworts, and mosses)
Phylum Psilophyta (psilophytes)
Phylum Lycopodophyta (club mosses)
Phylum Arthrophyta (horsetails)
Phylum Pterophyta (ferns)
Phylum Cycadophyta (cycads)
Phylum Coniferophyta (conifers)
Phylum Anthophyta (flowering plants)

Kingdom Fungi (fungi)

Phylum Myxomycophyta (slime molds)
Phylum Eumycophyta (true fungi)

Kingdom Animalia (Animals)

Phylum Mesozoa (mesozoans)
Phylum Porifera (sponges)
Phylum Cnidaria (coelenterates)
Phylum Ctenophora (comb jellies)
Phylum Platyhelminthes (flatworms)
Phylum Nemertina (ribbon worms)
Phylum Rotifera (rotifers)
Phylum Gastrotricha (gastrotichs)
Phylum Kinorhyncha (kinorynchs)
Phylum Nematoda (roundworms)
Phylum Nematomorpha (horsehair worms)
Phylum Acanthocephala (spiny-headed worms)
Phylum Entoprocta (peudocoelomate polyzoans)
Phylum Gnathostomulida (gnathostomulidans)
Phylum Ectoprocta (sea mosses, or moss animals)

Table 4-2. (Continued)

Phylum Brachiopoda (lamp shells)
Phylum Phoronida (phoronid worms)
Phylum Mollusca (mollusks)
Phylum Annelida (segmented worms)
Phylum Sipunculida (peanut worms)
Phylum Echiurida (spoon worms)
Phylum Priapulida (priapulidans)
Phylum Onychophora (onychophorans)
Phylum Tardigrada (water bears)
Phylum Pentastomida (tongue worms)
Phylum Arthropoda (arthropods)
Phylum Chaetognatha (arrow worms)
Phylum Pogonophora (beard worms)
Phylum Echinodermata (echinoderms)
Phylum Hemichordata (acorn worms)
Phylum Chordata (chordates)

menclature is the binomial system in Linnaeus's *Systema Naturae* of 1758 (tenth edition).

A species is referred to by citing both its generic and specific names; both names are latinized and are written in italics, with the first letter of the generic name capitalized. No two species can be assigned the same combination of generic and specific names.

The original specimen(s) from which a new species is described is called the **type specimen.** A representative species is similarly declared for each genus, and, likewise, representatives for higher taxa are assigned. The **law of priority** governs the naming of species. After a species has been initially named, and its description published, any subsequent identification and naming usually takes second place. However, it sometimes happens that a species is initially named in some zoological journal and then accidently renamed in a well-read journal. Zoologists use the second name and become used to it until someone runs across the original name. The original is supposed to be used, but many zoologists will stick to the more familiar name. In such cases the law of priority is subservient to the **law of continuity.** Hence, throughout the literature there may be several **synonyms** for members of the same species. Also, different investigators working on the same group of animals may arrive at independent conclusions concerning their evolutionary status, and as a result their classification schemes will be more or less synonymous.

Zoological family names are formed by adding -idae to the stem of the name of the type genus; subfamily names end with the suffix -inae. More details of zoological taxonomy can be found in Mayr's *Methods and Principles of Systemic Zoology* (see Selected References).

One commonly refers to a particular organism by using only the binomial nomenclature, as in *Canis familiaris* (the dog). Sometimes, if the genus about which one is writing has been identified recently in the text, an organism may be referred to by simply indicating the first letter of the genus followed by the entire specific name, for instance, *C. familiaris*.

Sometimes a species may be subdivided into two or more **subspecies**. Subspecies contain members that resemble each other in certain characteristics and differ somewhat from other populations of the same species. Reproductive isolation between subspecific populations is generally incomplete, arising primarily from geographical isolation, but some subspecies will eventually evolve into distinct species. A **trinomial nomenclature** is employed to describe subspecies. For example, two very similar types of robin are found in the southern and eastern parts of the United States. The eastern type is called *Turdus migratorius migratorius;* the southern type is called *Turdus migratorius achrustera*, the subspecies name *achrustera* meaning "duller color."

A rather complete classification of a species of crayfish follows. It should not be memorized, as it merely serves as an example of the complexity of taxonomy.

Kingdom:	Animalia
Grade:	Bilateria
Division:	Eucoelomata
Subdivision:	Schizocoela
Phylum:	Arthropoda
Subphylum:	Mandibulata
Class:	Crustacea
Subclass:	Malacostraca
Series:	Eumalacostraca
Superorder:	Eucarida
Order:	Decapoda
Suborder:	Reptantia
Section:	Macrura
Superfamily:	Nephropsoidea
Family:	Astacidae
Subfamily:	Cambarinae
Genus:	*Cambarus*
Species:	*bartoni*

Discriminating Characteristics

Animals are generally considered to be closely related if they share particular traits in common. A multitude of characteristics must be considered in determining taxonomic relationships. The most important characteristics used to discriminate zoological taxa are those that follow.

Homology and Analogy

Analogous structures perform similar functions. For example, butterfly and bird wings are analogous as they both function in flight. The anatomies of butterfly wings and bird wings are very different, however. Insect wings are supported by a framework of tubular thickenings, the tracheal air passageways, whereas vertebrate wings contain an internal framework of bones.

Homologous traits have similar embryonic developmental patterns. For example, the forelimbs of a bird and those of a bat are homologous; that is, the embryonic development of these vertebrate structures is similar. Homologous structures need not perform similar (analogous) functions, however. A bird wing, human arm, and cow leg are all embryologically homologous, yet they do not serve analogous functions.

Homology is an important basis for uncovering evolutionary relationships. Analogy is not, as it often reflects **convergent evolution**—similar adaptations by different types of animals to similar environmental demands.

Symmetry

Symmetry refers to the arrangement of parts in relation to planes and central axes—the regularity of form. The body plan of a symmetrical animal can be subdivided by one or more planes into halves with essentially equivalent geometric designs. *Most animals have symmetrical body plans.*

Asymmetry

The amoeba is an example of an **asymmetrical** organism; no plane will consistently divide it into equal parts. The amoeba's body has no definite form or static arrangement of parts. It lacks symmetry just as a

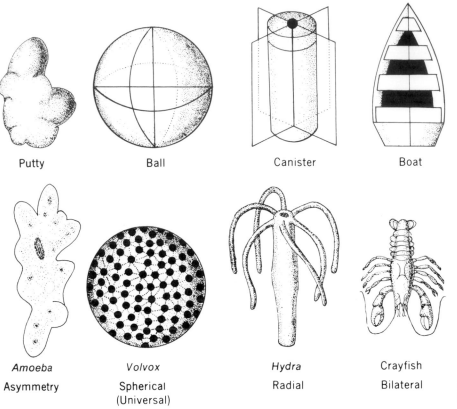

Putty   Ball   Canister   Boat

Amoeba   *Volvox*   *Hydra*   Crayfish
Asymmetry   Spherical   Radial   Bilateral   **Figure 4-1.** Patterns of symmetry
                   (Universal)   in animals.

lump of putty does (Fig. 4-1). Many other protozoans and most sponges are asymmetrical.

Spherical Symmetry

Spherical symmetry takes the shape of a ball, with every point on the animal's surface roughly equidistant from an imaginary center point. Such a body can be divided into similar halves by a cut through the center in any direction. Most spherical animals are free floating, such as some planktonic protozoans (e.g., radiolarians).

Radial Symmetry

A **radially symmetrical** animal is cylindrical in form and may be divided into a number of similar halves only around a central **longitudinal axis.** A longitudinal axis is an imaginary line extending from one pole of an animal to another pole at the opposite end. It is possible to draw a number of planes through such an axis that divide the body of a radially symmetrical animal into equal parts (called **antimeres**). A cylindrical soup can is a simple example of radial symmetry. (Fig. 4-1).

Some simple sponges and the majority of cnidarians (hydras, jellyfishes, sea anemones, etc.) and adult echinoderms (starfishes, sea urchins, etc.) are radically symmetrical. Radial symmetry seems best suited to **sessile** (attached or at least immobile) life-styles, as their similar antimeres enable these animals to obtain food or repel enemies from all sides of the longitudinal axis.

A variation of radial symmetry occurs when two mutually perpendicular planes divide a **biradially symmetrical** animal into symmetrical halves. The first plane passes through the longitudinal and **sagittal axes,** and the other is set crosswise (transverse) to the first. Note that each plane produces a different set of body halves. Biradial symmetry may be found in ctenophores (comb jellies) and in some sea anemones, which are best suited to a floating existence.

Bilateral Symmetry

Animals with **bilateral symmetry** generally have their main organs arranged in pairs on either side of the sagittal axis—in the plane extending longitudinally (anteroposteriorly) and dorsoventrally in the midline. The sagittal axis in the only plane that will divide the body of a bilaterally symmetrical animal into equal halves. A dorsal surface and a ventral surface are recognized, as are right and left sides. Bilateral symmetry is a general characteristic of motile animals, including all vertebrates and most invertebrates (Fig. 4-1). Bilateral symmetry favors locomotion and the differentiation of a head end.

It is doubtful that perfect symmetry is to be found anywhere in the animal kingdom. Animals said to show spherical symmetry usually only approach a spherical form. Traces of bilateral symmetry occur in many radially symmetrical animals. Although the human body is considered a good example of bilateral symmetry, the right and left sides are not identical. The human heart, for instance, is located to the left of the midline. Nevertheless, the concept of symmetry is of great importance in the classification of animals, as will become evident when the different phyla are studied.

Metamerism

Metameric animals have segmentally arranged bodies, composed of more or less similar parts, or with organs arranged in a linear series along the longitudinal axis. Each part is called a **metamere, somite,** or **segment.** The earthworm provides a good example of metamerism. Its body is made up of a great number of very similar segments each contain segmentally arranged chambers of the body cavity, blood vessels, muscles, nerve ganglia, and excretory organs (Fig. 16-2).

The earthworm exhibits **homonomous segmentation,** as it metameres are structurally and functionally very similar to one another. Each metamere acts as a semi-independent unit (having its own excretory organs, for example); however, the metameres are se-quentially connected and cannot function in isolation.

The crayfish, on the other hand, exhibits **heteronomous segmentation,** as the metameres in different regions of its body are dissimilar (Fig. 17-3). Heteronomous segmentation occurs in three phyla: Annelida, Arthropoda, and Chordata. Metamerism is not always apparent in adult external structures. This is true of most vertebrates, which nevertheless have metamerically arranged internal vertebrae, ribs, and nerves.

Cephalization and Polarity

Animals with bilateral symmetry and a few others are **polar;** that is, they have an anteroposterior differentiation of body parts. They have a definite anterior end, called the **oral** end because it bears the mouth and a posterior, or **aboral** end. A concentration of nervous tissue (brain and sense organs) at the oral end distinguishes the **head;** this condition is known as **cephalization.**

Cephalization and polarity are adaptations that represent the most efficient means of sampling most environments. As animals increased in complexity, they tended to move unidirectionally. It is advantageous for a motile animal to have its major sense organs and neural integration centers concentrated at its anterior end (the pole that meets the oncoming environment first).

Internal Body Cavities

The presence or absence of a body cavity is another means of distinguishing animal groups and tracing their evolutionary histories.

Acoelomate animals (such as flatworms) do not have an internal body cavity other than the lumen of the digestive tract; the space between their body wall and visceral organs is filled with mesenchymal tissue, a type of undifferentiated tissue consisting of scattered cells and extracellular material.

Most animals have an internal body cavity, of which two types are distinguished by their origins and constructions. A **pseudocoel** is an internal body cavity that is derived from the cavity (blastocoel) contained within the embryonic blastula (see Chapter 39). A pseudocoel represents simply a dispersion of mesenchyme and is not lined with a **peritoneal membrane.**

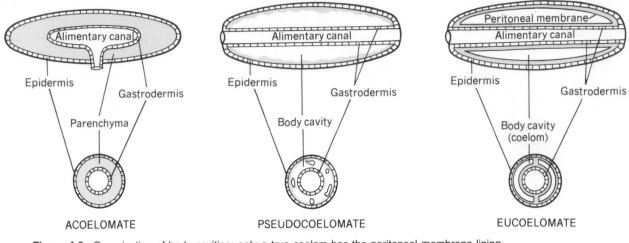

Figure 4-2. Organization of body cavities; only a true coelom has the peritoneal membrane lining.

Nematodes (roundworms) and rotifers are examples of pseudocoelomate animals.

A **coelom** forms later in development, after the disappearance of the blastocoel, and represents a true cavity within the mesoderm that is lined by a cellular peritoneum. The coelomate condition characterizes most animal phyla, including the largest and most diverse groups. Our own coelom is subdivided by the diaphragm into the thoracic and abdominal cavities. The manner of formation of the coelom during development is an important criterion used to classify invertebrates. Figure 4-2 compares the variations in body cavities.

An internal body cavity affords several advantages to an organism; for example, it can accommodate elongated and winding organ systems. Such cavities are filled with fluid, and the circulation of this coelomic fluid can assist the internal transport of nutrients, metabolic wastes, and respiratory gases. Coelomic fluid is also very important as a **hydrostatic skeleton,** supporting the body and assisting in locomotion.

Other Means of Distinction

Cell Number

Cell number is a quick indication of gross similarity or difference. Kingdoms Monera (bacteria and blue-green algae) and Protista (protozoans) are composed exclusively of unicellular organisms. Multicellular animals are called **metazoans.** Unicellular and multicellular organisms usually employ rather different methods to carry out their life processes. For example, whereas the amoeba may regulate its fluid content by pinocytosis and water-expulsion vesicles, humans must employ much more complex organ systems to achieve the same end.

Also, certain organs or even the entire bodies of some species or roundworms and rotifers contain a constant number of cells.

The Digestive Cavity

A digestive tract may take several forms, another gross indication of taxonomic similarity. Sponges have no digestive cavity at all, but most other multicellular animals have a **lumen** or **gut.** In some, such as the jellyfish, the gut is a blind sac with a single opening (mouth) serving for both the ingestion of food and the elimination of particulate wastes. Most multicellular animals have a complete digestive tract, with both a mouth and anus.

Embryonic Development

A most helpful aid to classification is similarities in embryonic patterns of development. The **biogenetic law,** first stated by Haeckel, has been in disfavor until its recent reconsideration in the light of modern embryology and evolution theory; it states that *ontogeny recapitulates phylogeny.* This means that evolutionary history that took hundreds of thousands of years (**phylogeny**) to accomplish is represented during the relatively short time span in which an organism develops from a fertilized ovum into a new individual (**ontogeny**). The human fetus, for example, has rudimentary gill slits and a tail for a short time during its prenatal development. Actually, ontogenic development is only roughly parallel to phylogenetic history, because many gaps, inclusions, and omissions occur. Initially, in prenatal development a multicellular animal undergoes changes that become progressively more individualized.

A major embryological point of dissimilarity among animals is the site at which the mouth forms on the **gastrula** (the ball of cells that forms early in development; see Chapter 38). In the **protostomes** (Gr. *protos*, first + *stoma*, mouth), the mouth forms from the **blastopore,** an opening in the gastrula. This is the case in annelids, arthropods, and flatworms. In **deuterostomes** (Gr. *deuteros*, second + *stoma*, mouth), the mouth develops elsewhere, whereas the anus develops from the blastopore. The deuterostomes include the echinoderms and chordates.

The pattern of cell divisions, or **cleavage,** following fertilization is also quite variable. The zygotes of animals that share similar evolutionary histories generally tend to cleave in similar fashion: **holoblastic cleavage,** in which the egg is completely divided, and **meroblastic cleavage,** in which the division is incomplete, form an index of egg yolk concentration. In general, the higher the concentration of yolk in the egg, the greater the probability of meroblastic cleavage in the zygote.

The number of tissue layers, or **germ layers,** is also important. Either two or three germ layers form during the development of multicellular animals. Sponges and cnidarians are **diploblastic,** having two germ layers (endoderm and ectoderm), whereas more complex multicellular animals are **triploblastic,** with three tissue layers (endoderm, ectoderm, and mesoderm between them).

Appendages

Any considerable projection from the body of an animal is called an **appendage.** Examples include tentacles, antennae, legs, fins, wings, and arms. Different types of appendages serve different purposes.

Locomotor, feeding, and sensory functions are usually performed by appendages. The number and morphology of appendages that an animal possesses (or lacks) are good means of characterization, especially among the arthropods. For example, whereas spiders, scorpions, and their allies have eight legs, majority of the insects have six.

Skeletal System

A skeleton is a supporting framework of an animal body. Skeletons composed of hard parts may be located inside (endoskeleton) or outside (exoskeleton) the body. Endoskeletons are characteristic of sponges, echinoderms, and chordates. Exoskeletons distinguish mollusks and arthropods.

As noted, fluid-filled internal body cavities may act as hydrostatic skeletons, aiding in support and locomotion.

Sexual Characteristics

Animals vary considerably in their sexual characteristics. **Dioecious** animals are unisexual—both sexes are present, but each individual ordinarily develops only one set of gonads, either male or female. **Monoecious,** or **hermaphroditic,** animals have both male and female gonads developed in each mature adult. Monoecious species are often closely related to other monoecious species.

Another reproductive attribute, the formation of dormant spores, is often important in the taxonomy of protozoans and very small invertebrates.

Larval Forms

One or several larval forms may occur during development from the fertilized egg (**zygote**) to the adult form. The importance of larval forms for taxonomic purposes cannot be overemphasized, as similar larvae suggest evolutionary affinities.

A group of chordates, the ascidians, were at one time classified as mollusks. The adult ascidian, upon which this determination was based, is sessile, lacks a notochord, and has a primitive nervous system. In 1866, Kowalevsky directed his attention to the motile larvae of ascidians and found that they had a well-developed central nervous system, a notochord, and meta-

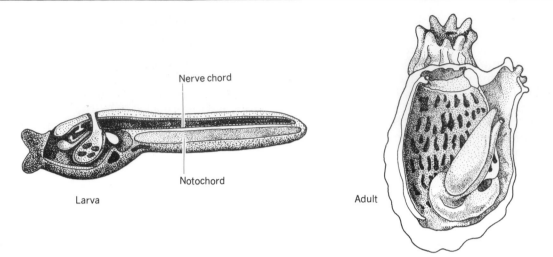

Figure 4-3. Larval and adult ascidian. Left: Larval form (note nervous and skeletal systems). Right: Adult form (lacking nervous and differentiated skeletal system).

meric bands of muscles (Fig. 4-3). The ascidian is now classified as a chordate on the basis of these larval characters.

Other discriminating bases exist, but they are too numerous and specific to mention here. Many will be described throughout later chapters. Now let's consider the kingdoms of living organisms and the major protozoan and animal phyla.

Kingdoms of the Living World

Our conception of the kingdoms in the living world have been revised several times over the years. Originally, everything living was just "living." Later, it was found convenient to divide the living world into two major categories: the kingdoms Plantae and Animalia. That scheme was acceptable for some time, but refinements became necessary as more information was obtained. It was suggested as early as 1860 that the number of kingdoms be increased to three, to include a new one, Protista, that would contain all unicellular organisms.

With the accumulation of more detailed data, other schemes were presented. For example, it was discovered that bacteria and blue-green algae lack nuclear membranes and subcellular organelles. These primitive organisms are called **procaryotes.** The cells of all other living organisms (**eucaryotes**) contain a distinct nucleus, bounded by a membrane, plus mitochondria and other organelles.

In 1959, following the lead of a few earlier scientists, R. H. Whittaker proposed a five-kingdom system (Fig. 4-4). We will use much of this system in the present text. The five kingdoms are Monera (the procaryotic bacteria and blue-green algae), Protista (unicellular algae and protozoans), Plantae (multicellular plants), Fungi (nonphotosynthetic plantlike organisms), and Animalia (multicellular animals). Note that "protozoans" are considered a subset of protists that exhibit animal-like characteristics. Table 4-2 presents the classification system in more detail. We will concern ourselves with only the animal-like protists and the multicellular animals.

Synopsis of the Major Phyla

Throughout this text we will be considering the characteristics of the various animal phyla in some detail. The purpose of the following brief overview is to give you an understanding of the great diversity of animal life. The estimates of numbers of living species are conservative, for newly recognized species are being named constantly. A more detailed taxonomy is presented in the classification appendix.

Some Major Phyla of Kingdom Protista

Four phyla of protozoans, or animal-like protists, are included in this kingdom, which also includes the eucaryotic algae phyla. Some 50,000 species have been described, but this figure is much too low because most of these primarily unicellular organisms have yet to be discovered and categorized. The protistan phyla are distinguished by their methods of lo-

comotion, nutrition, reproduction, number of nuclei, and possession of special organelles.

Phylum Sarcomastigophora

This phylum includes both the flagellated and amoeboid protozoans. Sarcomastigophorans possess flagella or pseudopodia (or both) but never cilia. There is generally only one type of nucleus, although they may be multinucleated. Sexual reproduction,

Figure 4-4. The five-kingdom system and general evolutionary tree. Smaller diagram illustrates differences in food sources.

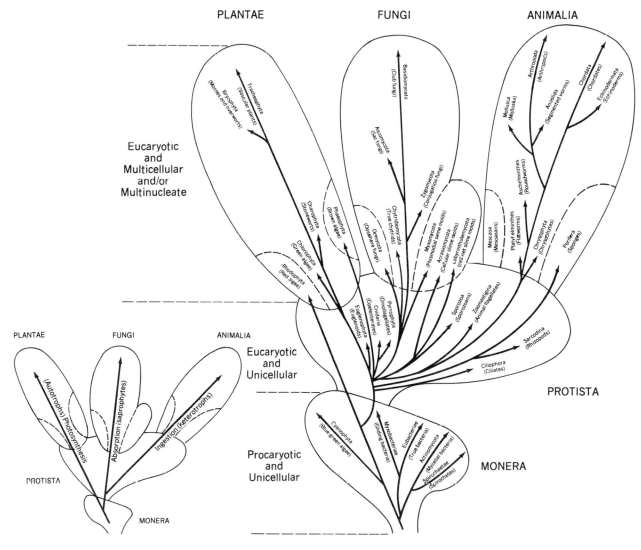

when it occurs, is by syngamy (union of gametes), not by spore formation or conjugation. Members of this phylum live in moist soils and aquatic habitats.

Within this phylum, zoologists recognize the class Phytomastigophorea; these photosynthetic organisms are separately classified by botanists and algologists (those who study algae). This is largely explained by the endosymbiotic origin of eucaryotes, as there are definitely phylogenetic affinites among many unicellular organisms whose only difference is the presence or absence of a chloroplast; one successful symbiotic invasion of an animal-like protozoan by a blue-green algae immediately creates a major evolutionary jump whereas the new organism retains many of the ancestor's characteristics. The equivalent taxonomic names given to various phytomastigophoran taxa by botanists are presented with the appropriate taxa in the classification appendix.

Phylum Sporozoa

These are all parasitic protozoans. They lack flagella or cilia, and any locomotion is accomplished by changing the shape of the cell. Sporozoans have a single type of nucleus. They are characterized by sporelike infective stages. The disease malaria is caused by a sporozoan.

Phylum Ciliophora

This phylum contains protozoans that bear cilia or ciliary organelles. Ciliates have two kinds of nuclei and may reproduce by asexual division or sexual conjugation. They lack spores, and most are free living in marine and freshwater habitats.

Major Phyla of Kingdom Animalia

Phylum Porifera

The phylum Porifera contains the sponges, of which some 10,000 species have been described. Some sponges exhibit radial symmetry, but most are asymmetrical. Their body forms vary, most being globular, vaselike, or treelike; many are colonial. The multicellular poriferans are diploblastic; that is, they possess only two germ layers. They have no actual tissues, much less organs. The body is riddled with ostia (pores), chambers, and canals that allow for water passage. Sponges have endoskeletons composed of spicules and/or proteinaceous fibers. The skeletal spicules are made of calcium carbonate, spongin, or silicon dioxide. Reproduction is sexual (by eggs and sperm) or asexual (by budding). Most sponges live in marine habitats; only a few species inhabit fresh water.

Phylum Cnidaria

The cnidarians or coelenterates are a diverse group of aquatic animals that are of considerable ecological importance. The phylum includes sea anemones, corals, jellyfishes, hydroids, and many others. Over 99 percent of the 9000 described species are marine. Symmetry is either radial or biradial. Their bodies may take the form of sessile polyps or free-swimming medusae. Cnidarians have two germ layers with a layer of mesoglea, a jellylike matrix, between them. They are the most primitive animals to exhibit a digestive system, composed of a saclike gastrovascular cavity with only a single opening through which materials are both ingested and egested. They also possess a nervous system of sorts, called a nerve net, and a primitive form of muscular system. Some possess an exoskeleton made of calcium carbonate or a glycoprotein called chitin. Reproduction may be asexual or sexual, with some cnidarians being monoecious. Cnidarians bear stinging cells each of which releases a projectile; these nematocysts are used in capturing food and defense.

Phylum Ctenophora

Ctenophores are transparent, gelatinous, monoecious animals that are commonly called comb jellies or sea walnuts. Less than 100 species are known, but they are quite common in ocean surface waters. They have biradial symmetry, which should indicate to you that they are usually spherical or ellipsoidal. Most are free swimming. Ctenophores usually possess eight rows of external comb plates that are used for locomotion. These organisms are the most primitive animals to exhibit a triploblastic body. They also have a primitive digestive tract. A nervous system consists of an anterior statocyst (a sense organ) and eight nerve strands beneath the comb plates that are used for locomotion. Reproduction is exclusively sexual.

Phylum Platyhelminthes

Platyhelminths are the most primitive bilaterally symmetrical animals. Phylum Platyhelminthes contains the free-living flatworms and parasitic flukes and tapeworms. Something less than 13,000 species have been described. Exhibiting essentially a triploblastic body, the platyhelminths are nevertheless acoelomates. Most have an incomplete digestive system and primitive excretory organs. The muscular system is also primitive, but a relatively advanced nervous system containing ganglia exists. Some platyhelminths have eyespots. They have no circulatory, respiratory, or skeletal systems. Most are monoecious. Platyhelminths inhabit the ocean, fresh water, land, and various living hosts.

Aschelminth Phyla

These are triploblastic, bilaterally symmetrical animals that possess a pseudocoel. The eight phyla commonly referred to as aschelminths include roundworms (nematodes), rotifers, and gastrotrichs, plus some minor groups. They are usually microscopic in size, but a few attain lengths of 1 m. The digestive system is usually complete. A nervous system is present, with several types of sense organs being well developed. Moderate cephalization is present. Sexes are usually separate. Most aschelminths are free living, but many nematodes parasitize plants and animals. Only about 12,000 species of this enormous assemblage have thus far been described.

Phylum Annelida

The phylum Annelida contains the true worms (9,000 known species) and is one of the most successful animal groups, having adapted to marine, freshwater, and terrestrial habitats. Annelids exhibit metameric segmentation. The phylum contains marine polychaetes, terrestrial and freshwater earthworms, and parasitic leeches. These worms have a coelom, divided by septa, within their triploblastic bodies. They exhibit bilateral symmetry. Annelida have a rather advanced, closed circulatory system as well as a complete, unsegmented digestive system. The excretory system includes one pair of nephridia within each metamere. The nervous system is situated ventrally, as it is in mollusks and arthropods. A sensory system, including tactile organs, photoreceptor cells, and sometimes eyes with lenses, is present. Annelids may reproduce hermaphroditically or with separate sexes, and some forms bud asexually.

Phylum Mollusca

This is one of the largest and most diverse of the animal phyla. Over 50,000 living species of snails, clams, squid, chiton, and other mollusks have been described. They are usually bilaterally symmetrical, with a triploblastic body, and most are unsegmented. Some have a dorsal mantle that secretes a calcium carbonate shell, which is a type of exoskeleton. Locomotion is generally accomplished with a ventral muscular foot, although various other mechanisms have evolved. An advanced circulatory system is present, complete with a heart and blood vessels. Digestive and respiratory systems are present, along with a complex nervous-sensory system including four distinct pairs of ganglia. Mollusks may be either monoecious or dioecious, with one or two gonads and a system of ducts. Fertilization may be external or internal. Mollusks are mostly marine, but freshwater and terrestrial species are quite common.

Phylum Arthropoda

Eighty percent of all animal species belong to the phylum Arthropoda, which includes insects, crustaceans (shrimp, crabs, etc.), centipedes, spiders, and ticks, among others. More than 800,000 species of arthropods have already been described, and some authorities estimate that there may be over 10 million species of insects alone! Arthropods exhibit bilateral symmetry with triploblastic, metameric bodies. The metameres are covered by a chitinous exoskeleton and are often consolidated into a head (indicating definite cephalization), thorax, and abdominal regions. Arthropods possess open circulatory systems. The muscular system is highly complex, composed primarily of striated muscles. The respiratory system may utilize the body surface, gills, tracheae, or book lungs. A well-developed excretory system is present, usually featuring Malpighian tubules. The nervous and sensory systems are well developed, including a dorsal brain, eyes, and tactile, auditory, equilibrium, and general sensory organs. Sexes are usually separate. Arthropods must periodically molt their exoskeletons.

Phylum Echinodermata

The echinoderms include about 6000 described species of starfishes, sea urchins, sea cucumbers, crinoids, and brittle stars. Their radial symmetry is characteristically expressed in pentamerous (five-way) divisions. They lack cephalization. These triploblastic animals possess a coelom and have an endoskeleton of calcareous ossicles. Locomotion is accomplished by tube feet that are controlled hydrostatically by a unique water vascular system. The tube feet also aid in respiration and feeding. The coelom has freely moving body cells known as amoebocytes in its fluid.

The nervous system usually consists of two or three networks located at different levels, although the sensory system is rather poorly developed. A few echinoderms are hermaphroditic, but the vast majority have separate sexes. These slow-moving or sessile organisms are all marine.

Phylum Chordata

Some 66,000 species of chordates have been described. Symmetry is bilateral, and the body is triploblastic. This phylum includes the most complex animals, and all possess a notochord (at least embryonically), the precursor of a vertebral column in higher chordates (see Fig. 33-7). Most chordates exhibit some form of metamerism and an endoskeleton. Organ systems reach their highest complexity in the phylum Chordata; these systems are so diverse that the student should consult the chapter on Chordata for details.

Characteristics of chordates that distinguish them from all other animals include pharyngeal gill slits (present in the embryos of all chordates, including humans); a dorsal nerve cord, at least partially hollow or tubular; and a notochord, a supporting rod between the dorsal nerve and the gut, which is the precursor to the vertebral column of vertebrates.

Chordates have successfully adapted to virtually all habitats in the biosphere. Phylum Chordata contains two invertebrate groups, the tunicates and lancelets, in addition to the backbone-bearing vertebrates. Vertebrate groups include the cyclostomes (lampreys and hagfish), the cartilaginous fishes (sharks and rays), the bony fishes, the amphibians (frogs, toads, salamanders), the reptiles (lizards, snakes, turtles, alligators), the birds, and the mammals (including *Homo sapiens*).

Summary

Taxonomy, the science of classification, today is allied with systematics in attempting to arrange living organisms according to their evolutionary relationships. Taxonomy and systematics are in a constant state of flux.

The binomial nomenclature system was presented as were the principal taxons, in decreasing order of generality, and kingdoms, phyla, classes, orders, families, genera, and species.

Some of the criteria that are considered in assessing taxonomic relationships are homology and analogy; symmetry (asymmetry, radial, biradial, and bilateral symmetries); cephalization and polarity; metamerism; body cavity presence and structure; cell number; digestive tract presence and variation; embryonic development; appendage presence and morphology; skeletal system presence and form; reproductive structures; and larval forms.

The major phyla of protozoans and animals were introduced. You are now ready to investigate these phyla in more detail. Look for patterns in the following chapters as these will help to unify your view of zoology.

Review Questions

1. About how many different animal species have been described at present? Do you expect that number to change?
2. Grouping together all animals living in tropical habitats would be an example of what type of classification system?

3. What is the difference between a coelom and a pseudocoel?
4. Do jellyfish exhibit cephalization? Explain.
5. Name three animal phyla that exhibit metamerism.
6. What is a species?
7. What kingdom(s) contain(s) animal-like organisms?
8. Which is more general and inclusive: a class or an order?
9. What is Linnaeus known for?
10. Name three animals that exhibit radial or biradial symmetry.
11. Of the discriminating bases listed in this chapter, which do you think are most important in taxonomy? Give your reasons.
12. What is the least complex phylum to show bilateral symmetry?
13. How many species of echinoderms live in freshwater habitats?
14. Name three characteristics of chordates that are not shared with any other animals.
15. Of the protistan phyla listed, which has two types of nuclei?
16. What is the function of a tube foot?
17. The phylum containing more species than all other phyla put together is phylum _____ .
18. What is another word for "hermaphroditic?"

Selected References

Boyden, Alan. *Perspectives in Zoology.* New York: Pergamon Press, 1973.

Blackwelder, R. E. *Guide to the Toxonomic Literature of Vertebrates.* Ames: Iowa State University Press, 1972.

Cain, A. J. *Animal Species and Their Evolution.* New York: Hutchinson's University Library, 1954.

CRM Books. *Biology Today, 3rd ed.*, Del Mar, California: Communications Research Machines, 1980.

Dodson, E. O. "The Kingdoms of Organisms." *Syst. Zool.*, **20**:265–281, 1971.

Margulis, L. "Whittaker's Five Kingdoms of Organisms: Minor Revisions Suggested by Considerations of the Origin of Mitosis." *Evolution.* **25**:242–245, 1971.

Frings, H., and M. Frings. *Concepts of Zoology.* Toronto: Macmillan Publishing Co., Inc., 1970.

Jahn, T. L. "Classifying Species by Computer," *New Scientist*, **29**:151–153, 1966.

Mayr, E. *Methods and Principles of Systematic Zoology.* New York: McGraw-Hill Book Company, Inc., 1969.

Mayr, E. (ed.). *The Species Problem.* Washington: American Association for the Advancement of Science, 1957.

Mayr, E. *Animal Species and Evolution.* Cambridge: Harvard University Press, 1966.

Savory, T. *Naming the Living World.* London: English Universities Press, Ltd., 1962.

Simpson, G. G. "The Principles of Classification and a Classification of Mammals," *Bulletin of the American Museum of Natural History*, **85**:1, 1945.

Simpson, G. G. *Principles of Animal Taxonomy.* New York: Columbia University Press, 1961.

Whittaker, R. H. "New Concept of the Kingdoms of Organisms." *Science*, **163**:150–159, 1969.

Introduction to the Protozoans—Kingdom Protista, Phylum Sarcomastigophora, Superclass Mastigophora

Protists represent an intermediate level of organization between the procaryotic bacteria and blue-green algae and the multicellular plants and animals. Kingdom Protista contains eucaryotic organisms that are generally unicellular; some form simple colonies of basically independent individuals that are never differentiated into tissues.

These one-celled organisms are studied for a variety of reasons. **Protozoans** are simple prototypes of more complex organisms; they suggest the pathways that have led to the gradual evolution of more complex living systems. Their microscopic size and rapid rate of reproduction make them convenient subjects for biological experiments. Also, some protists are of medical importance because they cause diseases, such as malaria and sleeping sickness.

It could be argued that a study of animal life should begin with organisms more familiar to us than the protists; for example, the earthworm, grasshopper, frog, or cat might be studied first. But this would be like starting an introductory math class with advanced calculus. Protists are relatively simple as compared with metazoans and, thus, serve as logical introductions to the multicellular animals. Progressing from the simplest animal-like protists (the protozoans) to the increasingly complex multicellular animals gives

one a feeling for the general course of evolutionary history.

Insofar as morphology is concerned, a single-celled protozoan is comparable, in some respects, to the individual cells of a multicellular animal body, but the physiology of the protozoan is comparable to that of the entire body of the multicellular animal. One of the intriguing things about protozoans is how a single cell can execute all the basic life processes.

Characteristics of Protozoans

One who examines a sample of pond water under a microscope for the first time discovers a new world. The Dutch scientist Antony van Leeuwenhoek was the first person to ever observe these "animalcules," as he called them, using a homemade microscope in 1674. Even today, relatively few people have seen protozoans. Although the number of individual protozoans exceeds that of all metazoans combined, protozoans are generally microscopic in size. Only a few (e.g., *Stentor*) are large enough to be visible to the unaided eye.

Active protozoans require moist habitats; they are omnipresent in freshwater ponds, lakes and streams, the ocean, and moist soils. Many live on or within the bodies of animals and plants. All protozoans are unicellular; some are colonial.

We noted in Chapter 4 that Linnaeus's division of all living organisms into plant and animal kingdoms was found to be too simplistic as more became known about the microbial world, especially about the consequences of symbiotic associations of animal-like and plantlike creatures. *Euglena*, for instance, is a one-celled organism that displays both animal-like and plantlike characteristics. Its motility suggests that it is an animal, but its green color and ability to photosynthesize organic compounds are characteristics of plants. Organisms of kingdom Protista that display animal-like traits will be discussed in the next four chapters under the descriptive term Protozoa. Four phyla of protozoans are commonly recognized:

Phylum 1. Sarcomastigophora. Protozoans without cilia (except for Opalinata); locomotion is by means of either pseudopodia (false feet) or flagella.

Phylum 2. Ciliophora. Protozoans that move by means of cilia or ciliary organelles.

Phylum 3. Sporozoa. Parasitic protozoans without motile organelles, but with a spore stage without polar filaments in their life cycle.

Phylum 4. Cnidospora. Parasitic protozoans with no cilia or flagella, but with a spore state having one or more polar filaments.

The major subdivisions of these phyla are as follows:

Phylum 1. **Sarcomastigophora**
  Superclass I. **Mastigophora**
    Class 1. **Phytomastigophorea**
    Class 2. **Zoomastigophorea**
  Superclass II. **Opalinata**
  Superclass III. **Sarcodina**
    Class 1. **Rhizopodea**
    Class 2. **Piroplasmea**
    Class 3. **Actinopodea**
Phylum 2. **Ciliophora**
  Class 1. **Ciliatea**

Phylum 3. **Sporozoa**
  Class 1. **Telosporea**
  Class 2. **Toxoplasmea**
  Class 3. **Haplosporea**
Phylum 4. **Cnidospora**
  Class 1. **Myxosporidea**
  Class 2. **Microsporidea**

Let us now see what some representatives of these groups look like, and how they meet the problems of survival.

The first phylum mentioned, Sarcomastigophora, contains three superclasses: Mastigophora (the flagellates), Opalinata (parasites with unusual cilialike organelles), and Sarcodina (the amoebas).

Superclass Mastigophora— The Flagellates

The mastigophorans are protists that usually possess one or more whiplike flagella. These organelles are used primarily for locomotion, but they may also function to capture food and as sensory receptors that explore the surrounding environment. Mastigophorans have a definite shape and are polar; one or more flagellae arise from the anterior end. These minute protists are abundant in puddles, ponds, and marine waters. Many are parasitic in animals, including humans and plants.

They are arranged in two distinct classes: Phytomastigophorea (plantlike flagellates with chloroplasts) and Zoomastigophorea (animal-like flagellates without chloroplasts).

Phytomastigophorans are an important component at the base of many aquatic food chains. They synthesize organic compounds photosynthetically, using carbon dioxide, water, and energy from sunlight that is captured by the green photopigment, chlorophyll. These green flagellates are eaten by very small crustaceans, which are in turn consumed by insect larvae and fish. Along with other algae, phytomastigophorans are a major source of several vitamins, including A and D, which are eventually utilized by the higher levels of the food chain.

One order of phytomastigophorans, the Volvocida, includes several colonial species that can be arranged

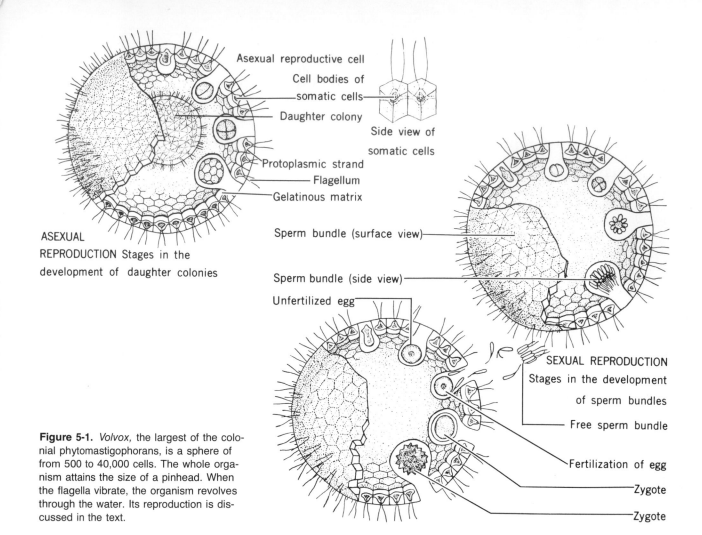

Asexual reproductive cell

Cell bodies of
somatic cells

Daughter colony

Side view of
somatic cells

Protoplasmic strand

Flagellum

Gelatinous matrix

ASEXUAL
REPRODUCTION Stages in the
development of daughter colonies

Sperm bundle (surface view)

Sperm bundle (side view)

Unfertilized egg

SEXUAL REPRODUCTION
Stages in the development
of sperm bundles

Free sperm bundle

Fertilization of egg

Zygote

Zygote

Figure 5-1. *Volvox,* the largest of the colonial phytomastigophorans, is a sphere of from 500 to 40,000 cells. The whole organism attains the size of a pinhead. When the flagella vibrate, the organism revolves through the water. Its reproduction is discussed in the text.

in a series from simple aggregations of cells (*Spindylomorum*) to very complex colonies (Fig. 5-1).

The euglenoid mastigophorans, while belonging to class Phytomastigophorea, exhibit most of the characteristics peculiar to the superclass Mastigophora, and so will be discussed in some detail here.

Euglena viridis—A Common Flagellate Habitat

Euglena viridis will serve as a representative example of the superclass Mastigophora. Euglenas are common in freshwater ponds, especially within clusters of vegetation. When present in sufficient num-

bers, they may even impart a greenish tinge to the water. They thrive in the laboratory in a jar of water that is exposed to plenty of indirect sunlight. Over 150 species of the genus *Euglena* have been described, differing from one another in size, behavior, and structural details.

Morphology

Euglena viridis is 0.1 mm or less in length, blunt at the anterior end, and pointed at the posterior end. Figure 5-2 presents its general morphology. Surrounding the cytoplasm is a thin elastic membrane, the **pellicle,** whose striated appearance is due to parallel spiral thickenings. The pellicle is rigid enough to maintain the shape of the body but sufficiently flexible to allow euglenoid movements. Near the anterior end is a funnel-shaped depression, the cell mouth (**cytostome**), that leads into the cell gullet (**cytopharynx**). The euglena does not ingest solid food, as these terms might imply. The cytopharynx enlarges posteriorly to form a vesicle called the **reservoir.** Next to this is a **water-expulsion vesicle** that regulates the water content of the cell. The water-expulsion vesicle was formerly referred to as the **contractile vacuole**, but this term is misleading because there is no muscular contraction; the vesicle is collapsed by cytoplasmic pressure. This organelle periodically collapses, discharging its fluid contents into the reservoir and from there out through the cytopharynx and cytostome.

Near the anterior end of the euglena is an orange-red **eyespot** (**stigma**), a light-sensitive organelle that helps to orient the photosynthetic protist toward sunlight. A **flagellum** arises from a basal body and extends anteriorly through the cytostome. Electron microscopic cross-sections reveal that the core of the flagellum contains the 9 + 2 microtubule arrangement of typical flagella (Fig. 2-21B).

Near the center of the euglena is an oval or spherical **nucleus** containing a central body, the **endosome,** consisting largely of genetic material. Also suspended in the cytoplasm are a number of green bodies, the **chloroplasts.** Their green color is due to the presence of chlorophyll. In *Euglena viridis* the chloroplasts are slender and radiate from a central point. In some species of euglena each chloroplast contains a **pyrenoid,** in which a starchlike storage material called **paramylum** is synthesized. It is produced by photosynthesis and represents reserve food material. Paramylum bodies may also be freely distributed throughout the cytoplasm in the form of discs, rods, and links.

Physiology

Nutrition

As do green plants, euglenas obtain food largely by photosynthesis, a process that takes place within the

Figure 5-2. *Euglena viridis.* (A) Photomicrograph. (B) Diagram of a stained specimen showing internal structure.

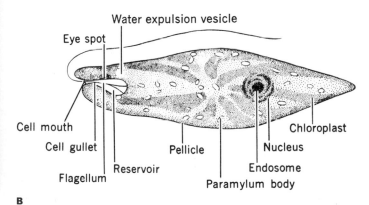

A

B

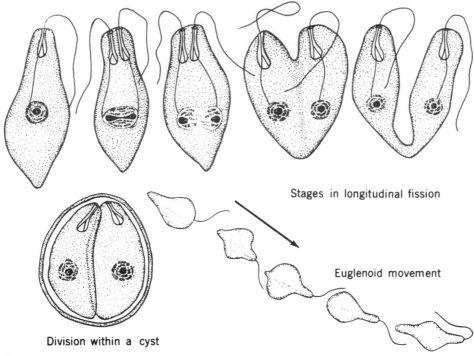

Stages in longitudinal fission

Euglenoid movement

Division within a cyst

Figure 5-3. Reproduction and euglenoid movement in *Euglena viridis*.

chloroplasts. Such **holophytic,** or **autotrophic, nutrition** involves the synthesis of complex organic compounds from simple inorganic molecules, using the energy of sunlight captured by means of chlorophyll. However, euglenas do require some materials that they themselves cannot manufacture. Vitamin B_{12}, for example, must be absorbed from the surrounding water. The euglena can survive in the dark by absorbing organic compounds that are dissolved in water, a process called **saprophytic** nutrition. When euglenas are maintained in the absence of light, their chloroplasts and pyrenoids degenerate and disappear. Although euglenas do not capture and eat other organisms (**holozoic** or **heterotrophic nutrition**), the animal-like flagellates of class Zoomastigophorea do consume protozoans, algae, and diatoms.

Locomotion

Swimming is accomplished with the aid of the whirling flagellum, which causes the mastigophoran to gyrate through the water in a helical path (shaped like the coil of a spring). When the flagellum con-

tracts, water is displaced, pulling the euglena forward. In essence, the flagellum is continually pushing off from a soft surface, the water. Although euglenas possess a definite shape, during locomotion they may employ wormlike movements, to which the term **euglenoid movement** has been applied (Fig. 5-3).

Reactions to Light

Euglenas (L. *eu,* good + Gr. *glene,* eyeball) are easily stimulated by changes in the direction and intensity of light. Laboratory experiments have shown that most species swim toward an indirect light source, such as that from a window. If a culture jar is examined, most of the protists will be found on the side toward the brightest light. This behavior is of distinct advantage to the holophytic euglena, because light is necessary for the photosynthetic process.

However, euglenas will swim away from direct sunlight, which will kill them after extended exposure. If a drop of water containing euglenas is positioned so that one half is in direct sunlight and the other half is shaded, the euglenas will avoid both extremes of shade and light, indicating that both are unfavorable to them. They will congregate in a small band between the two extremes, where the light intensity is best suited for them.

By shading various portions of an individual euglena, it has been found that the eyespot is especially sensitive to light. Note that this receptor is located at the anterior end of the body, which encounters new environments first.

Reproduction

Reproduction in euglenas is by **binary fission** (Fig. 5-3); the protist splits longitudinally into two new individuals. First, the nucleus divides in two by mitosis. Then, anterior organelles such as the reservoir are duplicated, and eventually the entire cell divides longitudinally. The old flagellum may be retained by one of the two daughter cells while a new flagellum is developed by the other.

During unfavorable conditions, such as a summer drought or a winter cold spell, many protists surround themselves with a protective shell (a **cyst**) and drastically reduce metabolic activities until the environment improves. Euglenas encyst by secreting a gelatinous wall around themselves. In this condition, periods of drought or cold may be successfully passed, after which the organisms again become active. Cysts are usually present in laboratory cultures on the sides of the culture dish. Before encystment the flagellum is discarded and a new one is produced when activity again resumes. Longitudinal division often occurs within the cyst. One cyst usually contains two euglenas, although further multiplication by longitudinal division may produce 4, 8, 16, or 32 young euglenas within a single cyst.

Other Mastigophorans

Chilomonas (Fig. 5-4) is a mastigophoran genus whose species are very common in nature. They are about 35 μ long and have two flagella at the anterior end. Chloroplasts are absent, and feeding is by saprophytic absorption of dissolved organic compounds through the pellicle.

Noctiluca (Fig. 5-4) is a genus of marine flagellates whose populations sometimes occur in such enormous numbers that, owing to their orange color, the seawater looks like tomato soup. Under such bloom conditions a quart of seawater may contain more than 3 million individuals. Species of *Noctiluca* are **bioluminescent;** that is, they glow with a bluish or greenish light when agitated. This gives parts of the ocean a striking appearance when one travels over it at night or walks along the wave wash.

Gymodinium (Fig. 5-4) is a dinoflagellate genus of which one species (*G. brevis*) may occur in such great numbers as to cause periodic red tides in temperate and tropical coastal waters. Actually, "red tide" is a misnomer for the brownish-amber discoloration of seawater caused by tremendous populations of this microscopic flagellate. Dinoflagellate blooms have been correlated with periods of heavy rainfall that leach fertilizer (nitrates and phosphates) from farmlands. Freshwater runoff enriches near-shore waters, fostering dinoflagellate growth. During a bloom, the density of dinoflagellates may quickly increase from about 1000 per liter to 80 or 90 *million* per liter. Toxins are released into the seawater by this teeming population in quantities great enough to kill fish and other sea life. Sometimes the adjacent shoreline becomes littered for miles with dead fish. Dinoflagellate toxins are among the most powerful natural poisons; experiments have shown them to be ten times more powerful than strychnine, a standard rat poison.

Another dinoflagellate, *Gonyaulax*, also causes seawater to appear rusty red at times because of its great numbers. *Gonyaulax catenella* is known to have been the cause of disastrous poisoning in humans. Mussels and other shellfish along the Pacific Coast feed on these dinoflagellates and accumulate the toxin in their tissues, thus making the shellfish poisonous for human consumption.

The genus *Mastigamoeba* (Fig. 5-4) contains species that live in fresh water or in moist soil. These unusual mastigophorans not only possess a flagellum but also form pseudopodia with which they ingest food particles much as amoebas do.

Many mastigophorans are very complex in structure, notably certain species that live in the intestines of termites and derive most of their nutriment from cellulose in the wood that the termites consume. In fact, the termites would starve if the flagellates were removed from their gut, because the insects cannot synthesize an enzyme that digests cellulose. Such an

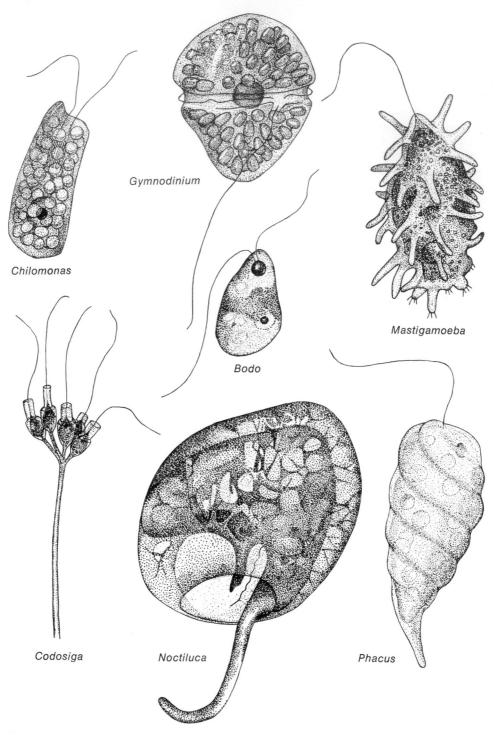

Chilomonas

Gymnodinium

Mastigamoeba

Bodo

Codosiga

Noctiluca

Phacus

Figure 5-4. Representatives of various genera of *Mastigophora*.

association of two species that is beneficial to both partners is called **mutualism.**

Mastigophorans in Water Supplies

Public health officials monitor drinking water for the presence of pathogenic (disease-causing) protozoans from sewage. Also of concern are various free-living mastigophorans that, when they occur in high densities, may render the water unpalatable. This is especially likely to occur when the water is confined in a quiet, open reservoir before being filtered for use. Representatives of three genera of colonial mastigophorans are illustrated in Figure 5-5. *Uroglenopsis* forms spherical colonies, with individuals embedded in the periphery of a gelatinous matrix. *Dinobryon* (Fig. 5-5) forms a branching colony and occurs most commonly in alkaline waters. Colonies of *Synura* (Fig. 5-5) consist of from 2 to 50 individuals arranged in radial fashion.

Among the several protozoans known to make water unpalatable for drinking, *Uroglenopsis* is perhaps the most undesirable because it imparts a fishy odor, reminiscent of cod-liver oil, to the water. Similar odors result from the presence of *Eudorina, Pandorina, Volvox,* and *Glenodinium.* Both *Synura* and *Pelomyxa* produce an odor resembling a salt marsh, whereas a culture of *Peridinium* smells like clam shells. The fishy odor of *Dinobryon* is somewhat like that of seaweed. *Ceratium* produces a vile stench. *Chlamydomonas* and *Mallamonas* are less objectionable, as their odors have an aromatic quality. Waters with blooms of *Cryptomonas* have been described as smelling like "candied violets."

All these odors are due to various aromatic oils produced by the flagellates during growth and are liberated when they die and undergo decomposition. Treatment of the water reservoir with copper sulfate is standard procedure for the control of these bad smelling populations, but such techniques appear to illustrate the cliché that "the cure is worse than the disease."

Some Parasitic Mastigophorans

It is important to become familiar with parasites to gain a complete picture of the biological world. It is estimated that there are more species of parasites than of free-living animals. This seems logical when one considers that every free-living host provides a variety of habitats for invasion and that parasites themselves are often parasitized in turn. The study of **parasitology** fosters an appreciation of the immense role that these organisms play in limiting the populations of their

Figure 5-5. Colonial mastigophorans that may render water unpotable.

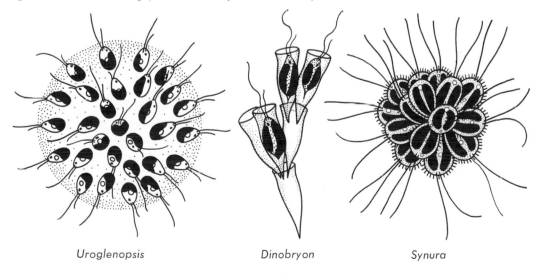

Uroglenopsis   Dinobryon   Synura

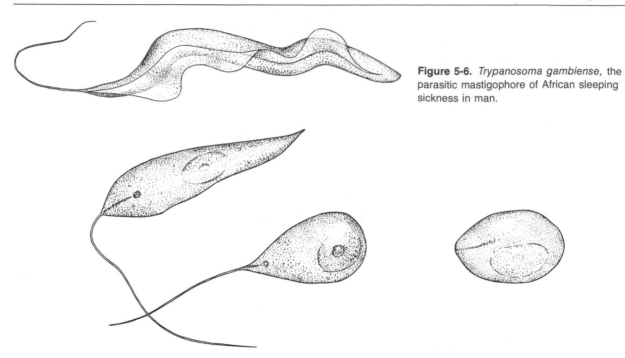

Figure 5-6. *Trypanosoma gambiense,* the parasitic mastigophore of African sleeping sickness in man.

Figure 5-7. *Leishmania donovani,* the parasitic flagellate of kala-azar in Asia. Top: Stage in man. Bottom: Stages in the sandfly, *Phlebotomus.*

hosts (in much the same way that predators do). One can also note how parasites have adapted to different hosts and, thus by comparison with free-living organisms, obtain a better understanding for the limits and potentials of living systems.

The most important mastigophoran parasites of blood and tissues are of two types; the **trypanosomes** and **leishmanias.**

The genus *Trypanosoma* is widespread in nature. These parasites live in the blood of many common mammals, birds, reptiles, fishes, and amphibians. Humans host three well-known species: *Trypanosoma gambiense* (Fig. 5-6) and *T. rhodesiense,* which cause two forms of African sleeping sickness, and *T. cruzi,* the cause of Chagas' disease in Texas and Latin America.

African sleeping sickness is transmitted by the bite of either sex of the bloodsucking **tsetse** fly. As an infected insect takes its blood meal, the slender, flagellated trypanosomes pass via of the fly's proboscis into the human bloodstream. Early stages of the disease are characterized by symptoms of fever and swelling of lymph nodes as the parasites multiply in the blood. Subsequent invasion of the nervous system results progressively in drowsiness, coma, emaciation (through inability to eat), and finally death. Treatment with drugs is effective if the disease is diagnosed in its early stages, before the central nervous system is invaded. Wild game serve as a reservoir from which tsetse flies can transmit the parasites to humans and domesticated animals. It is estimated that 25 percent of the total area of Africa is barred to agricultural development by **trypanosomiasis.**

South American trypanosomiasis, or Chagas' disease, produces quite different symptoms. The nervous system is not invaded, but the heart is damaged in acute stages of the disease. These trypanosomes are transmitted by bloodsucking kissing bugs (so called because they generally bite the face). Infection takes place by way of the bug feces, which the host unknowingly rubs into the irritating puncture wound. The parasites enter the bloodstream and are carried to tissues throughout the body. The muscular tissue of the heart is particularly susceptible to infestation, and death from cardiac failure commonly results. Kissing bugs also feed on other mammals; the armadillo is an important reservoir of the disease.

Leishmania infections are also of more than one type. Kala-azar, a disease common in Asia and parts of South America, is caused by *L. donovani* (Fig. 5-7). This parasitic flagellate is transmitted by the bite of sandflies. The parasites invade certain leukocytes and then the liver, spleen, kidneys, and bone marrow. Persons suffering from kala-azar usually show enlargement of both the spleen and liver, general emaciation, and a peculiar darkening of the skin. Various antimony compounds are used in treatment. Oriental sore, caused by *L. tropica,* occurs in northern Africa, southern Asia, and southern Europe. **Espundia,** caused by *L. brasiliensis,* is limited to Central and South Amer-

ica. All types of leishmaniasis are transmitted by sandflies.

There are four mastigophorans that live in the human digestive tract: *Trichomonas tenax* lives in the tartar of the mouth; *Chilomastix mesnili, T. hominis* (Fig. 5-8), and *Giardia lamblia* (Fig. 5-8) inhabit the intestine. *Chilomastix* is a pear-shaped protist with four conspicuous flagella at its broad anterior end. Its cysts are lemon shaped. *Trichomonas* is of similar contour but bears from three to five anterior flagella plus a longitudinal undulating membrane; there is also a central **axostyle** that protrudes caudally as a distinct tail. *Trichomonas* never forms cysts.

Giardia lamblia (Fig. 5-8) is an odd-looking mastigophoran with two anterior nuclei and three or four pairs of posteriorly directed flagella. The protozoan has one flat side by which it may adhere to an intestinal epithelial cell. Populations of *Giardia* may cover the

Figure 5-8. Intestinal mastigophorans of man.

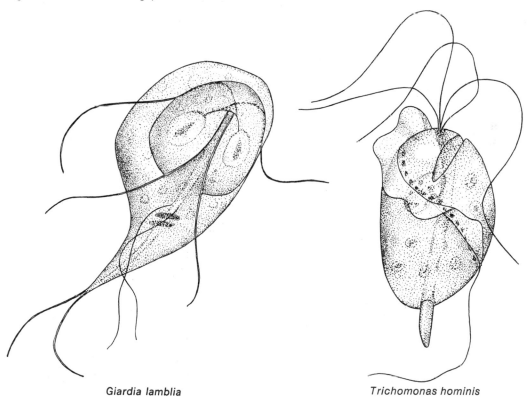

Giardia lamblia

Trichomonas hominis

entire inner surface of the upper small intestine and thereby mechanically interfere with absorption, particularly of fats; the presence of large amounts of unabsorbed fats in the intestine causes diarrhea. The cyst of *Giardia* is elongate, with four nuclei grouped at one end. Transmission from host to host occurs when the cysts are ingested in feces-contaminated food or water. Atabrine is an effective drug in the treatment of **giardiasis.**

Trichomonas vaginalis primarily inhabits the female urinary tract and vagina, where it may cause severe inflammation. Males may also harbor these parasites, which are transmitted by sexual intercourse. It closely resembles *T. hominis* in shape (Fig. 5-8).

Trichomonas foetus is an important cause of abortion in cattle, and *T. gallinae* produces a fatal infection in pigeons and other birds. *T. gallinarum* inhabits the lower intestine of chickens and turkeys and, in chronic cases, invades the liver.

Giardia infection may cause severe diarrhea in both rabbits and dogs. Blackhead in turkeys is caused by the mastigophoran *Histomonas meleagridis*, which is transported by a nematode parasite, *Heteralais gallini.*

Trypanosome diseases of animals are common in tropical areas. Horses, camels, cattle, pigs, dogs, and monkeys are susceptible to **nagana,** an African disease, caused by a trypanosome, as is **mal de Caderas,** a South American venereal disease of horses. *T. evansi* is responsible for **surra,** a serious disease of horses, dogs, elephants, camels, and other mammals. *T. lewisi* is a parasite found almost universally in rats. Bats, mice, sheep, goats, and other mammals harbor other trypanosome parasites.

Summary

Protozoans are members of kingdom Protista that display at least some animal-like traits. Protozoans are generally unicellular, yet they carry on all the basic processes necessary for life. The euglena, a motile mastigophoran with green chloroplasts, has a light-sensitive eyespot that enables it to distinguish light intensities. Protozoans are important because they

pervade the environment and affect nearly everything in it. Increased knowledge about these microscopic organisms has shown us that the living world is made up of a continuum of life forms, of which the protozoans represent the smallest.

Review Questions

1. Where might you go to find a euglena within your neighborhood?
2. In what ways do euglenas affect your life?
3. Imagine that you are conversing with a neighbor who is not as knowledgeable in zoology as you are. Explain by using examples why some organisms do not fit neatly into either plant or animal categories.
4. How is locomotion of a euglena affected when its flagellum is perpendicular to the axis of the cell?
5. Explain how a stigma enhances the survival of a euglena.
6. Discuss why protozoans are an important component of the food chain.

Selected References

Allen, W. E. "Red water in La Jolla Bay (California) in 1945," *Amer. Micr. Soc.*, **55:**149–153, 1946.
Biology Today, 3rd ed. Del Mar, California: Communications Research Machines, Inc., 1980.
Buetow, D. E. (ed.) *The Biology of Euglena*, New York: Academic Press, 1968.
Honigberg, B. M., *et al.* "Classification of Protozoa," *The Journal of Protozoology*, **11:**7–20, 1964.
Jahn, T. L., and J. J. Votta. "Locomotion in Protozoa," *Annual Review of Fluid Mechanics*, 4:93–101, 1972.
Manwell, R. D. *Introduction to Protozoology*, 2nd ed. New York: Dover Publications, Inc., 1968.
Sleigh, M. A. *The Biology of Protozoa*. New York, American Elsevier Publishing Co., Inc., 1973.
Storer, T. I., and R. L. Usinger, *General Zoology*. New York: McGraw-Hill Book Co., 1965.
Vickerman, K., and F. E. G. Cox, *The Protozoa*. Boston: Houghton Mifflin Co., 1967.

Kingdom Protista, Phylum Sarcomastigophora, Superclasses Opalinata and Sarcodina

In Chapter 5 we considered the flagellated protozoans that belong to the superclass Mastigophora of the phylum Sarcomastigophora. In this chapter we shall describe representatives of the remaining two superclasses, Opalinata and Sarcodina (Fig. 6-1). For quick summary of the characteristics of each superclass refer to Table 6-1.

Superclass Opalinata

These creatures are named opalinids because of their opalescent (iridescent) color. All members of this superclass are **symbiotic** (i.e., they live in intimate association with other organisms), mostly in the alimentary tract of frogs, toads, turtles, and salamanders. Opalinids are apparently **commensals,** benefiting from but not harming their hosts. Only one order, Opalinida, belongs to this superclass.

Morphology

Opalinids are unicellular, oblong-shaped protists that are distinguished by the presence of many oblique rows of cilia about 10 to 14 μ in length.* A

*The opalinids were once considered protociliates (rather than true ciliates) because macronuclei and micronuclei were not differentiated. As more information accumulated concerning their biol-

typical opalinid may measure 1.2 mm in length. The cytoplasm may contain anywhere from two to hundreds of nuclei, all of which are the same size and shape (monomorphic). No cytostome or water expulsion vesicle is present (Fig. 6-2).

Locomotion

It was once thought that the locomotor organelles of *Opalina* beat with a forward and return stroke (like cracking a whip). However, observation with high-speed microcinematography has revealed that these organelles beat in a continuous, helical-shaped manner that generates the appearance of traveling waves on the cell surface.

Reproduction and Life Cycle

Opalinids spend their active adult lives within the large intestines of amphibians and turtles. *Opalina*

ogy, particularly details of reproduction, it became increasingly clear that they should not be included with the ciliates, but placed in a group of their own.

The sarcomastigophoran superclass Opalinata was therefore established. Sexual reproduction of opalinids involves syngamy of the anisogamete rather than conjugation, which is characteristic of the ciliates. Division of opalinids occurs between the rows of the locomotor organelles. This alone establishes a close affinity with the mastigophorans.

Table 6-1. Comparison of Some Features of the Sarcomastigophora*

	Locomotion	Reproduction	Number of Nuclei	Nutrition	Cytostome
Mastigophora	Flagella	Asexual	One	Varied	Yes
Opalinata	Cilia	Asexual, sexual	Many	Saprozoic	No
Sarcodina	pseudopodia, flagella	Asexual	One	Holozoic	No

* All three groups have in common no spores and one type of nucleus.

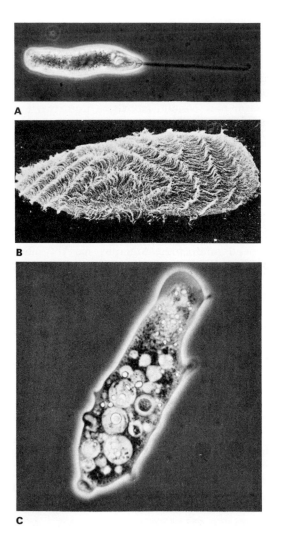

A

B

C

Figure 6-1. Photographs of Sarcomastigophora. (A) Mastigophora. (B) Opalinata. (C) Sarcodina. Note the differences due to variations in the predominant organelles.

ranarum, for example, is a European species that occurs regularly in the rectum of frogs and toads. Because the host does not live forever, the life cycle of the parasite must include an infective stage that can be transmitted to new hosts (Fig. 6-3).

From summer through early spring the multinucleated *Opalina* reproduce by fission, populating the rectum of their host with similar multinucleated individuals. Late in the spring a change occurs; the rate of fission increases, yielding smaller opalinids with only one or several nuclei. These smaller opalinids pro-

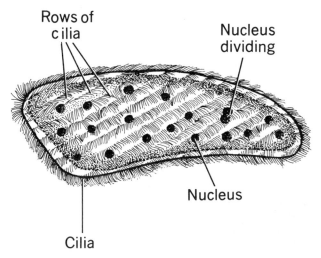

Rows of cilia

Nucleus dividing

Nucleus

Cilia

Figure 6-2. Structure of *Opalina.*

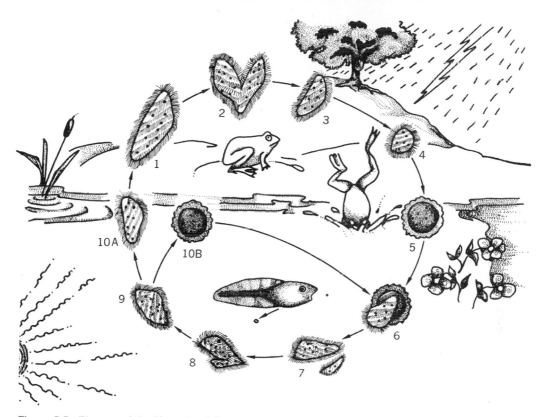

Figure 6-3. Diagram of the life cycle of *Opalina ranarum*. 1–3, trophozoite stages; 4–5, formation of cysts; 6, hatching of cyst; 7, formation of gametes; 8, copulation; 9, zygote; 10a, encystment of zygote and hatching from cyst of young asexual individuals in gut of young worm; 10b, cyst discharging into water and being swallowed by tadpole.

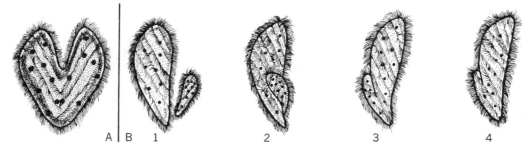

Figure 6-4. Reproduction in *Opalina*. (A) Fission. (B) Syngamy (fusion of gametes).

duce cysts that are expelled with the feces of the amphibian into the pond water, where they sink to the bottom. An encysted opalinid remains dormant until it is ingested by a tadpole feeding on bottom debris. Within the tadpole's intestine the opalinid emerges from its cyst and divides into **anisogametes** (gametes of unequal size). The gametes unite in pairs (syngamy) and form diploid **zygotes** (Fig. 6-4). A zygote may encyst and be discharged into the water to potentially infect another tadpole, or it may become a **trophozoite,** the active feeding stage of the organism.

You might wonder how this life cycle is regulated.

Consider the low probability of infecting new hosts (tadpoles) if cyst formation occurred during the fall or winter. It just so happens that cysts form during the late spring, when frogs are breeding and young tadpoles are developing. How is the onset of cyst formation triggered? Is it by some intrinsic "biological clock," or do the opalinids respond to some change in their external environment? Two obvious changes occur in the environment of opalinids just prior to the onset of cyst formation. The weather warms, and the host undergoes hormonal changes in preparation for spawning. Experiments have tested the roles of each of these environmental changes.

It was found that, when frogs were injected with pregnancy urine (which contains reproductive hormones) or pituitary extract,* cyst formation could be induced as early as January or February. Direct injection of one of the pituitary hormones initiated encystment in about one half the opalinid populations from September to December. However, frogs that were made sexually mature by transplanting pituitary glands into them fostered no encystment between January and March.

During the winter, maintaining frogs at 15–22°C, well above the ambient temperature encountered by frogs in nature, for one month did not induce encystment. But, after having been kept at these temperatures for two months, the frogs began to spawn in March, and the opalinids formed cysts. If the temperature was increased to 28–30°C for two weeks, the opalinids encysted even in the winter.

Cultures of *Opalina* maintained outside their hosts at 1°C during the winter appeared similar to those in frogs at the same season. Spring temperatures increased the temperature of the cultures, and cyst formation took place in late April—the same season that opalinids living in their natural hosts would encyst.

The fact that encystment can be induced prior to spring would seem to indicate that an intrinsic "biological clock" is not involved. That encystment was induced by temperature changes in cultures outside the frog indicates that the potential exists for a temperature-dependent mechanism that is independent of hormonal control.

---

* The pituitary is an endocrine gland at the base of the brain that secretes several hormones, some of which induce the onset of reproductive activity (see Chapter 35).

However, the injection experiments seem to contradict this hypothesis (assuming that the injections did not raise the temperature of the frogs). Obviously, more experimentation needs to be done to better understand the regulatory process in opalinid encystment.

Superclass Sarcodina

The superclass Sarcodina contains protozoans with pseudopodia. All sarcodines employ pseudopodia for feeding, and motile forms use them for locomotion. The familiar **amoeba** and its allies are naked, with ever-changing asymmetrical body shapes. But most sarcodines are housed within concentric shells of chitin, silica, or cemented foreign materials. We shall consider the amoeba in some detail before surveying the shelled sarcodines.

Amoeba proteus—A Sarcodine

You *must* observe a live amoeba through a microscope! There is something very intriguing in the way in which it constantly changes shape as it moves and engulfs food. Undoubtedly the alien nature of this **amoeboid movement** captured the imagination of the filmmakers responsible for such science fiction movies as *The Blob* and *Son of Blob*. Both films were about giant creatures that moved and ate in the same manner as do amoebas. Can you think of any zoological reasons why the existence of a one-celled blob 10 feet in diameter is unlikely?

The naked amoebas live in a variety of moist habitats, such as fresh water, the sea, and the soil, and as parasites within animals (including humans). One common, relatively large freshwater species, *A. proteus*, is usually selected to introduce the superclass Sarcodina. The amoeba (Fig. 6-1) lives in freshwater ponds and streams and can often be found on the undersides of dead lily pads and other vegetation in shallow water.

If pond water containing amoebas is studied under a microscope, some of the activities and a little of the structure of these protists can be observed. By changing environmental conditions such as temperature and light, one can also note their behavior.

Morphology

Amoeba proteus (Fig. 6-5) is about 250 μ in length. It appears under the microscope as an irregular, grayish mass of animated jelly that continually changes its shape by pushing out and withdrawing fingerlike projections called **pseudopodia.** Two types of cytoplasm are recognizable in the amoeba: a thin outer layer of clear semi-rigid protoplasm called **ectoplasm,** or **plasmagel,** surrounds the more centrally located, granular, fluidlike protoplasm called **endoplasm,** or **plasmasol.** Although the ectoplasm is surrounded by only a very thin and elastic cell membrane, amoebas can crawl over one another and never fuse. The endoplasm contains several cytoplasmic in-clusions that are larger than ordinary granules. One of these, the **nucleus,** is easy to see in the living animal. The nucleus is disc shaped and filled with chromatin granules. It regulates fundamental activities of the cell such as maintenance, growth, and reproduction. If an amoeba is cut into two pieces, the part containing the nucleus may continue to live and divide, but the one without a nucleus cannot divide and soon dies.

A clear bubblelike body can often be found lying near the nucleus; this is known as the **water-expulsion vesicle** (Fig. 6-5). At more or less regular intervals it migrates to the surface, where it ruptures and expels its fluid contents from the cell. Food vacuoles (**phagosomes**) may also be seen in the endoplasm.

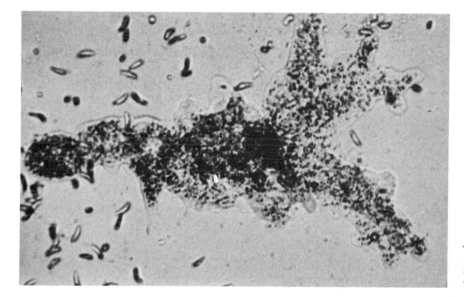

Figure 6-5. Top: Photograph of *Amoeba proteus.* Bottom: Diagrammatic representation of *Amoeba* showing main morphological features.

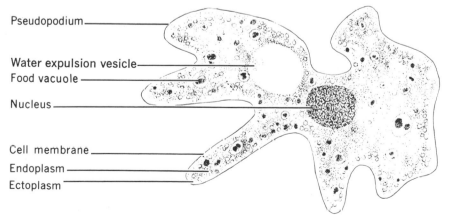

Pseudopodium

Water expulsion vesicle

Food vacuole

Nucleus

Cell membrane

Endoplasm

Ectoplasm

Physiology

An amoeba shares in common with more complex animals all the activities necessary for animal life—movement, the capture and digestion of food, the absorption and assimilation of the products of digestion, the elimination of waste, respiration, growth, reproduction, and response to changes in its environment.

Locomotion (amoeboid movement)

Amoebas move from place to place and ingest solid particles of food by means of fingerlike pseudopodia. These may arise at any point on the cell surface. The amoeba moves along by forming pseudopodia and then flowing into them. Ectoplasm at the rear of the cell is reconverted back into the fluid endoplasm. A blunt projection of ectoplasm first appears, then granular endoplasm flows into this tube of growing ectoplasm. The entire amoeba moves forward in the direction of the pseudopodium. Several pseudopodia often form simultaneously, but only one usually becomes dominant, and the others gradually shrink and disappear.* Amoebas have been observed to move at rates of about 2–3 cm per hour. The actual speed varies with temperature, increasing up to about 30°C, but ceasing at 33°C.

Feeding

The amoeba feeds primarily on smaller organisms, but small particles of organic matter are also ingested. Not every object encountered is phagocytosed; a distinct selection of food particles is evident (Fig. 6-6). The amoeba is able to capture rapidly swimming protists such as the ciliates *Tetrahymena* (Fig. 6-7) and *Paramecium*. A paramecium is sometimes held and actually cut in two by the pseudopodia before being ingested.

Food may be engulfed at any point on the surface of the amoeba, but it is usually taken by that part of the body being extended in the direction of locomotion—

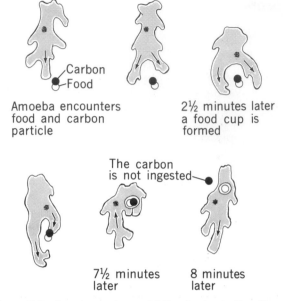

Amoeba encounters food and carbon particle

2½ minutes later a food cup is formed

The carbon is not ingested

7½ minutes later

8 minutes later

Figure 6-6. *Amoeba proteus* exhibiting food selection. The arrows indicate the direction of movement of the endoplasm in the pseudopodia.

the temporary anterior end. Food is usually ingested in the following version of **phagocytosis** (Fig. 6-6). Pseudopodia enclose the food particle from the sides; then thin sheets of cytoplasm cover the top and the bottom, thus entirely surrounding it. A vacuole (**phagosome**) is formed with walls that were formerly part of the cell membrane, the contents consisting of one or more food particles suspended in water. When the prey is active, a large food vacuole is often formed, and the prey is enclosed without being touched (Fig. 6-7); in this manner a dozen or more flagellates or ciliates may be ingested in one phagosome.

The whole process of phagocytosis occupies one or more minutes, depending on the type of food and the temperature (Fig. 6-6). The rate increases up to 25°C, but it decreases to zero at about 33°C.

Feeding occurs only when the amoeba is attached to some solid object. At certain times any organism that is not too large may be ingested, but food selection is evident when several species are present. For example, the small flagellate *Chilomonas* (Fig. 5-4) is engulfed more readily than is the larger ciliate *Col-*

---

* Many cells in multicellular organisms, including humans, exhibit typical amoeboid movements. For example, leukocytes (white blood cells) move through the walls of capillary blood vessels by means of pseudopodia. Leukocytes also engulf invading particles such as microorganisms by means of pseudopodia.

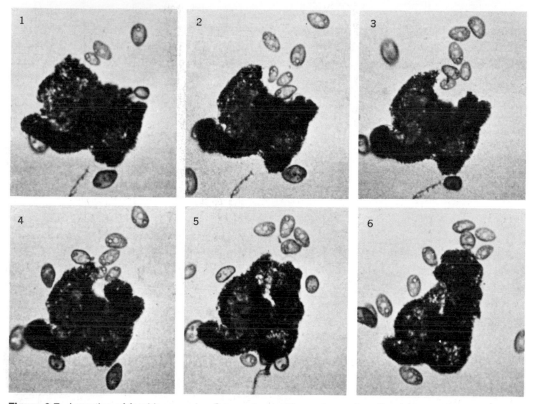

Figure 6-7. Ingestion of food by amoeba. Successive positions of pseudopodia of an amoeba capturing *Tetrahymena*, from a 16 mm motion picture. Magnification about 200 ×.

pidium, and the flagellate *Monas* is rarely taken if *Chilomonas* or *Colpidium* is available.

As many as 50 to 100 chilomonads may be ingested in a single day. When amoebas are fed exclusively on chilomonads, they grow and multiply for a few days but soon die, whereas, when fed exclusively on *Colpidium*, they grow large, become sluggish, and multiply slowly, but do not die. An amoeba may survive for 20 days or more without food, but it decreases in size until it is about 5 percent of its original volume.

Digestion

A food vacuole (Fig. 6-5) serves as a sort of temporary stomach. Digestive enzymes are secreted into it from cytoplasmic lysosomes. The contents of the vacuole are acidic at first and then become alkaline. In humans, as we shall see, food materials encounter an acid environment in the stomach and then an alkaline environment in the small intestine. Chilomonads

remain alive in food vacuoles for 3 to 18 minutes and are completely digested in from 12 to 24 hours. Proteins, fats, and starches are broken down into their smaller organic subunits. These small molecules diffuse through the vacuole membrane into the cytoplasm. The vacuole decreases in size until only indigestible materials remain. These wastes are then eliminated.

Egestion

Indigestible and sometimes partially digested particles are egested at any point on the surface of the amoeba, there being no special opening or organelle for elimination. Usually such particles are denser (heavier) than cytoplasm, and, as the amoeba moves forward, the wastes lag behind, finally passing out at the end away from the direction of movement. The amoeba flows away, leaving the indigestible solids behind (Fig. 6-8).

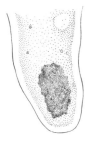
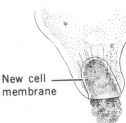

New cell
membrane

Figure 6-8. *Amoeba verrucosa.* Part of a specimen showing egestion of an indigestible particle; development of a new cell membrane prevents loss of endoplasm.

Assimilation

Some of the simple by-products of digestion that are absorbed are built up into the complex constituents of protoplasm by means of the process termed **assimilation,** or **anabolism,** which results in growth.

Dissimilation

The breakdown of complex molecules liberates energy that was stored in chemical bonds is known as **dissimilation** (or **catabolism**). Cellular respiration is a stepwise series of oxidations (Chapter 27) by which usable energy is derived from food nutrients. Mitochondria within the cytoplasm are the sites of cellular respiration, and the synthesis of ATP molecules is the outcome. The ATP molecules are transported throughout the cell, where their breakdown in turn provides energy for all cellular activities.

Secretion

Very little is known about secretion in the amoeba. Digestive fluids are undoubtedly secreted by lysosomes into the food vacuoles. Other substances of use in the life processes of the amoeba may also be secreted.

Excretion

The amoeba probably eliminates most of its soluble wastes, especially ammonia and carbon dioxide, through its plasma membrane. The water-expulsion vesicle may assist excretion, but its primary function is to regulate the water content of the cell. Some water enters the cell with the food, some is formed as a by-product of oxidation, and still more enters

through the cell membrane by osmosis. A water-expulsion vesicle is formed by the fusion of minute droplets of liquid. Its "wall" is a condensation membrane that disappears when the vesicle expels its contents. Water-expulsion vesicles form in various parts of the amoeba, often near the nucleus. They migrate to the posterior end of the amoeba, where their contents are discharged.

Respiration

The amoeba requires oxygen for cellular respiration and must eliminate carbon dioxide. Oxygen dissolved in water diffuses into the amoeba, and carbon dioxide diffuses out through the cell membrane. The water-expulsion vesicle may remove some dissolved carbon dioxide from the cytoplasm.

That oxygen is required by the amoeba can be demonstrated by replacing the air in the culture vessel with hydrogen gas. Amoeboid movements cease after 24 hours; if air is then reintroduced, the amoebas become active again; if not, death ensues.

Reproduction

When food is plentiful, an amoeba builds up protoplasm more rapidly than it breaks it down. When a critical size is attained, it reproduces by simply dividing into two. This asexual method of reproduction is called **binary fission** (Fig. 6-9). The nucleus divides by **mitosis; prophase** lasts 10 minutes, **metaphase** probably less than 5 minutes, **anaphase** about 10 minutes, and **telophase** about 8 minutes. During mitosis the amoeba becomes spherical and covered with small pseudopodia. In the telophase stage it elongates and separates into two daughter cells. The time required for the mitotic process (excluding interphase)

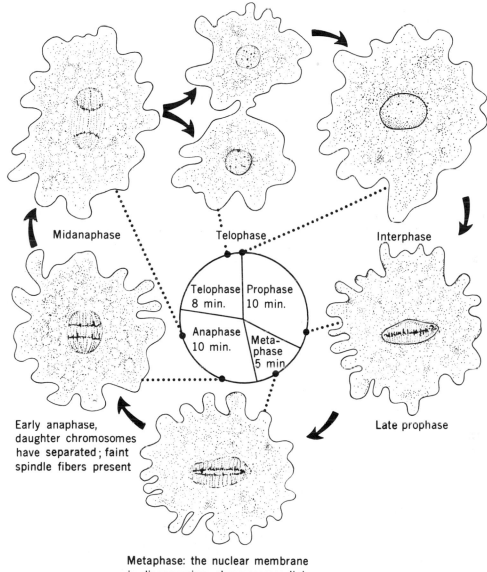

Midanaphase

Telophase

Interphase

| Telophase 8 min. | Prophase 10 min. |
| Anaphase 10 min. | Meta-phase 5 min. |

Early anaphase, daughter chromosomes have separated; faint spindle fibers present

Late prophase

Metaphase: the nuclear membrane is disappearing, chromosomes lining up at equatorial plate

Figure 6-9(A). The amoeba reproduces by binary fission and shows both external appearance and the division of the nucleus by mitosis. Begin study with the interphase stage and follow the arrows. The time in minutes for each stage is shown in the center.

depends on the temperature; at 24°C, it takes about 33 minutes. Under laboratory conditions, the amoeba divides every few days. Mitosis in many amoebas is somewhat unusual because of the formation of *several* **spindle poles** that accommodate the several hundred chromosomes (a remarkably large number) that may

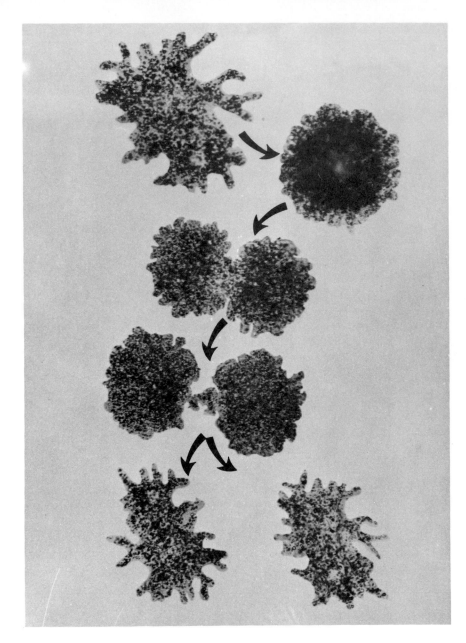

Figure 6-9(B). Composite photomicrograph of *Amoeba proteus* undergoing division. Compare stages depicted with the drawing (A).

contribute to their resistance to the influence of radiation (which generally destroys parts of chromosomes).

Development in the amoeba is simply a matter of growth. Growth is rapid just after division, and then gradually decreases until the critical size for division is once again reached, a process that averages about three days. An amoeba is potentially "immortal," for it reproduces by fission, and there is no death from old age. Death results from adverse environmental conditions or through predation by another organism (or from use in college laboratory experiments).

Behavior

The activities of the amoeba constitute its **behavior.** These include changes in shape, formation of

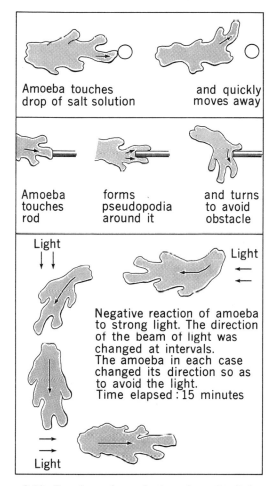

Amoeba touches
drop of salt solution

and quickly
moves away

Amoeba
touches
rod

forms
pseudopodia
around it

and turns
to avoid
obstacle

Light

Light

Negative reaction of amoeba
to strong light. The direction
of the beam of light was
changed at intervals.
The amoeba in each case
changed its direction so as
to avoid the light.
Time elapsed : 15 minutes

Light

Figure 6-10. Reactions of amoeba to various stimuli. Arrows show direction of movement.

pseudopodia, locomotion, and capture of food. These behaviors are responses to changes in the amoeba's external environment or to internal needs such as "hunger." The environmental change is called a **stimulus,** and the organism's reaction, a **response.** The amoeba responds to many types of stimuli, including changes in contact, light, temperature, chemicals, and electricity. Movement toward a stimulus is called a **positive response;** away from a stimulus it is a **negative response.**

Contact. When touched by a small rod, an amoeba will cease locomotion for a time and then move away,

thus exhibiting a delayed negative response to contact or mechanical shock (Fig. 6-10). However, when a floating amoeba touches a solid object, it will react positively.

Chemicals. Food selection by the amoeba undoubtedly involves responses to chemical stimuli; a positive response results in ingestion of a food particle and a negative response in its avoidance. Amoebas react negatively to various chemicals, such as table salt (sodium chloride), acetic acid, cane sugar, and methyl green (Fig. 6-10).

Light. The amoeba generally moves away from a strong light source (Fig. 6-10), but it may respond positively to a very weak light.

Temperature. As previously noted, the rate of locomotion depends on the temperature of the amoeba's environment. A local increase in temperature results in movements away from the stimulus, that is, in a negative response. If the temperature is decreased sufficiently, movements cease.

Conclusions. These examples of behavior show that the amoeba responds to a variety of stimuli. Its responses are advantageous to the individual and hence to the preservation of the species. Negative responses, elicited in most cases by injurious agents, such as strong chemicals, heat, and mechanical impacts, carry the organism away from danger.

The data thus far obtained indicate that factors are present in the behavior of the amoeba comparable to the habits, reflexes, and automatic activities of multicellular animals.

Other Sarcodines

The amoeba was first reported by Roesel in 1775, although which species he observed is not clear.

Our example, *A. proteus,* has been described as "a shapeless mass of protoplasm," but this is much too simplistic. Although the amoeba continually changes shape, it has definite characteristics such as a disc-shaped nucleus, blunt pseudopodia, and often longitudinal ridges on its surface. Furthermore, the amoeba contains all the cytoplasmic organelles (or endosymbionts) necessary for its metabolic activities. Many species of amoebas have been described from fresh and marine water and moist soils, and others are parasites in animals. Some have been placed in the genus *Amoeba,* and the rest are assigned to other genera.

Pelomyxa palustris is a large amoeba that may reach a diameter of 2 mm; it has many nuclei and flows along without definite pseudopodia. This species has drawn recent interest because of its primitive characteristics, such as the lack of spindle-aided mitosis and the absence of mitochondria. Its many nuclei divide by a bacterialike binary fission, whereas its energy needs are supplied by **endosymbiotic** aerobic bacteria. This certainly lends support for the theory of endosymbiotic origins of all mitochondria. Another large species of this genus, *P. carolinensis,* also known as *Chaos chaos* or the giant amoeba, contains about 50 to 100 times more volume than *A. proteus;* it may reach a length of from 2 to 5 mm and can be easily seen with the unaided eye. This species usually contains about 300 to 400 nuclei and 3 to 12 water-expulsion vesicles. Instead of dividing into two daughter amoebas, it generally divides into three.

Several types of common freshwater sarcodines are protected by shells. *Arcella* secretes its shell, whereas *Difflugia* builds a shell of minute sand grains. In both types, pseudopodia are thrust out through a circular opening in the shell; they serve, as in the naked amoebas, for locomotion and capture.

Heliozoans are freshwater amoebas that are often called "sun animalcules" because of their spherical bodies and radiating pseudopodia. The pseudopodia are not used for locomotion, but are covered with an adhesive ectoplasm that ensnares prey organisms. Each pseudopodium is supported by a central axial rod of protoplasmic fibers. These can change shape, allowing the pseudopodium to shorten, bend, or be resorbed into the cytoplasm. Most heliozoans lack a shell. *Actinophrys* is a common heliozoan genus.

Marine Sarcodines

Most of the roughly 8000 species of sarcodines live in the ocean. Foraminiferans ("foramen" usually refers to a hole), of which *Globigerina* is an example, construct a perforated shell of calcium carbonate, through which slender pseudopodia (Fig. 6-11) pro-

trude. Radiolarians generally build elaborate skeletons of glasslike silica.

Sarcodines and Geology

Although most protists leave no substantial remains after death, the skeletal structures of foraminiferans and radiolarians, and the organic walls of many protistan, may be preserved as fossils. Because species of this type have existed since very early times, a great number of distinct rock strata bear evidence of protistan life. Fossils of foraminiferans are important to geologists in analyzing the results of oil-drilling operations.

Today, as in the past, the skeletons of "forams" and radiolarians are continually sinking to the bottom of the ocean, where they form ever-growing layers of ooze. Calcareous ooze, composed of the calcareous skeletons of forams mixed with brown clay, is the most widespread sediment in the world today, carpeting over half the ocean floor. The greater portion of the Atlantic Ocean floor, an area of perhaps 20 million square miles, is covered with an ooze of foram skeletons of the genus *Globigerina.* Eventual compaction of such ooze results in the formation of chalks, limestones, and marbles. An enormous chalk deposit in Alabama and Mississippi, undoubtedly created in this way, is approximately 300 m thick in certain places. The chalk cliffs of Dover are composed mostly of foraminiferan shells. The stone in the Egyptian pyramids is composed of the skeletons of very large forams of the genus *Camerina* (formerly called *Nummulites*).

Siliceous ooze of radiolarian origin is especially abundant in deep equatorial waters. Approximately 1.4 million km² of the Pacific and Indian ocean basins are covered by this glasslike material. Radiolarian ooze may also eventually become sedimentary or metamorphic rock.

Soil Protists

Soil fertility is affected not only by bacterial action but also by the local protozoan fauna. Both high organic content and small particle size favor large populations of bacteria and protozoans in moist soil. There appears to be a definite relationship between the number of protists and the number of bacteria present, as the latter serve as food for many of the proto-

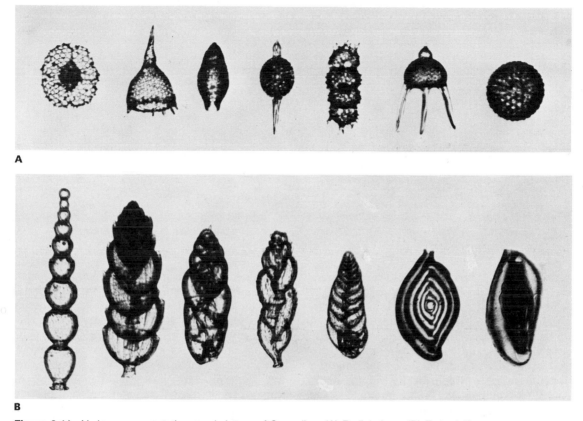

Figure 6-11. Various representative exoskeletons of Sarcodina. (A) Radiolarians. (B) Foraminiferans.

zoans. Some 200 to 300 protozoan species have been identified from soils, with small flagellates being most common and amoebas and ciliates correspondingly less so. Most soil-dwelling protists are found near the surface, the greatest concentration being at a depth of about 1 cm; a few are found at depths of about 50 cm. Very few are ever found in subsoil. Certain species show a surprisingly wide geographical distribution. *Amoeba proteus*, for example, has been found in soils from almost every part of the world. *Trinema* also has a cosmopolitan distribution.

Parasitic Sarcodines

Relatively few sarcodines are medically significant, and those few are endoparasites that generally infest parts of the digestive tract. *Entamoeba* (formerly *Endamoeba*) *gingivalis* lives in the human mouth, where it sometimes contributes to **pyorrhea,** a gum disease, although it is not considered the primary cause of that condition. Kissing is the most common means of transmission from host to host. Over 50 percent of the general population is infected.

In the intestine, especially in the colon, a number of amoebas may occur. *Entamoeba histolytica* (Fig. 6-12), which causes **amoebic dysentery (amoebiasis),** is the only serious disease-producing parasite of this group. About 10 percent of the general population is infected, especially in tropical regions, but fortunately most hosts are merely carriers; that is, the entamoebas are present but do no damage, and hence no symptoms appear. Occasionally, however, entamoebas may invade the intestinal wall and form abscesses, which later rupture and become persistent

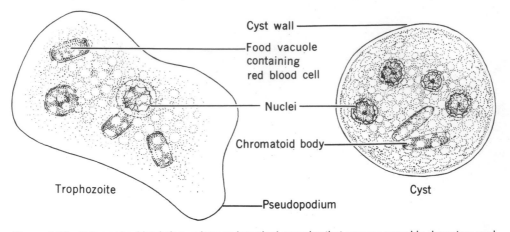

Figure 6-12. *Entamoeba histolytica,* a human intestinal amoeba that causes amoebic dysentery and amoebic liver abscess.

ulcers, resulting in diarrhea and dysentery. Occasionally, they invade the bloodstream and are carried to other parts of the body, such as the brain or liver, where abscess formation may cause death. Infected carriers pass encysted amoebas in their feces. These cysts may gain access to new hosts through contaminated food or water, soiled hands, or the activities of flies. Though more prevalent in the tropics, amoebiasis is fairly common in the temperate zones, and local outbreaks have even been reported in arctic regions. Certain drugs can effectively treat amoebic dysentery, once the presence of the parasites has been diagnosed by microscopic examination of stool samples.

Other intestinal amoebas are less pathogenic, such as *Entamoeba coli,* which is a harmless species often found associated with *E. histolytica. Dientamoeba fragilis,* which is characterized by the presence of two nuclei and does not form cysts, causes mild diarrhea.

Origins and Relationships between Sarcodines and Mastigophorans

Many zoologists theorize that the mastigophoran flagellates are the most primitive protozoans and that they probably share ancestry with the green algae. Many green mastigophorans, such as *Volvox,* can hardly be distinguished from other green algae. A close relationship between mastigophorans and sarco-

dines is indicated by the observation that certain species contain both amoeboid and flagellated stages in their life cycle. Indeed, some species can be induced to transform from one type of locomotion to the other by altering their environment; for example, a high salt concentration could give rise to an amoeboid form and a low salt concentration to a flagellated condition. Also, certain sarcomastigophorans (e.g., *Mastigamoeba*) possess both a flagellum and pseudopodia simultaneously. These lines of evidence suggest that the amoebas and flagellates evolved from common ancestors but have since undergone an immense evolutionary radiation. The primitive characteristics discussed concerning *Pelomyxa palustris* support a primitive amoeboid ancestor. The recent wider acceptance of endosymbiotic origins of eucaryotic protists, especially theories including *multiple* instances of organelle origin by endosymbiotic "invasions," presents added confusion to systematists, as caution must be exercized when organellar characteristics are used to derive evolutionary relationships.

Summary

Opalina commensals depend upon a cyst stage in their life cycle for transmission from host to host. Experiments to uncover the factors responsible for encystment were described in this chapter to illustrate the scientific approach to a problem.

Amoeba proteus is a sarcodine that lives in wet places and moves about with pseudopodia. Morphological, physiological, and behavioral adaptations enhance the amoeba's survival.

Review Questions

1. Suppose for a moment that a giant amoeba existed (it developed numerous invaginations of its cell membrane so that diffusion was no longer a problem). If such an amoeba attacked you, how might you repel it? (Name three ways).
2. Would you expect a marine amoeba (whose cytoplasm is isotonic to salt water) to possess water-expulsion vesicles?
3. Would you expect to find ATP in an amoeba? Where?
4. What would happen to a freshwater amoeba that for some reason was unable to form water-expulsion vesicles?
5. What observations suggest that members of the superclasses Sarcodina and Mastigophora are closely related?
6. Describe the scientific approach to a problem and illustrate with an example (other than *Opalina* encystment induction).

Selected References

Calkins, G. N., and F. M. Summers (eds.). *Protozoa in Biological Research.* New York: Columbia University Press, 1941.

Cheung, A. T., *et al. IV Int. Congress of Protozoology.* Clermont-Ferrand, France, 1973.

Cushman, J. A. *Foraminifera, Their Classification and Economic Use.* Cambridge: Harvard University Press, 1948.

Dogiel, V. A. *General Protozoology,* 2nd ed. Oxford at the Clarendon Press, 1965.

Hagelstein, R. *The Mycetozoa of North America.* Published by R. Hagelstein, Mineloa, New York, 1944.

Hall, R. P. *Protozoology.* Englewood Cliffs, N. J.: Prentice-Hall, Inc., 1953.

Hunter, S., and A. Lwoff (eds.). *Biochemistry and Physiology of the Protozoa.* II. New York: Academic Press, Inc., 1955.

Jahn, T., and F. F. Jahn. *How to Know the Protozoa.* Dubuque: Wm. C. Brown, 1949.

Jahn, T., and J. Votta. "Capillary Suction Test of the Pressure Gradient Theory of Amoeboid Motion," *Science,* **177**:636–637, 1972.

Jeon, K. W. (ed.) *The Biology of Amoeba.* New York: Academic Press, 1973.

Kirby, H. *Materials and Methods in the Study of Protozoa.* Berkeley: University of California Press, 1950.

Kudo, R. R. *Protozoology,* 5th ed. Springfield: Charles C Thomas, 1966.

Lwoff, A. *Problems of Morphogenesis in Protozoa.* New York: John Wiley & Sons, Inc., 1950.

Perkins, D. L., and T. L. Jahn. "Amoeboflagellate Transformations & the Gibbs-Donnan Ratio," *The Journal of Protozoology,* **17**(2):168–172, 1970.

Pitelka, D. R. *Electron-microscopic Structure of Protozoa.* New York: Pergamon Press, Inc., 1963.

Sleigh, M. A. (ed.) *Cilia and Flagella.* New York: Academic Press, 1974.

Sonneborn, T. M. "Breeding Systems, Reproductive Methods, and Species Problems in Protozoa," in *The Species Problem.* Washington: American Association for the Advancement of Science, 1957.

Wigg, D., E. Bovee, and T. Jahn. "The Evacuation Mechanism of the Water Expulsion Vesicle ("Contractile Vacuole") of *Amoeba Proteus,*" *The Journal of Protozoology,* **14**(1):104–108, 1967.

Kingdom Protista, Phylum Ciliophora

When compared with other protozoans, the ciliates are a relatively large and complex group. The ciliates are distinguished from other protozoans in possessing cilia, two types of nuclei, and a unique form of reproduction called conjugation. Most ciliates are free-living aquatic forms, but some are important parasites of humans and other animals.

The ciliates play important roles in aquatic food chains. Many ciliates ingest dissolved nutrients and smaller microorganisms and in turn serve as food for small multicellular animals, which are then eaten by larger animals.

Paramecium caudatum—A Freshwater Ciliate

Paramecia are common in pond water among decaying vegetation (Fig. 7-1). These ciliates were among the first microorganisms observed through the microscope following its invention in the seventeenth century. The biology of paramecium has been investigated through the years, and recently it and related ciliates have been used for studies of nutrition, respiration, cancer, heredity, reproduction, behavior, and ecology.

The body of *Paramecium caudatum* superficially resembles a twisted bedroom slipper. This representative ciliate is easy to obtain, and its relatively large size (about 0.3 mm) makes it easy to study. Many other species of ciliates may be found in cultures made from pond weeds and decaying plant and animal infusions.

Morphology

The ten well-known species of paramecium differ somewhat in size, shape, and structure. The following description applies principally to *P. caudatum* (Fig. 7-2) and *P. aurelia*. The former ranges from about 0.15 to 0.3 mm in length and the latter from about 0.12 to 0.2 mm.

The anterior end is blunt and the posterior end more pointed. The greatest width is behind the center of the body.

A depression called the **oral groove** extends from the anterior end obliquely downward and posteriorly into the endoplasm. The oral groove leads to the cell mouth (**cytostome**). The side containing the oral groove is designated **oral** and the opposite side **aboral.** As in the amoeba, the cytoplasm can be divided into an outer, comparatively thin and clear, **ectoplasm** and an inner, granular, **endoplasm.** In addition, a distinct elastic membrane, the **pellicle,** covers the outer surface of the ectoplasm. Fine hairlike **cilia** are regularly arranged on the surface of the pellicle. A large **water-expulsion vesicle** (Fig. 7-2) is usually situated near each end of the body, close to the aboral surface, and a variable number of **food vacuoles (phagosomes)** are present in the endoplasm. Two different nuclei are present: a large **macronucleus** concerned with vegetative (nonreproductive)

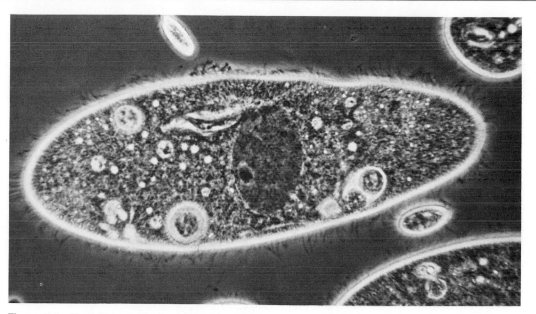

Figure 7-1. Photomicrograph of a paramecium. 250 ×.

functions and a smaller **micronucleus** that is important in reproduction. These nuclei are suspended within the endoplasm near the oral groove. A temporary opening, called the **cell anus,** can be observed only when undigested particles are discharged. It always reforms at the same point posterior to the oral groove.

The endoplasm occupies the central part of the body. Most of the larger granules contained within it have been shown by microchemical assays to be reserved food particles. The granules flow from place to place, indicating that the endoplasm is fluid. The ectoplasm does not contain any large granules, as its greater density prevents their entrance. If a drop or two of 35 percent alcohol is added to a drop of water containing paramecia, the pellicle in some specimens will be raised in the form of a blister. Under high power of the light microscope, the pellicle can then be observed to consist of a great number of hexagonal areas produced by ridges under the surface (Fig. 7-3). The distribution of the locomotor organelles, the cilia, corresponds to this arrangement of ridges, with one cilium projecting from the center of each hexagon. These hairlike organelles occur on all parts of the body, with those at the posterior end being slightly longer than elsewhere. The thousands of cilia have a complicated root system, called the **infraciliature.** Each cilium terminates in a **basal body;** the basal bodies are interconnected by fibers, and this infraciliary network helps to coordinate the activity of cilia, which is rather complicated within the oral groove, where cilia guide the food particles that are swept within their reach.

Physiology

Physiological processes similar to those described for the amoeba occur in the paramecium. Ciliates capture and ingest food, digest it, build up protoplasm, react to stimuli, carry on processes of respiration and excretion, and reproduce.

Locomotion

The cilia of the ciliates cover the body surface and are about 50 times shorter (5–10 μ) than are the flagella of mastigophorans. However, both cilia and flagella have the 9 + 2 fiber arrangement, and both normally beat in traveling helical waves. The coordinated

Anterior end

Cilium

Trichocyst

Food vacuole 3

Radiating canal

Water expulsion vesicle

Pellicle

Ectoplasm

Endoplasm

Food vacuole 2

Bacteria being ingested

Oral groove

Micronucleus

Macronucleus

Cytosome

Cytopharynx

Food vacuole forming

Cell anus

Radiating canal of water
expulsion vesicle

Food vacuole 1

Figure 7-2. Left: Spiral path of a free-swimming paramecium. Note that these animals rotate on their long axes at the same time that they are advancing. Right: Internal structure of *Paramecium caudatum*. Numbered food vacuoles show progress of digestion and absorption.

helical beats of all the cilia propel the ciliate forward (see Fig. 7-2). To reverse direction, the cilia do not change the pattern of the beat but, rather, change the direction of the axis of the helical waves. The ciliary motion describes a cone when the paramecium is stationary. As the paramecium swims forward, it spirals around a more or less straight course. This spiral locomotion results from two factors. There is rotation about the long axis of the body due to the oblique orientation of the cilia and, second, a swerving of the body away from the oral side caused by the oral groove's cilia beating faster than others. Another component of the spiral forward motion may be due to the so-called **metachronal waves** that sweep across the entire ciliary blanket.

Ciliary activity is apparently controlled and coordinated by the infraciliary network that interconnects the basal bodies of the cilia. It was once thought that the fibrils performed this function and were linked to a "neuromotor" that drove the entire system. However, this "neuromotor" was found to be a staining artifact.* How the fibrils function is not completely known. The beats of the cilia appear to be initiated by the presence of certain ions. Electrical stimulation or the action of cations can shift the axis of the helical waves, causing the ciliate to reverse its direction.

*The term "artifact" in zoology means an abnormality of cells or organisms that is *artificially* induced.

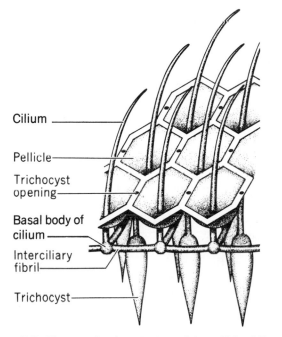

Cilium

Pellicle

Trichocyst
opening

Basal body of
cilium

Interciliary
fibril

Trichocyst

Figure 7-3. Diagram showing structure of the pellicle of *Paramecium*. The hexagonal areas are due to ridges; the cilia extend out from the center of the hexagonal areas, each being attached to a basal body; the basal bodies are located on longitudinal fibers. The carrot-shaped trichocysts are attached by delicate threads to the ridges.

Offense and Defense

The paramecium possesses carrot-shaped structures called **trichocysts** that are embedded in the ectoplasm just beneath the body surface (Fig. 7-3). These trichocysts are about 4 μ long and lie perpendicular to the body surface. A small amount of iodine or acetic acid, when added to a drop of water containing paramecia, causes discharge of trichocysts to the exterior. Each trichocyst is attached to a thread some 6 to 8 times longer than itself. After the explosive discharge, the organism is surrounded by a halo of long threads.

The function that trichocysts serve in *P. caudatum* is not known. The paramecium feeds principally on bacteria and on other protozoans that, as a rule, are smaller than itself. No special offensive weapons appear to be necessary for food capture. Paramecia themselves are attacked and preyed upon by other protozoans and by some small metazoans. For example, the ciliate *Didinium* uses its trichocysts offensively in capturing a paramecium. The trichocysts of *Didinium* contain an immobilizing toxin, whereas those of *Paramecium* do not. The paramecium in turn discharges its trichocysts, but these have no effect on the didinium, except perhaps to get in its way. Some zoologists have proposed that trichocysts serve to attach the paramecium to other objects. However, specimens lacking trichocysts are able to adhere to solid substrata. At times trichocysts are discharged with no apparent stimulus. It is possible that trichocysts secrete salts and may serve an **osmoregulatory** function.

Nutrition

Paramecia do not possess chlorophyll and hence are unable to manufacture food by photosynthesis as euglenas do. One species, however, *P. bursaria*, contains unicellular algae within its endoplasm. This species can grow and reproduce in a solution of inorganic salts if it is kept in the light. Paramecia are generally holozoic feeders that may consume as many as 5000 bacteria in a single day. Paramecia are selective feeders. For example, if different species of bacteria are present, they may feed on one species and not on another.

Bacteria, yeasts, small protozoans, and algae are captured with the aid of the cilia in the oral groove. These oral cilia drive a steady stream of water toward the cytostome. Food particles that are swept into the cytostome are carried down into the **cytopharynx** (cell gullet). Cilia in the cytopharynx gather the particles together near the end into a **food vacuole.** When the vacuole has reached a certain size, it is released into the surrounding cytoplasm, and the formation of another vacuole begins.

As soon as a vacuole containing the water and suspended food particles separates from the cytopharynx, it is swept away by the rotary streaming movement of the endoplasm, known as **cyclosis** (Fig. 7-4). This carries the food vacuole around a definite course that begins just behind the cytopharynx, passes posteriorly, then forward and aborally, and finally posteriorly to the vicinity of the oral groove. During this journey, the food is digested by enzymes

Anterior

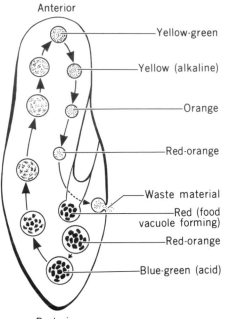

Yellow-green

Yellow (alkaline)

Orange

Red-orange

Waste material

Red (food
vacuole forming)

Red-orange

Blue-green (acid)

Posterior

Figure 7-4. *Paramecium caudatum.* Diagram illustrates cyclosis and the process of digestion. To indicate the path of the food vacuole within the paramecium, some yeast cells stained with Congo red should be placed near the animal. The yeast cells are taken into the body, where a food vacuole is formed. The arrows in the figure indicate the path of the food vacuole within the body (cyclosis). The change in color of the Congo red to blue-green indicates that the vacuole became acid soon after it was formed. As the vacuole with the yeast cells circulates through the body, it changes back to a red-orange color, indicating that the vacuole and its contents are becoming less acid. This type of experiment shows that, in digestion of food, the vacuole is first acid and then alkaline.

from the endoplasm, nutrients are absorbed, and the vacuole becomes smaller. The digested food is either stored—catabolized for metabolic energy—or converted into protoplasm. Undigested particulate wastes are eliminated through the cell anus.

Regulation of Water Content

The water-expulsion vesicles function primarily to remove excess water from the cell. Water is continu-

ally entering the cell by osmosis through the semipermeable pellicle, following the concentration gradient from the freshwater surroundings into the more concentrated, hypertonic cytoplasm.

Two water-expulsion vesicles are present in *P. caudatum,* occupying definite positions near each end of the body. Each communicates with the surrounding cytoplasm by means of a system of six to eleven **ampullae** (radiating canals) (Fig. 7-2). These ampullae fill with liquid, then discharge their contents into the swelling vesicle, which in turn expels the liquid through a permanent pore to the exterior. After each emptying, a new cycle of gradual expansion and sudden collapse commences.

Excretion

Most of the soluble waste products of paramecia (and other protozoans) diffuse through the pellicle into the external environment. Some nitrogenous wastes (ammonia) may also be excreted by the water-expulsion vesicles.

Respiration

Oxygen, dissolved in the water, diffuses through the surface of the body into the cytoplasm. Carbon dioxide primarily diffuses out across the general body surface, although the water-expulsion vesicles probably discharge some dissolved carbon dioxide as well as nitrogenous wastes.

Behavior

As in the amoeba and euglena, changes in the paramecium's environment serve as stimuli to which it responds in various ways.

One of the most common responses of the paramecium is known as the **avoiding reaction (negative response)** (Fig. 7-5). When a free-swimming paramecium encounters a harmful chemical, such as a strong salt solution, it responds with a characteristic series of behaviors that serve to remove it from the harmful stimulus. First, it swims backward for a short distance; then, its gyrations decrease in frequency, and it swerves toward the aboral side. The paramecium then pivots in a circle around its posterior end. Dur-

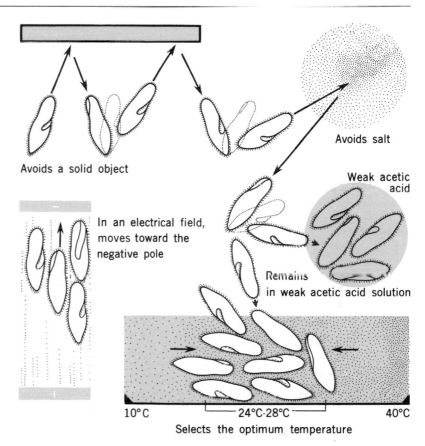

Avoids a solid object

Avoids salt

In an electrical field, moves toward the negative pole

Weak acetic acid

Remains in weak acetic acid solution

10°C   24°C-28°C   40°C

Selects the optimum temperature

Figure 7-5. Behavior of the paramecium in various environmental conditions.

ing this revolution, samples of the surrounding medium are drawn by ciliary currents into the oral groove. When the noxious chemical is no longer detected, the organism moves forward again. If this movement once more brings it into the region of the harmful chemical, the avoiding reaction is repeated. This behavior continues as long as the paramecium encounters a stimulus that elicits a negative response.

Chemicals of various sorts and concentrations have striking effects on the paramecium. As already discussed, a sodium chloride solution elicits a negative response. Weak acetic acid attracts the paramecium—that is, it responds positively—and mercuric chloride kills it (Fig. 7-5).

The paramecium reacts not only to chemicals but also to contact, changes in temperature, light, electric currents, and other stimuli (ions and electric fields can act directly on the cilia). If the anterior end of a paramecium, which is more sensitive than the other parts of the body, is touched with a glass rod, the avoiding reaction is evoked. Frequently, a slowly swimming paramecium comes to rest with its cilia in contact with an object; this **positive response** often brings it into an environment rich in food, as bacteria are more abundant on solid surfaces than dispersed in the water. Paramecia do not respond either positively or negatively to visible light, but they avoid ultraviolet rays.

The optimum temperature for the paramecium ordinarily lies between 24°C and 28°C. When several paramecia are placed on a slide that is heated at one end, they will swim about in all directions, giving the avoiding reaction when stimulated by heat, until they become oriented and independently move toward the

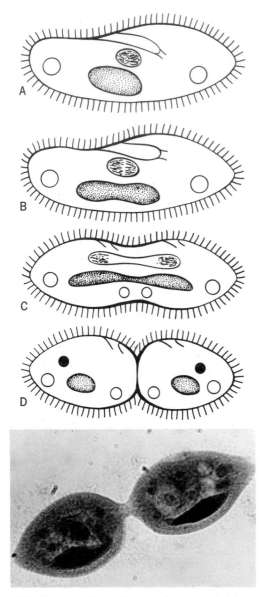

Figure 7-6. Binary fission of *Paramecium caudatum*.

specimen that is in contact with a solid surface is also affected by temperature, chemicals, heat, and other stimuli. The *physiological condition* of a paramecium determines the character of its response. This physiological state is a dynamic condition, changing continually with the process of metabolism going on within the organism.

Reproduction

The paramecium usually multiplies by simple **binary fission.** This asexual process is interspersed occasionally by a temporary union (**conjugation**) of two individuals that involves a mutual nuclear fertilization.

Binary Fission

Binary fission is an asexual process in which one fully grown individual divides transversely into two daughters (Fig. 7-6). First the micronucleus undergoes **mitosis,** producing two identical daughter micronuclei; these separate and migrate to either end of the body. Meanwhile the macronucleus elongates and then divides transversely by direct nuclear division without condensed chromosomes or spindle formation. The cytopharynx produces a bud that develops into another cytopharynx. Two new water-expulsion vesicles arise, and the number of cilia are roughly doubled. While these events are taking place, a constriction appears near the middle of the body; this cleavage furrow deepens until only a slender thread of protoplasm holds the two halves of the body together. This connection is finally severed, and the two daughter paramecia begin independent lives.

The entire process generally takes about 2 hours. The time, however, varies considerably, depending upon the temperature of the water, the quality and quantity of food, and other factors. The daughter paramecia increase rapidly in size, and at the end of 24 hours divide again if the temperature remains at 15–17°C. If the temperature is raised to 17–20°C, two divisions may take place in one day. A paramecium under optimum conditions may reproduce at the rate of 600 or more generations per year. If all the descendants of one individual were to live and reproduce at this rate, they would soon equal the earth in volume. Under natural conditions, however, popula-

cooler end. This is an example of trial-and-error behavior; that is, the organism tries all directions until one is discovered that allows it to escape from the harmful stimulation.

Frequently, a paramecium may be stimulated in more than one way at the same time. For example, a

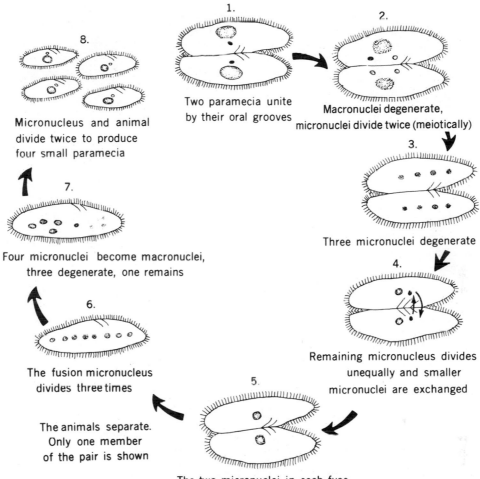

1. Two paramecia unite by their oral grooves

2. Macronuclei degenerate, micronuclei divide twice (meiotically)

3. Three micronuclei degenerate

4. Remaining micronucleus divides unequally and smaller micronuclei are exchanged

5. The two micronuclei in each fuse

6. The animals separate. Only one member of the pair is shown

The fusion micronucleus divides three times

7. Four micronuclei become macronuclei, three degenerate, one remains

8. Micronucleus and animal divide twice to produce four small paramecia

Figure 7-7(A). Conjugation in *Paramecium caudatum*. Not all stages are shown. In the interest of clarity, the macronuclei are omitted from the third stage; actually they do not disappear completely until after the conjugating animals have separated.

tion size is limited by lack of food, low temperatures, drought, predation by other animals, and other factors.

Conjugation

The paramecium ordinarily multiplies by binary fission, but at intervals this may be interrupted by the sexual process of conjugation (Fig. 7-7). Conjugation begins when two paramecia come together and touch along their oral surfaces, where a protoplasmic bridge forms between them (Fig. 7-8). During conjugation the pair continues to swim about. As soon as this union is effected, the macronuclei degenerate, and the micronuclei pass through a series of divisions in which the chromosomal number is reduced to one half (**haploid**) the normal complement; this has been likened to the meiotic maturation processes of metazoan germ cells (see Chapter 39). Three of the four haploid micronuclei disintegrate, and the fourth di-

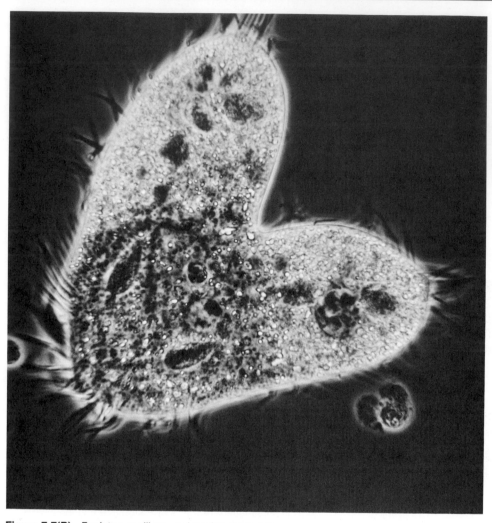

Figure 7-7(B). *Euplotes,* a ciliate, undergoing conjugation.

vides again to produce two gametic nuclei, one of which is smaller than the other. During conjugation there is a mutual interchange of micronuclei. The migratory micronucleus (Fig. 7-7) is smaller than the stationary micronucleus and may be considered comparable to the nucleus of a male germ cell. The migratory micronuclei are exchanged across the cytoplasmic bridge, and the stationary micronucleus of each fuses with the micronucleus of its neighbor. This fusion of haploid nuclei restores the normal (**diploid**) number of chromosomes and is therefore analogous to the fusion of gametes during other types of sexual reproduction. The ciliates then separate, and subsequent nuclear and cytoplasmic divisions restore the normal number of macronuclei and micronuclei.

Conjugation is similar to the fertilization of metazoans or other protozoans in that there is a mixture of nuclear materials (genes) from two individuals. However, after conjugation the ciliates continue to reproduce by asexual division in contrast to most metazoans

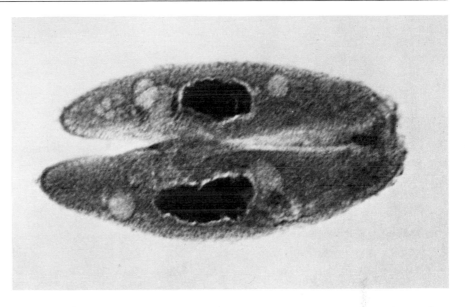

Figure 7-8. What is occurring in this photomicrograph?

in which there is asexual division during growth, but only sexual reproduction of the entire organism.

If paramecia are kept in a constant medium (e.g., hay infusion), they undergo a period of physiological depression about every three months, as evidenced by a decrease in their rate of division. Semiannual periods of physiological depression also occur, and recovery does not take place if the animals are kept under constant conditions or if conjugation is prevented. The protoplasm degenerates and becomes vacuolated, and the organisms become inactive and finally die. This suggests that conjugation is essential for continued fitness.

Experiments performed on one species suggest that in a varied environment neither conjugation nor death from old age necessarily occurs. In one experiment, a paramecium culture was carried through a period of over 25 years without conjugation by changing the character of the medium daily. During this time there were over 25,000 generations, and there was no evidence of a decline in vitality of the organisms, as indicated by the rate of division. The cycle may thus be prolonged by employing a varied culture medium. However, it is now known that the paramecium may undergo a process of self-fertilization called **autogamy** at regular intervals but that some other ciliates, over a period of years, do not undergo either conjugation or autogamy. Therefore, periodic nuclear reorganization may be essential in some ciliates but not in others.

Mating Types in Conjugation

At one time it was thought that there were no differences between the two paramecia that join in conjugation, but it is now recognized that there are distinct mating types within each species. For example, in *P. aurelia* there are at least sixteen varieties, most of which contain two mating types. Conjugation only occurs between individuals of the same variety but different mating types. *P. bursaria* contains six known varieties: one with two mating types, three with four, one with eight, and one in which conjugation has never been observed.

Other Ciliophorans

Ciliates are widespread in nature. Many species inhabit freshwater and marine habitats; others live in moist soil and as parasites upon or within the bodies of animals. Most ciliates are free living. Figure 7-9 shows representatives of some common free-living forms.

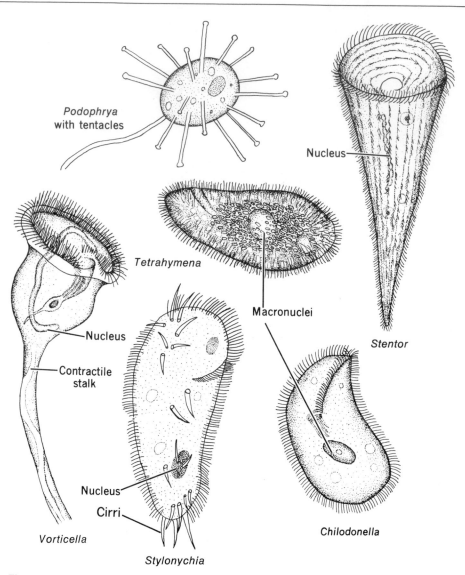

Figure 7-9. Some representative ciliophorans.

Tetrahymena, a small ciliate, can be cultured in a medium free from all other microorganisms and has about the same nutritional requirements as humans. Studies show that it must have a diet of vitamins, amino acids, salts, and sugar. This suggests that a single-celled organism may have metabolic processes nearly as complex as our own. This tiny organism is

playing an increasingly important role in physiological and genetic research.

Stentor is a trumpet-shaped ciliate, bluish in color, with a beaded macronucleus and spirally arranged cilia at its oral end; it may be free swimming or attached. *Stylonychia* has a flattened body and groups of cilia fused to form structures called **cirri,** which are used in creeping. *Vorticella* resembles an inverted bell attached to a contractile stalk.

Parasitic Ciliates

One well-known parasitic ciliate, *Balantidium coli,* is the largest protozoan (40–80 μ) found in the human intestinal tract. This large ciliate penetrates the membrane lining the colon, where it frequently causes ulcers. It is a definite disease-producing organism and may cause symptoms resembling acute amoebic dysentery. Infected persons pass large spherical cysts, which are easily identified by microscopic examination of their feces. *B. coli* is also common in pigs and in chimpanzees, both of which probably serve as reservoirs for human infection.

Many domesticated and wild animals, both terrestrial and aquatic, are parasitized by ciliates, most of which invade the digestive tract. One ciliate lives in the intestine of the earthworm. About 40 species of very complex ciliates live in the first and second stomach chambers of cattle; these may be commensals; that is, they neither benefit nor harm their host. Two common ciliates, *Kerona* and *Trichodina,* creep about on the bodies of certain aquatic animals and frequently occur on hydra; also of interest are species of *Podophyra,* which parasitize other ciliates.

Origin of Ciliphorans

The ciliates and mastigophorans are considered to be closely related to each other because of the presence in both groups of cilia and flagella, respectively, with their homologies of 9 + 2 fibers and basal bodies and certain biochemical similarities. Ciliates are more complex than mastigophorans in structure and function. Thus many investigators believe that the ciliates evolved from mastigophoran ancestors.

Summary

Protists belonging to the phylum Ciliophora possess cilia at some stage of their life cycle. Also, the nuclear material is separated into a large macronucleus and a smaller micronucleus, and reproduction by sexual conjugation occurs. Most ciliates are free living in fresh or salt water, but many are parasites of other animals. Structurally, the ciliates are the most complex protozoans.

Review Questions

1. What is taking place in Figure 7-8? Explain.
2. What is the function of trichocysts in *P. caudatum*? In *Didinium*?
3. How many kinds of organelles does *P. caudatum* have? What are they?
4. Describe the infraciliature of paramecium.
5. How are cilia and flagella similar? How do they differ?
6. What does a paramecium eat? How does the paramecium acquire its food?

Selected References

Corliss, J. O. *The Ciliated Protozoa: Characterization, Classification, and Guide to the Literature.* New York: Pergamon Press, Inc., 1961.

Jahn, T., and J. Votta. "Locomotion in Protozoa," *Annual Review of Fluid Mechanics,* **4,** 1972.

Jahn, T. L. "The Mechanism of Ciliary Movement. II. Ion Antagonism and Ciliary Reversal," *Journal of Cellular & Comparative Physiology,* **60**(3), 1962.

Jones, A. R. *The Ciliates.* New York: St. Martin's Press, 1974.

Kuznicki, L., *et al.* "Helical Nature of the Ciliary Beat of *Paramecium Multimicronucleatum,*" *The Journal of Protozoology,* **17**(1): 16–24, 1970.

Organ, A. E., *et al.* "The Mechanism of the Nephridial Apparatus of *Paramecium Multimicronucleatum,*" *The Journal of Cell Biology,* **37**(1): 139–145, 1968.

Sleigh, M. A. (ed.) *Cilia and Flagella.* New York: Academic Press, 1974.

Sonneborn, T. M. "Mating Types in *Paramecium Aurelia,*" *American Philsophical Society Proceedings,* **79,** 1938.

Wichterman, R. *The Biology of Paramecium.* New York: The Blakiston Company, 1953.

Kingdom Protista, Phyla Sporozoa and Cnidospora

All members of the protistan phyla Sporozoa and Cnidospora are endoparasites that can potentially invade the internal tissues of nearly every major animal group. Sporozoans and cnidosporans differ from other protozoans in two significant ways:

1. Most members of both phyla have resistant stages when they form **spores,** which are transmitted from host to host.*
2. Sporozoans and cnidosporans have no locomotor organelles in their adult (**trophozoite**) stages, whereas other protozoans do possess such organelles (e.g., flagella, cilia, pseudopodia).

Cnidosporans were once classified with the sporozoans in a single taxon. However, certain morphological features clearly distinguish the two. Cnidosporans produce spores with polar filaments. On this basis, as well as additional differences in their life cycles, Cnidospora and Sporozoa are now commonly differentiated into two separate phyla.

Phylum Sporozoa

These parasitic protozoans are found in hosts ranging in complexity from simple invertebrates to hu-

mans and other mammals. The life cycles of many sporozoans involve more than one host. For example, *Plasmodium vivax,* one of the sporozoans responsible for malaria in humans, spends part of its life cycle within certain mosquitoes *(e.g., Anopheles).*

Morphology

Sporozoans exhibit a variety of shapes, ranging from spherical to snakelike cells. A typical sporozoan has an **ectoplasm,** which serves as a spore wall, surrounding a granular **endoplasm.** One nucleus and numerous **carbohydrate storage granules** are found in the endoplasm of the adults (Fig. 8-1). Locomotor organelles are not present in the adult. Water-expulsion vesicles are also absent, as the parasite's cytoplasm is isotonic with the tissue fluids of its hosts and hence does not accumulate excess water by osmosis.

Most sporozoans are polar, possessing an anterior modification of the ectoplasm called a *mucron*. The mucron usually serves to attach the parasite to its host, but it may also be used in feeding (some unattached sporozoans possess mucrons).

Physiology

Locomotion

Although the trophozoite (adult) stage of sporozoans lacks locomotor organelles, these protozoans are sometimes observed to move. Such locomotion may involve longitudinal and transverse fibers that exist in

---

*This spore state appears similar to that of an encysted *Opalina.* The distinction between spore and cyst is somewhat hazy. In general, spores are much smaller than cysts. A spore refers to the capsule with its inhabitant, whereas a cyst often refers to the capsule alone.

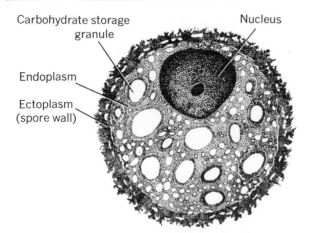

Carbohydrate storage granule

Nucleus

Endoplasm

Ectoplasm (spore wall)

Figure 8-1. Morphology of a typical sporozoan.

the ectoplasm. Some **sporozoite** stages (young infective sporozoans) possess locomotor organelles such as flagella or pseudopodia.

Feeding

Food is absorbed directly from the host through the cell membrane (saprozoic). Perhaps their limited mobility aids in feeding by increasing the mixing of molecules that are immediately adjacent to the cell membrane. The anterior mucron may also play a role in feeding.

Excretion and Respiration

Excretion and respiration take place by diffusion through the cell membrane (Fig. 8-2). Some accumulated waste materials are eliminated during fission.

Reproduction and Life Cycle

The life cycle of sporozoans is complex and often includes sexual and asexual stages. A generalized sporozoan life cycle involves three phases: (1) Following infection, the trophozoite reproduces asexually by **multiple fission (schizogony)** within the host; schizogony may be repeated several times, and large populations of these trophozoites, now called **merozoites,** result. (2) Eventually, by means of the process of **gamogony,** the asexual merozoites develop into sex-

ual **gametes,** and pairs of gametes unite to produce **zygotes.** (3) The zygote then multiplies to form the infective stage, the **spores,** that are encased within a protective spore envelope. Sometimes thousands of spores are produced from a single zygote through the process known as **sporogony.** This general sequence of schizogony, gamogony, and sporogony characterizes the phylum Sporozoa. Many variations have evolved as sporozoans have adapted to specific hosts.

The spore is often a spindle-shaped case containing many sporozoites (Fig. 8-3). Spores or their sporozoites serve as the infective stage in the sporozoan life cycle. They may pass out of one host in feces and then enter another host by means of contaminated food or drink; spores may also be transmitted by a bloodsucking insect (vector) that acts as an intermediate host.

Monocystis lumbrici—A Sporozoan Parasite of Earthworms

Monocystis lumbrici illustrates many characteristics of the phylum Sporozoa (Fig. 8-3). This parasite is almost invariably found in the seminal vesicles of the common earthworm (*Lumbricus terrestris*).

The life cycle of *Monocystis* is briefly outlined as follows: Mature spores in the soil are taken into the earthworm's digestive tract, where the sporozoites are released. The sporozoites penetrate the gut wall, enter the bloodstream, and eventually enter the seminal vesicles. The parasites, now called trophozoites, enter developing sperm cells and live at the sperm cell's expense. The sperm of the earthworm, which are deprived of nourishment by the parasite, slowly shrivel up, becoming tiny filaments on the surface of the trophozoite, making it resemble a ciliated organism. Pairs of older trophozoites come together in the sperm funnel and are surrounded by a cyst wall. Each then divides, producing a number of smaller cells, the gametes. Pairs of gametes unite to form zygotes. It is probable that gametes from the same trophozoite do not fuse with each other but rather cross-fertilize with those of the other trophozoite encosed within the same cyst. Each lemon-shaped zygote secretes a hard wall around itself and is now known as a spore. The nucleus of the spore divides successively into two, four, and finally eight daughter nuclei, each of which,

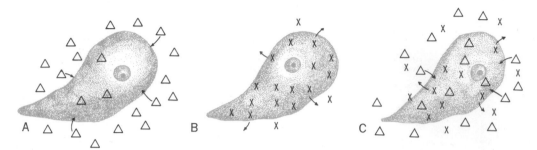

Figure 8-2. Osmoregulation in sporozoans. (A) As the sporozoan uses up food substances for growth and energy, the concentration of the food in the parasite becomes lower than the concentration outside the parasite, and more food substances diffuse in. (B) As food is used and respiration occurs, waste products build up in the parasite. These diffuse out into the host. (C) Because the total concentration of dissolved materials (food, waste, salts, etc.) is the same inside and outside the parasite, water is not accumulated or depleted.

together with a portion of the cytoplasm, becomes a sporozoite.

It is likely that the spores remain in the reproductive organs until the death of the worm, when they are released into the soil; or they may be transferred to the soil via the surface of the worm's egg cocoon. In either case the resistant spores remain dormant until they are ingested by a new earthworm host to start the cycle again.

Other Sporozoans

Some of the sporozoans of great importance to humans are described in the following paragraphs. Two types that are easily obtained for study are gregarines and coccidians.

Gregarines parasitize gut and body cavities of their invertebrate hosts, particularly annelids (segmented worms) and insects. They may be obtained from the intestines of grasshoppers, cockroaches, and mealworms. These insects ingest spores accidentally; once in the gut the spore envelope ruptures, and the sporozoites escape. The liberated sporozoites penetrate the cells of the intestinal wall and develop into trophozoites. After undergoing a period of growth, the trophozoites move into the lumen of the intestine, where they unite end to end (Fig. 8-4). The rest of the life cycle is similar to that of the gregarine *Monocystis*. The presence of an anchoring device in these

and other gregarines that live in the alimentary tract of their hosts seems to indicate a primitive form of cephalization. Attachment would be useful to an organism living in the gut, because fluid movements might otherwise eliminate it from the host's body.

Coccidians are intracellular parasites that infect both vertebrates and invertebrates, often with a complex life cycle that involves an alternation of vertebrate and invertebrate hosts. Coccidians are easily obtained for study from the rabbit. Oocysts may be found in the feces of most individuals. An oocyst consists of a single cell when eliminated, but four spores, each containing two sporozoites, develop after about three days.

Species of the haemosporidian *Plasmodium* cause **malarias.** Plasmodium trophozoites infect vertebrate red blood corpuscles where schizogony occurs, causing the erythrocytes to burst. The gamogony and sporogony stages occur in bloodsucking insects. Malaria is probably the most devastating human disease with regard to prevalence, mortality, sickness, and economic loss. Malaria is not yet under control, but there is hope that the concerted efforts of zoologists, physicians, engineers, educators, sociologists, and politicians may someday completely eradicate these parasites.

Three types of human malaria are each caused by a different species of *Plasmodium*. **Benign tertian malaria,** caused by *P. vivax*, is characterized by an attack of fever every 48 hours. **Quartan malaria,** caused by

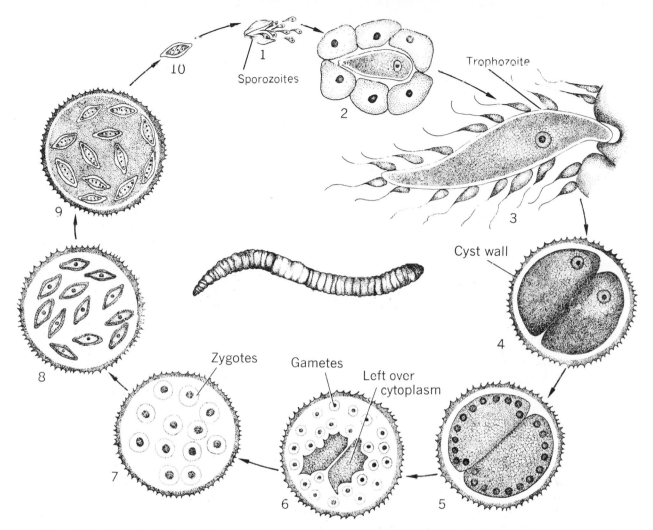

Figure 8-3. Life cycle of *Monocystis,* a sporozoan that lives in the common earthworm. 1, spores eaten by earthworm, sporozoites released; 2, young trophozoite in sperm morula; 3, trophozoite grows into worm, earthworm sperm develops from morula cells and attaches to trophozoite; 4, trophozoites associate in pairs and form a cyst wall (gametocytes); 5, nuclei of gametocytes divide; 6, gametes are produced; 7, gametes fuse (fertilization) to form zygotes; 8, zygotes secrete spore walls (sporocysts); 9, nuclei divide to form eight sporozoites in each spore; 10, spores liberated when earthworm dies. Spores may also be transferred to another worm during intercourse and infect testes of other worms. They may also be transferred to worm cocoon, entering the ground when the cocoon disintegrates.

P. malariae, usually produces fever every 72 hours. **Malignant tertian malaria,** caused by *P. falciparum,* manifests symptoms of irregular fever and causes the most human mortalities. The life cycles of all three species are essentially similar and are summarized below (Fig. 8-5).

Malaria is transmitted through the bite of a female mosquito (male mosquitoes lack piercing mouth parts and hence cannot suck blood). Animal malarias are carried by various specis of mosquitoes, but only those of the genus *Anopheles* harbor the human parasites.

As the insect's mouth parts pierce the skin, saliva containing **anticoagulants** is injected into the wound.

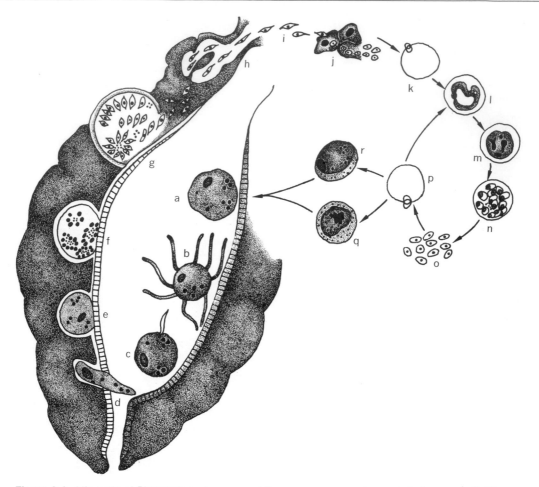

Figure 8-4. Life cycle of *Plasmodium vivax,* one of the sporozoans causing malaria in man. Left: Diagram of part of mosquito's body showing sexual stages. Right: Asexual stages that occur in the human blood vessel. Sexual stages: (A) egg in stomach, (B) sperm attached to outside of egg, (C) fertilization of egg, (D) ookinete, (E) young oocyst, (F) developing oocyst, (G) mature oocyst with sporozoites, (H) sporozoites in salivary gland. The sporozoites enter the human blood (I) while the mosquito feeds and migrate to liver tissues (J). From this reservoir, parasites enter red blood cells (K), and become trophozoites (L). Asexual stages include (M) the nuclear division in schizont, (N) the resulting merozoites, (O) the release of merozoites, (P) the merozoites entering an erythrocyte, which may result in (left) another trophozoite or a male (Q) or a female (R) gametocyte. If taken up by a feeding mosquito (Q) becomes an egg and (R) becomes a sperm.

This provides an opportunity for malarial parasites within the salivary glands of an infected mosquito to pass into the human bloodstream. This infective stage is the **sporozoite** (Fig. 8-4). The sporozoites do not invade the blood corpuscles directly but, rather, invade certain cells of the liver, where they grow and multiply. The persistence of this **exoerythrocytic** (outside the red blood corpuscles) stage provides a reservoir from which relapses of the disease occur. Within about two weeks **trophozoites** are released into the bloodstream, where they proceed to enter red blood corpuscles. Each parasitized erythrocyte contains a single trophozoite, which first assumes a ring shape and then an irregular form that soon fills the host cell. The mature trophozoite eventually divides asexually by a form of multiple fission (**schizogony**) into 12 or 18 merozoites. The corpuscle then ruptures, and the merozoites are released into the bloodstream. Each merozoite invades a new erythrocyte, where it too undergoes schizogony. Liberation of merozoites occurs every 48 hours in *P. vivax*, with

accompanying symptoms of chills and fever, since billions of erythrocytes are often ruptured simultaneously. In *P. malariae* the growth–schizogony cycles are synchronized in 72-hour periods, whereas those of *P. falciparum* are more variable.

Eventually, certain merozoites, instead of becoming trophozoites, develop into **gametocytes** (Fig. 8-5). These are potential gametes, but as long as they remain in the human host they undergo no further development. These cells are of no significance to the human host as they circulate harmlessly in the blood;

Figure 8-5. Various kinds of sporozoa. (A) Photomicrograph of two trophozoites (attached end to end) of *Gregarina blattarum* from cockroach intestine. (B) Oocyst of *Isospora hominis* from human intestine. (C) Schizont of *Plasmodium vivax* (encapsuling merozoites) in human red blood cell. (D) Trophozoite of *Leptotheca* from fish bladder. (E) Trophozoite of *Actinophilus* from centipede gut.

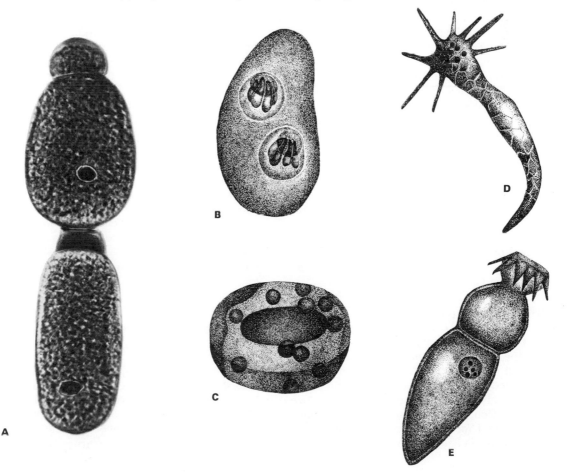

their survival depends upon their being sucked up by a mosquito of the appropriate type. Gametocytes that pass into a mosquito's stomach become active at once. There are two kinds: the female gametocyte, which develops into a single spherical **ovum** (or female gamete), and the male, which divides to form four to six male gametes (**sperm**). Union of a male gamete and an ovum produces a **zygote** (Fig. 8-5), which, because it has the ability to move about, is called an **ookinete.** The wormlike ookinete migrates through the intestinal wall and attaches onto the outer stomach wall. There its nucleus divides and redivides, resulting in the formation of a great number of sporozoites, which for a time, are contained within a swelling called an **oocyst** (Fig. 8-5). The stomach of an infected mosquito often harbors a considerable number of these oocysts, clearly visible on its exterior surface.

After 10 to 20 days, depending on temperature, the oocysts rupture, and the sporozoites are released into the insect's blood. Many of the motile sporozoites succeed in making their way to the mosquito's salivary glands. The next time that the mosquito takes a blood meal, sporozoites are injected into the wound along with the salivary anticoagulent.

Attempts to control human malaria have involved three approaches:

1. Treatment of infected humans with appropriate antimalarial drugs. This treatment has met with some success; however, in Southeast Asia and South America resistant strains of parasites have evolved, necessitating treatment with different combinations of antimalarial drugs.
2. Protection of uninfected individuals by the use of screens, nets, gloves, and the application of mosquito repellents.
3. Eradication of *Anopheles* mosquitoes by various means. One technique is to use poisonous insecticides, such as DDT. This effective short-term measure has serious drawbacks. Mosquitoes that survive the poisons may pass drug-resistant genes to the next generation of vectors, which can then be controlled only by higher doses of the poison, or a completely different insecticide may be necessary. In addition, some poisons take a long time to decompose and meanwhile are a threat to other organisms. More effective solutions include drain-

age and other engineering projects calculated to destroy the watery breeding places of the mosquitoes. Another promising approach is genetic control, which involves breeding a large population of mosquitoes, sterilizing them with chemicals or X-rays, and then releasing them to compete with the wild population for mates. *What effect would this have on the next generation of mosquitoes?*

Malaria is by no means confined to humans. Several species of *Plasmodium* infect monkeys, bats, squirrels, buffaloes, antelopes, and birds; at least 13 species occur in reptiles, and 2 have been reported in amphibians.

Also of interest is a malarialike disease called **redwater fever** or **Texas fever** that is caused by a sporozoan, *Babesia bigemina*, that parasitizes the red blood corpuscles of cattle. This species is transmitted among bovine hosts by a tick, *Boöphilus*. The parasites pass from the mother tick into her eggs and thus infect the larvae that emerge from them. Each generation of ticks is therefore able to transmit the disease to new bovine hosts. This was the first disease in which it was shown that an arthropod transmitted a protistan pathogen. This discovery, by Smith and Kilbourne in 1893, was a milestone in medical entomology.

Sporozoans are abundant in many animal hosts. Coccidial infections of the intestine are particularly destructive to rabbits and birds. *Nosema bombycis*, a sporozoan parasite of the silkworm caterpillar, once threatened the worldwide silk industry. Infested caterpillars that did not succumb developed into moths that laid infected eggs, and thus the infection continued. Pasteur studied the problem in the late 1860s and discovered that diseased eggs could be distinguished by microscopic examination.

Origins of Sporozoans

Parasitic protozoans no doubt ultimately evolved from free-living species, many by way of other parasitic species that had free-living ancestors. The phylum is presently subdivided into three classes with seven orders. The origin of the sporozoans is obscure, but it appears that they evolved from more than one ancestral source; that is, they are **polyphyletic.** Sporo-

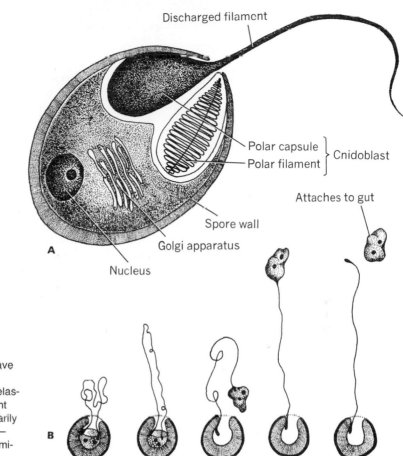

Figure 8-6. Phylum Cnidospora, *Myxidium* spore. (A) The Golgi apparatus appears to have a role in the formation of the polar capsules. Within each capsule is a tightly wound solid elastic polar filament. In the host the polar filament discharges (by straightening out) and temporarily attaches the spore to its host. (B) Discharge—emergence of filament and sporoplasm from microsporidean spore.

zoans with amoeboid sporozoites may have evolved from sarcodine ancestors, and those with flagellated sporozoites from mastigophorans.

Phylum Cnidospora

Cnidospora is the other phylum of spore-producing protozoan parasites. The spores of cnidosporans differ from those of sporozoans by the presence of one or more spirally wound **polar filaments.** Discharge of these spiral organelles apparently injects the sporoplasm from the spore into the intestinal cells of the host (Fig. 8-6B). Cnidosporans parasitize a wide variety of invertebrates and vertebrates. Some of their hosts, notably honey bees, silkworms, trout, and salmon are of considerable economic consequence.

Cnidosporans do not appear to infect any intermediate hosts. As in the case of Sporozoa, Cnidospora is probably a polyphyletic grouping; two classes are commonly recognized: **Myxosporidea** and **Microsporidea.**

The spores of myxosporideans are composed of several cells. The spore usually consists of two or three valves (Fig. 8-7), some **cnidocysts** (polar capsule + polar filament), and **sporoplasm** (internal amoeboid cells). Some biologists suspect that the myxosporideans may be degenerate metazoans because their cnidocysts closely resemble cnidarian nematocysts (Fig. 11-6) and their spores are made up of more than one cell. However, many others hypothesize that these parasites arose from unicellular ancestors, evolving cnidocysts as specialized adaptations for attachment and aggregating for mutual aid. From this

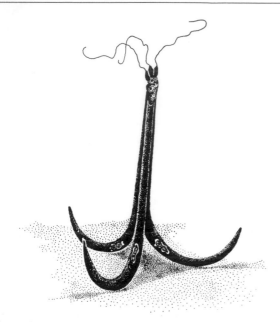

Figure 8-7. *Triactinomyxon ignotum* spore. Several cells make up this particular spore. Each valve is produced by one cell, each cnidocyst is produced by one cell, and the remaining cells form the sporoplasm. This spore gives the appearance of an anchor and may rest on underwater mud.

perspective, cnidocysts and nematocysts are considered to be roughly analogous structures. Myxosporideans parasitize fish almost exclusively, invading the gallbladder, urinary bladder, and other hollow organs.

The class Microsporidea contains intracellular parasites whose spores possess only a single polar filament. They most commonly infest the skin and muscle cells of fish or the gut epithelium and fat bodies of arthropods. Many other vertebrates and invertebrates are also parasitized by microsporideans, as are some protozoans (e.g., *Perezia* invades the cytoplasm of certain gregerines).

Summary

Sporozoans and cnidosporans are parasitic protozoans that form spores. They parasitize members of al-most every phylum in the animal kingdom. Great structural modifications must have occurred to adapt previously free-living organisms to parasitic life. This evolution has resulted in the absence of locomotor organelles, mouth, anal pore, and water-expulsion vesicle.

Review Questions

1. Some sporozoan spores may remain viable for years. How might such endurance be adaptive for these parasites?
2. Where in a host would you expect *not* to find a sporozoan that lacked an anterior anchoring structure? Where in the host might it reside?
3. What function does a polar filament play in the life cycle of a cnidosporan parasite?

Selected References

Chandler, A. C., and C. P. Read. *Introduction to Parasitology*, 10th ed. New York: John Wiley & Sons, Inc., 1961.

Gerberich, J. B. "An Annotated Bibliography of Papers Relating to the Control of Mosquitoes by the Use of Fish," *Amer. Midland Naturalist*, **36**:87, 1946.

Hocking, B. *Biology or Oblivion*. Cambridge, Mass.: Schenkman Publishing Company, Inc., 1965.

Levine, N. D. *Protozoan Parasites of Domestic Animals and of Man*. Minneapolis: Burgess Publishing Company, 1972.

Levine, N. D. and U. Ivens *The Coccidian Parasites (Protozoa, Sporozoa) of Rodents*. Urbana: University of Illinois Press, 1965.

Levine, N. D., and U. Ivens *The Coccidian Parasites of Ruminants*. Urbana: University of Illinois Press, 1970.

Olsen, O. W. *Animal Parasiter: Their Life Cycles and Ecology*, 3rd ed. Baltimore: University Park Press, 1974.

Mackie, T. T., G. W. Hunter, and C. B. Worth. *Manual of Tropical Medicine*. Philadelphia: W. B. Saunders Company, 1966.

Kingdom Animalia, Phylum Mesozoa

The last four chapters considered the way in which one-celled protozoans meet the problems of survival. Because each protist functions as a complete unit, like an organism composed of many cells, some biologists prefer to call them **acellular** (as opposed to unicellular) organisms. The protozoans are animal-like members of the kingdom Protista. We now enter the kingdom Animalia, the realm of multicellular animals (**metazoans**).

Metazoan Characteristics

All metazoans share three characteristics. First, they are all heterotrophic, multicellular organisms that develop from embryos. Heterotrophic organisms must derive their energy from preformed organic molecules; they cannot synthesize organic molecules from inorganic precursors as green plants do during photosynthesis. Second, metazoans are further qualified by having a body composed of more than one cell. This multicellular condition distinguishes the metazoans from the predominantly unicellular protozoans. Third, the fact that metazoans develop from embryos functionally distinguishes them from colonial protozoans (e.g., *Volvox*), for which there are no embryonic stages during the course of colony formation.

The number of cells poses a sharp distinction between kingdom Protista and the kingdom Animalia—one-celled organisms versus many-celled organisms.

However, as we have seen, there are increasing levels of complexity in the protozoans, culminating in the ciliates. These increasing levels of complexity continue in the animal kingdom, which is subdivided into more than 30 distinct phyla. Each phylum is characterized by a distinctive plan of body organization. It will be helpful to think of animals as falling somewhere on a spectrum, an organizational spectrum, as we investigate the different phyla. Keeping in mind that animals display a continuum of organizational complexity, we shall proceed to divide up kingdom Animalia and inspect one phylum at a time.

Phylum Mesozoa

The most simple animals, members of the phylum Mesozoa, are minute parasites. Their simplicity may be partly the result of adaptations for their parasitic existence. Most mesozoans (order **Dicyemida**) live in the excretory organs of squid and octopods. The rest (order **Orthonectida**) live in flatworms, polychaete worms, brittle stars, and other marine invertebrates (Fig. 9-1).

Morphology

The mesozoan body is wormlike, long and cylindrical, and bilaterally symmetrical and ranges in length from 0.5 to 10 mm. The body consists of one to several **axial cells** that function during reproduction sur-

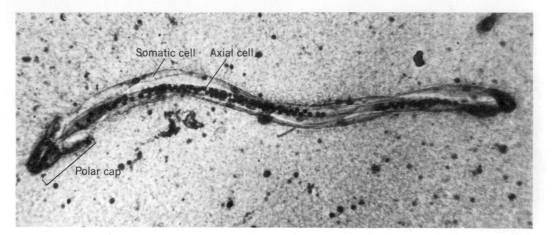

Figure 9-1. A representative mesozoan, *Dicyema,* a parasite of cephalopods.

rounded by a single layer of large ciliated **somatic** (body) **cells.** From 16 to 42 somatic cells are present, and anteriorly 8 of these form a **polar cap** that holds the dicyemid by means of apical cilia to the kidney of its host (Fig. 9-1).

Physiology

Dicyema must live **anaerobically,** that is, in the absence of oxygen, as there is little or no oxygen in cephalopod (squid and octopus) urine. Thus they derive ATP only from **glycolysis** (see Chapter 2).

It is likely that *Dicyema* absorbs nutrients directly from the host's urine. Electron microscope observations of *Dicyema* reveal numerous foldings of the somatic cell membranes and **pinocytotic vesicles** below these convolutions, suggesting that ingestion occurs in this manner. Adult dicyemids are adapted to live in cephalopod urine, but they cannot survive for long periods in seawater. The urine contains inorganic and organic molecules that may be used as energy sources and building materials by the parasite. If the parasite derived nutrients from the living tissues of the host, one would expect to see damage to the host's renal organs; but no such damage occurs.

Digestion and **excretion** appear to take place in each individual somatic cell.

Reproduction and Life Cycle

Adult dicyemids inhabit the renal organs (kidneys) of cephalopods, where they attach by means of the polar cap cilia, with the rest of the body drifting in the urine. Repeated asexual fission of the axial cell gives rise to new individuals that share the host. When the population of parasites in a host reaches a certain density, some adults give rise to hermaphroditic individuals (both male and female sex organs are present) that produce dispersal larvae. These ciliated larvae are discharged with the urine into the ocean, where their subsequent fate is unknown. Attempts to infect young cephalopods and other invertebrates with dicyemid larvae have repeatedly failed. It is assumed that the larvae infect some intermediate host, perhaps a bottom-dwelling filter feeder, and then develop into stages that can infect young cephalopods. These missing steps in the dicyemid life cycle remain a complete mystery, however.

Relations to Other Organisms

Mesozoans resemble some colonial protozoans by having external cilia, digestion that occurs in somatic cells, and special reproductive cells as in *Volvox.* Mesozoans are unlike typical metazoans in that their two cell layers are not comparable with ectoderm and endoderm, and they have no internal digestive tract. The name mesozoa implies that these organisms are intermediate between protozoans and metazoans. Mesozoans are either intermediate between unicellular and multicellular organisms or are possibly degenerate flatworms (phylum Platyhelminthes) that have adapted to a parasitic life style. Whatever the case may be, these small animals afford us a fascinating insight into the early workings of evolution.

Summary

Mesozoans are the simplest animals. They are small internal parasites (endoparasites) of some marine invertebrates. Their evolutionary affinities are unclear, but they certainly seem to be the most primitive metazoans that have survived to the present.

Review Questions

1. Because dicyemids are common parasites in the highly mobile cephalopods, what may be inferred about the fate of their dispersal larvae?
2. The lining of the human intestine is highly convoluted (folded). What does this suggest about the function of the intestine?

Selected References

Hadzi, J. *The Evolution of the Metazoa.* New York: Macmillan Publishing Co., Inc., 1963.

McConnaughey, B. H., and E. I. McConnaughey. "Strange Life of the Dicyemid Mesozoans." *Science Monthly,* **79:**277–284, 1954.

McConnaughey, B. II. *"The Mesozoa." In* M. Florkin and B. I. Sheer (eds.) *Chemical Zoology,* Vol. 2. New York: Academic Press, Inc., 1968.

Meglitsch, P. A. *Invertebrate Zoology.* New York: Oxford University Press, 1967.

Russell-Hunter, W. D. *A Biology of Lower Invertebrates.* New York: The Macmillan Company, 1972.

Phylum Porifera

The phylum Porifera (L. *porus,* pore + *ferre,* to bear) contains the sponges. For centuries sponges were considered to be plants, and it was not until 1825 that their animal nature was recognized. It is easy to imagine why they were mistaken for plants, as sponges are sessile and immobile, and some freshwater species are green, due to the fact that they contain photosynthetic algae. Sponges are thought to have diverged early from the main line of animal evolution and, next to the mesozoans, are the least complex metazoans. Sponges exhibit a multicellular organization that is somewhat intermediate between protists and typical metazoans.

Sponges are sessile animals in the adult stage. Dispersal to new habitats occurs in the larval stages, which either actively swim or are passively carried by water currents. Most of the poriferans, including the familiar bath sponges, are marine, but members of the family Spongillidae live in freshwater ponds.

Sponges are more complex than protists. Division of labor among somatic cells has resulted in cellular specialization, but there is no grouping and coordination of specialized cells to form definite tissues. And, without tissues, sponges also lack organs and organ systems. Hence, sponges have not advanced much beyond the cellular level of organization, even though they are multicellular animals.

Sponges exhibit great diversity in shape, size, structure, and geographical distribution; many are colonial. All sponges share, with variations, a simple body plan. Two things in particular characterize all sponges: (1) a system of canals and flagellated chambers and (2) skeletal elements consisting of calcareous or siliceous materials, the spicules, or proteinaceous fibers called spongin.

Leucosolenia—An Asconoid Sponge

Leuconsolenia (Fig. 10-1) exhibits the simplest and most primitive poriferan body plan. This small, whitish or yellowish sponge adheres to rocks along the seashore, just below the low-tide level, and consists of a group of slender, tubular individuals that project upward from a common network of horizontal tubes. The upright tubes have an opening, called the **osculum,** at their top. The inside of the tube contains a central cavity, or **spongocoel.** Buds may appear on the sides of the tubes. Many three-pronged (triradiate) **spicules** are interlaced in the soft body wall, thus serving to strengthen the body and hold it upright.

The body wall (Fig. 10-1) consists of two layers: (1) an outer epidermis consisting mainly of thin flat cells called **pinacocytes** and (2) an inner layer of flagellated **collar cells (choanocytes).** Between these two cell layers is a gelatinous extracellular substance known as **mesoglea,** which contains the spicules and many different kinds of amoeboid wandering cells (**amoebocytes).** The body wall is perforated by a multitude of **incurrent pores** that penetrate the **porocyte** cells and allow liquid to pass into the spongocoel. This rather simple type of canal system is known as the **asconoid** type (Fig. 10-2).

Note that the spongocoel is lined by a single layer of collar cells (Fig. 10-1); these choanocytes are in loose contact with one another and resemble certain

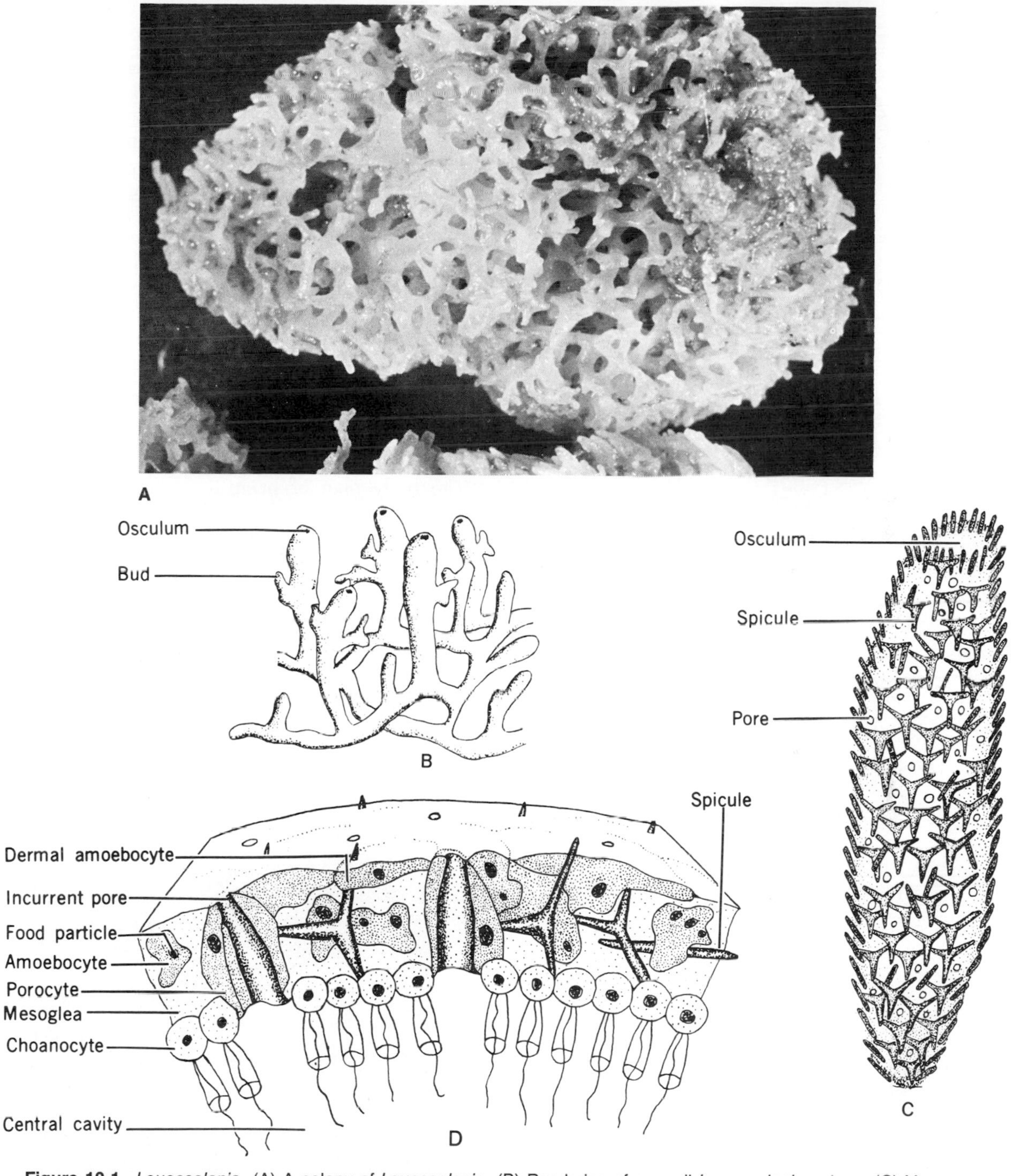
A
Osculum
Bud
B
Osculum
Spicule
Pore
C
Spicule
Dermal amoebocyte
Incurrent pore
Food particle
Amoebocyte
Porocyte
Mesoglea
Choanocyte
Central cavity
D

Figure 10-1. *Leucosolenia.* (A) A colony of *Leucosolenia.* (B) Rendering of a small *Leucosolenia* colony. (C) Young sponge of the ascon type, highly magnified. (D) Diagrammatic cross-section of *Leucosolenia,* showing the cellular structure of the body wall.

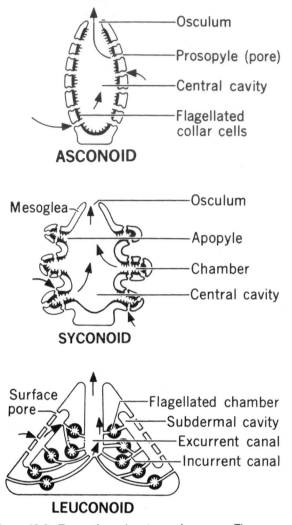

Figure 10-2. Types of canal systems of sponges. The syconoid type is derived, theoretically, by the evagination of the wall of the asconoid type of sponge into saclike chambers. Note how much each chamber of the syconoid type of sponge resembles the single chamber of a simple sponge. The leuconoid type, as in the case of the bath sponge, is more complex, with an elaborate system of canals and flagellated chambers. Arrows indicate the course of water through the various types of sponges.

mastigophorans in possessing a flagellum. The base of this single, long flagellum is surrounded by a collar of cytoplasmic tentacles. The constant beating of the

choanocyte flagella produce water currents that circulate through the incurrent pores into the spongocoel and out the osculum. Incoming water brings oxygen and food materials. The latter are trapped on the collar filaments, passed to the cell body of the choanocyte, and ingested. The food of sponges consists of minute suspended matter: unicellular algae, flagellates, bacteria, fine organic detritus, and possibly dissolved organic compounds. The small diameter of the incurrent pores limits entrance to particles less than 5 μ in diameter. Note that sponges are the only animals with many scattered mouths and one exit pore.

Scypha—A Syconoid Sponge
Morphology

The body plan of primitive sponges is generally elaborated in more advanced forms to increase the surface area covered by choanocytes. *Scypha*, also known as *Sycon* or *Grantia* (Fig. 10-3), is a slightly

Figure 10-3. *Scypha,* the comparatively simple sycon sponge.

more complex than *Leucosolenia*. As do other marine sponges, *Scypha* lives permanently attached by one end to rocks and other solid objects. It varies in length from 12 to 25 mm and resembles a slender vase that bulges slightly near the center. The osculum is surrounded by a ring of straight spicules. Smaller spicules protrude from other parts of its body.

Scypha has a folded body wall that looks almost as if the simple vase shape of *Leucosolenia* were repeatedly folded on itself. Choanocytes are restricted to the sac-shaped **flagellated,** or **radial, canals** that lie perpendicular to the central spongocoel.

Water currents are produced by the flagella of the choanocytes that line these canals. Water passes into the incurrent canals and then through inhalent openings called **prosopyles,** which lead into the flagellated canals. From there water passes out an opening, the **apopyle,** into the spongocoel and out the osculum. Sponges with this type of flow pattern are known as **syconoids** (Fig. 10-2).

Between the flagellated chambers lies mesoglea with wandering amoebocytes and skeletal spicules. Amoebocytes are the most functionally versatile cells in sponges. Many morphological variations including different sizes, shapes, and cytoplasmic inclusions are observed. This suggests the multitude of specialized functions that these generalized cells perform. Furthermore, the roles of individual amoebocytes do not appear to be fixed as they may potentially perform several functions during their life span.

Archaeocytes are amoebocytes capable of differentiating into other cells; they are usually progenitors to ova and sperm and are the chief sites of digestion in sponges. Specialized amoebocytes secrete the skeletal elements that support and protect the soft poriferan body; **sclerocytes** form spicules, whereas **spongiocytes** produce spongin. Other amoebocytes become pigment cells or food storage cells; some are meandering phagocytic cells, whereas others remain at a fixed location.

The skeleton of *Scypha* consists of a great number of spicules composed of calcium carbonate. Four varieties of spicules are always present: (1) long straight (**monaxon**) rods that guard the osculum, (2) short monaxon rods that surround the incurrent pores, (3) triradiate spicules (Fig. 10-4) that are embedded in the body wall, and (4) T-shaped spicules that line the spongocoel.

Physiology

Scypha consumes fine detritus particles and minute planktonic organisms that are drawn into it by the movement of each choanocyte flagellum. Some digestion occurs within the choanocytes. Most food particles, however, are transferred to amoebocytes for digestion. Digestion, as in protozoans, is intracellular. Distribution of digested nutrients throughout the sponge is accomplished by the diffusion of digested food from cell to cell, aided by the wandering amoeboid cells.

Excretion of particulate wastes is also provided by the amoebocytes. Waste particles are engulfed during **phagocytosis** and are discharged from the sponge when the cell is full. Dissolved wastes, primarily ammonia, diffuse from sponge cells directly into the surrounding water. Exchange of respiratory gases (oxygen and carbon dioxide) also occurs by diffusion across the general body surface.

Sponges are externally very quiet and sluggish, but internally they are among the most active and energetic of all animals, pumping water day and night through their porous bodies. The volume of water that flows through a sponge is tremendous: an average size sponge draws about 150 liters of water through its canal system in a single day.

True nervous tissue has not been demonstrated in sponges, and their behavior is what one would expect in the absence of nerves. Sponges are able to respond to certain stimuli, but the response, as in protozoans, is single celled. Instead of the porocytes of the asconoid sponges, the adult syconoid simply contains holes in the pinacocyte layer. These **prosopyles,** together with the oscula, are surrounded by contractile cells **myocytes** that are able to close these openings. Touching an osculum results in the immediate closure of the aperture, and a radiating wave of contraction travels from the site of irritation at a slow rate (0.25 mm per second in one report). Transmission of stimuli apparently passes through adjacent myocytes. The pinacocytes have a limited ability to contract when irritated and thereby change the shape of a sponge.

Reproduction

Scypha reproduces both sexually and asexually. In the latter case, a **bud** arises near the point of attach-

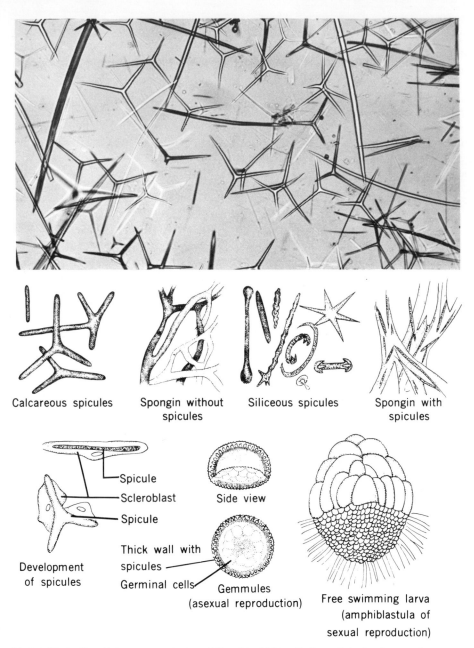

Figure 10-4. Top: Calcareous spicules of *Scypha.* 125 ×. Bottom: Spicules from various genera of sponges; gemmules of a freshwater sponge.

ment, eventually breaks free, and takes up a separate existence.

The sexually reproductive cells in sponges develop from amoebocytes in the mesoglea. Both ova and sperm occur in a single individual; that is, *Scypha* is **monoecious (hermaphroditic).** Sperm are liberated into the surrounding water and drawn into another sponge by its water currents. Ova are fertilized *in situ*

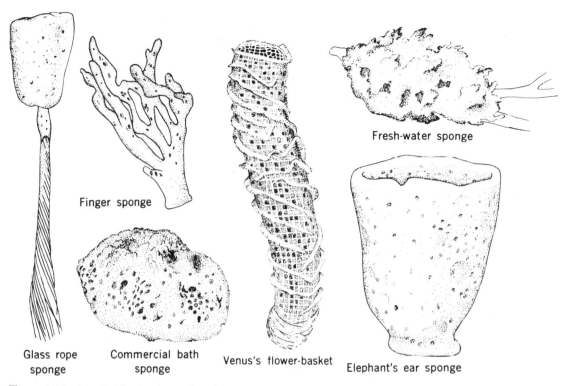

Finger sponge

Fresh-water sponge

Glass rope sponge

Commercial bath sponge

Venus's flower-basket

Elephant's ear sponge

Figure 10-5. Assorted freshwater and marine sponges.

and in a unique manner: incoming sperm are engulfed by amoebocytes or choanocytes, which then transfer the sperm to ova. Initial development also occurs within the parent sponge.

The fertilized **zygote** undergoes cleavage, eventually yielding a 16-cell stage that lies just below the maternal choanocytes. The layer of the embryo adjacent to the choanocytes is destined to become the outer epithelium, and the other layer will develop into choanocytes and amoebocytes. Eventually a flagellated **amphiblastula larva** develops. This motile larva escapes through the excurrent canals and osculum of its parent. After a free-swimming existence of several hours or days, the larva settles to the bottom and metamorphoses into a young adult sponge.

The embryology of sponges is peculiar. The flagellated cells of the larva do not become the outer (dermal) layer as do the ciliated cells of the larvae of certain higher animals; instead they produce the layer of choanocytes. In conjunction with this, the nonflagel-

lated cells produce the dermal layer as well as the mesoglea. No sponge has anything like an ectoderm and endoderm that characterize other metazoans.

Adaptive Modifications of Poriferans
Form, Size, and Color

Sponges may be simple, thin-walled, and tubular, like *Leucosolenia* (Fig. 10-1), or massive and more or less irregular in shape (Fig. 10-5). Many sponges conform to the contours of the rocks, shells, or plants to which they are attached; others are more regular in shape and attach to soft bottoms by means of rootlike masses of spicules. The form exhibited by the members of certain species may vary somewhat, depending on whether they grow in shallow or deep water. For example, *Microciona* in shallow water forms a thin encrustation on rocks, whereas in deeper water the colonies become massive and may reach heights of 15 cm. Some sponges are branched like trees; others are shaped like gloves, cups, or domes. The majority are irregular and amorphous, although a few are radially symmetrical. Sponges vary in size from species no larger than a pinhead to species that grow as big as barrels 2.5 m in diameter. Sponges are highly variable in color; many are white or gray, but others are yellow, orange, red, green, blue, purple, and velvety black.

Canal Systems

If it were not for the evolution of elaborate canal systems, sponges would have remained in the simple asconoid condition of *Leucosolenia* and could not have become massive in size; the complex foldings found in syconoids facilitates greater food and waste transport and, hence, larger size. The canal system furnishes an avenue for food through the body and for the elimination of wastes. Three types are usually recognized (Fig. 10-3), two of which have been described: (1) the simplest (or **asconoid**) type, as in *Leucosolenia;* (2) the **syconoid** type, as in *Scypha;* and (3) the **leuconoid** (**rhagonoid**) type, in which there is a great increase in the number of small flagellated canals lined with choanocytes. Water passes through an extensive labyrinth of canals and pores in the leuconoid grade of organization, which is exhibited by the majority of the 5000 described species of sponges. Because of their lack of a spongocoel, the largest excurrent canals often open directly to the outside; a number of oscula are frequently observed.

Skeletons

The skeletons of sponges consist of spicules of calcium carbonate (e.g., limestone), or silica dioxide (e.g., glass), or organic fibers of spongin (Fig. 10-4). Spongin is a flexible protein that is similar to collagen in composition. As discussed earlier, spongin fibers are secreted within the mesoglea by amoebocytes called **spongiocytes.** The fibers are fused to form an interconnected network. Spicules are secreted by **sclerocytes** in a complex process that generally involves the coordinated activities of several such amoebocytes. The concentration of dissolved silica in seawater is about 15 parts per 1,000,000. Therefore, to extract an ounce of skeleton, at least a ton of seawater must be drawn through the sponge.

Regeneration

In addition to asexual and sexual reproduction, sponges are able to **regenerate,** that is, to replace lost body parts by growth. In nature, fragments of sponges can regenerate into full-sized individuals. Some branching forms constrict the bases of their branches, pinching off buds that are distributed by currents to new habitats, where they regenerate.

The classic experiment that demonstrates regeneration involves forcing a sponge through a silk cloth sieve to separate it into individual living cells. The cells reaggregate a short time later to form several small sponges. If screened cells from two species are intermixed, the subsequent regenerations include segregation of the cells of the two species.

Classes of Sponges

Three classes of sponges are recognized: (1) **Calcarea,** comparatively simple shallow-water species with calcareous spicules and no spongin (e.g., *Leucosolenia*); (2) **Hexactinellida,** slightly more complex deep-sea sponges with siliceous spicules with six rays or hooks at the ends, (e.g., *Euplectella*); and (3) **Demospongiae,** the most common and most complex type of sponge with a skeleton composed of siliceous spicules without six rays or hooks, spongin fibers, or a combination of both. All demosponges display the leuconoid grade of body organization, and most form colonies; a few have no skeleton at all. The lack of absolute correlation between canal morphology and taxonomic classes suggests multiple independent origins of the more complex types from asconoid ancestors.

Freshwater Sponges

The majority of freshwater sponges belong to the family Spongillidae. Freshwater sponges are usually found in clear water, encrusted on stones, sticks, and plants, and are often yellow, brown, or green in color. These sponges may reproduce asexually by forming

Figure 10-6. Examples of sponges. (A) Venus's flower basket. (B) Common bath sponge. (C) Common encrusting demosponge.

gemmules, resistant stages that allow the sponge to pass through periods of environmental stress. A gemmule (Fig. 10-4) consists of a number of food-filled archaeocytes and other amoebocytes gathered into a ball and surrounded by a chitinous shell reinforced by spicules and/or spongin. Gemmules are formed as a regular part of the life cycle during the late summer and autumn; with the onset of winter the parent sponge disintegrates, but the gemmules' tough outer shells allow them to withstand adverse conditions, even freezing and some drying. With the return of more favorable conditions in the spring, the amoebocytes emerge from the gemmule and develop into an adult sponge.

More than 20 species of freshwater sponges occur in this country, and *Spongilla* is the most common genus. Some are green due to the presence of symbiotic algae within their amoebocytes.

Relations of Poriferans to Humans and Other Animals

Bath sponges, with which we once washed cars and cleaned kitchens (before modern synthetic sponges) and which the ancient Greeks and Romans used for drinking and painting, are actually the supple spongin skeletons of certain poriferans. These sponges live mostly near the shores of the Mediterranean and Caribbean seas and Australia. They are harvested from the ocean bottom, killed by trampling, and left to decay. The carcasses are then beaten and rinsed to remove all but the spongin skeleton.

Boring sponges live in shallow coastal waters throughout the world. They form irregular masses and are a bright yellow in color. Their name refers to their habit of attaching themselves to the shells of oysters, clams, and snails and boring them so full of holes that eventually the shells are entirely broken up.

Some sponges are poisonous, and human contact with them can produce skin irritations. Other sponges give off a stong unpleasant odor, and many contain sharp spiny spicules. Probably because of such protective devices sponges are seldom preyed upon. The nudibranch (Chapter 16) is the main predator of marine sponges, whereas insect larvae of the order Neuroptera (spongilla flies) live and feed on freshwater sponges. Sponges are sometimes used for purposes of concealment; certain species of crabs place sponges on their backs or on their legs. Other animals (e.g., small fishes, crustaceans, brittle stars) find the spongocoel of the sponge an excellent shelter in which to retreat for protection.

Of ornamental interest is a sponge known as Venus's flower basket, which constructs a beautiful skeleton of interconnected glassy spicules in the form of a cylinder about a foot long. Sponges of this type anchor in the mud of the ocean floor with a mass of long threadlike spicules at their basal end.

Origin of the Poriferans

Sponges are multicellular animals in which the somatic cells are somewhat differentiated for the performance of special functions. Although there is relatively little specialization of the somatic cells, the sponges are more complexly organized than colonial protists such as *Volvox*.

Despite the fact that sponges are multicellular and suggest the tissue level of organization, there are no organs as in most higher metazoans, and no digestive cavity is present. The many functions carried out by tissues, organs, and organ systems in most other metazoans are handled in sponges by individual cells. This is quite amazing, considering the structural complexity of many leuconoid sponges. Some polar sponges grow to a diameter of 4 m and have a dry weight of several tons! To organize the physiological functions of such a mass of cells without tissues and organs is a unique achievement in the animal kingdom.

Sponges are represented in the fossil record since the Cambrian period, 600 million years ago. Their evolutionary origin is puzzling because sponges are unusual in many ways. Sponges are the only animals with bodies built around a water canal system; a definite symmetry is lacking in many cases; and the complete absence of organs and the relatively low level of cellular differentiation and interdependence are relatively primitive characteristics.

It seems likely that sponges evolved from flagellated protozoans, because the structure of sponge choanocytes is almost identical to those of colonial mastigophorans known as choanoflagellates. Sponges apparently diverged from the mainstream of metazoan evolution quite early, or perhaps they evolved from protozoans by a completely different route than the other metazoans. Although a successful group, sponges are undoubtedly an evolutionary dead end.

Summary

Sponges are asymmetrical or radially symmetrical in form, and all are sessile animals in the adult stage. Their many cells are loosely arranged into two more or less definite layers, between which are amoeboid cells. Neither organs nor a mouth are present, and their cells act mostly independently. Sponges are usually supported by skeletons of spicules and/or spongin. The bodies of sponges contain pores, canals, flagellated chambers, and a central cavity through which flow currents of water. Collar cells, the choanocytes, line some of the body cavities. Most of the 5000 or more living species of sponges are marine; some 150 species inhabit fresh water. Typical body plans are illustrated in Figure 10-5.

Review Questions

1. If you cut a living sponge in half, would it be likely to feel pain? Explain as best as you can.
2. If you were to return the two halves of the sponge that were cut in the first question to their natural habitat, what would probably happen?
3. Explain the role of choanocytes.
4. Briefly distinguish among the three main types of canal organization in sponges.
5. While on a hike near a pond, a friend notices a live sponge under the water and remarks on the wonders of plants. You offhandedly remark that it is actually a primitive animal. How could you substantiate your statement? (Assume that you have any equipment necessary, except a zoology book.)
6. What are some of the characteristics common to both sponges and certain protozoans?

Selected References

de Laubenfels, M. W. "The Marine and Fresh Water Sponges of California," *U.S. National Museum, Proceedings*, **81**:1, 1932.

de Laubenfels, M. W. *A Guide to the Sponges of Eastern North America.* Coral Gables: University of Miami Press, 1953.

Dendy, A. "On the Origin, Growth, and Arrangement of Sponge Spicules," *Quart. J. Micr. Sci.*, **70**:1, 1926.

Fry, W. G. (ed.). *The Biology of the Porifera.* New York: Academic Press, 1970.

Harrison F. W., and R. R. Cowden (eds.). *Aspects of Sponge Biology.* New York: Academic Press, 1976.

Jewell, M. "Porifera," in Ward and Whipple, *Fresh-water Biology* (W. T. Edmondson, ed.), 2nd ed. New York: John Wiley & Sons, Inc., 1959.

Levi, C. "Ontogeny and Systematics in Sponges," *Syst. Zool.*, **6**:174, 1957.

MacGinitie, G. E., and N. MacGinitie. *Natural History of Marine Animals*, 2nd ed. New York: McGraw-Hill Book Company, 1967.

Moore, H. F., and P. S. Galstoff. "Sponges," in D. K. Tressler and J. McW. Lemon, *Marine Products of Commerce*, 2nd ed. New York: Reinhold Publishing Corporation, 1951.

Pennak, R. W. *Fresh-Water Invertebrates of the United States.* New York: The Ronald Press Company, 1953.

Rasmont, R., J. Bouillon, P. Castiaux, and G. Vandermeerche. "Utrastructure of the Choanocyte Collar-cells in Fresh-water Sponges," *Nature*, **181**:58, 1958.

Tuzet, O. "The Phylogeny of Sponges," in *The Lower Metazoa* (C. E. Dougherty, Z. N. Brown, E. D. Hanson, and W. D. Hartman, eds.). Berkeley: University of California Press, 1963.

Van Weel, P. B. "On the Physiology of the Tropical Fresh-water Sponge *Spongilla proliferans*: Ingestion, Digestion, and Excretion," *Physiol. Comp. Oecol.*, **1**:110, 1948.

Wilson, H. V., and J. T. Penny. "The Regeneration of Sponges (*Microciona*) from Dissociated Cells," *J. Exp. Zool.*, **56**:73–147, 1930.

11

Phylum Cnidaria

The cnidarians, or coelenterates, are a common and diverse group of aquatic animals that are of considerable ecological importance. Diversity within phylum Cnidaria is great: hydroids, jellyfishes, sea anemones, and many others are included. Cnidarians are entirely aquatic, and over 99 percent of the 9000 described living species are marine. Four characteristics distinguish the body plan of cnidarians: radial symmetry, a saclike gut, tissue grade of organization, and specialized stinging organelles called nematocysts.

An outstanding (but not unique) characteristic of cnidarians is their **radial symmetry** (Fig. 4-1). Note that an animal with radial symmetry has no head, no real anterior or posterior end, no left and right sides. The end with the mouth is called the **oral surface,** and the opposite end is the **aboral surface.** Individual cnidarians are radially (or sometimes biradially) symmetrical, but many cnidarians form complex colonies that may or may not show a symmetry of their own.

The basic cnidarian body plan is simply a sac composed of two epithelial layers, an outer **epidermis** and an inner **gastrodermis,** that are separated by a jellylike **mesoglea.** A single opening, called the **mouth,** opens into the **gut,** or **gastrovascular cavity.** The mouth is typically surrounded by one or more whorls of tentacles.

The basic cnidarian body plan takes two general forms: polyp and medusa (Fig. 11-1). A **polyp** is typically attached to a solid substratum at its aboral end, and its oral–aboral axis is elongated in the form of a cylinder. A **medusa** is typically a free-swimming jellyfish. Its oral–aboral axis is often short and flattened, to assist floating and swimming, and its oral surface is characteristically oriented downward.

Cnidarians exhibit a **tissue grade** of organization. There are many types of cells, and the cells are specialized for particular functions and organized into distinct tissues. These tissues form functional systems that are not as complexly organized as organ systems. Cnidarians have a digestive system, muscular system, nervous system, sensory system, and (sometimes) a skeletal system. Cnidarians lack distinct respiratory, circulatory, endocrine, and excretory systems.

Cnidarians are carnivorous animals that capture prey and defend themselves by means of specialized stinging organelles, called **nematocysts,** that reside in cells known as **cnidocytes.** Basically, a nematocyst consists of a hollow thread coiled inside a thin capsule. Upon discharge, the thread everts and penetrates, coils about, or sticks to the prey, immobilizing it with the aid of a toxin.

Phylum Cnidaria contains three classes: (1) **Hydrozoa,** which is represented by *Hydra*, the colonial *Obelia*, the hydrozoan jellyfish *Gonionemus*, and the colonial *Physalia* (Portuguese man-of-war); (2) **Scyphozoa,** or large jellyfish, typified here by *Aurelia;* and (3) **Anthozoa,** characterized by the sea anemone *Metridium* and the coral *Astrangia.*

Hydra—A Freshwater Hydrozoan

Hydras are simple cnidarians that are fairly common in freshwater ponds and streams. Nine known species occur in the United States. Colors generally range from white to brown, but *Chlorohydra viridissima* is green due to the presence of symbiotic algae within its gastrodermal cells.

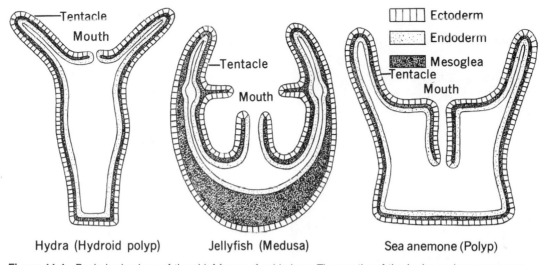

Ectoderm
Endoderm
Mesoglea

Hydra (Hydroid polyp)   Jellyfish (Medusa)   Sea anemone (Polyp)

Figure 11-1. Basic body plans of the chief forms of cnidarians. The mouths of the hydra and sea anemone are directed upward, but the jellyfish swims with mouth down. For purposes of comparison, however, the jellyfish is shown with its mouth up.

Hydras are minute animals that are often overlooked in their natural habitats. They usually range from 2 to 20 mm in length, and, seen with the naked eye, they resemble a short thread frayed at the unattached oral end.

External Morphology

The body of the hydra (Figs. 11-2 and 11-3) resembles an elastic tube that can be extended to lengths of about 2 cm. At the oral end, the **mouth** is surrounded by a circlet of usually six or seven **tentacles.** The tentacles can be contracted into small blunt projections or extended into very thin threads 7 cm or more in length (see Fig. 11-7). The tentacles move independently, capturing food and bringing it to the central mouth. The number of tentacles varies considerably, increasing with the size and age of the animal.

The aboral end, called the **foot** or **basal disc,** is usually attached to some solid object such as a submerged plant. The foot secretes a sticky substance that anchors the animal and also aids in locomotion. The foot may also secrete a gas bubble enclosed by a film of mucus. This bubble floats the animal to the water surface, where it may drift with its tentacles extended downward like a jellyfish (Fig. 11-7).

A conical elevation, the **hypostome,** lies between the tentacles and the mouth. The mouth contracts during rest or digestion into a minute circular pore, but, when swallowing prey, the mouth and the surrounding hypostome can dilate to a relatively large diameter.

Frequently, the body column of a hydra will be observed to possess **buds** in various stages of development (Fig. 11-3). When fully grown, each bud becomes detached and leads a separate existence. The **body stalk** contains the gastrovascular cavity.

Histology

Body Wall

The body wall of a hydra consists of two cellular layers: a thin outer **epidermis** and an inner **gastodermis,** which is about twice as thick as the epidermis (Fig. 11-4). Both layers are composed primarily of **epitheliomuscular cells.** A thin layer of noncellular jellylike material, the **mesoglea,** separates the epidermis from the gastrodermis. In some cnidarians (e.g., jellyfish) the mesoglea constitutes the bulk of the body, but in the hydra it is thin, especially toward the oral end of the body and in the tentacles, and at the center of the basal disc it is lacking altogether.

Figure 11-2. Photograph of a hydra showing a bud.

Mesoglea serves as a basement membrane for the epithelial cells and as an attachment site for their muscle processes.

The following outline lists the cellular elements of the three layes of the body wall:

1. Epidermis
   a. Interstitial cells
   b. Epitheliomuscular cells
   c. Germ cells (ova and spermatozoa)
   d. Sensory and nerve cells
   e. Mucus-secreting cells
   f. Cnidocytes (containing nematocysts)
2. Mesoglea. This middle, supporting layer is noncellular but is traversed by migrating cells and crossed by intercellular bridges and nerve cell processes.
3. Gastrodermis
   a. Nutritive-muscular cells (variously differentiated in different regions)
   b. Enzymatic gland cells
   c. Sparse interstitial cells
   d. Sensory and nerve cells

The primary components of both epidermis and gastrodermis are epitheliomuscular cells, which extend the full height of the epithelial layer and support the other cellular elements. The dimensions of the epitheliomuscular cells change with the expansion or contraction of the animal.

Epitheliomuscular cells have polygonal outer surfaces that are cemented together to form a continuous membrane over the animal that is interrupted only where stinging or sensory cells communicate with the external environment.

The nutritive-muscular cells of the gastrodermis line the entire gastrovascular cavity, which extends the length of the tentacles as well as the body stalk. The character of these cells varies in different regions, but, because all are concerned with digestion, they are called **nutritive cells. Flagella** at their free ends continuously circulate the fluid in the gastrovascular cavity. Food particles are engulfed by **pseudopodia.** The bases of the epidermal and gastrodermal cell types contain contractile fibers, which will be discussed later.

At the hypostome, the gastrodermal cells are either filled with small secretion granules or have a fine spongy texture, the two conditions alternating in adjacent cells. The **mucus-secreting gland cells** assist in swallowing food and preparing it for digestion. Contractile fibers at their bases form a sphincter that can close the mouth.

The nutritive-muscular cells rest on the mesoglea, but the **enzymatic gland cells** are wedged in between the free ends of the aforementioned cells. These pear-shaped gland cells are most abundant in the stomach and hypostome and are only sparsely distributed in the stalk and basal disc.

Interstitial cells are small rounded cells with a clear cytoplasm and a relatively large nucleus. Their undifferentiated cytoplasm lacks specialized structures, and they are frequently observed undergoing mitosis. The interstitial cells are very labile and can differentiate into any of the specialized cells of the hydra. As such, interstitial cells are the chief agents in the continuous process of reconstructing tissues in growth, budding, regeneration, and replacement of dead

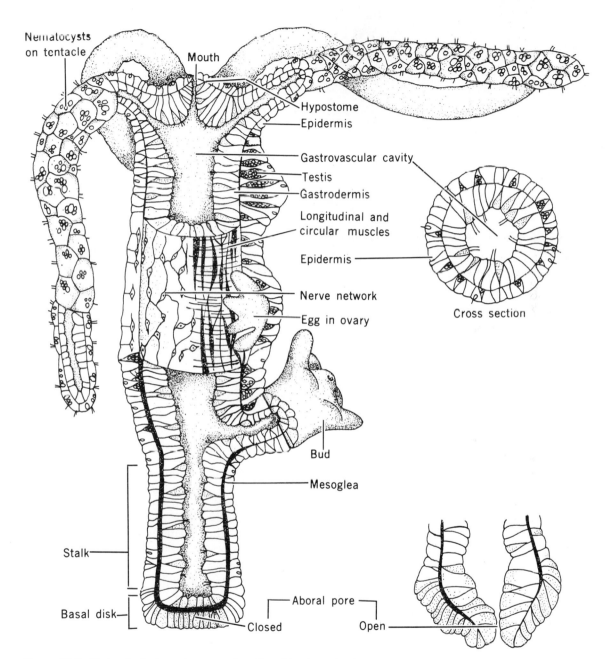

Figure 11-3. Parts of a hydra cut away and sectioned to show structure.

Nematocysts on tentacle

Mouth

Hypostome

Epidermis

Gastrovascular cavity

Testis

Gastrodermis

Longitudinal and circular muscles

Epidermis

Nerve network

Egg in ovary

Cross section

Bud

Mesoglea

Stalk

Basal disk

Closed

Aboral pore

Open

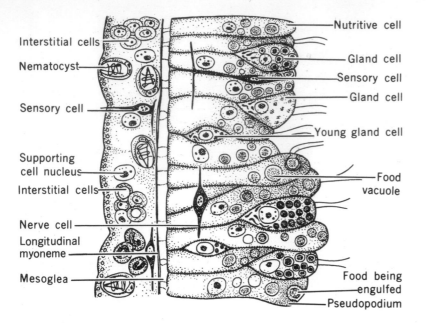

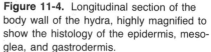

Interstitial cells

Nematocyst

Sensory cell

Supporting cell nucleus

Interstitial cells

Nerve cell

Longitudinal myoneme

Mesoglea

Nutritive cell

Gland cell

Sensory cell

Gland cell

Young gland cell

Food vacuole

Food being engulfed

Pseudopodium

Figure 11-4. Longitudinal section of the body wall of the hydra, highly magnified to show the histology of the epidermis, mesoglea, and gastrodermis.

cells. The interstitial cells also form the primordial germ cells of the **gonads.** Interstitial cells are critical to the vitality of the hydra, for, if these cells are selectively killed, the animal will die in a matter of days. Whether cnidocytes, germ cells, and gland cells arise from a common interstitial cell or from several types of interstitial cells is not known.

Muscular System

The muscular system of the hydra consists primarily of two contractile layers applied to opposite surfaces of the supporting mesoglea. The outer muscle layer is formed by longitudinally arranged contractile elements, the **myonemes,** of the epitheliomuscular cells. The inner muscle layer is within the nutritive-muscular cells and consists of circularly arranged contractile myonemes oriented perpendicular to the longitudinal axis of the stalk (Fig. 11-3). This circular muscle layer contracts slowly, but the longitudinally oriented myonemes that shorten the tentacles and body column are capable of rapid contraction.

Nervous System

The cnidarian nervous system is relatively primitive and consists of two general types of cells (Fig. 11-5): **motor nerve cells** and **sensory cells.** These are

distributed throughout the body to form networks in the epidermis and gastrodermis. The two nerve networks are interconnected by fibers that pass through the mesoglea. The **epidermal network** is more highly developed. The nerve cells of the epidermis lie just distal to the longitudinal muscle layer, beneath the bases of the epitheliomuscular cells. The processes of the nerve cells interlace to form a **nerve net** that extends throughout the entire body from the tip of the tentacles to the basal disc.

The nervous systems of more complex metazoans are usually unidirectional; that is, the nerve impulses are conducted in only one direction. In the hydra, nerve impulses can travel either way along a nerve process, as the synaptic contacts between adjacent nerve cells are nonpolar. The synapses tend to block nerve impulses unless the stimulus is strong and repeated. Therefore, slight stimuli elicit only local responses, but strong stimuli result in general responses that may involve the entire body. The velocity of nerve impulses in cnidarians is relatively slow (12–120 cm/sec) compared with that in humans (12,500 cm/sec).

The greatest concentration of nervous elements occurs around the mouth, where nerve fibers form a loosely organized **nerve ring** in the hypostome. A somewhat similar concentration of nerve fibers occurs in the basal disc.

Cnidarians have receptors for **touch, light,** and **chemoreception** ("taste" and "smell"). The sensory cells (Fig. 11-5) are slender, threadlike nerve cells that occur between the epitheliomuscular cells in both the epidermis and the gastrodermis. They frequently bear a hairlike process or some other specialized structure at the end that monitors the environment. The other end divides into two or more fibers that connect with the nerve net or with muscle fibers.

Nematocysts

Nematocysts (stinging capsules) are most numerous on the tentacles but are also present on all parts of the epidermis except the basal disc. Nematocysts are used in capturing prey and in locomotion. Each of these specialized organelles is formed inside a differ-

entiated interstitial cell called a **cnidocyte** (Fig. 11-6). On the body column, mature cnidocytes generally lodge between the epidermal cells, but on the tentacles and hypostome they lie invaginated within the epitheliomuscular cells, which serve as **host cells.** On the tentacles the host cells are large, and each contains a battery of several types of nematocysts.

Four types of nematocysts occur in the hydra: (1) The largest is the **penetrant,** which is surrounded by a pear-shaped capsule that almost fills the cnidocyte. Within the capsule is an inverted coiled tube, at the base of which are three large and many small spines. Three rows of minute spines spiral along the outside of the thread when discharged. Penetrant nematocysts inject paralyzing toxins into prey. (2) **Volvents** are small pear-shaped nematocysts that contain a thread that when discharged coils tightly around the

Figure 11-5. Principal cell types of the hydra.

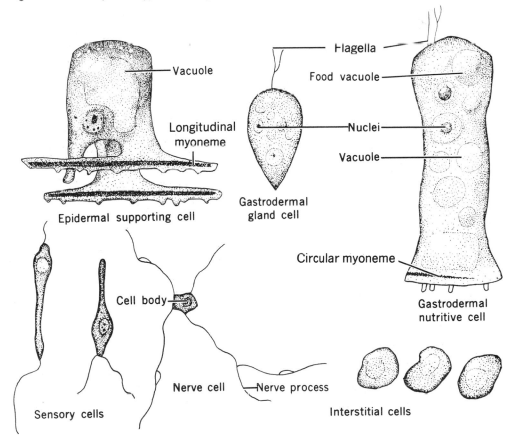

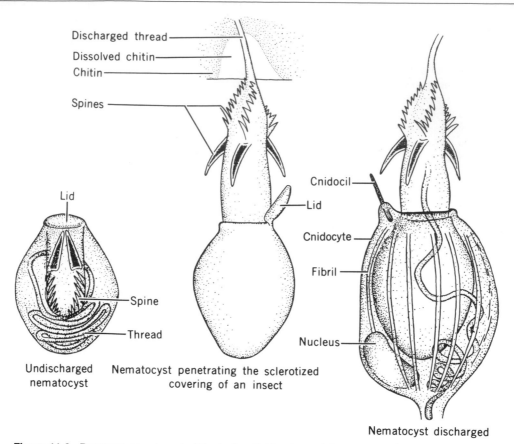

Figure 11-6. Penetrant nematocyst of the hydra. Cnidocyte on right.

hairs or bristles of the prey. The other two types of nematocysts secrete a sticky substance that is used in locomotion as well as in feeding. (3) The **oval glutinant** is large and has a long adhesive thread that bears minute spines. (4) The **small glutinant** bears a straight, unarmed thread.

Projecting from the outer surface of the cnidocyte is a hairlike process, the **cnidocil.** Nematocyst discharge results from a sequential combination of chemical and mechanical stimuli. A soluble, reduced chemical substance (**glutathione**) from the prey first sensitizes or lowers the firing threshold of the nematocyst, which then discharges upon mechanical contact with the cnidocil. Discharge is probably due to a change in water permeability of the capsule wall; water then suddenly enters the capsule, which becomes turgid,

and the coiled tube everts with explosive force. Note that nematocyst discharge is triggered directly by stimuli from the external environment, not as a result of nervous control; they are referred to as **independent effectors.**

Prey such as small crustaceans that contact a hydra's tentacles are immediately paralyzed and sometimes killed by a poison, called **hypnotoxin,** that is injected into it by the penetrant nematocysts. Hypnotoxin inhibits an enzyme in the Krebs cycle and thereby stops the metabolic production of ATP.

Cnidocytes develop from **interstitial cells** (Fig. 11-4) that are clustered in the epidermis near the base of the tentacles. As its nematocyst is developing, the cnidocyte migrates to its final site in the body, often the tentacles, where the organelle matures and develops a cnidocil. A cnidocyte is specialized solely for the formation of a single cnidocil. Once discharged, the tube of the nematocyst cannot be returned to its capsule, and the cnidocyte that houses that nematocyst perishes. However, neighboring interstitial cells soon give rise to new cnidocytes.

Differentiation of the Body Regions

Hypostome

The hypostome is richly supplied with sensory and motor nerve cells and is the most sensitive region of the body. Cnidocytes and a few interstitial cells are also located here. The gastrodermis of the hypostome is thrown into large deep folds when the mouth is contracted, so that a cross-section of the oral cone has a star-shaped configuration. The folds allow for distension of the mouth as prey are swallowed. The oral gastrodermis contains many mucus gland cells. Their secretions lubricate incoming food and initiate digestion. Food introduced directly into the stomach through a glass pipette, without first coming in contact with these mucus gland cells, is not digested.

Tentacles

The hollow tentacles can be rapidly elongated by pumping fluid from the cavity into them. A concentration of nutritive-muscular cells at the base of each tentacle acts as a sphincter that regulates the entrance or escape of this fluid. Contraction of epitheliomuscular cells shortens the tentacle, whereas contraction of nutritive-muscular cells helps to elongate it.

The Body Stalk Regions

The epidermis of the stomach part of the body column is about twice as thick as that of the hypostome and harbors most of the interstitial cells. As you will recall, the interstitial cells are responsible for the production of these structures. This is the region of nematocyst formation and of testes, ovaries, and buds. Epitheliomuscular cells form the supporting cells of the testis and ovary. The gastrodermis in this region is the chief digestive surface, effecting both extracellular and intracellular digestion. Many enzymatic gland cells are located here; their secretions extracellularly digest food into a broth of fine particles. The particles are then phagocytosed by nutritive-muscular cells where intracellular digestion completes the process in the food vacuoles. During the initial phases of digestion, the hydra confines large pieces of food within the cavity by muscular action.

Both muscle layers are well developed in the stalk, particularly the longitudinal layer in the epidermis. The stalk is primarily a region of extension and contraction and, therefore, is responsible for much of the flexibility that typifies the hydra. The mesoglea also reaches its greatest thickness in this region.

Basal Disc

The epidermis of the foot consists of columnar epitheliomuscular cells whose distal cytoplasm is filled with globules of the stocky mucus that these cells secrete. Their bases have muscle fibers that radiate from the center of the disc. Mesoglea is lacking in a small area near the center of the disc, and here epidermis and gastrodermis are in direct contact. This is the region of the **aboral pore,** which is opened as the hydra releases its hold on the substratum but is completely closed during attachment.

Physiology

Feeding

The food of the hydra consists primarily of small aquatic animals such as crustaceans, annelids, and insect larvae. Large hydras may consume young fish and tadpoles. Bits of meat may be ingested when offered to them in an aquarium.

The hydra normally rests with its basal disc attached to some object and its body column and tentacles extended into the water. In this position its radiating tentacles monitor a considerable volume of hunting territory. Any small animal that touches a tentacle is at once shot full of penetrants, affixed by glutinants, or grappled by volvents. It is common to find hydras that will not react to food when it is presented to them. This is because these animals will eat only after a certain interval of time has elapsed since their last meal. The physiological condition of the hydra, therefore, affects its response to the food stimulus.

The collision of an aquatic organism with the tentacle of a hungry hydra is not sufficient to cause nematocyst discharge, as it has been found that a chemical stimulus must also be present. This "feeding hormone," as mentioned earlier, has been identified by several investigators as reduced **glutathione.** When activated solely by a chemical stimulus, such as by crustacean secretions, a hungry hydra will initiate feeding movements such as extending and waving its tentacles.

The tentacle that has captured the prey bends slowly toward the **mouth.** Other tentacles help draw in the prey, and their nematocysts assist in quieting the victim. The mouth often begins to open before the food has reached it.

The edges of the mouth, lubricated with mucus, slip smoothly over the helpless prey as it slides slowly into the **gastrovascular cavity.** The body wall contracts behind the food and forces it downward. Frequently, organisms many times the size of the hydra are successfully ingested.

Digestion

Immediately after food is ingested, the gastrodermal gland cells show signs of great activity, while their nuclei enlarge and become granular signaling the appearance of proenzymes which are secreted into the gastrovascular cavity, where they begin at once to digest the food. The **flagella** of the **nutritive cells** create water currents that help to mix the enzymes with the food. This method of digestion differs from that of most protozoans and sponges in that it is *extracellular.* However, in the hydra extracellular digestion is followed by *intracellular* digestion. **Pseudopodia** of the gastrodermal cells engulf small particles of food (Fig. 11-4); and final digestion takes place within food vacuoles. The digested nutrients are absorbed and stored by the **gastrodermal cells.**

Earlier we noted that *Chlorohydra viridissima* is green in color. As in some freshwater sponges, this is due to the presence of unicellular algae, *Chlorella vulgaris,* within the animal cells. In hydra the algae populate the gastrodermal cells. Experimental evidence indicates that the algae photosynthesize and release considerable amounts of sugar to the hydra. In return, the algae receive protection and a constant supply of soluble nitrogenous compounds in the form of waste products of the animal's metabolism. This symbiotic relationship is intriguing because a carnivore is coupled directly with a primary producer, thus eliminating the herbivore link in the food chain.

All particulate wastes are egested from the mouth. A sudden contraction of the body wall expels the debris some distance from the hydra.

Respiration and Excretion

Oxygen diffuses into the cells from the water that surrounds the hydra, and carbon dioxide is eliminated by diffusion in the opposite direction. Soluble waste products of metabolism also diffuse through the general body surface.

Behavior

Spontaneous Movements

All movements of the hydra result from contraction of muscle fibers in the epidermis and gastrodermis. Body movements are elicited by two kinds of stimuli: internal (or spontaneous) and external. Spontaneous movements may be observed when the animal is attached and undisturbed. At intervals of several minutes, the body and/or tentacles contract suddenly and rapidly and then slowly expand in a new orientation. Hungry hydras display more spontaneous movements than do their well-fed counterparts.

Locomotion

Hydras can change location by several methods. One is a gliding movement, with the basal disc sliding slowly over the object to which it is attached. A second method is an inchworm type of movement (Fig. 11-7) in which the hydra bends over and attaches its tentacles by means of its adhesive glutinant nematocytes. It then slides its basal disc up close to them, after which it releases the tentacles and assumes an upright position. Another method is by turning somersaults. The animal attaches its tentacles, then releases its basal disc and flips its body column over to the other side of the tentacles. Such end-over-end movements may be repeated again and again. Infrequently, a hydra moves from place to place in an upside-down position by using its tentacles as legs. To rise to the water surface, a hydra may form a gas bubble in its basal disc; after detaching itself from the substrate, the hydra may float for some time.

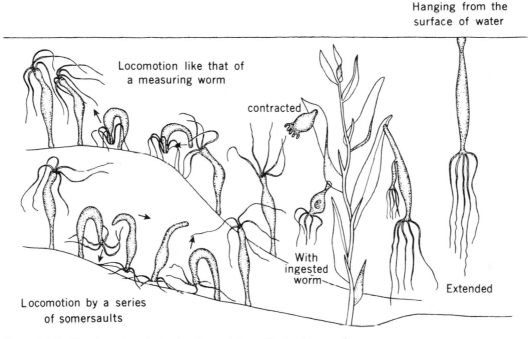

Figure 11-7. Sketches show hydra feeding and its methods of locomotion.

Contact

Hydras respond to mechanical stimuli in two ways. Mechanical shocks, such as sudden water movements, cause a rapid contraction of part or all of the animal. This is followed by a gradual extension until the original condition is regained. This is a nonlocalized response.

Local stimulation involves actually touching part of the body or tentacles with an object, such as a fine glass rod. Local stimulation of one tentacle may cause contraction of just that one tentacle, of all the tentacles, or of the tentacles and the body column. This indicates that there must be transmission of nerve impulses from one tentacle to another and to the body stalk. The structure of the **nerve net** makes such communication possible.

Reaction to Light

Hydras display a generalized response to light intensity. If a dish containing hydras is placed so that the illumination is not equal on all sides, the animals will tend to congregate in the brightest region. However, if the light is too strong, they will move to a place where the light is less intense. The hydra seeks an optimum light intensity by a method of trial and error. When put in a dark place, the hydra becomes restless and moves about in no definite direction; but, if white light is encountered, its locomotion becomes less rapid and finally ceases altogether. This behavior has considerable adaptive value as the small animals that serve as its prey are also attracted to well-lighted regions.

Other Stimuli

The response of hydras to changes in temperature is indefinite, although in many cases they move away from a heated region. Hydras do not seem to react to water currents.

The physiological condition of an animal determines to a great extent the degree of response to internal and external stimuli. Its physiological condition influences whether a hydra creeps upward toward light or sinks to the bottom, how it responds to chemicals or solid objects, and whether it remains stationary or moves about in an exploratory manner.

Reproduction

Reproduction in the hydra occurs both asexually by budding and sexually by production of sperm and ova. Asexual and sexual reproduction may occur simultaneously in an individual hydra.

Budding

Asexual reproduction by **budding** (Fig. 11-3) is a common occurrence in the hydra. Several buds are often found on a single animal. A bud is an outpocketing of the parent body wall, and it includes all three body layers and a gastrovascular cavity that is continuous with that of the parent. The bud first appears as a slight bulge in the body wall of the parent. This rapidly elongates and soon develops a circlet of blunt tentacles and a mouth. When fully grown, the bud detaches from the parent and becomes an independent individual. Budding may occur in almost any season and requires about two days for completion under favorable conditions.

Sexual Reproduction

Both ova and spermatozoa develop from interstitial cells. Some species of hydra form both kinds of **gametes** in one **monoecious** individual, but in **dioecious** species only one sex occurs. The sexual state can be induced in some species by lowering the water temperature; this may account for the appearance of sexual tissues in *Hydra oligactus* during the autumn and early winter. Stagnant water conditions in which wastes accumulate also induce sexual reproduction.

Spermatogenesis

The male gametes are formed in small conical or rounded elevations called **testes** that project from the surface of the upper body column (Fig. 11-3). There may be as many as 20 or 30 testes on one hydra. An indefinite number of interstitial cells congregate into a mass, causing the epidermis of the animal to bulge. Each of these interstitial cells becomes a **primordial germ cell;** it gives rise by mitosis to a variable number of **spermatogonia;** these undergo reduction divisions (**meiosis**), which reduce their chromosome number in half, to produce **spermatids** that transform into **spermatozoa.** The mature spermatozoa (**sperm**) swim about in the distal end of the testes and finally escape to the exterior through one or more small fissures in the protective covering. The testes of most hydras have definite nipples through which the sperm escape.

Oogenesis

An **ovum** is an interstitial cell that becomes large and spherical and possesses a large nucleus (Fig. 11-3). Several adjacent interstitial cells also begin to enlarge to form ova, but one finally incorporates the others. As the ovum grows it becomes scallop shaped, due to its confinement between the epidermal cells. When fully grown it becomes spherical but is still surrounded by epidermal cells, which remain rooted to the mesoglea but stretch enormously to cover the ovum. During maturation the number of chromosomes is reduced by meiosis from the somatic number of twelve to six. Now an opening appears in the epidermis and the ovum becomes free on all sides except where it is attached to the parent.

Fertilization

Fertilization usually occurs about as soon as the ovum is extruded. Several spermatozoa may penetrate the egg membrane, but only one enters the ovum itself. That sperm, containing six chromosomes, fuses with the nucleus of the ovum that also contains six chromosomes, and the somatic chromosome number is thereby restored.

Embryology

Cleavage (mitotic division) of the **zygote,** which now begins, is total and regular. A well-defined cavity is present at the end of the third cleavage division, the eight-cell stage. Cleavage produces the **blastula stage,** which resembles a hollow sphere with a single layer of epithelial cells composing its wall (Fig. 11-8). These cells are called the primitive **ectoderm.** By mitotic division they form endodermal cells that completely fill the central cavity. The early **gastrula stage**

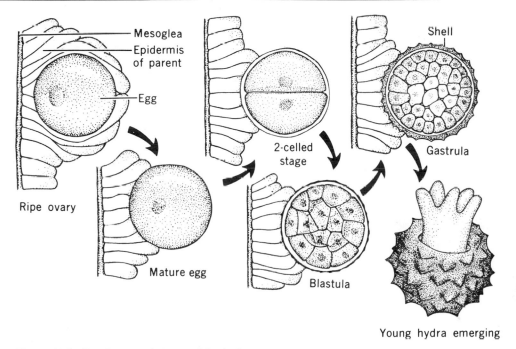

Figure 11-8. Developmental stages of the hydra.

is a solid sphere of cells differentiated into a single outer layer of ectoderm and an irregular central mass of **endoderm** (Fig. 11-8). The ectoderm secretes two envelopes around the gastrula: the outer envelope is a thick chitinous **shell** that may be covered with sharp projections; the inner envelope is a thin gelatinous membrane. Different species of hydra can be identified by the morphology of their shells.

Hatching

The shelled embryo now separates from the parent and falls to the bottom, where it remains dormant for several weeks. Then interstitial cells make their appearance. A subsequent resting period is followed by a breaking away of the chitinous shell and elongation of the hatched embryo. Mesoglea is now secreted between the ectodermal and endodermal cell layers, which differentiate to form the epidermis and gastrodermis. A circlet of tentacles arises at one end, and a mouth appears in their midst. The young hydra thus has no larval stage and grows directly into an adult.

Regeneration and Grafting

The ability of animals to regenerate lost parts was first described by Trembley in 1740. This Swiss naturalist found that, if hydras were cut crosswise into two, three, or four pieces, each part would regenerate into an entire animal. Trembley also observed regeneration when a hydra was split longitudinally into two or four parts. When the oral end is split in two and separated slightly, a "two-headed" animal results. When a hydra is turned inside out, the cells of the epidermis and gastrodermis migrate past each other through the mesoglea until they reestablish their original positions. Inverted hydras have also been observed to flip right-side out directly.

Regeneration occurs in many cnidarians and in at least some representatives of almost every metazoan phylum. The hydra has been used in many classic demonstrations of regeneration. Pieces of hydra that measure only 1.2×10^{-3} cm in diameter are capable of regenerating into entire animals, *provided that they contain all three body layers.*

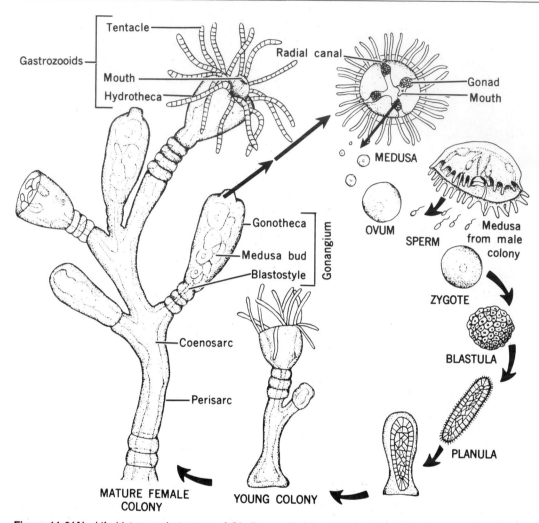

Figure 11-9(A). Life history and structure of *Obelia,* a colonial marine hydroid. The colony consists of polyps of two types, the feeding gastrozooids and the reproductive gonangia. Both gastrozooids and gonangia
are developed by asexual budding from stems attached to a branching rootlike tangle called the hydrorhiza.
Sperm and eggs are produced by medusae that bud from gonangia of different colonies; that is, the colonies
are dioecious. The embryo develops into a ciliated free-swimming planula larva; this attaches to the substratum and forms a new colony. The three kinds of individuals—the feeding polyps (gastrozooids), the asexual
reproductive polyps, and the sexual reproductive medusae—illustrate polymorphism.

Parts of one hydra may easily be grafted upon another; in this way many bizarre organisms have been
produced. Parts of two hydras of different species
have also been grafted successfully.

The ability to regenerate lost parts is of obvious
benefit to an animal. Limited regeneration takes
place continually in *all* animals; for example, new
human epidermal cells are produced constantly to

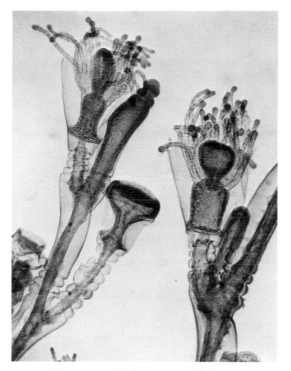

Figure 11-9(B). Mature *Obelia* gastrozooids.

replace those shed from the skin. In *Hydra* the interstitial cells replace cnidoblasts and other specialized cells that can no longer perform their functions.

Both internal and external factors influence the rate of regeneration and the character of the new part. Temperature, food, light, gravity, and mechanical contact are some of the external factors. In humans, various tissues are capable of regeneration, such as the skin, muscles, blood vessels, and bones. Lost body parts, however, cannot be regenerated in humans because the growing tissues do not coordinate properly. A decrease in regenerative ability seems to be correlated with an increase in complexity of animal body plans.

Other Hydrozoans
Obelia—A Colonial Hydroid

Obelia (Fig. 11-9) is a colonial hydrozoan that lives along the shores of the Atlantic and Pacific oceans to

depths of about 73 m. If you imagine a hydra that forms buds that remain attached to the parent, and that these buds specialize, some for feeding and others for reproduction, it is easy to understand the development and structure of *Obelia*. The specialized individuals of the colony are collectively called **zooids**. The colony is attached to rocks or algae by a rootlike mass (**hydrorhiza**), from which arise upright branches (**hydrocauli**). Hydralike feeding polyps (**gastrozooids**) and mouthless reproductive polyps (**gonangia**) protrude from each hydrocaulus, as shown in Figure 11-9. The hydrorhiza, hydrocauli, gastrozooids, and gonangia are all interconnected by a common gastrovascular cavity. The living tissues (**coenosarc**) of *Obelia* are protected by a nonliving chitinous covering (**perisarc**), which is ringed at intervals and which expands around the gastrozooids to form hydrothecae and around the gonangia to form gonothecae.

A gastrozooid resembles a hydra somewhat in structure and function. Gastrozooids are specialized for feeding and have about 30 solid tentacles. After the gastrozooids capture and ingest prey, extracellular digestion occurs in their gastrovascular cavities, and the fine particles of food that result are transported through the common **gastrovascular cavity** to all parts of the colony.

The gonangia are specialized for asexual reproduction. The central axis (blastostyle) of the gonangium gives rise to buds that develop into **medusae;** these escape through an opening in the end of the gonotheca. The free-swimming medusae are sexual individuals that produce either **ova** or **spermatozoa.** A fertilized ovum (zygote) develops into a ciliated free-swimming larva (**planula**) that soon attaches to a solid substratum and grows into a polypoid colony that reproduces by asexual budding.

In some cnidarians, such as *Obelia*, polypoid and medusoid stages alternate in the life cycle. The sessile hydroids, or polyps, generally reproduce asexually, and the free-swimming medusae reproduce sexually. Together a polypoid and medusoid stage forms a single cycle in such a cnidarian's life history; such a phenomenon of alternating sexual and asexual generations is called **metagenesis.** The polypoid and medusoid stages are not equally prominent in all hydrozoans. For example, the medusae in some species are inconspicuous (e.g., *obelia*) or degenerate,

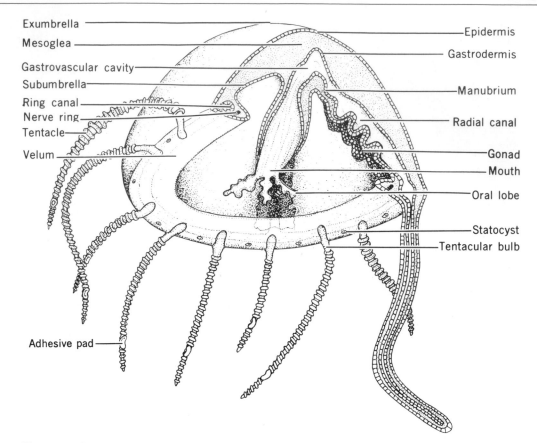

Exumbrella
Mesoglea
Gastrovascular cavity
Subumbrella
Ring canal
Nerve ring
Tentacle
Velum

Epidermis
Gastrodermis
Manubrium
Radial canal
Gonad
Mouth
Oral lobe
Statocyst
Tentacular bulb

Adhesive pad

Figure 11-10. Diagram of a hydrozoan medusa with part cut away to show the internal structure.

whereas in other species the polyp is only slightly developed (as in gonionemus).

Gonionemus—A Hydrozoan Medusa

Gonionemus is a common polypoid jellyfish found along the New England coastline. The medusa measures about 1–2 cm in diameter and bears around its margin from 16 to 80 hollow tentacles that bend at a sharp angle near the tip (Fig. 11-10). The brown gonads are clearly visible through the transparent bell-shaped body. The convex, aboral surface is called the **exumbrella,** whereas the concave, oral surface is the **subumbrella;** this is partly closed by a shelflike membrane, the **velum.** Water in the subumbrellar cavity is forced out through the central opening by contraction

of the body; this propels the medusa in the aboral direction by a sort of jet propulsion. Hanging down into the subumbrellar cavity is the **manubrium** with the **mouth** at its end, surrounded by four frilled **oral lobes.** The mouth leads to the **gastrovascular cavity** in the middle of the medusa, and four **radial canals** extend to a **ring canal** that circles the margin of the umbrella.

The two cellular layers are similar to those in the hydra, but the **mesoglea** is much thicker and gives the medusa a jellylike consistency. Suspended beneath the radial canals are the sinuously folded **gonads.** *Gonionemus* is **dioecious:** one individual produces either ova or spermatozoa. A free-swimming, ciliated **planula larva** develops from a fertilized ovum. The planula becomes fixed to some submerged object and matures to form a tiny, solitary polyp. This inconspicuous polyp then produces the medusae by asexual budding.

Physalia—A Polymorphic Colonial Hydrozoan

Physalia (Portuguese man-of-war) is a complex hydrozoan colony that is composed of many types of individuals. A colony containing two types of individuals is said to be **dimorphic**; one containing more than two kinds is **polymorphic.** *Physalia* (Fig. 11-11) consists of a gas-filled float (**pneumatophore**) from which a multitude of polyps is suspended. Some polyps are nutritive (**gastrozooids**), some are food capturing (**dactylozooids**), and others are reproductive (**gonozooids**) in function.

The surface of the float shimmers with beautiful iridescent colors—blues, pinks, violets, and purples—and the sail-shaped crest of the pneumatophore may glow a vivid carmine. The different types of zooids in *Physalia* arise from a single polyp by a budding process just beneath the float. This floating colony occurs in the Gulf Stream, and specimens are often cast upon Atlantic shorelines, especially during stormy periods. *Physalia* is passively carried from place to place by water currents and by winds blowing against the pneumatophore. The feeding dactylozooids, which may extend 60 feet into the water column, are armed with nematocysts powerful enough to inflict serious or fatal injury to human swimmers. The dactylozooids are able to catch large fish; they then contract and draw the prey up to the gastrozooids. These feeding polyps surround the prey and thereby enclose it in a digestive sac. Permanently attached medusoid zooids produce dispersal planula larvae.

Class Scyphozoa—The Jellyfish

Most of the larger jellyfish belong to the cnidarian class **Scyphozoa** (Gr. *skyphos*, cup + *zoön*, animal). Scyphozoan medusae are easily distinguished from hydrozoan medusa by their larger medusae, by the presence of usually eight sense organs along the margin of the umbrella, and by the absence of a distinct velum. Also, scyphozoans never form polymorphic

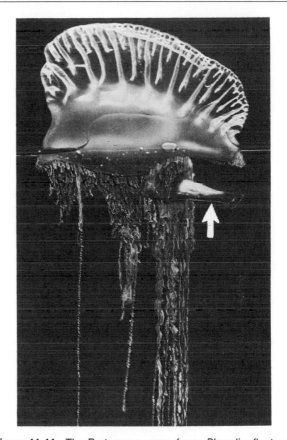

Figure 11-11. The Portuguese man-of-war, *Physalia,* floats on the surface of tropical seas. Hanging down from the float are long tentacles loaded with nematocysts. Many fish are captured by these streamerlike tentacles. The colony shown has just caught a fish; arrow points to fish. However, one species of fish of the genus *Nomeus* swims about among the tentacles of the Portuguese man-of-war with impunity. It appears to be immune to the poison of the stinging cells, possibly because it eats the tentacles of its host. These fish dart out to grasp a small food animal and hasten back amid the safety of the tentacles to devour it. The tentacles protect the fish, and particles of food not eaten by the fish are engulfed by the Portuguese man-of-war.

colonies, and the medusa is the predominate stage in the life cycle. Scyphozoan medusae usually range from 2.4 cm to 1 m in diameter, but a North Atlantic jellyfish, *Cyanea*, may grow to 2.3 m in diameter, with tentacles 36 m long and a wet weight of up to

A

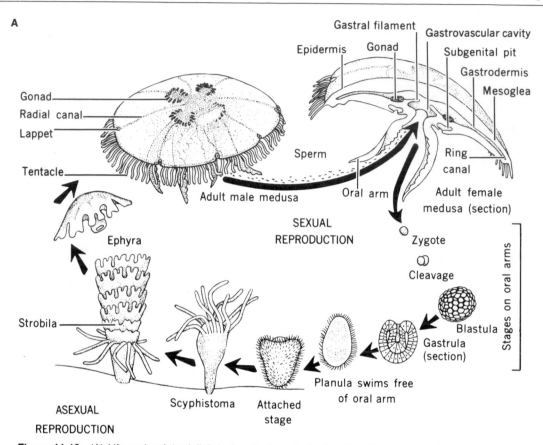

Gastral filament

Gastrovascular cavity

Epidermis · Gonad · Subgenital pit

Gastrodermis

Mesoglea

Gonad

Radial canal

Lappet

Sperm

Ring canal

Tentacle

Adult male medusa · Oral arm · Adult female medusa (section)

Ephyra

SEXUAL REPRODUCTION

Zygote

Cleavage

Stages on oral arms

Strobila

Blastula

Gastrula (section)

Scyphistoma · Attached stage · Planula swims free of oral arm

ASEXUAL REPRODUCTION

Figure 11-12. (A) Life cycle of the jellyfish, *Aurelia*. Longitudinal section through gastrula stage. Vertical section through adult. (B) An adult jellyfish, *Chrysaora* (see opposite page).

1 ton. Approximately 96 percent, however, is actually water. Scyphozoans float near the ocean surface; a very few species attach to rocks and seaweeds. **Metagenesis** generally occurs, but the asexual polypoid stage is relatively inconspicuous.

Aurelia (Fig. 11-12) is a common scyphozoan jellyfish that is white or bluish in color, with pink gonads. Four long **oral arms** hang down from the short **manubrium.** The **mouth** opens into a central **gastric cavity** from which four **gastric pouches** extend laterally. Within each gastric pouch is a gonad and a row of small **gastric filaments** bearing nematocysts. Numerous branching radial canals lead from the gastric cavity and pouches to a ring canal at the margin of the medusa. Eight sense organs (**rhopalia**) are regularly distributed around the umbrellar margin. The rhopalia contain receptors for equilibrium, light, and chemical stimuli.

The food of *Aurelia* consists of small organic particles and plankton that are trapped in mucus on the subumbrellar surface. The food-laden mucus is transported to the mouth by the oral arms. Ciliated grooves on these arms transport food to the gastric cavity. Preliminary digestion occurs in the gastric cavity and pouches, and flagellated **gastrodermal cells** produce currents that distribute the resulting broth throughout the **radial** and **ring canal systems.** The physiological processes in *Aurelia* are generally similar to those of the hydra.

The pleated **gonads** arise from gastrodermal inter-

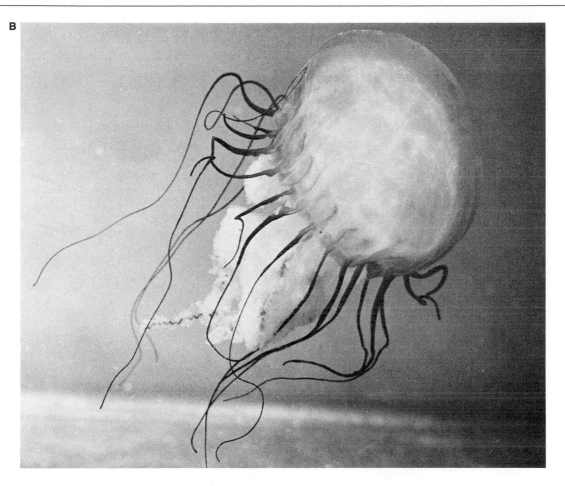

B

stitial cells in the floor of the gastric pouches. A fertilized ovum develops into a free-swimming **planula larva** that attaches to some object and develops into an elongated and deeply constricted polyp, known as a **scyphistoma** (Fig. 11-12). The scyphistoma divides transversely into discs until it resembles a pile of saucers; at this stage it is known as a **strobila.** Each disc develops tentacles and, separating itself from those below, swims away as a small medusa, called an **ephyra.** The ephyra grows directly into an adult jellyfish.

Pelagia is a pelagic (open-ocean) jellyfish that lacks a scyphistoma stage. In this species, the planula develops directly into an ephyra. Contact with *Chironex,* an Australian jellyfish, has been fatal to human swimmers. Accidental contact with their tentacles leads to

discharge of powerful nematocysts that inject paralytic toxins into the skin.

Class Anthozoa—Sea Anemones and Corals

Anthozoans (Gr. *anthos,* flower + *zoön,* animal) are sessile flowerlike animals. Whereas hydrozoans often alternate polypoid and medusoid stages in their life cyles, and scyphozoans emphasize the medusa (at times eliminating the polyp completely), the anthozoans are exclusively polypoid. Anthozoans do not form free-swimming medusae. Anthozoans are a large and diverse group, of which sea anemones and corals are familiar representatives.

Metridium—A Sea Anemone

Metridium dianthus (Fig. 11-13) is a **sea anemone** that fastens itself to wharf pilings and to solid objects in tidal pools along north Atlantic and Pacific coasts. It is a cylindrical animal with a crown of hollow **tentacles** arranged in several circlets around the slitlike **mouth.**

Its color varies from brownish to yellowish to whitish. The epidermis is tough. At either side of the **gullet** (**pharynx**) is a flagellated groove called the **siphonoglyph** that draws in water that inflates the body. The **gastrovascular cavity** below the pharynx consists of six **radial chambers** that are separated by six pairs of partitions called **primary septa** or **mesenteries.** Water passes from one radial chamber to another through pores (**ostia**) in these septa, and all are open below the gullet. **Secondary septa** project from the body wall into the gastrovascular cavity but do not reach the

Figure 11-13. (A) Morphology of the sea anemone, *Metridium,* a representative of the class Anthozoa. Left: Cross-section through the gullet shows the arrangement of the septa. Right: A part of the body has been cut away to show the internal structure. (B) Oral view of the sea anemone, *Xanthopleura* (see opposite page).

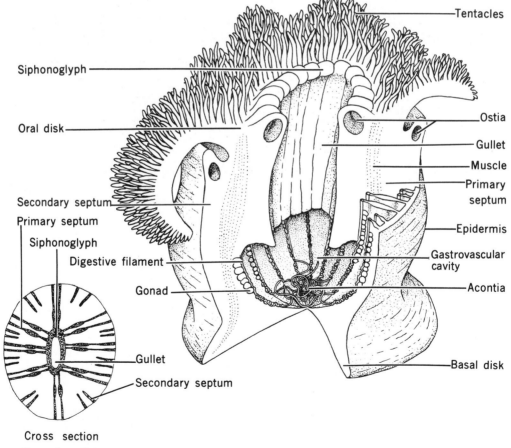

B

gullet. There is considerable variation in the number, position, and size of the septa. Their surfaces are covered with **gastrodermal cells,** and this greatly extends the digestive surface area of the gastrovascular cavity.

The septa below the gullet have thickened free margins called **digestive filaments.** These bear nematocysts, gland cells that secrete digestive enzymes, and flagellated tracts. The bases of these filaments extend into long delicate threads called **acontia,** which are armed with nematocysts and gland cells. Near the edge of the septa are the **gonads.**

Metridium is **dioecious.** Asexual reproduction may occur by **budding,** by fragmentation at the edge of the basal disc, or by longitudinal fission. Sexual reproduction takes place by fertilization of **gametes** in the surrounding seawater and the eventual formation of a

planula larva. After feeding in the plankton for a time, the larva attaches to a solid object and grows into an adult anemone.

Sea anemones are among the most beautiful and conspicuous inhabitants of tidal pools along the seacoast. When fully expanded, they form an underwater garden filled with flowerlike crowns of various colors, resembling more closely chrysanthemums or dahlias than the anemones after which they were named.

The habits of sea anemones are far from flowerlike, however! These carnivores trap any small animals that come within reach of their tentacles. They may be delicate in color, but they wield their batteries of nematocysts with paralyzing effect. The stunned prey is carried into the mouth, down the gullet, and into the gastrovascular cavity. There the food is digested by enzymes secreted by the gland cells on the digestive filaments, and the digestion products are absorbed by the general gastrodermis. Undigested wastes are ejected through the mouth.

Anemones are in turn preyed upon by some fishes, crabs, starfishes, and sea slugs. When disturbed or attacked, *Metridium* assumes a defense posture; it suddenly contracts and extrudes nematocyst-bearing acontia through minute pores in its body wall. Venom from these nematocysts contains two active proteins. One of the proteins stops nerve impulses by inhibiting active transport of ions in nerve cells. Together these proteins act synergistically to lyse red blood cells, in much the same way as occurs with bee and snake venom.

Astrangia—A Coral

Astrangia (Fig. 11-14) is a small colonial coral genus that inhabits northern temperate waters. *A. danae* lives along our mid-Atlantic shoreline, and *A. insignifica* is found along our Pacific Coast. Colonies are generally 5–8 cm in diameter and consist of a number of individuals living together on rocks near the shore. Each polyp looks like a small fragile sea anemone except that it rests within a calcareous exoskeleton that it has secreted. The "corals" on display in schools and museums are simply the skeletal remains of coral polyps.

Reef-Building Corals

In tropical seas coral colonies construct various types of reefs, atolls, and islands. Such reef-building corals are confined to marine waters that have a mean annual temperature of at least 20°C, principally between 30° N and 30° S latitude. The best-known **coral islands** are the Maldive Islands of the Indian Ocean, Wake Island, the Marshall Islands, the Fiji Islands of the Pacific Ocean, and those located in the Caribbean region. Bermuda is a coral island, and many of the houses there are built of limestone blocks mined from the surrounding reefs.

Coral reefs have been constructed by countless numbers of small polyps, each one secreting a cup-shaped skeleton around itself. As polyps die, new generations secrete new calcareous cups upon the old ones, and so only the surface of a coral mass is alive. Reef-building corals grow only in sunlit surface waters, at depths of 1 to 50 m, because the activities of symbiotic algae within their tissues are in some way necessary for calcification of the massive skeletons that form the reef.

Corals are carnivores that prey upon small planktonic animals, although they derive considerable nutriment from their algal symbionts. Corals are in turn preyed upon by some starfish and fishes.

There are three types of coral reef formations:

Figure 11-14. Colony of *Astrangia*. These corals secrete limestone cups into which the delicate polyps can retract.

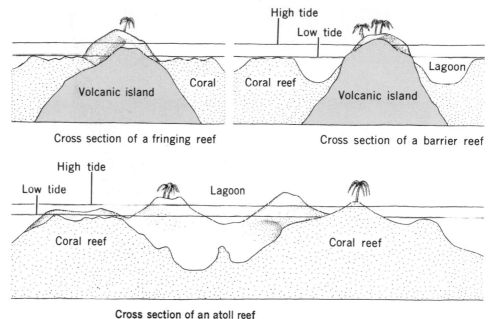

Figure 11-15. Development of a coral atoll. A fringing reef (top, left) is composed of coral that grows along the shore. The seaward side of the reef, high in nutrients, grows rapidly, leading eventually to a barrier reef (top, right). When the water rises, or the island sinks, the still-growing coral is all that is seen above the water (bottom). The resulting coral atoll may be a continuous reef, but it is usually divided by water channels extending through it from the ocean to the lagoon. Vegetation grows on accumulated debris.

Cross section of a fringing reef

Cross section of a barrier reef

Cross section of an atoll reef

(1) fringing reefs, (2) barrier reefs, and (3) atolls (Fig. 11-15).

A **fringing reef** is a ridge of coral built up from the sea bottom that is located so close to land that no navigable channel exists between it and the shore. Frequent breaks in the reef create irregular channels and pools that are inhabited by many different kinds of brilliantly colored animals.

A **barrier reef** is separated from the shore by a deep channel that is wide enough to afford passage for relatively large boats. The Great Barrier Reef is 810 km in length and lies 15 to 55 km from the northeast coastline of Australia. Its channel is from 20 to 50 m deep. Barrier reefs sometimes entirely surround tropical islands.

An **atoll** is a circle of reefs and islands composed of coral rubble that surround a central **lagoon**. Many coral atolls occur in the mid-Pacific Ocean, with lagoons of a few hundred meters to miles in diameter. Bikini Atoll has a lagoon area of 170 km² and a land area of only 1.72 km².

Relations of Cnidarians to Humans

Cnidarians are of considerable economic importance, though few are used as human food. Some scyphozoans are eaten in the Orient, and two species of anthozoans are consumed in Italy in a dish called ogliole. Precious corals (Fig. 11-16), usually bright red or pink, are crafted into necklaces and other jewelry.

The Horsehoe Atoll formation of Texas is the largest limestone petroleum reservoir in North America. This former atoll existed in the shallow seas that covered west Texas many millions of years ago but is now buried under thousands of feet of rock. It is from 40 to 55 km across and as much as 1000 m thick. More than 5000 wells have been drilled into the reef mass, and over 300 million barrels of oil have been recovered.

On tropical islands coral is used to build sidewalks, roads, landing strips, and houses.

Remote coral atolls have also been used to test the effects of atomic bombs and their resulting radioactive fallout on physical and biological systems.

The interstitial cells of hydras are providing an invaluable model for research in developmental biology.

Cnidarians have been used by other sciences as well. Astronomers have calculated that the earth's rotation is decreasing at a very gradual rate, on the order of 1 second every 50,000 years. If so, then days were shorter in the distant past. Because the time to travel around the sun is the same, there were more

Figure 11-16. Coral. Aerial views of a coral atoll in the Marshall Islands. Opposite page: From Gemini space craft. Bottom: From an aircraft. This page: Some representative corals.

days per year. Within the annual growth bands of certain coral exoskeletons, lines corresponding to daily growth can be observed and counted. Corals from 400 million years ago have over 400 of these lines per annual cycle, indicating that the year was then over 400 days long.

Origins of Cnidarians

Fossilized remains of medusae and polyps reveal little about cnidarian origins, other than that the phylum is very old (some formed almost a billion years ago). Fossil corals date from the Ordovician period (450 million years ago).

Many biologists believe that cnidarians evolved from colonial protozoans. Hydrozoans are considered to be the most primitive cnidarians because they have the simplest body plan. The scyphozoans and anthozoans probably evolved independently from hydrozoan ancestors, as did the ctenophores (Chapter 12).

Another group of biologists (probably a minority) consider the cnidarians to have evolved from flatworms (Chapter 13) because of the existence of the flatwormlike planula. Cnidarians have much simpler body plans than do flatworms, however. A popular recommendation is the divergence of all groups from planulalike ancestors that evolved previously from colonial protozoans.

Summary

As exemplified by the hydra, cnidarians are generally diploblastic and possess nematocysts. Contractile fibers are present in a more or less concentrated condition. Nerve cell processes (fibers) and sensory cells are characteristic structures; they may be a few in number and scattered or numerous and concentrated. The two principal cnidarian body forms are the polyp and the medusa. These are fundamentally similar in structure but are modified for different habitats. Although medusae may, upon superficial examination, appear to be very different from polyps, both are constructed on the same general plan. Both are radially symmetrical and have similar parts, the most noticeable difference being the relatively enormous quantity of mesoglea in the medusa. The water content of a medusa is very high, about 96 percent by weight.

Digestion in cnidarians is both extracellular and intracellular; enzymes are discharged into the gastrovascular cavity for preliminary chemical breakdown of prey organisms. The digested particles are transported to various parts of the gastrovascular cavity by flagellary currents and are then phagocytosed by the gastrodermal cells. Both respiration and excretion involve diffusion across the general surface of the epidermis and gastrodermis. Some cnidarians, notably stony corals, secrete an exoskeleton of calcium carbonate (limestone). Sense organs and nerve tissue provide for reception of various kinds of stimuli and for conduction of nerve impulses from one part of the body to another. Cnidarians are generally sensitive to changes in light intensity, temperature, mechanical stimuli, chemical stimuli, and gravity. Reproduction is both asexual, by budding and fission, and sexual, by the production of ova and spermatozoa, which unite to produce a zygote that develops into a ciliated planula larva.

Review Questions

1. Why might cnidarian interstitial cells be of interest to developmental biologists?
2. What are the functions of nematocysts? How do they work?
3. What is a medusa? Which cnidarians have a medusoid stage in their life cycles?
4. What are some major differences between primitive and advanced cnidarians?
5. Compare the eating habits of members from each cnidarian class.
6. Describe three types of coral reefs.
7. What functional explanation could be offered regarding the shape of medusae?
8. What are some differences between an individual cnidarian and a polymorphic colony? Give an example of each.

Selected References

Batham, E. J., *et al.* "The Nerve Net of the Sea Anemone, *Metridium senile*, the Mesenteries and Collu," *Quart. J. Micr. Sci.*, **101**:487, 1960.

Berrill, N. J. "The Indestructible Hydra," *Scientific American*, **197**:118, 1957.

Brien, P. "The Fresh-water Hydra," *American Scientist*, **48**:461, 1960.

Brunett, A. L. "Hydra: An Immortal's Nature,"*Natural History*, **68**:498, 1959.

Chapman, G. "Studies on the Mesoglea of Coelenterates," *Quart. J. Micro. Sci.*, **94**:155, 1953.

Chapman, G. B., and L. G. Tilney. "Cytological Studies of the Nematocysts of Hydra," *J. Biophys. Bichem. Cytol.*, **5**:69–78, 1959.

Crowell, S. "Behavioral Physiology of Coelenterates," *Amer. Zool.*, **5**:335, 1965.

Darwin, D. R. *The Structure and Distribution of Coral Reefs*, 3rd ed. New York: Appleton-Century-Crofts, 1896.

Florkin, M., and B. T. Scheer (eds.). *Chemical Zoology*, Vol. II: *Porifera Coelenterata and Platyhelminthes*. New York: Academic Press, 1968.

Fraser, C. McL. *Hydroids of the Pacific Coast of Canada and the United States*. Toronto: University of Toronto Press, 1937.

Fraser, C. McL. *Hydroids of the Atlantic Coast of North America*. Toronto: University of Toronto Press, 1944.

Gardiner, J. S. *Coral Reefs and Atolls*. London: Macmillan & Company, Ltd., 1931.

Hand, C. "On the Origin and Phylogeny of the Coelenterates," *Syst. Zool.*, **8**:191, 1959.

Hardy, A. C. *The Open Sea*. Boston: Houghton Mifflin Company, 1956.

Jones, C. S. "The Control and Discharge of Nematocysts in *Hydra*," *J. Exp. Zool.*, **105**:25, 1949.

Jones, O. A. and R. Eudean (eds.) *Biology and Geology of Coral Reefs*. New York: Academic Press, 1973.

Josephson, R. K. "The Coordination of Potential Pacemakers in the Hydroid *Tubellaria*," *Amer. Zool.*, **5**:483, 1965.

Lane, C. E. "The Portuguese Man-of-war," *Scientific American*, **202**:158, 1960.

Lenhoff, H. M. "Activation of the Feeding Reflex in *Hydra littoralis*," in Lenhoff and Loomis, eds., *The Biology of Hydra*. Coral Gables: University of Miami Press, 1961.

Lenhoff, H. M., and W. F. Loomis (eds.). *Symposium on the Physiology and Ultrastructure of Hydra and Some Other Coelenterates*. Coral Gables: University of Miami Press, 1961.

Lenhoff, H. M., L. Muscatine, and L. V. Davis. *Experimental Coelenterate Biology*. Honolulu: University of Hawaii Press, 1971.

Lentz, T. L., and R. J. Barnett. "Fine Structure of the Nervous System of Hydra," *Amer. Zool.*, **5**:341, 1965.

Loomis, W. F. "The Sex Gas of Hydra," *Scientific American*, **200**:145, 1959.

Mackie, G. O. "The Structure of the Nervous System in *Velella*," *Quart. J. Micr. Sci.*, **101**:119, 1960.

Mayer, A. G. *Medusae of the World*. Washington: Carnegie Institution of Washington, 1910.

Muscatine, L., and H. M. Lenhoff. *Coelenterate Biology: Reviews and New Perspectives*. New York: Academic Press, 1974.

Pantin, C. F. A. "Capabilities of the Coelenterate Behavior Machine," *Amer. Zool.*, **5**:581, 1965.

Rees, W. J. *The Cnidaria and Flies' Evolution*. New York: Academic Press, 1966.

Robson, E. A. "Nematocysts of *Corynactis*: The activity of the filament during discharging," *Quart. J. Micro. Sci.*, **94**:229–235, 1953.

Smith, F. G. W. *Atlantic Reef Corals*. Coral Gables: University of Miami Press, 1949.

Uchida, T. "Two Phylogenetic Lines the Coelenterates from the Viewpoint of Their Symmetry," Washington, *Proceedings of the Sixteenth International Congress of Zoology*, **1**:24, 1963.

Wiens, H. J. *Atoll Environment and Ecology*. New Haven: Yale University Press, 1962.

Yeatman, H. C. "Ecological Relationship of the Ciliate, *Kerona* to Its Host Hydra," *Turtox News*, **43**:226–227, 1961.

Yonge, C. M. "Ecology and Physiology of Reef Building Corals," in *Perspectives in Marine Biology* (A. A. Buzzati-Traverso, ed.). Berkeley: University of California Press, 1958.

Yonge, C. M. *A Year on the Great Barrier Reef*. New York: Putnam, 1930.

Phylum Ctenophora

The phylum Ctenophora (Gr. *ktenos*, comb + *phoros*, bearing) contains approximately 100 species of exclusively marine animals that resemble jellyfish (Fig. 12-1). Ctenophores are widely distributed in all oceans, being especially abundant in warm seas. They are transparent, gelatinous animals that are commonly called sea gooseberries or sea walnuts, because of their shape, or comb jellies, because of the comb-like locomotor organs (**comb plates**) that are arranged in eight rows from pole to pole. Ctenophores swim weakly through the water by the coordinated beating of these rows of long fused cilia. Ctenophores are biradially symmetrical; although most of their body parts are radially arranged, there exists a median longitudinal plane defined by the gut and tentacles (Fig. 12-2). The mouth is situated at one end and a sense organ at the opposite aboral end. Ctenophores generally swim with their oral end directed forward or down.

Pleurobrachia pileus is white or rose colored, ovoidal, and about 2 cm long, with tentacles about 15 cm long; it occurs in the North Atlantic from Long Island to Greenland, on European shores, and in the North Pacific. *Mnemiopsis leidyi* is a transparent, luminescent species, about 10 cm long, that lives along the east coast of the United States. It is often parasitized by a minute (1.5 mm long) sea anemone. Some bizarre body shapes occur among the ctenophores. For instance, *Cestus veneris* (Venus's girdle) may be only 6 cm tall but over 1 m long; transparent, but showing green, blue, and violet colors, it swims by muscular movements of its ribbonlike body as well as by beating its comb plates. It lives in tropical seas, but is sometimes carried north along our Atlantic Coast by the Gulf Stream.

Morphology and Histology

Most ctenophores possess two solid, contractile **tentacles** that emerge from blind pouches on either side of the body. The tentacles possess adhesive cells (**colloblasts**), which are modified epidermal cells used to snare prey. The colloblast is attached to the tentacle by a basal coiled spring. Adhesive mucus on the colloblast firmly anchors prey to the tentacles.

As in cnidarian jellyfishes, most of the ctenophore body is composed of transparent jellylike **mesoglea.** In the ctenophore, this middle layer contains a loose type of connective tissue, the **mesenchyme,** and other cells that are largely of ectodermal origin. A thin **epidermis,** derived from the ectoderm, covers the exterior and lines the **pharynx.** A **gastrodermis,** derived from the endoderm, lines the **stomach** and **gastrovascular canals.**

Physiology

Ctenophores are carnivores. Their food consists of small crustaceans, planktonic larvae, and other small animals. Feeding and digestion are much like that in the cnidarians, except that small **anal pores** help to eliminate particulate wastes in ctenophores. A series of canals branch out from the gut and supply the most metabolically active parts of the body.

The aboral sense organ (**statocyst**) is a unique balancing device in which a small granular particle

Figure 12-1. Phylum Ctenophora; *Mnemiopsis,* a comb jelly found in the North Atlantic.

known as the **statolith** sits on top of four tufts of cilia. These **balancer cilia** supply ciliary grooves that fork and communicate with the comb rows. When the ctenophore is tilted, the statolith presses differentially upon the balancer cilia, and this shift is transduced into neural impulses that affect the beating rates of the corresponding comb rows, thus initiating a righting response.

Reproduction

Ctenophores are **hermaphroditic.** Ova and spermatozoa are formed on the walls of the digestive canals just beneath the ciliated bands. The **gametes** pass by way of the mouth into the surrounding seawater, where fertilization occurs. The **zygote** develops into a free-swimming larval form, which develops directly into an adult.

Origin of the Ctenophores

Ctenophores probably evolved from cnidarian medusoid ancestors. The two groups were once con-

Figure 12-2. The structure of a typical ctenophore. Left: Aboral view of *Pleurobrachia.* Right: Side view of *Hormiphora.*

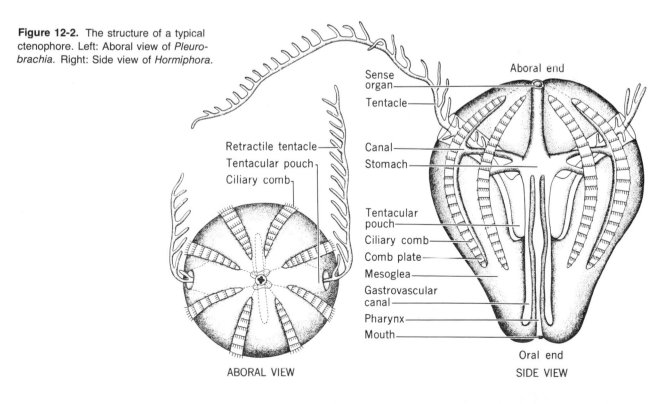

Retractile tentacle
Tentacular pouch
Ciliary comb

Sense organ
Tentacle
Aboral end
Canal
Stomach
Tentacular pouch
Ciliary comb
Comb plate
Mesoglea
Gastrovascular canal
Pharynx
Mouth
Oral end

ABORAL VIEW

SIDE VIEW

sidered as one phylum, **Coelenterata.** The features of the ctenophores are now considered sufficiently distinctive to warrant their classification in a separate phylum.

Summary

Ctenophores are biradially symmetrical carnivores with colloblasts and sense organs. Although they resemble cnidarians in some ways, ctenophores are a bit more complex.

Ctenophores differ from cnidarians in having comb plates, an aboral sense organ, mesenchyme, muscles, more complex organization of the digestive system with the anal pores, pronounced biradial symmetry, no nematocysts (except in one genus, *Euchlora*), and no planula larvae.

Review Questions

1. Would you rather be attacked by a cnidarian or a ctenophore? Explain.
2. How does a tilted ctenophore right itself?

Selected References

Harvey, E. N. *Bioluminescence*. New York: Academic Press, 1952.
Hickson, S. J. "Ctenophora," in *The Cambridge Natural History* (S. F. Harmer and A. E. Shipley, eds.). London: Macmillan and Company Ltd., 1906.
Mayer, A.G. *Ctenophores of the Atlantic Coast of North America*. Washington, Carnegie Inst. Pub. 162, 1912.

Phyla Platyhelminthes and Nemertina

Phylum Platyhelminthes— The Flatworms

The phylum Platyhelminthes contains the free-living and parasitic **flatworms.** Over 10,000 platyhelminth species have been described, and it is estimated that three times as many are yet to be cataloged. The platyhelminths are subdivided into three great groups: free-living flatworms, parasitic **flukes,** and parasitic **tapeworms.**

As do the cnidarians and ctenophores, platyhelminths display an **acoelomate** body organization; that is, they lack an internal body cavity other than the gut. But, whereas the cnidarians exhibit radial symmetry, the platyhelminths are **bilaterally symmetrical.** Their left and right sides are identical, but their anterior and posterior ends differ, as do their dorsal and ventral surfaces. Bilateral animals are typically motile and are characterized by **cephalization,** a condition reflecting a higher degree of development of receptors and nerve tissue at the anterior end of the body. Furthermore, the platyhelminths possess a third embryonic tissue, called **mesoderm.** Hence they are **triploblastic;** that is, their tissues are derived from ectoderm, endoderm, and mesoderm. The mesoderm gives rise to all tissues (except nervous tissue) between the epidermis and gastrodermis.

Mesoderm is perhaps anticipated by the mesoglea of cnidarians. Mesoglea is mostly a nonliving protein matrix with few cells. The mesoderm of platyhel-minths, on the other hand, is a loose connective tissue called **mesenchyme** that is derived from the endoderm during early development.

Platyhelminths are the simplest animals to exhibit a well-defined organ system level of complexity. Recall that, just as a tissue is a group of cells banded together to perform a particular function, an organ is a group of tissues cooperating in a common task; and an organ system is a group of organs cooperating to perform an even more general body function.

Phylum Platyhelminthes is subdivided into three classes: Turbellaria, Trematoda, and Cestoda. **Class Turbellaria** contains some 3000 described species of mostly free-living flatworms. Most turbellarians are aquatic, and the vast majority are marine. Most aquatic flatworms live on or within bottom debris. Some turbellarians have successfully invaded moist terrestrial habitats—"residing" beneath logs and in damp, moldy leaves.

Class Trematoda contains over 5000 described species of parasitic flukes that infest the body surface and internal organs of many vertebrates and invertebrates (especially mollusks and arthropods). Flukes suck blood, cell fragments, and tissue fluids from their hosts.

Class Cestoda contains over 200 described species of parasitic tapeworms. These endoparasites inhabit the intestine of vertebrates and absorb predigested nutrients from the host's gut through their general body surface.

Flukes and tapeworms are major parasites of eco-

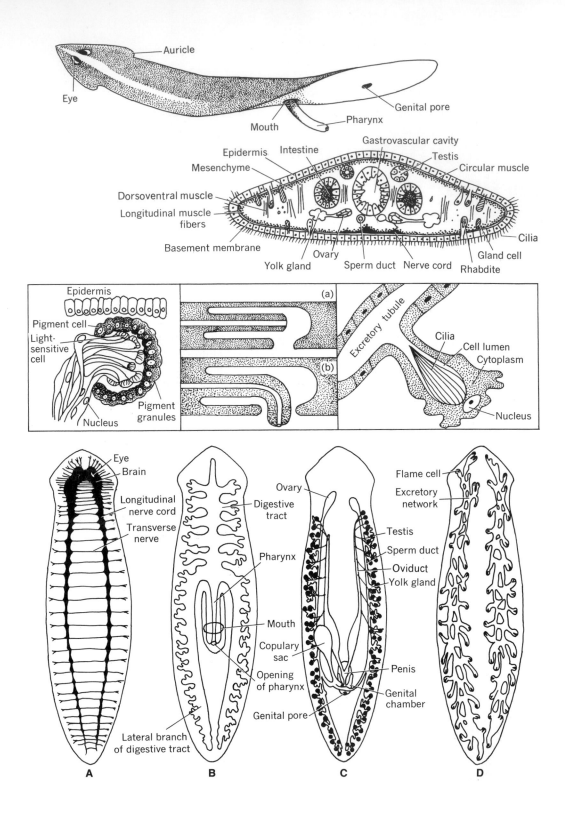

Auricle

Eye

Mouth

Pharynx

Genital pore

Epidermis

Intestine

Gastrovascular cavity

Testis

Circular muscle

Mesenchyme

Dorsoventral muscle

Longitudinal muscle fibers

Basement membrane

Yolk gland

Ovary

Sperm duct

Nerve cord

Cilia

Gland cell

Rhabdite

Epidermis

Pigment cell

Light-sensitive cell

Nucleus

Pigment granules

(a)

(b)

Excretory tubule

Cilia

Cell lumen

Cytoplasm

Nucleus

Eye

Brain

Longitudinal nerve cord

Transverse nerve

Ovary

Digestive tract

Pharynx

Mouth

Copulary sac

Opening of pharynx

Genital pore

Lateral branch of digestive tract

Flame cell

Excretory network

Testis

Sperm duct

Oviduct

Yolk gland

Penis

Genital chamber

A

B

C

D

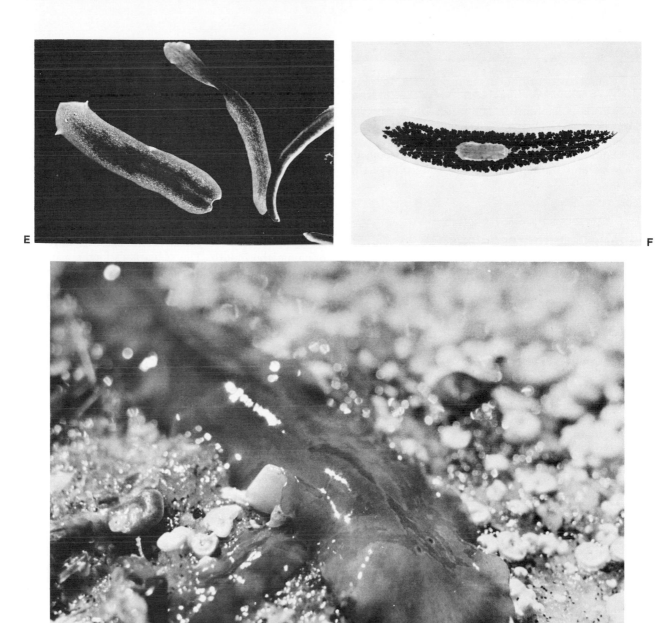

E

F

G

Figure 13-1. Morphology and anatomy of a planarian, *Dugesia tigrina.* Top, left: Planarian with pharynx extended. On facing page cross-section through the pharyngeal region is shown. Middle, left: Section through the eye. Incoming light is restricted to the opening in the cup of pigmented cells enabling the planarian to distinguish light direction. Middle, center: Pharynx (a) at rest in pharyngeal chamber and (b) protruded through the mouth. Middle, right: Single flame cell attached to a portion of duct in a freshwater planarian. Bottom: (A) nervous system, (B) digestive system, (C) reproductive system, and (D) excretory system. Above, (E) Adult planarian, *Dugesia,* (F) Injected digestive system of *Dugesia,* (G) A marine planarian, *Notoplana.*

nomic importance to humans and domesticated animals. Their study, combined with that of the parasitic roundworms (Chapter 14), comprises the science of **helminthology.** The parasitic platyhelminths have evolved complex adaptations for their special modes of life, and so the free-living turbellarians are considered to exhibit the more primitive body plan of the phylum.

Planaria—A Representative Turbellarian Flatworm

A common freshwater turbellarian in the United States is the planarian **Dugesia tigrina** (Fig. 13-1), which is often used in classroom study and zoological research. This flatworm lives on water plants and bottom debris in ponds, lakes, and rivers and in small streams under stones. Its dorsal surface is black, brown, or mottled and irregularly spotted with white; its ventral surface is white or grayish. The body, which may attain lengths of 15 to 18 mm, is broad and blunt at the anterior end and pointed at the posterior end.

External Morphology

The anterior or head end of this flatworm bears a pair of lateral projections, called **auricles,** that contain receptors for dissolved chemicals, touch, and detection of water currents. A pair of light-sensitive **ocelli** lie on the dorsal surface of the head between the auricles (Fig. 13-1). The **mouth** is not located on the head but rather opens on the midventral surface near the middle of the animal. The mouth communicates with the gut through a muscular tube, the **pharynx,** which is extended some distance out of the mouth during feeding. Posterior to the pharynx lies a smaller opening for the reproductive tract, the **genital pore.**

The flattened ventral surface of the body is covered with **cilia** and **mucus-secreting glands.** Mucus lubricates the interface between the animal and the substratum and leaves a distinctive slime trail in the animal's wake. Small turbellarians glide along on their cilia, but larger flatworms adopt muscular, crawling movements during which time they stretch out, attach the head, and then pull the tail forward.

Internal Morphology and Physiology

The space between the body wall and gut in turbellarians is filled with a loose network of **mesenchyme.** These tissues of mesodermal origin consist of a fibrous mesh with embedded fixed cells whose processes **anastomose** (join together) and free cells that can move about in amoeboid manner.

The **digestive system** of *Dugesia* consists of a **mouth,** a **pharynx,** and an **intestine** with three main trunks and many smaller lateral branches (Fig. 13-1). As in cnidarians, the **gut** is a blind sac with only one opening, the mouth. The planarian's food consists of insect larvae, annelid worms, crustaceans, and other small animals, living or dead. Receptors in the **auricles** are used to locate food; when the auricles are removed, the flatworm cannot find food. The prey is subdued with mucus and adhesive secretions. The **pharynx** is protruded into the prey's body, and, aided by the secretion of proteolytic enzymes that help digest the tissues, the mobile pharynx then sucks up the liquified food. Final digestion occurs within cells that line the planarian intestine. There is only one opening to the digestive cavity, and particulate wastes are eliminated through the mouth. Platyhelminths lack distinct circulatory and respiratory systems. Nutrients are circulated within the branches of the digestive system and fluid-filled spaces in the mesenchyme. Respiratory gases diffuse across the general body surface.

Soluble nitrogenous wastes (ammonia) diffuse directly into the external environment. Most turbellarians also have excretory organs called **protonephridia** whose primary function is to remove excess water from the body (**osmoregulation**). These are a complex network of small tubules on each side of the flatworm, from which branch many ciliated chambers called **flame cells** (Fig. 13-1).The flickering **cilia** create a current that forces the collected fluid from the tubules through **nephridiopores,** or opening on the body surface. In one turbellarian species, *Gyratrix hermaphroditis,* the freshwater form has pronounced flame cells, whereas the marine form, which looses water osmotically, does not, thus suggesting their importance in removing excess body water.

The **muscles** of the body wall are arranged in three layers. An outer circular layer is located just beneath the epidermis, followed by a middle layer of diagonally arranged fibers and then an inner longitudinal layer. Sometimes dorsoventral fibers also traverse the mesenchyme.

The well-developed **nervous system** lies below the body wall musculature (Fig. 13-1). It consists of an inverted V-shaped mass of tissue, the **brain,** and two ventral **longitudinal nerve cords** connected by **transverse nerves.** Nerves link the various anterior receptors to the brain. The pigmented **ocelli** are sensitive to light but do not form an image. Turbellarians show a very limited ability to **learn** that is, to modify their behavioral response to a particular stimulus on the basis of past experience (Chapter 31).

Turbellarians reproduce both asexually and sexually. Flatworms commonly reproduce asexually by **transverse fission.** Turbellarians are **hermaphroditic,** possessing both male and female **gonads** (Fig. 13-1), but cross-fertilization between different individuals is the rule. The **reproductive systems** are highly complex. Development is direct, without a larval stage, as the embryo flattens and develops directly into a young adult.

Turbellarians have remarkable powers of **regeneration** (Fig. 13-2). If an individual is cut transversely into two parts, the anterior fragment will regenerate a new tail and the posterior piece will develop a new head. A section cut from the middle of the body will regenerate both a head at the anterior end and a tail at the posterior end. Pieces of one flatworm can also be grafted onto another.

Figure 13-2. Planarian reproduction and regeneration. Top: Sexual reproduction. (A) Copulation. (B) Cocoon under rock. (C) Emergence and growth of young planarian. Middle: Asexual reproduction. The planarian divides by fission and the pieces develop and grow into adults. Bottom: Regeneration. (A) Specimen cut into two parts; the head regenerates another tail and the tail regenerates another head—both lengthen into normal specimens. (B) A split head regenerates a new head. (C) If the auricles are removed, the planarian cannot locate food (until new auricles regenerate).

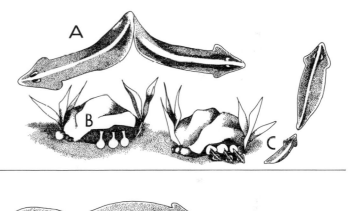

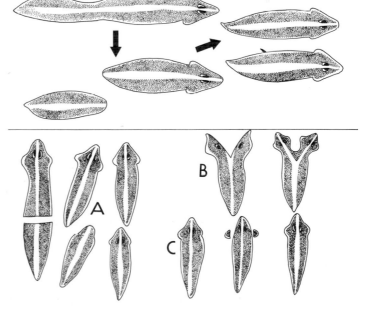

Fasciola hepatica—A Trematode (Fluke)

The class **Trematoda** contains the parasitic **flukes.***
Trematodes are generally small, leaf-shaped platyhelminthes with complicated life cycles that involve one
or more hosts. Most flukes are less than a few centimeters in length. The **mouth** is anterior and is surrounded by an **oral sucker.** One or more **ventral
suckers** and anchoring **spines** or **hooks** serve to attach
the fluke to its host. Flukes suck blood and tissue
fluids from their hosts. *Ecto*parasitic flukes attach to
the exterior of the host or to the walls of body cavities
with external openings such as the mouth, gills, or
urogenital tract, whereas *endo*parasitic flukes infest
the host's internal organs. Indeed, virtually every
organ system may be potentially infected by flukes
specialized for that particular habitat. This specificity
of parasite to host is further reflected in the complex
life cycles of many flukes.

Fasciola hepatica is known as the sheep liver fluke.
The adult flukes inhabit bile passages in the livers
of sheep, cows, pigs, and many other herbivores.
Human infestation by *F. hepatica* is relatively rare,
but this is probably due to infrequent exposure to the
parasite rather than to the parasite's inability to colo-
nize the human body. Humans may become infected
by eating aquatic plants (notably watercress) infested
with encysted fluke larvae. This occurs sporadically in
many parts of the world, including the continental
United States, Hawaii, Cuba, and Europe.

Flukes lack the ciliated epidermis that characterizes turbellarian flatworms. Instead, flukes are covered with a **cuticle,** a cytoplasmic **syncytium** that
serves as the site of diffusion of respiratory gases and
soluble metabolic wastes. In gut-dwelling endoparasites, the cuticle also protects the fluke from the host's
digestive enzymes.

The anterior mouth of *Fasciola* is surrounded by a
muscular disc, the oral sucker (Fig. 13-3). A ventral
sucker (**acetabulum**) also serves as an organ of attachment. Located between the two suckers is a single

---

* **Flukes** are sometimes confused with **leeches,** which they superficially resemble in form and method of feeding. Leeches, however,
are segmented worms that belong to the phylum Annelida; see
Chapter 16.

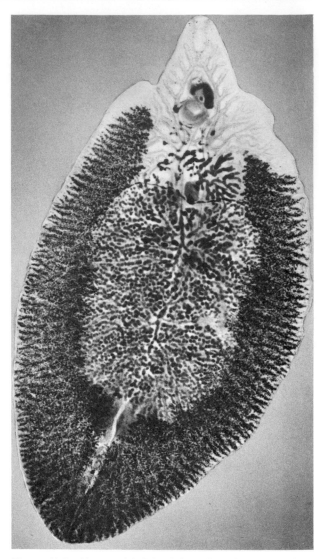

Figure 13-3. Above, adult fluke. Life cycle and structure of
the liver fluke, *Fasciola hepatica.* (A) Eggs are carried out of
the sheep body in the feces. (B) In water the eggs develop
into miracidia that burrow into a snail. (C) Sporocyst from
snail. (D) Redia larvae from sporocyst. (E) Cercaria larvae
from redia leave the snail and encyst on vegetation. (F) The
encysted metacercaria (here dissected from cyst). If the vegetation is eaten by a sheep, the metacercaria infect the sheep
and develop into adults. (G) Adult showing the digestive system on the left and the excretory system on the right.
(H) Adult showing the reproductive system.

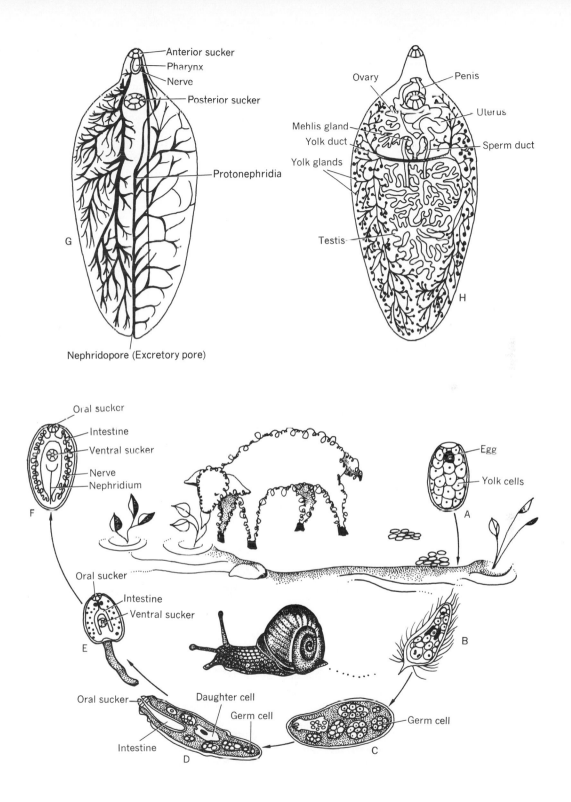

Anterior sucker
Pharynx
Nerve
Posterior sucker
Protonephridia
G
Nephridopore (Excretory pore)

Ovary
Penis
Mehlis gland
Uterus
Yolk duct
Yolk glands
Sperm duct
Testis
H

Oral sucker
Intestine
Ventral sucker
Nerve
Nephridium
F

Egg
Yolk cells
A

Oral sucker
Intestine
Ventral sucker
E

B

Oral sucker
Daughter cell
Germ cell
Intestine
D

Germ cell
C

midventral **genital pore.** An **excretory pore** opens at the extreme posterior end of the body.

The **digestive system** consists of a **mouth,** muscular **pharynx,** short **esophagus,** and **intestine** with two main branches. The host's tissues are injured by the oral sucker, cuticular spines in the pharynx, and proteolytic enzymes. The muscular pharynx pumps cell fragments and tissue fluids through the mouth and into the intestine. Particulate wastes are generally eliminated through the mouth (into the host's tissues). The flukes progressively destroy the liver tissue, causing a debilitating disease known as "liver rot."

The **excretory system** is a complex system of **flame cells** and collecting ducts that drain into two main **longitudinal ducts.** These empty into a common posterior **bladder** that communicates with the excretory pore.

The **nervous system** consists of a small **ganglion** at the anterior end of the body that is joined to a series of **longitudinal nerves** with interconnecting bridges (**commissures**). Sense organs are almost completely lacking.

Most flukes are **hermaphroditic;** that is, both male and female gonads are present in every adult fluke. The **reproductive system** is extremely well developed and quite complex. One liver fluke may produce as many as 500,000 **ova,** and, because the liver of a single sheep may contain more than 200 adult flukes, 100 million eggs may be produced within a single parasitized host.

Fertilization and early development occur within the **ootype** of the fluke. The shelled embryos pass through the bile duct into the sheep's intestine and then are eliminated with the feces (Fig. 13-3). Eggs that encounter water develop for about 15 days (depending on temperature) before hatching into ciliated larvae called **miracidia.** These swim about until they either die of exhaustion or encounter the appropriate species of freshwater snail, into which they burrow. Within the snail host, the miracidium changes into a simple saclike body, or **sporocyst,** in which several germ balls or **embryos** develop. Each embryo develops over the course of two or three weeks into a second kind of larva, the **redia.** These usually give rise asexually to one or more generations of daughter rediae. These asexual generations increase (up to 10^6

times) the number of larvae that may develop from a single egg. Eventually the redia produces a third kind of larva with a long tail, known as a **cercaria.** The cercariae leave the body of the snail, swim about in the water for a short time, and then encyst on a leaf or blade of grass at the shoreline. The encysted cercaria is now called a **metacercaria.** If vegetation bearing the encysted flukes is eaten by sheep, the metacercariae hatch in the small intestine and migrate through the bile duct into the liver, where they develop into mature flukes in about six weeks.

The tremendous number of eggs produced by a single fluke is necessary because most do not survive. Many eggs do not reach water; most miracidia do not find the particular species of snail necessary for their further development; and any metacercariae have little chance of being devoured by a sheep.

Opisthorchis sinensis (formerly *Clonorchis sinensis*) is an important human parasite in Japan and China. The adult flukes lives in the bile ducts of humans, cats, dogs, and other mammals. The life cycle of this parasite requires two intermediate hosts. The eggs are passed in the feces, and the miracidia invade certain snails. The cercariae enter various species of freshwater fish, where they form metacercariae. Humans and other mammals are infected by eating raw parasitized fish. Cooking the fish before ingestion kills the encysted metacercariae. Certain drugs can effectively eliminate adult flukes once they have become established in the liver.

Taenia solium—A Representative Cestode (Tapeworm)

Cestoda is the third and most specialized class of phylum Platyhelminthes. The vast majority of cestodes are **tapeworms,** endoparasites that inhabit the intestines of vertebrates. The flattened tapeworm body is divided into three general regions: scolex, neck, and strobila. The anterior **scolex** is armed with **hooks** and **suckers** that attach the parasite to the inner wall of the intestine. Behind the scolex is a short **neck region,** which buds off segments called **proglottids.** These contain the **reproductive organs** and form a ribbonlike chain (**strobila**). The strobila can be quite long; lengths exceeding 9 m have been reported. Tapeworms lack a mouth and digestive system; predigested nutrients are absorbed from the host's gut through the **cuticle** that covers the parasite's body. The cuticle also protects the cestode from the host's digestive enzymes. Adult tapeworms are generally well adapted to their specific hosts, and little harm appears to result, even in heavy infestations, other than their depriving the host of nutrition.

No digestive tract is present. The **nervous system** is similar to that of the turbellarians and flukes but not as well developed. Longitudinal and cross-connecting

Figure 13-4. Life cycle and structure of the pork tapeworm, *Taenia.*
(A) Embryo from ripe proglottid in feces of man. Pig ingests the embryo with food. (B) Bladderworm (cysticercus from muscle of pig). (C) When raw or partially cooked pork eaten by man, cysticercus everts, attaches to intestine, and grows into (D) adult.
(E) Mature proglottid with reproductive organs. (F) Gravid proglottid.

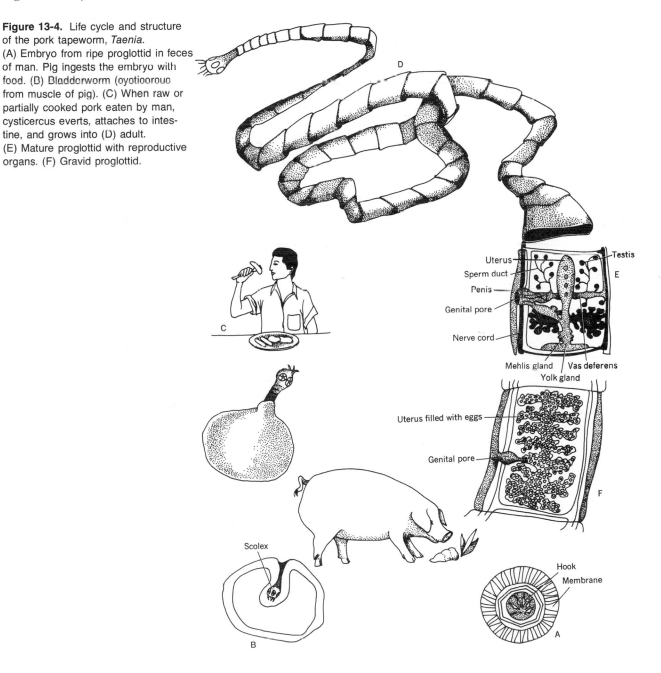

excretory canals drain fluid from flame cells in each proglottid.

All tapeworms are hermaphroditic, and a complete reproductive system occurs in each mature proglottid. A mature proglottid is almost completely filled with male and female gonads. Fertilization can occur within the same proglottid or between two moglottids of the same strobila (Fig. 13-4). Each fertilized egg develops into an oncosphere larva while still within the proglottid. The terminal proglottids are eliminated from the host in feces. If they are then eaten by a pig, the oncospheres hatch and bore through the wall of the intestine into the bloodstream. Oncospheres are carried in the blood to all organs of the body, especially muscles, where they develop into an encysted larval stage, called the cysticercus or bladder worm. The life cycle is completed when humans eat raw or poorly cooked infected pork. Upon entering the human small intestine, the cysticercus everts a scolex, attaches to the intestinal wall, and develops into a mature adult tapeworm.

Humans may also become "intermediate hosts" of this tapeworm if fertilized eggs are ingested in food and/or water that has been contaminated with feces. Then the oncospheres enter the bloodstream and encyst in various organs, including the brain, sometimes with fatal consequences.

Taenia solium, the pork tapeworm, lives as an adult in the human small intestine, where it attains lengths of 9 m. The parasite clings to the inner wall of the intestine by means of hooks and suckers on its scolex. Behind the scolex is a short neck followed by a string of 800 or 900 proglottids that gradually increase in size. Because the proglottids are budded off from the neck, those at the posterior end are the oldest and most mature. Two other tapeworm life cycles are shown in Figure 13-5.

Relations of Platyhelminths to Humans

The free-living flatworms are of little direct importance to humans, but the trematodes and cestodes are dangerous parasites. The scientific names, intermediate hosts, geographical distributions, and definitive hosts of some platyhelminth parasites of humans are presented in Table 13-1.

Certain fluke cercariae (blood flukes) have been incapable of surviving in humans, but they cause a severe swimmer's itch (a form of dermatitis) when they penetrate the skin of bathers who have become sensitized by repeated exposure. The first symptom of swimmer's itch is a prickly sensation followed by the development of very itchy pimples, which sometimes become pustular. This condition is common in the north central states and in southern Canada, as well as in some other parts of the United States, Europe, and India. Contact with trematode cercariae may be avoided by toweling off immediately. The parasite can be controlled in small bodies of water by adding copper sulfate to kill the snails that serve as intermediate hosts.

Hymenolepis nana, the dwarf mouse tapeworm, is the most common tapeworm in humans, with about 1 percent of the population being infested worldwide. Mice and rats serve as alternate reservoir hosts, and cockroaches and flour beetles are known to be intermediate hosts.

One tapeworm (*Taenia saginata*) was once sold in encysted form as a reducing aid to overweight people. The cyst was swallowed, and the resulting tapeworm absorbed much of the host's food. Clinical diseases sometimes resulted.

Origin of the Platyhelminths

The flatworms, especially some primitive turbellarians, resemble the cnidarians in certain respects: A single opening generally serves for both ingestion of food and egestion of wastes; the nerves may be arranged in a nerve net; and the mesenchymal connective tissue is reminiscent of the cellular mesoglea in higher cnidarians and ctenophores. Furthermore, some primitive (acoel) flatworms are remarkably similar to planula larvae; thus a popular phylogeny includes a planuloid ancestor for all metazoans higher than Porifera and possibly Mesozoa. Some biologists consider the ancestors of flatworms to be cnidarians, others point to ciliates or even annelids. Trematodes and cestodes are generally considered to have

Figure 13-5. Above, the adult human liver fluke, clonorchis. Below, tapeworm life cycles. Left: Life cycle of the dog tapeworm, *Echinococcum granulosus.* (A) Hydatid cyst develops in the tissues of a herbivore. (B) Carnivorous mammals (dogs) eat these tissues and adult; tapeworms develop in the intestine. (C) Eggs passed in feces are accidentally eaten by various herbivorous mammals (sheep, swine, and man); eggs then develop into hydatid cysts in the liver and other organs (A). Right: Life cycle of the broad tapeworm of man, *Dibothriocephalus laatus.* (A) Eggs passed in feces enter water and become ciliated larvae that are eaten by copepods. (B) Larva that develops in copepod. (C) copepods are eaten by fish and larvae encyst in fish muscles. (D) Raw or partially cooked infected fish eaten by man and other animals contain larvae that develop into adults.

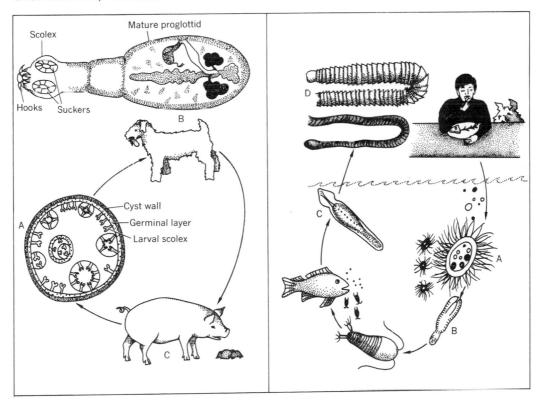

Table 13-1A. Some Common Trematodes of Man

	Scientific Name	Developmental Stages	Geographical Distribution	Definitive Hosts
Intestinal flukes	*Fasciolopsis buski*	In snails and on water-grown vegetables	China, Formosa, Sumatra, India	Man and pigs in China, Formosa, Thailand
	Heterophyes heterophyes	In snails and freshwater fish	Egypt, China, Japan	Man, cats, dogs in Egypt
Liver flukes	*Opisthorchis sinensis*	In snails and freshwater fish	China, Japan, Korea, Philippines	Man, cats, dogs, and fish-eating mammals
	Opisthorchis felineus	In snails and freshwater fish	Europe, Siberia	Man, cats, dogs
Lung flukes	*Paragonimus westermani*	In snails and freshwater crabs	Japan, China, Philippines, America	Man, tigers, dogs, cats, mustelids, and pigs
Blood flukes	*Schistosoma haematobium*	In snails	Africa, Near East, Portugal, Australia	Man
	Schistosoma mansoni	In snails	Africa, West Indies, N. and S. America	Man, certain rodents, and baboons
	Schistosoma japonicum	In snails	Japan, China, Philippines	Man, cats, dogs, pigs, etc.

Table 13-1B. Some Common Tapeworms of Man

	Scientific Name	Developmental Stages	Geographical Distribution	Definitive Hosts
	Diphyllobothrium latum	In copepods and freshwater fish	Widespread	Man, dogs, cats, bears, and foxes (intestine)
	Echinococcus granulosus	In liver, brain, and lungs of man, pigs, and sheep	Widespread	Dogs and other carnivores (intestine); not found in man except in larval–hydatid stage
	Taenia saginata	Muscles of cattle	Worldwide	Man
	Taenia solium	In muscles or central nervous system of pigs and man	Worldwide	Man

evolved from similar ancestors, but independently from the turbellarians.

Phylum Nemertina (Rhynchocoela)— Ribbon Worms

Nemerteans and platyhelminths are the only bilateral metazoans with no body cavity except the gut.

The 600 known species of phylum Nemertina are called ribbon worms because they are long and dorsoventrally flattened (Fig. 13-6). Nemerteans range in length from less than 2 cm to over 30 m. Most species live along the seashore, where they burrow into sand or mud, but a few inhabit freshwater and humid terrestrial habitats.

The distinguishing anatomical features of nemerteans (Fig. 13-7) are: (1) a long eversible **proboscis** that

Figure 13-6. Photograph of a ribbon worm.

is separate from the **gut** and enclosed in the tubular **proboscis sheath;** (2) a **circulatory system,** usually consisting of a median dorsal and two lateral **blood vessels;** and (3) a complete **digestive tract,** with both **mouth** and **anus.** The cavity surrounding the probos-

Figure 13-7. Structure of a nemertean. (A) Digestive, excretory, and circulatory systems. (B) Reproductive and nervous system. (C) Cross-section. (D) Details of the retracted (top) and extended (bottom) proboscis.

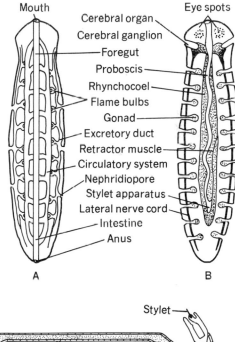

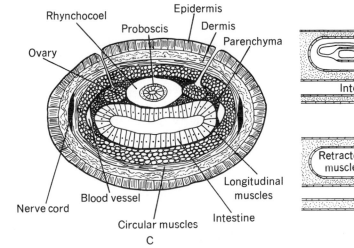

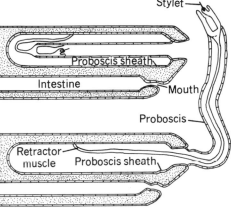

cis is called the **rhynchocoel;** it is not a coelom at all and is formed by an internal folding of ectoderm.

Nemerteans feed on other animals, both dead and alive. Their long proboscis, when used to capture prey and for defense, everts rapidly, wrapping its sticky surface around the prey. A terminal **stylet** (Fig. 13-7D) sometimes injects a toxin into the victim.

Ribbon worms may be found coiled in burrows in mud and sand or under stones and seaweed. Locomotion is effected by the cilia that cover the body surface, by contractions of the body muscles, or by attachment of the proboscis and a subsequent drawing forward of the body. Sexual reproduction involves release of gametes into the seawater, where fertilization occurs; the sexes are separate in most species.

Nemerteans have great powers of regeneration, and some reproduce in warm weather by fragmentation of the body. *Lineus socialis,* which is only 10 cm in length, can be cut into as many as 100 pieces, and each piece will regenerate a minute worm within four or five weeks. These minute worms may again be cut into pieces that regenerate, and these in turn may be cut up, and so on, until miniature worms result that are less than 1/200,000 of the volume of the original worm.

Nemerteans appear to be an evolutionary offshoot of the free-living platyhelminths. The nemertineans resemble the free-living flatworms in being bilaterally symmetrical; in having protonephridia, unsegmented flattened bodies, a ciliated epidermis, and similar nervous systems and sense organs; and in lacking a body cavity and respiratory system. However, they differ from the platyhelminths by having a complete digestive tract, a functional circulatory system, and a less complex reproductive system.

Summary

The platyhelminths are characterized as being unsegmented (except secondarily in cestodes), triploblastic, and bilaterally symmetrical. The body is flattened dorsoventrally. No internal body cavity other than the gut is present. Platyhelminths have no skeletal, circulatory, or respiratory systems, but they do have a protonephridial excretory system. Free-living flatworms have a head with sense organs, and all classes have a central nervous system that consists of a brain, longitudinal nerve cords, and various connecting branches. Most platyhelminths are hermaphroditic. The basic body plan of the free-living turbellarians has been modified considerably in the parasitic trematodes and cestodes.

The second acoel phylum discussed, Nemertina, consists of the free-living, burrowing ribbon worms. In contrast to the platyhelminths, ribbon worms have more advanced digestive and circulatory systems but less complicated reproductive systems.

Review Questions

1. Name some structures that are of mesodermal origin.
2. What are some differences between turbellarians, trematodes, and cestodes?
3. Describe the life cycle of one platyhelminth parasite in humans, and discuss ways of avoiding infection.
4. Redesign a turbellarian so that it could parasitize humans. Explain your modifications and relate them to the parasite's environment.
5. What characteristics of nemerteans suggests that they are more advanced than the flatworms?

Selected References

Beauchamp, R. S. A. "The Rate of Movement of *Planaria alpina,*" *J. Exp. Biol.,* **12:**271, 1935.

Bronsted, H. V. "Planarian Regeneration," *Biol. Rev.,* **30:**65, 1966.

Brychowsky, B. E. *Monogenetic Tremades, Their Systematics and Phylogeny.* J. W. Hargis, Jr. (ed.), trans. P. C. Oustinoff. Washington, D.C.: American Institute of Biological Sciences, 1961.

Cameron, T. W. M. *Parasites and Parasitism.* New York: John Wiley & Sons, Inc., 1956.

Caullery, Maurice. *Parasitism and Symbiosis,* trans. A. M. Lysaught. London: Sidgwick & Jackson, Ltd., 1952.

Chandler, A. C., and C. P. Read. *Introduction to Parasitology,* 10th ed. New York: John Wiley & Sons, Inc. 1961.

Cheng, T. *The Biology of Animal Parasites.* Philadelphia: W. B. Saunders Company. 1964.

Erasmus, D. A. *The Biology of Trematodes.* New York: Crane, Russak, 1972.

Faust, E. C., P. C. Beaver, and R. C. Jung. *Animal Agents and Vectors of Human Disease.* Philadelphia: Lea & Febiger, 1966.

Faust, E. C., P. Russel, and R. C. Jung. *Craig and Faust's Clinical Parasitology,* 8th ed. Philadelphia: Lea & Febiger, 1971.

Florkin, M., and B. T. Scheer (eds.). *Chemical Zoology,* Vol. II, *Porifera, Coelenterata and Platyhelminthes.* New York: Academic Press, 1968.

Hoare, C. A. *The Trypanosomes of Mammals.* Oxford: Blackwell Scientific Publications, 1972.

Hyman, L. H. "Platyhelminthes," *The Invertebrates,* II. New York: McGraw-Hill Book Company, Inc., 1951.

Hyman, L. H., and E. R. Jones. "Turbellaria," in W. T. Edmondson (ed.), Ward and Whipple, *Fresh-Water Biology,* 2nd ed. New York: John Wiley & Sons, Inc., 1959.

Jenkins, M. M., and H. P. Brown. "Sexual Activities and Behavior in the Planarian *Dugesia*" *Amer. Zool.,* **2:**418, 1962.

Jennings, J. B. "Studies on Feeding, Digestion, and Food Storage in Free-living Flatworms," *Biol. Bull.* **112:**63, 1957.

Jennings, J. B. "Further Studies on Feeding and Digestion in Triclad Turbellaria," *Biol. Bull.,* **123:**571, 1962.

Noble, E. R., and G. G. Noble. *Parasitology. The Biology of Animal Parasites,* 3rd ed. Philadelphia: Lea & Febiger, 1971.

Riser N. W., and M. P. Mouse (eds.) *Biology of Turbellaria.* New York: McGraw-Hill Book Company, 1974.

Schell, S. C. *How to Know the Trematodes.* Dubuque, Iowa: W. C. Brown Co., 1970.

Schmidt, G. D. *How to Know the Tapeworms.* Dubuque, Iowa: W. C. Brown Co., 1970.

Smyth, J. D. "the Physiology of Tapeworms," *Biol. Rev.,* **22:**214, 1947.

Swartzwelder, C., G. W. Hunger, III, and C. B. Worth. *Manual of Tropical Medicine.* Philadelphia: W. B. Saunders Company, 1966.

Swellengrebel, N. H., and M. N. Sterman. *Animal Parasites in Man.* New York: D. Van Nostrand Company, 1961.

Wardle, R. A., J. A. McLeod and S. Radinovsky. *Advances in the Zoology of Tapeworms.* Minneapolis: University of Minnesota Press, 1974.

Pseudocoelomate Phyla

Multicellular animals (metazoans) exhibit three basic grades of body organization: acoelomate, pseudocoelomate, and coelomate. **Acoelomate** metazoans do not have an internal body cavity other than the gut. Acoelomates are relatively simple animals: cnidarians, ctenophores, platyhelminths, and nemerteans. **Pseudocoelomate** metazoans have an internal body cavity in addition to the gut. This **pseudocoel** is not a "false cavity," as its name suggests, but a fluid-filled chamber that is derived from the embryonic blastocoel (Chapter 39). The pseudocoelomates include a diverse array of small marine and freshwater animals of uncertain origins and relationships. All other metazoans possess another type of body cavity called a coelom, which is derived from embryonic mesoderm.

The pseudocoel is located between the body wall and the internal organs; it lacks the cellular lining (**peritoneum**) of true coelomates. Possession of such an internal body cavity is advantageous for several reasons: (1) The pseudocoel may act as a **hydrostatic skeleton,** a hydraulic system in which muscles exert pressure upon a water-tight compartment. The shape of the pseudocoel changes according to the way in which the muscular pressure is applied. Internal pressure (turgor) supports the body, and changes in shape may enhance locomotion. (2) The **perivisceral** ("around the internal organs") **fluid** that fills the pseudocoel serves as a medium for the internal circulation of gases, nutrients, and wastes. Pseudocoelomate animals lack a distinct circulatory system. (3) The pseudocoel also allows the internal organs, now freed from continuous contact with the body wall, to lengthen and assume shapes that enhance their physiological functions.

The pseudocoelomate animals are difficult and confusing to classify. Eight types are commonly recognized. Five of these were once considered as classes of the phylum **Aschelminthes.** The aschelminths include **rotifers, gastrotrichs, kinorhynchs, nematodes,** and **nematomorphs.** The three other pseudocoelomate groups, **Acanthocephala, Entoprocta,** and **Gnathostomulida** are considered as separate phyla.

Phylum Rotifera
External Morphology

The animals in the class Rotifera are commonly known as wheel animals because they possess an anterior crown of cilia, called a **corona,** that appears to rotate like a wheel when viewed under the microscope. The corona is used in locomotion and feeding (Fig. 14-1). Rotifers are among the smallest metazoans; in fact, most are microscopic (0.1–0.4 mm). Because rotifers often possess fascinating body forms and a transparent body wall, they seldom fail to attract the attention of amateur microscopists. Most rotifer species live in fresh water, especially in shallow ponds and bogs, but some are marine, and a few live among terrestrial mosses and lichens. The tail-like **foot** is often

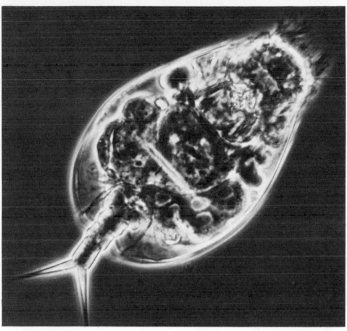

Figure 14-1. Phylum Rotifera. Top: Photomicrograph of a rotifer. Bottom: General structure of a female rotifer—left, dorsal view; right, lateral view.

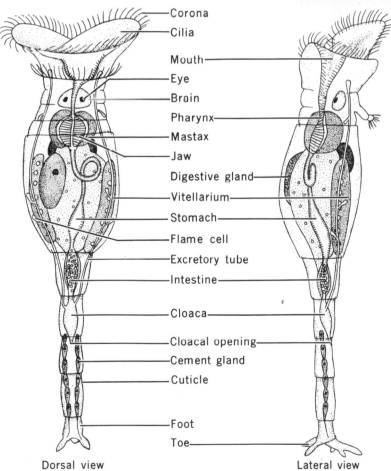

Corona
Cilia
Mouth
Eye
Brain
Pharynx
Mastax
Jaw
Digestive gland
Vitellarium
Stomach
Flame cell
Excretory tube
Intestine
Cloaca
Cloacal opening
Cement gland
Cuticle
Foot
Toe

Dorsal view

Lateral view

forked and contains **pedal glands** that allow the rotifer to temporarily attach to underwater plants and other objects. The body is usually cylindrical in shape and is covered by a proteinaceous **cuticle.**

Internal Morphology and Physiology

Rotifers eat minute organisms and organic debris that are swept by the coronal cilia into the **mouth.** The food then passes through the **pharynx** into a grinding organ called the **mastax.** Here chitinous **jaws,** which are constantly at work, break up the food. The movements of these jaws in the mastax clearly distinguish a living rotifer from other animals. A short **esophagus** then leads to the **stomach,** where food is digested. Nutrients are absorbed through the walls of the **intestine,** and wastes pass into the **cloaca** and out of the **cloacal opening** (anus). The cloaca also receives materials from the excretory and reproductive systems.

The **excretory system** consists of two **protonephridia,** one on either side of the body. **Flame cells** are drained by two tubules that empty into a common **bladder.** The bladder is emptied through the cloaca at rates of one to four times per minute. Such a large volume of discharge indicates that the protonephridia serve primarily as an **osmoregulatory organ** to remove excess water from the body, as in the platyhelminths.

The **nervous system** includes a **brain** (a bilobed **ganglion**) and two main **lateral nerves** that connect to muscles, an eyespot (**ocelli**), **antennae,** and various **chemoreceptors.** There is no respiratory system; gas exchange occurs by diffusion.

Reproduction

Reproduction in rotifers is remarkable. The sexes are separate, but males are known in only a few species; and, where found, males are usually smaller than the females and are often degenerate. The **eggs** of species in which males have never been observed develop **parthenogenically** (i.e., without fertilization). Some rotifers produce two kinds of eggs. During the summer, only females are produced from thin-shelled eggs by parthenogenesis. With the onset of winter, males are produced, and sexual reproduction produces fertilized eggs that are protected by thick shells. When favorable conditions return, these winter eggs hatch into female rotifers.

Another peculiarity of rotifers is their power to resist desiccation. Certain species, if dried slowly, secrete gelatinous envelopes that resist dehydration. In this dormant condition they survive seasons of drought and temperature extremes.

Some common rotifers are *Epiphanes senta* (formerly called *Hydatina senta*), a species used widely in experimental research; *Asplanchna,* which often occurs in enormous numbers in the plankton of the Great Lakes; *Floscularia,* which lives in a transparent tube and has a beautiful corona with five knobbed lobes; *Melicerta,* which constructs a tube of spherical pellets; and *Philodina,* characterized by a slender rose-colored body.

Rotifers eat microscopic organisms and in turn serve as food for small aquatic crustaceans and other animals. These are preyed upon by fish, which may be harvested for human food. Thus rotifers serve as important intermediate links in freshwater food chains.

Origins of Rotifers

The resemblance between certain rotifers and the trochophore larvae of some mollusks and annelids is quite striking and has led some investigators to theorize that rotifers are closely related to the ancestors of the mollusks and annelids. However, some of the most competent investigators believe that the resemblance to certain trochophore larvae is purely coincidental, a result of similar adaptations to similar environments and life-styles. If so, the rotifers probably evolved from primitive turbellarian ancestors.

Phylum Gastrotricha

These microscopic animals live in both fresh and salt water and are often abundant in bottom debris. They range in length from 0.06 to 1.5 mm, and some 400 species are known. *Chaetonotus* (Fig. 14-2) is a typical gastrotrich. The gastrotrich body is indistinctly divided into **head, neck, trunk,** and **toes.** Locomotion is accomplished by longitudinal bands of **cilia** on the ventral surface. The anterior **mouth** is sur-

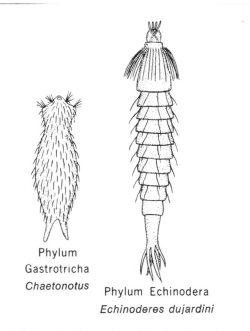

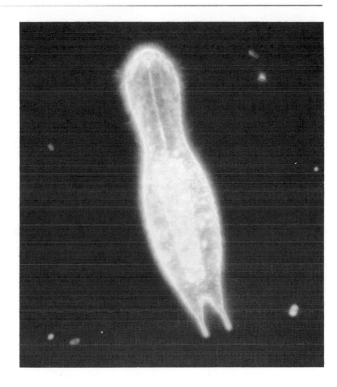

Phylum
Gastrotricha
Chaetonotus   Phylum Echinodera
Echinoderes dujardini

Figure 14-2. Phylum Gastrotricha, *Chaetonotus,* a free-living species. Phylum Kinorhyncha, *Echinoderes dujardini,* a marine species. Photo on right of a living gastrotrich.

rounded by **oral bristles,** and the dorsal surface bears many slender **spines.** Bacteria, protozoans, and organic debris are sucked into the **pharynx.** A straight intestine leads to an **anus** near the posterior end of the body. Gastrotrichs, unlike other aschelminths, are **hermaphroditic.** The eggs are very large and develop without any larval stage. The other organ systems do not differ significantly from the rotifer body plan.

Phylum Kinorhyncha

Kinorhynchs are minute animals that inhabit shallow, muddy ocean bottoms. About 100 species of kinorhynchs have been recognized. Kinorhynchs are somewhat larger than most rotifers and gastrotrichs, but they never exceed 1 mm in length. The body, which consists of a **head, neck,** and **trunk,** is covered with a spiny **cuticle** that is divided into 13 segments. Kinorhynchs burrow through mud, feeding on organic debris. Many questions about these obscure

animals remain unanswered. *Echinoderes dujardini* (Fig. 14-3) is reddish in color and lives in mud and among algae in the north Atlantic Ocean.

Phylum Nematoda—The Roundworms

Nematodes are the largest aschelminth group and the third largest (after the arthropods) group of metazoans. Over 10,000 species have been described. Nematodes are found in marine, freshwater, and terrestrial habitats and also as parasites in animals and plants. Nematodes have adapted to an incredible range of physical conditions; they are found in the ocean depths, high mountains, polar regions, deserts, and hot springs. They are often present in enormous numbers. A single decomposing apple yielded approximately 90,000 nematodes belonging to several species.

Nematodes are called **unsegmented roundworms** to distinguish them from the flatworms and segmented annelids (e.g., earthworms). Roundworms

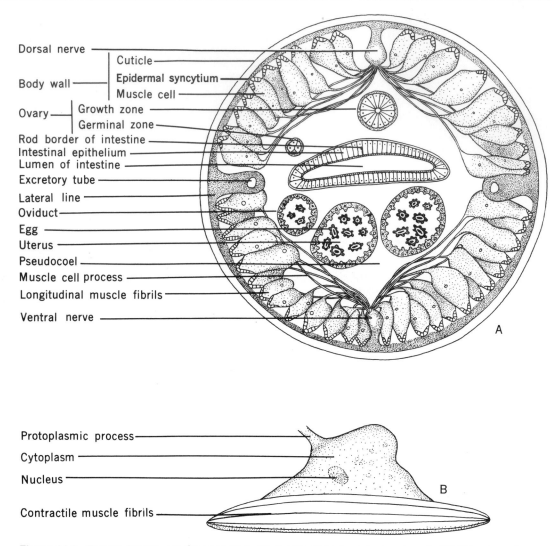

Dorsal nerve

Cuticle

Body wall

Epidermal syncytium

Muscle cell

Ovary

Growth zone

Germinal zone

Rod border of intestine

Intestinal epithelium

Lumen of intestine

Excretory tube

Lateral line

Oviduct

Egg

Uterus

Pseudocoel

Muscle cell process

Longitudinal muscle fibrils

Ventral nerve

A

Protoplasmic process

Cytoplasm

Nucleus

Contractile muscle fibrils

B

Figure 14-3. Female *Ascaris.* (A) Cross-section. (B) Details of an *Ascaris* muscle cell. (C) Side view of a specimen. (D) Dissection to show the internal organs.

are typically slender, cylindrical animals, tapered at one or both ends, with usually a smooth glistening surface.

Many of the parasitic nematodes undergo amazingly complex life histories. A brief description of *Ascaris* (which has a relatively simple life cycle) is presented below.

Ascaris lumbricoides—A Representative Roundworm

External Morphology

This common roundworm parasitizes the intestine of humans and pigs. It is estimated that one third of

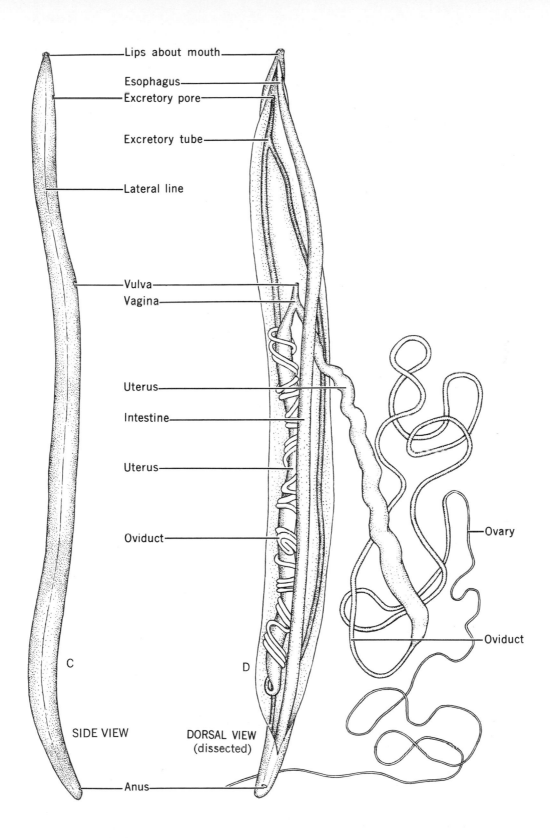

Lips about mouth

Esophagus
Excretory pore

Excretory tube

Lateral line

Vulva
Vagina

Uterus

Intestine

Uterus

Oviduct

Ovary

Oviduct

C

D

SIDE VIEW

DORSAL VIEW
(dissected)

Anus

all the people in the world are infected with *Ascaris*. This roundworm does not require an intermediate host, and transmission is by ingestion of the parasite's eggs in food or water contaminated with human feces.

Ascaris exhibits the elongated cylindrical shape that is characteristic of nematodes, but it is exceptionally large for a roundworm. Females are considerably larger than males, measuring up to 25 cm in length and 6 mm in diameter. Males are distinguished by a sharply curved posterior end. Narrow dorsal and ventral white lines running the entire length of the body mark the location of longitudinal **nerve cords,** and broader lateral lines denote **excretory tubules** on either side. The tough chitinous **cuticle** is smooth but marked with fine striations. The **mouth** at the anterior end is surrounded by one dorsal and two lateroventral **lips.** Near the posterior end of the male is the **cloacal** opening, or **anus,** from which extend two chitinous rods, the **penial spicules,** that assist sperm transfer during copulation. In the female, the **genital pore** (**vulva**) is located on the ventral surface about one third of the body length from the anterior end.

Internal Morphology and Physiology

The body contains a straight digestive tract and other visceral organs (Fig. 14-3). Between the intestine and the body wall is a fluid-filled body cavity, the **pseudocoel.** The digestive tract is very simple. The small **mouth** opens into a muscular **esophagus,** or **pharynx,** which is from 10 to 15 mm long. The esophagus pumps fluids from the intestine of the host. The nutrient is then absorbed through the wall of the parasite's long intestine. The posterior portion of the intestine is known as the **rectum** in the female, and wastes are discharged directly through the **anus.** But in the male the intestine and reproductive system open into a common passageway, the **cloaca,** which leads to the exterior.

Nematodes contain a dual excretory system. Two longitudinal **excretory tubes,** one in each lateral line (Fig. 14-3), empty through a single ventral **excretory pore** near the anterior end of the body. One or two **renette glands,** which are single cells that empty near the excretory pore, can also be seen.

A ring of nervous tissue surrounds the esophagus and gives off two large **nerve cords,** one dorsal and the other ventral, and several smaller nerves.

Reproduction and Life Cycle

The male **reproductive system** is a single coiled tube. A threadlike **testis** at the blind end produces amoeboid **spermatozoa.** These **gametes** mature as they pass through the **sperm duct** (**vas deferens**) into the wider **seminal vesicle,** where they are stored. During copulation sperm are discharged through a short muscular **ejaculatory duct** that opens into the cloaca.

The female (Fig. 14-4) has a Y-shaped reproductive system. Each coiled branch of the Y consists of an **ovary, oviduct,** and **uterus.** The two uteri unite into a short muscular tube, the **vagina,** which opens through the **vulva** to the exterior.

Fertilization occurs within the oviduct. Each **zygote** is then surrounded by a thick, rough-surfaced shell and is passed out through the vulva. A mature female *Ascaris* may contain as many as 27 million ova at one time, and about 200,000 shelled eggs per day are passed into the host's gut.

The eggs are then eliminated in the host's feces. *Ascaris* eggs are very resistant to desiccation; if depos-

Figure 14-4. Internal structure of the free-living vinegar eel, *Turbatrix aceti.* Parts of digestive and reproductive systems are shown. Natural size 2 mm in length.

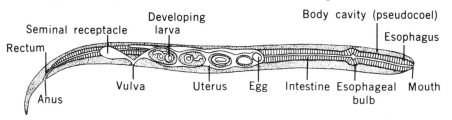

ited in soil, the **embryos** remain alive for many months or years. Infection with *Ascaris* results from ingesting eggs in food or water contaminated by feces or soil.

The eggs pass through the stomach and hatch in the small intestine. The larvae burrow through the wall of the small intestine and enter the bloodstream, which transports them to the liver, then to the postcaval vein to the heart, then to the lungs. Next they bore into the air passages and migrate up the trachea (windpipe) to the throat, where they are swallowed. This journey through the host requires about 10 days. Their second visit to the host's intestine lasts for the duration of their adult lifespan of about one year. Sexual maturity is attained after about two months in the intestine.

Ascaris worms found in humans and pigs are difficult to distinguish morphologically, but it is thought that they must differ physiologically because cross-infection between the two hosts has rarely been observed. The mature worms do not generally harm the intestine of their host, but considerable nutriment is robbed to support their massive reproductive output. The wandering larvae sometimes lodge in organs other than the lungs, with pathological results.

Other Nematodes
Turbatrix—The Vinegar Eel

The vinegar eel, *Turbatrix (Anguillula) aceti,* is a free-living nematode, easily obtained from the bottom of a barrel of cider vinegar. These roundworms are about 2.5 mm in length and are visible to the unaided eye when held before a bright light. As do other nematodes, *T. aceti* exhibits distinctive side-to-side lateral undulations of the body. These **thrashing movements** result from the fact that the body musculature of roundworms contains only **longitudinal fibers,** with no circular muscles.

Females are almost twice as long as males. The major anatomical features of a female vinegar eel are shown in Figure 14-4. The **ova** are fertilized within the body of the female and also develop there. The thin egg membrane ruptures in the **uterus,** and the young worms hatch before being released from the mother's body; this is called **ovoviviparous reproduction.**

Other Free-Living Roundworms

Free-living nematodes are mostly small, rarely exceeding 1 cm in length. Many possess an **oral spear** with which they puncture the root cells of various plants, including economically valuable species. Roundworms live in almost every moist habitat. Most species are widely distributed, but some occur only in very specialized habitats. *Turbatrix silusiae*, a relative of the vinegar eel, is found in the felt mats under beer mugs in Germany.

Nematode Parasites of Plants

More than 1,000 species of nematodes are known to parasitize plants. The damage that these tiny parasites do to cultivated crops is staggering. The common garden roundworm *Meloidogyne* invades the roots of over 1,700 species of plants. Eggs are laid in the roots or in the nearby soil. Certain nematodes stimulate plant tissue to form knotlike galls on the roots. Others enter leaves and move about eating the contents of cells. Some roundworms stay on the surface of the plant, bury their anterior end into the tissue, and feed upon the sap.

These parasites undermine the plant's vigor, open the way for invasion by bacteria and fungi, and injure growing points. Partial control of roundworm parasites can be achieved by crop rotation, soil sterilization, and the development of resistant varieties of plants.

Nematode Parasites of Animals

Four of the many roundworms that parasitize vertebrate animals are described in the paragraphs that follow.

The **dog ascarid,** *Toxocara canis,* is especially prevalent in puppies, which became infected as fetuses when larvae migrate through the mother's placenta. Mature dogs become infected by swallowing the eggs. The larvae migrate through the body as do the larvae of *Ascaris* in humans. Dogs acquire a degree of resistance to the infection, and after three or four months the worms are passed out and susceptibility to further infection is reduced.

The **cecum worm of chickens,** *Heterakis gallinae* (Fig. 14-5), lays eggs that are passed in the feces of infected birds. If the eggs are swallowed by other birds, they hatch in the small intestine, and the young nematodes move into the **ceca** (two blind pouches that branch from the junction of the small and large intestines). They do not seriously injure domestic fowl, but they carry with them a protist parasite, *Histomonas meleagridis,* that causes a disease in turkeys known as **blackhead.**

The **urinary worm of rats,** *Trichosomoides carssicauda,* has a most unusual male–female relationship. The male is much smaller than the female and lives in her vagina. Infective eggs are released into the urine. If ingested by another rat, the eggs hatch in the stomach. The larvae burrow into the bloodstream, and those that reach the urinary tract develop into mature adults.

The **horse roundworm,** *Strongylus vulgaris,* has a worldwide distribution but is especially prevalent in warm climates. This parasite lives in the cecum or colon of horses, attached by its mouth to the mucosal lining, and sucks blood. Loss of blood due to heavy infestation results in anemia. The eggs of *Strongylus* are voided in the feces, and infective larvae develop

in the soil. When ingested by a horse, the larvae penetrate the intestinal wall. Some larvae enter the bloodstream and travel to the lungs, then migrate up the trachea, are swallowed, and reenter the intestine, where they develop to maturity. Other larvae lodge in the wall of the mesenteric arteries that supply the intestines; there they cause fibrous clots that obstruct the flow of blood or dilated aneurysms that weaken the arterial wall.

Nematode Parasites of Humans

Humans are parasitized by at least 45 species of roundworms, some of which cause widespread suffering and death.

Ascaris

Ascaris lumbricoides (described earlier) is a prevalent human parasite in the tropics. Migration of large numbers of ascarid larvae through the lungs causes inflammation and can result in pneumonia. The adults may populate the intestine in such large numbers that

Figure 14-5. Examples of Phyla Nemertinea (A,B) class Nematomorpha (C).

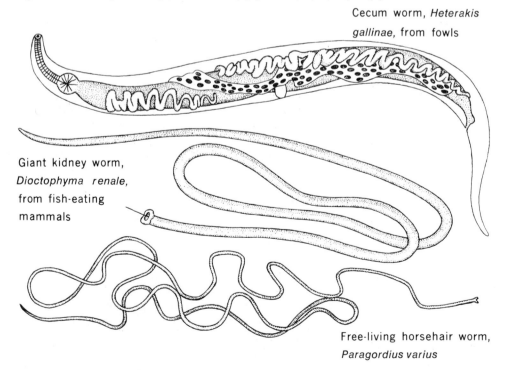

Cecum worm, *Heterakis gallinae,* from fowls

Giant kidney worm, *Dioctophyma renale,* from fish-eating mammals

Free-living horsehair worm, *Paragordius varius*

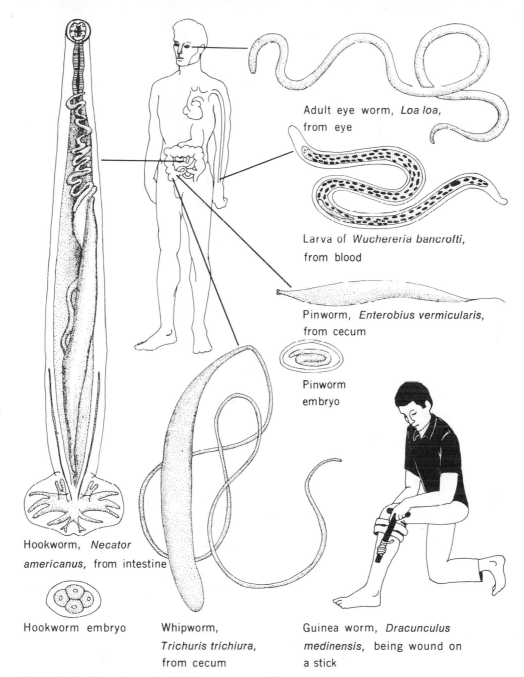

Adult eye worm, *Loa loa*, from eye

Larva of *Wuchereria bancrofti*, from blood

Pinworm, *Enterobius vermicularis*, from cecum

Pinworm embryo

Hookworm, *Necator americanus*, from intestine

Hookworm embryo

Whipworm, *Trichuris trichiura*, from cecum

Guinea worm, *Dracunculus medinensis*, being wound on a stick

Figure 14-6. Several roundworms that parasitize man.

a fatal intestinal obstruction results. Thousands of worms have been recorded in a single patient, but even a hundred can block the passage of food through the intestine. A frequent complaint of patients suffering from **ascariasis** is abdominal pain or discomfort. Nervous symptoms such as headaches or convulsions

183

may result from the toxic substances that are secreted by the worms. Fortunately, several drugs are available that safely and effectively remove the worms.

The degree of ascarid infection is a rough measure of local sanitation, as transmission is by fecal contamination. Ascariasis is also common in some parts of the United States. Infection occurs more frequently in children than in adults because of careless sanitary habits. Infection can generally be prevented by increasing sanitary conditions.

Hookworms

The hookworms *Ancylostoma duodenale* and *Necator americanus* (Fig. 14-6) are widespread nematode parasites of humans. *N. americanus* is more common in the United States. Other hookworms parasitize domesticated and wild animals. Hookworm larvae develop in moist, feces-contaminated soil and usually enter the body of their human host by boring through the skin of the foot, especially between the toes. They enter the bloodstream and are carried to the lungs. Then the larvae make their way through the air passages and are swallowed. The hookworms attach themselves by means of **oral hooks** to the walls of the intestine and feed by suction, ingesting blood and tissue fluids. In the case of the dog hookworm (and probably also the human hookworm), blood is continuously being ingested by the worm, and droplets of fluid consisting mainly of red blood corpuscles are constantly being eliminated from its anus. A single worm may withdraw blood from its host at the rate of 0.8 ml per day. Infection by 500–1000 hookworms may result in a daily loss of one pint to one quart of blood. Victims of severe hookworm infections are anemic and prone to other diseases.

Hookworm disease is not as widespread in this country as has been true in the past, thanks to concerted efforts by public health officials to improve sanitation at the turn of this century. Hookworm disease is prevalent in much of the tropics, however, where soil and climate favor these parasites. Hookworm disease can be cured by several drugs. The most important preventive measure is to dispose of human feces in ways that do not contaminate the soil, thus giving the eggs no opportunity to hatch and develop into infective larvae. Wearing shoes also helps to prevent infection.

Trichina Worms

Trichinella spiralis causes the disease **trichinosis** in humans, pigs, rats, and many wild mammals. Human trichinosis is especially common in north temperate countries, and about 20 percent of the U.S. population is estimated to be infected. The parasites enter the human body when inadequately cooked meat from an infected pig or bear is eaten (Fig. 14-7). The larvae soon mature in the human intestine, and each mature female worm bears about 15,000 **ovoviviparous** young. These larvae burrow through the intestinal wall into the bloodstream, where they are transported throughout the body. They eventually encyst in muscle tissue, the tongue and diaphragm being favored. Heavily infected muscles may contain as many as 15,000 encysted parasites per gram of tissue.

Pigs acquire the disease chiefly by eating restaurant meat scraps and slaughterhouse garbage; rats may also become infected in this way. The incidence of trichinosis among hogs fed raw garbage is almost 20 times higher than is that among other hogs. The U.S. government does not inspect pork for the trichina worm. Heat kills the encysted larvae, so your only protection is to avoid eating uncooked pork. Pink color in freshly cooked pork is evidence of inadequate cooking. Humans are usually a dead end for this parasite, whose transmission depends upon ingestion of raw flesh that contains encysted larvae.

The presence of encysted larvae causes local inflammation and may impair muscle function. The severity of symptoms seems to depend upon the number of larvae that are ingested. Of the estimated 60 million human hosts in the United States, only about 600 cases are clinically diagnosed per year. There is no known cure, but treatment with cortisone helps relieve inflammation and allergic symptoms. The incidence of trichinosis in the United States is decreasing because federal and state laws now require that garbage be cooked before being fed to hogs.

Pinworms

The human pinworm, *Enterobius vermicularis* (Fig. 14-6), measures 9 to 12 mm in length and lives in the upper part of the large intestine. Pinworm infection has a worldwide distribution and is especially prevalent in temperate, industrialized nations, where

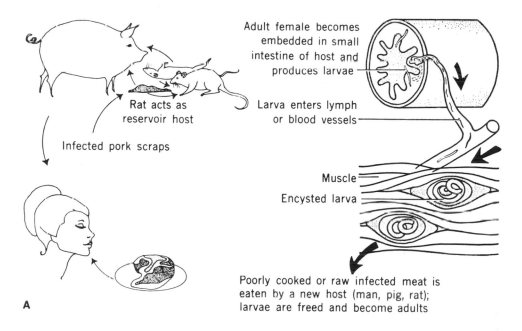

Figure 14-7. (A) Life cycle of *Trichinella*. (B) Encysted *Trichina* in pork. (C) Living *Trichinella* larva digested out of cyst.

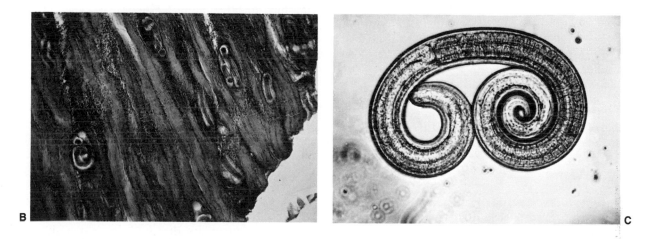

perhaps 50 percent of the population is infected. Pinworms are well-adapted parasites that cause little obvious damage to their hosts. The adults feed on tissue fluids in the intestinal mucosa. Gravid females migrate to the anus, where eggs are deposited. The eggs cause local inflammation, and subsequent scratching spreads the eggs to clothes and food. If ingested, the eggs hatch in the small intestine, and the pinworms mature in the large intestine. Other pinworm species with basically similar life cycles parasitize wild and domesticated mammals.

Whipworms

Trichuris trichiura (Fig. 14-6) inhabits the human intestine. Its anterior end is drawn into a long, slender, whiplike process that burrows into the intestinal mucosa. Whipworms feed on blood and tissue fluids.

There is no intermediate host; eggs escape in the host's feces and mature in the soil. If swallowed, the eggs hatch in the small intestine, and the larvae develop into adults as they migrate to the large intestine. Heavy infestations may cause abdominal discomfort, anemia, and bloody stools. It has been estimated that 400 million people worldwide are infected with whipworms. Sanitary disposal of human excrement interrupts the life cycle and prevents the spread of this parasite.

Filaria Worms

Filaria roundworms are parasites found in the tropics; their life cycles involve an alternation of mammalian and insect hosts. *Wuchereria bancrofti* is widespread in tropical countries. The minute larvae (**microfilariae**) circulate in the human bloodstream and then invade lymph vessels where they mature into sexual adults. Obstruction of the lymph vessels sometimes results in a tremendous enlargment of limbs or genitals (**elephantiasis**). The larvae can be killed with drugs, but surgery is required to bypass obstructed lymph vessels. Mosquitoes act as intermediate hosts and vectors that transmit the parasite from one human to another.

Another filarial roundworm is the eye worm, *Loa loa*, of western and central Africa (Fig. 14-6). The adults migrate through the subdermal connective tissues of the human host and sometimes pass across the eyeball. Local swellings accompany these migrations. Tabanid flies act as transmitting vectors.

Guinea Worm

Dracunculus medinesis, the guinea worm, is a common human parasite in arid regions of Africa, Arabia, India, South America, and the West Indies. The adult female, which may reach lengths of over 1 m, migrates through the subcutaneous tissues of the arms, legs, and shoulders. It eventually penetrates the skin, discharging thousands of active larvae when that part of the body is submerged in fresh water. The larvae may bore into a small crustacean *(Cyclops)*, and humans become infected by swallowing them while drinking water. The method of extracting the parasite, practiced by natives for hundreds of years, is to roll the guinea worm up gradually on a stick, a few

turns each day, until the entire worm has been drawn from the body (Fig. 14-6). Serious infection may occur if the worm is broken while still in the body.

Nematode Parasites of Insects

Researchers have discovered several roundworms that attack insects, and there is some hope that these nematodes can be used in pest control. One of these carries a type of bacterium that quickly kills many insects. The nematode acts as a microsyringe that introduces the bacteria into the infected insect's body cavity. This bacterium has proved lethal to codling moths and at least 35 other kinds of insects, including the corn earworm, boll weevil, pink bollworm, vegetable weevil, cabbage worm, and white fringed beetle.

Origin of the Nematoda

Nematodes seem to occupy a rather isolated side branch of the evolutionary tree, with few clear relationships to other animal groups. In some respects they resemble the platyhelminths in that they are unsegmented and possess excretory canals (but no flame cells). However, nematodes generally lack cilia, have a complete digestive tract with a mouth and an anus, and are **dioecious** (separate sexes). It is evident that the parasitic roundworms evolved from free-living roundworms.

Phylum Nematomorpha— Horsehair Worms

The name "horsehair" comes from a medieval superstition that this roundworm developed spontaneously from horsehairs that fell into water. Indeed, their thin elongated bodies are shaped like horsehairs, and their dark brown or black color further supports the resemblance. *Paragordius varius* is a common species in North America (Fig. 14-5). The 230 described species of nematomorphs are a widely distributed group of parasites; most live in fresh water during part of their life cycle, but a few are marine.

Their life cycle is as follows: The sexual adults live in water and lay their eggs there. The larvae that hatch penetrate the body of certain insects or crustaceans and migrate to the host's body cavity. There they mature until they escape as young adults. Nematomorphs have been found in the body cavities of beetles, crickets, and grasshoppers, and it has been suggested that these terrestrial insects ingest the larvae in drinking water. Infested crickets appear to migrate to the edge of water. If a cricket is caught by a wave, *Paragordius* emerges within 20 to 50 sec. The adults do not feed and apparently perform only reproductive functions; they mate and lay eggs.

Nematomorphs resemble nematodes in that they have a similar body form, a **cuticle,** a simple musculature, and the absence of segmentation. However, nematomorphs also display distinctive characteristics such as a body cavity that is nearly filled with **parenchyma (mesenchyme),** a degenerate **digestive tract** in adults, a single **nerve cord,** and a **cloaca.** No circulatory, respiratory, or excretory organs are present. The physiology of this group is not well understood.

Phylum Acanthocephala—The Spiny-Headed Worms

Acanthocephalans are parasitic pseudocoelomates that, in their adult stages, infest the intestines of vertebrates. A distinctive feature of the group in an anterior **proboscis** that is covered with recurved **spines** (Fig. 14-8). These hooks are used to attach the parasite to the intestinal wall, sometimes causing signifi-

cant damage to the mucosa. There are about 400 species of spiny-headed worms, and they range in length from 2 to 50 cm. The body is generally elongate, flattened, and capable of extension. No digestive tract is present at any stage in their life cycle; predigested food is absorbed directly from the host's intestine. The sexes are separate, and the **reproductive systems** are complex. Their life cycle requires two hosts. The definitive hosts are vertebrates, especially fish and birds, whereas intermediate hosts are always arthropods, either insects or crustaceans. Acanthocephalan species have been reported in the United States to reside in fish, turtles, birds, rats, mice, pigs, squirrels, dogs, and humans. *Macracanthorhynchus hirudinaceus,* found in the intestines of hogs, is of some economic importance.

The acanthocephalans differ from the aschelminths in the absence of a digestive tract and presence of an armed proboscis, circular muscles, and certain peculiarities of their reproductive organs.

Phylum Entoprocta

Another pseudocoelomate group is the phylum Entoprocta, a group of free-living, sessile animals that inhabit coastal marine waters throughout the world. There are about 90 species; entoprocts are rather common but are often overlooked due to their small size.

The entoproct body consists of a **stalk** that bears a

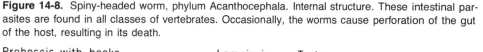

Figure 14-8. Spiny-headed worm, phylum Acanthocephala. Internal structure. These intestinal parasites are found in all classes of vertebrates. Occasionally, the worms cause perforation of the gut of the host, resulting in its death.

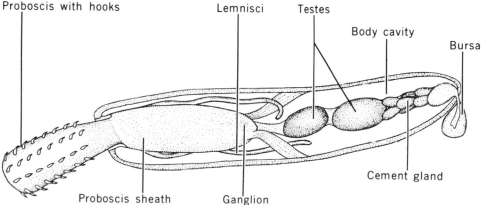

Proboscis with hooks   Lemnisci   Testes   Body cavity   Bursa

Proboscis sheath   Ganglion   Cement gland

tentacle-crowned **head,** or **calyx** (Fig. 14-9A). The basal end of the stalk is attached to a rock, algae, or some sessile animal. In colonial entoprocts the zooids are interconnected by horizontal **stolons.** The calyx contains the viscera and bears 8 to 30 ciliated **tentacles** that surround the **mouth** and **anus.** Movement is generally restricted to a characteristic flicking motion, which has suggested their common name of "nodding heads."

The phylogenetic affinities of these organisms are poorly understood. They were once thought to be bryozoans (ectoprocts), but their lack of a true coelom resulted in their status as a pseudocoelomate, although they are not very similar to the other groups discussed.

Phylum Gnathostomulida

Members of this problematic, relatively recently discovered phylum of small (0.5 to over 3mm) organisms are actually acoelomates (Fig. 14-9B). Gnathostomulidans live in large numbers in fine-grained marine sand habitats, where they graze on algae and bacteria with specialized **jaws** and **mouth.** Their oral configuration is similar to the jaw structure of rotifers; this, with other traits, such as the similarity of **sensory cells** (in the form of **ciliary pits**) among gnathostomulidans, rotifers, and gastrotrichs warrants discussion of Gnathostomulida with the pseudocoelomates. The **excretory system** (**protonephridia**), ciliated epithelial layer (no cuticle), acoelous condition, and lack of an anus ally these slender creatures with tubellarians; they are most likely intermediate between the acoel and pseudocoel organisms.

Reproduction is by the transfer of sperm in a mucous ball from the male **copulatory organ** to the **bursa** (sperm sac) of another worm, where it fertilizes the eggs transferred from the single ovary of the recipient. Both separate sexes and hermaphroditic organisms have been observed. **Circular** and **longitudinal muscle** fibers aid in movement, as well as the locomotion provided by the ciliated epithelium. There are no circulatory or respiratory systems.

Summary

Aschelminths are bilaterally symmetrical animals with a pseudocoel. Many are parasitic and of great

Figure 14-9. Structures of typical members of phyla Entoprocta (A) and Gnathostomulida (B). Both are partially cut away to show internal anatomy.

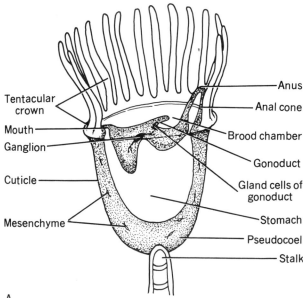
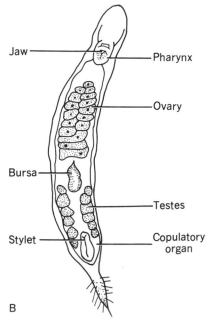

medical and agriculture interest. Acanthocephalans, or spiny-headed worms, are so specialized for parasitism that their evolutionary relationships to other groups are difficult to establish. Entoprocts are somewhat intermediate between the pseudocoelomates and the true coelomates, whereas gnathostomulidans are actually acoelomates, but more closely resemble the pseudocoels in other traits.

This assemblage of diverse animals is typified by their possession of a pseudocoel, a body cavity that is derived from the blastocoel and not lined with a cellular peritoneum. Their origins are obscure, especially because their soft bodies are not well represented in the fossil record. Perhaps these pseudocoelomate groups are remnants of an archaic fauna that dwindled due to competition from animals with a more efficient coelomate body plan.

Review Questions

1. Name an aschelminth that is in our food chain (that is, eaten by successively larger animals that are, in turn, eaten by humans).
2. One of the greatest boons to humans was the use of fire. What survival value might cooking food have?
3. Compare and contrast a typical flatworm, nemertean, aschelminth, and acanthocephalan using drawings.
4. Name three ways that one might avoid parasites, and name the particular parasite that would be thus avoided.
5. In what ways do these metazoans parasites resemble protist parasites? How could you account for such similarities?

Selected References

Baer, J. G. *Ecology of Animal Parasites*. Urbana: University of Illinois Press, 1952.

Cameron, T. W. M. *Parasites and Parasitism*. New York: John Wiley & Sons, Inc., 1956.

Chitwood, B. G., and M. B. Chitwood. *An Introduction to Nematology*, Section 1, 2nd ed. Baltimore: Monumental Printing Company, 1950.

Croll, N. A., and B. E. Matthews. *Biology of Nematodes*. New York: Wiley, 1977.

Crompton, D. W. T. *An Ecological Approach to Acanthocephalan Physiology*. New York: Cambridge University Press, 1970.

Dougherty, E. C. (ed.). *The Lower Metazoa*. Berkeley: University of California Press, 1963.

Goodey, T. *Soil and Freshwater Nematodes*. New York: John Wiley & Sons, Inc., 1951.

Harris, J. E., and H. D. Crofton. "Structures and Function in the Nematodes: Internal Pressure and Cuticular Structure in Ascaris," *J. Exp. Biol.*, **34**:116, 1957.

Hull, T. G., *et. al. Diseases Transmitted from Animals to Man*. Springfield, Ill.: Charles C Thomas, 1946.

Hyman, L. H. *The Invertebrates: Acanthocephala, Aschelminthes, and Entoprocta*, III. New York: McGraw-Hill Book Company, 1951.

Moore, D. V. "Acanthocephala," in *McGraw-Hill Encyclopedia of Science and Technology*, I. New York: McGraw-Hill Book Company, 1960.

Nicholas, W. L. *The Biology of Free-living Nematodes*. Oxford: Clarendon Press, 1975.

Olsen, O. W. *Animal Parasites: Their Biology and Life Cycles*. Minneapolis: Burgess, 1962.

Rogers, W. P. *The Nature of Parasitism*. New York: Academic Press, 1962.

Sasser, J. N., and W. R. Jenkins (ed.). *Nematology*. Chapel Hill: University of North Carolina Press, 1960.

Shipley, A. E. "Nemanthelminthes," *Cambridge Natural History*, II. London: Macmillan Company, Ltd., 1896.

Thorne, G. *Principles of Nematology*. New York: McGraw-Hill Book Company, 1961.

Yorke, W., and P. A. Maplestone. *The Nematode Parasites of Vertebrates*. Philadelphia: The Blakiston Company, 1926.

Introduction to the Coelomates

The coelomate phyla comprise the majority of metazoans and include the largest and most diverse animal groups. Unlike a **pseudocoel,** which is a persistent **blastocoel,** a **coelom** is a secondarily derived body cavity. A **coelom** (Fig. 15-1) represents an internal space within the mesoderm that is completely surrounded by a cellular epithelial layer, the **peritoneum.** The coelom completely fills the body of many coelomates, separating the body wall from the digestive tract. One part of the mesoderm thereby becomes closely applied to the body wall and the other to the gut. This separation has allowed for the independent development and elaboration of **somatic** (referring to the body wall) and **visceral** (digestive tract) tissues and organs.

The coelom also provides room for elongated and winding organs, and its **coelomic fluid** cushions the viscera from the effects of body wall movements. The fluid is important in the internal transport of metabolites and wastes, even in coelomates that have a separate circulatory system. The coelomic fluid may also serve as a medium by which nitrogenous wastes and gametes are discharged from the body. Finally, the fluid-filled coelom may also act as an internal **hydrostatic skeleton** that can be used both to support the body and to facilitate body movements.

Coelomates may be divided into two great branches, **protostomes** and **deuterostomes,** according to how their coelom is formed (see also Chapter 37). In protostomes, which includes the annelids, mollusks, and arthropods, the coelom forms as a series of splits in the embryonic mesoderm (**schizocoely**). Deuterostomes, such as the echinoderms and chordates, are characterized by a coelomic body cavity that develops **enterocoelously,** that is, by outpocketings of the primitive gut. As they enlarge, the pouches pinch off and become separated from the gut. The final result of enterocoely is remarkably similar to that of schizocoely in protostomes: the space between the body wall and gut contains a fluid-filled coelom that is lined with peritoneum and divided into lateral compartments by dorsal and ventral **mesenteries** that suspend the gut. Deuterostomes include echinoderms and chordates. Other embryological distinctions between protostomes and deuterostomes and discussed in Chapter 38.

Some Minor Coelomate Phyla

Following are descriptions of five minor and somewhat enigmatic phyla that are especially interesting because of their evolutionary significance. All probably diverged during the earliest adaptive radiations that primarily left the major protostome and deuterostome lines of evolution. This is supported by the mixtures of protostome and deuterostome characteristics observed in the so-called **lophopharate phyla (Ectoprocta, Phrororida, Brachiopoda)** and the phylum **Pogonophora.** The phylum **Chaetognatha** is much less problematic; it is very likely the most primitive deuterostome phyla represented today.

Ectoprocts, phoronids, and brachiopods are rarely seen or recognized by most people due to their small sizes and subtidal habitats, but these three phyla of small marine animals have interested and puzzled scientists for years. They share some features that are tantalizingly similar to and yet different from other animals. All three phyla are characterized by a crown of ciliated tentacles, called the **lophophore,** an extended coelomic structure that serves to capture food. A few other features are also similar, but generally there is no close anatomical correlation between them. This is undoubtedly a reflection of their long and specialized evolutionary histories. Although considered protostomes, the lophophorates probably di-

Figure 15-1. The relationship of body cavities to germ cell layers. The cross section of the earthworm is depicted to aid in the identification of a true coelom.

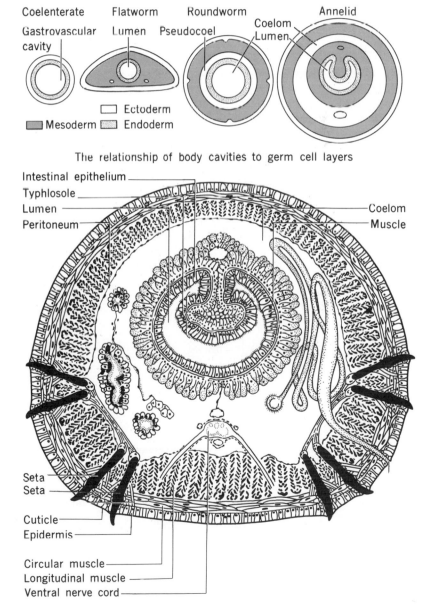

The relationship of body cavities to germ cell layers

verged near the protostome–deuterostome split in the phylogenetic tree. The typically deuterostome style of coelom formation (enterocoel) is exhibited in brachiopods; this, and the modification of cleavage patterns observed during the development of brachiopods and ectoprocts, contributes to the difficulty of assigning these groups to either protostomes or deuterostomes. These early branches of the animal kingdom are intriguing as living showpieces of some directions that evolution took eons ego.

Phylum Ectoprocta (Bryozoa)

The ectoprocts are a major phylum of small aquatic animals. Some 2600 extant (i.e., living today) species have been described, and another 4000 are known from the fossil record. Adult bryozoans are all sessile, most are colonial, and many resemble hydrozoans in form, but they are more advanced in internal structure (Fig. 15-2). The majority of bryozoan species live in the ocean, but a few inhabit fresh water. Marine ectoprocts have chitinous or calcareous exoskeletons,

Figure 15-2. Phylum Ectoprocta. Internal structure of a typical marine ectoproct.

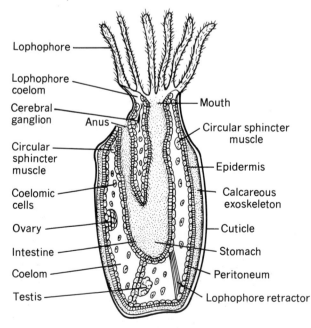

but those of freshwater species contain no hard parts. *Bugula* is a common marine genus, and *Plumatella* is a common freshwater genus. A colony of *Plumatella* is primarily made up of cylindrical, branched tubes. All the individuals (**zooids**) in a colony are derived by asexual budding from one settled larva. The anterior of each zooid has a rounded ridge, the **lophophore**, which bears a horseshoe-shaped double row of **tentacles.** The 40 to 60 tentacles are hollow and ciliated. When the tentacles are spread out in the water, the cilia cause currents that sweep planktonic organisms into the **mouth.** Internal organs of a typical ectoproct are indicated in Figure 15-2. Between the digestive tract and the body wall is a true coelom that is lined with a **peritoneum.** There are no respiratory, circulatory, or excretory organs. Ectoprocts are **hermaphroditic.** The larvae of some bryozoans resemble a **trochophore** (see Chapter 16); this suggests an ancient alliance with higher protostomes.

Certain freshwater ectoprocts produce disc-shaped buds, called **statoblasts** that secrete a hard chitinous shell. These survive when the parent dies in the autumn or during a drought and start new colonies in the spring or when the wet season returns.

Certain freshwater ectoprocts are known to foul water pipes. They form thick crusts inside pipes, and dead colonies sometimes break loose, clogging small pipes and meters.

Since their first appearance in the Cambrian period (about 600 million years ago), the ectoprocts have made substantial contributions to layers of calcareous rock during every geological period.

Phylum Brachiopoda—Lamp Shells

Brachiopods are marine animals whose approximately 250 species live within a calcareous **bivalve shell** (Fig. 15-3). They are all **sessile** or **sedentary;** some attach to an object by a muscular stalk called the **peduncle;** others cement their shell directly to rock; and some inhabit burrows in mud. Because of their shell, they were long regarded as mollusks. The **valves** of the brachiopod shell are unequal in size, however, and are dorsal and ventral instead of lateral as in the bivalve mollusks. Their common name, lamp shell, refers to the resemblance of their shells to the oil lamps of ancient Romans. Within the shell is a con-

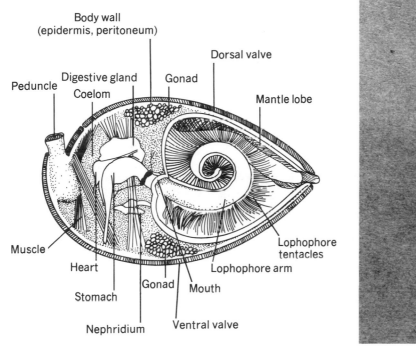

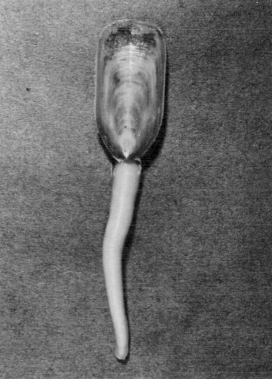

Figure 15-3. Phylum Brachiopoda. (A) Internal anatomy of a brachiopod. (B) Photograph of *Lingula,* a brachiopod considered to be a "living fossil."

spicuous **lophophore** that consists of two coiled ridges that bear ciliated **tentacles.** Food is drawn into the **mouth** by the lophophore. A true **coelom** is present, within which lie the **stomach, digestive gland,** a short blind **intestine, nephridia,** and a poorly defined **circulatory system.**

Phylum Brachiopoda is extremely old, dating from Cambrian times. Although found in all oceans today, brachiopods were formerly much more numerous and exhibited a much greater variety of forms than is the case at present. The fossil record contains about 30,000 recognized species, compared with the several hundred present today. Some such as *Lingula* (Fig. 15-3) are apparently much the same today as they were in the Ordovician period, over 400 million years ago. *Lingula* is thought to be the oldest animal genus with modern representatives. It is called a "living fos-

sil" because it has changed very little during long geological periods.

Phylum Phoronida

Phoronida (Fig. 15-4) is a small phylum of wormlike marine lophophorates with about 20 species distributed throughout the world's oceans. Phoronids are elongated animals that typically live in clusters. The many **tentacles** on the double-rowed **lophophore** bear cilia that create water currents. Planktonic organisms are trapped in mucus on the tentacles and passed by cilia to the **mouth.**

The body cavity is a true **coelom.** The **nervous system** is simple. A definite **circulatory system** is present as are a pair of **nephridia,** and **reproduction** is by

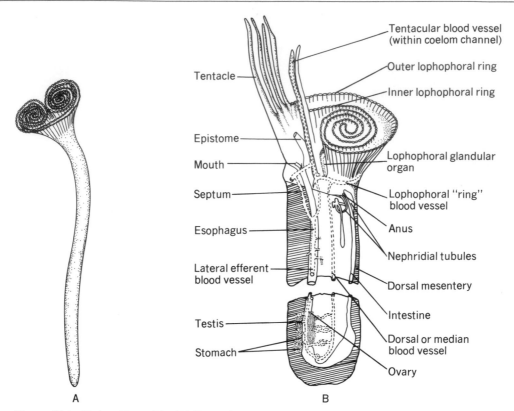

Figure 15-4. Phylum Phoronida. (A) External view. (B) Internal anatomy.

asexual and sexual means. A free-swimming larva, called an **actinotroch,** somewhat resembles a trochophore.

Phylum Pogonophora—Beard Worms

The pogonophorans are extremely thin animals; their bodies may exceed 30 cm in length yet measure only 2 mm in diameter. They dwell in deep ocean waters along the continental shelves, inhabiting chitinous tubes in muddy sediments. Pogonophorans seem to have a worldwide distribution but were not discovered until 1900, when the first specimens were recognized. Before that time, any that were recovered from deep ocean trawls were discarded under the mistaken impression that they were frayed pieces of hemp rope!

Pogonophorans are crowned by a mass of as many as 200 **tentacles** (Fig. 15-5); hence their common name "beard worms." Pogonophorans are remarkable because they are the only free-living metazoans that have no trace of an alimentary canal; there is no mouth, gut, or anus! How they feed is not precisely known because they have rarely been observed in their deep-sea habitats.

Pogonophorans have a well-developed **circulatory system** and an epidermal **nervous system.** Their **coelom** has been reported to be of the **enterocoel** type, as in deuterostomes; this and general body form suggest affinities to the hemichordates (see Chapter 23). But the discovery of the segmented anchoring portion (**opisthoma**) of its body has led others to place the group with the segmented annelids (see Chapter 17). A third point of confusion is the likeness of the crown of tentacles to a lophophore. The larvae of this

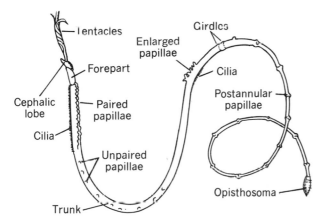

Figure 15-5. Phylum Pogonophora. External anatomy of a pogonophoran worm removed from its tube.

phylum have not been studied; certainly more detailed studies are needed to completely understand the phylogenetic relationships of pogonophorans.

Phylum Chaetognatha—Arrow Worms

Chaetognaths are common members of most marine plankton communities. There are only about 50 species, but several are cosmopolitan in distribution. Chaetognaths are known as arrow worms because of their distinctive shape (Fig. 15-6) and rapid, darting locomotion. They are important predators of other planktonic animals, including copepods and fish larvae. Arrow worms average about 3 cm in length and are virtually invisible underwater because their body wall and visceral organs are transparent.

Their **digestive tract** is a relatively simple tube. The **nervous system** of these active predators is complex and centralized, and cephalic sense organs include **eyes** and receptors for chemicals and water currents. Arrow worms lack circulatory, respiratory, and excretory systems. Chaetognaths are **hermaphroditic,** and development is direct. The **coelom** develops **enterocoelously,** attesting to deuterostomous affinities. Chaetognaths have such a unique constellation of characters that they probably represent a very early offshoot from the deuterostome line that led to the echinoderms and chordates.

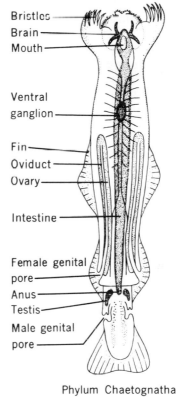

Phylum Chaetognatha
Sagitta hexaptera

Figure 15-6. Phylum Chaetogognatha, *Sagitta hexaptera*, an arrow worm 6 mm in length.

Summary

The presence and method of formation of the coelom is an important characteristic used for the classification of animals. This fluid-filled cavity is lined with a cellular peritoneum; diffusion of nutrients, wastes, and other metabolites, protection of visceral organs, and bodily support (hydrostatic skeleton) are the coelom's most common functions. Almost all true coelomates may be divided into two major lines of evolution—the protostomes and the deuterostomes. The five minor coelomate phyla described probably represent the most primitive coelomates. Except for the deuterostomic chaetognaths, these organisms exhibit both protostome and deuterostome traits, suggesting evolutionary divergences from the common ancestors of the protostomes and deuterostomes.

Review Questions

1. Distinguish between schizocoely and enterocoely, protostomes and deuterostomes
2. What characteristics of chaetognaths support its assignment to deuterostomes?
3. What information would be useful to clarify the taxonomic position of Pogonophora?

Selected References

Bassler, R. S. "The Bryozoa, or Moss Animals," *Smithsonian Inst. Ann. Rep.*, 1920.

Hickman, C. P. *Integrated Principles of Zoology*. St. Louis: The C. V. Mosby Company, 1970.

Johnson, W. H., L. E. Delanney, E. C. Williams, and T. A. Cole. *Principles of Zoology*, 2nd ed. New York: Holt, Rinehart and Winston, 1977.

Lynch, W. "Factors Influencing Metamorphosis of *Bulgula* Larvae," *Biol. Bull.*, **103**:369, 1952.

MacGinitie, G. E., and N. MacGinitie. *Natural History of Marine Animals*, 2nd ed. New York: McGraw-Hill Book Company, 1967.

Marsden, J. "Regeneration in *Phoronis Vancouverencis*," *J. Morphol.*, **101**:307, 1957.

Michael, E. L. *Classification and Vertical Distribution of the Chaetognatha of the San Diego Region*. University of California Publications, Zool. 1911.

Rattenbury, J. "The Embryology of *Phoronopsis Viridis*," *J. Morphol.*, **95**:289, 1954.

Russell-Hunter, W. D. *A Biology of Higher Invertebrates*. New York: Macmillan Publishing Co., Inc., 1971.

Schuster, R. O., and A. A. Grigarick. *Tardiguada from Western North America, with Emphasis on the Fauna of California*. Berkeley: University of California Press, 1965.

Stephen, A. C., and S. J. Edmonds. *The Phyla Sipunculida and Echiura*. London: British Museum (Natural History), 1972.

Phylum Mollusca

Phylum Mollusca is one of the greatest and most diverse phyla of the animal kingdom (Fig. 16-1). Over 80,000 living species of mollusks have been described. Mollusks have successfully adapted to almost every available habitat in the biosphere. They are found in all oceans, from the abyssal depths to the surface waters, and in many freshwater habitats. Mollusks are also one of the few animal groups, along with the arthropods and chordates, that have successfully exploited land, where they frequent deserts to Arctic–Alpine environs.

Mollusca is commonly divided into six classes, four of which contain the majority of living species: **class Bivalvia** includes clams, scallops, and oysters; **class Gastropoda,** the snails and slugs; **class Cephalopoda,** the squids and octopods; and **class Amphineura,** the chitons. The remaining two classes provide valuable insights into molluscan evolution: the members of **class Monoplacophora** are peculiar "living fossils," and **class Scaphopoda** contains mollusks with distinctive tusk-shaped shells.

At first glance, these diverse animals do not appear to have much in common, but closer examination reveals an array of similar features. Mollusks are generally **bilaterally symmetrical** and unsegmented. Most secrete a **shell** of calcium carbonate that protects and supports their soft tissues. The molluscan body is organized into three general regions: head, foot, and visceral hump. The **head** bears the mouth and sense organs. The muscular **foot** is primitively used for locomotion, crawling in snails and digging in clams, but in squid it is modified into tentacles for capturing prey. The **visceral hump** contains the digestive tract and other visceral organs. A fold of the body wall, called the **mantle,** generally covers the visceral hump. The mantle secretes the shell. The **mantle cavity** between the visceral hump and the mantle houses the **gills** and receives openings from the reproductive, excretory, and digestive systems. Figure 16-2 indicates how this basic molluscan body plan has become modified during the evolution of the various classes.

Mollusks are generally slow-moving, bottom-dwelling (**benthic**) animals. Most mollusks are small, averaging perhaps 2–5 cm in length, although giant pelagic squids attain lengths of at least 17 m! Most mollusks are herbivores, but some (e.g., squid) are active predators. Mollusks typically employ a feeding organ called a **radula** (Fig. 16-3), which is armed with rows of chitinous teeth. The radula is protruded from the mouth and worked back and forth to rasp the food into fine particles.

The sexes are usually separate, although certain groups are **hermaphroditic.** Fertilization is generally external in aquatic mollusks, and an enormous number of gametes may be shed into the the water. An oyster, for example, may produce 500 million ova in a single season. Development in most mollusks includes a **trochophore larvae** (Fig. 16-4) and a subsequent **veliger larva,** so called because of a band of cilia, the **velum,** in front of its mouth. The velum is a locomotor organ, but water currents are the major factor in the dispersal of molluscan larvae. The presence of a trochophore stage in the life histories of many mollusks and annelids suggests that these two groups evolved from a common ancestor.

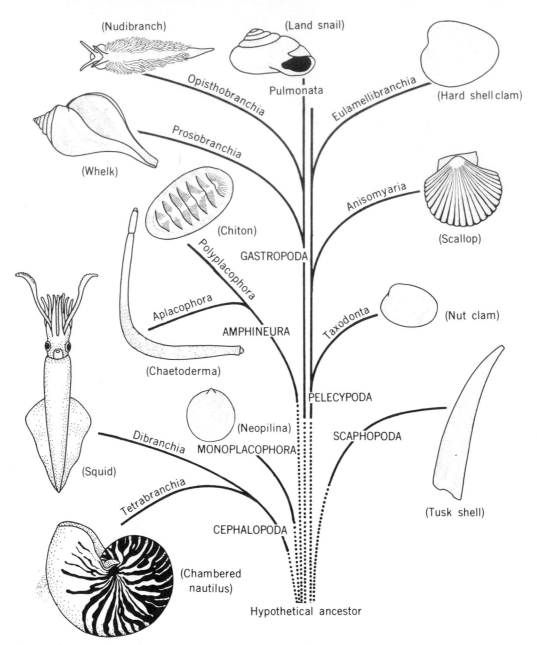

Figure 16-1. Representative mollusks. The lines suggest possible relationships.

Class Bivalvia—The Clams and Related Forms

The **Bivalvia** (**Pelecypoda**) includes some 11,000 described species of **clams, mussels, oysters, scal-** lops, and related **bivalve mollusks.** All bivalves are aquatic, and the vast majority are marine, with few freshwater species. Nearly all bivalves are bottom-dwelling animals that either burrow through sand and mud or live attached to rocky substrata; a few (e.g., scallops) are capable of swimming.

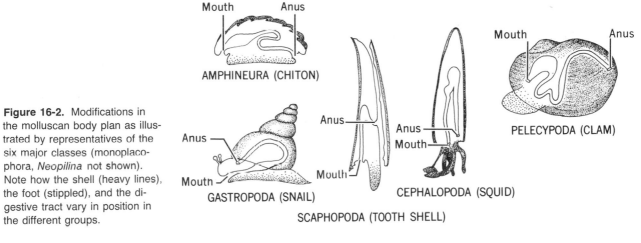

Mouth   Anus

AMPHINEURA (CHITON)

Figure 16-2. Modifications in the molluscan body plan as illustrated by representatives of the six major classes (monoplacophora, *Neopilina* not shown). Note how the shell (heavy lines), the foot (stippled), and the digestive tract vary in position in the different groups.

Anus

Mouth

GASTROPODA (SNAIL)

Anus

Mouth

SCAPHOPODA (TOOTH SHELL)

Anus
Mouth

CEPHALOPODA (SQUID)

Mouth   Anus

PELECYPODA (CLAM)

Radula

Mouth

Figure 16-3. Radula of a gastropod. Left; Diagram on anterior longitudinal section of a snail. Right; Dorsal view of a few teeth from radula of a snail; that at extreme left is a central tooth. The radula is pressed against food and moved rapidly back and forth, rasping off small particles.

Figure 16-4. (A) Two stages in the development of a mollusk; both are free-swimming larva. Both annelids and mollusks have a trochopore larval stage. (B) Photomicrograph of trocophore larva.

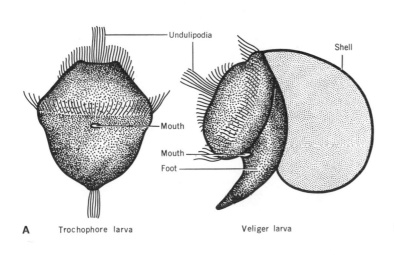

Undulipodia

Mouth

Mouth

Foot

Shell

A   Trochophore larva   Veliger larva

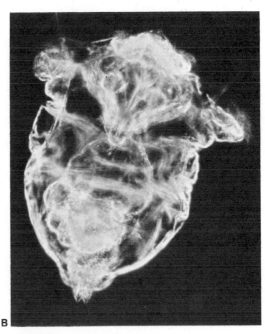

B

Anodonta—A Freshwater Clam
External Morphology

Clams usually lie partly buried in muddy or sandy sediments. They burrow from place to place by means of their muscular **foot** (Fig. 16-5), which can be extended from the anterior end of the bivalve shell. Water carrying oxygen and suspended food is drawn into the mantle cavity through a ventral **incurrent siphon** at the posterior end of the **shell.** A dorsal **excurrent siphon** exhausts deoxygenated water and wastes from the mantle cavity.

The bivalve shell consists of two generally symmetrical halves called **valves.** Concentric ridges called **growth lines** etch the exterior of each valve; these represent intervals of rest between successive periods of growth, with annual lines being most conspicuous. The dorsal **umbo** is the oldest part of the valve.

The outer epithelium of the mantle secretes the shell, which consists of three layers (Fig. 16-6): (1) a thin outer proteinaceous layer, the **periostracum,** which is often pigmented, that protects the underlying mineral layers from dissolution by acids in the water; (2) a middle **prismatic layer** that consists of polygonal calcium carbonate crystals ($CaCO_3$); and (3) an inner **nacreous layer (mother-of-pearl)** that is composed of many horizontal sheets of calcium carbonate that produce an iridescent sheen.

The valves are joined together at the umbo by a tough elastic **hinge ligament** that tends to spring the valves apart. This tendency is opposed by anterior

Figure 16-5. Freshwater clam *Anodonta*. (A) Inside view of a valve showing point of attachment of mantle and points of attachment of muscles. (B) External view of a living animal with foot protruding from shell and arrows showing direction of water flow through siphons.

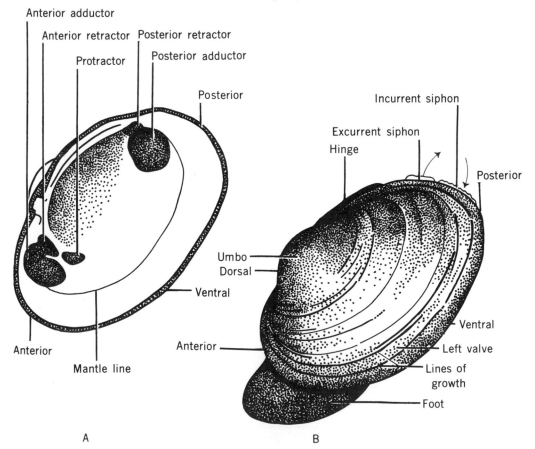

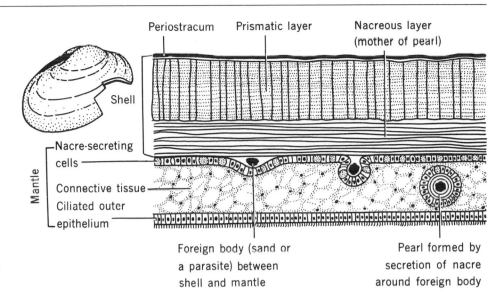

Periostracum Prismatic layer Nacreous layer (mother of pearl)

Shell

Mantle

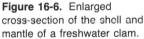

Nacre-secreting cells

Connective tissue
Ciliated outer epithelium

Foreign body (sand or a parasite) between shell and mantle

Pearl formed by secretion of nacre around foreign body

Figure 16-6. Enlarged cross-section of the shell and mantle of a freshwater clam.

and posterior **adductor muscles,** which contract to close the shell.

The two folds of the mantle that secrete the valves enclose the mantle cavity, which contains the gills, foot, and visceral mass (Fig. 16-7).

Internal Morphology and Physiology

Digestive System

Water containing minute plants, animals, and debris enters the mantle cavity through the incurrent siphon. Suspended particles are trapped in the mucus that covers the gills. The food-laden mucus is transported by cilia to the ventral edge of the gill and from there anteriorly to the **labial palps.** These flaps of tissue surround the **mouth** and serve as a sorting mechanism that selects food particles and rejects sediment grains. Food enters the mouth and passes through the short **esophagus** into the bulbous **stomach.**

Bivalves have a unique structure, the **crystalline style,** a gelatinous rod that projects into the stomach from the intestine. The style is composed of an acidic mucus that contains digestive enzymes. It is rotated by cilia against a chitinous **gastric shield.** The style is thereby slowly worn away, releasing its enzymes and

mixing the stomach contents as it does so. The enzymes accomplish preliminary digestion by breaking down food into fine particles. These are passed through numerous tubules into the **digestive gland** that surrounds the stomach. Here the food particles are engulfed by amoeboid cells, and the remaining stages of digestion occur intracellularly.

The **intestine** receives undigested wastes and compacts them into feces that are expelled into the exhalent water current. The intestine emerges from the stomach, loops through the visceral mass, and penetrates the pericardial cavity and heart, after which it becomes the **rectum.** Its opening, the **anus,** is positioned near the excurrent siphon.

Circulatory System

The circulatory system consists of a dorsal heart, blood vessels, and open spaces called **sinuses.** The **heart** lies in a protective sac, the **pericardium** (Fig. 16-7). The **pericardial cavity** is part of the **coelom,** which also contains the kidneys and gonads. The heart has two **atria** that receive oxygenated blood from the gills and mantle. The blood is passed to the single **ventriole** and is pumped through the **anterior** and **posterior aorta** into a system of large vessels that open into sinuses that bathe the tissues. Contractions

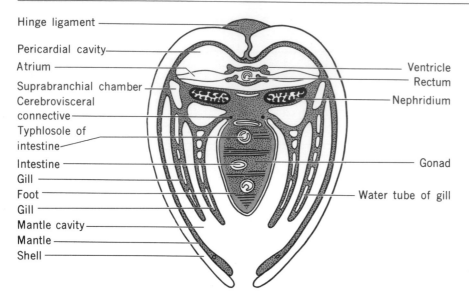

Hinge ligament

Pericardial cavity

Atrium

Suprabranchial chamber

Cerebrovisceral connective

Typhlosole of intestine

Intestine

Gill

Foot

Gill

Mantle cavity

Mantle

Shell

Ventricle

Rectum

Nephridium

Gonad

Water tube of gill

Figure 16-7. Cross-section of a freshwater clam through the region of the heart.

of the foot, mantle, and siphons help to move blood through the sinuses. Deoxygenated blood from the tissue sinuses is collected into vessels that deliver it to the gills.

Respiratory System

The general body surface in contact with water functions in respiration, but most of the oxygen and carbon dioxide exchange occurs in the **gills (ctenidia).** A pair of gills extends into the mantle cavity on either side of the foot (Fig. 16-7). Each ctenidium consists of two plates, or lamellae, made up of a large number of vertical **gill filaments** (Fig. 16-8) that are strenthened by chitinous rods and interconnected by horizontal bars. **Interlamellar partitions** that cross between the two lamellae form many vertical water tubes within each gill. The **water tubes** of each gill pair join dorsally in a common **suprabranchial chamber.**

Cilia on the gill filaments draw water through the incurrent siphon and over the gills. At the gill surface water passes through many microscopic openings, the **water pores,** into the water tubes, where respiratory gases are exchanged across the highly vascularized epithelium. The water then passes dorsally to the suprabranchial chambers and from there to the excurrent siphon.

Excretory System

Two **kidneys (nephridia)** lie just beneath the pericardial cavity. Each nephridium is differentiated into glandular (dark spongy mass) and bladderlike portions. One end of the nephridium opens into the pericardial cavity, and the other end opens into a **suprabranchial chamber.** Nitrogenous wastes filter from the blood through the heart wall into the pericardium. There the wastes are transported into the coelomic fluid to the kidneys. Some wastes are absorbed directly from the blood by the glandular nephridium. The urine is excreted into the exhalent water of the suprabranchial chambers.

Nervous System and Sense Organs

Three pairs of interconnected **ganglia** are present (Fig. 16-9): **cerebropleural ganglia** near the anterior adductor muscle, **pedal ganglia** in the foot, and **visceral ganglia** near the posterior adductor muscle. Sense organs include light receptors on the margins of the siphons. A small vesicle (**statocyst**) containing a calcareous concretion (**statolith**) in the foot is an organ of equilibrium. A thick patch of yellow epithelial cells (**osphradium**) covers each visceral ganglion. The osphradia are thought to be useful in detecting for-

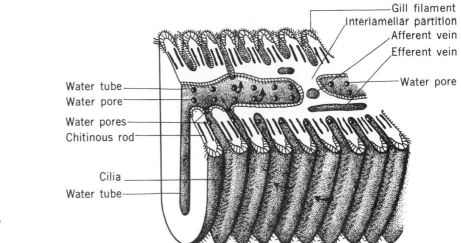

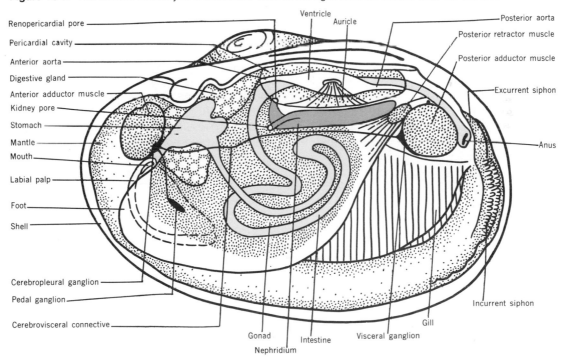

Figure 16-8. Respiratory system of a freshwater clam. Horizontal section through a gill showing arrangement of gill filaments, blood vessels, and water tubes. Arrows indicate direction of water currents.

Figure 16-9. The internal anatomy of a freshwater clam in the right valve as viewed from the left side.

eign materials in water. The edges of the mantle are provided with numerous sensory cells that are probably stimulated by tactile contact and light.

Reproduction

Most bivalves are **dioecious,** but a few are **hermaphroditic.** The **gonads** are situated in the visceral mass (Fig. 16-9). The **gametes** are generally shed into the surrounding water, where fertilization occurs. Marine bivalves generally pass through **trochophore** and **veliger** larval stages.

Many freshwater clams have life histories in which the young parasitize fish. Their eggs develop into a peculiar larva known as a **glochidium,** a modified veliger (Fig. 16-10). In *Anodonta* the eggs are usually fertilized in August, and the glochidia are retained in the gills of the mother throughout the winter. The following spring the larvae are released, and, if they chance to contact a fish, they seize hold of its skin by closing their valves. The skin of the fish grows around the larval clams, forming "worms" or "blackheads." After parasitizing the tissues of the fish for three to many weeks, a young clam emerges, settles to the bottom, and grows into a free-living adult. Such parasitic habits ensure the wide dispersal of many freshwater clams.

Other Bivalves

Not all bivalves burrow through soft sediments (Fig. 16-11). Adult **oysters,** for example, live permanently attached by their left valve to some solid object. Oysters feed in much the same way as do clams. A single oyster may produce half a billion eggs in one season; these develop into ciliated larvae, commonly called spat, that are dispersed by water currents.

Most **pearls** come from oysters that are not closely related to the edible oyster. Pearls result when a foreign particle becomes lodged between the mantle and the shell. This stimulates the deposition of layers of **nacre** (mother-of-pearl) around the irritating particle, forming a pearl. Otherwise bivalves can also produce pearls (Fig. 16-12).

Scallops are able to swim by clapping their valves together, thereby forcing jets of water from their mantle cavity. Enormous populations of scallops migrate from time to time, disappearing and then reappearing for unknown reasons.

Other bivalves can bore through hard substrata such as wood or certain rocks. These include the **shipworms** that cause extensive damage to submerged timbers (Fig. 16-20).

Class Gastropoda—The Snails and Slugs

Gastropoda is the largest class of mollusks, and its members have been most successful in adapting to different habitats. Although popularly identified only as snails, gastropods are a very diverse group (Fig. 16-13); over 35,000 living species have been discribed. Marine gastropods crawl over all types of bottom substrata, and there are also **pelagic** and **planktonic** gastropods in the ocean. Many freshwater

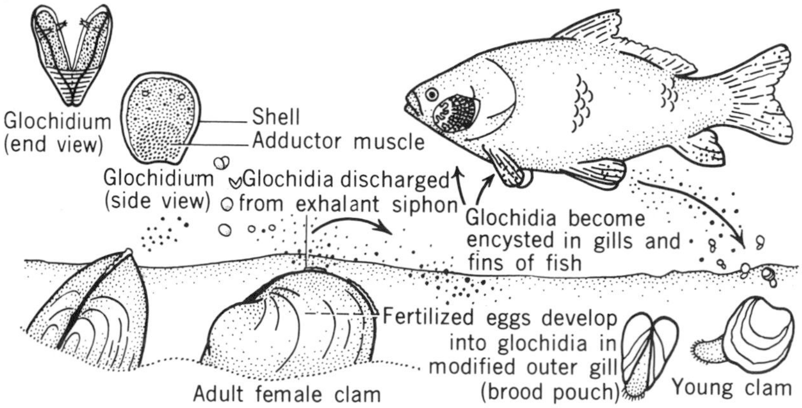

Figure 16-10. Life cycle of some freshwater clams. The larval clam or glochidium passes out of the excurrent siphon of a female and attaches itself to the gills of a fish; finally, the glochidium drops from the fish host to take up a free-living existence as a young clam.

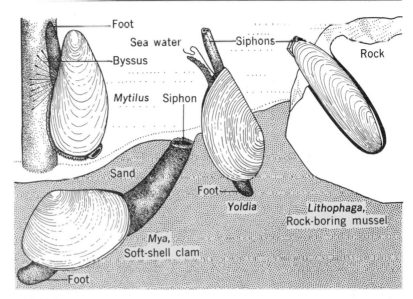

Figure 16-11. Some marine bivalves showing modes of life. *Mytilus,* the edible mussel, attached by byssal threads to a wooden pier. *Mya,* the mud clam, a burrowing form with a long siphon. Contrary to a general misconception, the "head" end of a clam is the end opposite the siphons. The walrus is said to feed almost entirely on this clam. *Yoldia* is capable of leaping through the water; note the united retractable siphons. *Lithophaga,* the rock-boring mussel, is said to secrete an acid to dissolve the rock into which it burrows.

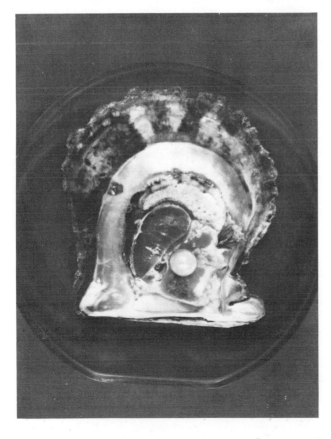

habitats have been invaded by gastropods, and even dry terrestrial communities are home to these adaptable mollusks. One major adaptation of terrestrial gastropods is loss of the **ctenidia** and conversion of the mantle cavity into an air-breathing **lung.**

A characteristic feature that distinguishes all gastropods from other mollusks is **torsion.** This has nothing to do with the spiral coiling of the shell. Rather, torsion is a 180° rotation (with reference to the head and foot) of the visceral hump and mantle that occurs during the veliger larval stage of all gastropods. Some consequences of torsion in the adult snail are illustrated in Figure 16-14: The mantle cavity now faces anteriorly, allowing **retractor muscles** to pull the head into the shell and block the entrance with the foot or, in some gastropods, the **operculum** (Fig. 16-14). The **digestive tract** is twisted into a U-shaped loop, with the **anus** opening just behind the head.

Other general characteristics of gastropods include a broad and flat muscular **foot** that is used for crawling. The **head** is relatively large and cephalized and bears one or two pairs of **tentacles** that contain a number of **chemoreceptors** and **photoreceptors.** Gastro-

Figure 16-12. A pearl in a marine oyster. Pearls are protective secretions made of the same substance (nacre) that lines the bivalve shell.

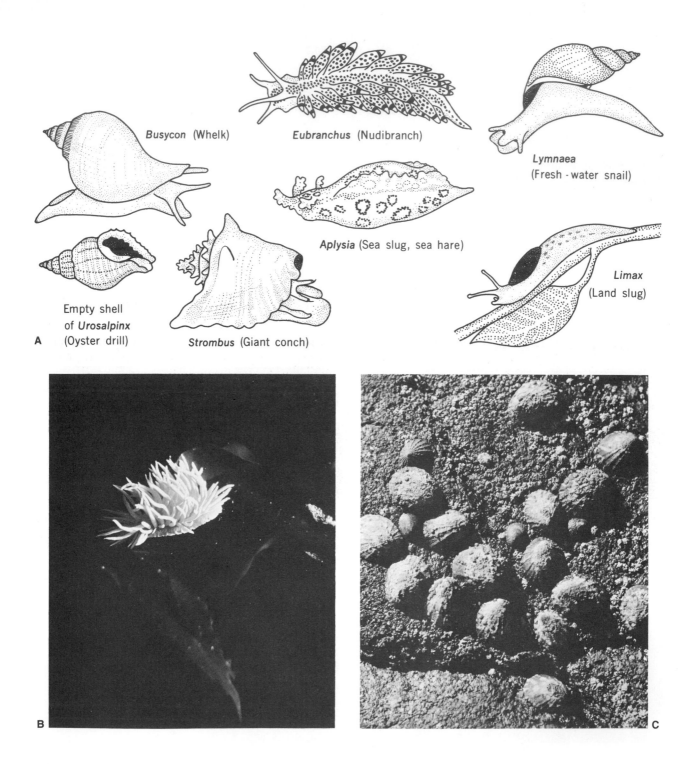

Busycon (Whelk)

Eubranchus (Nudibranch)

Lymnaea
(Fresh - water snail)

Aplysia (Sea slug, sea hare)

Empty shell
of *Urosalpinx*
(Oyster drill)

A

Strombus (Giant conch)

Limax
(Land slug)

B

C

Figure 16-13. (A) Representative gastropods. (B) The nudibranch, *Hopkinsia*. (C) The limpet, *Acmea.* (D) The garden snail, *Helix.*

pods usually have a well-developed **radula** that has been modified in various ways to accommodate different diets. The shell is often **asymmetrical** and spirally coiled.

Whelks and **periwinkles** are among the most common marine snails. The largest whelk off our shores is the **queen conch,** which lives along the Florida Keys and the West Indies. Its shell may be 40 cm long and it may weigh as much as 2 kg. Periwinkles crawl over rocks in the upper intertidal zone. The discarded

shells of these and other dead snails provide mobile homes for hermit crabs.

All gastropods do not have spiral shells. **Limpets** are gastropods with dome-shaped shells that cling tenaciously to rocks in the surf (**intertidal**) zone. Limpets feed on microscopic algae and bacteria that they scrape from the rocks with their radulae.

Land slugs (**pulmonates**) are related to land **snails** but have completely lost their shells. To avoid dehydration, slugs seek moist microhabitats such as

Figure 16-14. External view (left) and internal morphology (shell removed) of a marine gastropod, genus *Busycon.*

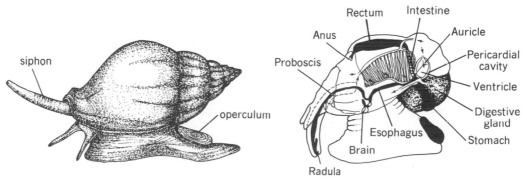

beneath ground litter; at night, they emerge to feed on vegetation. **Sea slugs** are shell-less gastropods that live below low-tide level; many are brilliantly colored.

Sea butterflies (**pteropods**) are planktonic gastropods that live in the surface waters of the open ocean. They may be so abundant in some regions that their delicate shells contribute significantly to the bottom sediments.

Class Cephalopoda—The Squids and Octopods

Class Cephalopoda contains the most specialized mollusks, including squids, octopods, cuttlefish, and nautiloids. About 500 living species have been recognized, and some 10,000 more are known from the fossil record. Of the modern cephalopods, only the **chambered nautilus** has an external shell; in others, the shell has been reduced and enclosed within the body or lost entirely.

The basic molluscan body plan has been modified in cephalopods by a great dorsoventral elongation of the **visceral hump.** This brought the head and foot together, and they became fused. Part of the foot formed a funnel-shaped **siphon** through which water is expelled from the mantle cavity, and the head is surrounded by a ring of **tentacles.**

The foot of a **squid** (Fig. 16-15) is elaborated into ten arms and a siphon. The arms bear **suckers** that are used for capturing and handling fish and other prey. Locomotion is brought about by contraction of the muscular mantle cavity, which forces a jet of water through the siphon. When the siphon is directed forward, the jets of water propel the animal backward. If the siphon is directed backward, the squid is thrust forward. Located posteriorly, lateral **fins** act as stabilizers and can push the squid slowly forward or backward, or enable it to hover by undulatory movements.

The shell of the squid is a noncalcified horny **pen** that is concealed beneath the skin of the back (anterior surface). The true **head,** the short region between the arms and the mantle collar, bears two large **eyes.** The **digestive system** includes a muscular **pharynx, esophagus** with **salivary glands, stomach** surrounded by the **liver, pancreas, cecum, intestine, rectum,**

and **anus.** There are two powerful horny **jaws** in the pharynx, and a **radula** is also present. The stomach acts as a **gizzard** that crushes and grinds the food. The food particles are mixed with digestive enzymes from the liver and pancreas. The semiliquid food thus formed enters the cecum for further digestion and then passes into the intestine, where nutrients are absorbed. The anus opens next to the siphon, and feces are flushed from the mantle cavity with the exhalent water.

Above the rectum is the **ink sac,** with a duct that opens near the anus. When the squid is attacked, it emits a cloud of inky fluid through its siphon. This "smoke screen" interferes with the visual and chemoreceptors of the predator and thereby helps the squid to escape.

In conjunction with their active lives, most cephalopods have evolved a **closed circulatory system** in which the blood is confined within blood vessels. **Capillary networks** replace the open sinuses of other mollusks, and **accessory hearts** help maintain blood flow through the gills. The blood contains a respiratory pigment, **hemocyanin,** as in all gastropods.

Two **gills** and two **nephridia** are present. The **nervous system** is highly developed, with a number of **ganglia** fused into a collar around the esophagus. Sense organs include two very highly developed **eyes,** two **statocysts** (for equilibrium), and two **olfactory organs.** The eyes are large and superficially similar to those of vertebrates (Fig. 16-16). In squids the sexes are separate, and fertilization is internal following copulation.

Squids and octopods are noted for their color changes. Their skin contains many pigment cells, called **chromatophores,** that are filled with blue, purple, red, and yellow pigments. Tiny muscles attach to the periphery of these cells. When the muscles contract the chromatophores are expanded to display their pigments. The color displays are used for camouflage and communication. Squid and octopods have color vision.

Near the coast of Newfoundland, **giant squids** are known to reach 17 m or more in total length, with arms as thick as a person's leg and suckers as big as teacups. Giant squids are food for sperm whales, which may dive a mile beneath the ocean surface in their pursuit.

Cuttlefish also have ten arms, but unlike squid they

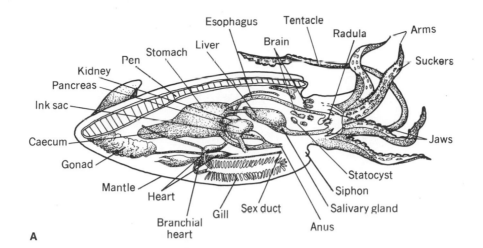

A

B

C

Figure 16-15. (A) Internal structure of *Loligo, a typical squid.* (B) A population of *Loligo,* (C) Closeup of *Loligo.* Note chromatophores, siphon, eyes, and position of tentacles.

Figure 16-16. Diagram of a section through the eye of a squid (right) and a vertebrate (left). These eyes are constructed on the same principle as a camera; light enters through an opening (pupil) in the diaphragm (iris) and passes through a lens that focuses an image on the film (retina). The squid's eye is remarkably similar to the vertebrate eye but differs in having the photosensitive cell layer of the retina directly exposed to the image, devoid of interfering nerves and cell bodies.

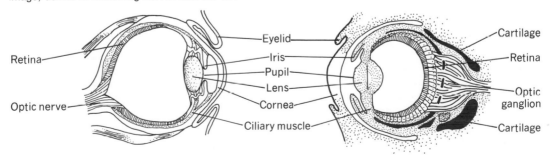

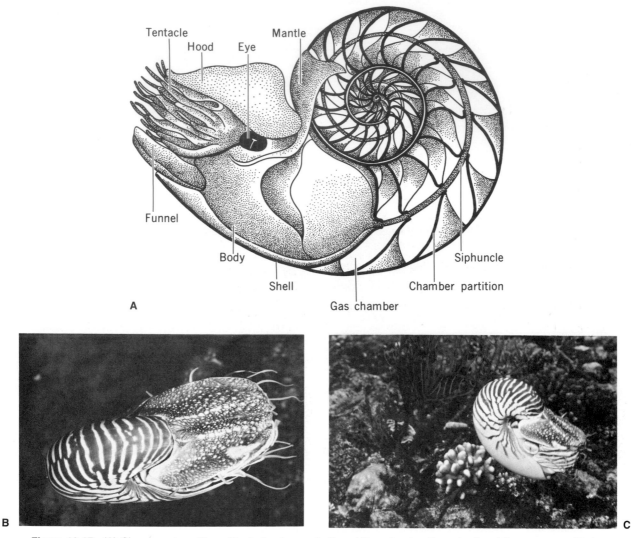

Figure 16-17. (A) Chambered nautilus with shell cut away to the midline showing the animal and the many gas-filled chambers in which it lived during successive stages of growth. The siphuncle is a limy tube that encloses a cord of living tissue that extends from the last to the original chamber. These animals live in the south Pacific Ocean. Natural size is up to 25 cm in diameter. (B) View of *Nautilus* in natural habitat. Note extended sensory tentacles. (C) A side view of same animal. Observe eyes, shell, and hood.

have an internal calcareous skeleton (**cuttlebone**). Dried cuttlebones are used as bill sharpeners and calcium sources for caged birds. Cuttlefish ink has provided the dark reddish-brown sepia pigment used by artists for centuries.

The **chambered nautilus** (Fig. 16-17) is the only modern cephalopod with an external shell. Its shell is coiled like a watch spring and is divided by cross-walls into a series of successively larger compartments. Each chamber is built when the old compartment is outgrown, and a new wall is secreted behind it. The empty chambers are partially filled with gas, which

Figure 16-18. The common octopus is a cephalopod without a shell. It pulls itself over the rocks with its arms; it can also move by expelling water from its funnel. The sucker-bearing arms are used to seize the animals on which it feeds.

helps buoy the animal in the water. The head bears 60–90 tentacles without suckers.

Octopods (Fig. 16-18) are bottom-dwelling cephalopods that have lost all trace of a shell. The octopus creeps about on its tentacles or jets itself through the water. Octopods feed primarily on crabs and other crustaceans, mollusks, and occasionally fish. The octopus has become a popular animal for scientific studies relating to their nervous system, learning, and behavior.

Class Amphineura—The Chitons

Chitons (Fig. 16-1) are mollusks with a shell of eight dorsal plates arranged in a linear series, each overlapping the one behind like shingles on a roof. Chitons typically cling to surf-swept rocks by means of their broad muscular **foot.** When detached, a chiton rolls up like an armadillo, with its soft parts protected by the hard shell. They feed on microscopic life that is scraped from the rocks with their radulae. About 600 species of these primitive mollusks are recognized.

Class Scaphopoda—The Tusk or Tooth Shells

Scaphopods are a small group (200 living species) of mollusks with a curved conical shell that is open at both ends. All scaphopods are marine, living partly embedded in sand or mud and ranging from the subtidal zone down to very deep waters. The **foot** is elongated and modified for digging. Tentacular **filaments** select organic material from the sand and convey it to the **mouth.** Water enters and leaves the hole of the top of the shell, and respiratory gas exchange occurs across the wall of the **mantle.** Development includes a **trochophore** larval stage.

Class Monoplacophora

In 1952, ten "living fossils," now identified as *Neopilina galatheae,* were dredged from a depth of 3,591 m. *Neopilina* (Fig. 16-19) is a living representative of Monoplacophora. The class name was originally proposed for a group of extinct **Paleozoic** (600 to 270 million years ago) mollusks that were known only from their fossils. *Neopilina,* which is radically differ-

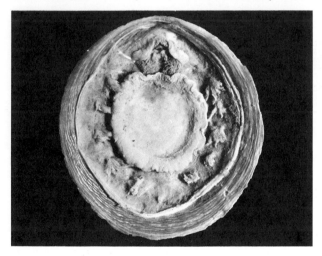

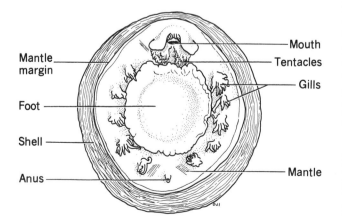

ent from all other mollusks, has a single dome-shaped shell, much like that of a limpet. There is no evidence of the torsion of internal organs that distinguishes gastropods. The **foot** is much like that of a chiton. The **heart** has two **auricles** and two **ventricles.** The **digestive system** includes both a **radula** and a crystalline **style.**

The most remarkable characteristic of *Neopilina* is the **metameric** arrangement of certain organ systems. There are six pairs of **retractor muscles,** five pairs of **gills,** and six pairs of **nephridia.** These organ systems are **serially arranged** and coincide remarkably. Also, unlike other gastropods, the **coelom** is spacious.

Relations of Mollusks to Humans

The value of mollusks as scavengers is little appreciated. **Clams,** for example, are continually filtering organic particles from the water in which they live. Some freshwater **mussels** are used as indicators of environmental conditions, such as detection of pesticide residue and toxic pollutants.

Mollusks, especially the bivalves, have long been used as food. Evidence of their popularity with native Americans is found in the enormous piles of oyster shells (kitchen middens) that mark the location of their coastal campsites. **Oysters** exceed in commercial value any other seafood. About 30 million bushels are gathered annually along U.S. coastlines. Oyster culture is also being carried on with success to increase production. The spat are allowed to settle on suspended platforms that keep the oysters beyond the reach of predatory snails and starfish. The **soft-shell** or **long-neck clam,** *Mya arenaria,* and the **hard-shell clam,** *Mercenaria mercenaria,* are both widely used as food. One edible mussel, *Mytilus edulis,* is eaten extensively in Europe but infrequently in this country. Only the large adductor muscles of **scallops,** *Pecten* (Fig. 16-1), are generally consumed. In certain parts of Europe **snails** (**escargots**) are considered a delicacy, as is the **abalone** on the western coast of the

Figure 16-19. Photograph and diagram of a segmented mollusk, *Neopilina galatheae,* a primitive mollusk that represents a class that probably originated over 450 million years ago. The largest specimen collected measured about 37 mm in length, 33 mm in width, and 14 mm in height.

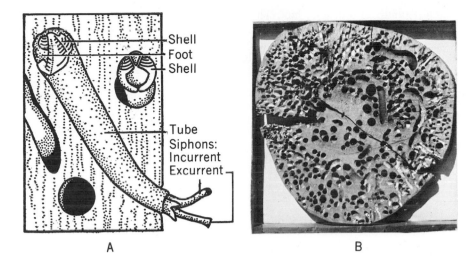

Shell
Foot
Shell

Tube
Siphons:
Incurrent
Excurrent

A   B

Figure 16-20. (A) Section of wood with a shipworm, *Teredo,* in its burrow. *Teredo* is a bivalve, but the two shells that are used for boring enclose only a very small part of the anterior end of the body. The shipworm feeds on wood particles and minute organisms. (B) Damage done to a wharf pile section, driven in March 1944 and removed August 1945.

United States. **Squids** (Fig. 16-14), **cuttlefishes,** and **octopods** are esteemed by palates in southern Europe and in the Orient.

The **giant clam,** *Tridacna,* of the tropics may reach a length of 1 m and weigh 227 kg. The large shells are used as baby cradles by natives of the East Indies.

Among the valuable materials derived from mollusks are **pearls** and pearl buttons. Pearls are formed especially by pearl oysters, *Pinctada,* off the coasts of Ceylon, India, Japan, and northern Australia. The Japanese produce cultured pearls by inserting a foreign body into the mantle of the bivalves. Pearls also occur in the common oyster and in clams, but these are not considered valuable. The shells of freshwater mussels are used to manufacture pearl buttons, and many tons are collected annually from the Mississippi and Ohio river systems. Overharvesting of mussels for buttons has seriously depleted their populations, and pollution has limited their range. As a result, many species of the world's richest freshwater molluscan fauna, those of the great Mississippi River system, are threatened with extinction.

Not all mollusks are considered beneficial to human pursuits. **Slugs** attack crops in greenhouses and gardens. **Shipworms,** *Teredo* (Fig. 16-20), burrow through wooden boat hulls, wharves, and piles, weakening and destroying them. The most important destructive role of mollusks is played by the freshwater **snails** that host trematode parasites of humans and domesticated animals. First among these are probably the intermediate snail hosts of schistosomes blood flukes of humans, cattle, and other animals. Worldwide, some 200 million humans are believed to be parasitized by schistosome flukes.

Origin of the Mollusks

The discovery of metameric organs in *Neopilina* has sparked some controversy concerning the evolutionary origin of the mollusks. Some researchers believe that *Neopilina* is a very primitive mollusk and that its metamerism indicates that the mollusks evolved from the segmented annelids. Other investigators (perhaps the majority) consider *Neopilina* to be a primitive mollusk that evolved metamerism after diverging from the other mollusks. This school holds to the traditional view that the mollusks and the annelids evolved independently from unsegmented, trochophorelike ancestors.

In any event, the great evolutionary plasticity of the molluscan body plan (Fig. 16-1) has been a principal factor in the great success of this phylum.

Summary

Mollusks are one of the most highly successful animal groups. Mollusks possess some unique features including a soft body with a shell-secreting mantle and a cephalized muscular foot. The basic body plans of mollusks consists of a head, visceral mass, and foot. This plan has been modified considerably among the different classes, and, as a result, the mollusks have successfully invaded many habitats and adapted to many different ways of life.

Review Questions

1. Describe how the basic molluscan body plan is modified in the major classes.
2. Describe feeding and digestion in a clam.
3. Give some examples of the diversity of mollusks that permits occupation of different habitats.

Selected References

Abbott, R. T. *American Sea Shells*. New York: D. Van Nostrand Company, 1954.

Coker, R. T., and A. F. Clark. "Natural History and Propagation of Fresh Water Mussels," *Bull. U.S. Bur. Fisheries*, **37**:77, 1921.

Duncan, C. J. "The Life Cycle and Ecology of the Freshwater Snail *Physa fontinalis*," *J. Anim. Ecol.*, **28**:97, 1959.

Grave, B. H. "Natural History of the Shipworm, *Teredo navalis*, at Woods Hole, Mass.," *Biol. Bull.*, **55**:260, 1928.

Keen, A. M. *Marine Molluscan Genera of Western North America*. Stanford: Stanford University Press, 1963.

Kline, G. "Notes on the Stinging Operation of *Conus*," *Nautilus*, **69**:76, 1956.

Lemche, H. "A New Living Deep-sea Mollusc of the Cambro-Devonian Class Monoplacophora," *Nature*, **179**:43, 1957.

Malek, E. A. *Laboratory Guide and Notes for Medical Malacology*. Minneapolis: Burgess Publishing Company, 1962.

Mead, A. R. *The Giant African Snail: A Problem in Economic Malacology*. Chicago: The University of Chicago Press, 1961.

Morris, P. A. *A Field Guide to the Shells of Our Atlantic Coast*. Boston: Houghton Mifflin Company, 1947.

Morris, P. A. *A Field Guide to Shells of the Pacific Coast and Hawaii*. Boston: Houghton Mifflin Company, 1952.

Morton, J. E. *Mollusks*. London: Hutchinson University Library, 1958.

Mozley, A. *An Introduction to Molluscan Ecology*. London: H. K. Lewin, 1954.

Nichols, J. T., and P. Bartsch. *Fishes and Shells of the Pacific World*. New York: Macmillan Publishing Co., Inc., 1945.

Peterson, R. P. "The Anatomy and History of the Reproductive System of *Octopus Bimaculoides*," *J. Morph.*, **104**:61, 1959.

Pilsbry, H. A. *Land Mollusca of North America (North of Mexico)*. Philadelphia: Academy of Natural Sciences, Monograph 3, Vol. 1, Parts 1 and 2, 1939 and 1940; Vol. 2, Parts 1 and 2, 1946 and 1948.

Pratt, H. S. *A Manual of the Common Invertebrate Animals*. Philadelphia: Blakiston, 1932.

Purdion, R. D. *The Biology of the Mollusca*, 2nd. ed. New York: Pergamon Press, 1977.

Russell-Hunter, W. D. *A Biology of Lower Invertebrates*. London: The Macmillan Company, Ltd., 1972.

Stiles, K. A., and N. R. Stiles. "The Pearl, a Biological Gem," *Bios.*, **14**:3, 1943.

Solem, G. A. *The Shell-makers: Introducing Mollusks*. New York: Wiley, 1974.

Tressler, D. K. *Marine Products of Commerce*, 2nd ed. New York: Reinhold, 1951.

Yonge, C. M. *Oysters, (The New Naturalist)*. London: William Collins Sons & Co., Ltd., 1950.

Young, J. Z. "Learning and Discrimination in the Octopus," *Biol. Rev.*, **36**:32, 1961.

Web, W. F. *Handbook for Shell Collectors*, 9th ed. St. Petersburg, Fla., 1951.

Wells, M. J. *Brain and Behavior in Cephalopods*. Stanford, Cal.: Stanford University Press, 1962.

Wilbur, K. M., and C. M. Yonge (ed.). *Physiology of Mollusca*, Vol. 1. New York: Academic Press, 1964.

Annelida and Related Minor Phyla

The phylum Annelida contains most well-known worms and is one of the largest and most successful animal groups. An outstanding characteristic of the annelids is **segmentation,** or **metamerism:** The body consists of a linear series of essentially similar units, called **segments, somites,** or **metameres.** Metamerism evolved twice in the animal kingdom, once in the annelid–arthropod line and again in the chordates. In both cases metamerism was an adaptation for locomotion: for burrowing among early annelids and for undulatory swimming among chordates. In both cases the principal metameric structures are the body wall muscles.

Also included in this chapter are three minor worm phyla, Sipunculida, Echiurida, and Priapulida. Sipunculids and echiurids are unsegmented worms that show clear phylogentic origins near the annelid and mollusk branches of evolution. Priapulida is one of the most problematic phyla to relate to other animals; because of its coelom and wormlike nature, it is included here, with a recognition of its uncertain origins.

Annelids inhabit virtually all marine environments and have successfully invaded many freshwater and moist terrestrial habitats. Three principal classes of annelids are recognized (Fig. 17-1). Class **Oligochaeta** includes the familiar earthworms. Nearly all oligochaetes live in moist terrestrial or freshwater habitats. Class **Polychaeta** is the largest and most diverse annelid group, containing 4000 of the 7000 known species. Polychaetes typically inhabit marine and brackish waters. Class **Hirudinea** includes the carnivorous and bloodsucking leeches, which are found in all three major environments. A fourth group, **Archiannelida,** contains minute worms that live between sand grains, and another aberrant group of parasitic annelids comprise the **Myzostomids.**

Class Oligochaeta, *Lumbricus terrestris*—The Earthworm

The common earthworm *Lumbricus terrestris* is used to illustrate the principal characteristics of the annelids. Figure 17-2 shows structural features of this segmented worm.

Earthworms are soft and naked and, to avoid dehydration, must live in moist earth from which they emerge chiefly on damp nights. The burrows usually extend less than 1 m underground. Earthworms can force their way through soft earth, but they must eat their way through harder soil. The earth that passes through the digestive tract is deposited at the mouth of the burrow as castings.

External Morphology

The cylindrical body of *Lumbricus* varies in length from about 15 cm to 30 cm. The dorsal surface is darker than the slightly flattened ventral surface. The approximately 100 segments (somites) are clearly marked by grooves that extend around the body. At

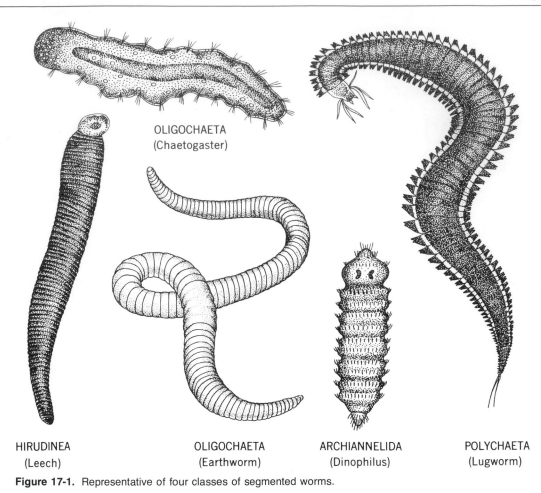

OLIGOCHAETA
(Chaetogaster)

HIRUDINEA   OLIGOCHAETA   ARCHIANNELIDA   POLYCHAETA
(Leech)   (Earthworm)   (Dinophilus)   (Lugworm)

Figure 17-1. Representative of four classes of segmented worms.

the anterior end a fleshy lobe, the **prostomium** (Fig. 17-2), projects over the mouth; the prostomium is not considered a true segment. It is customary to number the segments, beginning at the anterior end, to facilitate recognition of the organs that they house. Segments from 31 or 32 to 37 are notably swollen in mature worms, forming a saddle-shaped enlargement called the **clitellum,** which is used during reproduction. Every segment, except the first and last, bears four pairs of chitinous bristles called **setae** (Fig. 17-3). The setae help the worm to grip to the walls of its burrow and may be manipulated by **retractor** and **protractor muscles.** In mature worms, the setae on segment 36 are modified as **reproductive claspers** that grip the partner during copulation.

The body is covered by a thin transparent **cuticle** that is secreted by underlying epidermal cells. The cuticle protects the body from physical and chemical injury. It contains numerous pores that pass secretions from unicellular mucous glands that keep the cuticle moist. The cuticle is also marked with fine striations, causing its surface to appear iridescent.

External openings of various sizes allow the entrance and exit of materials: (1) The crescent-shaped **mouth** opens ventrally in the first segment, the **peristomium** (see Fig. 17-5); the mouth is overhung by the prostomium. (2) The oval **anus** opens in the last segment. (3) The openings of the **sperm ducts** are situated on each side of segment 15 (Fig. 17-2); these openings have swollen lips, and a slight ridge extends

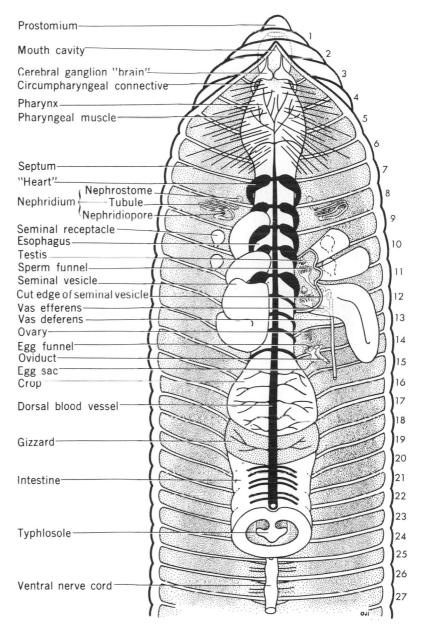

Prostomium

Mouth cavity

Cerebral ganglion "brain"
Circumpharyngeal connective

Pharynx
Pharyngeal muscle

Septum

"Heart"

Nephridium { Nephrostome
— Tubule
Nephridiopore

Seminal receptacle

Esophagus

Testis

Sperm funnel

Seminal vesicle

Cut edge of seminal vesicle

Vas efferens

Vas deferens

Ovary

Egg funnel

Oviduct

Egg sac

Crop

Dorsal blood vessel

Gizzard

Intestine

Typhlosole

Ventral nerve cord

Figure 17-2. Anterior part of an earthworm, *Lumbricus,* with dorsal body wall removed to show the internal organs.

from them posteriorly to the clitellum. (4) The openings of the **oviducts** are small round pores, one on each side of segment 14; fertilized ova pass through them. (5) The openings of the **seminal receptacles** that receive sperm during copulation appear as two pairs of minute pores concealed within the grooves between segments 9 and 10 and segments 10 and 11.

(6) A pair of **excretory nephridiopores** (Fig. 17-3) open on the ventrolateral surfaces of every segment except the first three and the last. (7) The fluid-filled **coelom** communicates with the exterior by means of dorsal **pores,** one at the anterior middorsal edge of each segment from 8 or 9 to the posterior end of the body.

217

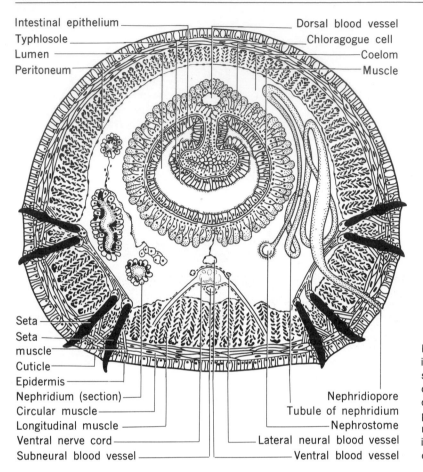

Intestinal epithelium
Typhlosole
Lumen
Peritoneum

Dorsal blood vessel
Chloragogue cell
Coelom
Muscle

Seta
Seta
muscle
Cuticle
Epidermis
Nephridium (section)
Circular muscle
Longitudinal muscle
Ventral nerve cord
Subneural blood vessel

Nephridiopore
Tubule of nephridium
Nephrostome
Lateral neural blood vessel
Ventral blood vessel

Figure 17-3. Cross-section of an earthworm illustrating the advances in complexity of structure, correlated with the appearance of a coelom and the development of systems of organs. Left side of drawing shows sectioned parts of nephridium as they actually appear; right side shows an earthworm nephridium as it appears in a dissection. Rarely does a cross-section show all four pairs of setae.

Internal Morphology and Physiology

An earthworm may be dissected along its length by a mid-dorsal incision passing through the body wall to obtain a general view of its internal structures (Fig. 17-2). The annelid body is essentially a double tube; the cylindrical body wall is separated from the tubular digestive tract by the coelomic cavity. The coelom is divided into compartments by means of partitions, called **septa,** that underlie the grooves in the external body wall (Fig. 17-2). The straight digestive tract is suspended by the septa in the center of the coelom. The coelom is lined with an epithelium, called the **peritoneum** (Fig. 17-3), that is derived from mesoderm.

The coelomic cavity is filled with a colorless fluid. A large perforation is present in the midventral part of each septum. **Sphincter muscles** around this opening control the flow of coelomic fluid from one compartment to another.

Muscular System

The body wall contains two layers of muscles. Beneath the epidermis lies a layer of **circular muscles.** These muscle fibers are long and spindle shaped; when the contract, the diameter of body becomes smaller and the worm elongates. Beneath the circular layer is a thick longitudinal layer of muscle fibers lying parallel to the length of the worm; when these contract, the worm shortens and the diameter of the body becomes greater.

Digestive System

The **digestive tract** (Fig. 17-2) consists of (1) a mouth (buccal cavity) in segments 1 to 3, (2) a thick pharynx in segments 4 and 5, (3) a narrow, straight tube, the esophagus, that extends through segments 6 to 14, (4) a thin-walled enlargement, the crop, in segments 15 and 16, (5) a thick-walled muscular gizzard in segments 17 and 18, and (6) a thin-walled intestine extending from segment 19 to the terminal anus. The cavity of the intestine is not a simple cylindrical tube; its dorsal wall is infolded, forming an internal longitudinal ridge, the **typhlosole,** that increases the digestive surface. Surrounding the digestive tract and dorsal blood vessel is a layer of yellowish **chloragogue** tissue (Fig. 17-3). These cells play important roles in intermediary metabolism; they synthesize and store glycogen and fats, break down proteins, and synthesize urea. Three pairs of **calciferous glands** lie alongside the esophagus in segments 10 to 12. Their primary function is the excretion of excess calcium and carbonate ions from the blood.

The food of the earthworm consists principally of pieces of leaves and other vegetation, particles of animal matter, and organic humus in soil. Food is gathered at night when the worms are most active. They crawl out on the surface of the ground, holding fast to their burrows with their posterior ends, and explore the neighborhood. Food particles are pumped into the **buccal cavity** by contractions of the pharynx. Earthworms also digest the organic matter from the soil that they ingest while burrowing.

In the **pharynx,** the food is mixed with a salivary secretion that contains mucus and a protease. Food then passes through the **esophagus** to the **crop,** where it is temporarily stored. The **gizzard** is a grinding organ where food is broken into minute fragments by being squeezed and rolled about. Solid particles such as sand grains, which are frequently ingested, probably aid this grinding process. The food then passes into the **intestine,** where most of the digestion and absorption of nutrients occurs.

Digestion in the earthworm is very similar to that in higher animals. Enzymes that aid in the breakdown of food include **amylase,** which digests carbohydrates; **cellulase,** which digests cellulose; **pepsin** and **trypsin,** which digest proteins; and **lipase,** which digests fats. The nutrients that result from digestion are absorbed through the wall of the intestine, assisted by the amoeboid activity of some of the cells, into the bloodstream. Absorbed nutrients also enter the coelomic fluid, which distributes them directly to the tissues that it bathes. No circulatory system is necessary in unicellular protists and such simple metazoans as hydra, planaria, and ascaris because food is either digested within the cells or comes into direct contact with them. In large complex animals, however, a special system of organs is required to distribute digested food to all cells of the body.

Circulatory System

The **blood** of the earthworm is contained in a system of tubular vessels that supply all parts of the body. There are five longitudinal **blood vessels.** These main vessels and their connectives are shown in Figure 17-4 and are as follows: the **dorsal vessel** above the gut; the **ventral (subintestinal) vessel** just below the gut; the **subneural vessel** below the ventral nerve cord; and two **lateral neural vessels,** one on either side of the ventral nerve cord. Blood flows anteriorly in the dorsal vessel and posteriorly in the ventral vessels.

The longitudinal vessels are interconnected by **lateral vessels** in every somite. These connective vessels supply the somatic and visceral tissues. The dorsal vessel receives blood from the lateral vessels, which drain the body wall and digestive tract, and carries it anteriorly. The ventral vessel delivers blood posteriorly to the lateral vessels.

The dorsal vessel functions as a **heart.** Blood is forced forward by wavelike contractions of the dorsal vessel, beginning at its posterior end and traveling quickly forward. Such wavelike contractions are called **peristaltis.** One-way **valves** (Fig. 17-4) in the dorsal vessel prevent backflow of blood from the anterior end. In segments 7 to 11, the blood passes from the dorsal vessel through the **aortic arches** into the ventral vessel. These so-called "hearts" regulate blood pressure; they receive spurts of blood from the dorsal vessel and then contract in such a way that the ventral vessel receives blood at a steady pressure. Valves in the aortic arches prevent backflow. From the ventral vessel, the blood passes to the body wall, the intestine, and the nephridia. In their target organs the connective vessels ramify into exceedingly

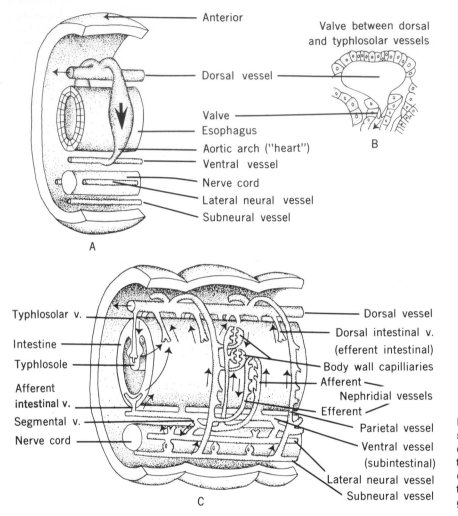

Figure 17-4. Earthworm circulatory system. (A) One pair of "hearts" and other vessels. (B) A section to show the structure of a valve. (C) A three-dimensional view of an opened section of an earthworm to show the general scheme of the circulation.

thin tubules, the **capillaries.** Blood that is carried to the body wall receives oxygen and eliminates carbon dioxide through the cuticle before being circulated back to the dorsal vessel.

The blood consists of fluid **plasma** in which are suspended a great number of colorless amoeboid cells (**corpuscles**). The blood's red color is due to a respiratory pigment, **hemoglobin,** that is dissolved in the plasma. Hemoglobin binds with oxygen and thus increases the amount of oxygen that can be transported in the blood. In the vertebrates, hemoglobin is located within red blood corpuscles.

The exchange of materials between the capillary blood and the tissue cells takes place across minute tissue spaces. Blood plasma and a few corpuscles pass through the capillary walls into the tissue spaces, and this tissue fluid bathes the cells. An interchange of oxygen and carbon dioxide occurs, and waste products of cellular metabolism are collected and transported into the bloodstream for disposal.

Respiratory System

The earthworm possesses no distinct respiratory system. Oxygen and carbon dioxide diffuse through the moist skin. Many capillaries lie just beneath the

Figure 17-5. (A) Pair of earthworms in copulation. Because this usually occurs at night, it is not often observed. [Courtesy of General Biological Supply House, Inc., Chicago] Diagrams B, C, and D illustrate regeneration. (B) A new anterior end (dotted) regenerates in place of one which has been removed. (C) A new posterior end (dotted) regenerates in place of a missing part. (D) Three separate worm parts have been grafted together to yield a longer worm.

cuticle, and oxygen diffuses into the blood and combines with hemoglobin. Earthworm hemoglobin is a relatively inefficient oxygen carrier compared with its counterpart in birds and mammals. Carbon dioxide diffuses in the opposite direction, going from the blood through the skin into the external environment.

Excretory System

The excretory system of oligochaetes consists of pairs of **nephridia** (Figs. 17-2 and 17-3) in every segment except the first three and the last. Each nephridium occupies the coelomic compartments of two successive segments, penetrating the septum between them. In the anterior somite the nephridium opens into a ciliated funnel, the **nephrostome,** that communicates with a tubule that passes through the septum. The preseptal, tubular portion is typically short, whereas the postseptal tubule is highly coiled. The **cilia** on the nephrostome draw in soluble wastes from the coelomic fluid; other wastes are received directly from the blood vessels that surround the tubule. Nitrogenous wastes (urea and ammonia) are eliminated through a **nephridopore** that opens on the ventrolateral surface of the postseptal somite. **Chloragogue cells** may temporarily store excretory wastes prior to release into the coelomic fluid. Under normal conditions of adequate water supply, the urine of earthworms is generally abundant and more dilute (hypotonic) than the blood.

Nervous System

The nervous system is illustrated in Figures 17-2 and 17-6. A bilobed mass of nerve tissue, the **brain** or **cerebral ganglia,** lies dorsal to the pharynx in the

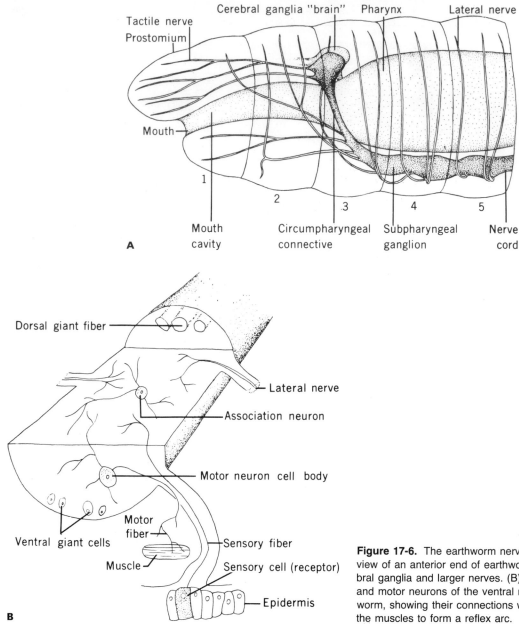

Tactile nerve
Prostomium
Cerebral ganglia "brain"
Pharynx
Lateral nerve

Mouth

1
2
3
4
5

Mouth cavity
Circumpharyngeal connective
Subpharyngeal ganglion
Nerve cord

A

Dorsal giant fiber

Lateral nerve

Association neuron

Motor neuron cell body

Ventral giant cells

Motor fiber

Muscle

Sensory fiber

Sensory cell (receptor)

Epidermis

B

Figure 17-6. The earthworm nervous system. (A) Side view of an anterior end of earthworm showing the cerebral ganglia and larger nerves. (B) Diagram of sensory and motor neurons of the ventral nerve cord of an earthworm, showing their connections with the epidermis and the muscles to form a reflex arc.

third segment. The brain is connected by a pair of **circumpharyngeal connectives** to two **subpharyngeal ganglia** beneath the pharynx. A **ventral nerve cord** extends posteriorly from the subpharyngeal ganglia. The ventral nerve cord lies within the body wall musculature and enlarges into a ganglion in every seg-

ment. Each ganglion represents two ganglia fused together. Three pairs of **lateral nerves** leave the nerve cord in each segment. The brain and ventral nerve cord constitute the **central nervous system;** the nerves that supply each somite represent the **peripheral nervous system.**

The functions of nervous tissue are reception, conduction, and stimulation. These functions are performed by nerve cells called **neurons.** Each nerve cell consists of a **cell body** from which extend several to many processes called **nerve fibers.** The branching fibers of one neuron juxtapose with the fibers and cell bodies of many other neurons, and in this way impulses can be communicated throughout the nervous system. Bundles of nerve fibers held together by connective tissue are called **nerves,** whereas a collection of nerve cell bodies is called a **ganglion.**

Peripheral neurons are either motor or sensory. **Motor nerve fibers** (Fig. 17-6) extend from motor nerve cells in the ganglia of the central nervous system and pass out to the muscles and glands of the body wall and viscera. Nerve impulses sent along motor nerve fibers stimulate their target organs. **Sensory nerve fibers** originate from receptor cells that monitor the external and internal environments of the animal. Receptor cells interpret environmental stimuli as nerve impulses that are sent through sensory fibers to the central nervous system.

Sensitivity of the earthworm to light, chemicals, and touch is due to the presence of numerous epidermal **sensory receptors** (Fig. 17-7). **Photoreceptor** sensory cells in the inner epidermis are distributed over the dorsal surface of the body, especially at the anterior and posterior ends of the worm. Earthworms are negatively phototaxic. Scattered clusters of **chemoreceptor** neurons extend through the cuticle and monitor the presence of chemicals (e.g., food or toxins) in the external environment. Free nerve endings in the skin are activated by tactile stimuli. All these receptors send nerve fibers into the ventral nerve cord and brain.

Reproduction

Earthworms do not reproduce asexually, although they can regenerate lost body parts. Earthworms are all hermaphroditic (Fig. 17-2). The **gonads** are restricted to a relatively few somites at the anterior end of the worm.

The female system includes a pair of **ovaries** in segment 13 and a pair of **oviducts.** Each of these tubules opens by means of a ciliated funnel into segment 13, penetrates the septum, and passes to the exterior through a **genital pore** in segment 14. A pair of **egg sacs** are small outpocketings (**diverticula**) of the septum that the oviducts penetrate and are the sites of ova maturation in some worms. Two pairs of **seminal receptacles,** which receive sperm during mating, are found in segments 9 and 10.

The male gonads are two pairs of minute, glove-shaped **testes** in segments 10 and 11. The testes and a ciliated sperm funnel are contained in three pairs of **seminal vesicles,** conspicuous saclike structures that surround the testes and in which the **spermatozoa** mature. The **sperm funnel** collects the gametes from each seminal vesicle and delivers them through tiny ducts, the **vas efferens.** The two ducts on each side of the body connect to form a common **vas deferens,** which leads to the outside through a male **genital pore** in segment 15.

Self-fertilization does not occur. Mating takes place at night and requires two to three hours for completion. Spermatozoa are transferred from each worm to the other during **copulation.** Two worms come together along their ventral surfaces, each facing the opposite direction (Fig. 17-5). Then spermatozoa from the seminal vesicles of each worm are discharged through sperm funnels, sperm duct, and male pores into **seminal grooves.** Muscular contractions of the body wall propel the sperm along the grooves and into the seminal receptacles of the other worm. After each has received sperm from the other, the worms separate.

A few days later, the clitellum secretes a band-shaped **cocoon,** which consists of mucus enveloping a tough chitinous membrane. The cocoon is forced for-

Figure 17-7. Diagram of the epidermis of the earthworm showing sense organs.

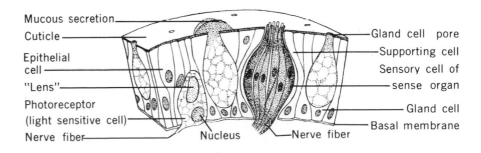

Mucous secretion
Cuticle
Epithelial cell
"Lens"
Photoreceptor (light sensitive cell)
Nerve fiber
Nucleus
Nerve fiber
Gland cell pore
Supporting cell
Sensory cell of sense organ
Gland cell
Basal membrane

ward by movements of the worm. Eggs are discharged into the cocoon as it passes over the female genital pores. Then sperm are deposited in the cocoon from the seminal receptacles. As the front and end of the cocoon slip over the prostomium, the ends close and seal. Inside the cocoon, which is about the size of an apple seed, the fertilized eggs develop over a two- to three-week period into minute worms. As a result, earthworms have no swimming larval stages as do the marine polychaetes (discussed in a later section of this chapter).

Behavior

Earthworms respond to light, chemicals, and mechanical stimuli.

Responses to Light

Earthworms are very sensitive to light; sudden illumination at night will generally cause them to retreat quickly into their burrows. Photoreceptor cells occur in the dorsal epidermis of every somite and are especially abundant in the most anterior and posterior somites. Each light-sensitive cell contains a transparent lens that focuses light on the neurofibrils that ramify throughout the cell. Very slight differences in light intensity can be distinguished by means of these photoreceptor cells. Earthworms generally avoid strong illumination. A positive reaction to faint light has been demonstrated for the manure worm, *Eisenia foetida;* this reaction may correlate with their emergence from their burrows at night. It is an interesting fact that, although earthworms react negatively to sunlight, they respond positively to red light and may be collected at night with the use of such a light.

Responses to Chemicals

Responses to chemicals in its environment may bring the earthworm into regions where food is available or turn it away from toxic substances. Earthworms respond positively to the presence of moisture, which is necessary for respiration. Darwin explained the earthworm's preference for certain kinds of food by supposing that discrimination between edible and inedible substances was possible only when they were in direct contact with the body. Such reception re-

sembles the sense of taste in more complex animals. Positive responses seem to depend upon contact stimulation, but negative responses (e.g., retracting into the burrow) may occur when certain unpleasant chemicals are still some distance from the body.

Reactions to Mechanical Stimuli

Mechanical stimulation, if continuous and not too strong, elicits a positive reaction. This makes sense when one considers that earthworms live in almost continuous contact with the walls of their burrows. Response to sounds is not due to a sense of hearing but, rather, to mechanical stimuli produced by soil-borne vibrations. Darwin showed that musical tones produced no response, but, if a flowerpot containing earthworms was placed upon a piano and a note was struck, the worms immediately drew back into their burrows.

Learning in Earthworms

Whether or not learning occurs in protists or in such simple metazoans as sponges and hydras is uncertain. But experiments have indicated that the earthworm, like the cephalopod mollusks, is capable of what psychologists call "latent memory," or the storing of impressions until a later time when they may be useful.

In one experiment, worms could escape from a lighted chamber by entering a branched T-shaped passageway of glass tubing. If the worms turned to the right at the "T," they entered a dark chamber filled with damp earth and moss, a favorable environment. If they turned left, they received a mild electric shock. In early trials, the worms were observed to turn left as often as right. But at the end of 20 days, they turned to the left only 5 times out of 20. By the end of 40 days they were turning left only once in 20 trials.

Regeneration and Grafting

Earthworms have considerable capacity for regeneration. An anterior piece of worm regenerates a new tail, but a headless tail will not regenerate if the anterior 15 or so segments have been cut off. A posterior piece may in certain cases regenerate a second tail.

Such a double-tailed worm slowly starves to death.

Grafting experiments involve holding pieces of worms together until they unite. Three pieces from several worms may be united to make a long worm. Two pieces may fuse, forming a worm with two tails, or an anterior piece may be united with a posterior piece to make a short worm.

Other Oligochetes

The 3100 known species of oligochaetes include the terrestrial earthworms plus many freshwater and a few marine forms. Oligochaetes are specialized for a burrowing existence, with streamlined bodies. Aquatic oligochaetes are generally unnoticed but are found in all freshwater habitats, especially in shallow waters. Most burrow through bottom muds and debris, ingesting organic matter as do their terrestrial relatives. Gas exchange generally occurs across the general body surface, but parts of the body wall may be elaborated into leaflike or tubular **gills.** A few aquatic oligochaetes can survive periods of desiccation or cold by **encystment.**

Among the interesting aquatic oligochaetes are species of *Aeolosoma* (Fig. 17-8), which are only 1 mm long and spotted due to red oil globules in their skin. These worms live among algae, consist of from 7 to 10 segments, and reproduce asexually by transverse fission. *Nais* is light brown in color, 2 to 4 mm long, and consists of from 15 to 37 segments. It also lives among algae and may reproduce by budding. *Tubifex tubifex* is a common bottom-dwelling freshwater annelid, reddish in color, and about 4 cm long. It lives in a tube from which the posterior end of the worm projects and waves back and forth through the water. *Tubifex* favors waters polluted with sewage. Large populations that often occur in patches on muddy bottoms look like masses of tiny reddish hairs.

Among the smallest of the oligochaetes are species of *Chaetogaster*, which may be only 0.44 mm long. The largest earthworms are known from southern Australia, where *Megascolides australis* may attain lengths of 3 m. The number of segments in oligochaetes varies from 7 in *Aeolosoma* to over 600 in *Rhinodrilus*.

Class Polychaeta

The polychaetes consist largely of free-living marine worms that exhibit typical annelid characteristics. The body tends to be long and wormlike. It generally consists of a distinct **head** region and a **trunk. Segmentation** is well marked both internally and externally. The outer **cuticle** is usually soft and requires a wet environment to prevent desiccation. The **digestive system** consists of a straight tube with an anteroventral **mouth** and a posterodorsal **anus.** The **circulatory system** has a **dorsal vessel** in which **blood** moves anteriorly and a **ventral vessel** in which it flows posteriorly, with **transverse vessels** interconnecting these longitudinal vessels. The **nervous system** has a dorsal **brain** in or near the **prostomium** and paired ventral **ganglia** in a ladderlike arrangement. **A giant nerve fiber** system, which is usually present, consists of longitudinal fibers that mediate rapid responses.

Figure 17-8. Asexual reproduction in annelids.

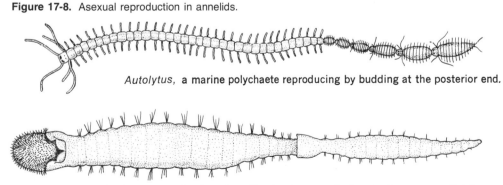

Autolytus, a marine polychaete reproducing by budding at the posterior end.

Aeolosoma, a fresh-water oligochaete reproducing by transverse division.

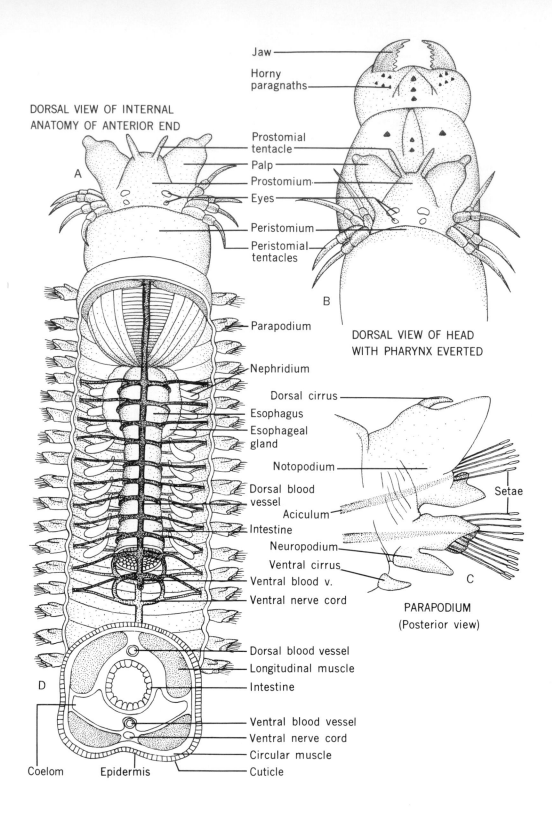

DORSAL VIEW OF INTERNAL
ANATOMY OF ANTERIOR END

A

Jaw
Horny
paragnaths

Prostomial
tentacle
Palp
Prostomium
Eyes

Peristomium

Peristomial
tentacles

B

DORSAL VIEW OF HEAD
WITH PHARYNX EVERTED

Parapodium

Nephridium

Esophagus
Esophageal
gland

Dorsal blood
vessel
Intestine

Ventral blood v.
Ventral nerve cord

Dorsal cirrus

Notopodium

Aciculum

Setae

Neuropodium
Ventral cirrus

C

PARAPODIUM
(Posterior view)

Dorsal blood vessel
Longitudinal muscle

Intestine

Ventral blood vessel
Ventral nerve cord
Circular muscle
Cuticle

D

Coelom   Epidermis

226

Nephridia are present in most body segments. Most polychaetes are **dioecious** (in contrast to the earthworm), with the two sexes resembling each other externally. **Gonads** may arise in many segments, and **gametes** are generally produced in enormous numbers. Development includes a free-swimming **trochophore larva.** Some polychaetes reproduce asexually. *Autolytus,* for instance, produces buds at its anterior end, thus forming a linear row of offspring, each of which develops a head before separating from the parent (Fig. 17-8).

A distinctive polychaete feature are **parapodia** (Fig. 17-9), which are fleshy extensions from the lateral body walls of most somites. They are usually conspicuous and provided with various structures such as **cirri, scales,** and **gills.** Bundles of **setae** project from the parapodia, which function in locomotion, tube building, food gathering, and other important ways.

Neanthes—The Sandworm

The principal characteristics of polychaetes are exhibited by the sandworm, *Neanthes virens* (Fig. 17-9), a common polychaete that lives in sand or mud along the seashore. By day it rests in its burrow, but at night it ventures forth in search of food.

The body is flattened dorsoventrally and may reach a length of 50 cm or more, with 100 to more than 200 segments. The **head** is well developed. The **prostomium** above the mouth (Fig. 17-9) bears a pair of **tentacles,** two pairs of simple **eyes,** and a pair of lateral **palps.** The **peristomium** bears four tentacles on either side. Small animals are captured by a pair of strong chitinous **jaws** that are everted with part of the **pharynx** when *Neanthes* is feeding. The somites behind the peristomium each bear a pair of lateral **parapodia** that are used for swimming and crawling.

The body wall consists of an outer **cuticle,** secreted by epidermal cells just beneath it, and several underlying **muscle** layers. The **digestive, circulatory, excretory,** and **central nervous system** follow the basic annelid plan, as exhibited by the earthworm. The

Figure 17-9. *Neanthes virens,* the sandworm. (A) Anterior end of the body with dorsal wall removed. (B) Dorsal view of head with pharynx everted. (C) Posterior view of a parapodium. (D) Diagrammatic cross-section of *Neanthes* showing internal structure.

sexes are separate. **Ova** or **spermatozoa** are shed into the sea water, where fertilization occurs. A planktonic **trochophore larva** develops from the fertilized egg (Fig. 17-10).

Other Polychaetes

Polychaetes have adapted to a great diversity of niches, and the general body plan described in the preceding paragraphs has been correspondingly modified in many ways. The common names for polychaete families suggest the variations in shape and structure that have evolved: **sea mice, scale worms, fire worms, glass worms, proboscis worms, bamboo worms, gold crowns, gooseberry worms, lugworms, feather dusters,** and **shield worms.** The structural diversity of polychaetes reflects the wide range of marine habitats to which they have adapted.

Most polychaetes are free living, but some are **commensals,** and a very few are parasites. Most are marine, but many live in brackish water, some live in freshwater, and a few are terrestrial. Polychaetes are most abundant in the upper 60 m of the ocean. Metamerism may be **homonomous** (with the somites generally alike), but usually there is considerable departure from this primitive condition. In *Chaetopterus* (Fig. 17-11), parts of the anterior somites and parapodia are modified to function as suction discs, as water-pumping fans, and as specialized feeding organs. In feather duster worms, the peristomium is elaborated into a feathery **tentacular crown,** or food-gathering organ, plus a plug-shaped operculum that closes the mouth of its tube when the worm is withdrawn.

The tubes of polychaetes are also highly variable. These shelters may consist of spun threads (modified setal secretions, as in some scale worms), transparent horny tubes, tough leathery tubes, calcareous tubes, or clear glasslike tubes (some serpulids). Extraneous materials such as sand grains or shell particles are frequently cemented together, sometimes after being selected with regard to size, color, and weight. The following brief descriptions of two polychaetes hint at the great diversity of these annelids.

The **Pacific palolo worm,** *Eunice* (Fig. 17-12), was first encountered in the Samoan Islands, where it is eaten by the natives. Enormous numbers of these marine worms appear periodically in certain localities

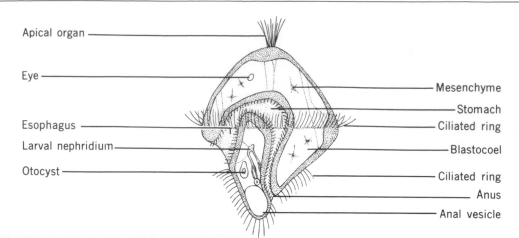

Apical organ ——————————

Eye ————————————

Esophagus ————————

Larval nephridium ————

Otocyst ————————————

—————— Mesenchyme

—————— Stomach

—————— Ciliated ring

—————— Blastocoel

—————— Ciliated ring

—————— Anus

—————— Anal vesicle

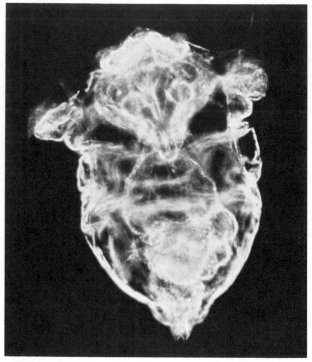

Figure 17-10. Drawing and photo of a side view of the trochophore larva of a polychaete, *Eupomatus*.

for only a few hours. These reproductive swarms occur almost invariably during the months of October and November, usually within the third quarter of the moon. During the night of the swarm, the posterior halves of the worms break off and swim to the ocean surface. Ova and sperm are shed into the sea during the early morning, sometimes in such enormous numbers that the ocean surface has been likened to a thick noodle soup. The fertilized eggs rapidly develop into **trochophore larvae,** which in three days sink to the bottom. Other palolo worms occur in different parts of the world, particularly in warm seas. The Atlantic palolo swarms in June and July.

Chaetopterus is a highly specialized tube-dwelling

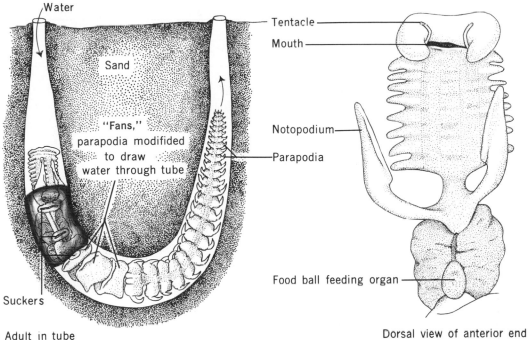

Water

Sand

"Fans,"
parapodia modifided
to draw
water through tube

Suckers

A Adult in tube

Tentacle

Mouth

Notopodium

Parapodia

Food ball feeding organ

Dorsal view of anterior end

B

Figure 17-11. (A) The marine polychaete, *Chaetopterus,* in its tube.
Arrows indicate the direction of water currents. (B) An intact *Chaetopterus* removed from its tube.

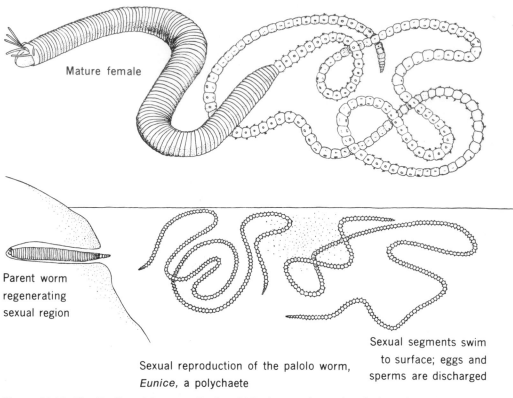

Mature female

Parent worm regenerating sexual region

Sexual reproduction of the palolo worm, *Eunice,* a polychaete

Sexual segments swim to surface; eggs and sperms are discharged

Figure 17-12. The Pacific palolo worm, *Eunice viridis,* burrows in coral reefs; it produces many posterior segments filled with eggs or sperm that are periodically cast off.

polychaete (Fig. 17-11). When fully grown it may reach 15 to 30 cm. The body is highly bioluminescent and consists of three distinct regions. *Chaetopterus* secretes a U-shaped, opaque, parchmentlike tube, up to 50 cm long, that lies completely buried in mud or sand except for its two open ends. The worm maintains its position within the spacious tube by means of long anterior **notopodia** and three **ventral suckers** that represent highly modified parapodia. Seawater is drawn through the tube by three **muscular fans.** Another remarkably modified parapodial structure is an organ that constructs a **mucous net** that strains food particles from the circulating seawater. This polychaete has a worldwide distribution, usually colonizing broad sand flats where there is little current. It is found along the Pacific Coast and the Atlantic Coast from North Carolina to Cape Cod.

The Archiannelids

Archiannelids comprise about 50 species of aberrant marine annelids that exhibit many primitive features. These microscopic worms inhabit the spaces between sand grains in shallow marine waters. They typically lack parapodia and setae, and their internal organs are reduced to various degrees. Whether archiannelids are primitive or degenerate is not known. Many taxonomists consider them to be highly specialized polychaetes.

Polygordius appendiculatus (Fig. 17-13) lives along sandy shorelines of the Atlantic and Mediterranean. It is about 1 inch long and indistinctly segmented externally. The prostomium bears a pair of **cephalic tentacles,** and the posterior end bears two **anal tentacles.** A pair of **ciliated pits,** one on either side of the **pros-**

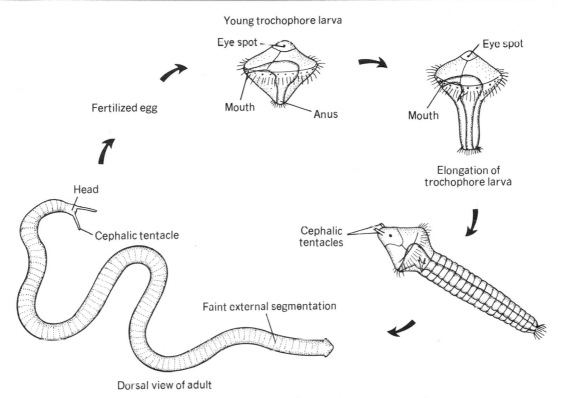

Figure 17-13. Stages in development of *Polygordius appendiculatus,* one of the archiannelids.

tomium, probably serve sensory functions. The development of *Polygordius* includes a larval **trochophore** stage. The trochophore elongates from its posterior end, which becomes segmented. New segments are added from the posterior, just anterior to the **anus,** as the larva transforms into an adult. Such **teleoblastic growth** is characteristic of all annelids. As a consequence, the segments are formed posteriorly: the "younger" segments are at the anterior end.

The Myzostomids

Another aberrant group of annelids, **Myzostomida,** is often considered as a separate class. All are parasites of echinoderms, notably sea lilies (crinoids). In size they range from 0.5 to 9 mm, and the flattened, disc-shaped body bears five pairs of **parapodia** on its ventral surface. Individuals are **protandric;** that is, the younger ones function as males, but later, with increase in size and age, they become females; cross-fertilization is thus ensured. The zygote gives rise to a swimming **trochophore.**

Class Hirudinea—The Leeches

Hirudinea contains the **leeches.** Most leeches are bloodsucking parasites, but about 25 percent of the 500 known species are predatory carnivores. Most species inhabit fresh water, but some are marine and a few are terrestrial. Leeches range in size from 1 to 30 cm in length, and their bodies are capable of great contraction and extension. The dorsoventrally flattened body is typically tapered at the anterior end. **Metamerism** is relatively reduced in leeches, and the number of segments is fixed at 34. Secondary annulations of the cuticle obscure the original segmentation however, and internal septa are reduced or absent. Figure 17-14 illustrates the principal structures of a leech.

The anterior and posterior segments are modified to form **suckers.** The anterior sucker generally sur-

rounds the mouth. The **digestive tract** is adapted to accommodate infrequent blood meals. The **mouth** is provided with three **jaws** armed with chitinous **teeth** that bite through the skin of the host. Then the muscular **pharynx** pumps blood from the wound. A short **esophagus** leads from the **pharynx** into an enormous **crop,** which has 11 pairs of lateral branches. Here the blood is stored until digested in the small globular **stomach.** A tremendous amount of blood may be in-

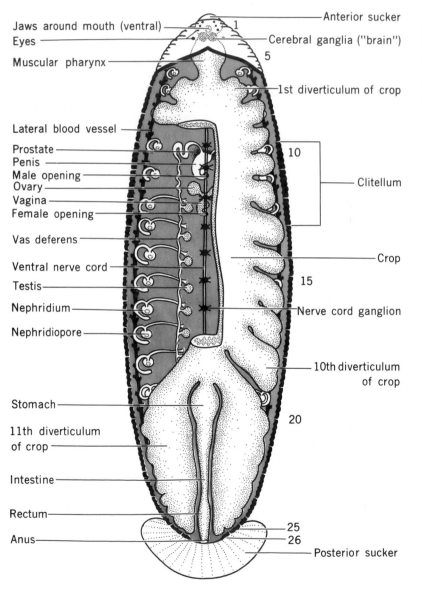

Jaws around mouth (ventral)
Eyes
Muscular pharynx
Anterior sucker
Cerebral ganglia ("brain")
1st diverticulum of crop
Lateral blood vessel
Prostate
Penis
Male opening
Ovary
Vagina
Female opening
Vas deferens
Ventral nerve cord
Testis
Nephridium
Nephridiopore
Clitellum
Crop
Nerve cord ganglion
10th diverticulum of crop
Stomach
11th diverticulum of crop
Intestine
Rectum
Anus
Posterior sucker

Figure 17-14. Dorsal view showing the segmentation and internal anatomy of the leech. Part of the crop is cut away on the left side to show the ventral nerve cord and reproductive organs. The numbers on the right indicate the internal segmentation or somites as shown by the nerve ganglia.

gested, up to ten times the leech's body weight. And, because it may take as long as nine months to digest such a meal, leeches can feed infrequently.

Respiration occurs mainly through the general body surface. Waste products are removed from the blood and coelomic fluid by up to 17 pairs of **nephridia.** Leeches are **hermaphroditic.** Copulation and cocoon formation are similar to that in earthworm reproduction. Some leeches do not secrete a cocoon, but carry the eggs on their body or deposit them on stones.

Relations of Annelids to Humans

Of the segmented worms, earthworms and leeches most directly influence human welfare. Charles Darwin demonstrated, by careful observations extending over a period of 40 years, the great ecological importance of earthworms. One m² of ground may contain over 20 earthworms. Their feces (castings) are scattered over the soil surface; castings are especially noticeable in the early morning. Darwin estimated that more than 4.5 kg of soil may be carried to the surface as castings in a single year on 1 m² of ground; and in 20 years, a layer 8 cm thick would be transferred from the subsoil to the surface. In addition to this turnover, continuous honeycombing with earthworm burrows makes the soil more porous and ensures better penetration of air and moisture. This in turn accelerates bacterial decomposition of organic matter, which releases nutrients required by plants.

Earthworms may serve as intermediate hosts of parasitic flatworms and roundworms. For example, they are intermediate hosts in the life cycles of a cestode (*Amoebotaenia*) of chickens and a nematode **lungworm** (*Metastrongylus*) of pigs. Earthworms are passive carriers of the nematode *Syngamus trachea*, which infests the windpipe of fowls.

As a transporter of soil, the marine polychaete *Arenicola*, the **lugworm** (Fig. 17-1), is even more effective than the earthworm. The amount of sand brought to the surface on 19 measured sites was 20.4 castings per square meter; the average amount of sand brought to the surface each year on these areas was estimated to be 472 kg per square meter, or a layer about 33 cm deep. As is true for earthworms, lugworms are widely used by fishermen as bait.

Oyster pests include polychaete worms of the genus *Polydora*, which cause mud blisters in the inner nacreous layers of the shells and may weaken or kill the oyster. Oyster growers call this condition "worm disease." Because of *Polydora* invasions, in some regions where oyster cultures once flourished, farming had to be discontinued, or different methods had to be introduced, such as rearing the spat (young oysters) on elevated surfaces. Other bivalve mollusks may also be attacked.

Sedentary polychaetes contribute to the fouling of ship hulls, dikes, and harbor installations. They not only cause destruction of the building materials, but they also add to the weight and drag of a vessel so that its speed may be substantially reduced. Periodic drydocking of ships is required to clear the hulls of fouling organisms. Some tube-building polychaetes construct sand or limestone reefs that may modify shore contours and trap sediments.

Use of certain polychaetes as food, such as the palolo, has already been mentioned. The harvested worms consist almost entirely of yolk-laden eggs and make a highly nutritious meal.

The widespread occurrence of polychaete **fireworms** along tropical shores is of interest because of the injuries that they can inflict. The worms are sometimes as long as 30 cm, covered with striking color patterns. Fireworms crawl conspicuously over rocky surfaces, but the unwary person who picks one up is startled by a severe burning sensation. The injuries are produced by harpoonlike bristles that penetrate the skin.

The former use of the **leech** in medicine was based on the theory that many illnesses were due to "bad blood." **Bloodletting** by leeches was thus considered a cure for many ailments, whereas modern doctors commonly transfuse blood into the body. However, leeching was so common during the last few centuries that doctors were often called "leeches." *Hirudo medicinalis* and other species were used in various parts of the world. The red stripe on a barber's pole indicated that the barber was also a bloodletter (the white stripe signified the bandage). Medicinal bloodletting by leeches is now extremely rare in this country.

Leeches, especially in tropical regions where land

leeches live among dense vegetation, may attach themselves in large numbers to humans and other vertebrates. Such leeches caused much discomfort to Napoleon's soldiers when they invaded North Africa.

The salivary glands of leeches produce a substance called **hirudin,** which prevents blood clotting while the leech is feeding. For this reason, a wound made by a leech may bleed for some time after the parasite has detached itself. Hirudin was formerly extracted from leech salivary glands for use as an anticoagulant during surgical operations.

Origin of the Annelids

Annelids certainly originated in marine waters. Polychaetes gave rise to the oligochaetes, perhaps after freshwater habitats were invaded. Oligochaetes exhibit many adaptations for burrowing through soil: the body has become streamlined by loss of parapodia and cephalic tentacles; also, fertilization is internal, and the embryos are protected by a waterproof cocoon. The leeches subsequently evolved from the oligochaetes and have become highly specialized as parasites.

Annelids share the trochophore larval stage with the mollusks, echiuroids, and sipunculids. The nervous system of the trochophore resembles that of platyhelminths, and the adult annelidan nervous system resembles that of arthropods. This continuum of characters firmly places the annelids within the mainstream of protostome evolution.

Phylum Sipunculida—The Peanut Worm

Phylum sipunculida contains about 300 species of bottom-dwelling marine worms that generally range in length from 18 to 36 cm. Most inhabit intertidal and subtidal zones in warmer waters. Sipunculids derive their common name "peanut worm" from their shape when contracted. The sipunculid body is divided into an anterior **introvert** and a posterior **trunk.** The introvert generally bears one or more rows of **tentacles** and can be retracted into the trunk when the animal is disturbed (Fig. 17-15).

Many sipunculids burrow through sand and mud by extending the introvert into the substratum, expanding and anchoring it, and then retracting it to pull the trunk forward. Some sipunculids feed by entrapping suspended organic debris and plankton on their mucus-covered tentacles; others ingest sediments as they burrow, much as earthworms do.

Sipunculids have a U-shaped **digestive tract,** as do many sedentary and inactive animals. Their **anus** opens dorsally on the trunk near its junction with the introvert. The **coelom** is spacious, and its fluid serves internal transport functions; distinct respiratory and circulatory systems are lacking. Excretion is by two large **nephridia.** The sipunculid body wall and **nervous system** are quite similar to those of annelids, except that sipunculids do not exhibit metamerism. Development includes a **trochophore larval** stage, and coelom formation is **schizocoelous.** Sipunculids are considered to have been derived from early

Figure 17-15. Phylum Sipunculida. (A) Internal anatomy. The position of the introvert retracted (B) and extended (C) is shown.

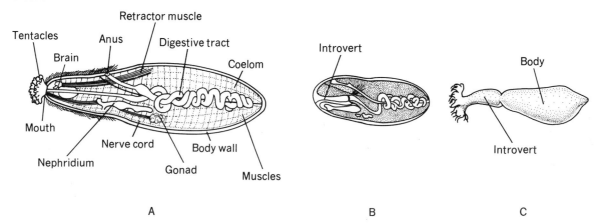

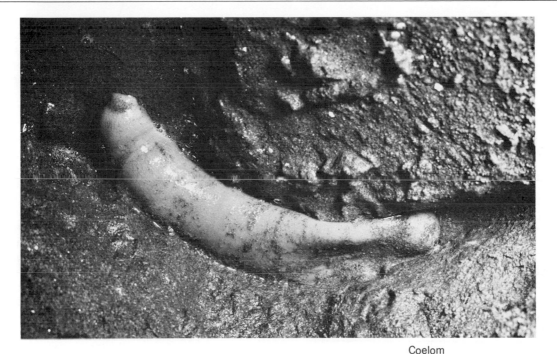

Figure 17-16. Phylum Echiuroidea, *Urechis caupo*, an echiur-oid common in California. Inside its burrow it spins a slime tube that catches food pumped in with water. Then it ingests both the food and the tube. Internal structure is shown at right.

annelidan stock before metamerism became established. They have since evolved secondary characters, such as a U-shaped gut, that adapt them for rather sedentary life-styles.

Phylum Echiurida

Phylum Echiurida contains about 60 species of marine worms that are rather similar to sipunculids in size and habitat. The echiuroid body is divided into two parts: an anterior **proboscis** and a sausage-shaped **trunk.** The proboscis is highly extensible but lacks tentacles and cannot be withdrawn into the trunk.

Echiuroids typically inhabit semipermanent U-shaped, or blind, burrows. Some are ciliary-mucoid feeders that trap suspended debris and plankton on their mucus-covered proboscis. Others are deposit feeders that sweep the algae and bacterial surface film from the substratum surrounding their burrows. *Urechis* (Fig. 17-16) constructs a mucus net inside its U-shaped burrow (much like the polychaete *Chaetop-*

terus); peristaltic body wall contractions pump water through the burrow, and suspended food is trapped in the funnel-shaped net.

The **anus** opens at the posterior tip of the trunk. The spacious and thin-walled hindgut also serves as a respiratory surface. The coelomic fluid contains **hemoglobin.** Most echiuroids have a simple **circulatory system** that corresponds to the basic annelid plan. The **nervous system** is also like that in annelids, except that there are no ganglia. Excretion is by **nephridia.** Development includes a free-swimming **trochophore** larval stage, and their embryology exhibits similarities to both annelids and mollusks. Although no metamerism is evident in the adult, larvae show transitory segmentation that disappears during metamorphosis to the adult, again supporting a relationship to the annelids. Echiuroids were probably another evolutionary offshoot from near the base of the branch that gave rise to the early polychaetes.

Phylum Priapulida

This phylum of unsegmented, true coelomates does not show clear affinities to the other phyla discussed; some characteristics, such as the eversible pharynx, suggest relationships to the pseudocoelous acanthocephalons. Until more information is obtained about this group of marine worms that inhabit the muds and sands of cold marine shores, their phylogenetic status will remain uncertain.

Growing as large as 12 cm, a priapulid buries in sediments and feeds by predation on other soft-bodied worms, including other priapulida. Figure 17-17 shows the internal anatomy, which includes an **eversible pharynx** lined with **spines** (**teeth**). The rest of the **digestive system,** paired **urogenital organs, circular** and **longitudinal muscle** layers, and other features are also illustrated. The external anatomy reveals an anterior **proboscis** or **presoma,** a superficially segmented **trunk,** and one or two **caudal appendages.** The latter serve for respiration; gases diffuse directly into the **coelom,** which serves as the sole method of circulation. The **nervous system** includes a **nerve ring** around the pharynx and a **ventral nerve chord.** Reproduction occurs with separate sexes by external fertilization; direct development into an adultlike larva proceeds, but it remains encased in a chitinous cyst for two years before the first of its several **molts** occurs. More information is needed concerning their

Figure 17-17. Phylum Priapulida. External appearance (left) and internal anatomy (right).

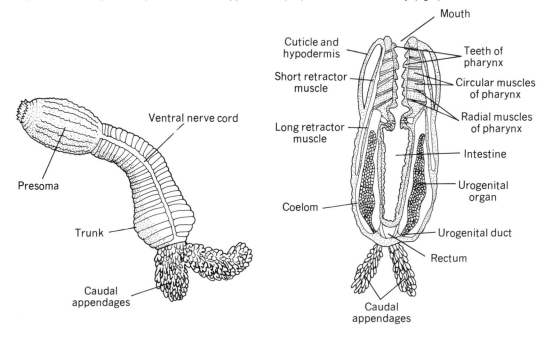

early embryology, such as the method of coelom formation.

Summary

Annelids possess a basic architecture from which more complex organisms could evolve. The body is divided into a linear series of similar segments. Definite organ systems are present to accommodate the increased size and specialization of the organism. The coelom is well developed; the nephridia drain excretory wastes from the coelomic fluid and blood. Respiratory parapodia anticipate the gills of higher organisms. The nervous system has become centralized, with a dorsal brain, a pair of ventral nerve cords, and a pair of ganglia in each segment. The digestive tract is a straight tube with an anterior mouth and posterior anus; the body wall musculature consists of outer circular and inner longitudinal layers. The gametes are derived from the mesoderm. Cleavage of the egg is spiral unless obscured by excessive yolk. Three classes of annelids are generally recognized, and two groups (Archiannelida and Myzostomids) are probably related to polychaetes.

Class Oligochaeta. Terrestrial or freshwater; without parapodia and few setae; head not well developed; hermaphroditic; no trochophore larva. *Lumbricus terrestris* (Fig. 17-1).

Class Polychaeta. Marine; parapodia well developed and provided with setae that are variously modified; prostomium and first few segments sometimes highly cephalized; sexes usually separate; larva typically a trochophore. *Neanthes virens* (Fig. 17-9).

Class Hirudinea. Parasitic or predacious; mostly freshwater or terrestrial; without parapodia or setae; body with 33 segments plus prostomium; posterior and often an anterior sucker; hermaphroditic; coelom reduced by encroachment of connective tissue. *Hirudo medicinalis* (Fig. 17-1).

Phyla Sipunculida and Echiurida are marine worms that share many characteristics with annelids; lack of metamerism (except in the larvae of echiuriods) suggests early evolutionary divergence from the annelids.

Phylum Priapulida consists of predacious marine worms that share fewer characteristics with annelids than do the above groups. They are a relatively unknown and less studied group, and as such their relations to other organisms are currently uncertain.

Review Questions

1. What is the significance of metamerism?
2. What is a trochophore, and which organisms are associated with it?
3. How does teleoblastic growth compare with the growth of tapeworm proglottids?
4. Describe the circulation of blood through an earthworm.
5. Contrast reproduction in the annelid classes and correlate it with their environments.
6. What fostered the evolution of a circulatory system in annelids?
7. Describe the ecological importance of burrowing annelids.
8. Give evidence that supports the annelid relationships of Sipunculida and Echiurida.
9. What feature of the priapulid caudal appendage supports its role as a respiratory organ (see Fig. 17-17)?

Selected References

Bahl, K. N. "Excretion in the Oligochaeta," *Biol. Rev.*, **22**:109, 1947.

Ball, R. C., and L. L. Curry. "Culture and Agricultural Worth of Earthworms," *Bulletin 222*. East Lansing: Michigan State University, 1956.

Barnes, R. D. *Invertebrate Zoology.* Philadelphia: W. B. Saunders Company, 1963.

Barrett, T. J. *Harnessing the Earthworm.* Boston: Bruce Humphries, Inc., 1947.

Beddard, F. E. *Earthworms and Their Allies.* Cambridge: Cambridge University Press, 1901.

Bell, A. W. "The Earthworm Circulatory System," *Turtox News*, **25**:89, 1947.

Brinkhurst, R. O., and B. G. M. Jamieson. *Aquatic Oligochaeta of the World.* Toronto: University of Toronto Press, 1971.

Dales, R. P. *Annelids.* London: Hutchinson University Library, 1963.

Edwards, C. A., and J. R. Lofty. *Biology of Earthworms*. London: Chapman and Hall, 1972.

Fauchald, K. *The Polychaeta Worms: Definitions and Keys to the Orders, Families, and Genera*. Natural History Museum of Los Angeles County, 1977.

Gates, G. E. "On Segment Formation in Normal and Regenerative Growth of Earthworms," *Growth*, **12**:165, 1948.

Grove, A. J. "On the Reproductive Processes of the Earthworm *Lumbricus Terrestris*," *Quart. J. Micro. Sci.*, **69**:245, 1925.

Johnson, W. H., L. E. Delanney, E. C. Williams, and T. A. Cole. *Principles of Zoology*, 2nd ed. New York: Holt, Rinehart and Winston, 1977.

Krivanek, J. O. "Habit Formation in the Earthworm, *Lumbricus Terrestris*," *Physiol. Zool.*, **29**:241, 1956.

Lavarack, M. S. *The Physiology of Earthworms*. New York: Macmillan Publishing Co., Inc., 1963.

Mann, K. H. *Leeches (Hirudinea), Their Structure, Physiology, Ecology and Embryology*, Vol. II. New York: Pergamon Press, 1962.

McConnaughey, B., and D. L. Fox. "The Anatomy and Biology of the Marine Polychaete *Throacophelia Mucronata*," *Univ. Calif. Publ. Zool.*, **47**:319, 1949.

Moment, G. B. "On the Way a Common Earthworm, *Eisenia Fotida*, Grows in Length," *J. Morph.*, **93**:489, 1953.

Moore, J. P. "Annelida," *Encyclopaedia Britannica*. Chicago: Encyclopaedia Britannica, Inc., 1956.

Sawyer, R. T. *North American Freshwater Leeches, Exclusive of the Piscicolidae, With a Key to All Species*. Urbana: University of Illinois Press, 1976.

Introduction to Arthropoda and Other Related Phyla

Arthropods are the most successful group of animals. More than 800,000 species of living arthropods have been described, and there are certainly many, many more. Some authorities estimate that there may be 10 million species of insects alone! There are more species of arthropods than of all other metazoans put together (Fig. 18-1). In fact, about 80 percent of all animal species belong to phylum Arthropoda, and of these 95 percent are insects. Yet insects are certainly not the only type of arthropod. Also included in this great phylum are such diverse groups as the crustaceans, millipedes, centipedes, spiders, sea spiders, horseshoe crabs, and the extinct trilobites.

A second criterion for measuring the success of arthropods is the sheer numbers of individuals. It is estimated, for example, that there are more copepods in the world ocean than there are individuals of all other metazoans combined.

A third criterion is the great number of habitats that have been successfully invaded by large numbers of their species. Arthropods live almost everywhere: marine, freshwater, and terrestrial habitats; at all depths, altitudes, and latitudes; and both as free-living animals and as parasites.

Origin of the Arthropods

Few contemporary biologists would question that the arthropods are closely related to the annelids.

But many consider the arthropods to be a **polyphyletic** group, one that possibly represents several independent lines of evolution from ancestral annelidan stock.

Onychophorans are a little-known but interesting group of terrestrial protostomes that have many features that suggest evolutionary ties with both the annelids and the arthropods. Onychophorans probably diverged very early from the annelid–arthropod stem; some onychophorelike fossils are known from the **Cambrian period** (600 to 500 million years ago). Modern onychophorans have retained many primitive features that give us an insight into the workings of evolution at that early age. Two other phyla, **Tardigrada** and **Pentastomida**, also show affinities to the arthropod and annelid protostomes; as with onycophora, they are probably surviving offshoots of the adaptive radiations dominated by the annelids and arthropods.

Phylum Onychophora

There are 65 described species of onychophorans, and all are confined to restricted localities in the tropics and south temperate latitudes. *Peripatus* (Fig. 18-2) is a typical genus. Onychophorans are very susceptible to dehydration and are active only at night. They seek humid microhabitats, such as beneath logs and ground litter in tropical rain forests.

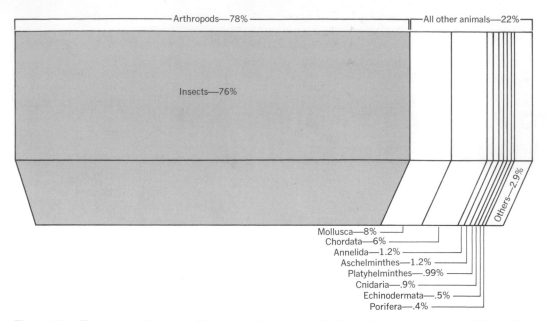

Figure 18-1. There are well over a million known living species in the entire animal kingdom. Of these, the vast majority are insects.

Figure 18-2. *Peripatus sp.,* an onychophoran. Top: drawing to show external structure Bottom: Photo of living animal. It is a walking wormlike animal that is neither an annelid nor a typical arthropod. Because it has both annelid and arthropod characteristics, it is the only living animal that comes near to being a common relative to annelids and arthropods. Therefore it is considered by some as a connecting link between the two phyla. [Photo reproduced by permission from L. J. and M. J. Milne, *The Biotic World and Man,* p. 48. Copyright 1952, by Prentice-Hall, Inc., Englewood Cliffs, N.J.]

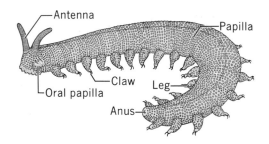

External Morphology

The external anatomy of onychophorans is suggestive of both polychaetes and centipedes. The roughly conical body ranges from about 2 to 15 cm in length. The head region bears two large **antennae** and a ventral **mouth.** A short **oral papilla** and a clawlike **mandible** flank the mouth on either side (Fig. 18-3). The antennae, papillae, and mandibles represent the appendages of three **cephalic somites** that became fused during development, as in arthropods.

Fourteen to 43 pairs of **legs** also attest to **metameric** organization. Each leg bears two terminal **claws;** hence the name onychophora ("claw bearer").

The onychophoran body wall is distinctive: thin and flexible as in annelids, yet covered with a **chitinous cuticle** as in arthropods. The cuticle is only 1μ thick and is thus extremely flexible and permeable. It is not composed of external plates but is raised into minute tubercles that cover the body. Lack of a solid exoskeleton allows the body to be distorted to an amazing degree; onychophorans can squeeze through openings as small as one ninth of their resting diameter. Crevices inside rotting logs are favorite retreats; there they are safe from desiccation and predation.

Their thin cuticle makes onychophorans suscepti-ble to predation, and they meet this threat with a unique spitting apparatus. When confronted by a predator (e.g., a scorpion, spider, or centipede), the onychophoran squirts a sticky slime through a pore on the tip of each oral papilla. Body wall contractions squirt the fluid as a stream and the adhesive material quickly hardens. Distances of 0.5 m may be achieved, and large prey can be immobilized.

Except for the thin chitinous cuticle, the onychophoran body wall is similar to that of annelids. The **coelom** is greatly reduced however, and the principal body cavity is a **hemocoel.** Vestiges of the coelom persist around the gonads and nephridia, as in arthropods and mollusks.

Internal Morphology and Physiology

Onychophorans are generally carnivores, and their prey includes insects, snails, and worms. The straight **digestive tract** consists of a **mouth, pharynx** with **salivary glands, esophagus, intestine,** and terminal **anus.**

The **circulatory system** is similar to that of arthropods. A dorsal **heart** pumps the colorless **blood** forward into a hemocoel that consists of interconnected **sinuses.** Lateral **ostia** in each segment communicate with the heart. The **nephridia** are metamerically arranged, with a **nephridiopore** opening at the base of each leg.

A bilobed **brain** overlies the pharynx. Two **ventral nerve cords** that are connected by **lateral commissures** extend the length of the body. Sense organs include a pair of tactile **antennae,** a **pigment-cup ocellus** at the base of each antenna, **chemoreceptors** around the mouth, **sensory bristles** on the body wall tubercles, and **hygroreceptors** located on the antennae and body surface that monitor humidity.

Onychophorans respire through a **tracheal system** that is somewhat similar to that of insects. Openings called **spiracles** are widely distributed over the body; each communicates with a series of **tracheal tubes** that permeate the body and carry air directly to the tissues.

The sexes are separate, and fertilization is internal. **Eggs** may be deposited in moist soil, or they may develop within the **uterus** of the mother. Embryonic development is quite similar to that of the terrestrial arthropods.

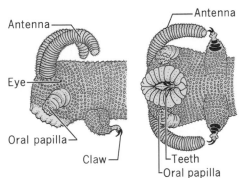

Figure 18-3. *Peripatus sp.* Left: Lateral view, the "head" bears two extensible antennae near the base of which are a pair of simple eyes. The numerous papillae that cover the whole body give it a velvety texture. Right: Ventral view, the legs are stubby and unlike those of typical arthropods, but the claws are arthropodlike. In each jaw are embedded two backward-pointing, clawlike teeth.

Phylum Tardigrada—The Water Bears

Phylum Tardigrada contains about 180 species of tiny, highly specialized animals. They have a worldwide distribution and are not uncommon, but they are rarely noticed due to their minute size and cryptic habits. Most tardigrades are less than 0.5 mm in length. The majority inhabit thin films of water that cover terrestrial mosses and lichens; others live interstitially in marine sediments. Most tardigrades feed on the juices contained within plant cells; others prey upon rotifers and nematodes.

Tardigrades have plump, **bilaterally symmetrical** bodies without clearly demarcated head or tail regions (Fig. 18-4). They crawl about awkwardly on four pairs of stubby **legs** that extend from the ventrolateral body wall; each leg terminates in four to eight **claws.** Their clumsy antics when viewed under a microscope no doubt suggested their common name, "water bears."

The body is covered by a cuticular exoskeleton that is periodically **molted** to accommodate growth. The body wall musculature is broken up into discrete strands with definite origins, insertions, and actions. The **coelom** of tardigrades is confined to the cavity of the single **gonad;** their main body cavity is a **pseudocoel** that acts as a **hemocoel.** Like many minute aquatic metazoans, tardigrades lack distinct circulatory and respiratory systems. Their **nervous system** has metamerically arranged **ganglia** along a double **ventral nerve cord** (Fig. 18-4).

Tardigrades are able to survive periods of desiccation by entering a state of suspended animation (cryptobiosis) that is characterized by loss of body water, shriveling of the body, and extremely low rates of metabolism. Tardigrades have emerged from periods of cryptobiosis lasting seven years. The onset and release from cryptobiosis are fascinating to observe with a microscope.

Tardigrades are **dioecious,** but **parthenogenesis** is common. The eggs are generally surrounded by thick **shells.** Development is direct, with no intermediate larval stages. Coelom formation is unique: five pairs of metameric coelomic pouches arise as outpocketings of the gut. The four anterior pairs then disintegrate, and the posterior pair fuses to form the single gonad. Such an **enterocoelous** method is a **deuterostome** characteristic, but the **metameric** arrangements of the coelomic pouches and nervous system suggest affinities to the annelid–arthropod line. The evolutionary origin of tardigrades is hence somewhat enigmatic.

Phylum Pentastomida— The Linguatulid Worms

One phylum of lesser protostomes remains. The pentastomids are a small group of specialized parasites that exhibit many arthropodlike characters (Fig. 18-5). The hosts of adult pentastomids are predominantly tropical reptiles (crocodiles and snakes), but hosts also include carnivorous mammals and birds.

Pentastomids have short wormlike bodies that range in length from 2 to 13 cm. Two pairs of short anterior **appendages** bear **claws** that attach the para-

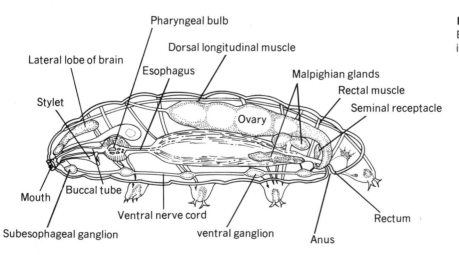

Figure 18-4. Phylum Tardigrada. External view (top photograph) and internal anatomy (bottom).

Pharyngeal bulb

Dorsal longitudinal muscle

Lateral lobe of brain

Esophagus

Malpighian glands

Rectal muscle

Stylet

Seminal receptacle

Ovary

Mouth

Buccal tube

Rectum

Subesophageal ganglion

Ventral nerve cord

ventral ganglion

Anus

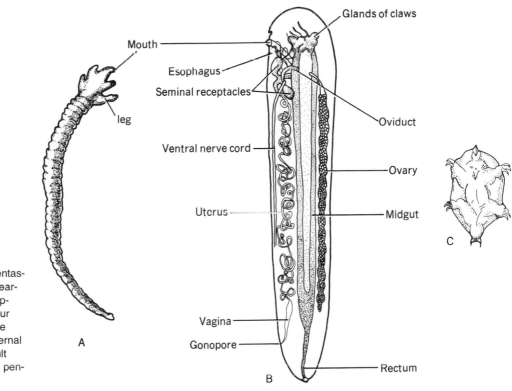

Mouth

Esophagus

Seminal receptacles

Ventral nerve cord

Uterus

Vagina

Gonopore

leg

Glands of claws

Oviduct

Ovary

Midgut

Rectum

A

B

C

Figure 18-5. Phylum Pentastomida. (A) External appearance, showing the five appendages (mouth and four clawed legs) that give the phylum its name. (B) Internal anatomy of a typical adult worm. (C) Structure of a pentastomid larva.

site to the lungs and nasal passages of its host. The parasites suck the blood of their host into a simple tubular **gut.** They lack circulatory, respiratory, and excretory organs. Their **nervous system** corresponds to the basic annelid–arthropod plan, with three pairs of **metameric ganglia** along the **ventral nerve cord.** Their **cuticle** is also similar to that of arthropods, being composed of **chitin** and requiring periodic **molting** during larval growth.

Pentastomids have complex life cycles that require an intermediate host (e.g., fishes when crocodiles are the definitive host, rabbits when the sexual adults infest dogs). Development of the larva (Fig. 18-5C) is initiated inside the intermediate host and completed if the definitive host eats the intermediate. Humans may become parasitized in this manner.

Characteristics of the Arthropods

With this introduction we can now consider arthropod characters in more detail. Ten main features set the arthropod body plan apart from those of other animals.

(1) One outstanding arthropod feature is **heteronomous metamerism;** their segments are specialized to perform specific functions. Arthropod segments and appendages exhibit a division of labor and a tendency to differentiate and specialize. Somites tend to fuse into groups that form more or less distinct body regions, or **tagmata.** Primitive arthropod groups show little **tagmosis** or heteronomy, but more specialized groups clearly exhibit a reduction in the number of body segments and their incorporation into distinct tagmata (e.g., the **thorax** and **abdomen** of insects). Associated with heteronomy is a tendency toward differentiation and specialization of the head region (**cephalization**).

Heteronomy is most clearly seen in the appendages. Annelid parapodia are generally all alike, but the appendages of an arthropod are usually specialized to perform distinct functions (e.g., sensory antennae, various mouthparts, walking or grasping appendages, and reproductive appendages).

243

(2) Arthropods are covered by a hardened **exoskeleton** composed of nitrogenous **polysaccharides** secreted by the epidermis. Around the joints the exoskeleton is soft and flexible, but over most of the body it is stiffened by proteins or lime salts.

(3) **Jointed appendages** are required where the exoskeleton is stiff, as no movement is possible in a jointless armor. The bodies of most arthropods are also jointed between segments or tagmata.

(4) A dermomuscular tube (as in annelids) would be useless inside a relatively inflexible exoskeleton. Arthropod **muscles** therefore tend to be broken up into discrete bundles or strands that perform specific movements. The exoskeleton provides an extensive and firm inner surface to which muscles can attach in a tremendous variety of ways.

(5) With the development of a hard exoskeleton and the breakup of an originally tubular musculature into discrete bundles, the arthropods had a body with great evolutionary potential for diverse specializations. One disadvantage of an exoskeleton, however, is that an animal encased in such an armor cannot grow continuously. The exoskeleton must be periodically shed or **molted,** and rapid growth occurs during the short interval before a new exoskeleton hardens. Growth of arthropods is thus stepwise and discontinuous.

(6) Another arthropod feature is the presence of **striated muscles.** In annelids striated muscles are generally lacking, but the main body musculature of arthropods is striated. Striated fibers allow for more rapid movements and responses than do smooth muscles.

(7) Arthropods lack cilia and flagella. As a consequence, all movements depend exclusively upon muscular contraction. Also, fertilization is generally achieved by copulation or by active transfer of sperm packets (**spermatophores**), as the gametes are typically nonmotile.

(8) The arthropod **coelom** is greatly reduced, and its remnants serve only excretory and reproductive functions. The stiff exoskeleton would prevent the coelom from functioning as a hydrostatic skeleton, as it does in annelids.

(9) The major body cavity of arthropods is an extensive **blood sinus,** or **hemocoel.** The **circulatory system** is an **open** system, with the dorsal heart and major vessels opening into the hemocoel.

(10) A tenth feature that sets the arthropods apart from their annelidan ancestors is their embryology. Arthropod eggs tend to be relatively yolky, and primitive spiral cleavage of other protostomes tends to be replaced by superficial cleavage patterns. Also, arthropod development is generally direct, without any simple larval stages like a trochophore. Arthropod **larvae,** when present in the life history, are complex little animals with at least three pairs of appendages.

Resumé of Arthropod Classification

Phylum Arthropoda is divided into three subphyla: Trilobitomorpha, Chelicerata, and Mandibulata.

The extinct **trilobites** comprise the subphylum **Trilobitomorpha** (Fig. 18-6). These primitive arthropods left abundant fossils in rocks strata that date from 600 to 225 million years ago.

Subphylum **Chelicerata** contains three classes. Class **Merostomata** includes the **horsehoe crabs;** these living fossils inhabit soft bottoms along the northwestern Atlantic and Caribbean coasts and also

Figure 18-6. Fossil trilobite. Such extinct arthropods lived in warm primeval seas from about 550 to 200 million years ago.

the Pacific Coast from Korea and Japan to the Philippines and East Indies. Class **Pycnogonida** contains the **sea spiders,** a curious group of minute yet cosmopolitan marine carnivores. The class **Arachnida** includes the familiar terrestrial **scorpions, spiders, harvestmen, ticks,** and **mites.**

Subphylum **Mandibulata** includes the vast majority of living arthropods. Class **Crustacea** comprises a diverse group of primarily marine and freshwater forms, although a few are terrestrial. **Crabs, lobsters,** and **crayfish** are some familiar crustaceans. Class **Insecta** is the largest and most successful group of animals, having adaptively radiated into every terrestrial and many shallow aquatic habitats. The mandibulates also include the **centipedes** (class **Chilopoda**), the **millipedes** (class **Diplopoda**), and two minor groups. The general appearances of five major classes are compared in Figure 18-7.

The next three chapters will survey these major groups of arthropods in some detail. Chapter 19 will

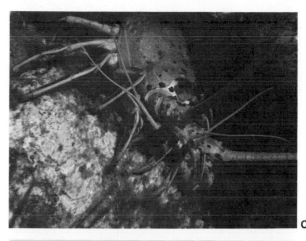

Figure 18-7. Representatives of the major classes of arthropods, showing body divisions and appendages. (A) Chilopoda (Centipede). (B) Diplopoda (Millipede). (C) Crustacea (Spiny lobster). (D) Arachnoidea (Spider). (E) Insecta (Fly).

cover the chelicerates, and the mandibulates will be discussed in Chapters 20 and 21.

Summary

Onychophorans display characteristics that are somewhat intermediate between the annelidan and arthropod body plans. Tardigrades and pentastomids also exhibit likenesses to the annelid–arthropod protostomes. The arthropods are the largest and most successful metazoan phylum. Distinctive characteristics include heteronomous metamerism, tagmosis, and a hardened exoskeleton with jointed appendages. A host of allied features enhance the potential adaptability of the arthropod body plan.

Review Questions

1. Discuss three criteria by which the arthropods are considered evolutionarily successful.
2. Why does *Peripatus* need hygroreceptors?
3. How does the exoskeleton affect the basic body plan of arthropods?
4. Summarize the annelid and arthropod characteristics exhibited by Onycophora, Tardigrada, and Pentastamida.
5. Outline the major arthropod taxa.

Selected References

Clarke, K. U. *The Biology of the Arthropoda*. New York: American Elsevier Publishing Co., Inc., 1973.

Dakin, W. J. "The Anatomy of West Australian *Peripatoides*," *Pro Zool. Soc.*, London, 1920.

Dakin, W. H. "The Eye of *Peripatus*," *Quart*.

Johnson, W. H., L. E. Dolanney, E. C. Williams, and T. A. Cole. *Principles of Zoology*, 2nd ed. New York: Holt, Rinehart and Winston, 1977.

Sedgwick, A. "A Monograph of the Development of *Peripatus Capensis*," *Quart. Jour. Mic. Sci.*, 1885–1888.

Snodgrass, R. E. "Evolution of the Annelida, Onychophora, and Arthropoda," Washington: The Smithsonian Institution, 1938.

Snodgrass, R. E. *A Textbook of Arthropod Anatomy*. Ithaca: Comstock Publishing Associates, 1952.

Southward, A. J. "Feeding of Barnacles," *Nature*, **175**:1124, 1955.

Tiegs, O. W., and S. M. Manton. "The Evolution of the Arthropoda," *Biol. Rev.*, **33**:255, 1938.

Phylum Arthropoda, Subphylum Chelicerata

Subphyla Chelicerata and Mandibulata represent the two evolutionary lines that have produced the modern arthropods. Subphylum chelicerata includes three classes (Fig. 19-1): **Merostomata** (horseshoe crabs), **Arachnida** (spiders, scorpions, etc.), and **Pycnogonida** (sea spiders). These three classes all share the following constellation of characters.

The body of chelicerates is divided into two regions: an anterior **cephalothorax** and a posterior **abdomen.** Chelicerates lack antennae. Their most anterior appendages are feeding structures called **chelicerae,** which are anterior to the mouth. The first pair of postoral appendages, called **pedipalps,** are variously modified for specific functions in the different classes.

Class Merostomata—The Horsehoe Crabs

Merostomates are the closest living relatives of the extinct trilobites (subphylum Trilobitomorpha), and the genus *Limulus* is at least 200 million years old. Only three genera and four species represent this class today. The **horseshoe,** or **king, crab,** *Limulus polyphemus* Fig. 19-2), lives along sandy Atlantic shores from Maine to the Gulf of Mexico.

The horseshoe crab's body is divided into an anterior **cephalothorax** and a posterior **abdomen.** The cephalothorax has a hard dorsal **carapace** formed by the fusion of eight embryonic segments, the posterior seven of which bear **uniramous appendages** (i.e., having only one branch). The first pair are three-jointed **chelicerae** that end in a small **pincer.** A pair of **pedipalps** and four pairs of **walking legs** each have six joints and also terminate with pincers. All chelicerates lack jaws, but the walking legs of horseshoe crabs have grinding **gnathobases** that crush the sandworms and soft-shelled clams that compose their diet. The eighth cephalothorax segment bears a seventh pair of appendages of unknown function. The dorsal surface of the cephalothorax bears two anterior **ocelli** and two lateral **compound eyes.**

The fused cephalothorax is hinged to the abdomen, which is composed of six fused segments and a posterior spine or **telson.** Each abdominal segment bears a pair of **biramous appendages** (i.e., having two branches). The most anterior pair are fused to form a **genital operculum;** two **genital pores** are located on its underside. The last five pairs are called **book gills** because their broad, overlapping, outer branches serve as respiratory surfaces.

Class Merostomata also contains the extinct **eurypterids** (Fig. 19-1), known only from the fossil record. One such fossil species was almost 3 m in length! This is amazingly large for arthropods, whose periodically molted exoskeletons seem to limit their size. Eurypterids invaded fresh waters and may have become amphibious and even terrestrial. At some point before their extinction, they gave rise to the arachnids.

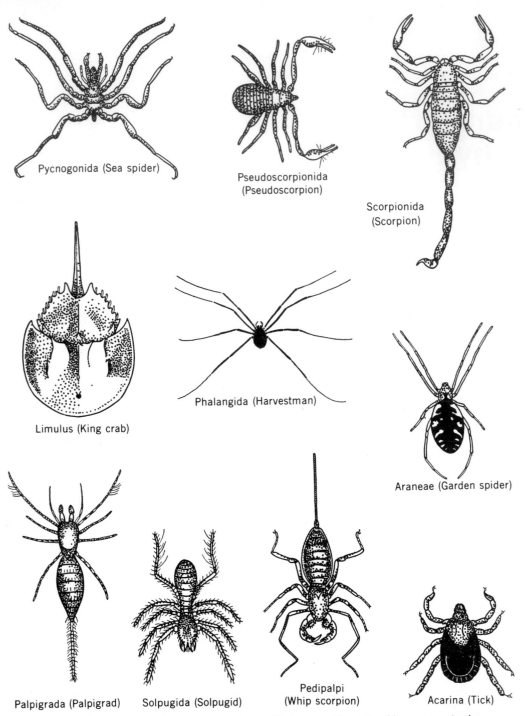

Figure 19-1. Some representatives of the subphylum Chelicerata. The sea spider represents class Pycnogonida; *Limulus* is of class Merostomata; and the remaining examples are included in class Arachnida.

Pycnogonida (Sea spider)

Pseudoscorpionida (Pseudoscorpion)

Scorpionida (Scorpion)

Limulus (King crab)

Phalangida (Harvestman)

Araneae (Garden spider)

Palpigrada (Palpigrad)

Solpugida (Solpugid)

Pedipalpi (Whip scorpion)

Acarina (Tick)

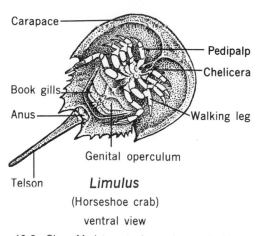

Figure 19-2. Class Meristomata, horseshoe crab, *Limulus polyphemus* (formerly *Xiphosura*), ventral view. Natural size about 37 cm long. It is called a living fossil because it has undergone little change during long geological periods.

Class Arachnida—The Spiders and Relatives

Class Arachnida is the largest and most successful group of modern chelicerates, comprising well over 60,000 described species. This class is divided into ten orders, of which the **scorpions (Scorpiones)**, **spiders (Araneae)**, **harvestmen** or "daddy longlegs" **(Opiliones)**, and **mites (Acarina)** are the most abundant and familiar. Scorpions are uhndoubtedly the most primitive arachnids, as they possess the complete series of segments and appendages that generally characterize the chelicerates. The remaining orders are characterized by a progressive reduction of **metamerism.** This involves a reduction in the number of abdominal segments and sometimes the fusion of the cephalothorax with the abdomen to form a single body region. Internal modifications include the fusion of nerve ganglia to form a concentrated **brain** region and a reduction in the number of **heart ostia** and **respiratory** structures.

The vast majority of arachnids are terrestrial, although a few have secondarily become aquatic. The conquest of land necessitated some major anatomical and physiological adaptations. Most notable was conversion of the merostomate book gills into air-breathing **book lungs** and **tracheae.** And appendages became modified for terrestrial locomotion. Two distinct evolutionary potentials were also exploited within the class: **silk production** (in spiders, pseudoscorpions, and some mites) and **poison glands** (in scorpions, spiders, and pseudoscorpions). The spiders will be considered in some detail in the following discussion.

Order Araneae—Spiders

External Morphology

The spider body (Fig. 19-3) consists of a **cephalothorax** and an **unsegmented abdomen.**

The cephalothorax bears six pairs of appendages: one pair of **chelicerae**, one pair of **pedipalps**, and four pairs of **walking legs.** The chelicerae have a terminal **claw** or **fang** through which fluid from **poison glands** is injected into prey. The poison may be **neurotoxic** (poisonous to nerve tissue) or **hemolytic** (causing tissue decomposition of red blood cells). The pedipalps have grinding surfaces at their bases that are used to crush the prey into small pieces. The **mouth** is a minute opening between the bases of the pedipalps; it serves for the ingestion of liquids only. The pedipalps of mature males are modified into **copulatory organs** that transfer sperm packets to the female during mating.

The eight walking legs each consists of seven parts: (1) **coxa**, (2) **trochanter**, (3) **femur**, (4) **patella**, (5) **tibia**, (6) **metatarsus**, and (7) **tarsus.** The distal tarsus bears two or three claws and often a pad of hairs, the **claw tuft,** that enables the spider to adhere to vertical and upside-down surfaces.

The abdomen bears no appendages. It is reduced to four segments; the anterior two house the **book lungs,** or **tracheal bundles,** and the posterior two form the **spinnerets.** The somites of both the cephalothorax and the abdomen are fused, and these two body regions are joined by a narrow waist, the **peduncle.** Near the anterior end of the abdomen, on the ventral surface, is the **genital opening,** which in some female spiders is covered by a flat plate, called the **epigynum.** On either side of the epigynum is a slitlike opening of the respiratory organs or book lungs. Some spiders also possess **tracheae** that open through **spiracles** near the posterior end of the ventral abdomen (Fig. 19-3). Just posterior to the spiracles are three pairs of **spinnerets** (Figs. 19-3 and 19-4), which are

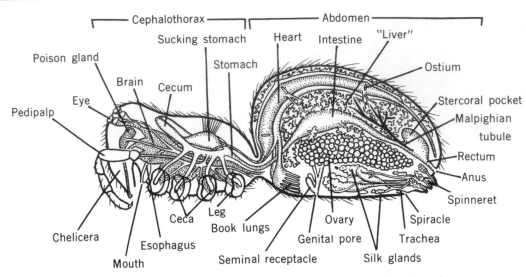

Figure 19-3. Internal structure of a spider as seen with the left side of the body removed.

Internal Morphology and Physiology

Feeding and Digestion

Spiders feed mainly on insects, although larger species may capture small vertebrates. Because only liquid food can be ingested, the tissues of the prey are first liquified by powerful **digestive enzymes** that are regurgitated into or over the victim. Spiders ingest food in two ways: (1) Those with weak chelicerae puncture the body of the prey with their fangs and then alternately inject digestive fluid through this hole and suck out the liquified tissues until only the exoskeleton remains; (2) those with strong chelicerae crush the insect into small pieces between their chelicerae and pedipalps as salivary digestive enzymes are regurgitated. The liquified tissues are sucked up by contractions of a posterior enlargement of the esophagus called the pumping stomach.

The **digestive system** consists of a **mouth** and **pharynx,** followed by a horizontal **esophagus** that leads into the **pumping stomach,** which in turn opens into the **true stomach.** Five pairs of blind pouches

called **ceca** branch from the stomach in the cephalo-thorax. The **intestine** passes almost straight through the abdomen but is enlarged midway where ducts enter from the **digestive gland** that secretes a digestive fluid and also absorbs nutrients. Near its posterior end, another enlargement, called the **cloacal chamber,** receives wastes. A short **rectum** then terminates in the **anus.**

Respiratory System

Respiration is carried on by **book lungs** or **tracheae,** or both. There may be one or two pairs of book lungs. Each **book lung** consists of an air-filled chamber whose dorsal surface is elaborated into leaflike lamellae that are filled with blood. Oxygen and carbon dioxide are exchanged by diffusion across the extensive surface area of the lamellae. The book lung communicates to the outside through the **spiracle,** a slit in the body wall. Muscular contractions ventilate the book lung by expanding or contracting its volume. In most spiders the posterior pair of book lungs have evolved into **tracheae.** These are systems of tubules that open through a spiracle and then ramify throughout the body tissues. Some small spiders have completely replaced book lungs with tracheal tubules.

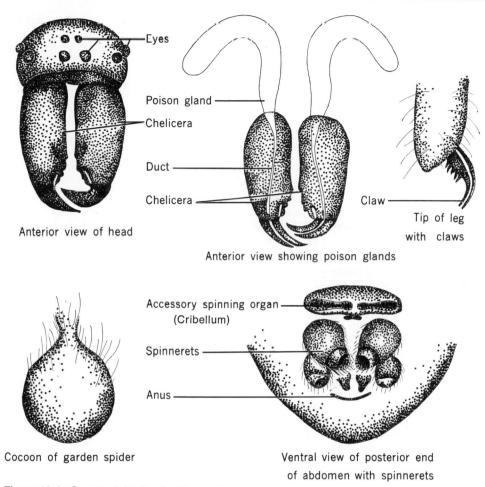

Eyes

Poison gland

Chelicera

Duct

Chelicera

Claw

Anterior view of head

Anterior view showing poison glands

Tip of leg
with claws

Accessory spinning organ
(Cribellum)

Spinnerets

Anus

Cocoon of garden spider

Ventral view of posterior end
of abdomen with spinnerets

Figure 19-4. Structural details of spiders and a cocoon.

Circulatory System

The circulatory system consists of a heart arteries, veins, and a number of open sinuses. The **heart** is situated in the abdomen dorsal to the digestive tract. It is a muscular contractile tube within a large blood sinus called the hemocoel. Two to five pairs of openings, called **ostia,** allow blood to flow from the pericardium into the heart. The heart pumps blood into an **anterior aorta,** which branches and supplies the cephalothorax, and into a **posterior aorta,** which irrigates the abdominal tissues. The **blood,** which is

colorless and contains mostly **amoeboid corpuscles,** passes from the **arteries** into **sinuses** among the tissues and is carried to the book lungs where it is oxygenated. The blood is then returned through **veins** to the pericardial sinus, completing its circuit.

Excretory System

The excretory organs are paired **Malpighian tubules** that open into the cloacal chamber and also one or two pairs of **coxal glands** in the floor of the cephalothorax. Both organs function by absorbing, and concentrating dissolved wastes for excretion; coxal glands absorb directly from the blood, whereas Malpighian tubules absorb wastes from cells in the posterior region of the intestine. The coxal glands are sometimes degenerate, and their openings are difficult to find; these glands are homologous with the green glands of the crayfish (see Fig. 20-4).

Nervous System and Sense Organs

The nervous system consists of a bilobed **brain** of fused **ganglia** above the esophagus, a large **subesophageal ganglion,** and **nerves** that extend to various organs. **Sensory hairs** that cover the body and appendages are important sense organs. There are usually eight **simple eyes** (Fig. 19-4) arranged in two rows along the anterior dorsal margin of the carapace. The eyes of only a few families of hunting spiders are capable of forming distinct images; in web builders and others, the eyes function primarily for perception of degrees of illumination and detection of moving objects. The sense of **smell** is well developed, and an organ of **taste** is located in the pharynx.

Reproduction

The sexes are separate, and the **testes** or **ovaries** form a network of tubes in the abdomen. During mating, the **sperm** are ejaculated onto a minute sperm web, the **spermatophore,** and are then picked up by the male's specialized pedipalps and transferred to the **seminal receptacles** of the female. Complex courtship behaviors often precede mating. Sometimes the female kills and eats the male after mating. The sperm move from the seminal receptacles to fertilize the **eggs** as they leave the female's body. The eggs are laid in a **silk cocoon,** which is attached to the web or other substratum, or are dragged behind the female. The young spiderlings remain in the cocoon until after their first **molt.** Up to 15 molts occur before sexual maturity.

Silk Production

The silk spinning organs of spiders are three pairs of modified appendages called **spinnerets** (Figs. 19-3 and 19-4). These are pierced by hundreds of microscopic tubes through which a fluid secreted by ab-

dominal **silk glands** is extruded. Spider silk is formed from a **scleroprotein** that hardens as it is extruded; it is elastic and stronger than a steel thread of the same diameter. Some spiders play out a safety line, which acts much like a mountaineer's safety rope. Spiders also use silk to build snares, spin webs, and construct cocoons. Spiderlings disperse themselves by spinning a long thread on which they are carried by the wind.

Many spiders possess an accessory spinning organ, the **cribellum** (Fig. 19-4), in addition to the spinnerets. A sticky kind of silk used to snare prey is emitted from this organ.

Not all spiders build webs, but those that do construct some of the most remarkable edifices seen in the natural world. An **orb web** (Fig. 19-5) is spun in the following manner: A thread is stretched across the span selected for the web; then, from a point on this thread near its center, other threads are drawn out and attached in radiating lines. These threads all become dry and smooth. On this foundation a spiral of sticky silk is spun. The spider stands in the center of the web or retires to a nest at one side and waits for an insect to become entangled in the sticky thread. It

Figure 19-5. The web of an orb-weaving spider.

then rushes out and spins thread about its prey until the struggling ceases. **Poison glands** are often used to kill prey.

Male spiders do not spin webs, but wander about capturing insects or lie in wait for them in some place of concealment.

Some Spiders of Special Interest

More than 30,000 species of spiders are known, and about 3000 species live in the United States. Only the **black widow, brown recluse,** and **tarantula** (Fig. 19-6) are potentially harmful to humans residing in North America.

Crab spiders (Fig. 19-7) are so named because of their habit of walking sideways. Some are white or yellow and seem to favor flowers of similar color, waiting for an unwary insect to land. **Jumping spiders** (Fig. 19-7) are noted for their elaborate mating displays. **Tarantulas** (Fig. 19-6) are the giants of the spi-

der world, reaching a body length of 6 cm and a leg span of 20–25 cm. Spiders of this size are able to capture mice, small lizards, and small birds. Tarantulas have been maintained in captivity for over 20 years, but most other spiders in the wild probably live less than one year. The **trap-door spider** digs a burrow in the ground about 9 cm deep and closes it with a hinged door. **Wolf spiders** are large spiders that actively hunt for prey. They care for their young by transporting them on their backs.

Other Arachnids
Scorpions

Scorpions (Fig. 19-1) are the most primitive arachnids, and their body plan most closely resembles that of their merostomate ancestors. The scorpion cephalothorax is composed of eight fused segments that are covered by a dorsal carapace. The **cephalothoracic**

A

B

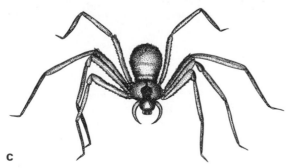

C

Figure 19-6. Several potentially harmful arachnides. (A) Photograph of black widow spider *Latrodectus* sp on web strands. The female may kill and eat the male after mating. (B) Photograph of a tarantula. Tarantula is the largest of the spiders. The tarantula does not spin a web, but stalks its prey like a lion. Although feared by many people, experiments on species in the United States have shown that their "bites" are no more harmful than a bee sting. However, it is true that some South American tarantulas are more injurious. (C) Drawing of a brown recluse, a poisonous spider.

Jumping spider
(Salticus)

Crab spider
(Theridion)

House spider
(Misumena)

Figure 19-7. Some common spiders.

appendages are a pair of short **chelicerae,** a pair of large, clawed **pedipalps,** and four pairs of **walking legs.** The pedipalps and first two pairs of walking legs have grinding structures at their bases that are used to crush prey. The **abdomen** consists of 12 unfused segments and a terminal **telson.** The first abdominal segment bears **gonopores,** and the next four segments have ventrolateral **spiracles** that open into the **book lungs.** The food of scorpions consists mainly of insects and spiders that are grasped by the pedipalps and injected with a paralyzing poison by a sting on the telson.

Scorpions measure 1–20 cm in length. They live in tropical and subtropical regions, hiding during the daytime but hunting actively at night. Only two spe-

cies in the United States are considered dangerous to humans. Both are species of *Centruroides,* small, straw-colored desert scorpions found in Arizona and New Mexico.

Mites and Ticks

In mites and ticks (Fig. 19-8), all the segments and body divisions are fused together. Mites are generally smaller than ticks. Both groups are notorious as parasites, although many mites are free living. In fact, mites are among the most cosmopolitan of animals, being widely distributed from polar regions to desert hot springs.

The free-living species have clawed **chelicerae,**

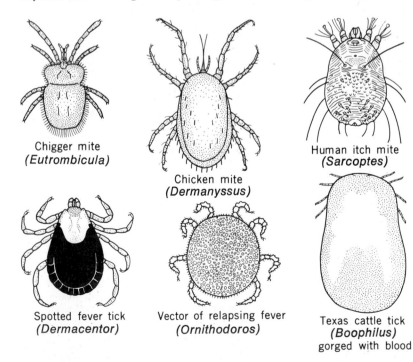

Chigger mite
(Eutrombicula)

Chicken mite
(Dermanyssus)

Human itch mite
(Sarcoptes)

Spotted fever tick
(Dermacentor)

Vector of relapsing fever
(Ornithodoros)

Texas cattle tick
(Boophilus)
gorged with blood

Figure 19-8. Some parasitic ticks and mites.

tactile **pedipalps,** and four pairs of **legs** spaced widely apart. The chelicerae of parasitic species are usually modified for cutting and sucking, and their pedipalps are often hook shaped for attaching to the host. These parasites do considerable damage to crops and spread several diseases, including **Rocky Mountain spotted fever,** among vertebrate hosts. Like all arthropod parasites, they must temporarily detach from their host during the molting process.

Class Pycnogonida—The Sea Spiders

These peculiar chelicerates are commonly called **sea spiders** because of their marine habitats and spiderlike forms. Pycnogonids are usually minute and their bodies are "all legs," on which they move quite slowly (Fig. 19-1). For these reasons they are often overlooked by the casual naturalist. Sea spiders are actually quite common marine animals, however. Some 500 species have been recognized, and they are found in all oceans at all depths. Most crawl slowly over the bottom, where they feed on corals, hydrozoans, and bryozoans.

The pycnogonid body is composed essentially of a **head region** with its **appendages,** as the **abdomen** is reduced to a tiny posterior protrusion. The anterior **chelicerae** bear **pincers** that are used to tear open their minute prey. A **proboscis** is then inserted through the gash, and tissue juices are pumped into the **pharynx.** The remaining appendages include a pair of sensory **pedipalps,** a pair of small, egg-bearing (**ovigerous**) legs that have ten joints and are used by males to carry the developing **eggs,** and four pairs of **walking legs.** The walking legs have eight joints and are quite long and thin. Respiratory gas exchange occurs across the extensive body surface.

Relations of The Chelicerates to Humans

Most arachnids are not harmful to humans. Many feed on insect pests and so are indirectly beneficial. A few species damage food plants; others parasitize humans and domesticated animals; and a few transmit disease.

The **red spider,** *Tetranychus,* is destructive to many species of plants and may become a pest in greenhouses and cotton fields. The **clover mite** injures clover and fruit trees, and the **pear-leaf blister mite** damages pears and apples.

Scorpions are noted for their poisonous sting. Statistics from the Arizona State Department of Health show that during a 20-year period there were 64 deaths from scorpion stings. The state of Durango, Mexico reports more than 1700 deaths from scorpions over a 41-year period. Scorpions are mostly confined to the tropics, and some are not very poisonous.

The black widow and the brown recluse (Fig. 19-6) are the only spiders in the United States whose bites need be feared. Two species of **black widow** spider occur in the United States: *Latrodectus mactans* in the south and *L. variolus* in the north. The poisonous adult female can be recognized by its glossy black color, globose abdomen, and a distinctive reddish spot on the undersurface of its abdomen, usually shaped like an hourglass but sometimes like one or two triangles. The **brown recluse,** *Loxosceles reclusa,* can be recognized by a violin-shaped mark on its cephalothorax. It is 1 cm or less in length and varies in color from light to dark chocolate brown. Its range includes most of the central United States; recent reports include occurrences on the east and west coasts of the United States.

The **chigger mites,** or **red bugs** (Fig. 19-8), produce a very distressing itch that may continue for several days, and some transmit microscopic pathogens. One of the itch mites, *Sarcoptes* (Fig. 19-8), causes **mange** in dogs, cats, humans (**scabies**), and other mammals. Domesticated animals are infested by several species of mites and ticks. The **chicken mite** (Fig. 19-8) is a serious pest of poultry. **Scab mites** attack horses and dogs among others. The **sheep scab mite** may seriously injure sheep. It causes intense irritation, loss of wool, and decreased vitality.

Several ticks are important vectors of disease. **Rocky Mountain spotted fever** is associated with *Dermacentor,* which transmits rickettsial pathogens from rodents and larger mammals to humans; the female ticks pass the disease-producing microorganisms to their offspring through the egg. An eastern variety of spotted fever occurs in rural districts from New York to Florida; cases have also occurred in the Middle West. A vaccine has been developed that confers immunity for about one year. The **Rocky Mountain wood tick** is also responsible for **tick paralysis** (apparently due to a poisonous salivary secretion) that may prove fatal to children.

Another disease known to be transmitted by ticks as well as by certain insects is **rabbit fever,** or **tularemia.** Relapsing fever, characterized by alternating periods of elevated and normal temperatures, occurs in various parts of the world, including, in the United States, Texas, Kansas, Montana, Utah, and California. A different species of tick of the genus *Ornithodoros* apparently transmits the disease microorganisms (spirochetes) in each locality. **Texas fever** (now eradicated from the United States) in cattle is transmitted from diseased animals to healthy ones by the **cattle tick** *(Boöphilus)* (Fig. 19-8); females pass the pathogenic sporozoan to their offspring through the egg.

Summary

Subphylum Chelicerata includes the marine merostomates (horseshoe crabs) and pycnogonids (sea spiders) and the terrestrial arachnids (scorpions, spiders, mites and ticks, and others).

Chelicerates differ from other arthropods in having no antennae or true mandibles. The body usually has two divisions, the cephalothorax and the abdomen. The cephalothorax bears six pairs of appendages: chelicerae, pedipalps, and four pairs of walking legs.

Review Questions

1. What characteristics do all chelicerate arthropods share in common?
2. How do spiders feed?
3. Construct a chart outlining how the cephalothoracic appendages have become modified for different functions in the various chelicerate groups.

Selected References

Baker, E. W., and G. W. Wharton. *Introduction to Acarology.* New York: Macmillan Publishing Co., Inc., 1952.

Comstock, J. H., and W. J. Gertsch. *The Spider Book.* New York: Doubleday, 1940.

Gertsch, W. J. *American Spiders.* New York: D. Van Nostrand Company, Inc., 1949.

James, M. T., and R. F. Harwood. *Herm's Medical Entomology,* 6th ed. New York: Macmillan Publishing Co., Inc., 1969.

Kaston, B. J., and E. Kaston. *How to Know the Spiders.* Dubuque: W. C. Brown Company, 1953.

King, P. E. *Pycnogonids.* London: Hutchinson Publishing Group, 1973.

Price, P. W. (ed.). *Evolutionary Strategies of Parasitic Insects and Mites.* New York: Plenium Publishing Corporation, 1975.

Savory, T. H. *The Biology of Spiders.* London: Sedgwick & Jackson, 1928.

Savory, T. H. *Introduction to Arachnology.* London: Muller, 1974.

Phylum Arthropoda, Subphylum Mandibulata, Class Crustacea

Subphylum Mandibulata

Despite the relative success of the arachnids, the subphylum Mandibulata contains the vast majority of the arthropods. In contrast to the chelicerates, the mandibulates possess at least one pair of preoral **antennae.** They are also characterized by a pair of appendages called **mandibles** that flank the mouth. Two pairs of accessory mouthparts, called the **maxillae,** are also typically present.

Subphylum Mandibulata contains four great arthropod groups: the predominantly aquatic **crustaceans** and the typically terrestrial **insects, centipedes,** and **millipedes.** The crustaceans are the subject of this chapter; the primarily terrestrial mandibulates are the subject of Chapter 21.

Class Crustacea

The crustaceans are distinguished from the terrestrial mandibulates by the possession of two pairs of preoral antennae, whereas the terrestrial classes have only a single pair. Another major difference is that the appendages of crustaceans are **biramous** (i.e., composed of two branches), whereas those of insects, centipedes, and millipedes are **uniramous.** Also, the second maxillae of crustaceans are not fused together to form a lower lip or labium (as they are in insects). Finally, the crustaceans are the only large arthropod group living today that is primarily aquatic. Over 26,000 species of crustaceans have been described, and the class is commonly divided into eight subclasses.

The crustacean body may be divided into two regions, the **head** and the **trunk.** The head varies little within the class and typically bears five pairs of appendages: a pair of small preoral first antennae (**antennules**), a second pair of longer **antennae,** a pair of **mandibles** that flank the **mouth,** and postoral **first maxillae** and **second maxillae.**

The morphology of the trunk and its appendages is quite variable and reflects the adaptive trends that have shaped the different crustacean groups. In very primitive crustaceans (e.g., fairy shrimp), the trunk is composed of many distinct and separate segments, each of which bears similar appendages. In most crustaceans, however, **tagmatization** has generally produced two differentiated trunk regions: an anterior **thorax** and a posterior **abdomen.** The thorax is generally covered by a shieldlike dorsal **carapace.** A terminal extension (**telson**) of the last abdominal appendage bears the **anus.**

The number of trunk segments and the types of trunk appendages vary widely from group to group. General tendencies toward reduction of segments and specialization of appendages has culminated in the body plans of advanced crustaceans such as shrimp, crabs, lobsters, and crayfish.

The Crayfish—A Specialized Crustacean

The crayfish is a familiar crustacean that inhabits freshwater streams and ponds throughout the world. The genus *Cambarus* is common in our central and eastern states, and *Pacifastacus* lives in the western United States. The anatomy of the Atlantic lobster, *Homarus americanus,* differs from the crayfish only in minor details.

External Morphology

The body is covered by a hard **cuticle** containing **chitin** and impregnated with lime salts. This **exoskeleton** (Fig. 20-1) is thinner and flexible at the joints, al-

lowing movement. A typical body segment is surrounded by a convex dorsal plate (**tergum**), a ventral transverse plate (**sternum**), and plates (**pleura**) projecting down the sides (Fig. 20-1).

The body consists of two distinct regions: a fused anterior cephalothorax and a flexible posterior abdomen. The **cephalothorax** consists of segments 1–12, which are enclosed dorsally and laterally by a shieldlike carapace. A beaklike **rostrum** projects anteriorly between the eyes. The **mouth** is situated ventrally near the posterior end of the head region; it is partly obscured by the feeding appendages. A Y-shaped groove separates the carapace into three parts: a middorsal strip, the **areola,** and two large convex flaps, the **branchiostegites,** that protect the gills on either side.

The **abdomen** contains six unfused segments and a terminal extension, the **telson,** that bears the **anal opening** on its ventral surface.

Every segment bears a pair of jointed appendages. They are all variations of a prototype consisting of a basal stem, the **protopodite,** that bears two branches;

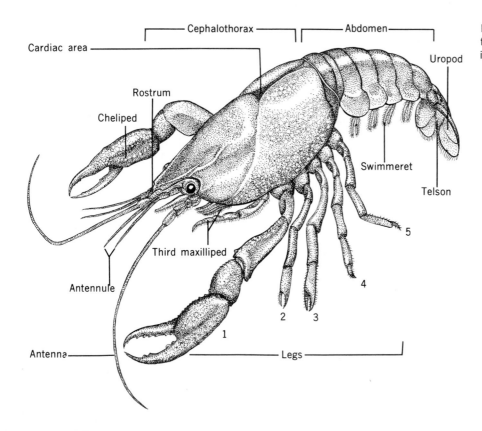
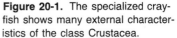

Figure 20-1. The specialized crayfish shows many external characteristics of the class Crustacea.

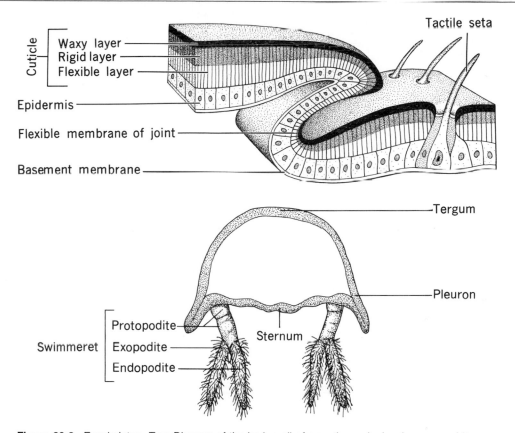

Figure 20-2. Exoskeleton. Top: Diagram of the body wall of an arthropod, showing some of its modifications. The rigid layer is replaced by a flexible membrane in places where movement occurs. Bottom: Diagram of a cross-section of the third abdominal segment of the crayfish.

an inner **endopodite;** and an outer **exopodite** (Fig. 20-2).

Beginning at the anterior end, the appendages are arranged as follows (Fig. 20-1): the sensory **antennules** and **antennae** precede a pair of **mandibles,** behind which are the first and second **maxillae.** The thoracic region bears the first, second, and third **maxillipeds,** the **chelipeds** (**pincers**), and four other pairs of **walking legs.** The abdomen bears five pairs of ventral **swimmerets** plus a pair of highly flattened appendages, called **uropods,** on the sixth abdominal segment. The different types of appendages are adapted to perform specialized functions.

Three general kinds of appendages can be distinguished in the adult crayfish: (1) **foliaceous,** the second maxilla; (2) **biramous,** the swimmerets; and (3) **uniramous,** the walking legs. All have probably been derived from a simple biramous prototype as seen in Figure 20-3, the modifications reveal the complementarity of structure and function. The uniramous walking legs pass through a biramous stage during their embryological development, and biramous embryonic maxillipeds are made foliaceous by expansion of their basal segments.

Structures that share similar fundamental organization and embryonic derivation, regardless of function, are said to be **homologous.** When homologous structures are repeated in a series, the condition is known as **serial homology.** The highly specialized appendages of the crayfish are serially homologous but have

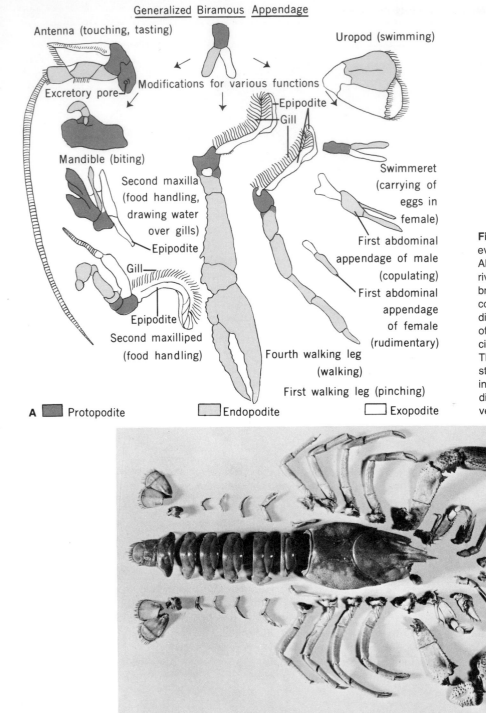

Generalized Biramous Appendage

Antenna (touching, tasting)

Excretory pore

Mandible (biting)

Modifications for various functions

Second maxilla (food handling, drawing water over gills)

Epipodite

Gill

Epipodite

Second maxilliped (food handling)

Uropod (swimming)

Epipodite

Gill

Swimmeret (carrying of eggs in female)

First abdominal appendage of male (copulating)

First abdominal appendage of female (rudimentary)

Fourth walking leg (walking)

First walking leg (pinching)

A ▨ Protopodite ▢ Endopodite ☐ Exopodite

Figure 20-3. (A) Homology and evolution of crayfish appendages. All are believed to have been derived from a generalized two-branched (biramous) appendage consisting of protopodite, endopodite, and exopodite. This basic plan of structure has been modified (specialized) for the various uses noted. The appendages demonstrate in a striking way the changes that occur in the evolution of structures. (B) A disarticulated crayfish skeleton, revealing the appendage.

B

become specialized for different functions (as discussed in subsequent paragraphs).

Internal Morphology and Physiology

The crayfish (Fig. 20-4) contains all the major organ systems that characterize higher animals in general. The **coelom,** as in other arthropods, is restricted to cavities enclosing the gonads and excretory organs. Certain internal organs such as the nervous system retain their primitive metameric arrangement; others such as the excretory organs have been consolidated.

Nutrition

The crayfish's diet consists principally of living animals such as snails, tadpoles, insect larvae, small fishes, frogs, and other crayfish, but decaying organic matter is also scavenged. They feed at night and are most active at dusk and daybreak. Their method of feeding may be observed in the laboratory if a little fresh meat is offered to them. The **maxillipeds** and **maxillae** manipulate the food as it is torn and crushed into small pieces by the **mandibles.**

Digestive System

The digestive tract of crustaceans can usually be divided into three regions: a foregut, midgut, and hindgut. The foregut and hindgut represent invaginations of ectoderm, and their hardened linings are shed at each molt. Only the midgut is lined with endodermal tissues that are secretory and absorptive.

The **foregut** of *Cambarus* consists of the **mouth,** a short **esophagus,** and most of the **stomach,** which is divided into two chambers. An anterior **cardiac chamber** contains chitinous teeth (**gastric mill**) that grind the food into minute particles that enter the smaller **pyloric chamber.** Here fine combs and bristles strain the liquified food so that only fine particles enter the **midgut,** which comprises the posterior portion of the pyloric stomach.

A duct from the paired **digestive glands (hepatopancreas)** enters each side of the midgut. Digestive enzymes are squeezed through these ducts into the midgut, where preliminary extracellular digestion occurs. The resulting nutrient broth is then drawn into the digestive glands for absorption.

The **hindgut,** or **intestine,** which is lined with cuticle, serves only to eliminate undigested wastes. The tubular intestine passes through the abdomen and opens at the **anus** on the ventral side of the telson.

Circulatory System

The **blood** transports nutrients throughout the body, oxygen from the gills to the body tissues, carbon dioxide from the tissues to the gills, and metabolic waste products to the excretory organs.

The blood plasma is an almost colorless liquid that contains a dissolved respiratory pigment, **hemocyanin.** Various types of **amoebocytes** are suspended in the plasma. Some are involved in **phagocytosis,** whereas others function in **blood clotting.**

The **heart** is a muscular saddle-shaped sac lying within the **pericardial sinus** in the middorsal thorax. It may be considered a dilation of the dorsal vessel, resembling that of the earthworm. Blood enters the heart from the pericardial sinus through three pairs of **valves** called ostia—two dorsal, two lateral, and two ventral.

The circulation of blood through the crustacean body involves an anterior flow (1) from the dorsal heart into the **aorta,** then (2) through major **arteries** into the **blood sinuses** or **hemocoel,** where the tissues are bathed, then (3) through **afferent channels** to the **gills,** where oxygen and carbon dioxide are exchanged, and finally (4) through **efferent gill channels** to the pericardium. Rhythmical contractions of the heart propel the blood through the arteries. Movements of the body and appendages help to push the blood through the open blood sinuses. One-way valves in the heart and arteries regulate the direction of blood flow.

Respiratory System

Respiratory gas exchange in crustaceans occurs across the body surface in regions where the exoskeleton is thin. The endopodites of the crayfish's thoracic appendages are elaborated by numerous vascularized filaments into plumelike **gills.** The **branchial chambers** containing the gills lie between the lateral plates of the carapace (branchiostegites) and the body wall (Fig. 20-4).

Water enters the branchial chamber through open-

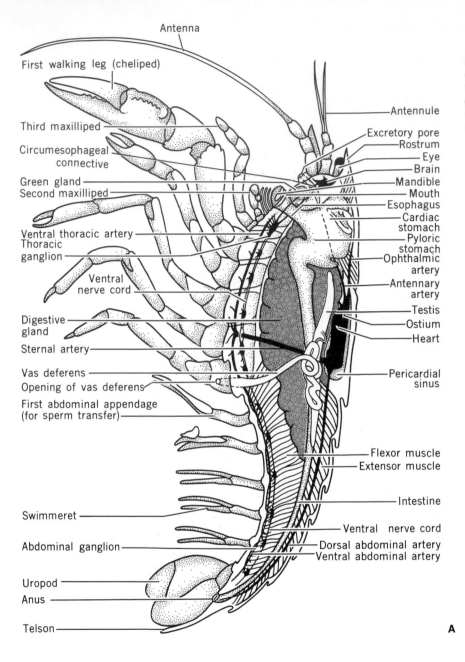

Antenna

First walking leg (cheliped)

Third maxilliped

Circumesophageal
connective

Green gland
Second maxilliped

Ventral thoracic artery
Thoracic
ganglion

Ventral
nerve cord

Digestive
gland

Sternal artery

Vas deferens
Opening of vas deferens

First abdominal appendage
(for sperm transfer)

Swimmeret

Abdominal ganglion

Uropod
Anus

Telson

Antennule

Excretory pore
Rostrum
Eye
Brain
Mandible
Mouth
Esophagus
Cardiac
stomach
Pyloric
stomach
Ophthalmic
artery
Antennary
artery
Testis
Ostium
Heart

Pericardial
sinus

Flexor muscle
Extensor muscle

Intestine

Ventral nerve cord
Dorsal abdominal artery
Ventral abdominal artery

Figure 20-4. Internal structure of a crayfish. (A) A female crayfish. (B) Cross-section through heart region of a female crayfish, showing arrangement of the gills and other organs (see opposite page).

A

ings at the base of the legs. Specialized parts of two appendages ventilate the gills; the paddlelike **epipodite.** The **gill bailer** of the second maxilliped is primarily responsible for pumping water anteriorly through the branchial chamber on each side of the body.

Excretory System

The excretory organs are a pair of rather large coelomic bodies, called **green glands** (Fig. 20-4), situated in the ventral part of the head anterior to the esophagus. Each green gland consists of a glandular portion

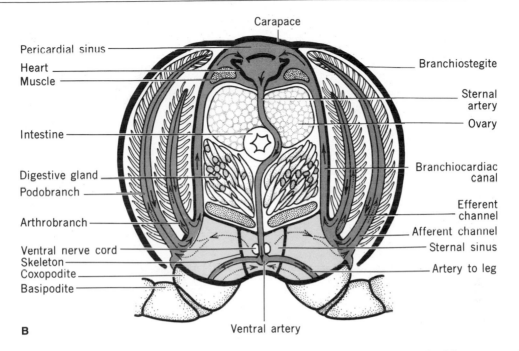

that is green in color, a thin-walled dilation called the bladder, and a duct opening to the exterior through an excretory pore on the basal segment of an antenna. The principal nitrogenous waste is ammonia, which also diffuses across the gills. A major purpose of the green gland in crayfish is to maintain ionic and osmotic conditions of body fluids.

Nervous System

The nervous system (Fig. 20-4) of the crayfish is in many respects similar to that of the earthworm, but it is further concentrated and developed in the head and thorax.

The **central nervous system** includes a dorsal ganglionic mass, the **brain (supraesophageal ganglia)**, and two **circumesophageal connectives** that pass to the **subesophageal ganglia.** These are the most anterior ganglia of the double **ventral nerve cord** that intercommunicates six pairs of **ganglia** in the thorax and six more pairs in the abdomen. The brain is a compact mass, larger than that of the earthworm, that sends nerves to the eyes, antennules, and antennae. The ganglia and connectives of the ventral nerve cord are more intimately fused than in the earthworm; it is

difficult to make out the double nature of the cord except between the fourth and fifth thoracic ganglia, where the sternal artery passes between them.

The **visceral nervous system** consists of three **nerves,** one from the posterior surface of the brain and one from each circumesophageal connective, that form a branching network or **nerve plexus** on the pyloric stomach.

Sense Organs

The crayfish has more highly developed sense organs than do the annelids. The sense of touch aids them in finding food, in avoiding obstacles, and in many other ways. **Tactile receptors** are located in specialized hairlike bristles, or **setae** (Fig. 20-2), on various parts of the body. These are especially abundant on the mouthparts, chelipeds and other chelae (claws), and edge of the telson. Vision in the crayfish is especially valuable in detecting moving objects. Receptors for sound have not been observed in crayfish, but pressure waves can probably be detected by the tactile receptors. In aquatic animals it is so difficult to distinguish between taste and smell that both are included in the term **chemical sense. Chemore-**

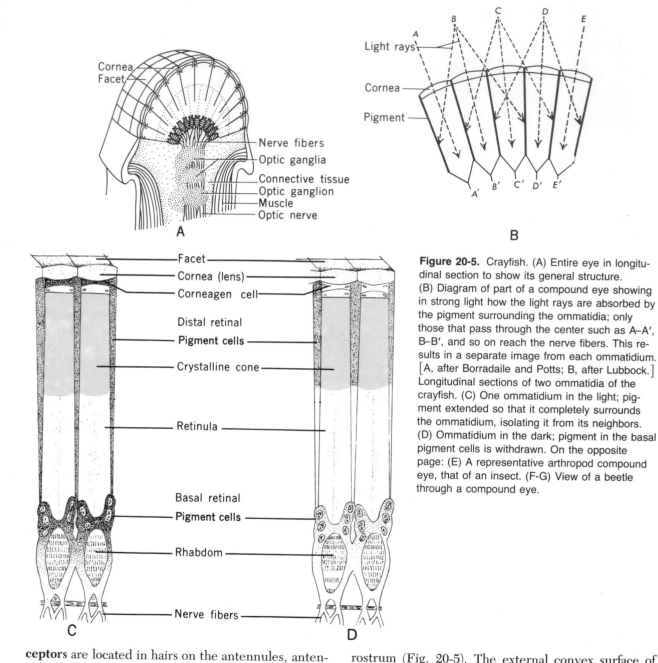

Figure 20-5. Crayfish. (A) Entire eye in longitudinal section to show its general structure. (B) Diagram of part of a compound eye showing in strong light how the light rays are absorbed by the pigment surrounding the ommatidia; only those that pass through the center such as A–A′, B–B′, and so on reach the nerve fibers. This results in a separate image from each ommatidium. [A, after Borradaile and Potts; B, after Lubbock.] Longitudinal sections of two ommatidia of the crayfish. (C) One ommatidium in the light; pigment extended so that it completely surrounds the ommatidium, isolating it from its neighbors. (D) Ommatidium in the dark; pigment in the basal pigment cells is withdrawn. On the opposite page: (E) A representative arthropod compound eye, that of an insect. (F-G) View of a beetle through a compound eye.

ceptors are located in hairs on the antennules, antennae, mouthparts, and other places.

Two **compound eyes** are situated at the end of movable stalks that extend from under each side of the rostrum (Fig. 20-5). The external convex surface of the eye is covered by a modified portion of transparent exoskeleton called the **cornea.** This window is subdivided by a large number of fine lines into four-

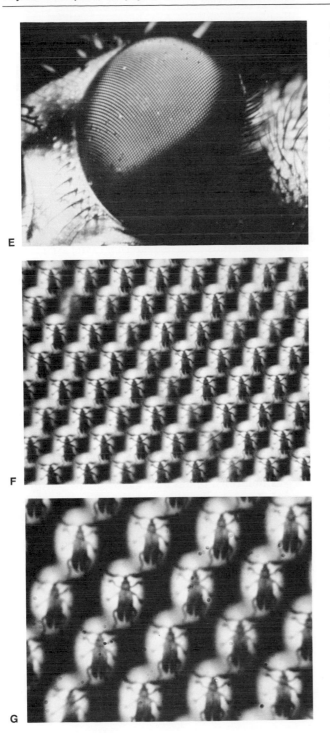

E

F

G

sided panes termed **facets**.* Each facet is the external window of a long slender visual rod known as an **ommatidium**.

Each compound eye is constructed of similar ommatidia lying side by side, each separated from its neighbors by a layer of dark **pigment cells** (Fig. 20-5). The average number of ommatidia in a single crayfish eye is 2500; the eye of an Atlantic lobster may possess 14,000 ommatidia.

Two ommatidia are diagrammed in Figures 20-5C and 20-5D. Beginning at the outer surface, each ommatidium consists of the following parts: (1) a **cornea**, (2) two **corneagen cells** that secrete the cornea after each molt, (3) a crystalline cone formed by four **cone cells**, (4) two distal **pigment cells** surrounding the crystalline cone, (5) several **retinular cells** that form a central **rhabdom** where they meet, and (6) a number of black **basal pigment cells** around the base of the retinular cells. The **cornea** and **crystalline cone** act as lenses; the **rhabdom** and **retinular cells** are the receptor element (**retinula**) of the ommatidium. Fibers from the **optic nerve** enter the base of the ommatidium and communicate with the inner ends of the retinular cells.

The eye of the crayfish is supposed to produce a **mosaic**, or **apposition, image.** This is illustrated in Figure 20-5B, where the ommatidia are represented by A–E and the fibers from the optic nerve by A′–E′. Most rays of light from any point (A, B, or C) will encounter the dark pigment cells surrounding the ommatidia and be absorbed. Only rays that pass directly through the center of the cornea will penetrate to the retinula and stimulate nerve impulses in the fibers of the optic nerve.

Thus the retinula of each ommatidium receives a single impression from the light that reaches it at each instant. Neighboring ommatidia are oriented in slightly different directions and so will receive different stimuli. Thus the ommatidia in a compound eye collectively monitor the external environment within the field of their numerous optic axes. Each visual unit responds to a fragment of the total field, and the resulting fragmentary images are subsequently integrated into a single composite picture. Compound eyes are particularly good at detecting moving objects

* the facets (and, thus, ommatidia) of some other crustaceans are hexagonal.

and function best at short distances; they do not distinguish form and size very clearly.

In weak light the proximal and distal pigments are retracted toward the ends of the ommatidia (Fig. 20-5D). This allows the eye to collect more light. When this occurs the ommatidia no longer act independently; light can pass from one ommatidium to another, and a continuous image is thrown on the retinular layer. The resulting **superposition image** is much less distinct than the apposition image formed at strong light intensities.

The **statocyst** is an organ of equilibrium. The statocysts of *Cambarus* are chitin-lined sacs situated in the basal segment of each antennule (first antenna). The base of the statocyst contains a ridge with many fine **sensory hairs** that are innervated by a single nerve fiber. Attached to these hairs are a number of large sand grains, the **statoliths,** that the animal places there after each molt. Beneath the sensory cushion are glands that secrete a substance that glues the grains to the hairs.

The statocysts monitor the orientation of the crustacean with respect to the gravitational vertical, as any departure will cause the statoliths to pull on their sensory hairs in different ways. Information from these receptors is integrated within the central nervous system, which formulates appropriate righting responses.

The statocysts are shed with the exoskeleton when the crayfish molts; thus individuals that have just molted are unable to orient themselves. Some elegant experiments have elucidated statocyst function. In one, shrimps that had just molted were placed in filtered seawater. When supplied with iron filings, the animals filled their statocysts with them. A magnet was then held near the statocyst, and the shrimp oriented themselves in a position that compromised between the two pulls, that of gravity and that of the magnet.

Endocrine Glands

In each eyestalk there is a neural **X-organ** that produces several hormones that pass along nerve axons to the **sinus gland** located in the base of the eyestalk. These hormones appear to control the spread of pigment granules in epidermal **chromatophores** and the retraction of retinal pigments. They also govern, to a greater or lesser extent, metabolic rate, growth, and viability. They regulate the frequency of molting and are necessary for normal deposition of calcium salts in the exoskeleton. Reproduction is also regulated by them to some extent. In the maxillae a small gland called the **Y-organ** produces another hormone that induces molting and sexual maturation. These hormones are distributed to their target organs by the bloodstream, as in the vertebrates.

Muscular System

The musculature of the crayfish is complex. The muscles are all attached to the inner surface of the exoskeleton, instead of constituting part of the body wall as in annelids or being external to the skeleton as in the vertebrates. The largest muscles in the crayfish are situated in the abdomen (Fig. 20-4) and are used to flex the tail forward upon the ventral surface of the thorax, resulting in a backward swimming locomotion. Other muscles extend the abdomen in preparation for another stroke. Muscles of considerable size are also situated in the thorax and within the tubular appendages, especially the chelipeds.

Reproduction and Development

Crayfish are almost without exception **dioecious.** The male **gonads** consist of two white **testes** partially fused into three lobes that lie in the thorax and anterior abdomen. On each side a long coiled sperm duct, the **vas deferens,** opens at the base of the fifth walking leg. Germ cells within each testis pass through two maturation divisions and then develop into **spermatozoa.** As many as 2 million sperm may be stored in the testes and sperm ducts until copulation occurs.

In the female, the two **ovaries** resemble the testes in form and location in the body (Fig. 20-4). A short **oviduct** leads from each ovary to an opening in the base of the third walking leg. The **seminal receptacle** is a cavity in a fold of cuticle between the fourth and fifth pairs of walking legs.

Sperm are transferred from the male to the seminal receptacle of the female during **copulation,** which usually occurs between early spring and autumn. The **ova** are fertilized as they are laid, several weeks to several months after copulation. The eggs are fastened to the female's swimmerets with a sticky substance and are aerated by being moved back and forth through the water.

Cleavage of the egg is superficial, and the embryo appears first as a thickening on one side. The eggs hatch in 2 to 20 weeks, and the larvae cling to the egg stalk. The young stay with the mother, clinging to her swimmerets with their legs, until they can care for themselves. After 2 to 7 days they **molt;** they molt at least six times during the first summer. The cuticle of the crayfish loosens and drops off. Meanwhile, the epidermal cells have secreted a new covering. Molting is necessary before growth can proceed, as the exoskeleton is hard and nonelastic. As soon as the exoskeleton is shed, the body increases in bulk due to intake of water through the gills. The shed exoskeleton is generally eaten to recycle calcium salts. It takes several weeks for the new exoskeleton to completely harden.

Sometime before each molt, a quantity of calcium is absorbed from the old exoskeleton and distributed by the blood to the stomach where it is deposited in calcareous bodies called **gastroliths.** The formation of gastroliths is under the endocrine control.

Regeneration

The crayfish and many other crustaceans are able to regenerate lost parts to a much more limited extent than are cnidarians, platyhelminths, and annelids. Regeneration occurs if the second and third maxillipeds, the walking legs, the swimmerets, or the eyes are lost. Growth of regenerated tissue is more frequent and rapid in young crayfish than in adults.

As in the earthworm, the rate of regeneration depends upon the amount of tissue removed. If one cheliped is amputated, a new one regenerates less rapidly than if both chelipeds and some of the other walking legs are removed.

The regenerated appendage is not always like that of the one lost, a phenomenon called **heteromorphosis.** For example, when the region containing the seminal receptacle of an adult is experimentally removed, another is regenerated; although this new structure is as large as that of the normal adult, it is comparable in complexity only to that of an early larval stage. A more remarkable phenomenon is the regeneration of an apparently functional antennalike organ in place of a degenerate eye that was removed from the blind crayfish *Cambarus pellucidus*. In this case a nonfunctional organ was replaced by a functional one of a different character.

An interesting aspect of crayfish regeneration is the presence of definite breaking points near the bases of the walking legs. If a cheliped is grasped or injured, it is reflexively lost by the crayfish at this breaking plane. The other walking legs, if injured, are also thrown off at the free joint between the second and third segments.

This reflexive breakage of body parts at a definite point is known as **autotomy,** a phenomenon that also occurs in some hydroids, polychaetes, bivalve mollusks, and echinoderms. A special (**autotomizer**) muscle flexes the leg and amputates it at the breaking point. A membrane on the basal stub then constricts, minimizing blood loss. A new leg as large as the lost one regenerates from the end of the stump in the course of several molts. Autotomy is a unique adaptation that prevents undue loss of blood when a leg or other body part is injured or lost to a predator.

Behavior

Crayfish are more active between dusk and dawn than during the daytime. At night they move about in search of food. When at rest, the crayfish usually faces outward from a place of concealment and extends its antennae. In this position it may learn the nature of an approaching object without being detected.

Crayfish elicit behavioral responses to a variety of stimuli. Positive responses to contact are exhibited to a marked degree by crayfishes that seek to place their bodies in contact with a solid object if possible. The normal resting position of the crayfish is to bring its side or dorsal surface in contact with the walls of a hiding place.

Light of various intensities generally causes the crayfish to retreat. Colored lights are preferred to white. Negative response to light is adaptive, as crayfish seek dark hiding places, such as under stones.

Responses to food are partly due to chemical sense. Positive responses result from stimulation by food substances. For example, if meat juice is placed in the water near an animal, the antennae become animated and the mouthparts perform vigorous chewing movements. The meat juice causes general restlessness and movement toward the source of stimulation, but crayfish seem to depend chiefly on touch for accurate localization of food. Acids, salts, sugar, and other chemicals produce a sort of negative response, indicated by the animal scratching the carapace, rubbing the chelae, or pulling at the part stimulated.

It has been shown by simple experiments that crayfish are able to form **habits** and to modify them. Maze experiments, in which the animals are presented with a choice of paths, have been used in these studies. They learn by experience and modify their behavior slowly or quickly, depending upon their familiarity with the situation. The chief factors in the formation of habits are the chemical sense, touch, sight, and the muscular sensations resulting from the direction of turning. Experiments show that the animals are able to learn a path even when the possibility of following a scent is eliminated.

Locomotion

Locomotion is accomplished by either walking or swimming. Crayfish are able to walk in any direction —forward usually, but also sideways, obliquely, or backward. The fourth pair of legs is most effective in walking and bears nearly the entire weight of the animal; the fifth pair serves as props and pushes the body forward; the second and third pairs are less efficient for walking because they are modified as grasping organs. Crayfish generally swim only when frightened. The crayfish then extends its abdomen, spreads out the uropods and telson, and by sudden contractions of the flexor abdominal muscles flips the abdomen down and thus darts backward.

Cave Crayfishes

There are at least 12 different species of cave crayfishes in the United States, some of which are restricted to the waters of a single cave, such as Mammoth Cave, Kentucky.

Interesting modifications adapt these crayfish to their dark, subterranean environments. All are blind; the eyes are atrophied, and the eyestalks are more or less undeveloped. Pigmentation is absent, and the body is pale. They are mostly small in size, with chelae not well developed. The antennae are long and highly specialized as tactile organs.

Other Crustaceans

The crustaceans (complementing their insect cousins on land) have adapted to virtually every aquatic habitat and have even made a few successful forays into terrestrial environments. The familiar **crabs,** for example, have adapted to abyssal, subtidal, intertidal, estuarine, and semiterrestrial ways of life. An unusual representative is the **coconut crab,** *Birgus,* which climbs palm trees on tropical islands to harvest coconuts. Most crustaceans are carnivores or carrion feeders, however. **Euphausids,** or **krill,** are among the most important links in marine food chains. Large, circumpolar euphausids prey upon smaller crustaceans and are in turn the primary food of many plankton-feeding (baleen) whales. The majority of crustaceans are mobile, but **barnacles** are not; following planktonic larval development these crustaceans settle on hard substrata and secrete a permanent calcareous house. Many crustaceans are planktonic, such as the water flea, *Daphnia,* which is a major herbivore in many freshwater communities. There are also numerous parasitic crustaceans, such as the barnacle, *Sacculina,* that infests the tissues of crabs.

These few examples suggest the great diversity of the crustaceans (Fig. 20-6) that have adaptively radiated and become specialized for many different environments and niches. Only a few representatives of the following five major subclasses will be surveyed here:

CLASS CRUSTACEA

Subclass	Branchiopoda	Fairy shrimp, water fleas
	Ostracoda	Ostracods
	Copepoda	Copepods
	Cirripedia	Barnacles
	Malacostraca	Euphausids, shrimp, lobsters, crayfish, crabs

Subclass Branchiopoda

Branchiopods are primitive crustaceans with elongated bodies composed of generally similar somites that bear relatively unspecialized appendages. They are generally small, averaging only several millimeters in length, and nearly all are filter feeders. Most

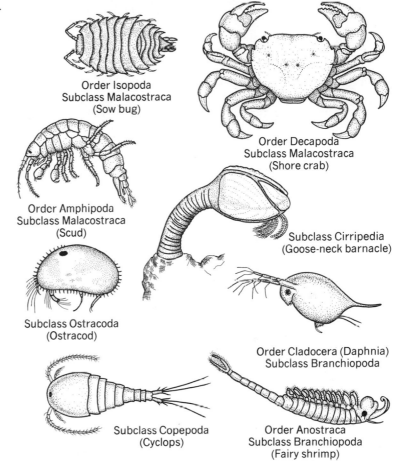

Figure 20-6. Representative crustaceans.

Order Isopoda
Subclass Malacostraca
(Sow bug)

Order Decapoda
Subclass Malacostraca
(Shore crab)

Order Amphipoda
Subclass Malacostraca
(Scud)

Subclass Cirripedia
(Goose-neck barnacle)

Subclass Ostracoda
(Ostracod)

Order Cladocera (Daphnia)
Subclass Branchiopoda

Subclass Copepoda
(Cyclops)

Order Anostraca
Subclass Branchiopoda
(Fairy shrimp)

branchiopods live in fresh water, but brine shrimp, *Artemia,* inhabit salt lakes.

Fairy shrimp, *Branchinecta,* are common in freshwater pools. They are semitransparent and swim on their backs. Eggs laid during the summer become buried in the bottom mud and are able to withstand drying and winter cold. **Water fleas,** *Daphnia,* are also quite common in freshwater habitats. A laterally compressed carapace surrounds the thorax and abdomen but not the head. Enlarged second antennae are used for swimming, and the trunk appendages are specialized for filtering microscopic algae from the water.

Subclass Ostracoda

Ostracods are a curious group of marine and freshwater crustaceans whose bodies are completely en-

cased within a carapace that forms a bivalved shell. Ostracods generally scurry over the substratum by extending their abdomen from the shell and flexing it to push the animal along. Ostracods are generally filter feeders. They are important members of shallow-water communities, where they are often enormously abundant. A few species inhabit moist soils in tropical forests.

Subclass Copepoda

Copepods are some of the most important animals in aquatic ecosystems. It has been estimated that there are more individual copepods than all other metazoans put together. Many copepod species live in the ocean, some in fresh water, and a relatively large number parasitize fish and many invertebrates. The copepod body is divided into an anterior **meta-**

some and a narrow posterior **urosome.** The median larval eye usually persists as the adult eye.

Planktonic copepods, such as *Calanus*, are generally filter feeders that consume phytoplankton (microscopic algae) and other small drifting animals. These grazing copepods, in turn, form the principal food of some sharks, fishes, whales, and numerous other marine animals.

Cyclops, a common copepod in freshwater ponds, sometimes serves as an intermediate host for the broad fish tapeworm (cestode) and the guinea worm (nematode) that infect humans.

Subclass Cirripedia

The **barnacles** are perhaps the most highly modified crustaceans. In fact, for a long time they were considered to be mollusks! The familiar barnacles are sessile animals that are typically encased within calcareous shells. The shell is secreted by a carapace that envelops the body as a mantle (Fig. 20-7). When a barnacle molts, it only sheds the exoskeleton on its torso and appendages; it never sheds the covering of the mantle, which grows continually. The elongated thoracic appendages bear fine plumes of setae. When feeding, these appendages are thrust through the shell aperture, and planktonic organisms are raked in toward the mouth.

The **acorn** and **gooseneck barnacles** are familiar to most people because they commonly attach to rocks and wharf pilings in the intertidal zone. Gooseneck barnacles attach to the substrata by means of a long flexible stalk. Acorn barnacles, *Balanus*, possess a thick shell but no stalk. When the tide recedes they close the six plates of their shells to avoid dehydration.

Many barnacles are parasitic. *Sacculina* attaches itself to marine crabs and sends rootlike processes throughout the host's tissues (Fig. 20-7).

Subclass Malacostraca

The largest crustacean subclass is **Malacostraca,** which includes about 18,000 species. Included here are most of the animals that people generally think of when "crustacean" is mentioned. Most malacos-

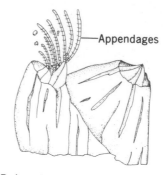

Balanus (acorn barnacle)

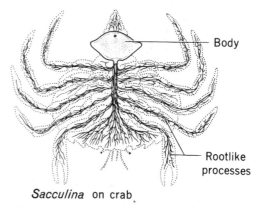

Sacculina on crab.

Figure 20-7. Top: Rock barnacles are common marine crustaceans that live permanently attached. One is shown with appendages extended and feeding; the other is withdrawn into its shell. Bottom: *Sacculina,* a barnacle, is a curious marine crustacean that in the adult stage is parasitic on a crab. The crab is represented by dotted lines; the parasite body from which the roots penetrate the tissues of the host is represented by solid lines.

tracans have a carapace and display shrimplike, crablike, or lobsterlike body forms.

Isopods are dorsoventrally flattened malacostracans that have successfully invaded freshwater and terrestrial habitats. The familiar **pill bugs** that inhabit moist retreats under logs and stones are isopods. Many pill bugs develop a **tracheal respiratory system** in their abdominal appendages. Their young develop in a fluid bath within a broad pouch and breathe by means of gills.

Amphipods are laterally compressed malacostracans that lack a carapace. Among the extremely cosmopolitan amphipods are the common beachhoppers that live in sandy beaches. These scavengers feed on decaying organic matter, especially seaweed that is washed onto the shore. They have even been found in the waters beneath the ice shelves of Antarctica.

Euphausids have elongated shrimplike bodies and are especially abundant in cool marine waters. These **krill** feed primarily on phytoplankton and in turn are consumed in immense numbers by blue, humpback, and other baleen whales.

Shrimps and **prawns** are of great commercial importance. Many species are gregarious and congregate in large schools—a behavior that is capitalized upon by shrimp fishermen, who trawl long V-shaped nets to catch them.

The **crabs** are a very large assemblage with many specialized members. They are generally characterized by an extreme reduction and underfolding of their abdomen. Familiar examples include the edible **rock crab** (*Cancer*), the **blue crab** (*Callinectes*), the **fiddler crabs** in estuaries, the **hermit crabs** that live inside abandoned gastropod shells, and **sand crabs** that live in the surf zone of sandy beaches.

The **crayfish** and **lobsters** are specialized for bottom-dwelling modes of life. They are among the most specialized crustaceans.

Relations of Crustaceans to Humans

Crustaceans are of immense value as human food, both directly and indirectly. The smaller species are often present in enormous populations and constitute important links in the food chains of freshwater and marine communities. Commercially, shrimp are the most important crustaceans as human food; crabs, lobsters, and crayfish follow in that order. Blue or other edible crabs are eaten extensively; they are called **soft-shelled crabs** just after molting. Soft-shelled crabs may be kept soft for a week or more on ice. Low temperature slows metabolism, and so the new exoskeleton develops slowly.

Crayfish are used for food, especially in Europe, Louisiana, and on the Pacific Coast. The large abdominal muscles of shrimps and prawns are sold fresh or canned. The shrimp industry in the United States is centered in the Gulf of Mexico and in California.

In some southern states, crayfish damage cotton and other crops by devouring the plants. They also occasionally burrow into the supports of levees and weaken them.

Although some crustaceans are parasites of aquatic animals, none parasitize humans or other terrestrial animals. Copepods serve as intermediate hosts of several parasitic worms of humans; for example, certain species of *Cyclops* host larvae of the guinea worm and certain tapeworms. Crayfish and crabs are potential intermediate hosts for lung flukes (*Paragonimus*).

Summary

Crustaceans, insects, centipedes, and millipedes are all mandibulate arthropods. Mandibulates are distinguished from the chelicerates by possessing at least one pair of preoral antennae, by chewing mandibles that flank the mouth, and by two pairs of postoral maxillae.

Crustaceans are the only large arthropod group that are primarily aquatic. The crustaceans are distinguished from the terrestrial mandibulates by the possession of biramous appendages and two pairs of preoral antennae.

Review Questions

1. What major characteristics distinguish a horseshoe crab from an edible rock crab?
2. List the head appendages of a typical crustacean and describe their functions.
3. Describe feeding and digestion in the crayfish.
4. How do you suppose the freshwater *Cyclops* got its name?
5. What do crustaceans have in common with annelids? What advances have the crustaceans made?
6. Describe how cave crayfish have adapted to their environment.
7. Barnacles lack compound eyes and hearts. How can these deficiencies be correlated with their sessile way of life?

Selected References

Barnard, J. L., R. J. Menzies, and M. C. Bacescu. *Abyssal Crustacea*. New York: Columbia University Press, 1962.

Carlisle, D. B., and F. Knowles. *Endocrine Control in Crustaceans*. New York: Cambridge University Press, 1959.

Cohen, M. J. "The Function of Receptors in the Statocyst of the Lobster *Homarus Americanus*," *J. Physiol.*, **130**:9, 1955.

Cushing, D. J. "The Vertical Migration of Planktonic Crustacea," *Biol. Rev.* **26**:158, 1951.

Dahl, E. "Main Evolutionary Lines Among Recent Crustacea," in *Phylogeny and Evolution of Crustacea*, H. B. Whitington and W. D. I. Rolfe (eds.). Cambridge: Museum Comparative Zoology, 1963.

Green, J. A. *A Biology of Crustacea*. London: H. F. and G. Witherby, Ltd., 1961.

Hansen, H. J. *On the Comparative Morphology of the Appendages in the Arthropoda. A. Crustacea*. Copenhagen: Gylenda, 1925.

Huxley, T. H. *The Crayfish, an Introduction of the Study of Zoology*. London: Kegan, Paul, Trench, Trubner, 1880.

Marshall, S. M., and A. P. Orr. *The Biology of Calanus Finmarchicus*. London: Oliver & Boyd, 1955.

Passano, L. M. "The Regulation of Crustacean Metamorphosis," *American Zool.*, **1**:89, 1961.

Pennak, R. W. *Freshwater Invertebrates of the United States*. New York: Ronald Press, 1953.

Smith, W. L., and M. H. Chanley. *Culture of Marine Invertebrate Animals*. New York: Plenum Publishing Corporation, 1975.

Southward, A. J. "Feeding of Barnacles," *Nature*, **175**:1124, 1955.

Waterman, T. H. (ed.). *The Physiology of Crustacea; Metabolism and Growth*, Vol. 1. *Sense Organs, Integration and Behavior*, Vol. 2. New York: Academic Press, 1960–1961.

Young, D. *Developmental Neurobiology of Arthropods*. New York: Cambridge University Press, 1973.

Phylum Arthropoda, Subphylum Mandibulata, Class Insecta, Class Diplopoda, Class Chilopoda

Over 750,000 species of insects have already been described, yet perhaps ten times as many remain to be recognized. It has been estimated that, for every star that can be seen with the unaided eye on a clear night, there are 100 species of insects on the earth! Obviously such an assemblage cannot be adequately covered in this general textbook.

The study of insects comprises a distinct discipline of biology called **entomology,** and entomology textbooks and courses treat the insects in considerable detail. We shall only discuss some general aspects of insect anatomy and physiology, using the grasshopper as an example, and we shall then consider some special adaptations of their versatile body plan. Insect taxonomy is covered in some detail in the classification appendix; Figure 21-1 illustrates the variety of orders of the class **Insecta.**

Insects are primarily terrestrial arthropods with bodies consolidated into three **tagmata:** head, thorax, and abdomen. The **head** bears four pairs of **appendages:** one pair of **antennae,** one pair of **mandibles,** and two pairs of **maxillae.** A shelflike extension of the **exoskeleton** called the **labrum** forms an **upper lip** over the **mouth.** A posterior **labium** derived from the

fusion of the second maxillae forms a **lower lip.** The **thorax** is composed of three **somites,** each of which bears a pair of **legs.** One or two pairs of **wings** are generally borne on the posterior thoracic somites. The **abdomen** contains 10 or 11 somites. Abdominal appendages are restricted mainly to **reproductive** and **tactile organs** at the posterior tip.

Class Insecta, *Romalea*—The Grasshopper

The grasshopper is one of the least specialized insects and therefore exhibits the essential features of the class. Special adaptations of the grasshopper that depart from the general insect body plan include leathery forewings, enlarged hindlimbs, structures for making sounds, and auditory organs.

External Morphology

As is true for other arthropods, the grasshopper is covered by a hard **exoskeleton** (Fig. 21-2), or **cuticula,** that is divided into a linear row of plates known as

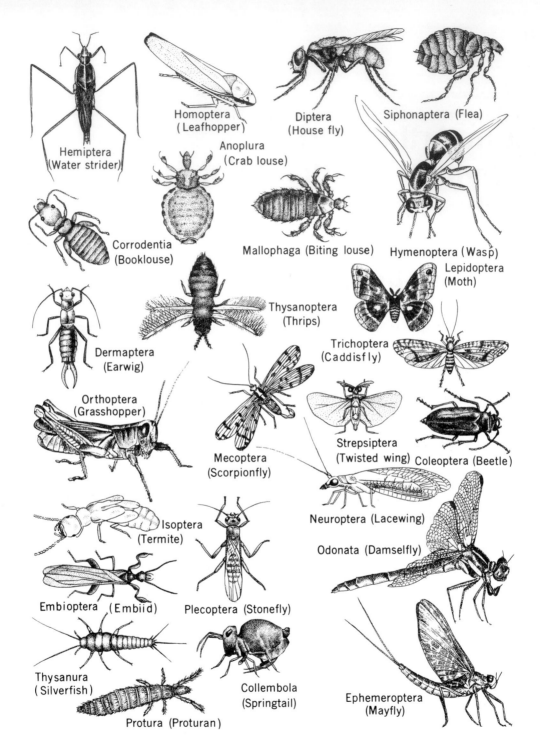

Figure 21-1. Representatives of some orders of insects.

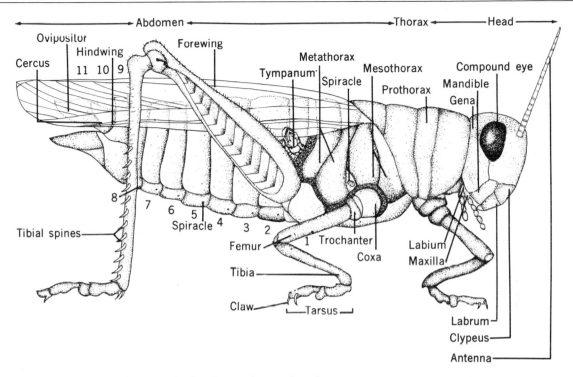

Figure 21-2. External features of a female grasshopper *Romalea*.

sclerites. A groove called a **suture** is often seen where the sclerites meet, but sometimes the sutures between sclerites are indistinguishable. Sutures are often accompanied by an infolding of the cuticula that reinforces the exoskeleton and provides sites for muscle attachment. The **tracheae** are formed in a similar manner. The exoskeleton is thin and flexible in intersegmental regions to allow movement of the appendages and body.

Head

The fused segments of the **head** are indistinguishable in the adult but may be observed in the embryo. The dorsal region of the head is known as the **vertex** (Fig. 21-3); the front portion is called the **frons,** and the sides are the **cheeks,** or **genae.** The rectangular sclerite below the frons is the **clypeus.** Each side of the head bears a compound eye. Three simple eyes (**ocelli**) are also present, two on top of the head and one median between the compound eye.

The food of the grasshopper consists of vegetation that it bites off and macerates by means of chewing mouthparts (Fig. 21-3). An **upper lip,** or **labrum,** is attached to the ventral edge of the clypeus. Beneath the labrum is a membranous tonguelike organ, the **hypopharynx,** that is derived from the floor of the head anterior to the mouth. On either side of the **mouth** is a hard **jaw,** or **mandible,** with a toothed surface adapted for grinding. Behind the mandibles are a pair of **maxillae,** consisting of several parts and with lateral **sensory palps.** The **labium,** or **lower lip,** also bears sensory palps. The labrum and labium serve to hold food between the mandibles and maxillae, which move laterally, thus grinding the food between their serrated edges. The maxillary and labial palps are supplied with **chemoreceptors** that serve to distinguish different kinds of food.

Each of the two **compound eyes** (Fig. 21-3) is covered by a transparent part of the cuticula, the **cornea,** which is divided into a large number of **hexagonal facets.** Each facet is the outer end of a photoreceptive

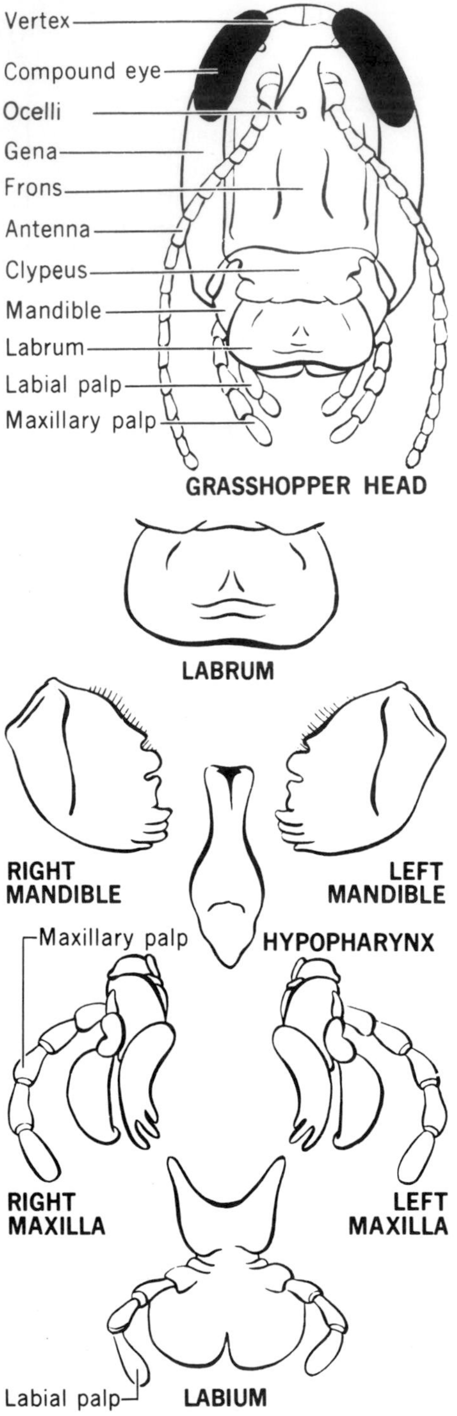

Vertex
Compound eye
Ocelli
Gena
Frons
Antenna
Clypeus
Mandible
Labrum
Labial palp
Maxillary palp

GRASSHOPPER HEAD

LABRUM

RIGHT MANDIBLE　　**LEFT MANDIBLE**

Maxillary palp　　**HYPOPHARYNX**

RIGHT MAXILLA　　**LEFT MAXILLA**

Labial palp　**LABIUM**

Figure 21-3. Grasshopper head and mouthparts.

ommatidium. Intergration of all the ommatidia gives mosaic vision, as described for the crayfish (Chapter 19).

Each of the three **ocelli** (Fig. 21-4) consists of a transparent **lens,** a modification of the cuticula; pigmented **epidermal cells;** and **photoreceptive retinal cells.**

The threadlike, multisegmented **antennae** are supplied with numerous **tactile hairs** and **olfactory pits;** their location on the highly movable antennae allows a larger area to be explored.

Thorax

The thorax is separated from the head and abdomen by flexible joints. It consists of three segments; an anterior **prothorax,** a middle **mesothorax,** and a posterior **metathorax.** Each segment bears a pair of legs, and the mesothorax and metathorax both bear a pair of wings. On either side of the mesothorax and metathorax is an opening (**spiracle**) into the respiratory system.

A typical thoracic segment consists of a dorsal **tergum** composed of four fused **sclerites** in a row, two lateral **pleurons** made up of three sclerites, and a ventral **sternum** composed of one sclerite.

Each **leg** (Fig. 21-2) is composed of the following series of segments: the **coxa** articulates with the body, followed by the small **trochanter** fused with the **femur,** the **tibia,** and the **tarsus.**

The hind legs of grasshoppers are modified for jumping. The elongated metatarsal legs have very muscular femurs. The anterior two pairs of legs are specialized for landing. The tarsal segments are equipped with pads that act as shock absorbers and with spines that are used to cling to the substratum.

Insect **wings** are outgrowths of the exoskeleton and so are covered with cuticle on their upper and lower surfaces. **Veins** that ramify through the wings carry blood vessels, **respiratory trachioles,** and **sensory nerves** (Fig. 21-5). The veins are fortified with cuticle and thus help to strengthen the wings. The veins vary in number and arrangement in different species of insects and are very useful for classification.

Each wing articulates at a movable hinge, formed

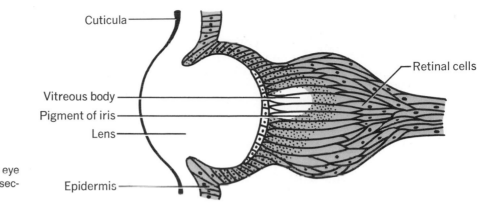

Figure 21-4. Ocellus or simple eye of the honeybee in longitudinal section.

Cuticula

Vitreous body

Pigment of iris

Lens

Epidermis

Retinal cells

by the dorsal thoracic tergum and a dorsal process of the lateral plate (**pleuron**). This process serves as a fulcrum upon which the wings move up and down as a lever. Six to twelve pairs of **primary flight muscles** do *not* directly move the wings themselves but rather move the tergum up and down. Half the flight muscles are orientated vertically and half horizontally. When the vertical muscles contract, the tergum is depressed and the wings are raised. Then contraction of the longitudinal muscles bows the tergum outward, causing a downstroke of the wings.

The **mesothoracic wings** of the grasshopper are leathery and not folded; they serve as protective covers for the delicate metathoracic wings, which are thin and folded like a fan. The metathoracic wings allow a gliding flight when the grasshopper jumps.

Abdomen

The slender abdomen consists of 11 segments, with those at the posterior extremity being modified for copulation or egg laying. Along the lower sides of the

Figure 21-5. Insect wing. Left: Generalized insect wing showing the chief veins. Right: Enlarged cross-section of wing showing a vein that consists of the outer surface of the wing, blood space, and trachea.

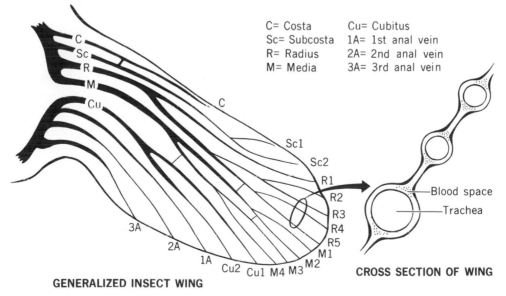

C= Costa
Sc= Subcosta
R= Radius
M= Media

Cu= Cubitus
1A= 1st anal vein
2A= 2nd anal vein
3A= 3rd anal vein

Blood space

Trachea

GENERALIZED INSECT WING

CROSS SECTION OF WING

abdomen there are eight pairs of small **spiracles** through which the animal breathes. In the grasshopper the ventral sternum of the first abdominal segment is fused with the **metathorax;** on either side of this segment is an oval **tympanic membrane** covering an **auditory sac.** Segments 2 to 8 are unmodified. In the male, the **sternum** of segment 9 is elongated ventrally, giving an upward twist to the abdomen. The end of the female abdomen is more tapered than is that of the male and forms the egg-laying **ovipositor.**

Internal Morphology and Physiology

The internal organ systems (Fig. 21-6) lie within a **hemocoel** that is filled with **blood (hemolymph).** All the systems characteristic of higher animals are represented.

Muscular System

The muscles are **striated.** They are very soft and delicate, yet strong and capable of rapid contraction. The muscles are segmentally arranged in the abdomen, but they are disbanded into more discrete units in the head and thorax. The most conspicuous muscles are those that move the mandibles, wings, metathoracic legs, and the ovipositor.

Digestive System

The principal parts of the **digestive tract** (Fig. 21-6) are the foregut, midgut, and the hindgut.

The **foregut** begins at the **mouth,** on either side of which opens a duct from **salivary glands** that produce saliva and digestive enzymes. A tubular **esophagus** enlarges into a **crop,** a storage compartment that lies within the mesothoracic and metathoracic segments. This leads into a grinding organ called the **proventriculus (gizzard).**

The **midgut,** or **ventriculus (stomach),** extends posteriorly into the abdomen. Six double cone-shaped pouches, the **gastric ceca,** secrete **digestive enzymes** into the midgut. The products of digestion are absorbed through the stomach wall into the surrounding hemolymph, which distributes the nutrients throughout the body.

The **hindgut** is made up of the **ileum** and the **colon,** which expands into the **rectum** and opens through the

Figure 21-6. Internal organs of a grasshopper viewed with the left side of the body wall removed; tracheae not included.

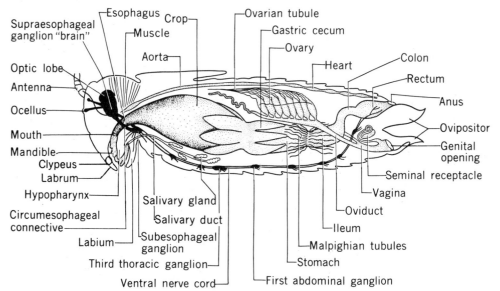

anus. The anterior end of the ileum receives excretory products from the Malpighian tubules. Because both the foregut and the hindgut are lined with cuticle, little absorption takes place in them.

Circulatory System

As in other arthropods, the insect circulatory system is an **open system,** with the dorsal **heart** and major vessels opening into the **hemocoel,** or body cavity. The heart is located in the middorsal abdomen, just under the body wall. It is divided into a row of chambers; into the base of each opens a pair of **ostia.** These openings are closed by **valves** when the heart contracts. The **pericardial sinus,** in which the heart lies, is filled with blood, which enters the heart from the pericardial sinus through the ostia and from there is pumped anteriorly through the **aorta** into the hemocoel, where it bathes all the tissues.

The **blood (hemolymph)** serves chiefly to transport food and wastes, as there is a separate respiratory system. The blood consists of a clear plasma in which **phagocytes** are suspended; these amoeboid cells phagocytose foreign organisms and plug wounds to prevent bleeding.

Respiratory System

The respiratory system (Fig. 21-7) consists of a network of air-filled tubes, the **trachae,** that communicate with every part of the body. The tracheae consist of a single layer of cells lined with cuticula, which is thickened at intervals by spiral rings that prevent the tracheae from collapsing. The tracheae open through from eight to ten **spiracles** on each side of the body: one above the mesothoracic and/or metathoracic legs and one on each lateral surface of the first seven or eight abdominal segments.

As the tracheae near their target organs, they subdivide into the smallest branches, called **tracheoles,**

Figure 21-7. Diagram of the tracheae in the body of a grasshopper. (A) The tracheal system consists of air-filled tubes that branch into others. Arrows indicate that the grasshopper inhales through spiracles located in the anterior part of the body and exhales through those limited to its abdomen. (B) A large tracheal trunk and some of its branches.

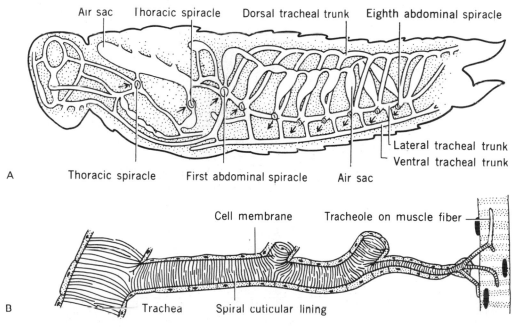

which are less than 1μ in diameter. The ends of the tracheoles branch into a network over the tissue cells. Their tips are filled with liquid, and the actual exchange of oxygen and carbon dioxide occurs by diffusion through this liquid medium.

In the grasshopper and certain other insects, some of the tracheae become expanded into thin-walled **air sacs.** Contraction and expansion of the abdomen ventilates the tracheal system by enlarging and compressing the air sacs, which act much like bellows.

Excretory System

The organs of excretion (Fig. 21-6) are the **Malpighian tubules** that coil about in the hemocoel and open into the anterior end of the hindgut. The final nitrogenous waste product of amino acid catabolism in insects is **uric acid,** a compound that is much less toxic than ammonia and that can therefore be excreted in a more concentrated form.* Uric acid is formed within the body tissues and passes into the blood, from which it is selectively absorbed across the walls of the Malpighian tubules. Water is reabsorbed into the hemolymph, and the uric acid is condensed into crystals and discharged into the hindgut for elimination through the anus.

Nervous System

The nervous system (Fig. 21-8) of insects conforms to the basic arthropod pattern but is relatively consolidated, with a specialized brain and an anterior migration and fusion of the **ventral nerve cord ganglia.** The **brain (supraesophageal ganglion)** is located dorsally in the head and consists of three pairs of fused ganglia. These ganglia supply the eyes, antennae, and other cephalic organs. The brain is joined by two connectives around the esophagus to the **subesophageal ganglion.** This ganglion consists of the three anterior pairs of ventral nerve chain ganglia fused together and innervates the mouth parts. The **ventral nerve chain** continues with a pair of large ganglia in each thoracic segment. The ganglia in the metathoracic segment are particularly large, as they represent the fusion of

---

* Birds and reptiles are other groups of terrestrial animals that excrete uric acid to conserve water; mammals excrete **urea,** with less efficient, but nevertheless satisfactory, results.

ganglia from both this segment and the first abdominal segment. Five pairs of ganglia are present in the abdomen. The pair in the second abdominal segment represents the ganglia from the second and third abdominal segments fused together; and the pair in the seventh segment represents the seventh to the eleventh ganglia combined. Also connected to the brain are the ganglia of the **sympathetic (autonomic) nervous system,** which controls the "involuntary" activities of the digestive tract, heart, aorta, and reproductive system.

Sense Organs

Grasshoppers possess sense organs of sight, hearing, touch, taste, and smell. The **compound eyes** and **ocelli** (Fig. 21-3) have already been noted. Vision by means of compound eyes in the crayfish has been described in Chapter 20. The ocelli are considered to be primarily organs of light perception, as they apparently cannot form images.

The pair of **auditory organs** are located on the sides of the tergum of the first abdominal segment. Each consists of a **tympanic membrane (tympanum)** stretched across an almost circular sclerotized ring. Airborne sound vibrations set the tympanic membrane in motion, and this activates a slender point beneath the membrane that is connected to sensory nerve fibers. Some insects hear sounds beyond the range of the human ear. Grasshoppers produce sound by rubbing the rough-surfaced tibia of the hindleg against a wing vein, causing the latter to vibrate **(stridulation).**

The **antennae** are supplied with the principal organs of smell. Organs of taste are located on the **sensory palps** of the mouth parts. Hairlike **touch receptors** are present on various parts of the body, particularly the antennae.

Reproductive System

Female grasshoppers can easily be distinguished from males by the presence of the **ovipositor** (Fig. 21-2). In the female there are two **ovaries.** Each consists of several tapering egg tubules called **ovarioles** that lack central cavities (Fig. 21-9). The ovarioles contain the developing **ova** arranged in a linear series, along with **nurse cells** and other accessory cells. The

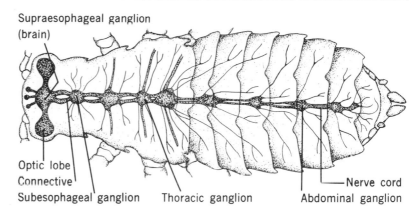

Figure 21-8. The grasshopper nervous system in dorsal view. [Redrawn from R. E. Snodgrass, *Principles of Insect Morphology.* New York: McGraw-Hill, 1935.]

oocytes grow and mature as they proceed posteriorly down the ovariole; hence the ovariole becomes gradually larger toward its posterior end. The ovarioles of each ovary communicate with an **oviduct** into which the eggs are discharged. The two oviducts unite to form a short **vagina** that leads to the single **genital opening** between the plates of the ovipositor. A tubular **seminal receptacle (spermatheca)**, which connects with the dorsal wall of the vagina, receives the spermatozoa during copulation and releases them when the eggs are laid.

The male has two testes in which **spermatozoa** develop (Fig. 21-9). The **gametes (sperm)** are discharged into a **vas deferens.** The two **vasa deferentia** unite to form an **ejaculatory duct** that enters the **penis,** opening ventrally on the eighth abdominal segment. Accessory glands that secrete the **seminal fluid** in which the sperm are transported are present at the anterior end of the ejaculatory duct.

The ova are fertilized as they are deposited by penetration of sperm through an opening (**micropyle**) in one end of the eggshell. Grasshoppers undergo gradual metamorphosis (as discussed in the next section).

Life Cycles of Insects

The life cycles of various insects involve different kinds of **metamorphosis,** or transition, from larval to adult stages. Metamorphosis may be lacking, gradual, incomplete, or complete (Fig. 21-10).

No Metamorphosis

The dipluran, *Campodea staphylinus* (order **Aptera**) is a primitive **wingless insect** (Fig. 21-10) about 3 mm long that lives under leaves and other forest litter. The young that hatch from the eggs look exactly like

Figure 21-9. Diagrams of the reproductive systems of insects in general. Left: Female. Right: Male. Some species deviate from this general plan in lacking one or more organs.

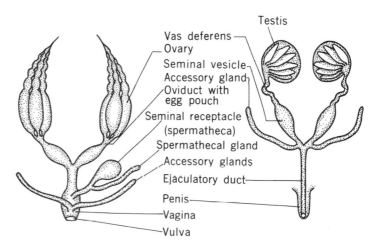

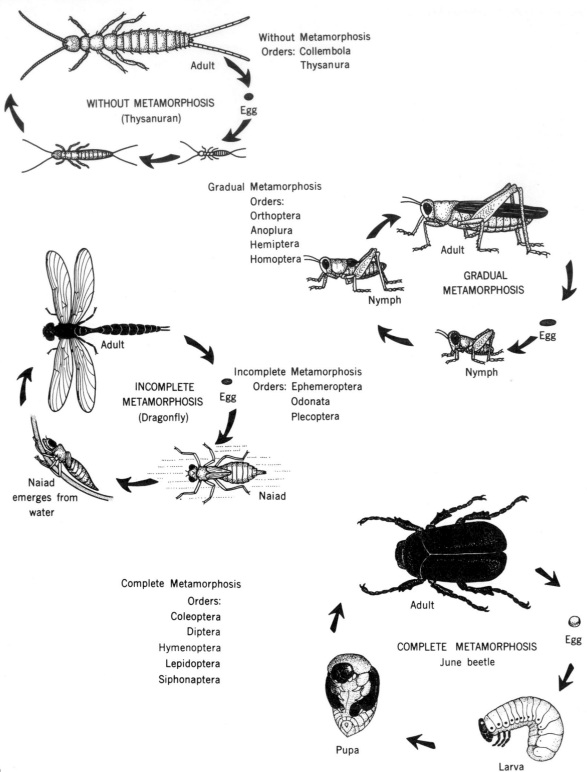

Without Metamorphosis
Orders: Collembola
Thysanura

WITHOUT METAMORPHOSIS
(Thysanuran)

Adult

Egg

Gradual Metamorphosis
Orders:
Orthoptera
Anoplura
Hemiptera
Homoptera

Adult

GRADUAL
METAMORPHOSIS

Nymph

Egg

Nymph

Adult

INCOMPLETE
METAMORPHOSIS
(Dragonfly)

Egg

Incomplete Metamorphosis
Orders: Ephemeroptera
Odonata
Plecoptera

Naiad
emerges from
water

Naiad

Complete Metamorphosis
Orders:
Coleoptera
Diptera
Hymenoptera
Lepidoptera
Siphonaptera

Adult

Egg

COMPLETE METAMORPHOSIS
June beetle

Pupa

Larva

miniature adults. They molt several times before reaching sexual maturity.

Gradual Metamorphosis

Grasshoppers, cockroaches, true bugs, leafhoppers, and some other insects undergo a gradual metamorphosis. The young of these insects resemble the adults upon hatching, except that their wings and reproductive organs are underdeveloped. The adult condition is attained gradually with successive molts. Each intervening stage is called a nymph. The wings develop gradually from small external buds of the exoskeleton.

Incomplete Metamorphosis

This type of metamorphosis is exhibited by insects such as **dragonflies, stoneflies,** and **mayflies** in which the **nymphs,** or **naiads,** are aquatic and the adults are aerial. Incomplete metamorphosis involves changes in body form that are greater than in gradual metamorphosis but less dramatic than in a complete metamorphosis.

Metamorphosis of the dragonfly (order **Odonata**) is incomplete. The naiad of the dragonfly has **rectal gills** that are ventilated by alternately drawing water into and expelling it from the hindgut.

Metamorphosis of the mayfly is also incomplete. Mating takes place during flight, after which the female lays several masses of 80 to 300 eggs on a stone in the water. The eggs hatch in about a month, and the naiads live underwater, breathing by means of tracheal gills and feeding upon minute plants (diatoms and algae). Maturation is accompanied by as many as 45 molts (an unusually high number for insects) and requires from 6 to 9 months. When ready to venture into the air, a naiad swims to the surface; a split appears along its back, and a gauzy-winged adult emerges and flies to a nearby object where it rests for 18 to 24 hours. Then it molts once again and is ready to fly into the air and find a mate. The adult life of both males and females last only a few hours or days.

Figure 21-10. Four types of life cycles found in insects: no metamorphosis, gradual metamorphosis, incomplete metamorphosis, and complete metamorphosis.

The scientific name of their order, **Ephemeroptera,** means "living but a day."

Complete Metamorphosis

Butterflies, moths, beetles, flies, bees, and other advanced insects undergo complete metamorphosis, which is characterized by a succession of egg, larval, pupal, and adult stages.

Pieris rapae (order **Lepidoptera**) is the common white **cabbage butterfly.** Its **larvae,** or **caterpillars** (Fig. 21-11), feed on the leaves of cabbage, turnip, mustard, horseradish, and radish plants. The larvae will die if they hatch on the wrong type of plant. The butterfly probably distinguishes one plant from another by means of olfactory sense. A few eggs are laid, one by one, on a suitable host plant. The bullet-shaped eggs are fastened to the leaf by their flat end. The larva eats its way out of the distal end of the egg shell and then devours the rest of the shell. It then begins to chew holes in the leaves with its mandibles. The caterpillar molts when it can grow no larger within its cuticular covering. A split appears in the dorsal exoskeleton near the anterior end, and the larva crawls out and expands before a new exoskeleton hardens. The caterpillar's legs are of two kinds: three pairs of jointed legs on the thorax and five pairs of unjointed, temporary **prolegs** on the abdomen. It possesses six small simple eyes on the head but no compound eyes. **Green cabbage caterpillars** mimic very closely the leaves on which they feed and are therefore difficult for birds and other predators to detect.

When full grown (about 3 cm in length), the caterpillars attach themselves with a thread from their **silk glands** to a leaf or some other object. Butterfly caterpillars do not spin silk cocoons as do many moth caterpillars. After a time the body becomes shorter and thicker, the skin splits down the back and is pushed off at the posterior end, revealing a greenish **pupa,** or **chrysalis** (Fig. 21-11).

The pupa is a quiescent stage in which adult structures are formed. The digestive system changes from one fitted for solid food to one that can utilize liquid food, and the muscular system of the crawling larva becomes modified for flight. The nervous system is reorganized, wings grow out from pads of larval tissue, and the reproductive organs grow to maturity.

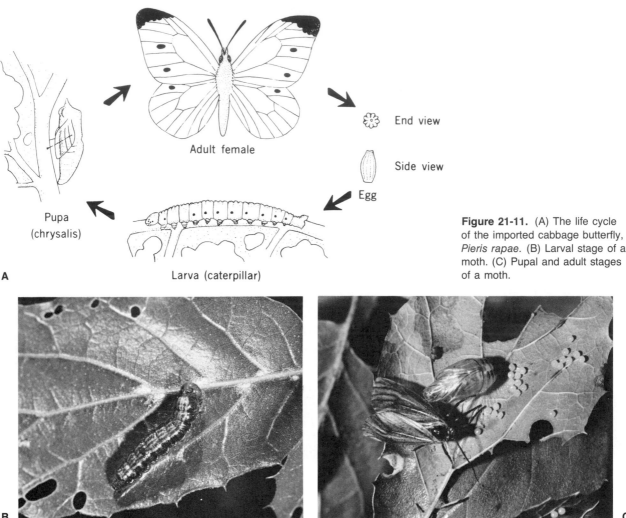

Pupa
(chrysalis)

Adult female

End view

Side view

Egg

A

Larva (caterpillar)

Figure 21-11. (A) The life cycle of the imported cabbage butterfly, *Pieris rapae.* (B) Larval stage of a moth. (C) Pupal and adult stages of a moth.

B

C

This complete metamorphosis is fueled by fat that was stored during the larval stage. After about ten days, the pupa splits open, and an adult butterfly emerges, expands, and spreads its wings; after the wings dry and harden, the butterfly flies away. The adult butterfly possesses a long coiled **proboscis** that can be extended to probe flowers for nectar.

Many variations occur in the four types of metamorphosis just described. For example, among the **social termites** only one caste contains nymphs that are sexually mature, whereas the nymphs of the "sev-

enteen-year locust" live underground for 13 to 17 years before attaining adulthood. **Aphids** may be **ovoviviparous** or **oviparous,** and their eggs may be fertilized or develop parthenogenically.

On other points of interest, the larvae of many beetles are called **grubs,** and fly larvae are called **maggots.** Many **moth caterpillars** spin **cocoons** in which they pupate; the **cockroach** secretes an egg case to protect her eggs; and certain larvae, especially certain **hymenopterans** (e.g., **wasps**), stimulate the formation of **plant galls.**

Adaptive Modifications of Class Insecta

The great success of the insects rests upon an array of anatomical and physiological adaptations that have allowed them to endure and exploit terrestrial habitats. We shall now examine some aspects of the basic body plan of insects that contribute to their success.

Wings

Insects are the only invertebrates that are capable of flight, and this capacity has certainly enhanced their exploitation of terrestrial habitats. The mobility afforded by flight confers advantages in food gathering, predator avoidance, and active dispersal into favorable habitats.

Most insects bear two pairs of wings. Certain primitive insects (e.g., **Thysanurans,** Fig. 21-1) never possess wings in their life histories; others (e.g., **lice** and **fleas,** Fig. 21-27) are adapted to parasitic life, and their wings are secondarily lost. Various modifications are exhibited among the winged orders of insects. The **flies (Diptera)** have a pair of reduced hindwings (**halteres**) that are important in maintaining equilibrium during flight. In **beetles (Coleoptera,** Fig. 21-1), the forewings are sheathlike and are called **elytra;** some stiffened forewings are used like aircraft wings, whereas the hindwings propel the animal forward. The forewings of **grasshoppers (Orthoptera,** Fig. 21-1) are leathery and are known as **tegmina** (sing., **tegmen.**).

The frequency of wing beats differs according to the species. Yellow **swallow-tailed butterflies** average about 6 beats per second, **dragonflies** about 30, **honeybees** about 400, and **houseflies** about 1000. In contrast, the wings of hummingbirds make about 750 beats per second during forward flight, and an accomplished pianist can perform a fast trill at about 10 notes per second.

Legs

Insect legs are highly modified for special functions. Running insects, such as the **ground beetle,** possess long, slender legs (Fig. 21-21). The **preying** mantis has forelegs adapted for grasping; the hindlegs of the **grasshopper** are used in leaping (Fig. 21-2) the forelegs of the **mole cricket** are modified for digging, and the legs of the **water bug** are specialized for swimming.

The legs of the **honeybee** (Fig. 21-12) are remarkably adapted for a variety of functions. The **prothoracic legs** possess two useful structures, the **pollen brush** and the **antenna cleaner.** The **femur** and **tibia** of this pair are clothed with branched hairs used for gathering pollen. The surface of the **first tarsal joint** is covered with **bristles,** constituting the cylindrical pollen brush that is used to brush up and collect pollen within reach of the front legs. On the distal edge of the tibia is a flattened movable spine, the **velum,** that fits over a curved indentation in the proximal tarsal segment. This entire structure is the antenna cleaner, and the row of teeth that lines the indentation is known as the **antenna comb.**

The last tarsal joint of every leg bears a pair of notched **claws** that enable the bee to obtain a foothold on rough surfaces. Between the claws is a fleshy glandular lobe, the **pulvillus;** its sticky secretion allows the bee to cling to smooth objects. **Tactile hairs** are also present.

The **mesothoracic legs** are also provided with a pollen brush, but, instead of an antenna cleaner, a **spur** is present at the distal end of the tibia. This spur is used to dislodge wax from the wax pockets on the ventral side of the abdomen and to remove pollen from the pollen basket.

The **metathoracic legs** possess three very remarkable structures: pollen basket, pollen packer, and pollen combs. The **pollen basket** consists of a concavity in the outer surface of the tibia with rows of curved bristles along the edges. Pollen is stored in this basketlike structure, allowing the bee to transport a larger load on each trip from the hive. On the inner edge of the distal end of the tibia is a row of stout bristles, the **pecten.** The opposing proximal end of the metatarsus bears a plate or lip, the **auricle.** When the joint between these structures is flexed backward, the auricle glides over the outer surface of the pecten and presses against the oblique outer surface of the end of the tibia (the pollen basket). The auricle and the pecten, working together, constitute the **pollen packer,** as their manipulations force the sticky pollen

Leg segments C = Coxa
Tr = Trochanter
F = Femur
Ti = Tibia
Ta = Tarsus

Fore- and hindwings hooked together

Antenna cleaner (comb and velum) in use

Pollen basket

C
Tr
F
C
C
F
Tr
Ti

Eye-brush

Pecten
Auricle
Pollen packer

Ti
Tr
F

Velum

Pollen comb

Spur

Ta

Pollen brush

Wax plate

Ta

Sting

Antenna comb

Pollen brush

Ti
Ta

Prothoracic legs

Ta

Pollen brush

Mesothoracic legs

Metathoracic legs

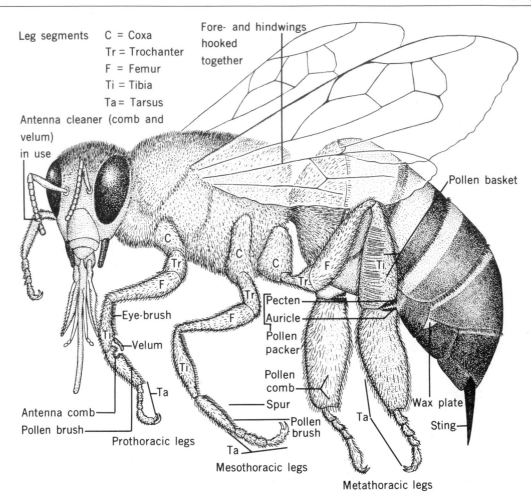

Figure 21-12. Bee adaptations. Leg of the honeybee *Apis mellifera* worker shows many of the structural modifications for gathering pollen, manipulating wax, and cleaning the antennae.

masses into the pollen basket. In loading the pollen baskets, the pollen brushes of the mesothoracic legs collect pollen from other parts of the body and are themselves cleaned by being drawn between the **pollen combs** of the hindlegs. Each pollen comb is then scraped over the pecten of the opposite leg, the sticky pollen being deposited on the outer surface of the pecten or falling on the upper surface of the auricle. The leg is then flexed backward at this joint, the auricle squeezing the pollen outward and upward and thus packing it into the pollen basket.

Mouthparts

Insects have adapted to many different diets and methods of feeding. The mouthparts of insects are in most cases modified for either chewing (**mandibulate**) or sucking (**suctorial**). The grasshopper posseses typical mandibulate mouthparts (Fig. 21-13). The mandibles of insects that consume vegetation are adapted for crushing; those of carnivorous species are usually sharp and pointed, suitable for cutting and tearing. Suctorial mouthparts are adapted for piercing the tis-

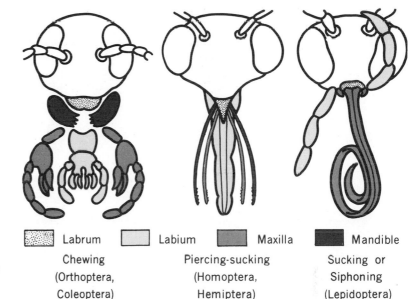

Labrum   Labium   Maxilla   Mandible

Chewing   Piercing-sucking   Sucking or
(Orthoptera,   (Homoptera,   Siphoning
Coleoptera)   Hemiptera)   (Lepidoptera)

Figure 21-13. Some modifications in the fundamental mouthparts of insects. Note the high degree of specialization that adapts the animals to different types of food.

sues of plants or animals and sucking juices. The specialized mouthparts of the **honeybee** are suctorial. In the female **mosquito** (Fig. 21-14), the labrum and hypopharynx are combined to form a **sucking tube,** and the **mandibles** and **maxillae** are **piercing organs.** The hypopharynx carries saliva, and the labium constitutes a sheath in which the other mouthparts lie when not in use. The **proboscis** of **butterflies** and **moths** is a sucking tube formed by the maxillae (Fig. 21-13).

Antennae

The antennae of insects are usually **tactile, auditory,** or **olfactory** in function. They differ in form and structure (Fig. 21-15), and in a given species the antennae of the male often differs from those of the female (**sexual dimorphism**).

Interesting experiments by von Frisch (see Selected References) led him to conclude that bees can distinguish certain odors. For example, they can select an odor derived from orange peel from among 43 other odors. He also demonstrated that bees can find places through a sense of smell. Experimental removal of antennae indicated that the olfactory sense organs are located on these appendages.

The antennae also play a role in insect communication by means of **pheromones.** A pheromone is a chemical substance secreted by an animal that affects the behavior of other animals of the same species. Ants, for example, maintain food trails by means of a pheromone secreted from the abdomen. After worker ants have located a food source, they lay down a scent trail back to the nest. Other ants from the nest then follow the pheromones to the food.

Digestive Tract

Of the internal organs of insects, the digestive tract and respiratory system are of particular interest. The digestive tract is modified according to the character of the food; that of the grasshopper is typical of **herbivorous** insects. **Suctorial** insects, for example, butterflies and moths, possess a muscular pumping **pharynx** and a **crop** that stores juices (Fig. 21-16).

Respiratory System

Insect respiratory systems are generally like that of the grasshopper, but modifications occur in many species, especially those with aquatic larvae. Naiads often lack spiracles; many obtain oxygen by means of

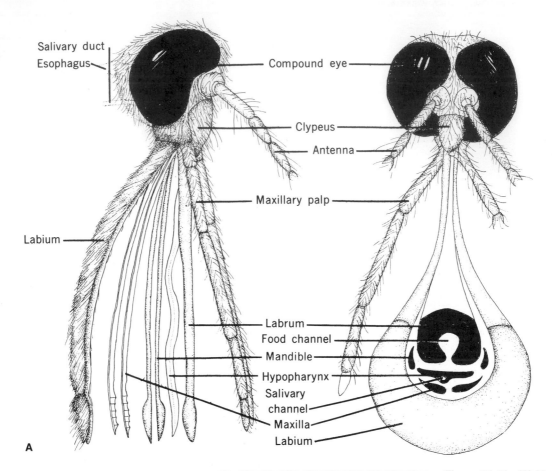

Salivary duct
Esophagus
Compound eye
Clypeus
Antenna
Maxillary palp
Labium
Labrum
Food channel
Mandible
Hypopharynx
Salivary channel
Maxilla
Labium

A

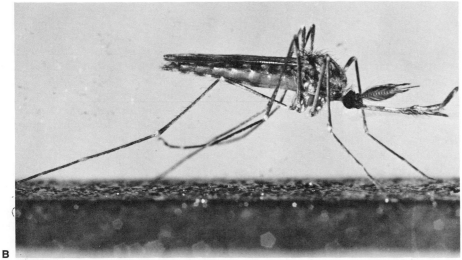

B

Figure 21-14. (A) Mouthparts of female mosquito showing the modifications for piercing and sucking. The mouthparts that are shown in solid black are those used in penetration of the skin. (B) An adult mosquito perched for action.

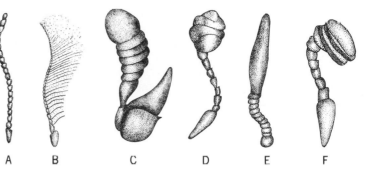

Figure 21-15. Antennae of various families of Coleoptera showing variations within a single order of insects. (A) Elateridae, click beetle; (B) Lampyridae, firefly; (C) Gyrinidae, whirligig beetle; (D) Silphidae, *Necrophorus*, burying beetle; (E) Curculionidae, weevil; (F) Scarabaeidae, lamellicorn beetle.

threadlike or leaflike outgrowths, called **tracheal gills,** at the sides or posterior end of the body.

Hormones

Much of the classic work on insect hormones was done by V. B. Wigglesworth using the bloodsucking bug, *Rhodnius.* It is now well established that hormones play important roles in the regulation of insect growth and development and other processes.

Three major **endocrine organs** are known to be involved in insect **molting** and **metamorphosis.** These hormone-secreting organs are the **brain, prothoracic glands,** and **corpora allata.** The brain secretes **activation hormone,** which stimulates the secretion of **ecdysone** (molting hormone) by the prothoracic glands. Ecdysone has been shown to be involved in many growth and developmental processes in addition to molting. **Juvenile hormone** is produced by the corpora allata, associated with the nerves innervating the foregut. If these glands are experimentally removed from insects in their first or second immature stage, pupation occurs at the next molt, and a midget adult results. Conversely, if the corpora allata from a

young larva are implanted into one that is about to undergo a final molt, metamorphosis does not occur, and instead another immature stage results.

Coloration

Many insects are brilliantly colored, especially the butterflies, moths, and beetles. Coloration of some insects varies with the season, and successive broods may have different color patterns. Males and females are often colored differently; they are sexually dimorphic, as, for example, in the swallow-tailed butterflies of the genus *Papilio.* Some insects are colored like their habitual surroundings; such **cryptic coloration** conceals them from predators. Others advertise their toxicity or distastefulness with **warning coloration** that contrasts to their surroundings. Still others derive some degree of protection by mimicking the warning coloration of other species. A well-known example of this is the **viceroy butterfly**'s **mimicry** of the coloring of the distasteful **monarch.**

In some insects, the colors displayed result from pigments imbedded in the cuticula. In others they are due to the structure of the cuticula, which diffracts

Figure 21-16. Digestive tract of a lepidopteron (with siphoning mouthparts). Diagram of the internal organs from the left side.

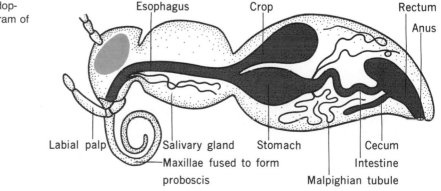

Esophagus — Crop — Rectum — Anus — Labial palp — Salivary gland — Maxillae fused to form proboscis — Stomach — Malpighian tubule — Intestine — Cecum

sunlight into its constituent colors by means of numerous fine scales, as on the wings of butterflies.

Other Interesting Adaptations

One has only to study the structure, physiology, and behavior of an insect to discover adaptive modifications. A few interesting examples follow: (1) The **walking stick** (Fig. 21-17) not only resembles a dead twig but has the habit of feigning death when disturbed. (2) The male **cricket** possesses a specialized sound-producing apparatus consisting of a **file** (Fig. 21-17) on the base of the hindwing that scrapes over thick veins on the forewing (**stridulation**), producing the pleasant call that attracts mates. The frequency of

calls increases with ambient temperature. The significant part of the sound produced by some classes of insects has so high a frequency that it is inaudible to the human ear. (3) **Dragonfly naiads** breathe by means of **rectal gills** that line the enlarged posterior end of the digestive tract; oxygen is removed from the water that is drawn into and expelled from this cavity. (4) The labium of this naiad (Fig. 21-17) is much elongated and can be extended rapidly from its folded resting position beneath the head so as to impale prey on its terminal hooks. (5) Aquatic **mosquito larvae** obtain air through a tube that is thrust through the water surface. (6) **Water striders** (Fig. 21-17) have long, slender legs that do not break through the surface film as they skim over the water. (7) **Fireflies**

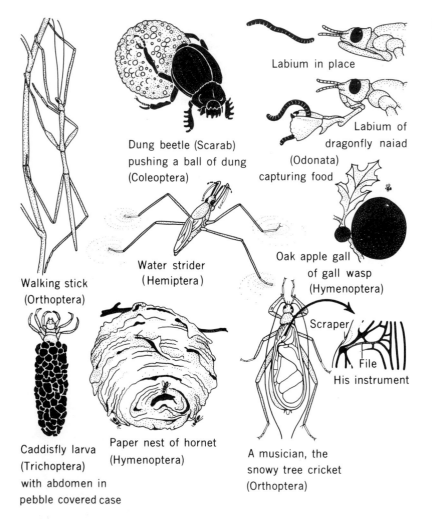

Figure 21-17. Special structural and physiological adaptations of insects.

Dung beetle (Scarab)
pushing a ball of dung
(Coleoptera)

Labium in place

Labium of
dragonfly naiad
(Odonata)
capturing food

Walking stick
(Orthoptera)

Water strider
(Hemiptera)

Oak apple gall
of gall wasp
(Hymenoptera)

Scraper

File

His instrument

Caddisfly larva
(Trichoptera)
with abdomen in
pebble covered case

Paper nest of hornet
(Hymenoptera)

A musician, the
snowy tree cricket
(Orthoptera)

have a **bioluminescent organ** capable of emitting light; their larvae, known as **glowworms,** are also luminescent. (8) **Dung beetles,** including the sacred **scarab** of the Egyptians (Fig. 21-17), roll up balls of dung that feed their larvae. (9) The aquatic larvae of most **caddis flies** (Fig. 21-17) build portable protective cases of sand grains or vegetable matter fastened together with silk. (10) Certain **hornets** build nests of wood pulp (Fig. 21-17). (11) **Gall wasps** stimulate plants to develop abnormal growths called galls, presumably caused by growth-stimulating substances secreted by the insect.

Social Insects

Various types of intraspecific associations occur among animals. Many protists live in colonies, as do many cnidarians whose polymorphic zooids may differ conspicuously from one another. In many cases animals of the same species band together for breeding purposes or are attracted to a limited area because of the presence of food. Birds are often gregarious during seasonal migrations. Certain mammals congregate in herds partly for protection; for example, male bison, when attacked, form a circle around the cows and calves.

More complex animal societies involve **division of labor;** the principal activities are reproduction, obtaining food, and defending the colony. Many of the most complex examples of social life occur among the insects, especially the wasps, bees, ants, and termites.

Wasps and bees may be solitary or social. Solitary wasps and bees dig nests in soil or wood or construct a nest of mud. Wasps provision their nests with caterpillars or other arthropods that have been paralyzed by their sting; bees provide pollen ("bee bread"). After laying an egg in the nest, they close the entrance and provide their offspring with no more parental care.

Bumblebees and Honeybees

Bumblebees and **honeybees** are familiar types of social bees. The inseminated **queen** bumblebee survives through the winter. In the spring she lays a few eggs in a cavity in the ground. The **worker bees** that hatch are infertile females that carry on all activities of the colony except laying eggs. At the end of the summer, males (**drones**) and fertile females (**queens**) hatch from some of the eggs. These fertile bees mate, and the sperm receptacles of the queens are filled with sperm. The workers and drones then die, and the race is maintained during the winter by the queens alone. Honeybees exhibit an even more complex social organization (Fig. 21-18).

Observers of honeybees have long realized that they had some system of communicating with each other, but the Austrian zoologist Von Frisch (1973 Nobel prize winner for physiology and medicine) clearly demonstrated that their main method of broadcasting a source of nectar or pollen to others in the hive depends on rhythmic movements and odors. For a discussion of this dance, see Chapter 31.

Do bees have color vision? Honeybees can apparently distinguish four colors: blue-green, yellow-green, blue-violet, and ultraviolet (which is invisible to humans). Bees can also visually distinguish geometric figures (e.g., a solid triangle versus three parallel lines).

Something is known about taste in bees. They can distinguish salt, sour, sweet, and bitter just as we can. Honeybees are further able to determine different degrees of sweetness.

Ants

Many ants live a complicated social life. An ant colony, unlike a colony of bees, may contain several fertile **queens** and at certain periods fertile **drones.** Infertile females may be of several types: (1) **soldiers** to guard the colony, (2) **workers** to gather food, and (3) workers to care for the eggs and young. These different types of individuals are morphologically different; that is, the species are **polymorphic.**

Social ants have developed effective solutions to survival problems, and some of these parallel important advances in human cultures, notably **agriculture** and the **domestication of animals.** Agricultural ants maintain crops of fungi that they use for food. Other ants have domesticated aphids and use their wastes for food. Some complex ant societies migrate great distances, overcoming prey and physical barriers by working together (for instance, certain ants will form a bridge of their bodies to permit the colony to cross a

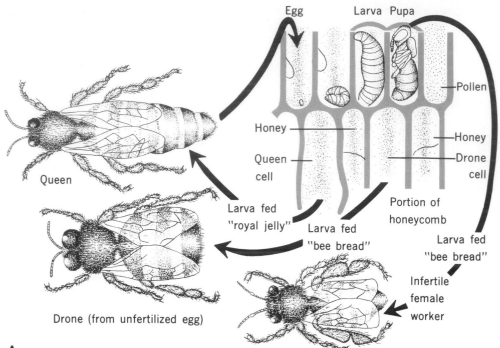

Egg   Larva   Pupa

Pollen

Honey

Queen cell

Honey

Drone cell

Portion of honeycomb

Queen

Larva fed "royal jelly"

Larva fed "bee bread"

Larva fed "bee bread"

Drone (from unfertilized egg)

Infertile female worker

A

B

Figure 21-18. (A) Life cycle of the honeybee *Apis mellifera* showing growth stages and three adult castes: the worker, drone, and queen. In nature, cells of the honeycomb are arranged parallel to the floor of the hive. (B) Beehive. The large queen bee is surrounded by the workers.

stream). There are some species in which the queen will invade host ant nests and take over either by force (large vicious ones) or by infiltration (reduced in size and not noticed by the host workers).

Termites

The most complex social life of all insects is that of the termites (Fig. 21-19). The colony contains three principal member types: **sexuals** (**kings** and **queens**), **workers,** and **soldiers.** The sexually reproductive individuals may possess functional wings, small nonfunctional wings, or no wings at all. Winged kings and queens leave the colony, mate, lose their wings, and start a new colony.

The second caste consists of the male and female workers; the third is male and female soldiers. All lack wings and functional sex organs. The workers are more numerous than are any other caste. They care for the eggs and young, feed and tend the queen, obtain food, cultivate fungus in special chambers (certain species), excavate tunnels and galleries, construct mounds, and perform other duties.

The soldiers are the most highly specialized. Two types may be present: one has a large body, strong head, and huge mandibles for killing intruders; the other **nasute** carries on chemical warfare by means of pore in the head through which a repellent fluid may be ejected.

Relations of Insects to Humans

Many insects of importance to human welfare have been mentioned in this chapter. Some of them are beneficial, and others are injurious. Among the bene-

Figure 21-19. Castes and life cycle of the termite. It feeds on dead wood.

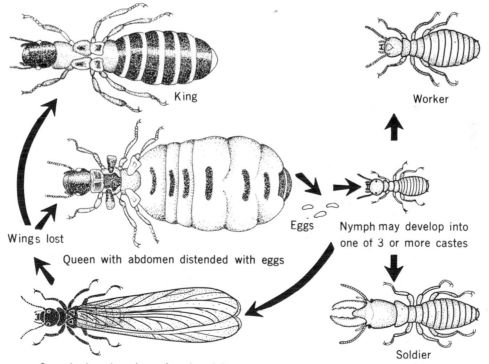

King

Worker

Eggs

Nymph may develop into one of 3 or more castes

Wings lost

Queen with abdomen distended with eggs

Soldier

Sexual winged male or female adult

ficial insects are those that produce honey, wax, silk, shellac, and cochineal; those that cross-fertilize (pollenize) flowers; and those that destroy injurious insects by either devouring or parasitizing them. Injurious insects include farm and household pests and those that transmit diseases.

Beneficial Insects

About 200 million kg of **honey** are recovered from the hives of honeybees in the United States every year. Some 5 million kg of **beeswax** are also collected annually. **Silkworms** (Fig. 21-20) spin about 300 m of thread to make each cocoon, and some 25,000 cocoons are necessary to manufacture one pound of silk. Annual world production of silk is about 25 million kg. **Lac** insects (family **Coccidae**) secrete a wax known as shellac. The red dye known as **cochineal** is made from the dried bodies of certain tropical American **scale insects** that live on cactus; it is no longer widely used, because aniline dyes are now largely substituted instead.

Bees are the most important insects that pollenate flowers. For example, apples, pears, blackberries, raspberries, and clover depend upon them for flowering and fruit formation. Certain moths and other insects also pollenate flowers as they gather nectar. Note that many of our important food plants could not survive without insects. An interesting example is the Smyrna fig, which would not bear fruit in California until a minute fig insect was introduced to pollenize its flowers.

Predatory Insects

Predatory insects are generally beneficial because they devour vast numbers of other insects, most of which are injurious. **Ground beetles, tiger beetles, antlions,** and **lady beetles** (all shown in Fig. 21-21) are common examples. The **Australian ladybird beetle** was introduced into California because it was known to devour the fluted scale insect that was attacking orange and lemon trees there. Other predatory species have also been introduced with favorable results.

Insects that parasitize (Fig. 21-22) other insects are also beneficial to humans. They usually lay their eggs in the host larvae, and the young that hatch slowly devour and finally kill their host before it becomes an adult. Some parasitic insects exhibit **polyembryony:** each of their eggs produces not one but many larvae; as many as 3,000 have been reported from a single egg of a small parasitic wasp. Sometimes parasitic insects parasitize other parasitic insects, and these in turn may be parasitized, and so on, a condition known as **hyperparasitism.** There is certain truth in the following humorous lines:

Big fleas have little fleas
upon their backs to bite 'em.
And little fleas have lesser fleas,
and so *ad infinitum.*

Many insects are scavengers, and vast quantities of dead animal and vegetable matter are eaten by them, thus preventing putrefaction. **Blowflies** are especially effective scavengers because they lay enormous num-

Figure 21-20. Life cycle of the silkworm *Bombyx mori.*

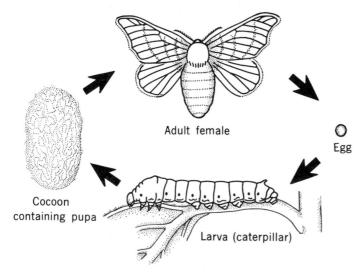

Adult female

Egg

Cocoon containing pupa

Larva (caterpillar)

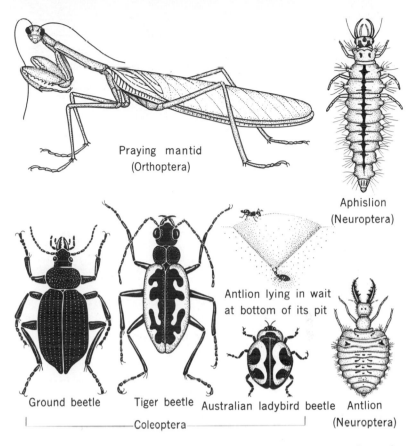

Figure 21-21. Predacious insects that now benefit man.

Praying mantid
(Orthoptera)

Aphislion
(Neuroptera)

Antlion lying in wait
at bottom of its pit

Ground beetle   Tiger beetle   Australian ladybird beetle   Antlion
|————————Coleoptera————————| (Neuroptera)

bers of eggs in animal carcasses, and the larvae that hatch from them are extremely voracious. As Linnaeus remarked, flies can devour the carcass of a horse more quickly than a lion can. **Houseflies** (Fig. 21-23), **water scavenger beetles, burying beetles,** and **dung beetles** (Fig. 21-17) are some other noteworthy scavengers.

Other insects have great potential for human benefit. Insects that feed on crops can sometimes be controlled by the local release of predatory insects (biological control) instead by insecticides. **White ants** (worker termites) are tasty, nutritious, and consume paper among other things. Mass culture of such insects might transform some of our wastes into food.

Insects Harmful to Plants

The U.S. Department of Agriculture estimates that insects annually cause about $4 billion in damage to farm crops, forests, stored foodstuffs, and domesticated animals. Some are native species, such as the **potato beetle** (Fig. 21-24), **Rocky Mountain locust,** and **army worm;** others were introduced from foreign parts, such as the **San Jose scale, European corn borer, cotton boll weevil, Japanese beetle,** and **Mediterranean fruit fly.** Sucking insects that are farm pests include the **aphids, scale insects, stink bugs, Hessian fly maggots,** and **leafhoppers.** The **chinch bug** is especially injurious to corn and small grains. Chewing insects of importance include **wireworms, white grubs, European corn borers, flat-headed borers, bark borers, alfalfa weevils, corn-ear worms,** and **cotton boll weevils.** Stored grain is destroyed in large quantities by beetles of various kinds, especially weevils, and by certain moth caterpillars.

Insects Injurious to Domesticated Animals

Domesticated animals are often seriously injured by insects. **Biting lice,** such as the **chicken louse** (Fig.

Hymenopteran ovipositing in egg of codling moth (Lepidoptera)

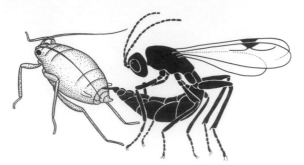

Hymenopteran ovipositing in an aphid (Homoptera)

Pyrgota fly (Diptera) ovipositing in abdomen of June beetle (Coleoptera)

Cocoons (pupae) of a braconid wasp (Hymenoptera) on the larva (hornworm) of the tomato sphinx moth (Lepidoptera)

Figure 21-22. Some insects that show a parasite–host relationship. It is estimated that there are roughly 11,000 species of parasitic insects in North America.

Figure 21-23. Life cycle of the housefly, *Musca domestica*. This fly, as is true of insects in general, has a tremendous reproductive potential. It has been said that, if this potential were unchecked, the descendants of a pair of flies in ten years would weigh more than the earth.

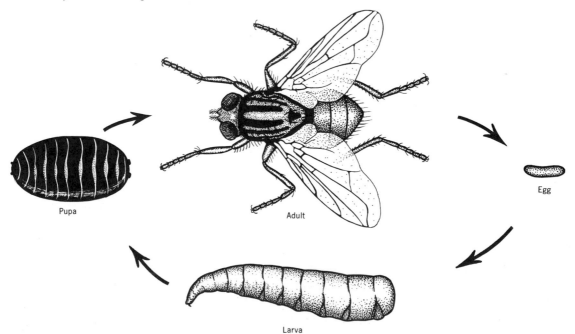

Pupa

Adult

Egg

Larva

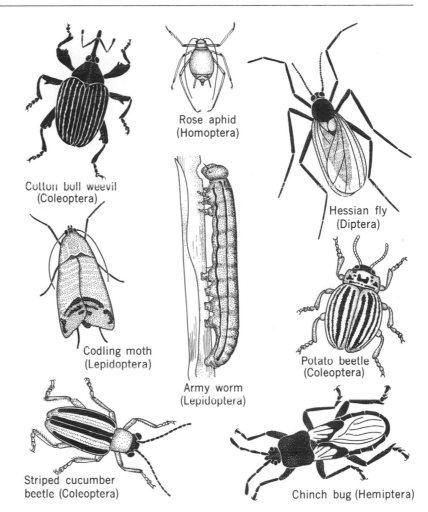

Figure 21-24. Some herbivorous insects.

Cotton boll weevil
(Coleoptera)

Rose aphid
(Homoptera)

Hessian fly
(Diptera)

Codling moth
(Lepidoptera)

Army worm
(Lepidoptera)

Potato beetle
(Coleoptera)

Striped cucumber
beetle (Coleoptera)

Chinch bug (Hemiptera)

21-25), may feed on skin and feathers, and their activities cause constant irritation. **Sucking lice** have mouthparts adapted for sucking the blood and tissue fluids of mammals. The **horn fly** is a serious bloodsucking pest of cattle. **Botflies** spend their larval lives inside the stomachs of horses (Fig. 20-26). The maggots of the **ox warble fly** (Fig. 20-26) produce holes in the skin of cattle and damage the hide.

Household Insect Pests

Insects that invade the home (Fig. 21-26) are sometimes only annoying, but they may also be destructive. Food may be spoiled or contaminated by **cock-** **roaches, ants, fruit flies,** and **weevils.** Clothing, carpets, furs, and feathers may be damaged by **clothes moths** and **carpet beetles.** Among the piercing insects that are annoying are **stable flies, bedbugs,** and **mosquitoes.**

Insects That Transmit Human Diseases

Insects that carry diseases (Fig. 21-27) are one of the greatest enemies of human welfare. Noteworthy vectors include **houseflies** that spread the bacteria that cause **typhoid** and **summer diarrhea; mosquitoes** that transmit the microorganisms that cause **malaria,**

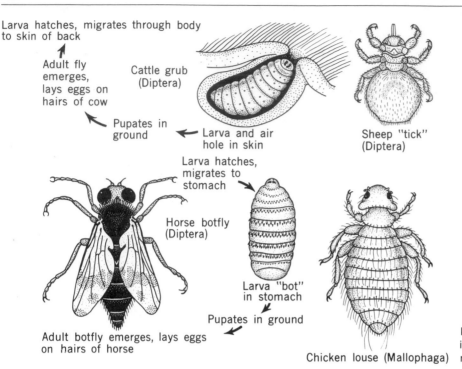

Larva hatches, migrates through body to skin of back

Adult fly emerges, lays eggs on hairs of cow

Cattle grub (Diptera)

Pupates in ground

Larva and air hole in skin

Sheep "tick" (Diptera)

Larva hatches, migrates to stomach

Horse botfly (Diptera)

Larva "bot" in stomach

Pupates in ground

Adult botfly emerges, lays eggs on hairs of horse

Chicken louse (Mallophaga)

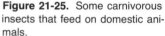

Figure 21-25. Some carnivorous insects that feed on domestic animals.

yellow fever, and **filariasis; fleas** that convey the bacteria of **bubonic plague** from rats and ground squirrels to humans; **body lice** (cooties) that transmit **typhus fever;** and **tsetse flies** that are vectors of **African sleeping sickness** in humans and **nagana** and other diseases in domesticated animals.

Our federal and state governments and educational institutions all recognize the necessity of controlling injurious insects, and hence **economic entomology** has become a most active field of investigation. Health departments devote considerable funds and effort to the control of insects, especially houseflies and mosquitoes.

Other Terrestrial Mandibulates

In addition to the insects, the phylum Arthropoda also includes two other major classes of terrestrial mandibulates: the millipedes and centipedes. Also included with the mandibulates are two minor centi-

pedelike classes, **Pauropoda** and **Symphyla,** that will not be discussed here. These four groups are sometimes called **myriopods** ("numberless feet"), a collective term that has no taxonomic significance.

The myriopods are all terrestrial animals that appear to have evolved independently from an ancient mandibulate stock. They have a worldwide distribution but are especially abundant in humid tropical and subtropical environments. Myriopods are often overlooked by the casual observer because these reclusive animals live under the ground cover or in the soil.

Myriopods are generally characterized by a **head** and a long **trunk** composed of many **segments.** The head bears a single pair of **antennae;** mouthparts typically include one pair of **mandibles** and two pairs of **maxillae.** Each of the many trunk segments typically bears one or more pairs of **legs.**

The visceral organs are somewhat similar to those of primitive insects. The tubular **heart** extends dorsally throughout the trunk, with paired **ostia** opening in each segment. The **nervous system** is relatively uncentralized. Respiratory gas exchange is generally

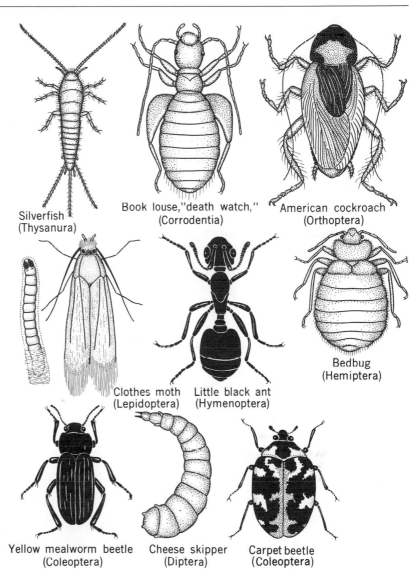

Silverfish
(Thysanura)

Book louse,"death watch,"
(Corrodentia)

American cockroach
(Orthoptera)

Clothes moth
(Lepidoptera)

Little black ant
(Hymenoptera)

Bedbug
(Hemiptera)

Yellow mealworm beetle
(Coleoptera)

Cheese skipper
(Diptera)

Carpet beetle
(Coleoptera)

Figure 21-26. Household insects.

assisted by **tracheae,** and **Malpighian tubules** are the principal excretory organs.

Class Chilopoda

The mandibulate class Chilopoda contains **centipedes** (Fig. 21-28). Their bodies are usually dorsoventrally flattened and consists of from 15 to 173 segments, depending on the species. Each trunk appendage bears one pair of legs, except the first, which has poison claws, and the last two, which usually lack appendages. Centipedes are carnivores. Their prey consists of insects, worms, mollusks, and other small animals that they kill with their **poison claws** and then chew with their **mandibles.** The **antennae** are long, consisting of at least 12 segments, and some of these

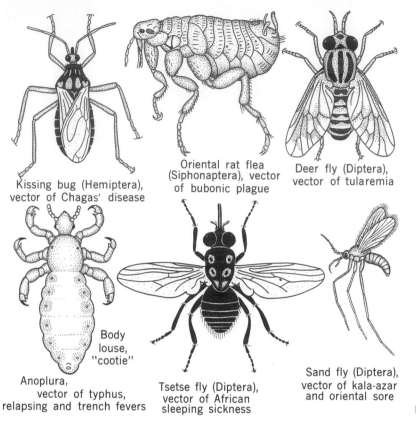

Kissing bug (Hemiptera),
vector of Chagas' disease

Oriental rat flea
(Siphonaptera), vector
of bubonic plague

Deer fly (Diptera),
vector of tularemia

Body
louse,
"cootie"

Anoplura,
vector of typhus,
relapsing and trench fevers

Tsetse fly (Diptera),
vector of African
sleeping sickness

Sand fly (Diptera),
vector of kala-azar
and oriental sore

Figure 21-27. Some species that transmit human diseases (vectors).

Figure 21-28. Centipedes have many legs. In the East Indies there is a giant centipede nearly 30 cm long.

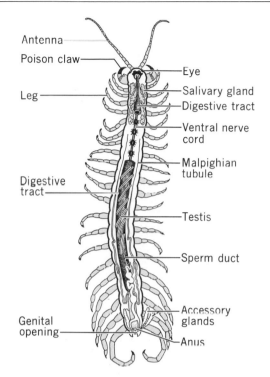

Figure 21-29. Drawing of centipede showing the internal organs and external features of *Scolopendra*.

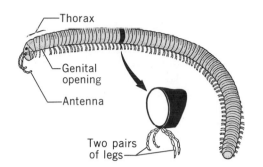

Figure 21-30. Millipedes are shy animals that hide in dark places, avoiding the light. They can easily be distinguished from centipedes by the subcylindrical body and the two pairs of legs on most segments.

predators have **compound eyes.** The internal anatomy of a common centipede is shown in Figure 21-29.

Centipedes are swift-moving creatures. Many live under the bark of logs or under stones. Some of the poisonous centipedes of tropical countries belong to the genus *Scolopendra*. These may reach 30 cm in length, and their bite is painful and even dangerous to humans. The common house centipede (*Scutigera*) has 15 pairs of very long legs and lives in damp places, such as basements. It is not only harmless but actually beneficial as it feeds on insects.

Class Diplopoda

The mandibulate class Diplopoda comprises the **millipedes.** Their bodies are more or less cylindrical and consist of from about 25 to more than 100 segments (Fig. 21-30). Each **trunk** segment (except the first three) bears two pairs of **legs.**

The mouth parts are a pair of **mandibles** and a pair of **maxillae.** One pair of short **antennae,** which bears **olfactory receptors,** and clumps of **simple eyes (ocelli)** are usually present. **Scent glands** in most segments secrete a noxious fluid that is used in defense. In fact, a Micronesian millipede ejects for several centimeters a highly irritating fluid that can cause temporary blindness. The **tracheal tubes** are usually unbranched; they develop from pouches that open just in front of the legs.

Millipedes move very slowly despite their numerous legs. Some are able to coil themselves into a ball when disturbed, thereby shielding their relatively soft underbodies. They live in dark, moist microenvironments and feed principally on decaying vegetable matter and sometimes on living plants.

The sexes are separate, and eggs may be laid in a nest of damp earth. The young have six segments and only three pairs of legs when they hatch and resemble wingless insects. Other segments are added just in front of the anal segment in successive molts during growth.

Common millipedes include species of *Narcens* and *Julus;* the latter occur in meadows and gardens throughout the United States.

Summary

Arthropods are remarkable for their successful invasion of terrestrial habitats that have proved limiting or insurmountable to most other groups of invertebrates. Protists, nematodes, and earthworms have populated moist soils, and some leeches, snails, onychophorans, and crustaceans have adapted to humid habitats. But only the chelicerate arachnids and the mandibulate insects, millipedes, and centipedes are fully terrestrial. Of these, the insects have been by far the most successful. The evolution of wings opened many new niches to them. Insects have been able to exploit virtually every terrestrial habitat in which the temperature can support activity by poikilothermic (cold-blooded) animals. Some insects have secondarily invaded fresh waters as juveniles or adults and have become important to the ecology of these communities as well. Only the open ocean has rebuffed the adaptive radiation of the insects.

Review Questions

1. Describe the functions of tracheae.
2. If a hypothetical insect eats cars, describe some possible modifications of its mouthparts (use homologies). If some species began to specialize on certain parts of the car (the seats, tires, glass, etc.), what further modifications might one expect?
3. Insects are the only invertebrates that fly. What general insect attributes made flight possible?
4. Explain the reasons for the extraordinarily large number of insect species.
5. Distinguish among the different kinds of insect metamorphosis and give examples.

Selected References

Auerbach, S. L. "The Centipedes of the Chicago Area with Special Reference to Their Ecology," *Ecol. Monogr.*, **21**:97, 1951.

Belkin, J. N. *Fundamentals of Entomology*, Part 1. Baltimore: Bio-Rand Foundation, Inc., 1972.

Brues, C. T., A. L. Melander, and F. M. Carpenter, *Classification of Insects*, rev. ed. Cambridge, Mass.: Museum of Comparative Zoology Bulletin, Vol. 108, 1954.

Burtt, E. "Exudate from Millipedes, with Particular Reference to Its Injurious Effects," *Trop. Dis. Bull.*, **44**:7, 1947.

Carpenter, F. M. "The Geological History and Evolution of Insects," *American Scientist*, **41**:256, 1953.

Causey, N. B. "Studies on the Life History and the Ecology of the Hothouse Millipede, *Orthromorpha gracilis*," *Amer. Midl. Nat.*, **29**:670, 1943.

Chamberlin, R. B. "On Mexican Millipedes," *Bull. Univ. Utah*, **34**:1, 1943.

Cloudsley-Thompson, J. L. *Spiders, Scorpions, Centipedes, and Mites*. New York: Pergamon Press, 1958.

Dethier, V. G. *The Physiology of Insect Senses*. New York: John Wiley & Sons, Inc., 1963.

Goetsch, W. *The Ants*. Ann Arbor: University of Michigan Press, 1957.

Graham, S. A. *Forest Entomology*. New York: McGraw-Hill Book Co., 1952.

Herms, W. B. *Medical Entromology*, 5th ed., rev. by M. T. James. New York: Macmillan Publishing Co., Inc., 1961.

Hocking, B. *Biology or Oblivion*. Cambridge, Mass: Schenkman Publishing Company, Inc., 1965.

Horn, D. L. *Biology of Insects*. Philadelphia: W.B. Saunders Company, 1976.

James, M. T., and R. F. Harwood. *Herm's Medical Entomology*, 6th ed. New York: Macmillan Publishing Co., Inc., 1969.

Jaques, H. E. *How to Know the Insects*, 2nd ed. Dubuque: William C. Brown Company, 1947.

Johnson, C. G. "The Aerial Migration of Insects," *Scientific American*, **209**:132, 1963.

Lees, A. D. *The Physiology of Diapause in Arthropods*. Cambridge: Cambridge University Press, 1955.

Linsenmaier, W. *Insects of the World:* drawings, photos and text. New York: McGraw-Hill Book Company, 1972.

Metcalf, C. L., and W. P. Fling. *Destructive and Useful Insects*, 4th ed., rev. by R. L. Metcalf. New York: McGraw-Hill Book Co., 1962.

Michener, C. D., and M. H. Michener. *American Social Insects*. Princeton: D. Van Nostrand Company, 1951.

Oldroyd, H. *Collecting, Preserving and Studying Insects*. London: Hutchinson & Company, Ltd., 1958.

Price, P.W. (ed.). *Evolutionary Strategies of Parasitic Insects and Mites*. New York, Plenum Publishing Corporation, 1975.

Pringle, J. W. S. *Insect Flight.* New York: Cambridge University Press, 1957.

Roeder, K. D. (ed.). *Insect Physiology.* New York: John Wiley & Sons, 1953.

Ross, H. H. *A Textbook of Entomology,* 2nd ed. New York: John Wiley & Sons, 1956.

Saunders, D. S. *Insect Clocks.* New York: Pergamon Press, Inc., 1976.

Snodgrass, R. E. *Principles of Insect Morphology.* New York: McGraw-Hill Book Company, 1935.

Steinhaus, E. A., and R. F. Smith. *Annual Review of Entomology.* Stanford: Annual Reviews, Inc., 1956.

Steinhaus, E. A. (ed.). *Insect Pathology,* Vols 1 and 2. New York: Academic Press, 1963.

Thorpe, W. H. "Orientation and methods of communication of the honeybee and its sensitivity to the polarization of light," *Nature,* **164:**11, 1949.

von Frisch, K. *Bees, Their Vision, Chemical Senses, and Language.* Ithaca: Cornell University Press, 1950.

von Frisch, K. *The Dancing Bees.* New York: Harcourt Brace Jovanovich, 1955.

Wenner, A. M. "Sound communication in honeybees," *Scientific American,* **210:**116, 1964.

Wigglesworth, V. B. *The Physiology of Insect Metamorphosis.* New York: Cambridge University Press, 1954.

Wigglesworth, V. B. "Metamorphosis and differentiation," *Scientific American,* **200:**100, 1959.

Wigglesworth, V. B. *The Principles of Insect Physiology,* 5th ed. London: Methuen & Company, Ltd., 1959.

Wigglesworth, V. B. *The Life of Insects.* Cleveland: The World Publishing Company, 1964.

Wigglesworth, V. B. *Insects and the Life of Man.* New York, John Wiley & Sons, Inc., 1976.

Williams, C. B. *Insect Migration.* New York: William Collins Sons and Company, 1958.

Wilson, E. O. *The Insect Societies.* Cambridge: Harvard University Press, 1971.

Phylum Echinodermata

Echinoderms are conspicuous marine animals that have been known since ancient times. The term "echinoderm" was developed in 1734 and is derived from the Greek words meaning "spiny skin." But all the echinoderm classes were not recognized until 1801. Living representatives include (Fig. 22-1) starfishes (**Asteroidea**), brittle stars and basket stars (**Ophiuroidea**), sea urchins and sand dollars (**Echinoidea**), sea cucumbers (**Holothuroidea**), and sea lilies and feather stars (**Crinoidea**). Echinoderms are completely marine, and they inhabit all depths of the ocean.

Echinoderms are distinctly different from other metazoans in many ways. They exhibit **pentamerous radial symmetry,** with the body typically divided into five equal parts. Echinoderms are uncephalized, and the oral–aboral axis is perpendicular to the pentamerous plane. A **calcareous endoskeleton** is typically present, as is a complete **digestive tract.** The **nervous system** is complex and contains both central and peripheral components. There are generally no discrete excretory or respiratory systems. An outstanding characteristic of the echinoderms is their complex **coelom** and its derivatives: the **water vascular systems, perivisceral coelom,** and **hemal system.** The water vascular system is a unique organ system that functions in locomotion, feeding, respiration, and excretion.

Class Asteroidea

About 2000 extant species of **starfishes,** or **sea stars,** have been described. These familiar echinoderms live in marine waters throughout the world, especially in coastal waters. The asteroid body is composed of a central **disc** and radiating **arms** (**rays**). Many starfishes have 5 rays, but some (e.g., *Heliaster,* the sun star) have as many as 45 rays. Asteroids crawl over rocky, sandy, or muddy bottoms. Most are active carnivores, but some are suspension feeders.

Asterias—A Representative Starfish

A visit to a rocky coastline will reveal large numbers of starfish inhabiting shallow tide pools. The starfish has long been considered the most typical example of the echinoderm body plan and of their general biology.

External Morphology

On the lower, or **oral,** surface is the **mouth,** centrally situated on the animal, and five **ambulacral grooves,** one in each ray, from which extend two or four rows of **tube feet.** The rays may be flexed slowly by a few muscle fibers in the body wall. On the upper, or **aboral,** surface are many spines of various sizes, **pedicellariae** and dermal branchiae at the base of the spines, a **madreporite** that is the entrance to the water vascular system, and the **anal opening.**

Internal Morphology and Physiology

Skeletal System

The **endoskeleton** of **calcareous plates,** or **ossicles** (Fig. 22-2), is bound together by muscle and connec-

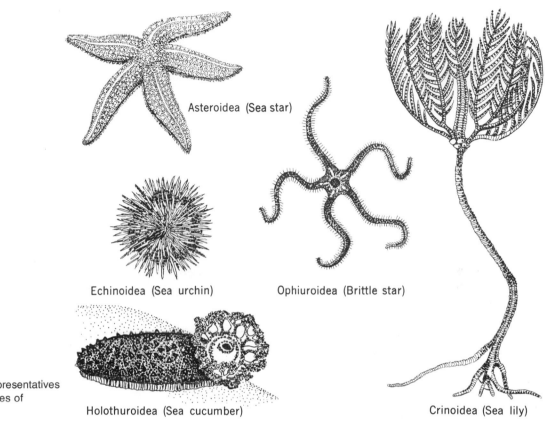

Asteroidea (Sea star)

Echinoidea (Sea urchin)

Ophiuroidea (Brittle star)

Figure 22-1. Representatives of the living classes of echinoderms.

Holothuroidea (Sea cucumber)

Crinoidea (Sea lily)

tive tissue fibers. The aboral **calcareous spines** are short and blunt and covered by **epidermis.** Around their bases are many modified spines, the **pedicellariae,** which resemble tiny jaws or scissor blades mounted on a stalk. When stimulated mechanically and chemically, the pedicellariae may be opened and closed by several sets of muscles. Their function is to keep the aboral surface clean, to aid in capturing food, and to protect the delicate **dermal branchiae.** The latter, also known as **papulae,** are thin-walled, hollow extensions of the perivisceral coelom that aid in excretion and facilitate gas exchange with the surrounding seawater.

Water Vascular System

The water vascular system (Fig. 22-3) is a division of the coelom that is peculiar to echinoderms. It is essentially a system of hollow, interconnected tubes filled with water. The water within the system seems to be under a very slight positive pressure. Water probably enters through the madreporite on the aboral disc, perhaps due to the action of cilia that line the entire system. The terminal organs of the water vascular system are the **tube feet (podia)** in the grooves on the oral surface.

Beginning with the madreporite, the **stone canal** runs downward and enters the **ring canal,** which encircles the mouth. From the ring canal, five **radial canals,** one in each ray, pass outward just above the ambulacral grooves. The radial canals give off side branches from which arise the **ampullae** of the tube feet (Fig. 22-3).

The water vascular system is a hydraulic pressure system. The starfish moves by means of its tube feet. A podium elongates as muscles surrounding its am-

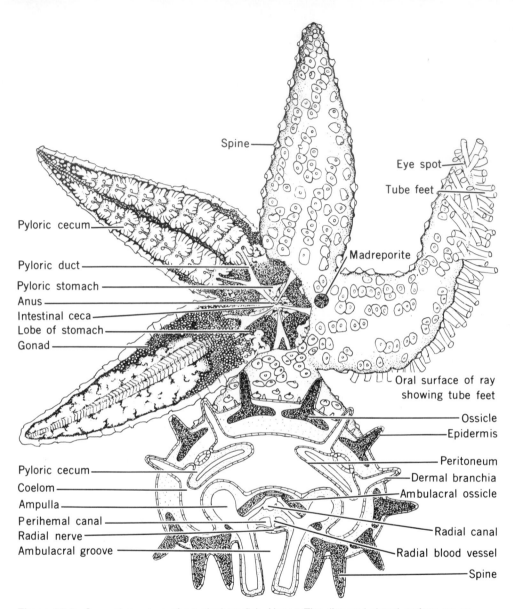

Figure 22-2. General structure of a typical starfish. Upper: The disc and aboral surface are removed from two rays. Lower: One ray is shown in cross-section, without gonads.

pulla contract, thus forcing fluid into the foot. To prevent a ballooning effect, a series of annulated rings extend the length of the tube foot and limit the expansion of its diameter. As the bottom of the podium, the **sucker,** becomes pressed against the substratum, contraction of local muscles elevates its center, creating an adhesive suction. Sticky secretions from **gland** cells at the tip of the podium also aid adhesion. **Longitudinal muscles** then contract, shortening the podium and helping to pull the starfish forward. The combined effects of the forward thrusts of many coordinated tube feet produces locomotion in the starfish. After each tube foot completes a forward thrust, its sucker is detached from the substratum; the foot then

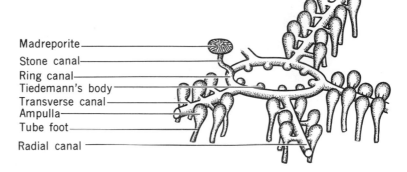

Madreporite
Stone canal
Ring canal
Tiedemann's body
Transverse canal
Ampulla
Tube foot
Radial canal

Figure 22-3. Diagram of starfish water vascular system. One of the radial canals is cut off at the base; the other four are cut off near the base.

contracts and points forward to start the process again. All the tube feet act in a coordinated way by extending in the same direction, but not all at the same time. The starfish advances slowly, at about 15 cm per minute.

The tube feet are used not only for locomotion but for clinging to surf-swept rocks, for capturing and handling food, and for respiration and excretion. There are also nine small spherical swellings, called **Tiedemann's bodies,** on the inner wall of the ring canal. Their function has not been established with certainty, but they may be involved in amoebocyte production.

Digestive System

The digestive tract (Fig. 22-2) is short and highly specialized. The **mouth** opens into a very short **esophagus** that leads into a thin-walled sac, the **stomach.** The stomach consists of two parts: a large **cardiac** portion and a smaller, aboral, **pyloric** portion. From the pyloric stomach a tube passes into each ray and then divides into two branches called **hepatic ceca,** which possess many lateral pouches. They are green in color. Above the stomach is the slender **intestine,** which opens through the **anus** on the aboral surface. Two branched pouches, brown in color, arise from the intestine and are known as **rectal,** or **intestinal, ceca.**

The food of the starfish consists of almost any animal matter, such as fishes, oysters, mussels, barnacles, clams, snails, worms, and crustaceans. Small food particles may be passed to the mouth by the pedicellariae or the tube feet. Bivalves are straddled and partially opened by the starfish; then the cardiac

stomach is everted between the gape in the shell (Fig. 22-4). **Digestive enzymes** are released, and the prey's tissues are reduced to a soupy broth. Food particles and fluid are swept by cilia into the hepatic ceca, where further digestion may proceed either extracellularly or intracellularly. Nutrients are stored in the hepatic ceca and released into the hemal system or coelomic fluid for distribution. The rectal ceca may also have a digestive function.

Internal Transport Systems

The principal medium of internal transport in echinoderms is the fluid in the **perivisceral coelom.** Wandering amoebocytes (**coelomocytes**) assist in the transport of nutrients and wastes and also function in clotting.

The water vascular system and the perivisceral coelom are the two most developed derivatives of the echinoderm coelom. The third division, the **hemal system,** is usually poorly developed except in the **holithuroids (sea cucumbers).** The principal function of the hemal system appears to be distribution of food materials that enter it via amoebocytes from the digestive tract.

Respiration

Respiratory gas exchange also occurs across the surfaces of the **dermal branchiae** and **podia.** As mentioned earlier, the dermal branchiae are outpocketings of the coelom that pass through minute openings in the skeleton. They are covered with **cilia** on both their internal and external surfaces. The external cilia

Figure 22-4. Starfish, *Pisaster ochraceus,* eating a mussel. The starfish attaches its tube feet to the two shells of the mussel and, by a continuous pull, virtually at right angles to the surface of each shell, eventually opens it enough to allow the stomach of the starfish to evert through its own mouth, coming in contact with the soft parts of the clam. Thus the bivalve is actually ingested in its own shell.

keep a current of oxygenated water passing over the outside, and the internal cilia circulate the coelomic fluid into the branchiae.

Excretion

Ammonia is the principal soluble nitrogenous waste, and it diffuses through thin areas of the body surface, especially the tube feet and dermal branchiae. Particulate wastes are picked up by wandering **amoebocytes** within the coelom; these amoebocytes are subsequently eliminated through the walls of the **dermal branchiae.**

Nervous System and Sense Organs

The nervous system of asteroids consists of interconnected peripheral and central components. The **peripheral nervous system** is a subepidermal **nerve net** that extends throughout the body, including the tube feet. This system controls local responses of the pedicellariae, spines, and dermal branchiae to specific stimuli. **Lateral tracts** conduct neural impulses at higher speeds to and from the **central nervous system.** This consists of a **nerve ring** around the mouth plus five large **radial nerves** that run parallel to the radial canals of the water vascular system.

Most echinoderms respond to **touch, gravity, light,** and **chemical stimuli,** yet sense organs are poorly developed. The primary receptors are **sensory cells** located in the epidermis, especially on the podia and the margins of the ambulacral grooves. At the end of each asteroid ray is a small **tactile tentacle** and a light-sensitive **eyespot** composed of 80–200 **ocelli.** Most asteroids are **negatively phototaxic;** they are generally nocturnal and seek the shade of algae and crevices during the day.

Asteroids show a very limited capacity to learn, but they can be conditioned to modify some responses to particular stimuli. They are dull students, however, slow to learn, and rapidly forgetful.

Reproduction and Regeneration

The sexes of the starfish are separate. A pair of branched **gonads** lie in the perivisceral coelom at the base of each arm (Fig. 22-3). Female starfish have

been known to release as many as 2.5 million **eggs** in two hours, and 200 million eggs may be produced in a single breeding season. Male starfish produce many times that number of **sperm.** Fertilization and development generally occur externally, in the seawater.

Echinoderm development includes a ciliated larval stage called a **bipinnaria** (Fig. 22-5) that has **bilateral symmetry** before it metamorphoses into a **radially symmetrical** young starfish. Echinoderms are clearly **deuterostomes,** as their coelom develops **enterocoelously.**

Asteroids are noted for their ability to **regenerate** a piece of their body into an entire individual. Some asteroids and ophiuroids (brittle stars) foster such asexual regeneration **fissiparously;** that is, the disc divides into pieces with one or more arms.

Class Ophiuroidea—Brittle Stars and Basket Stars

Class Ophiuroidea is the second largest group of living echinoderms, with at least 1700 known species. Ophiuroids (Fig. 22-6) exhibit two general body forms that are commonly called brittle stars and basket

Figure 22-5(A). The life cycle of the common starfish *Asterias vulgaris*. Note that in the later larval stages the starfish has bilateral symmetry before it attains radial symmetry.

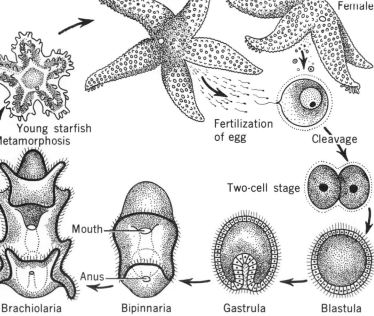

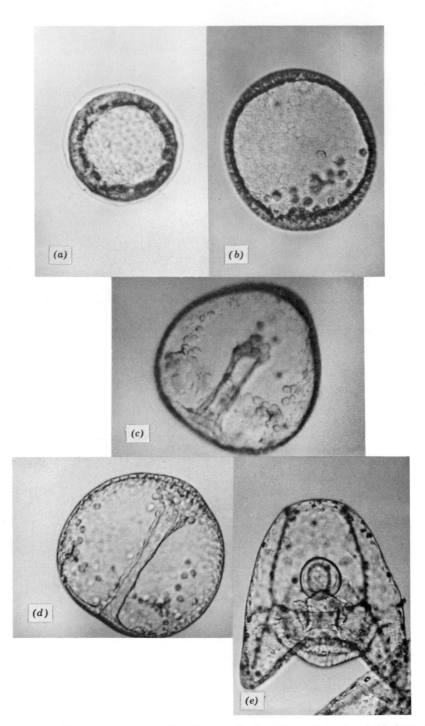

Figure 22-5(B). Life cycle of the common sea urchin. Compare the five stages shown here with those of the starfish.

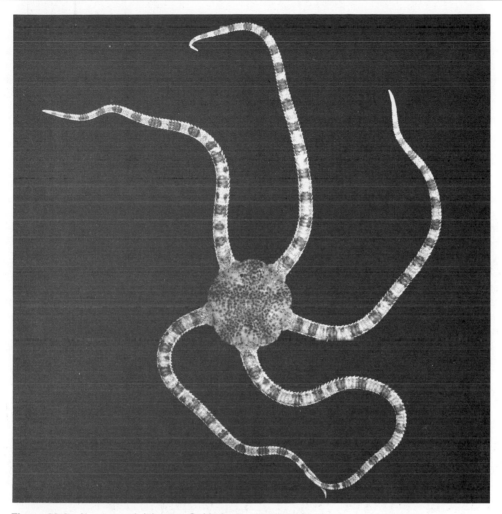

Figure 22-6. A common brittle star, *Ophioderma panamensis*.

stars. **Brittle stars,** so named because of their ability to self-amputate (**autotomy**), typically have only five arms, but in **basket stars** the five primary arms branch repeatedly to form a mass of tentaclelike coils.

Several differences from asteroids are readily apparent. The thin **arms** are distinctly set off from the **central disc,** and the **ambulacral grooves** on the oral side are covered by **plates; tube feet** protrude through small holes (as in sea urchins). There are no suckers on the tube feet, which are used for sensory and food gathering functions instead of locomotion.

The **madreporite** is ventral (oral); there are no pedicellariae, and there is no anus.

The arms of ophiuroids are capable of great lateral mobility, and, as a result, another common name for them is **serpent stars.** Snakelike movements of the arms allow ophiuroids to crawl rapidly over the substratum. Ophiuroids feed on detritus and small animals and plants. Tube feet near the mouth may shove in mud along with food, or prey may be grasped by the arms. The ophiuroid **mouth** is equipped with crushing **jaws** with **teeth.** Some capture fine detritus in mucus secreted on the arms; podia and flagella sweep food toward the mouth. Ophiuroids are found throughout the ocean, especially in protected microhabitats and are often abundant on soft bottoms of the abyssal depths.

Class Echinoidea—Sea Urchins and Sand Dollars

The name **echinoidea** means "like a hedgehog," which aptly describes these spine-covered **sea urchins, sand dollars,** and **heart urchins.** Echinoids lack rays or arms. The **ambulacral** areas bearing the **tube feet** extend from near the mouth in five bands up the sides to near the anus on the aboral surface. *Arbacia punctulata* is a purple-colored sea urchin that lives from Cape Cod to southern Mexico (Fig. 22-7). Its Pacific Coast counterpart is the purple urchin, *Strongylocentrotus purpuratus.* These urchins are roughly globular in shape. The **test (shell)** is made up of **calcareous plates** that are solidly fused together. Both the ambulacral and interambulacral areas bear movable **spines** about 25 mm long. The spines are moved by muscles around little knoblike elevations called **tubercles.** Several types of **pedicellariae** are present.

Food consists of plant and animal matter that is in-gested by means of a complex chewing apparatus, called **Aristotle's lantern,** that has five calcareous **teeth** (Fig. 22-7). The **water vascular system** consists of an aboral **madreporite,** a **stone canal,** a **ring canal,** and five **radial canals** that extend meridionally and connect with the ampullae of the podia via **transverse canals** (Fig. 22-7). **Respiration** takes place in sea urchins through five pairs of **peristomial** ("around the mouth") **gills** and the podia.

Sea urchins are variously colored; they may be white, purple, green, yellowish green, gray, black, and so on. Some species possess very long **poisonous spines,** such as the *Diadema* of Florida and the West Indies. Heart urchins (e.g. *Echinocardium*) represent variations of the echinoid body plan.

Class Holothuroidea—Sea Cucumbers

Sea cucumbers (Figs. 22-1, 22-8) differ from other echinoderms by having elongated, cylindrical bodies and by lying on their sides.

The **secondary radial symmetry** shared by all echinoderms is modified in the holothuroids by an elongation of the oral–aboral axis, thus producing a **tertiary bilateral symmetry.** Some sea cucumbers are long, slender, and wormlike; others are fat and sausage shaped. About 500 living species of holothuroids are

Figure 22-7(A). Sea urchin, *Arbacia,* showing both external and internal structures.

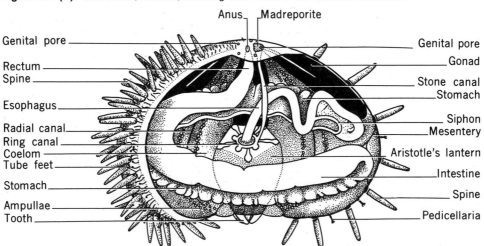

Figure 22-7(B). A population of sea urchins, *Strongylocentrotus purpuratus*.

Figure 22-7(C). Sea urchin test with spines removed.

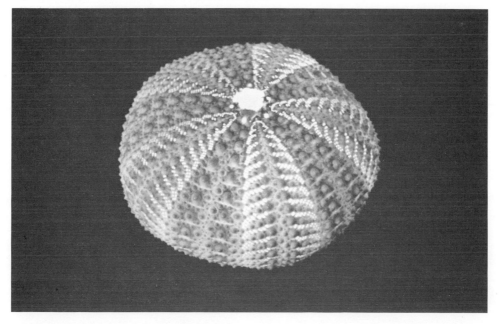

A

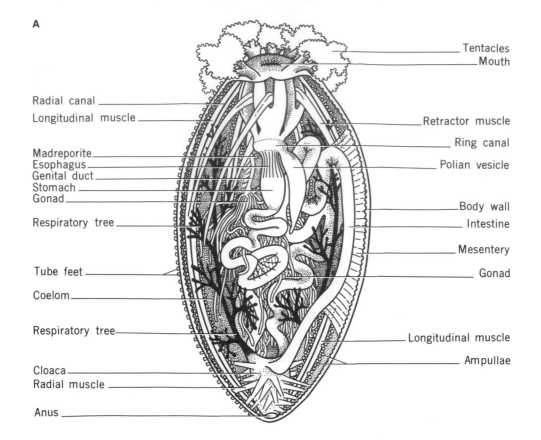

Tentacles
Mouth
Radial canal
Longitudinal muscle
Retractor muscle
Ring canal
Madreporite
Esophagus
Polian vesicle
Genital duct
Stomach
Gonad
Body wall
Respiratory tree
Intestine
Mesentery
Tube feet
Gonad
Coelom
Respiratory tree
Longitudinal muscle
Ampullae
Cloaca
Radial muscle
Anus

B

Figure 22-8. (A) Sea cucumber, *Cucumaria,* showing the tube feet and internal structures. (B) The giant sea cucumber, *Thelonotus,* 1 m long.

recognized, most of which inhabit tropical seas. Their colors are varied: brown, yellowish, reddish, whitish, black, pink, purplish, and so on. One Puget Sound species, *Psolus chitinoides,* has an orange-colored body and crimson-colored neck and tentacles.

Sea cucumbers live sluggishly on the ocean floor; some burrow in mud or sand, leaving only the ends exposed. Their **endoskeleton** is greatly reduced, and the soft but muscular body wall contains numerous microscopic **calcareous ossicles.**

From 10 to 30 of the **tube feet** surrounding the mouth are modified as food-gathering **tentacles.** Most sea cucumbers feed on organic matter digested from the sand or mud that they shovel into their mouths with the aid of their tentacles as they move about. The **digestive tract** (Fig. 22-8) consists of a **mouth,** a short **esophagus,** a small muscular **stomach,** and a long, looped **intestine** that expands into a muscular **cloaca** that opens through the **anus.**

Respiration is performed in the **cloaca.** Connected to the cloaca are two branching series of tubules, called **respiratory trees,** within the coelom on either side of the digestive tract. These tubules are filled by slow, pumping contractions of the cloaca and then emptied by an abrupt contraction of the tubule walls.

Class Crinoidea—Sea Lillies and Feather Stars

Class Crinoidea contains the most ancient and primitive of the living echinoderms. There are about 4350 recognized species of fossil crinoids but only 650 living species. Of these, about 80 species known as **sea lilies** (Fig. 22-9) are attached to the sea bottom by a long, jointed **stalk.** The others, called **feather stars,** have secondarily lost the stalk and have adopted a free-living existence. The stalked sea lilies live in deep water, often on mud substrata, whereas the feather stars are most abundant in shallow, rocky habitats.

Figure 22-9. The crinoid, *Comanthus,* a common inhabitant of shallow tropical waters.

The crinoid body may be conveniently divided into three regions: the body proper (**crown**), which encloses the viscera; the **stalk (stem)**; and attachment devices (if any). The crown bears several to the many **arms** that extend freely into the water. There are primitively five arms, but in most crinoids each arm forks at least once to produce ten in all. This branching process may be repeated up to eight or nine times. The arms bear many smaller branches, called **pinnules,** giving them a feathery appearance. Crinoids feed on small planktonic organisms that are trapped in mucus that covers their arms.

Autotomy in Echinoderms

Ophiuroids, many crinoids, and some asteroids exhibit **autotomy.** Ophiuroids derive their common name "brittle stars" from their tendency to cast off (autotomize) one or more arms when roughly handled. The autotomized arm wiggles for a while and then dies. This presumably diverts the attention of a predator and allows the ophiuroid to escape. The missing arm is soon regenerated. A starfish with regenerating arms is often encountered in nature.

Under unfavorable conditions, holothuroids may rupture and expel their viscera: respiratory trees, digestive tract, and gonads. Eviscerated sea cucumbers generally regenerate the missing organs. There are some indications that **evisceration** occurs under natural conditions during the warmer months.

Relations of Echinoderms to Humans

Echinoderms are of considerable importance to humans. In the Orient, **sea cucumbers** are dried in the sun and sold as trepang, or *bêche de mer*, for culinary use, especially in soups. The gonads of sea urchins are also eaten in certain Asian and tropical countries.

The dried skeletons of echinoderms have been crushed and used as a fertilizer because of their high content of calcium and nitrogenous compounds.

Predatory starfishes are very destructive in oyster beds; a sea star has been observed to consume ten oysters or clams in a day. Several control measures are in general use: (1) In some regions, a moplike tangle of threads is dragged across the oyster beds, and the starfishes grab onto these with their pedicellariae. The mop is then removed and plunged into hot water, killing the animals. (2) An effective method of killing starfishes is to spread quicklime over the oyster beds are a concentration that is harmless to the oysters but fatal to the starfish. (3) Bivalves are often cultured on platforms suspended above the bottom and out of reach of predatory asteroids and snails.

A coral-eating starfish, *Acanthaster planci*, has caused much excitement because of its recent spread in the tropics. One theory holds that the phenomenon is a natural periodic occurrence (with a period of about 100 years); another view is that it was precipitated by the building of harbors. There is much speculation on the subject, but no one really knows yet what caused the sudden population boom.

Origin of the Echinoderms

In the preface to one of her comprehensive volumes on the invertebrates, Dr. Libbie Hyman states, "I also here salute the echinoderms as a noble group especially designed to puzzle the zoologist." Today, this statement is as true as when it was made over two decades ago. As more scientists investigate echinoderm biology, the problems that they study are becoming increasingly complicated. Although the echinoderms appear anatomically simple, the results of many studies demonstrate that they are physiologically very complex.

Echinoderms and cnidarians, because of their radial symmetries, were at one time placed together in a group called **Radiata.** The anatomy of the adults and larvae, however, show that these phyla really occupy widely separate positions in the animal kingdom.

Adult echinoderms are so distinct that they cannot be compared with any other group of animals, and we must look to their larvae for signs of relationship. The bilateral bipinnaria larva is either a modification for free-swimming dispersal or a vestige of the ancestral body plan. The latter view is accepted by most zoologists. The ancestors of echinoderms were very likely bilateral, wormlike animals that secondarily became radial and took up sessile habits. Contemporary free-living echinoderms have probably been derived from fixed ancestors whose symmetry they still reflect. The oldest echinoderm fossils are crinoidlike filter feeders that attached to the substratum by their aboral surface. The tube feet used for feeding by these sessile ancestors became modified for locomotion during the course of evolution as the echinoderms became unanchored and inverted. Perhaps a crinoidlike ancestor that lay on its side began to locomote like the holothuroids, suggesting an evolutionary pattern for that group. Mutations that permitted this type to turn farther over (until the oral surface was down) gave rise ultimately to the other echinoderms.

Echinoderms have several features common to the chordates and are considered their closest relatives. Common features include the presence of an ectodermal neural system, an endoskeleton, and some embryological similarities.

Summary

Echinoderms are radially symmetrical as adults but bilaterally symmetrical as larvae. There is no segmentation. The body wall usually contains calcareous plates that form an endoskeleton. The nervous system is close to the ectoderm and has an oral nerve ring and radial nerves. Sexes are usually separate. A unique water vascular system, including tube feet, is usually present.

Review Questions

1. Describe the possible consequences to a starfish that for some reason lost its entire water vascular system.
2. What would be the probable fate of a starfish born without pedicellariae?
3. In the sea urchins, the coelomic fluid has the ability to coagulate (like our blood), which leads one to believe there is some functional significance in keeping this fluid inside the coelom. What is the significance?

4. Chesapeake Bay oyster fishermen at one time attempted to eliminate predatory starfishes by cutting them in half and tossing them overboard. Why do you suppose they discontinued this method of disposal?

Selected References

Binyou, John, *Physiology of Echinoderms* New York: Pergamon Press, Inc., 1972.

Boolootian, R. A. The Coagulation of Echinoderm Body Fluids. Doctoral dissertation. Stanford University, 1957.

Bury, H. "The Metamorphosis of Echinoderms," *Quart. J. Micr. Sci.*, 38:45, 1895.

Chia, F., and A. H. Whiteley. *Developmental Biology of the Echinoderms*. Thousand Oaks, California: American Society of Zoologists, 1975.

Clark, A. M. *Starfishes and Their Relations*. London: British Museum (Natural History), 1962.

Coe, W. R. *Echinoderms of Connecticut*. Connecticut State Geological and Natural History Survey, Bull. 19, 1912.

Feder, H. M. "On the Methods Used by the Starfish, *Pisaster ochraceus*, in opening three types of bivalved mollusks," *Ecology*, 36:764, 1955.

Fell, H. B. "echinoderm Embryology and the Origin of Chordates," *Biol. Rev.*, 23:81, 1948.

Harvey, E. G. *The American Arbacia and Other Sea Urchins*. Princeton: Princeton University Press, 1956.

Hyman, L. H. "Echinodermata," in *The Invertebrates*, Vol. 4. New York: McGraw-Hill Book Company, 1955.

Jennings, H. D. "Behavior of the Starfish *Asterias forreri de loriol*," *Univ. Calif. Pub. Zool.*, 4:53, 1907.

Millot, N. (ed.) *Echinoderm Biology*. New York: Academic Press, 1967.

Nichols, D. *Echinoderms*. London: Hutchinson & Company, Ltd., 1962.

Phyla Hemichordata and Chordata

Phylum Hemichordata

Hemichordates include two groups of wormlike marine animals (**acorn worms,** Fig. 23-1, and **pterobranchs**) that live in tubes within bottom muds or among rocks and algae. The hemichordates (Gr. *hemi,* half + L. *chorda,* chord) may be considered modern representatives of an ancient "link" between the echinoderms and the chordates because they possess features common to both groups. Their echinodermlike features include a partially **subectodermal nervous system,** similar **larval stages, water vascularlike coelomic cavities,** a similar marine habitat, and similar patterns of filter feeding among some groups. Chordatelike features include a partial **dorsal hollow nerve chord, gill pouches** (used primarily for feeding), and a structure that was once thought to be a **notochord** (Fig. 23-2). Many zoologists now consider the hemichordates to lack a true notochord, and so these little-known animals are now considered as a separate phylum. The evolutionary origin of the hemichordates is unclear, but they share basic **deuterostome** characters with the **chaetognaths, echinoderms,** and the **primitive chordates;** all perhaps evolved from a common ancestor.

Phylum Chordata

Phylum Chordata, to which we ourselves belong, is one of the largest and most diverse groups of animals.

There are more than 40,000 species of chordates, of which the bony fishes account for about one half. The chordates, arthropods, and mollusks are the three metazoan groups that have successfully adapted to marine, freshwater, and terrestrial habitats. The great evolutionary success of these phyla is interestingly coupled with their positions on the summits of the two great branches of metazoan evolution: the arthropods and mollusks on the protostome branch and the chordates on the deuterostome branch.

Chordata is commonly associated principally with the backbone-bearing vertebrates, but it also includes two groups of invertebrate animals. All chordates are characterized by three features that are present in some stage of their life histories: (1) **pharyngeal gill slits,** or **pouches;** (2) a **dorsal nerve cord,** at least partially hollow or tubular; and (3) a supporting rod, the **notochord,** between the dorsal nerve and the digestive tract.

The various chordates differ widely from one another, and it is customary to separate them into **subphyla.** Some zoologists also include the **Hemichordata** as a subphylum. Commonly accepted as subphyla of phylum Chordata are:

1. **Urochordata. Tunicates (sea squirts)** and several similar marine forms.
2. **Cephalochordata.** Small fish-shaped animals called **lancelets.**
3. **Vertebrata.** Cranium and vertebrae present. In-

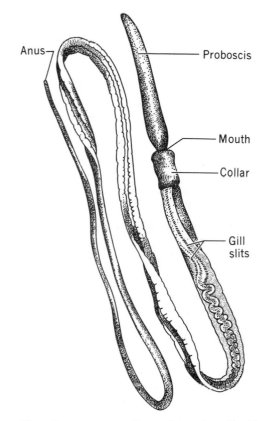

Figure 23-1. An acorn worm, *Saccoglossus kowalevskii,* a species that lives on sand flats from Massachusetts Bay to Beaufort, North Carolina. Natural size about 17 cm long.

cludes **fishes, amphibians, reptiles, birds,** and **mammals** (Fig. 23-3).

Subphylum Vertebrata includes most of the chordates, but the other two subphyla of **protochordates** are of considerable interest as they are more primitive and hence give us some idea of the animals from which the vertebrates possibly evolved.

The **coelom** of all chordates develops **enterocoelously,** as outpocketings of the primitive gut, and this firmly allies them with the echinoderms, chaetognaths, and pogonophorans. Echinoderms possess an endoskeleton, bilateral larvae, and many embrological similarities to the chordates.

It is significant to note that **metamerism** evolved in the enterocoelous vertebrates independently of the schizocoelous annelids and arthropods. The embryos of all three groups are clearly metameric, but later development obscures the serial arrangement of organs in arthropods and vertebrates, as the metameres are consolidated and specialized to perform distinct functions.

A discussion of the primitive chordate subphyla and further introduction to vertebrates follows. Chapters 24 to 30 consider the classes within subphylum Vertebrata in detail. Figure 23-3 shows the evolutionary origins and abundances of the nine vertebrate classes, and the classification appendix contains a detailed phylogeny of the chordates. When reading the next six chapters, the student is urged to re-

Figure 23-2. Acorn worm, *Balanoglossus.* (A) Midsagittal section showing some internal structures. Some zoologists call the short anterior structure, which has long been called a notochord, a "stomochord," but this new name does not rule out the possibility of its being homologous with the notochord of the vertebrates. (B) Cross-section through the trunk region.

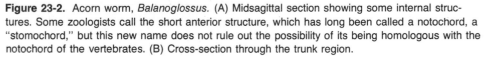

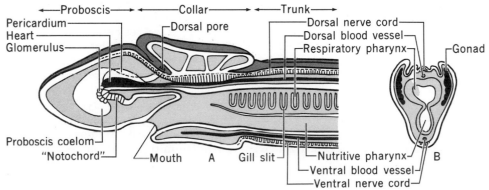

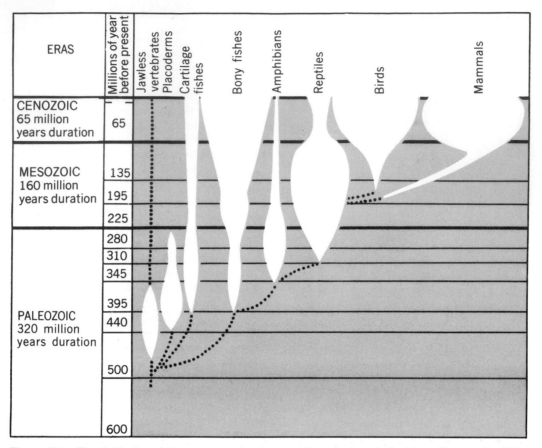

ERAS	Millions of years before present									
		Jawless vertebrates	Placoderms	Cartilage fishes	Bony fishes	Amphibians	Reptiles	Birds	Mammals	
CENOZOIC 65 million years duration	65									
MESOZOIC 160 million years duration	135									
	195									
	225									
PALEOZOIC 320 million years duration	280									
	310									
	345									
	395									
	440									
	500									
	600									

Figure 23-3. The distribution of the major vertebrate groups throughout geological time. Changes in width of the white areas indicate the relative abundance of each group through time; broken lines suggest possible sources and time of origin for certain groups.

fer to this diagram and to Figure 23-3 to understand the evolutionary relationships of the numerous taxa discussed.

Subphylum Urochordata— The Tunicates or Ascidians

The urochordates are entirely marine. They are either sessile or free swimming and are widely distributed throughout the oceans of the world, from near the surface to depths of over 5 km. They may live as solitary individuals, or they may form **colonies** by **asexual budding.** Some are brilliantly colored or **bioluminescent.** Only the sessile **ascidians,** or **sea squirts,** will be considered here; see the classification appendix for a summary of other urochordates.

Adult ascidians are saclike animals that received the common name "sea squirt" because, when irritated, they may eject water through two openings in their unattached end. The term **tunicate** was formerly applied to the group because their body wall is enclosed by a leathery epidermal secretion, known as a **tunic** that contains **cellulose.**

The chordate characteristics of tunicates were not recognized until their early development and larval metamorphosis were fully investigated. It was then discovered that their typical tadpole-shaped **larva** (Fig. 23-4), which is about 5 mm long, possesses (1) a distinct **notochord;** (2) a **dorsal neural tube** in the tail, enlarging in the trunk and ending in a **cephalic vesi-**

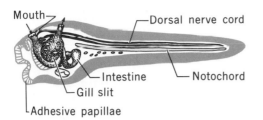

Figure 23-4. The free-swimming larva of a tunicate, showing all three chordate characteristics: notochord, dorsal nerve cord, and gill slits.

cle, which is considered the forerunner of the **brain** of the vertebrates; and (3) a **pharyngeal sac** that opens to the exterior by a multitude of **ciliated gill slits.**

The tail propels the larva forward by lateral strokes. After a short free-swimming existence, the larva be-

comes attached to some object by adhesive **papillae** on its anterior end. It then undergoes a **retrogressive metamorphosis** during which the tail with the notochord disappears and the nervous system is reduced to a single **ganglion.**

A typical adult tunicate (Fig. 23-5) is permanently attached to the substrate by a **base** or **stalk** surrounded by a thick, tough, elastic **tunic.** This is composed of a celluloselike substance, a material rarely found in animals but common in plants. The tunic is lined by a membranous **mantle** (or body wall) that contains **epidermis, muscle fibers,** and **blood vessels.**

At the unattached end of the animal are two openings: the **incurrent siphon (branchial opening),** into which a current of water passes, and the **excurrent siphon (atrial opening)** through which water is expelled. This flow of water brings food into the **diges-**

Figure 23-5. Internal structure of a tunicate, *Molgula.* Tunic, mantle, and pharynx removed from left side. Arrows indicate flow of water currents through the animal.

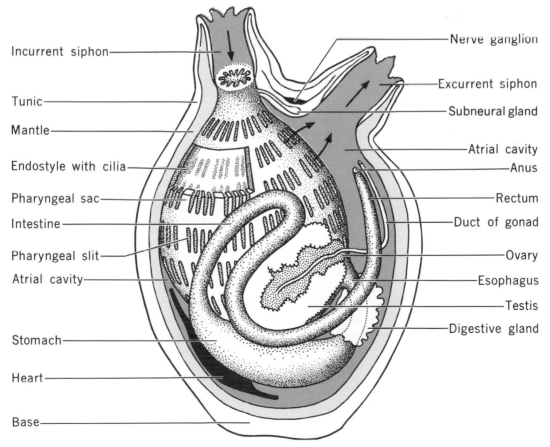

tive tract, furnishes oxygen for respiration, and carries away wastes and **gametes.** Water first enters the saclike **pharynx** and then passes through the numerous **gill slits** that perforate its walls. Suspended food materials are trapped in mucus that lines the gill slits, and respiratory gases are exchanged across the vascularized surface of the pharynx. **Cilia** drive water through the gill slits into the **atrial chamber,** which almost completely surrounds the pharynx. The anus and duct of the single **gonad** empty into the atrium near the excurrent siphon, and so their products are eliminated with the expelled water.

Microscopic plants and animals are trapped in mucus and are transported by cilia into the **esophagus,** which leads to the **stomach.** A **digestive gland** connects by a duct to the stomach. The **intestine** bends back upon itself before opening at the **anus.** A single **nerve ganglion** lies between the two siphons and supplies nerves to all parts of the body. Located below this ganglion is a **subneural gland** that seems to be homologous with the posterior lobe of the vertebrate pituitary gland. Hormone extracts from ascidian subneural glands have been shown to elicit pituitary like responses (e.g., uterine contractions) when administered to small mammals.

The **circulatory system** consists of a tubular **heart** with a large vessel connected to each end, and each vessel gives off branches into open **blood sinuses.** A unique feature of ascidian circulation is that the direction of blood flow is reversed at short intervals. The significance of this reversal of heartbeat is unknown.

Nearly all ascidians are **hermaphroditic,** but self-fertilization is probably uncommon. Some species also reproduce asexually by **budding.**

Subphylum Cephalochordata, *Amphioxus*—A Representative Cephalochordate

Subphylum Cephalochordata contains about 30 species of small marine animals with uniformly elongated, fish-shaped bodies. *Amphioxus*, a typical cephalochordate, is cosmopolitan in distribution but is confined to pure sand and gravel bottoms of shallow marine waters, such as tidal flats. Cephalochordates are of special interest because they exhibit chordate characteristics (a notochord, gill slits, and a dorsal tubular nerve cord) in a simple condition that may be reminiscent of the ancestors of the vertebrates.

Amphioxus measures about 5 cm long. The semitransparent body is pointed at both ends and is laterally compressed. It burrows in clean sand with its head or tail, concealing all but the anterior end. **Amphioxus** can swim through sand or water by means of lateral undulations of its body. When it ceases to move, it falls on its side.

External Morphology

Amphioxus (Fig. 23-6) is shaped like a fish but lacks lateral fins and a distinct head. Along the middorsal line, a low **dorsal fin** extends the entire length of the

Figure 23-6. *Amphioxus,* an animal that illustrates the three fundamental chordate characteristics. An adult with part of the body wall removed from left side to show the general structure. Natural size is about 5 cm long.

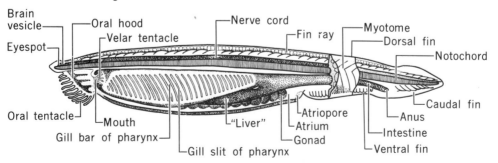

body and widens into a **caudal fin** at the posterior end. The caudal fin extends forward on the ventral surface to form the short **ventral fin.** The fins are strengthened by rods of connective tissue called **fin rays.** Anterior to the ventral fin, each side of the integument is expanded into a **metapleural fold** (Fig. 23-7).

The body wall is divided into V-shaped muscle segments, called **myotomes,** that are separated from one another by connective tissue. The myotomes on one side of the body alternate with those on the other side. The muscle fibers that they contain are arranged longitudinally and attach to the connective tissue partitions. As a result, they are able to produce the lateral body movements in swimming.

The protruding **mouth** is surrounded by an **oral hood** from which project many ciliated **oral tentacles** (**cirri**). The **anus** opens on the left side of the body near the base of the caudal fin. The **atriopore,** an opening through which water used in respiration passes to the exterior, is located just anterior to the ventral fin.

The funnel-shaped **vestibule** is the cavity of the oral hood anterior to the mouth. The 22 ciliated tentacles that project from the edge of the oral hood are provided with **sensory cells.** The mouth is surrounded by a membrane of circular muscle fibers, the **velum,** that may close its opening. Twelve sensory oral or **velar tentacles** act as a strainer, preventing entrance of coarse particulates.

Figure 23-7. A cross-section of *amphioxus* in the pharyngeal region, showing various internal structures, including some of the coelomic cavities.

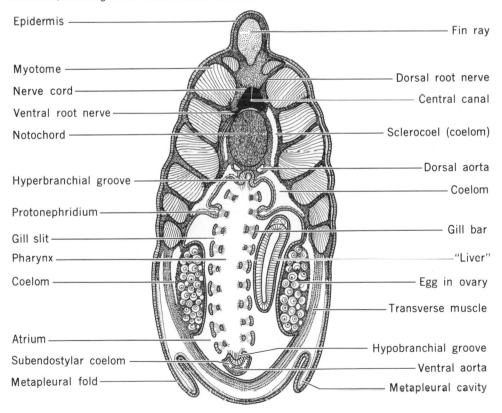

Internal Morphology and Physiology

Skeletal System

Amphioxus has a well-developed **notochord** (Figs. 23-6 and 23-7) that acts as the main skeletal support of the body. The notochord is a rod of connective tissue lying dorsal to the digestive tract and extending almost the entire length of the body. It is composed of vacuolated cells that are made turgid and rigid by their fluid contents. Other skeletal structures are the connective tissue rods that form the fin rays and similar structures that support the oral cirri and gill bars.

Digestive System

Food of *Amphioxus* consists of minute organisms that are carried into the mouth in a water current produced by cilia on the gills.

The **digestive tract** is complete. The **mouth** opens into a spacious **pharynx** whose walls are pierced with almost 180 pairs of **gill slits;** these sieve food particles from water that is drawn into the mouth. A ventral ciliated groove, called the **endostyle,** secretes mucus. **Cilia** convey the mucus upward over the gill bars, where it entraps food particles. Cilia then pass the food-laden mucus to the **stomach.** A blind **hepatic cecum** (liver) emerges from the ventral side of the stomach, whose main body communicates with the straight **intestine.** The **anus** opens ventrally, just anterior to the caudal fin.

Circulatory System

The cephalochordate circulatory system is generally similar to those in higher chordates such as fishes except that it lacks a heart and contains **open sinuses** in the tissues. The **subintestinal vein** collects nutrient-laden blood from the intestine and carries it forward into the **hepatic portal vein** and then to the **hepatic cecum.** The hepatic vein leads from cecum to the **ventral aorta.** Blood is pumped by means of rhythmic contractions of the **ventral aorta** into **afferent branchial arteries** that supply the gill bars and then through **efferent branchial arteries** into the paired **dorsal aortae.** The blood is oxygenated during its passage through the gill slits. Oxygenated blood passes back into the **median dorsal aorta** and finally, by way of **intestinal capillaries,** into the **subintestinal**

vein. The direction of the blood flow—posterior in the dorsal and anterior in the ventral vessel—is the same as that in the vertebrates but the reverse of that found in invertebrates such as annelids.

Respiratory System

The **pharynx** (Fig. 23-7) is attached dorsally and hangs down into a cavity called the **atrium.** The atrium is lined with an ectodermal epithelium and thus represents an enclosed part of the external environment. Water is drawn through the mouth into the pharynx, from which it goes through the **gill slits** into the atrium, exiting through the atriopore. The gill slits are framed by ciliated **gill bars** that are supported by rods of connective tissue. Respiratory gas exchange occurs as water flows through the gill slits.

Coelom

The **coelom** is reduced in adult cephalochordates to cavities around the digestive tract. The position of the coelomic cavities in the pharyngeal region is shown in Figure 23-7.

Excretory System

The excretory organs are simple **nephridia (protonephridia)** situated dorsolaterally to the pharynx. Each bears several clusters of flagellated **solenocytes (flame cells).** About a hundred pairs of protonephridia connect the dorsal coelom with the atrial cavity. These protonephridia are not homologous to the tubules of the vertebrate kidney, which presents an evolutionary enigma that has not yet been solved.

Nervous System

Amphioxus possesses a central **nerve cord** that lies dorsal to the digestive tract (Fig. 23-7), in contrast to the ventral nerve cords of annelids and arthropods. The nerve cord rests on the notochord and is almost as long. A minute **central canal** traverses its entire length and widens at the anterior end to form a **cerebral vesicle,** which marks the location of the **brain.** Two pairs of **sensory nerves** arise from the cerebral vesicle and supply the anterior region of the body. The rest of the nerve cord gives off nerves on opposite

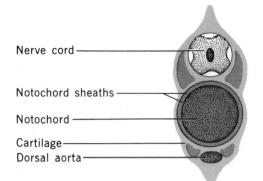

Figure 23-8. Notochord of a young dogfish shark *Squalus*. Cross-section shows nerve cord and sheaths of notochord.

sides that alternate with one another. These nerves are of two kinds: **dorsal nerves,** with sensory functions that pass to the skin, and **ventral nerves,** with motor functions that enter the myotomes. Sense organs include an **olfactory pit** in young individuals, **epidermal sensory cells** on the oral and velar tentacles, and a row of very **simple eyes,** consisting of one **ganglion** and one **pigment cell,** which appear as black spots on the neural tube.

Reproductive System

In *Amphioxus* the sexes are separate. The paired **gonads** (Fig. 23-6) project into the atrium. **Eggs** and

sperm are discharged into the atrial cavity and reach the exterior through the **atriopore.** Fertilization takes place externally in the seawater.

Subphylum Vertebrata

Subphylum Vertebrata contains chordates having a **segmental backbone,** or **vertebral column.** They also possess an axial **notochord** at some time in their lives. The notochord persists into adulthood in some of the lower vertebrates, becoming modified by a buildup of cartilage (Fig. 23-8) that becomes segmented and forms the vertebral column. In higher vertebrates the vertebral column is made up of a series of bones called **vertebrae,** and the notochord disappears before the adult stage is reached.

The vertebrates resemble the urochordates and cephalochordates in their **metamerism** and **bilateral symmetry,** in the possession of a **coelom,** a single tubular **dorsal nerve cord,** and a **notochord** and **gill slits** at some stage in their lives. They differ from other chordates and resemble one another in the possession of cartilaginous or **bony vertebrae,** usually two pairs of **appendages,** an internal and **jointed skeleton,** a ventral **heart** with at least two chambers, and **red blood cells (erythrocytes).**

The body of a vertebrate may be divided into a **head,** usually a **neck,** and a **trunk.** In many vertebrates the trunk extends posteriorly into a **tail.** Two

Figure 23-9. A diagrammatic longitudinal section of a generalized vertebrate, showing the plan of structure.

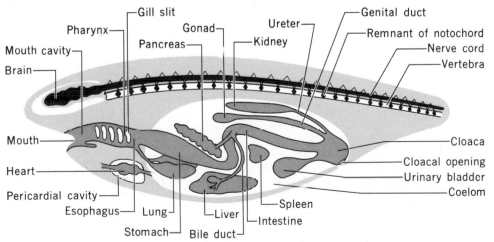

Table 23-1. Taxonomy of subphylum Vertebrata

Subphylum	Class
Pisces	1. Ostracodermi—extinct armored fishes
	2. Agnatha—lampreys and hagfishes
	3. Placodermi—extinct fishes with primitive jaws
	4. Chondrichthyes—sharks, rays, and skates
	5. Osteichthyes—bony fishes
Tetrapoda	6. Amphibia—frogs, toads, salamanders
	7. Reptilia—turtles, lizards, and snakes
	8. Aves—birds
	9. Mammalia—mice, whales, and humans

pairs of lateral appendages are generally present, **thoracic** (e.g., pectoral fins, forelimbs, wings, or arms) and **pelvic** (e.g., pelvic fins, hindlimbs, or legs). These appendages serve in locomotion, body support, and other special functions.

The basic vertebrate body plan can be presented most clearly with diagrams showing longitudinal sections and cross-sections through the body (Fig. 23-9). As in *Amphioxus*, the nerve cord is dorsal, but now it extends anteriorly to the notochord and enlarges into a **brain.** The notochord becomes invested by the vertebrae. The **digestive canal** forms a more or less convoluted tube that lies in the spacious **coelom.** The **liver, pancreas,** and **spleen** are situated near the digestive canal. In the anterior trunk region are the **lungs** and **heart.** The **kidneys** and **gonads** lie above the digestive canal. Table 23-1 summarizes the taxonomic scheme for vertebrates; see the Classification Appendix for more details.

Summary

The chordates have several features in common, including gill pouches with associated gill slits, a dorsal hollow nerve chord, and an endoskeleton consisting of an axial notochord that, in the higher chordates, is replaced by vertebrae. The chordates can be pictured as "upside-down" invertebrates due to the dorsal position of the nervous system and the ventral heart. There is much evidence, chiefly embryological, to link the echinoderms with the chordates; phy-

lum Hemichordata represents this evolutionary transition.

Review Questions

1. What three characteristics are shared by all chordates?
2. What features distinguish the vertebrates from other chordates?
3. List the seven extant classes of living vertebrates and cite some representatives of each.

Selected References

Alexander, R. M. *The Chordates.* New York: Cambridge University Press, 1975.

Barrington, E. J. W. *The Biology of the Hemichordata and Protochordata.* San Francisco: W. H. Freeman, 1965.

Barrington, E. J. W., and R. P. S. Jefferies *Protochordates.* New York: Academic Press, 1975.

Berrill, N. J. "Metamorphosis in Ascidians," *J. Morphol.* **81**:249, 1947.

Berrill, N. J. *The Origin of Vertebrates.* Oxford: Clarendon Press, 1955.

Bullock, T. H. "The Anatomical Organization of the Nervous System of Enteropneusta," *Quart. J. Micr. Sci.,* **86**:55, 1945.

Carson, R. L. *The Sea Around Us.* New York: Oxford University Press, 1951.

Colbert, E. H. *Evolution of the Vertebrates.* New York: John Wiley & Sons, Inc., 1955.

Conklin, E. G. "The Embryology of Amphioxus," *J. Morphol.,* **54**:69, 1932.

Morgan, T. H. "The Development of *Balanoglossus,*" *J. Morphol.,* **9**:1, 1894.

Romer, A. S. *The Vertebrate Story.* Chicago: University of Chicago Press, 1959.

Smith, H. M. *Evolution of Chordate Structure.* New York: Holt, Rinehart and Winston, 1960.

Stirton, R. A. *Time, Life, and Man.* New York: John Wiley & Sons, Inc., 1959.

Webster, D., and M. Webster. *Comparative Vertebrate Morphology.* New York: Academic Press, 1974.

Phylum Chordata, Subphylum Vertebrata, Class Agnatha, Ostracoderms and Cyclostomes

The next seven chapters will survey the vertebrates. Eight classes of vertebrates are generally recognized. Representatives of four of these classes are aquatic and are popularly known as **fishes.** The major characteristics that distinguish these classes will be described in this and subsequent chapters. The following general definition encompasses all their representatives: Fishes are aquatic, **poikilothermic** (cold-blooded) vertebrates that breathe with **gills** and bear **fins** that are supported by an inner skeleton of rodlike **fin rays.** This general definition applies alike to the jawless **lampreys** and **hagfishes,** to the cartilaginous **sharks** and **rays,** and to the many species of **bony fishes.**

Taxonomy and Evolution of Jawless Fishes

The fishes include the classes **Agnatha** (primitive jawless fishes), **Placodermi** (extinct primitive jawed fishes), **Chondrichthyes** (cartilaginous fishes), and **Osteichthyes** (bony fishes).

Class Agnatha may be subdivided into the extinct, jawless, and armored fishes (subclass **Ostracodermi**) and the jawless, unarmored derivatives (subclass **Cyclostomata**) that are represented today by the lampreys and hagfishes.

Subclass Ostracodermi—Extinct, Armored, Jawless Fishes

The characteristic features of this ancestral stock can be determined by fossil reconstructions and by examining the form and function of their contemporary descendants. These lines of evidence suggest that the earliest vertebrates were probably aquatic, swimming animals that were equipped with a dorsal **spinal cord** and a flexible **notochord;** their **mouth** was a simple opening without jaws; and their **pharynx** bore a series of paired **gills slits** through which water exited after passing over the **gills.**

The fossil record indicates that the first vertebrates were freshwater agnaths of the subclass **Ostracodermi,** now extinct. The agnaths evolved over 450 million years ago and were abundant in the ocean for a period of 100 million years. The head of an ostracoderm was encased by a solid **bony shield,** and the rest of the body was covered by **bony plates** (Fig. 24-1). These first vertebrates were filter feeders that trapped small food particles on their gills.

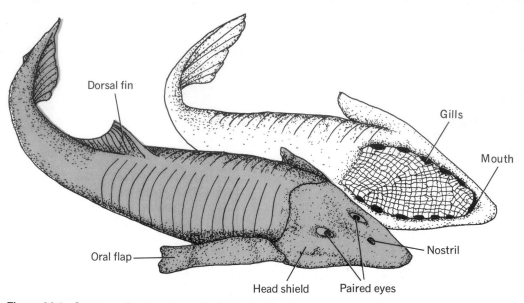

Figure 24-1. One type of ostracoderm, *Cephalaspis,* dorsal and ventral views.

Subclass Cyclostomata—The Lampreys and Hagfishes

Only two modern agnathan groups persist, the **lampreys** and the **hagfishes,** which together comprise about 50 species. Both have lost the bony skeletons that characterized their ancestors and have evolved long, eel-shaped bodies. Cyclostomes lack jaws and paired fins; they have only one **olfactory pit** and feed on the blood and tissues of other fishes that they attack with their **rasping mouths.** Some cyclostomes probably migrated into the ocean, but they were still tied to fresh water for reproduction. The modern lampreys inhabit both marine and freshwater habitats, but the marine species return to freshwater breeding streams. The hagfishes are entirely marine.

Petromyzon marinus—A Representative Cyclostome

Petromyzon marinus, a sea lamprey (Fig. 24-2), lives along the North Atlantic Coast and in the Great Lakes. It swims near the bottom by undulations of its eel-shaped body. During the spring, adult sea lam-preys ascend freshwater rivers and streams to spawn. During these migrations they rest in fast-moving water by adhering onto submerged boulders with their sucker mouths.

External Morphology

The sea lamprey (Fig. 24-2) reaches a length of nearly 1 m. Landlocked populations such as those in the Great Lakes attain a maximum size of only 60 cm. The body is nearly cylindrical, except at the posterior end where it is laterally compressed. There are two **dorsal fins** and one **caudal fin.** There is no exoskeleton. The soft, mottled, greenish-brown skin is covered with a slimy secretion produced by **epidermal glands.** A row of segmental **sensory pits,** the **lateral line,** is located along each side of the body and the head. The **mouth** (Fig. 24-3) lies within a suctorial disc, the **oral (buccal) funnel,** and is held open by a ring of cartilage. Around the oral funnel are a number of **sensory papillae** and horny "teeth." At the apex of the oral funnel is the mouth, which bears horny **teeth** and a protruding pistonlike **tongue.** Each side of the head bears an **eye** and seven **gill slits.** Between the eyes on the dorsal surface is a single **nasal opening.** The **anus** opens on the ventral surface near the poste-

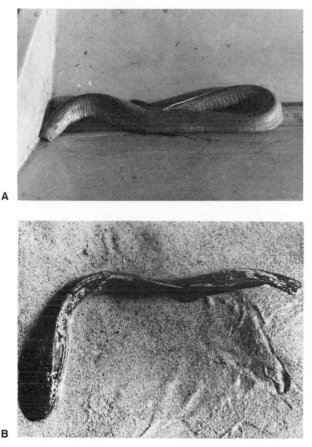

A

B

Figure 24-2. (A) Adult sea lamprey *Petromyzon marinus,* a jawless vertebrate, about 50 cm long, showing the characteristically mottled back of a sexually mature adult. The sea lamprey is an eel-like exoparasite of fish. (B) Adult hagfish.

rior end; just behind it is the **urogenital opening** in the end of a small papilla.

Internal Morphology and Physiology

Skeletal System

The **notochord** of *Petromyzon* persists as a well-developed structure in the adult (Fig. 24-4). In the trunk region it is supplemented by small cartilaginous **neural arches.** The organs in the head are supported and protected by a cartilaginous cranium. A cartilaginous **branchial basket** supports and protects the gill structures.

Muscular System

The muscles in the walls of the trunk and tail are **segmentally** arranged, as they are in higher fishes. The **rasping tongue** is moved by large **retractor** and

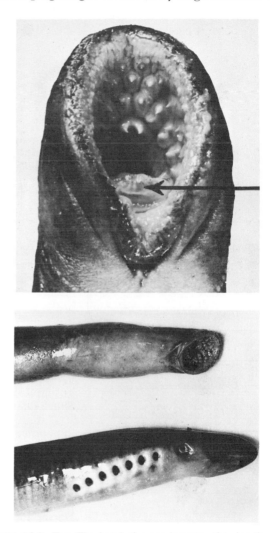

Figure 24-3. Top: The head of a sea lamprey showing the oral funnel, which serves as a suction cup by which it attaches itself to its prey. It is by means of the sharply pointed horny teeth inside the oral funnel and the rasplike (arrow) tongue that it can penetrate through the scales and flesh of its prey. Bottom: Note the laterally placed eye, behind which are gill slits. [Courtesy of Institute for Fisheries Research, Michigan Department of Natural Resources.]

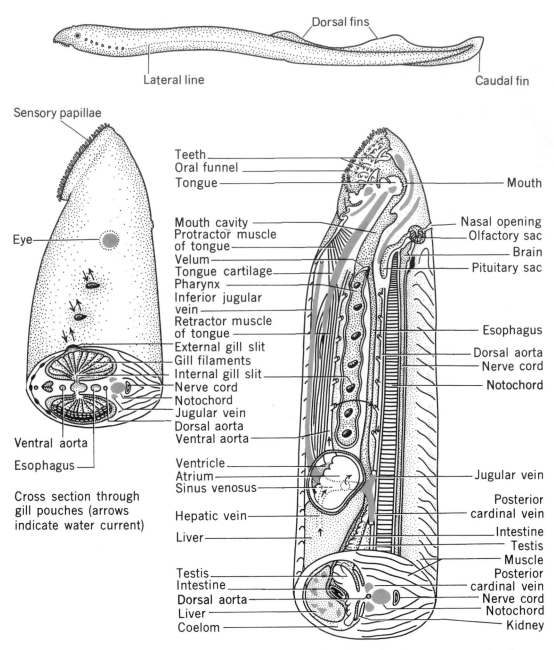

Figure 24-4. Anterior portion of an adult lamprey, *Petromyzon*. Right: Lateral portion cut away to show internal structures related to feeding. Left: Cross-section through gill pouches (arrows indicate water current).

smaller **protractor** muscles. The buccal funnel is operated by a number of **radiating muscles.**

Digestive System

The adult sea lamprey feeds chiefly on the blood of fishes. Expansion of the **oral funnel** (Fig. 24-3) causes the **mouth** to act like a sucker and enables the lamprey to fasten itself to the integument of fishes (Fig. 24-5). Prey include shad, sturgeon, cod, and mackerel in the ocean and lake trout, whitefish, walleye, and carp in the Great Lakes. **Horny teeth** on the top of the lamprey's tongue are used to rasp away the scales and flesh of its victim and to suck out the blood. The mouth cavity opens posteriorly into two tubes, an upper **esophagus** and a ventral **pharynx** (Fig. 24-4). The esophagus leads to the **digestive tract,** whereas the pharynx is part of the respiratory system. A fold called the **velum** at the anterior end of the pharynx prevents food from entering the respiratory system.

There is no distinct stomach. A **valve** separates the posterior end of the esophagus from the straight **intestine.** An inner fold, called the **typhlosole,** partitions the intestine in a spiral manner, increasing its absorptive surface area. The digestive tract terminates at the **anus.** A **liver** is present, but there is usually no bile duct in the adult.

Circulatory System

The lamprey possesses a heart, a number of arteries and veins, and many lymphatic sinuses. The **heart** (Fig. 25-5) lies in the **pericardial cavity** and consists of an **atrium,** which receives the blood from the **veins,** and a **ventricle,** which pumps blood into the **arteries.** A **lymphatic system** is present.

Respiratory System

Respiration is carried on by means of seven pairs of **gill pouches** that open to the **pharynx** internally and through **gill slits** to the external environment. Each gill pouch has numerous **gill filaments** that contain many **capillaries.** Respiratory gas exchange occurs between the blood in the filament capillaries and the water that passes through the pouch.

In a larval lamprey, the water used in respiration passes in through the mouth and out the gill slits, as in higher fishes. But in adult lampreys, water is drawn into the gill pouches through the gill slits (Fig. 24-4). This irrigation method, which is unlike that in the higher fishes, is necessary because the lamprey cannot pass water through the mouth when attached by its oral funnel to its prey.

Excretory System

The **excretory** and **reproductive systems** are so closely united in the lamprey and other fishes that they are commonly referred to as the **urogenital system.** The **kidneys** lie along the dorsal wall of the body cavity, and each sends urine through a **urinary duct** into the **urogenital sinus,** which opens to the outside through the **urogenital opening.**

Marine lampreys face some important problems when they enter freshwater streams to spawn. Seawa-

Figure 24-5. A lake trout showing a typical lamprey scar (arrow). [Courtesy of U.S. Fish and Wildlife Service.]

ter has salt concentrations that are slightly higher than lamprey blood and much higher than those in fresh water. As a result, lampreys lose body water by osmosis while in the ocean. This water loss is countered by the drinking of seawater and the subsequent active transport of excess salts from their gills. But the opposite problem is encountered when the lamprey enters fresh water. Then the blood is **hypertonic** to the water, and so water tends to diffuse into the body. A physiological change occurs (probably under endocrine control) so that the kidney retain salts and the lamprey excretes salt-free water; the gills may also actively take up salts from the freshwater environment.

Nervous System

The **brain** of the adult lamprey is very primitive. The vertebrate brain can be subdivided into many sections; their functional significance is discussed in more detail for higher vertebrates. The **forebrain** consists of a large pair of **olfactory lobes;** behind these are the small **cerebral hemispheres** attached to the **diencephalon;** ventral to the latter is a broad **infundibulum,** and above it a **pineal** structure. On the **midbrain** is a pair of large **optic lobes.** In the **hindbrain,** the rudimentary **cerebellum** is a small dorsal, transverse band, but the ventral **medulla oblongata** is fairly well developed. There are only 10 pairs of **cranial nerves** instead of the usual 12 in higher vertebrates. The **nerve cord** is flat and lies on the floor of the **neural canal.** The **autonomic nervous system,** which controls involuntary activities, consists only of an **intestinal plexus** linked with the brain.

Sense Organs

Organs of taste, smell, equilibrium, and sight are present in the lamprey. The receptors for taste are situated on the pharyngeal wall between the gill pouches. The organ of smell is an **olfactory sac** (Fig. 24-4) that lies in the **nasal capsule;** this sac communicates with the external environment through a single nasal opening located on the dorsal surface between the eyes. The olfactory sac gives off a ventral tube, called the **pituitary,** or **hypophyseal,** sac, whose function is unknown.

The **balancing organs** of *Petromyzon*, which lie in the **auditory capsule,** have two **semicircular canals,** whereas those of the hagfishes have only one.

The **eyes** of the adult lamprey, though primitive, are excellent visual organs. Posterior to the nasal opening, there is a well-developed median **pineal eye** with a clear lens and pigmented retina.

Some time ago, it was discovered that the eyes of marine fishes and terrestrial animals have one visual pigment (**rhodopsin,** a light-sensitive chemical based on vitamin A_1) and that freshwater animals have another (**porphyropsin,** a light-sensitive chemical based on vitamin A_2). Lampreys migrating from the ocean to freshwater habitats switch from rhodopsin to porphyropsin visual pigments. The mechanism and function for this switch are not fully known, but lampreys have both vitamins in their bodies.

Lampreys also produce electrical fields with which they can detect prey.

Endocrine System

Where the pituitary sac comes in contact with the infundibulum of the brain, it gives off numerous small follicles that become separated, forming the **pituitary gland.** Studies using radioactive iodine have shown that the **endostyle** of the larva, which aids in food getting, is the forerunner of the **thyroid gland** in the adult lamprey. Iodine is an important component of **thyroxin,** the hormone produced by the thyroid gland; it regulates the general metabolism of the animal.

Reproductive System

The sexes are separate in the adult; however, the **gonad** is **protandric,** first producing male gametes and then female gametes within the same individual. The single gonad fills most of the abdominal cavity at the time of sexual maturity. There is no genital duct; **eggs** or **sperm** are released into the **coelom,** after which they make their way through two **genital pores** into the urogenital sinus. From there, they pass out through the urogenital opening into the water, where fertilization occurs.

Sea lampreys become sexually mature in May or early June; then both sexes migrate into freshwater streams, sometimes using their suctorial mouths to "hitchhike" on a passing boat. They seek a gravelly bottom in a moderately fast-flowing stream. There they move stones to form a shallow rounded depression. The female then adheres to a stone in the nest, and the male attaches to the female by his oral funnel. Partly entwined, they move back and forth as eggs and sperm are discharged into the water, where fertilization takes place.

Each female sea lamprey produces from 24,000 to 107,000 eggs, depending on her size; the average female lays 62,500 eggs. Both sexes die soon after spawning. The eggs hatch out into small wormlike **larvae,** known as **ammocoetes,** after about 21 days. The blind larvae leave the nest and drift downstream into quiet water, where they burrow into the silt-covered bottom. The larval period lasts from 3 to 12 or more years, depending on water temperature and the amount of available food. Inspiration of water for respiratory purposes appears to be largely responsible for drawing bottom debris and food organisms into the mouth of the larva. An **endostyle** on the floor of the pharynx secretes mucus that entangles the food, as in the *Amphioxus.* Then cilia transport the food-laden mucus into the esophagus.

The ammocoete lies buried in mud and sand and probably keeps its skin free from bacteria, fungi, and other parasitic growths by means of an integumentary secretion. During a winter several years after burrowing, the ammocoete larva undergoes a **metamorphosis** into an adult lamprey and then migrates to the ocean; landlocked forms migrate into the Great Lakes. It is at this stage that the parasitic species begin to feed on fish.

Other Cyclostomes

One of the nonparasitic **brook lampreys** of North America, *Entosphenus lamottenii,* breeds in the spring. The adults move stones by means of their **oral funnels** until a space is cleared on the bottom where a number of them may congregate. A male clings to the head of a female, winds his tail about her body, and discharges **spermatozoa** over the **eggs** as they are extruded. Brook lampreys also undergo prolonged larval development. During the transformation to an adult, the **digestive system** degenerates, so that the adult must live on stored food reserves. Soon after **spawning,** the adults die.

Hagfishes have been aptly called the "vultures of the sea," for these scavengers subsist on dead and dying prey. Hagfishes are quite common along all coasts at depths from 30 to 300 m. Their small, jawless **mouths** are equipped with fingerlike toothed structures that arise from either side of the **buccal cavity.** These are used to rasp the soft tissues from dead and dying fish, leaving nothing but a bag of skin and bones.

Hagfishes are **protandric,** and development is direct, without a larval stage. Their blood is isotonic with seawater.

The chordate characters of the cyclostomes are obvious. Some, such as persistent notochord and many gill slits, are similar to the primitive characters of *Amphioxus.* Others are more advanced, such as a distinct head, cranium, better-developed brain, and cartilaginous neural arches. Cyclostomes are more primitive than are other fishes, as indicated by the absence of hinged jaws, paired limbs, true teeth, and complete vertebrae.

Relations of the Cyclostomes to Humans

Lampreys have been a popular food for centuries in many parts of Europe. Attempts have been made in the United States to market lampreys as palatable food or to find other commercial uses for them, but so far these efforts have met with little or no success. Larval lampreys sometimes serve as bait for commercial and sport fishermen. The **Atlantic lamprey** *(Petromyzon marinus)* has invaded all the Great Lakes, passing the Niagara Falls barrier by way of the Welland Canal. In the late 1930s and early 1940s, lampreys began entering the upper Great Lakes from Lake Ontario through man-made canals (constructed without ecological impact studies). In 1946, commercial fishermen took a catch of lake trout from Lake Michigan of about 5.5 million pounds, but the catch in 1953 was only 482 pounds. Both overfishing and the detrimental effects of the marine lamprey led to the demise of commercial fishing on the Great Lakes.

At present a chemical poison (TFN) that is somewhat selective for lampreys is used for lamprey control; electrical barriers (**weirs**) are also used to block migrations. This program, plus effective commercial fishing regulations have resulted in a comeback of the Great Lakes fisheries.

Summary

The first vertebrates, the jawless ostracoderms of some 400 million years ago, are known only from their fossil remains. They were heavily armored freshwater filter feeders.

Modern cyclostomes include the lampreys and the hagfishes. All hagfishes and some lampreys are marine, but lampreys must return to freshwater streams to reproduce. Lamprey eggs are not adapted to develop in salt water, which represents a hypertonic environment. Interestingly, it is the adults that must change their osmoregulatory processes as they return to spawn in the freshwater habitats of their agnathan ancestors. Lampreys and hagfishes are specialized parasites and scavengers and so have avoided, to some extent, competition with the jawed fishes for food resources.

Review Questions

1. Hagfishes reproduce on the ocean floor. What are some differences that you might expect to see between a hagfish and lamprey egg?
2. What is an ostracoderm?
3. How can an agnathan eat without a jaw?
4. Give some reasons for thinking that vertebrates first arose in fresh water. Can you think of an argument against this theory?
5. How would a lamprey osmoregulate in a tub of concentrated salt water?

Selected References

Applegate, V. C. *Natural History of the Sea Lamprey, Petromyzon marinus in Michigan.* Special Scientific Report, Fisheries No. 55, U.S. Fish and Wildlife Service, 1950.

Applegate, V. C., and J. W. Moffett. "The Sea Lamprey," *Scientific American,* **192**:36, 1955.

Dineley, D. L., and E. J. Loeffler. *Ostracoderm Faunas of the Delorme and Associated Siluro-Devonian Formations North West Territories, Canada.* London: Paleontological Association, 1976.

Gage, S. H. *The Lake and Brook Lampreys of New York.* Wilder Quarterly Century Book. Ithaca: Comstock, 1893.

Hardisty, M. W. and I. C. Potter (eds.). *The Biology of Lampreys.* New York: Academic Press, 1971–1972.

Hubbs, C. L. "The Life-cycle and Growth of Lampreys," *Papers Mich. Acad. Sci.,* **4**:587, 1924.

Johnson, W. H., L. E. Dolanney, E. C. Williams, and T. A. Cole. *Principles of Zoology,* 2nd ed. New York: Holt, Rinehart and Winston, 1977.

Lennon, R. E. "Feeding Mechanism of the Sea Lamprey and Its Effects on Host Fishes," U.S. Fish and Wildlife Service, *Fish. Bull.,* **56**:245, 1954.

Nicol, M. A. C. *The Biology of Marine Animals,* 2nd ed. London: Sir Isaac Pitman & Sons, Ltd., 1968.

Parker, P. S., and R. E. Lennon. *Biology of the Sea Lamprey in its Parasitic Phase.* U.S. Fish and Wildlife Service, Research Dept. 44.

Reynolds, T. E. "Hydrostatics of the Suctorial Mouth of the Lamprey," *Univ. Calif. Pub. Zool.,* **37**:15, 1931.

Romer, A. S. *The Vertebrate Body,* 4th ed. Philadelphia: W. B. Saunders Company, 1970.

Sawyer, W. H. "Endocrines and Osmoregulation Among Fishes," *American Zoologist,* **13**(3), 1973.

Scheer, B. T. *Animal Physiology.* New York: John Wiley & Sons, Inc., 1963.

Wood, D. W. *Principles of Animal Physiology.* New York: American Elsevier Publishing Company, Inc., 1970.

Phylum Chordata, Subphylum Vertebrata, Classes Placodermi and Chondrichthyes

Taxonomy and Evolution of Jawed Fishes

The lampreys and hagfishes are all that remain of the once-flourishing Agnatha. Some 400 million years ago, the **placoderms** (class **Placodermi**) started to become the predominate fishes of the time. These fishes had a set of movable **jaws** that were derived from a translocated set of **gill arches.** Thus equipped, the placoderms managed to outcompete the agnaths for food resources and gradually replaced them. Placodermi is the only vertebrate class that is extinct. This may have been due to competition from more efficient jawed forms—the **cartilaginous fishes** (class **Chondrichthyes**) and the **bony fishes** (class **Osteichthyes**). These two groups evolved from the placoderms independently: the cartilaginous fishes in marine waters and the bony fishes in fresh waters, although the latter group soon invaded, and later dominated, the oceans as well. Neither offshoot was encumbered by the heavy external armor of the placoderms, and this advantage was perhaps the key to their competitive success. By about 205 million years ago, all the placoderms were extinct, and the bony fishes started to diversify (Fig. 23-3).

Class Chondrichthyes is divided into subclasses **Elasmobranchii,** or **Selachii** (sharks, rays, and skates), and **Holocephali** (chimaeras or ratfishes). The taxonomy and evolution of the bony fishes are discussed in Chapter 26.

Class Placodermi—Extinct, Armored, Jawed Fishes

Placoderms are the first jawed vertebrates in the fossil record. They evolved first in fresh water, but some placoderms later successfully invaded the ocean. As with their ostracoderm ancestors, they were heavily armored (Fig. 25-1). There were many kinds of placoderms, but they all had one thing in common: a primitive **jaw.** In some cases the jaw was composed of **bony plates,** and in others it was derived from the first **gill bar** (supportive bone onto which the gill filaments are attached) of their agnathan ancestors. Possession of a movable jaw allowed placoderms to consume a wider variety of foods than could their filter-feeding ancestors, and many became carnivores. Speed and agility are obviously advantageous for an active predator, and these traits were realized through increased musculature, stabilizers (**fins**), and refinements of the nervous system and sense organs.

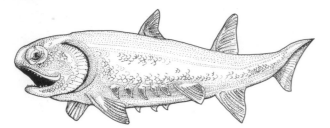

Figure 25-1. Drawing of fossil placoderm. The jaw, though primitive, enabled placoderms to become carnivorous chordates—a decisive evolutionary step.

Such adaptations allowed the placoderms to exploit many different niches in their aquatic environments.

Class Chondrichthyes— The Cartilaginous Fishes

The vertebrate class Chondrichthyes contains the **cartilaginous fishes:** the **sharks, rays, skates,** and **chimaeras.** These are the most primitive living vertebrates that have **complete vertebrae, movable jaws,** and **paired appendages.**

The cartilaginous fishes have much less massive armor than the placoderms, and this allows them much greater mobility. They also possess well-developed paired fins, and these provide greater stability and control in the water. These adaptations followed development of the placoderm jaw, thus allowing for a successful predatory existence.

The cartilaginous fishes exhibit a number of structural advances over the modern cyclostomes; they possess **paired fins,** a **lower jaw, gill arches,** and **placoid scales.** The features that distinguish the chondrichthyes from the bony fishes will be discussed in Chapter 26.

Squalus acanthias—A Representative Cartilaginous Fish

The common **dogfish shark,** *Squalus acanthias,* is abundant off the coasts of New England and northern Europe. It is a characteristic representative of the cartilaginous fishes. Furthermore, the dogfish shark pos-

sesses many basic vertebrate features in simple form, and its study thus helps to facilitate an understanding of the more complex systems of higher vertebrates.

External Morphology

The streamlined, spindle-shaped body is about 75 cm long (Fig. 25-2). There are two **dorsal fins,** one behind the other, each with a spine at its anterior end; two **pectoral fins;** and two **pelvic fins.** In the male the pelvic fins bear reproductive appendages known as **claspers.** Between the pelvic fins is the **cloacal opening,** sometimes called an **anus.** The tail, or **caudal fin,** is **heterocercal;** that is, it is asymmetrical, with the vertebral column extending into its dorsal portion (Fig. 26-14).

The **mouth** is a transverse slit on the ventral surface of the head. Above the mouth on either side is an **eye,** and each ventral **nostril** opens into a blind pouch, the **olfactory sac.** Anterior to each pectoral fin are six **gill slits,** the most anterior of which is modified as a **spiracle,** which is homologous to the middle ear of higher vertebrates.

The gray-colored skin is covered with **placoid scales** (Fig. 25-3). Each scale consists of a bony **basal plate** with a central spine of **dentine** covered with a hard enamel-like **vitrodentine.** Over the jaws, the placoid scales are modified into **teeth** (Fig. 25-3) that are directed backward and used for holding and tearing prey. Placoid scales are homologous with vertebrate teeth, as shown by their similar embryological development. There is apparently no enamel on the surface of placoid scales, but enamel-forming organs are present as in developing teeth. The homology of teeth and scales reflects the fact that the mouth lining in vertebrates is derived from inturned skin and hence possesses integumentary structures.

Internal Morphology and Physiology

Skeletal System

The skeleton is composed entirely of **cartilage.** The cartilaginous skeleton of the Chondrichthyes is probably a regressive, not a primitive, characteristic, as their placoderm ancestors had bony skulls and armor.

The **axial skeleton** consists of the **skull** and **vertebral** column. The cartilaginous skull is much more

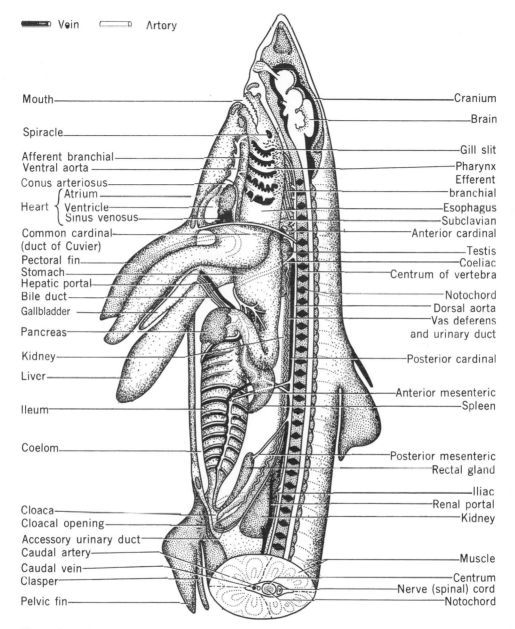

Vein ━━ Artery ━━

Mouth — Cranium
— Brain
Spiracle — Gill slit
Afferent branchial — Pharynx
Ventral aorta — Efferent
Conus arteriosus — branchial
Atrium — Esophagus
Heart { Ventricle — Subclavian
Sinus venosus — Anterior cardinal
Common cardinal — Testis
(duct of Cuvier) — Coeliac
Pectoral fin — Centrum of vertebra
Stomach —
Hepatic portal — Notochord
Bile duct — Dorsal aorta
Gallbladder — Vas deferens
— and urinary duct
Pancreas —
Kidney — Posterior cardinal
Liver —
Anterior mesenteric
Ileum — Spleen
Coelom —
Posterior mesenteric
Rectal gland
Cloaca — Iliac
Cloacal opening — Renal portal
Accessory urinary duct — Kidney
Caudal artery —
Caudal vein — Muscle
Clasper — Centrum
— Nerve (spinal) cord
Pelvic fin — Notochord

Figure 25-2. The internal organs of the spiny dogfish shark *Squalus acanthias.*

highly developed than in the cyclostomes. It is composed of the **cranium,** or **brain case,** and two large anterior **nasal capsules** and two posterior **auditory capsules.** The vertebrae are hourglass shaped, and the **notochord** persists in the lens-shaped spaces between them. The **visceral skeleton** consists of the upper and lower **jaws** (Fig. 25-3), the **hyoid arch,** and five **branchial arches** supporting the gills. The **appendicular skeleton** consists of cartilages of the **pectoral** and **pelvic girdles** and the fins that they support.

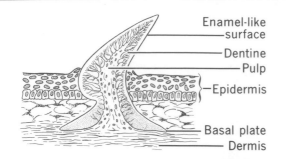

Enamel-like
surface

Dentine

Pulp

Epidermis

Basal plate

Dermis

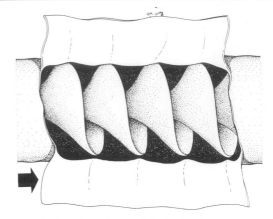

Figure 25-4. Intestine of a dogfish shark cut open to reveal shape of spiral valve. Food moving in direction indicated by arrow encounters a larger surface area (spiral valve) than it would if gut tube were hollow.

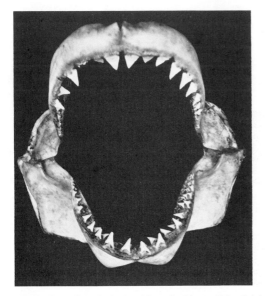

Figure 25-3. Top: Detail of placoid scale (dermal denticle) as seen in section. The pointed scales give the skin a rasplike texture. In addition, these scales probably give rise to teeth. [Modified after Kerr.] Bottom: Jaw and teeth of the great white shark, *Carcharodon carcharias.* This shark, about 5 m in length and weighing about 1,300 kg, bit a 10-kg piece out of a basking shark. It was caught off Moss Landing in Monterey Bay, California. [Photograph courtesy of Science Software Systems.]

Digestive System

The **digestive tract** is longer than the body length. The **mouth** (Fig. 25-2) leads into a large **pharynx** that is pierced by the **spiracles** and **gill slits.** The pharynx leads into the short, wide **esophagus** that opens into the U-shaped **stomach.** The junction of the stomach and **intestine** is marked by a circular sphincter muscle, the **pyloric valve.** The intestine terminates in the

cloaca that opens to the exterior through the **cloacal opening.**

Within the intestine is a spiral fold of mucous membrane, called the **spiral valve** (Fig. 25-4), that slows the passage of food and increases the absorptive surface area.

A slender fingerlike **rectal gland,** which excretes salt, attaches dorsally near the point at which the small and large intestines join.

The large **liver** consists of two long lobes; the **bile** that it secretes is stored in a **gallbladder** before passing through the **bile duct** into the intestine. A **pancreas** supplies an array of additional **digestive enzymes.**

Circulatory System

As in cyclostomes and most bony fishes, only venous blood enters the heart (Fig. 25-2). The **two-chambered heart** lies in a **pericardial cavity** that is separated from the viscera by a **transverse septum.** Deoxygenated blood returning from the body tissues passes through the **sinus venosus** into the thin-walled **atrium.** The atrium pumps the blood into the **ventricle,** whose thick, muscular walls then pump it into the **conus arteriosus.** From there the blood passes through the **ventral aorta** into the **afferent branchial arteries,** where it is oxygenated as it passes through the **capillaries** of the gills. Blood then passes into the efferent branchial arteries, which carry it to the dorsal aorta. The dorsal aorta supplies oxygenated blood to the various parts of the body through their capillary beds. Veins carry deoxygenated blood back to the sinus venosus, and the cycle continues.

A number of veins constitute the hepatic portal system, which transports blood from the digestive tract, pancreas, and spleen to the liver, from which hepatic sinuses return it to the sinus venosus. Another shunt, the renal portal system, conveys blood from the posterior part of the body to the kidneys. Blood leaves the kidneys by way of several renal veins, which empty into the posterior cardinal sinuses, and these return the blood to the sinus venosus.

Cartilaginous fishes have a unique method of osmoregulation in seawater. They maintain a high concentration of nitrogenous waste urea in their blood. This increases the osmotic concentration of the blood, making it slightly hypertonic to seawater. As a result, the Condrichthyes do not experience osmotic water loss.

Respiratory System

External respiration occurs by means of gills (Fig. 25-5). These are folds of mucous membrane well supplied with capillaries. The gills are supported by the hyoid arch, the first four branchial arches, and gill rays, to which are attached the gill filaments. Water entering the mouth passes between the branchial arches, bathing the gills and supplying oxygen to the branchial blood vessels; the water then passes out through the spiracles and gill slits.

Because the gills are impermeable to urea, no urea diffuses out.

Excretory System

The dogfish shark possesses two ribbonlike kidneys (Fig. 25-8), one on either side of the dorsal aorta. Urine from each kidney is carried through small ducts into a large ureter (urinary duct). These two ducts empty into a common urogenital sinus; from there, the urine is eliminated through the cloacal opening. Not much urea is excreted, however, as it is retained in the blood for osmoregulation. A series of yellowish bodies called adrenals are located on the medial border of each kidney. (see Chapter 37).

Nervous System

The brain (Figs. 25-6 and 25-7) is more highly developed than in the cyclostomes. There are two remarkably large olfactory lobes, a cerebrum of two hemispheres, a pair of optic lobes, and a cerebellum that projects posteriorly over the medulla oblongata. Eleven pairs of cranial nerves leave the brain and

Figure 25-5. Respiratory structures in the dogfish shark. The left side of the pharynx is cut lengthwise and is laid open to show the gills and other structures.

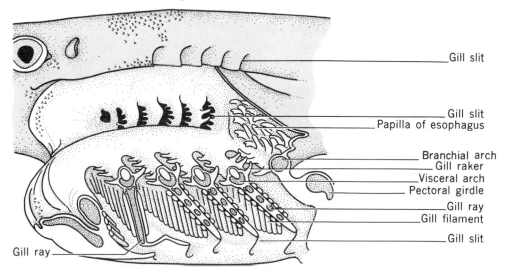

Gill slit

Gill slit
Papilla of esophagus

Branchial arch
Gill raker
Visceral arch
Pectoral girdle
Gill ray
Gill filament

Gill slit

Gill ray

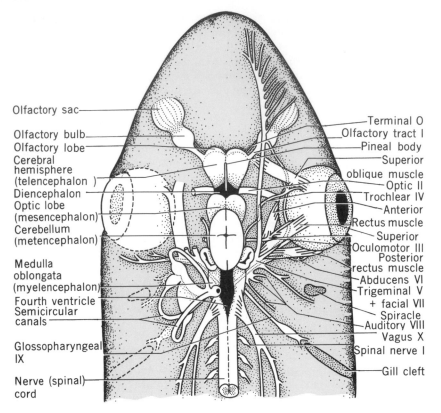

Olfactory sac
Olfactory bulb
Olfactory lobe
Cerebral hemisphere (telencephalon)
Diencephalon
Optic lobe (mesencephalon)
Cerebellum (metencephalon)
Medulla oblongata (myelencephalon)
Fourth ventricle
Semicircular canals
Glossopharyngeal IX
Nerve (spinal) cord

Terminal 0
Olfactory tract I
Pineal body
Superior oblique muscle
Optic II
Trochlear IV
Anterior
Rectus muscle
Superior
Oculomotor III
Posterior rectus muscle
Abducens VI
Trigeminal V + facial VII
Spiracle
Auditory VIII
Vagus X
Spinal nerve I
Gill cleft

Figure 25-6. Dorsal view of brain and cranial nerves of dogfish shark. Roman numerals signify cranial nerves.

Figure 25-7. Longitudinal section between first and second ventricles of dogfish shark brain showing structure.

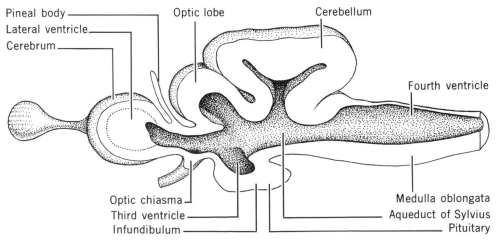

Pineal body
Lateral ventricle
Cerebrum
Optic lobe
Cerebellum
Fourth ventricle
Medulla oblongata
Aqueduct of Sylvius
Pituitary
Optic chiasma
Third ventricle
Infundibulum

innervate the head region. The **nerve (spinal) cord,** protected by the vertebral column, is a dorsoventrally flattened hollow tube. **Spinal nerves** arise from its sides in pairs.

Sense Organs

The two **olfactory sacs** are typically large. The **ears** are membranous sacs, each with three **semicircular canals,** that lie within the **auditory capsules.** The **eyes** are well developed. Along each side of the head and body is a longitudinal groove called the **lateral line;** it contains a canal with numerous openings to the surface. Inside the canal are **sensory hair cells** that are innervated by a branch of the **tenth cranial nerve.** The lateral line system seems to be a pressure-

sensitive organ that enables the fish to detect low frequency vibrations and other types of water movements.

The surface of the head also has **sensory canals;** these open into pores (**ampullae of Lorenzini**) containing **pit organs** with **sensory hairs.** It was once thought that this system was for heat reception; however, recent findings show that it is for **electroreception.** At close range this system possibly detects the electrical activity of the prey's nerves and muscles.

Reproductive System

The sexes are separate (Fig. 25-8). The **spermatozoa** of the male arise in two **testes** and are conveyed

Figure 25-8. Urogenital systems of dogfish shark.

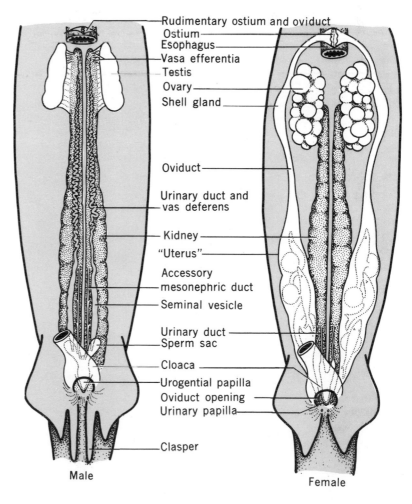

Rudimentary ostium and oviduct
Ostium
Esophagus
Vasa efferentia
Testis
Ovary
Shell gland

Oviduct

Urinary duct and vas deferens

Kidney
"Uterus"

Accessory mesonephric duct
Seminal vesicle

Urinary duct
Sperm sac

Cloaca
Urogential papilla
Oviduct opening
Urinary papilla

Clasper

Male

Female

A

B

Figure 25-9. Photographs and drawings of some cartilaginous fishes showing extreme variation in form. (A) Gray "killer" shark nearly 2 m in length. (B) Electric ray (torpedo) with electric organs in pectoral fins. (C) Ratfish. (D) Sawfish.

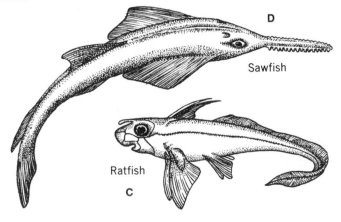

D

Sawfish

Ratfish

C

through the **vasa efferentia** to the convoluted **vasa deferentia,** which empty into the **urogenital sinus.** During **copulation,** the spermatozoa are transferred to the oviducts of the female with the aid of the **claspers** on the male's pelvic fins.

The **ova** of the female arise in paired **ovaries** that are attached to the dorsal wall of the abdominal cavity. The gametes break out into the coelom and enter a funnel-like opening, or **ostium,** that is common to both **oviducts.** The dogfish shark has been known since Aristotle's time to be **ovoviviparous:** the shelled eggs hatch inside the mother's body. An expanded anterior portion of each oviduct is called a **shell gland,** and the posterior oviduct is enlarged in the dogfish shark to form a "uterus" in which the young develop (Fig. 25-8). The oviducts have separate openings into the cloaca. **Oviparous sharks** (a minority of species) discharge eggs protected by rectangular horny **egg cases.**

Other Chondrichthyes

Class Chondrichthyes contains the sharks, rays, and chimaeras. Most of these cartilaginous fish inhabit tropical marine waters.

Sharks are primarily marine carnivores whose mouths are well armed with many replaceable triangular teeth. Shark teeth are prone to fossilization, and their geological record extends back more than 320 million years. Their present diversity of forms and numbers, estimated at some 300 species, appears to equal their abundance in the past.

With the exception of some whales, the sharks include the largest of all vertebrates (Fig. 25-9). The **whale shark** *(Rhineodon typicus)* is from 12 to 15 m long. The whale shark feeds on plankton, but the **great white shark,** *Carcharodon cacharias,* (Fig. 25-3), which reaches a length of 10 m, has earned the name of "man-eater" by its repeated attacks on human beings. One of the most peculiar sharks is the **hammerhead,** whose head is shaped like the head of a mallet, with an eye on each side.

Rays and **skates** differ from sharks in having dorsoventrally flattened bodies. These cartilaginous fishes generally live and obtain their food on the ocean floor. Bottom rays use their "wings" to create water currents that displace sediments and uncover clams and polychaete worms. These prey are then sifted from the sediments, chewed by crushing plates in the mouth, and ingested. Skates are more active forms that pounce upon bottom-dwelling crustaceans, mollusks, worms, and fish. Larger relatives of the skates include the **devil ray** *(Manta)* and **sawfish.** The sawfish (Fig. 25-9) is abundant in the Gulf of Mexico and reaches a length of 3 to 5 m. The saw is about 1.5 m long and is used for capturing prey; the "saw" is swung back and forth in a school of fishes to injure some of them sufficiently for capture. It is also used as a weapon of defense.

The **chimaeras,** or **ratfishes** (Fig. 25-9), are moderately large fishes, about 2 m long, that inhabit deeper waters along the continental margins. Their common name is derived from the fantastic chimaera monster of Greek mythology that had the head of a lion and the tail of a dragon. This allusion is supported by the erect dorsal and large lateral pectoral fins that frame their head and by their long slender tails. The dorsal and pectoral fins are used for locomotion. Chimaeras live near the ocean floor and feed on mollusks and other benthic invertebrates.

Relations of the Cartilaginous Fishes to Humans

Sharks feed chiefly on crustaceans, squids, fish, and other aquatic animals. Human beings are rarely prey, which is not what one might infer from some newspaper accounts. However, sharks have occasionally attacked swimmers and divers in temperate waters. Most of these attacks have occurred near beaches or reefs. The frequency of attacks has increased as an ever-increasing number of sports enthusiasts, equipped with fins, snorkels, and aqualungs, have invaded the sharks' domain. The **sting rays** are often encountered by bathers, and an occasional death results from the injury inflicted by its barbed spine.

Sharks compete with humans for food. They eat large numbers of lobsters, crabs, and commercial fishes. In certain parts of the world, especially the Orient, the smaller sharks and some skates are used for food. In America, much prejudice exists against using shark flesh for food, and so it is often sold fresh as "grayfish." Also, **dogfish sharks** are now being canned in the United States under a trade name. The fins of sharks and **rays** are considered a delicacy in certain Oriental countries. Sharkskin leather is used in the manufacture of shoes and handbags. Sharkskin, tanned with the scales on, is called shagreen and has been used by cabinetmakers as an abrasive. It is also used for binding books and as a covering for jewel boxes. During World War II, the liver of the dogfish shark was America's chief supplementary source of vitamin A, and many shark livers are still processed for this purpose. The pituitary gland (**hypophysis**) also provides an extract for medical use.

Summary

Placoderms were the first vertebrates with jaws. They lived in freshwater streams and eventually invaded the ocean. Before becoming extinct, the placoderms gave rise to the bony fishes in fresh water and the cartilaginous fishes in marine waters. Predation became possible with the evolution of a movable jaw, and this method of feeding provided selective pressures that fostered development of paired fins, reduction of bony armor, improvement of senses, and other adaptations.

Review Questions

1. What lines of evidence suggest that sharks evolved in a marine environment?
2. What are some predatory modifications seen in dogfish sharks that are not found in placoderms?
3. Compare the osmoregulatory problems of a freshwater cyclostome and a marine cartilaginous fish.
4. Sharks rely heavily on sight for finding prey. Cite anatomical evidence for this. (It was previously thought that olfaction was the main sense used, but shark repellents are successful because they include dyes that act as smoke screens, not because of foul-smelling chemicals.)

Selected References

Bigelow, H. B., and W. C. Schroeder, "Sharks, Rays, and Chimaeroids," in *Fishes of the Western North Atlantic*. New Haven: Yale University, Sears Foundation for Marine Research, Parts 1 and 2, 1948–1953.

Colbert, E. H. *Evolution of the Vertebrates*. New York: John Wiley & Sons, Inc., 1955.

Daniel, J. F. *The Elasmobranch Fishes*. Berkeley: University of California Press, 1934.

Gilbert, P. W. "The Behavior of Sharks," *Scientific American*, **207**:60, 1962.

Gilbert, P. W. (ed.). *Sharks and Survival*. Boston: D. C. Heath, 1963.

Gilbert, P. W., R. F. Mathewson, and D. P. Rall (eds.). *Sharks, Skates, and Rays*. Baltimore: John Hopkins Press, 1967.

Hyman, L. H. *Comparative Vertebrate Anatomy*, 2nd ed. Chicago: The University of Chicago Press, 1942.

Lineaweaver, T. H., and R. H. Backus. *The Natural History of Sharks*. Philadelphia: J. B. Lippincott, 1970.

Sawyer, W. H. "Discussion: Endocrines and Osmoregulation among Fishes," *American Zoologist*, **13**(3), 1973.

Scheer, B. T. *Comparative Physiology*. New York: John Wiley & Sons, Inc., 1948.

Schultz, L. P., and E. M. Stern. *The Ways of Fishes*. New York: D. Van Nostrand Company, 1948.

Spoczyuska, J. O. I. *An Age of Fishes: The Development of the Most Successful Vertebrate*. New York: Charles Scribner's Sons, 1976.

Young, J. Z. *The Life of Vertebrates*, 2nd ed. London: Oxford University Press, 1962.

Phylum Chordata, Sublcass Vertebrata, Class Osteichthyes, The Bony Fishes

Class Osteichthyes contains the highly diversified **bony fishes** (Fig. 26-1), with more living species (30,000) than any other vertebrate class. Familiar bony fishes include salmon, tarpon, herring, tuna, cod, bass, and perch. The great success of the bony fishes has been attributed to their relatively small sizes and diversified feeding habits. Whereas the cartilaginous fishes are relatively large animals that probably average about 2 m in length, the bony fishes are generally much smaller, averaging perhaps 15 cm in length. The bony fishes have herbivorous, omnivorous, and carnivorous representatives, whereas the cartiliginous fishes are exclusively carnivorous. Small size and varied feeding habits have allowed the bony fishes to exploit more niches, with consequently greater adaptive radiations. Another advantage of bony fishes is their possession of an air-filled **swim bladder,** with which they can adjust their buoyancy. And greater facilities for **osmoregulation** have allowed the bony fishes to fully exploit both marine and freshwater habitats.

Taxonomy and Evolution of Bony Fishes

Class Osteichthyes contains two subclasses. Subclass **Sarcopterygii** includes the **fleshy-finned fishes** whose fleshy fins are framed with internal bony skeletons. Subclass **Actinopterygii** includes the **ray-finned fishes** whose fins are supported by many soft or hard fin rays.

Subclass Sarcopterygii— Fleshy-finned Fishes

The evolution of the fleshy-finned fishes is marked by adaptations of their pectoral and pelvic fins into primitive walking organs and conversion of their swim bladder into an air-breathing lung. This group gave rise to the first amphibians that invaded terrestrial habitats about 350 million years ago. Modern sarcopterygians include the coelacanth (Fig. 26-2) and the lungfish. **Coelacanths** (order **Crossopterygii**) were known only from the fossil record until 1938, when a living specimen was recovered from deep waters off the coast of Madagascar. Several more of these living fossils have since been caught.

The **lungfishes** (order **Dipnoi**) survive today in shallow fresh waters in Africa, South America, and Australia. Their olfactory sacs communicate with the mouth cavity, and the swim bladder opens into the foregut. The lungfish swims to the surface of the water and swallows air into its swim bladder, which functions as a lung. As their freshwater pools dry up, some lungfishes dig into the bottom mud and

Figure 26-1. Representatives of the bony fishes. The figures are not drawn to scale.

Triglidae (Gurnard)

Exocotidae (Flying fish)

Amblyopsidae (Blind cave fish)

Ameiuridae (Bullhead)

Centrarchidae (Bass)

Percidae (Perch)

Characidae (Characin)

Cyprinidae (Carp)

Gobiidae (Gobies)

Ostraciontidae (Trunkfish)

Anguillidae (Eel)

Salmonidae (Trout)

Blenniidae (Blennies)

Lophiidae (Sargassumfish)

Amiidae (Bowfin)

Syngnathidae (Seahorse)

Neoceratodontidae (Lungfish)

Lepisosteidae (Alligator gar)

Acipenseridae (Sturgeon)

Coelacanthidae (Coelacanth)

Polypteridae (Lobefinned fish)

Figure 26-2. An ancient fish, the coelacanth, *Latimeria,* first caught off the coast of Africa in 1938; supposedly extinct for millions of years. It is a "living fossil," estimated to have lived in the Devonian period. Length about 1.5 m. Note the thick lobe-shaped fins.

await the rainy season in a quiescent, air-breathing state.

Subclass Actinopterygii— Ray-finned Fishes

The ray-finned fishes are the most abundant and familiar of the modern osteichthyes. Four superorders are represented today; the superorder **Paleonisci,** a group of freshwater and marine fishes that closely resembled and competed with the sharks, is extinct. Superorder **Polypteri** now consists of only two genera; *Polypterus* (Fig. 26-1) is characterized by **ganoid scales** (discussed below), the spiral intestinal valve, as in sharks, and the use of its swim bladder as a lung. The **chondosteans** (superorder **chondrostei**) flourished from about 400 to 150 million years ago but are represented today by the sturgeons and paddlefishes (Fig. 26-3). These are both curious-looking, somewhat degenerate, creatures that have lost the ganoid scales, most all the external bony armor (except for the sturgeon's bony plates), and bony endoskeletons of their ancestors (i.e., they are cartilagenous). Approximately 240 million years ago, the chondosteans gave rise to the **holosteans** (superorder

Holostei), which were quite successful during the late Mesozoic era but are now represented by only the **gar pikes** (Fig. 26-3) and **bowfins.** The holosteans in turn gave rise to the **teleosts** (superorder **Teleostei**), which have flourished since about 200 million years ago and now include more than 85 percent of the 30,000 known species of modern osteichthyes.

The teleost fishes are also ray finned, but they are distinguished from the chondrosteans and holosteans by their having fully ossified skeletons and symmetrical, **homocercal** tails. The teleosts are the dominant fishes in both marine and freshwater habitats. The taxonomy of superorder Teleostei is in a state of flux, but some 24 teleost orders may be recognized. Several of these orders contain about half the teleost species and include two basic body types.

The **clupeiform body type** is named after the order (**Clupeiformes**) that contains the **herring, tarpon, salmon,** and **pike.** Clupeiform teleosts generally have soft fin rays and smooth cycloid scales. Their swim bladder communicates with the foregut, and their pelvic fins are in the primitive posterior abdominal position. These teleosts exhibit an array of primitive characteristics that are intermediate between those of their holostean ancestors and the more advanced perciform teleosts.

Figure 26-3. (A) Shovel sturgeon, *Scaphirhynchus platorynchus,* a unique fish of great importance for food, seen lying under a large catfish. (B) Paddlefish, *Polydon spathula,* a fish easily identified by its elongated snout. This is one of the largest North American freshwater fishes, exceeding 150 pounds in weight. (C) Longnose gar, *Lepisoteus osseus.*

The **perciform body type** is named after the largest vertebrate order (**Perciformes**) and includes the **perch, tuna, mackerel, sea bass, rockfish, marlin, sailfish, bluefish,** and **jacks.** They are morphologically characterized by having hard and spiny fin rays, **ctenoid** scales with rough and spiny margins, air bladders that have become independent of the gut, and, most noticeably, pectoral fins that have been displaced dorsally and pelvic fins that have migrated so far forward that they now lie ventral and anterior to the pectorals.

These fishes dominate the coastal marine waters and account for much of the world fisheries catch. The perciform fishes diverged early from the primitive teleost line. Their sharp fin spines afford greater protection than do those of the softrayed clupeiform teleosts, and their fin positions allow for more versatile swimming movements. In addition, their jaws are more movable, thereby allowing them to exploit different food resources. All these traits have been used to good advantage during the successful adaptive radiations of the perciform teleosts.

Perca flavescens—A Representative Bony Fish

External Morphology

The **yellow perch,** *Perca flavescens* (Fig. 26-4), inhabits freshwater streams and lakes of the eastern United States. Its body attains lengths of approximately 30 cm and is divisible into **head, trunk,** and **tail.** There are two **dorsal fins,** a **caudal fin,** a single median **anal fin** just posterior to the ventral **anus,** two lateral **pelvic fins,** and two lateral **pectoral fins.** Along each side of the body is a **lateral line** sense organ. The head bears a **mouth** with well-developed **jaws** armed with **teeth,** a pair of lateral **eyes,** a pair of **external nares** in front of each eye, and **opercula** that cover the gills. The skin is provided with a protective but flexible covering of overlapping **scales.** Numerous glands in the skin produce the slimy mucus that makes the fish slippery.

The body of perch and of most other bony fishes are streamlined and spindle shaped. Bony fishes are able to remain stationary without much muscular exertion because the internal **air bladder** allows them to main-

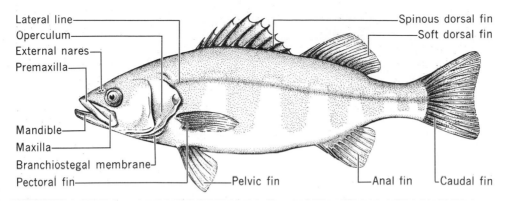

Figure 26-4. External features of the yellow perch, *Perca flavescens*.

tain their buoyancy; their average body weight is the same as the volume of water they displace. The principal locomotor organ is the tail. Alternating contractions of muscular bands on either side of the trunk and tail pivot the caudal fin, thus displacing water and propelling the fish forward.

The fins are integumentary expansions that are supported by spiny or soft **fin rays.** The paired pectoral and pelvic fins are used much like oars when the fish is swimming slowly. These lateral fins also aid the caudal fin in steering. Movement up or down results from holding the lateral fins in certain positions: obliquely backward with the anterior edge higher for ascent and obliquely forward for descent.

Fishes must maintain their equilibrium in some way as the muscular dorsal region is the heaviest part of the body and this tends to turn them over. The dorsal, anal, and caudal fins increase the vertical surface of the body and (much like the keel of a boat) help the animal maintain an upright position. The paired lateral fins also act as balancers. If one or two of the paired fins are removed, the fish soon learns to compensate for their loss.

Internal Morphology and Physiology

Skeletal System

The **exoskeleton** of the perch includes scales and fin rays. The **scales** develop within pouches in the dermis. They are arranged in oblique rows and overlap, thus forming a protective yet flexible covering that allows lateral body movements. The exposed poste-

rior edge of each scale is toothed and therefore rough to the touch. Scales of this kind are called **ctenoid** scales (Fig. 26-13). The **fin rays** support the fins. The rays in the spinous dorsal fin (Fig. 26-5) and in the anterior edges of the anal and pelvic fins are **hard spines.** The caudal, pectoral, pelvic, soft dorsal, and anal fins are supported by **soft fin rays.**

The bones of the **endoskeleton** are displayed in Figure 26-5. They include the **skull** and **vertebral column** (comprising the **axial skeleton**), **ribs, pectoral girdle, pelvic girdle,** and **interspinous bones,** the latter helping to support the unpaired fins. The body of the fish is, to a considerable extent, supported by the surrounding water, so its bones need not be as strong as those of terrestrial vertebrates.

The skull consists of a large number of parts—some bone, others cartilage. Also part of the **axial skeleton,** the **vertebrae** are simple and comparatively uniform in structure. Ribs are attached by ligaments to the abdominal vertebrae and form a framework protecting the body cavity. There is no sternum.

The **visceral skeleton** is composed of seven paired **branchial arches,** more or less modified. The first, or **mandibular arch,** forms the **jaws.** The second, or **hyoid arch,** is modified as a support for the opercula. Arches 3 to 7 support the gills and are known as **gill arches.** The first four gill arches bear spinelike ossifications, called **gill rakers,** that act as sieves to prevent solid particles from contacting the delicate gills. In many species, they also function in keeping small food particles from escaping between the gill arches.

The **appendicular skeleton** is represented in the perch by the **pectoral** and **pelvic girdles** with their

Figure 26-5. Skeleton of a perch *Perca flavescens.* The various fascia are artifact, and the ventral structure is also not part of the skeleton.

associated **fins** and by the **median fins** (Fig. 26-5). The simple pelvic girdle is not very typical in form, being degenerate or possibly primitive.

Muscular System

The principal **muscles** are those used in locomotion, respiration, and feeding. The lateral movements of the body employed in swimming are produced by four longitudinal bands of muscles: two heavy dorsal bands and a thinner lateral band on each side of both trunk and tail. These muscles are arranged segmentally as **myomeres.** Other sets of muscles move the gill arches, hyoid region, opercula, and jaws.

Digestive System

Aquatic insects, mollusks, and small fishes constitute a large part of the perch's diet. Such prey are captured by the **jaws** and held by the **teeth.** These many conical teeth are borne on the **mandibles** (lower jaw) and **premaxillae** (upper jaw) and on the roof of the mouth. They are used only for holding food, not for chewing. A rudimentary **tongue** projects from the floor of the mouth cavity; it is not capable of inde-

pendent movement, but it does function as a tactile organ. The **mouth cavity** leads into the **pharynx,** which is pierced on either side by four gill slits. Food passes through a short esophagus to the **stomach,** and then through the **intestine;** wastes are eliminated through the **anus.** The relative positions of the stomach, intestine, **liver, gallbladder, pyloric cecum,** and **anus** are shown in Figure 26-6. The **pancreas** is located in the first loop of the intestine, but it is so diffuse that it is not readily observed in gross dissection.

Circulatory System

The blood of the perch contains oval, nucleated **red blood cells** and amoeboid **white blood cells.** The two-chambered **heart** lies within the **pericardial cavity** beneath the pharynx. The circulatory pattern (Fig. 26-7) conforms rather closely to that already described for the dogfish shark (Chapter 25).

Respiratory System

The perch breathes with four pairs of **gills.** Each gill is supported by a **gill arch** that bears a double row of **gill filaments** that are abundantly supplied with capil-

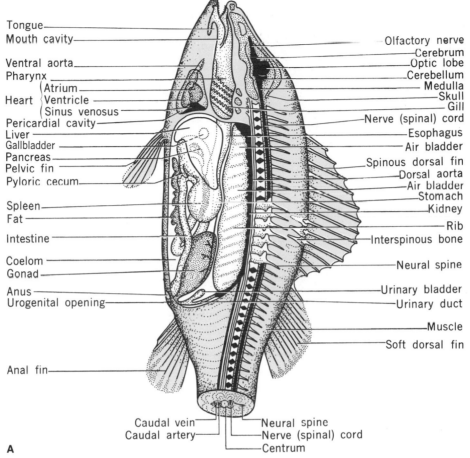

Tongue
Mouth cavity

Ventral aorta
Pharynx
Heart { Atrium
Ventricle
Sinus venosus
Pericardial cavity
Liver
Gallbladder
Pancreas
Pelvic fin
Pyloric cecum

Spleen
Fat

Intestine

Coelom
Gonad

Anus
Urogenital opening

Anal fin

Olfactory nerve
Cerebrum
Optic lobe
Cerebellum
Medulla
Skull
Gill
Nerve (spinal) cord
Esophagus
Air bladder
Spinous dorsal fin
Dorsal aorta
Air bladder
Stomach
Kidney
Rib
Interspinous bone
Neural spine
Urinary bladder
Urinary duct
Muscle
Soft dorsal fin

Caudal vein
Caudal artery
Neural spine
Nerve (spinal) cord
Centrum

A

Figure 26-6. (A) Gross internal anatomy of the yellow perch, *Perca flavescens.* (B) Dorsal view of perch brain.

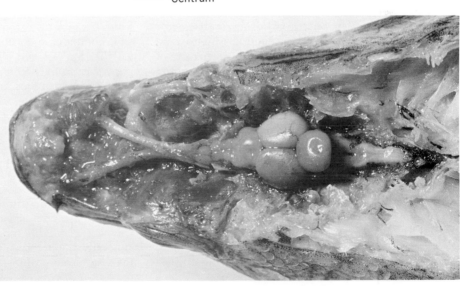

B

351

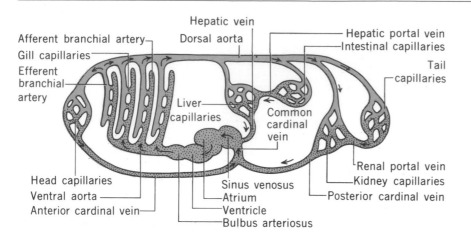

Figure 26-7. Diagram of the main blood vessels of a fish (yellow perch) as seen in lateral view.

laries. The **afferent branchial arteries** (Fig. 26-7) carry deoxygenated blood from the heart to capillaries in the gill filaments, where exchange of respiratory gases occurs. Carbon dioxide diffuses from the gill, and oxygen is gleaned from the continuous stream of water that enters the mouth and bathes the gills before exiting via the opened **opercula.**

The oxygenated blood is collected into the **efferent branchial arteries** and delivered to the dorsal aorta for delivery throughout the body. Respiration requires a constant supply of water passing over the gills, and a fish out of water generally dies of suffocation.

The **air bladder** is a comparatively large, thin-walled sac lying in the dorsal part of the body cavity. The air bladder is filled with gas and is a hydrostatic organ or "float." The gas that it contains is a mixture of oxygen, nitrogen, and carbon dioxide that is secreted from the blood vessels in its walls. The air bladder decreases the specific gravity of the fish, making its body weight equal to the weight of the water that it displaces. The fish is thus able to maintain a stationary position in the water column with minimal muscular effort. The amount of gas within the air bladder depends upon the pressure of the surrounding water and is regulated to some extent by the fish according to depth. If a fish is suddenly brought to the surface from a great depth, its air bladder, relieved of considerable pressure, expands, often forcing the stomach out of the mouth. In some fishes the air bladder may serve in sound production. In the lungfishes the air bladder can function as an air-breathing organ.

Excretory System

The **kidneys** lie between the spinal column and the air bladder. They extract ammonia and other waste products of cellular respiration from the blood. Two thin **urinary ducts (ureters)** carry the urine into a **urinary bladder** (Fig. 26-6), where it is stored before elimination through the **urogenital opening,** just posterior to the anus.

The ancestors of the modern bony fishes evolved in fresh water, and their kidneys were adapted to eliminate from their relatively hypertonic tissues the excess water that entered, by osmosis, the gill membranes and alimentary canal. As a result, freshwater fishes excrete a relatively large volume of urine each day, comparable to about 20 percent of their body weight. Special **salt-absorbing glands** in their gills glean salts from their aquatic environment and help to offset the loss of ions in the urine.

Marine bony fishes are equipped with kidneys similar to those evolved by their freshwater ancestors. These kidneys are adapted to excrete dilute urine, which is disadvantageous because water is continually lost into the hypertonic saltwater environment from the blood flowing through the permeable gills. To offset this osmotic water loss, marine teleosts habitually swallow seawater; the liquid component is retained while excess salts are actively eliminated by specialized salt glands in the gills. Also, their kidneys are generally smaller than those of their freshwater relatives, so urine elimination is reduced to only about 4 percent of the fish's body weight per day.

Nervous System

The **brain** of the perch (Fig. 26-6 and Chapter 36) is more highly developed than is that of a cyclostome or shark. The four principal divisions are well marked: the **cerebrum, optic lobes, cerebellum,** and **medulla.** The brain gives off ten pairs of **cranial nerves** to the sense organs and anterior portion of the body. The **spinal cord** passes through the neural arches of the vertebrae. **Spinal nerves** arise laterally from the spinal cord.

Sense Organs

The principal sense organs are **cutaneous receptors** (including the lateral lines), **olfactory sacs, ears,** and **eyes.**

The integument, especially near the lips, is supplied with **tactile receptors.** Barbels on some fishes, such as the bullhead (Fig. 26-1), also function as sensory organs for locating food. The **lateral line** contains sensory cells that serve to detect vibrations and currents in water.

The two **olfactory sacs** lie in the anterior part of the skull and communicate with external environment by means of a pair of **nares** (Fig. 26-4) in front of each eye. The olfactory sacs are not connected with the mouth cavity and play no role in respiration. Their inner surfaces are folded and contain many chemoreceptor cells that detect dissolved substances as water passes into the front opening and out the rear opening.

Each **ear** consists basically of three **semicircular canals** that function in maintaining equilibrium. Experiments indicate that goldfishes can hear, suggesting that the ear serves as an organ of both hearing and equilibrium. Water-borne sound waves are transmitted by the bones of the skull to the fluid within canals, as in cyclostomes and chondrichthyes. A ventral enlargement of the canals, the **sacculus,** contains concretions of calcium carbonate called ear stones, or **otoliths,** that aid in detecting body tilt and acceleration.

The **eyes** of the perch differ in several respects from those of the terrestrial vertebrates. Movable **eyelids** are absent in bony fishes, as their aquatic environment keeps the eyeball moist. The **cornea** is flattened and has about the same refractive index as the water. The **lens** is almost spherical. The pupil diameter is usually larger than that of terrestrial vertebrates, which allows the entrance of more light rays. This is necessary, because semidarkness prevails at even moderate depths. Many fishes are nearsighted. When at rest the eye focuses clearly at about 40 cm, but it can detect moving objects much farther away. To focus on distant objects, the lens is pulled backward. (The cartilaginous sharks that pursue rapidly moving prey have lenses that are set for distance vision.)

The "four-eyed" fish, *Anableps* (Fig. 26-8), is an interesting example of a very specialized visual sys-

Figure 26-8. "Four-eyed" fish, *Anableps.* Upper pupil allows vision above the water line; lower pupil allows underwater vision. Light coming into the upper pupil strikes the lower portion of the retina, and illumination of the lower pupil strikes the upper retina.

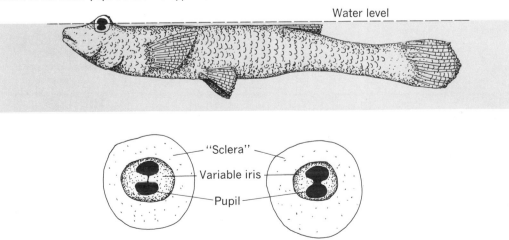

tem. Double lenses permit this fish to see both above and below the water simultaneously. This adaptation allows part of the eye to remain above the water, watching for insect food.

Reproductive System

The sexes are separate (**dioecious**). The single **ovary** or **testes** lie in the posterior region of the body cavity. The ovary in the female probably results from fusion of two ovaries in the embryo. The **gametes** pass through the reproductive ducts and out the **urogenital opening.** During the spring perch migrate from the deep waters of lakes and ponds, where they have spent the winter, to shallow waters along the shore. The female lays many thousands of **eggs** in a long ribbonlike mass. The male fertilizes the eggs by depositing **sperm** (milt) over them. Very few eggs de-

velop because of the lack of parental attention, disease, predation, and other factors.

The young hatch in approximately a week or two, depending greatly upon water temperature. Most of the egg consists of yolk. During development (Fig. 26-9) the young fish lives upon the stored food in the yolk sac but is soon able to obtain food from its environment. Vertebrate embryology is discussed more fully in Chapter 39.

Adaptive Modifications of Bony Fishes
Body Shape

The bodies of most fishes are **fusiform** (spindle shaped, Fig. 26-10) and laterally compressed, as in the perch. Such streamlining minimizes resistance to progress through the water. Variations in form are

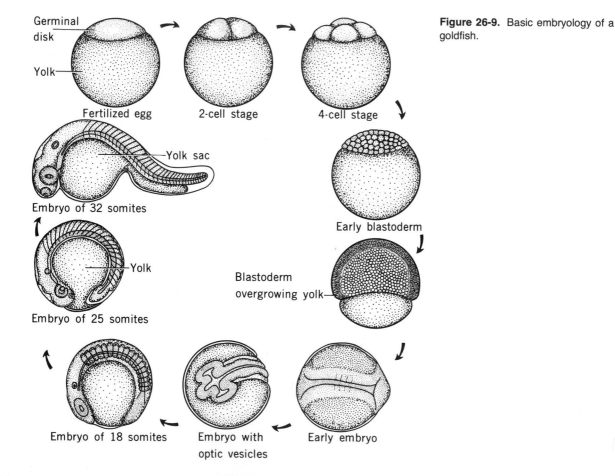

Figure 26-9. Basic embryology of a goldfish.

Germinal disk

Yolk

Fertilized egg

2-cell stage

4-cell stage

Yolk sac

Embryo of 32 somites

Early blastoderm

Yolk

Blastoderm overgrowing yolk

Embryo of 25 somites

Embryo of 18 somites

Embryo with optic vesicles

Early embryo

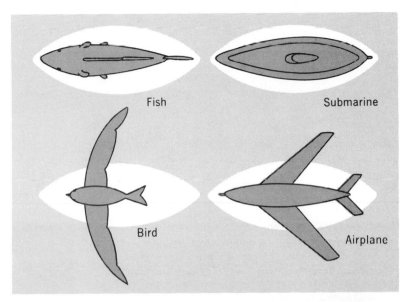

Figure 26-10. Diagrams of organisms that have a fusiform shape and machines that have been modeled after them. This shape is characteristic of birds capable of rapid flight and fast-moving objects in air and water. The flightless fossil birds differed about as much in form from the fast-flying birds of today as the first airplanes differed in shape from a modern jet.

correlated with the habits and habitats of certain fish. For example, **flounders** (Fig. 26-1, 26-11) have flat bodies that are adaptations for life on the ocean floor; **eels** (Fig. 26-1) have long cylindrical bodies that enable them to enter holes and crevices; and certain fishes (Fig. 26-12) bear leaflike body extensions that tend to conceal them among seaweed.

Figure 26-11. Adaptive coloration in the flounder *Pleuronichthys*. The flounder changes pattern and color to blend with its surroundings. Dispersion or concentration of pigment bodies is due to the chromatophores. Even when placed on such an unnatural background as that on the right, the resemblance is striking.

Figure 26-12. Australian sea dragon, *Phyllopteryx,* the most bizarre of all seahorses. The postures and antics of these fishes are as distinctive and grotesque as their appearance. The leaflike extensions from the body tend to conceal the fish among seaweed. Natural size up to 30 cm long.

Fins and Tails

Fins are generally considered to arise in the fish embryo as continuous median and lateral folds of the integument (Fig. 26-13). Parts of these folds later disappear, isolating the dorsal, caudal, anal, pelvic, and pectoral fins. All these fins vary considerably in shape and position.

The shape of the tail and associated caudal fin differs in the main groups of fishes and is therefore important in fish classification. The five main types of tails and caudal fins found in fishes are heterocercal, diphycercal, homocercal, protocercal, and hypocercal (Fig. 26-14).

The **heterocercal** tail is two lobed, with the vertebral column extending into the larger dorsal lobe. The stroke of the asymmetrical heterocercal tail forces the anterior part of the body downward. This tail is consequently characteristic of fishes that have ventrally situated mouths, many of which feed on the bottom (e.g., sturgeon and many primitive osteichthyes). Heterocercal tails also characterize the cartilaginous sharks.

In **diphycercal** tails, the vertebral column extends straight to the posterior tip of the body, with the spear-shaped caudal fin developed symmetrically

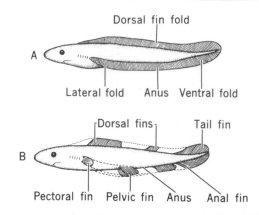

Figure 26-13. Diagrams illustrating the finfold theory of the origin of fins. (A) The continuous folds of the paired and unpaired fins in the embryo. (B) Parts of the continuous folds disappear to form the permanent fins. Photograph shows flying fish in flight after breaking the surface of the water.

above and below. Lungfishes have tail fins of this type.

The **homocercal** tail is externally and internally symmetrical, but the fin is better developed than the diphycercal type. The vertebral column stops at the base of the caudal fin. The stroke of the homocercal tail forces the fish straight forward. It is characteristic of fishes with a terminal mouth and is the type possessed by most bony fishes.

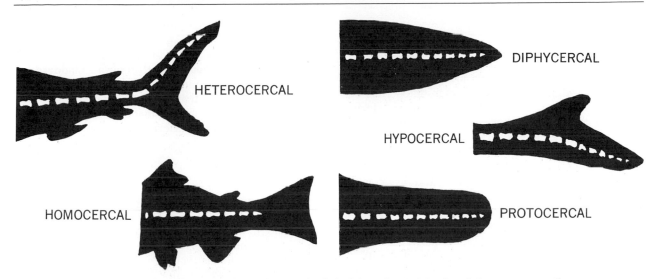

Figure 26-14. General types of tails in fishes. See text for descriptions.

The **protocercal** type is straight and occurs in some larval fishes. The notochord, or vertebral column, forms the axis of the protocercal tail and extends to the posterior tip of the rounded caudal fin.

The **hypocercal** tail is like an inverted heterocercal fin. This type is characteristic of very primitive types and was present in the now extinct ichthyosaurs, marine dinosaurs that swam in Mesozoic seas.

Scales

Fish scales form a protective exoskeleton. There are three principal types of scales: ganoid, cycloid, and ctenoid (Fig. 26-15). **Ganoid** scales are remnants of the heavily ossified scales of the placoderms. Ganoid scales occur in only a few primitive fishes such as the **gar pike** (Fig. 26-3). These heavy, rhomboid-shaped scales consist of a thick layer of shiny enamel (**ganoine**) deposited over a basal portion of bone.

Cycloid and ctenoid scales represent modifications of this primitive condition in which the scales are reduced to a layer of thin bone. **Cycloid scales** are nearly circular with concentric rings around a central point. **Ctenoid scales** are similar to cycloid scales, but their exposed posterior surfaces bears small spines. In many fishes the scales develop into large spines or fuse to form bony plates that are protective. Some fishes (e.g., catfishes) are scaleless.

Figure 26-15. The different types of scales on bony fishes. Ctenoid scale shows winter growth rings (numbered) that are used to determine age in years. The winter rings show slower growth periods that result from low food supply. [Photomicrographs of cycloid and ctenoid scales courtesy of Institute for Fisheries Research, Michigan Department of Natural Resources.]

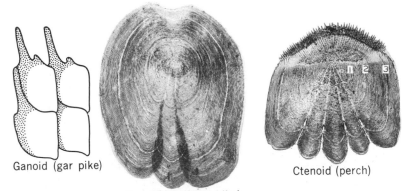

Ganoid (gar pike)

Cycloid (northern pike)

Ctenoid (perch)

Coloration

The general impression is that fishes are not brightly colored, but this may be because their colors quickly fade after death. Many fishes, especially those in tropical waters, are brillantly pigmented. The colors are due either to pigments within special dermal cells, called **chromatophores,** or to reflection and iridescence from guanine crystals within the scales. Chromatophore pigments are red, orange, yellow, or black, but other colors may be produced by their combination; for example, yellow and black when blended produce brown. The colors are usually arranged in a definite pattern consisting of transverse or longitudinal stripes and spots of various sizes. Coral reef fishes have long been famous for their brilliant colors, and many temperate freshwater fishes (e.g., **rainbow trout**) exhibit bright hues distributed in striking and intricate patterns. Fishes that inhabit the sunless abyssal ocean depths are typically black.

Certain fishes can disperse or concentrate the pigment in their chromatophores. These changes are mediated through the fisher's eyes and nervous system. The chromatophores are manipulated in such a way that the color of the immediate environment is mimicked, thus helping to conceal the fishes (Fig. 26-11). Change in color is slow in some fishes, but flounders and some others take only a few minutes.

Male fishes are often more brightly colored than are females of the same species, especially at the time of spawning. Territorial displays often involve an intensification of color by the antagonists. The outcome of such confrontations is often marked by the fading of the loser's color.

Sound Production

A surprisingly large number of fishes can produce sounds audible to humans; these noises are used either to bring the sexes together or to warn or startle enemies. Fishes make sounds in various ways. Certain **sculpins** vibrate their gill covers against the sides of their heads to produce a humming tone. The **hogfishes** grunt by gnashing their pharyngeal teeth. The **sea robins** produce a grunt by means of special muscles in their air bladders. The **croakers,** edible fishes that live along our southern Atlantic Coast, probably make more noise than any other kind of fish; organisms as deep as 10 m can be heard at the water surface. The croakers produce sounds by contracting special muscles associated with their air bladders.

Since the development of sonar, undersea sounds have acquired more than a purely academic interest, because some of the noises made by fishes (and marine mammals) can interfere with submarine detection.

Relations of Bony Fishes to Humans

Although some fishes destroy valuable food fishes and other aquatic animals, many are of use to humans by serving either as food or as source of recreation. Among the freshwater game fishes are various species of **trout, bass, pike,** and **perch** (Fig. 26-1); marine game fishes include **tarpon, sea bass,** and **tuna** (Fig. 26-16).

Marine food fish are of great commercial value. Vast quantities of **herring** (Fig. 26-16) are smoked, salted, and pickled. **Anchovies** off the Peruvian coast are the main food of vast numbers of aquatic birds responsible for guano (excrement) deposits that are used as fertilizer. **Mackerel** are caught in enormous numbers. The flounder family (Fig. 26-11) contains **halibuts, soles, plaice,** and **turbots. Codfishes** (Fig. 26-16) are especially valuable, along with other members of their family, such as **pollocks, haddocks,** and **hakes.** The average annual catch of codfish is over 2 billion pounds. Cod-liver oil is also a good source of vitamins A and D. Rivaling the codfish in value are the **salmon** of the Pacific Coast, which are dried and canned in large quantities. Freshwater food fishes include the **whitefish, lake trout, catfish,** and **perch.** The eggs of **sturgeons** and certain other fishes are consumed as caviar, especially in Russia. Processed fish meal (ground, dried fish) has become important in the fertilizer and pet food industries and holds great promise as a human dietary supplement as well.

Sport fishing is becoming increasingly popular with millions of people. The money spent by individuals in pursuing their sport runs into many millions of dollars. Enormous sums are also spent by federal and state governments in rearing fish and stocking streams and lakes.

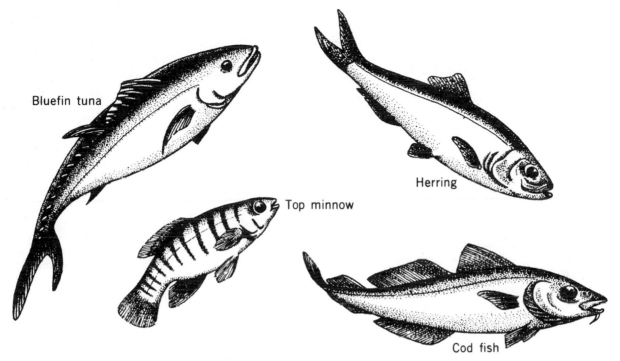

Figure 26-16. Some bony fishes important to man.

Many food fishes have decreased markedly in numbers due to overfishing, pollution by sewage and industrial wastes, and other causes. Federal and state governments have consequently placed certain restrictions on fishing and have also undertaken to propagate certain species artifically. These include freshwater whitefish, lake trout, pike, perch, and bass, and marine codfish, haddock, salmon, flounder, and the sardines.

Among the beneficial fishes should be mentioned the **top minnows**, *Gambusia* (Fig. 26-16), that feed voraciously on mosquito larvae. These fish are introduced into bodies of fresh water to prevent the breeding of mosquitoes that transmit malaria and yellow fever.

In recent years, considerable use has been made of fishes as experimental animals, especially in the fields of genetics, evolutionary biology, embryology, animal behavior, and pharmacology. Fishes, being the largest group of vertebrates, exhibit enormous variations and remarkable adaptations, all of great biological and evolutionary interest.

Summary

Perca flavescens, the yellow perch, has served as our representative member of class Osteichthyes. Its dorsal, caudal, anal, pelvic, and pectoral fins are characteristic of most bony fishes, although various modifications in shape and position occur.

The skeletal system is substantially ossified but is not as supportive as that of terrestrial vertebrates. The respiratory system utilizes gills, which are essentially concentrated capillary networks in constant contact with flowing water. The sense organs monitor tactile, olfactory, auditory, gustatory (assumed, because they reject some foods), and visual stimuli. The lateral line sense organ is unique to fishes (and some amphibians).

Tail and fin structures (heterocercal, diphycercal, homocercal, and protocercal) characterize the various groups of fishes and are useful clues in reconstructing their evolution. Ganoid, cycloid, and ctenoid scales also differentiate the three principal groups of bony fishes.

Review Questions

1. When a fish is swimming slowly, which fins might it use for propulsion?
2. What are the differences between the three scale types?
3. Where is the sternum in a bony fish?
4. What is the function of the tongue in bony fishes?
5. What is the relation between respiration and the external nares of bony fishes? (Ignore the few exceptions.)
6. How is the pupil of a bony fish's eye generally different from the pupil of terrestrial vertebrates?
7. The tail and caudal fin are differentiated into a number of types. Name and briefly describe each.
8. Which tail type(s) is (are) symmetrical?
9. Why do you suppose a hypocercal tail was advantageous to air-breathing ichthyosaurs?
10. Can any members of the class Osteichthyes breathe air directly? If so, how?
11. What other function do the gills have other than for respiration?
12. Describe briefly the function of the kidney of a freshwater and a marine fish.

Selected References

Alexander, R. M. *Functional Design in Fishes*. London: Hutchinson, 1967.

Berg, L. S. *Classification of Fishes Both Recent and Fossil*. Ann Arbor: Edwards, 1940.

Brown, M. D. *The Physiology of Fishes*, Vols. 1 and 2. New York: Academic Press, 1957.

Curtis, B. *The Life Story of the Fish: His Morals and Manners*. New York: Harcourt Brace Jovanovich, 1949.

Harden, J. F. R. *Fish Migration*. New York: St. Martin's Press, 1968.

Herald, E. S. *Living Fishes of the World*. New York: Doubleday & Company, 1961.

Hoab, W. S., and D. J. Randall (eds.). *Fish Physiology*. New York: Academic Press, 1971.

Lagler, K. F. *Freshwater Fishery Biology*. Dubuque: W. C. Brown Company, 1956.

Lagler, K. F., J. E. Bardach, and R. R. Miller. *Ichthyology*. New York: John Wiley & Sons, Inc., 1962.

Lanham, U. *The Fishes*. New York: Columbia University Press, 1962.

Norman, J. R. *A History of Fishes*, 5th ed. London: Ernest Benn, 1960.

Schults, L. P., and E. Stern. *The Ways of Fishes*. New York: D. Van Nostrand Company, 1948.

Spoczyuska, J. O. I. *An Age of Fishes: The Development of the Most Successful Vertebrate*. New York: Charles Scribner's Sons, 1976.

Sterba, G. *Freshwater Fishes of the World*, translated from German by D. W. Tucker. New York: The Viking Press, Inc., 1963.

Treassler, D. K. *Marine Products of Commerce*. New York: Chemical Catalogue Company, 1932.

Phylum Chordata, Subphylum Vertebrata, Class Amphibia—Frogs, Toads, Salamanders, and Caecilians

This chapter considers the vertebrate class that first successfully invaded terrestrial habitats (Fig. 27-1). The ancestors of modern amphibians evolved some 400 million years ago from air-breathing, freshwater, fleshy-finned fishes, the **Sarcopterygii.** As the name suggests, amphibians are partly aquatic and partly terrestrial, never venturing far from moist habitats. Most amphibians lay their eggs in water. The larvae, which respire by gills, are known as **tadpoles,** or **pollywogs.** Some amphibians are confused with reptiles, especially the lizards, because of general similarity of body form. However, almost all reptiles possess scales and are not slimy, whereas amphibians usually have a smooth, slimy skin without scales (except in a few rare species).

Modern frogs, toads, salamanders, and caecilians retain the basic **mesonephric kidney** of their aquatic ancestors. Constructed to eliminate excess osmotic water, this kidney is a hindrance on land, where water must be conserved. Water loss is further enhanced by their moist skin, which acts as an auxiliary respiratory surface that simultaneously provides an extensive surface for evaporative water loss.

An array of physiological and behavioral adaptations have allowed amphibians to invade the land, especially in humid tropical and subtropical habitats, where they outnumber all other terrestrial verte-
brates. Their large bladder acts as an internal water reservoir when the amphibian ventures onto land. Amphibians also exploit the spatial heterogeneity of their habitats, seeking shade and humid microhabitats to maintain their water balance at optimal levels. Another factor contributing to the success of amphibians is a relatively low metabolic rate, which allows them to go for long periods without eating and to survive periods of drought or cold in a dormant state.

One trait inexorably links amphibians to their ancestral aquatic habitats: Amphibians do not lay waterproof eggs, so they must breed in water. Even the highly terrestrial toads must return to fresh waters to reproduce.

Taxonomy and Evolution of Amphibians

The fossil record indicates that amphibians first appeared during a period of domination by fishes (the **Age of Fishes**) about 400 million years ago. Within 50 million years they increased so rapidly in numbers that the period is often spoken of as the **Age of Amphibians** (Fig. 23-3). The first amphibians are placed in the superorder **Labyrinthodontia,** so named because of the extensively folded dentine on the teeth of many. These salamanderlike animals probably

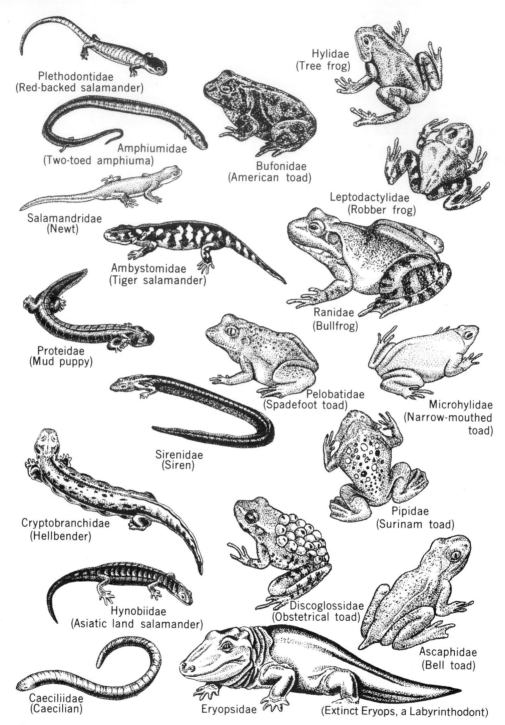

Figure 27-1. Some families of amphibians.

lived in fresh water or on land, reaching lengths of 3 m or more.

The **salamanders** resemble most closely the primitive amphibian body plan, whereas **frogs** and **toads** have become specialized for jumping. The wormlike **caecilians** are modified for burrowing and are not known from the fossil record.

Out of a total of ten orders of amphibians, there are only three living today consisting of about 3000 species, in a number very much smaller than in the other principal classes of vertebrates. The extant orders are:

1. Order **Anura,** containing about 1740 species of frogs and toads that are tailless in the adult stage.
2. Order **Caudata (Urodela),** containing some 200 species of amphibians with tails, commonly known as salamanders.
3. Order **Apoda (Gymnophiona),** containing about 60 species of legless amphibians, commonly called caecilians, that inhabit tropical and subtropical regions.

The United States is a paradise for students of amphibians because of the great numbers of amphibian species and individuals that live here. Because all amphibians require moisture, they are generally found in or near bodies of fresh water and in moist places such as under logs and stones in damp woods. As is true for birds, more amphibians are heard than are seen, and recordings are available that identify their calls.

Order Anura—The Frogs and Toads

The three major families of this order include the true frogs (**Ranidae**), tree frogs (**Hylidae**), and true toads (**Bufonidae**). The true frogs can be found everywhere except in Australia, New Zealand, and southern South America. About 19 species of the genus *Rana* live in the United States. Of these, the **leopard frog,** *Rana pipiens* (Fig. 27-2), is the most common. The **bullfrog,** *R. catesbeiana* (Fig. 27-1), is the largest frog in this country, often reaching a body length of 18 to 20 cm. Bullfrogs usually remain in or near water. They possess a deep bass voice. In the northern part of their range, the tadpoles do not become frogs during the first year, as do those of the leopard frog, but metamorphose into adults during the second or even the third year. Other true frogs include the green frog, *R. clamitans,* the eastern wood frog, *R. sylvatica,* and the pickerel frog, *R. palustris.*

Figure 27-2. *Rana pipiens*, the leopard frog.

Rana pipiens—The Leopard Frog

Frogs are frequently used for laboratory study because they are abundant and relatively easy to dissect. Investigation of this representative amphibian's morphology and physiology gives an understanding of the general vertebrate body plan and introduces the study of more complex vertebrates such as humans. The following account of frog biology applies generally to *Rana pipiens* (the leopard frog, Fig. 27-2), *R. catesbeiana* (the bullfrog), and most other common species.

The leopard frog lives in or near freshwater lakes, ponds, and streams throughout the North American continent except those on the Pacific slope. The frog swims in water and walks or leaps on land. The hindlimbs are elongated and muscular. When the frog is on land, the hindlimbs are folded to the sides of the body; they can be suddenly extended to propel the frog through the air. Likewise, while swimming the hindlimbs are alternately folded up and extended; during their backward extension, the webbed toes are spread apart, increasing the resistance to the water. Frogs frequently float with just the tip of their nose breaking the surface and their hindlegs loosely outstretched. If disturbed in this position the frog dives underwater: The hindlimbs are flexed, an action that submerges the frog, and the forelimbs direct the animal downward. Then the hindlimbs are extended in a propelling stroke.

Frogs croak mostly during the breeding season but also at other times of the year, especially in the evening and when the atmosphere is humid. Croaking may occur when the frog is either above or below the surface. Underwater croaking is achieved by circulating air from the lungs past the vocal cords into the mouth cavity, and then back again.

External Morphology

The body of the frog may be divided into **head** and **trunk** regions; there is no neck. The **eyes** normally protrude from the head but are drawn into their **orbits (sockets)** when the **eyelids** are closed. The frog, along with many other amphibians and reptiles, has a clear **nictitating membrane** that can be drawn laterally across each eye for additional protection (especially underwater). We humans have vestigial nictitating membranes, partially visible at the medial corner of each eye. Posterior and slightly ventral to each of the frog's eyes is a flat **tympanic membrane (eardrum).** A pair of **external nares (nostrils)** is situated dorsally near the end of the snout. The external nares lead to a **nasal cavity** that opens into the **mouth cavity** through a pair of **internal nares.** The **mouth** of the frog is relatively wide, extending from one side of the head to the other.

The two short **forelimbs** serve to support the anterior portion of the body. The **hands** possess four **digits** and a rudimentary **thumb (pollex).** In males, the pollex is thicker than the corresponding forefinger of females, especially during the breeding season. The muscular hindlimbs are folded together when the frog rests on land, and their five long **toes** are connected by **webs.** The **cloacal opening** (sometimes called the **anus**) is situated at the posterior end of the trunk.

The Integument

The smooth **integument (skin)** is loosely attached to the body. Extending behind the eyes along either side of the body is a ridge, called the **dorsolateral fold** (absent in the bullfrog), that may be distinctly pigmented. The skin is colored by scattered **pigment granules** in the epidermis and by pigment cells known as **chromatophores** in the dermis. There are two different kinds of chromatophores, containing black and brown or yellow and red pigments. There are also **interference cells** that contain whitish crystals. Green colors are produced by various combinations of these cells. The color of the dorsal and lateral surfaces is darker than the ventral surface, which is whitish. Such **protective coloration** camouflages the frog from terrestrial and aquatic predators.

The skin consists of two tissue layers, an outer epidermis and an inner dermis (Fig. 27-3), as in other vertebrates. The **epidermis** contains several layers of cells. The outer **stratum corneum** is horny and nonliving and consists of broad, thin, stratified **squamous epithelial cells.** Beneath this lies a transitional zone of **polygonal cells,** which in turn rest upon basal **columnar cells** of the **stratum germinativum.** Periodic **molting** involves shedding of the stratum corneum. New cells are continually being formed by the stratum ger-

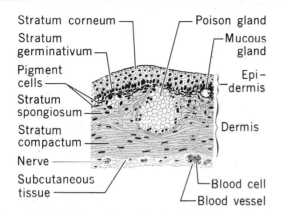

Figure 27-3. Section of frog skin showing microscopic structure (histology). Vertical section.

minativum, and the outward pressure of these growing cells results in the flattening of the surface layers.

The **dermis** consists of two layers: a loose outer **stratum spongiosum,** which contains glands and chromatophores, and a dense inner layer of connective tissue, the **stratum compactum,** which contains white and yellow fibers, a few smooth muscle cells, blood vessels, and nerves. Beneath the dermis is a loose subcutaneous connective tissue that attaches the skin loosely to the body wall.

The integument is richly supplied with two types of glands. **Mucus glands** are present in great densities, as many as 60 per square millimeter of surface. Each gland consists of epithelial secreting cells that are surrounded by muscle fibers and connective tissue; a duct leads to the surface of the integument. Mucus formed by the cells is discharged into the lumen of the gland and is forced through the duct by muscular contraction. The mucus moistens the frog's skin, thus facilitating some respiratory gas exchange to occur across the integument. **Poison glands** secrete a whitish fluid with a burning taste that serves as a means of protection against predators. Poison glands are larger and less numerous than are mucus glands.

Internal Morphology and Physiology

If the body wall of the frog is dissected along the ventral midline, the visceral organs in the **coelomic body** cavity will be exposed (Fig. 27-4). The heart lies within a saclike **pericardium** that is partially surrounded by the three reddish-brown lobes of the liver. The two lungs are located anteriorly. Coiled within the body cavity are the stomach and intestines. The kidneys are flat, reddish bodies attached to the dorsal portion of the coelom, lying distal to the thin membrane, the **peritoneum,** that lines the coelom. The digestive tract and reproductive organs are suspended by double layers of peritoneum called **mesenteries.** In males, the two testes are small ovoid organs suspended by membranes on either side of the digestive tract. In females, the egg-filled ovaries and oviducts occupy a large part of the body cavity during the breeding season.

Skeletal System

The supporting framework of the frog's body is an **endoskeleton** consisting largely of cartilage and bone. **Cartilage** (gristle) is a connective tissue in which the cells are embedded in a firm, tough, and elastic matrix that they secrete (Fig. 3-3). **Bone** is a connective tissue impregnated with mineral salts. About 58 percent of bone tissue is calcium phosphate, 7 percent is calcium carbonate, and 33 percent is organic matter such as living cells and gelatinous substances. The endoskeleton supports the soft tissues and organs of the body, furnishes points of attachment for the skeletal muscles, and protects such vital organs as the brain, eyes, spinal cord, heart, and lungs.

A careful study of the skeleton reveals many interesting interrelationships of structure and function. Each ridge, line, depression, or protuberance on a bone usually has some functional significance.

There are three main subdivisions of the skeleton: the axial skeleton, the visceral skeleton, and the appendicular skeleton. The **axial skeleton** comprises the skull, vertebral column, and sternum (breastbone); the frog has no ribs. The **visceral skeleton** is comprised of a series of bones in the jaws, throat, and ear area. The **appendicular skeleton** consists of the pectoral and pelvic girdles and the bones of the limbs that they support (Fig. 27-5).

Axial Skeleton. A large part of the **cranium** (Fig. 27-6) consists of cartilage that has undergone **ossification,** in which minerals replace cartilage, or that has

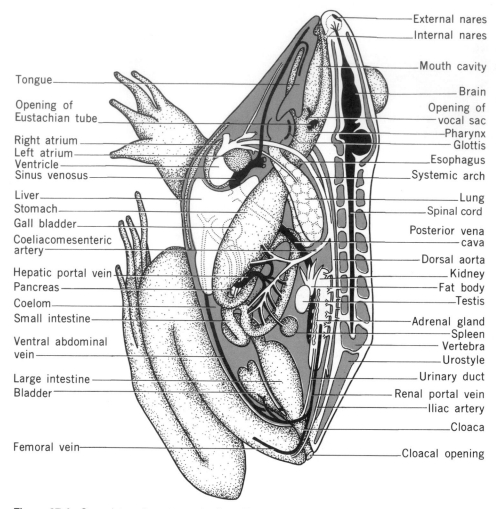

Tongue

Opening of
Eustachian tube

Right atrium
Left atrium
Ventricle
Sinus venosus

Liver
Stomach
Gall bladder
Coeliacomesenteric
artery

Hepatic portal vein
Pancreas
Coelom
Small intestine

Ventral abdominal
vein

Large intestine
Bladder

Femoral vein

External nares
Internal nares

Mouth cavity

Brain
Opening of
vocal sac
Pharynx
Glottis
Esophagus
Systemic arch

Lung
Spinal cord

Posterior vena
cava
Dorsal aorta
Kidney
Fat body
Testis
Adrenal gland
Spleen
Vertebra
Urostyle
Urinary duct
Renal portal vein
Iliac artery

Cloaca

Cloacal opening

Figure 27-4. Gross internal anatomy of a frog. Frog is vertical.

developed from connective tissue without passing through a cartilage stage. The spinal cord passes through a large opening, the **foramen magnum,** in the posterior end of the brain case. On either side of this opening are convexities of the **exoccipital** bones; these **occipital** condyles articulate with a pair of concavities on the first vertebra and enable the frog to pivot its head.

The **vertebral column** (Fig. 27-3), or backbone, consists of nine **vertebrae** and a posterior bladelike extension, the **urostyle.** The vertebrae are held together by **ligaments.** The vertebral column serves as a firm axial support that also allows bending of the body anterior to the urostyle. A frog has a sternum, but no ribs.

Visceral Skeleton. The upper and lower **jaws, hyoid,** and cartilages of the **larynx** all constitute the visceral skeleton; they are preformed in cartilage and then strengthened by ossification. The upper jaw

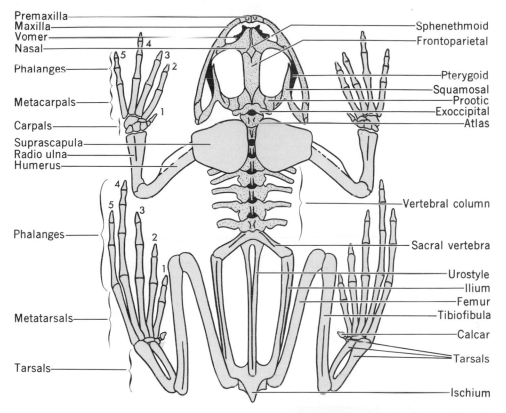

Premaxilla
Maxilla
Vomer
Nasal

Phalanges

Metacarpals

Carpals

Suprascapula
Radio ulna
Humerus

Phalanges

Metatarsals

Tarsals

Sphenethmoid
Frontoparietal

Pterygoid
Squamosal
Prootic
Exoccipital
Atlas

Vertebral column

Sacral vertebra

Urostyle
Ilium
Femur
Tibiofibula
Calcar
Tarsals

Ischium

Figure 27-5. Skeleton of the frog. Note axial (shaded) and appendicular skeletons.

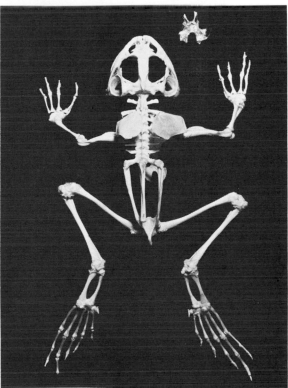

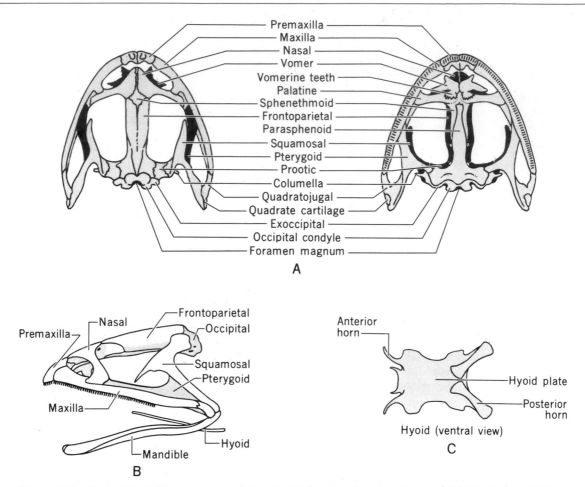

Premaxilla
Maxilla
Nasal
Vomer
Vomerine teeth
Palatine
Sphenethmoid
Frontoparietal
Parasphenoid
Squamosal
Pterygoid
Prootic
Columella
Quadratojugal
Quadrate cartilage
Exoccipital
Occipital condyle
Foramen magnum

A

Frontoparietal
Nasal
Occipital
Premaxilla
Squamosal
Pterygoid
Maxilla
Hyoid
Mandible

B

Anterior horn
Hyoid plate
Posterior horn
Hyoid (ventral view)

C

Figure 27-6. Skull of the bullfrog *Rana catesbeiana* in (A) dorsal and ventral views and (B) lateral view. (C) Ventral view of the hyoid bone.

(**maxilla**) consists of a pair of **quadratojugals.** The maxillae and premaxillae bear **teeth.** The lower jaw is the **mandible.** The **hyoid** and its processes (Fig. 27-6) aid in the support of the tongue and throat muscles.

Appendicular Skeleton. The pectoral girdle and sternum (Fig. 27-7), composed of bone and cartilage, support the forelimbs, serve as attachment sites for their muscles, and protect the organs lying within the anterior portion of the trunk. The principal parts of the **pectoral girdle** are the **suprascapulae, scapulae, clavicles, coracoids,** and **epicoracoids;** of the **sternum:** they are the **episternum, omosternum,** mesosternum, and **xiphisternum.** The suprascapulae lie dorsal to the vertebral column, and the rest of the girdle passes downward on either side and unites with the sternum in the ventral midline.

The **pelvic girdle** supports the hindlimbs. It consists of two sets of three parts each: the **ilium,** the **ischium,** and the **pubis.** The pubis is cartilaginous. The anterior end of each ilium attaches to one of the transverse processes of the ninth (**sacral**) vertebra. Where the parts of each half of the pelvic girdle unite, there is a concavity called the **acetabulum,** into which the head of the femur articulates.

The forelimbs and hindlimbs differ in size but have

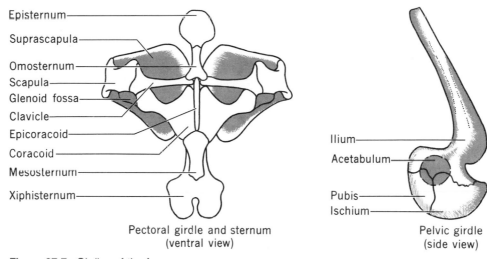

Episternum
Suprascapula
Omosternum
Scapula
Glenoid fossa
Clavicle
Epicoracoid
Coracoid
Mesosternum
Xiphisternum

Pectoral girdle and sternum
(ventral view)

Ilium
Acetabulum
Pubis
Ischium

Pelvic girdle
(side view)

Figure 27-7. Girdles of the frog.

similar component bones (Table 27-1). Each **forelimb** consists of a **humerus** that articulates at its proximal end with the **glenoid fossa** of the pectoral girdle and with the **radioulna** at its distal end (Figs. 27-6 and 27-8). The bone of the **forearm (radioulna)** consists of the **radius** and **ulna** fused. The **wrist** contains six small **carpal** bones arranged in two rows. The palm of the hand is supported by four proximal **metacarpal** bones. Digits 2 and 3 consist of two **phalanges** each, and digits 4 and 5 consist of three phalanges each. The rudimentary thumb is framed by the first metacarpal bone.

Each **hindlimb** consists of a **femur** (thighbone), a **tibiofibula** (**tibia** and **fibula** fused) or lower leg bone, four **tarsal** bones, the five **metatarsals** of the sole of

the foot, the **phalanges** of the digits, and the **prehallux (calcar)** of the accessory digit.

Muscular System

The muscular system is responsible for movement. Muscles (Figs. 27-8 and 27-9) are of three principal types: **smooth (visceral)**, **cardiac**, and **striated (skeletal)**. These differ in structure and function, as described in Chapter 3. Smooth muscle in the viscera and cardiac muscle in the heart are under **involuntary control.** Striated muscles, which move the bones of the skeleton, are under **voluntary** and **reflex control.**

A skeletal muscle that bends one body part upon another, as the leg upon the thigh, is called a **flexor;**

Table 27-1. Major Bones of the Frog

Forelimb (arm)	Number of Bones	Hindlimb (leg)	Number of Bones
Humerus (upper arm)	1	Femur (thighbone)	1
Radioulna (forearm)	2 fused	Tibiofibula (lower leg bone or shank)	2 fused
Carpals (wrist)	6	Tarsals (ankle)	4
Metacarpals (palm of hand)	5	Metatarsals (sole of foot)	5
Phalanges (fingers)	10	Phalanges (toes)	14 + prehallux

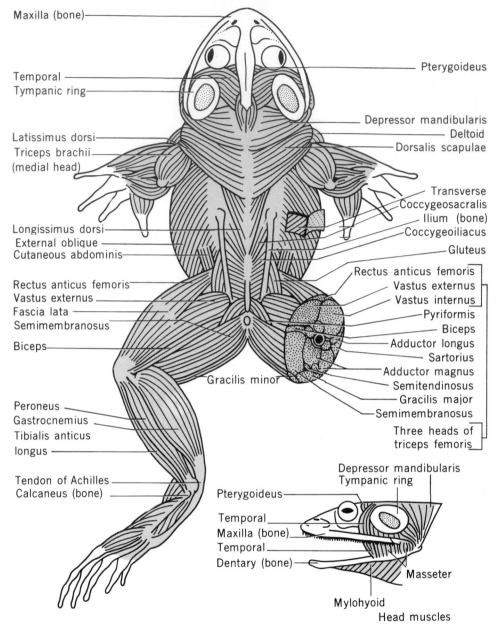

Figure 27-8. Muscles of the bullfrog *Rana catesbeiana,* dorsal view.

Labels (clockwise from upper left):

Maxilla (bone)

Temporal
Tympanic ring

Latissimus dorsi
Triceps brachii
(medial head)

Longissimus dorsi
External oblique
Cutaneous abdominis

Rectus anticus femoris
Vastus externus
Fascia lata
Semimembranosus

Biceps

Peroneus
Gastrocnemius
Tibialis anticus
longus

Tendon of Achilles
Calcaneus (bone)

Gracilis minor

Pterygoideus

Depressor mandibularis
Deltoid
Dorsalis scapulae

Transverse
Coccygeosacralis
Ilium (bone)
Coccygeoiliacus
Gluteus

Rectus anticus femoris
Vastus externus
Vastus internus
Pyriformis
Biceps
Adductor longus
Sartorius
Adductor magnus
Semitendinosus
Gracilis major
Semimembranosus

Three heads of
triceps femoris

Depressor mandibularis
Tympanic ring

Pterygoideus

Temporal
Maxilla (bone)
Temporal
Dentary (bone)

Masseter

Mylohyoid

Head muscles

one that straightens out a flexed part is called an **extensor;** one that draws a part toward the midline of the body is an **adductor;** one that moves a part away from the body midline is an **abductor;** one that lowers a body part is a **depressor;** one that raises a part is a **levator;** and one that rotates one part on another is a **rotator.** These movements depend on the **origin** (less movable end) and **insertion** (more movable end) of the muscle and the nature of the **articulations (joints)** of the bones that it connects.

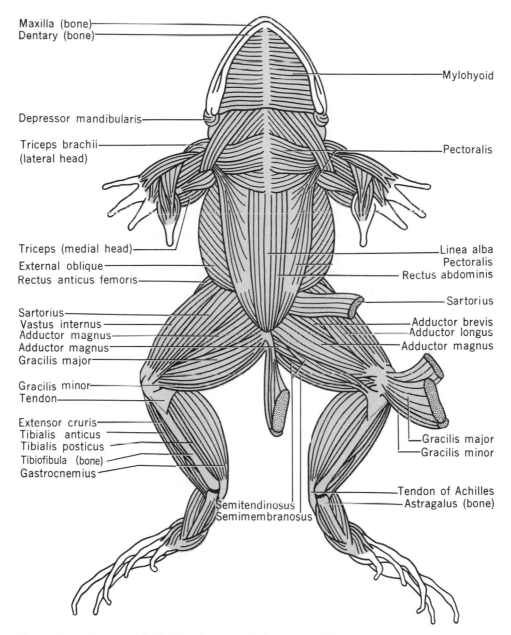

Maxilla (bone)
Dentary (bone)
Mylohyoid
Depressor mandibularis
Triceps brachii (lateral head)
Pectoralis
Triceps (medial head)
External oblique
Rectus anticus femoris
Linea alba
Pectoralis
Rectus abdominis
Sartorius
Vastus internus
Adductor magnus
Adductor magnus
Gracilis major
Sartorius
Adductor brevis
Adductor longus
Adductor magnus
Gracilis minor
Tendon
Extensor cruris
Tibialis anticus
Tibialis posticus
Tibiofibula (bone)
Gastrocnemius
Gracilis major
Gracilis minor
Semitendinosus
Semimembranosus
Tendon of Achilles
Astragalus (bone)

Figure 27-9. Muscles of the bullfrog *Rana catesbeiana,* ventral view.

Table 27-2 gives the name, origin, insertion, and action of some of the muscles of the hindlimb. A muscle may have more than one origin or insertion, such as the triceps femoris, which arises from three distinct "heads" (Fig. 27-8).

The following are a few major muscles of other parts of the body. The **rectus abdominis** extends longitudinally along the ventral side of the trunk; the **external oblique** covers most of the sides of the trunk; the **transverse muscle** lies beneath the external

Table 27-2. Name, Origin, Insertion, and Action of Some Muscles of the Hindlimb of the Frog

Name	Origin	Insertion	Action
Sartorius	Ilium, just in front of pubis	Just below head of tibia	Flexes leg, adducts thigh
Adductor magnus	Ischium and pubis	Distal end of femur	Adducts thigh and leg
Adductor longus	Ventral part of ilium	Joins adductor magnus to attach on femur	Adducts thigh and leg
Triceps femoris	From three heads: one from acetabulum, and two heads from ilium	Upper end of tibiofibula	Abducts thigh and extends leg
Gracilis major	Posterior margin of ischium	Proximal end of tibiofibula	Adducts thigh, flexes leg
Gracilis minor	Tendon behind ischium	Join tendon of gracilis major and tibiofibula	Adducts thigh, flexes leg
Semimembranosus	Dorsal half of ischium	Proximal end of tibiofibula	Adducts thigh, flexes leg
Biceps (iliofibularis)	Dorsal side of ilium	Tibiofibula	Adducts thigh, flexes leg
Gastrocnemius	Distal end of femur; tendon of triceps	Tendon of Achilles	Flexes leg, extends foot
Tibialis posticus	Posterior side of tibiofibula	Proximal end of astragalus	When foot is flexed, acts as an extensor
Tibialis anticus	Distal end of femur longus	Proximal end of astragalus and calcaneus (ankle bones)	Extends leg, flexes foot
Peroneus	Distal end of femur	Distal end of tibiofibula, head of calcaneus	Extends leg
Extensor cruris	Distal end of femur	Anterior surface of tibiofibula	Extends leg

oblique and serves to contract the body cavity; the **pectoralis major** moves the forelimbs; and the **mylohyoid** raises the floor of the mouth cavity during respiration.

The musculature of other vertebrates, particularly humans, is described in Chapter 32.

Digestive System

The principal functions of the digestive system are to receive, digest, and absorb food. The food of the frog consists primarily of living insects. These are usually captured by the specialized extensile **tongue.** The prey adheres to sticky secretions on the tongue, which is then withdrawn into the mouth. Large insects and other bulky prey are pushed into the mouth by the forefeet. No attention is paid to objects that are not moving; extension of the tongue is a reflex initiated by a small spheroidal object moving with a minimum velocity across the retinal field of vision. A captive frog will starve to death if placed among hundreds of dead flies.

The **mouth cavity** is large. The tongue lies on the floor of the mouth cavity, with its base attached to the anterior part of the mouth and its forked, free tip lying behind. The small, conical teeth (**vomerine teeth**) are borne on the upper jaw and on two bones called **vomers** in the roof of the mouth cavity. These teeth are used only for holding food and not for chewing. New teeth replace any that are lost.

The mouth cavity communicates through a horizontal slit with the **esophagus.** The inner surface of this tube bears longitudinal folds that allow distension as large, unchewed prey are swallowed. If the object swallowed proves undesirable, it can be regurgitated through the mouth. Histologically, the esophagus resembles the stomach.

The crescent-shaped **stomach** lies primarily in the left side of the coelom. Its anterior **cardiac end** is larger than the esophagus, but it decreases in size toward the posterior **pyloric end,** where it joins the small intestine. The stomach is held in place by dorsal and ventral folds of the **peritoneum.** The thick walls of the stomach consist of four layers of epithilial and muscular tissues.

The two largest digestive glands are the pancreas

and the liver (Fig. 27-10). The **pancreas** lies between the small intestine and the stomach. It is a much-branched tubular gland that secretes an alkaline digestive fluid into the **common bile duct.** The **liver** performs many metabolic functions, notably the filtration of toxic substances from the blood, the collection of dead erythrocytes (red blood cells), and the production of **bile,** an alkaline fluid that emulsifies fats. Bile is stored in the **gallbladder** until food enters the intestine, which stimulates its release through the common bile duct into the anterior portion of the small intestine, the **duodenum.**

The pyloric sphincter controls the rate at which food leaves the stomach and enters the duodenum. The duodenum leads to the coiled **ileum** that in turn widens abruptly into the **colon (large intestine).** The intestines are suspended in the coelom by a dorsal fold of the peritoneum. Six layers of cells make up the intestinal wall.

The digestive tract and the urinary and reproductive ducts all open into a common chamber called the **cloaca.** Wastes and gametes pass from the cloaca to the external environment through the **cloacal opening.**

Digestion in the vertebrates is discussed in Chapter 33.

Circulatory System

The circulatory system distributes blood throughout the body and maintains the interstitial fluid in about the same state (**homeostasis**) at all times as it delivers nutrients and removes wastes. The circulatory system of the frog consists of a heart, arteries, capillaries, and veins.

Blood. The liquid portion of the blood, called **plasma,** contains three kinds of cells: red cells (erythrocytes), white cells (leukocytes), and spindle cells (thrombocytes) (Fig. 27-11). The plasma carries nutrients and wastes in solution.

The **erythrocytes** are elliptical, flattened, nucleated cells that contain the respiratory pigment **hemoglobin.** Hemoglobin combines with oxygen in the capillaries of the respiratory organs and transports it to the tissues of the body. The **leukocytes** are of several types (Fig. 27-11 and Chapter 35); they vary in size, and most are capable of independent amoeboid movement. The **thrombocytes** are small spindle-shaped, nucleated cells. Thrombocytes are unstable: when they contact the rough edge of a torn blood vessel, they break down, releasing an enzyme called **thromboplastin (cephalin)** that is needed to form a **blood clot** (Chapter 35).

Blood corpuscles arise principally in the central **marrow** of the bones. They also increase in number by mitotic division while circulating in the blood. Some white cells are probably formed in the **spleen,** an organ in which worn-out red cells are destroyed.

Heart. The frog's heart is diagrammed in Figure 27-12. In birds and mammals the heart contains four chambers, two atria and two ventricles; but the amphibian heart has three chambers, with only one ven-

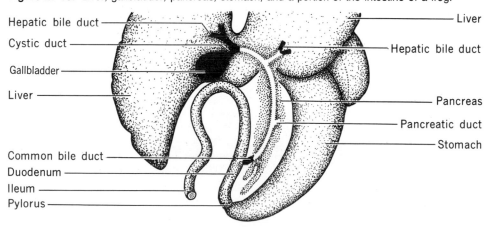

Figure 27-10. Liver, gallbladder, pancreas, stomach, and a portion of the intestine of a frog.

Hepatic bile duct

Cystic duct

Gallbladder

Liver

Common bile duct

Duodenum

Ileum

Pylorus

Liver

Hepatic bile duct

Pancreas

Pancreatic duct

Stomach

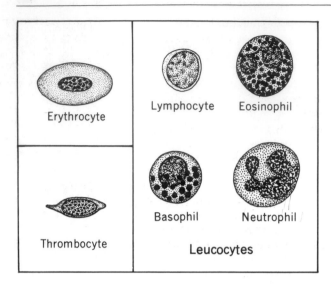

tricle. In the frog, the muscular **ventricle** is conical in shape. Anterior to the ventricle lie the two **atria,** whose walls are somewhat less muscular than the ventricle. Note that the right atrium seems to wrap around the area of the **conus arteriosus,** where the arteries carrying blood from the heart originate. The conus arteriosus extends anteriorly from the ventricle.

Arteries. The arteries (Fig. 27-13) carry blood away from the heart. The conus arteriosus divides near the anterior border of the atria into two vessels (Fig. 27-12). Each branch, called a **truncus,** gives rise to the following three arteries:

Figure 27-11. Types of blood cells in the frog. See Chapter 35 for details on different leukocytes.

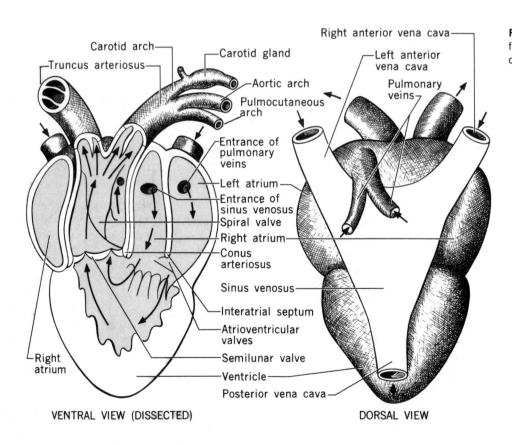

Figure 27-12. Heart of the frog. Arrows indicate direction of blood flow.

1. The **common carotid artery** divides into the **external carotid (lingual)**, which supplies the tongue and neighboring parts, and the **internal carotid,** which gives off other arteries to the tongue, the roof of the mouth, the eye, and the brain. Where the common carotid branches, there is a swelling called the **carotid gland;** this reservoir serves to equalize the blood flow, especially in the internal carotid artery.

2. The **pulmocutaneous artery** branches to form the **pulmonary artery,** which passes to the lungs, and the **cutaneous artery,** which supplies arteries to the other respiratory organ, the skin.

3. The **aortic (systemic) arches** pass outward on either side of the digestive tract and then unite to form the **dorsal aorta.** Before this union several branches are given off, including the **subclavian artery,** which arises at about the level of the shoulder and extends into either forelimb as the **brachial artery.** The dorsal aorta gives off the **coeliacomesenteric artery,** which supplies arteries to the stomach, pancreas, liver, intestine, spleen, and cloaca. Then the dorsal aorta gives off several **renal arteries,** which supply the kidneys. A small posterior **mesenteric artery** arises near the posterior end of the dorsal aorta and passes to the large intestine and, in females, to the uterus. The dorsal aorta finally divides into two **common iliac arteries,** which distribute blood to the ventral body wall, the rectum, bladder, the anterior part of either thigh (**femoral artery**), and other parts of the hindlimbs (**sciatic artery**).

Veins. The following describes the venous flow back to the heart (Fig. 27-13). Blood from the lungs is collected in the **pulmonary veins** and is poured into the left atrium. The **sinus venosus** is a triangular thin-walled chamber that collects deoxygenated (and carbon dioxide-rich) blood from three major veins (**venae cavae**) located dorsally and delivers it to the right atrium. The venae cavae consist of the two **anterior venae cavae** and the **posterior vena cava.** The anterior venae cavae receive blood from the tongue, thyroid, and neighboring parts; (2) the **innominates,** which collect blood from the head by means of the **internal jugulars** and from the shoulder by means of the **subscapulars;** and (3) the **subclavians,** which collect blood from the forelimbs by means of the **brachial vein** and from the side of the body and head by means of the **musculocutaneous veins.** The posterior vena cava receives blood from the kidneys by means of four to six pairs of **renal veins,** from the liver by two **hepatic veins,** and from others.

The veins that carry blood to the kidneys, where urea and other metabolic wastes are removed, constitute the **renal portal system.** The **renal portal vein** receives blood from the hindlimbs by means of the **sciatic** and **femoral** veins and from the body wall by means of the **dorsolumbar veins.** Although the circulatory systems of all vertebrates are built on the same general plan, there is no renal portal system in mammals.

The liver receives blood from the **hepatic portal system.** The femoral veins from the hindlimbs divide, and their ventral branches unite in the **anterior abdominal vein** and collect blood from the ventral body wall before entering the liver. The **hepatic portal vein** carries blood into the liver from the stomach, intestine, spleen, and pancreas. This shunt of blood from the digestive tract through the liver before entering the main circulation allows the liver to remove food nutrients from, or add substances to, the blood as physiological needs require.

Circulation. Circulation in the frog takes place in the following manner: The sinus venosus contracts, forcing nonoxygenated blood into the right atrium (Fig. 27-12). Oxygenated blood from the lungs simultaneously passes into the left atrium. Then both atria contract and force their contents into the single ventricle. Backflow of blood is prevented by means of **valves** (Fig. 27-12). Experiments have shown that the oxygenated and nonoxygenated bloods mix in the ventricle, and it is likely that mixed blood is pumped to all parts of the frog's body. Respiration through the skin of the frog, both in water and on land, is thought to compensate in part for the recycling of some nonoxygenated blood back into the general circulation.

The blood that is thus pumped through the arteries enters smaller and smaller blood vessels until the extremely narrow **capillaries** are reached. Here nutrients and oxygen are delivered to the tissues. Water, oxygen, and dissolved molecules permeate the thin capillary walls and form the **interstitial fluid** that directly bathes the body's cells. The passage of blood through the capillaries can be observed microscopic-

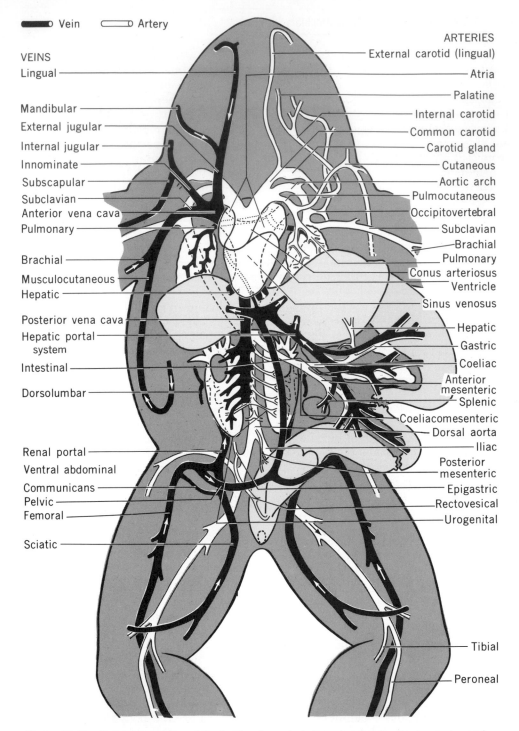

Vein ⬛▭   Artery ▭

VEINS

Lingual

Mandibular

External jugular

Internal jugular

Innominate

Subscapular

Subclavian

Anterior vena cava

Pulmonary

Brachial

Musculocutaneous

Hepatic

Posterior vena cava

Hepatic portal
   system

Intestinal

Dorsolumbar

Renal portal

Ventral abdominal

Communicans

Pelvic

Femoral

Sciatic

External carotid (lingual)

Atria

Palatine

Internal carotid

Common carotid

Carotid gland

Cutaneous

Aortic arch

Pulmocutaneous

Occipitovertebral

Subclavian

Brachial

Pulmonary

Conus arteriosus

Ventricle

Sinus venosus

Hepatic

Gastric

Coeliac

Anterior
mesenteric

Splenic

Coeliacomesenteric

Dorsal aorta

Iliac

Posterior
mesenteric

Epigastric

Rectovesical

Urogenital

Tibial

Peroneal

Figure 27-13. Circulatory system of the bullfrog in ventral view, showing the larger arteries and veins in relation to the internal organs. Arrows indicate blood flow.

ally in the web of the frog's foot or in the tail of the tadpole.

Lymphatic System

Much of the interstitial fluid is reabsorbed into the capillaries that merge into veins, but some is drained from the tissues by a separate type of the circulatory system, called the **lymphatic system.** This system of the frog includes many **lymph vessels** of various sizes that form networks alongside the blood vessels, but are difficult to see. The colorless, watery **lymph** contains **leukocytes** and various constituents of the blood **plasma.** Frogs and toads, unlike other vertebrates, have several **lymph spaces** between the skin and the body. Four **lymph hearts,** two near the third vertebra and two near the end of the vertebral column, force the lymph by pulsations into the internal jugular vein and a branch of the renal portal veins.

Chapter 35 contains a discussion of the circulatory and lymphatic systems.

Respiratory System

The primary functions of the respiratory system are to supply oxygen to the tissues and to eliminate carbon dioxide that forms as a by-product of cellular metabolism. Respiration involves both external and internal phases. **External respiration** involves the movement of air into the **lungs** (breathing) and the diffusion of respiratory gases between the lungs and the bloodstream. **Internal respiration** involves the transport of oxygen and carbon dioxide to and from the cells of the body.

External respiration is mediated by **gills** in most aquatic vertebrates (Chapter 26) and by lungs in terrestrial vertebrates. External respiration in the adult frog is carried on largely by the lungs but also occurs to a considerable extent through the **moist skin** and the **mucous membrane** lining the mouth cavity. As is shown in Figure 27-14, air passes through the **external nares** into the **nasal canals.** Air exits the canals through the **internal nares** into the **mouth cavity.** The external nares are then closed, and the floor of the mouth is raised, which forces the air through the **glottis** into the **larynx** (voice box), and then into two very short tubes, the **bronchi,** that enter the two lungs.

The **lungs** are ovoid sacs with thin elastic walls.

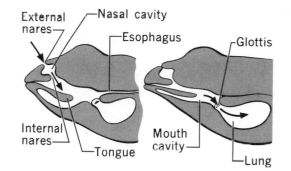

Figure 27-14. Respiratory movements of the frog. Left; External nares are open and air enters the mouth cavity. Right; External nares are closed, the floor of the mouth cavity is raised, and air is forced into the lungs. Labels have been omitted from the very short larynx and bronchus. Arrows show pathway of air to lungs.

Their inner surfaces are elaborated by a network of partitions into many minute chambers called **alveoli.** The alveoli are the functional units of the respiratory system of terrestrial vertebrates. Blood capillaries are numerous in the walls of the alveoli, and it is here that oxygen diffuses into the blood and carbon dioxide into the lung. Air is expelled from the lungs and respiratory passageways of the frog by contraction of body wall muscles.

The **larynx** is strengthened by cartilages. Across its lumen are stretched two elastic bands, the **vocal cords.** The croaking of the frog is produced by vibrations of the vocal cords due to their contraction as air is expelled from the lungs. Laryngeal muscles regulate the tension of the cords and hence the pitch of the sound. Males of many frog species have a pair of **vocal sacs** that open into the mouth cavity. Such sacs serve as resonators that increase the intensity of sound.

Respiration in the vertebrates is discussed more fully in Chapter 36.

Excretory System

The excretory system's functions are waste excretion and water regulation. A certain amount of nitrogenous waste is excreted by the skin, lungs, liver, and intestinal walls, but most is removed as urea from the blood by the two **kidneys.** The kidneys lie dorsal to the peritoneum in the **subvertebral lymph space.**

From each kidney the **urine** passes through a **urinary duct** (**ureter**) into the **cloaca.** It may be voided at once through the **cloacal opening** or stored in the **bladder** temporarily (Figs. 27-15 and 27-16). For a discussion of kidney function, see Chapter 34.

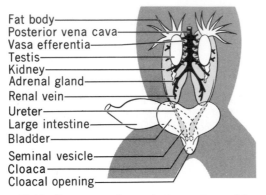

Fat body
Posterior vena cava
Vasa efferentia
Testis
Kidney
Adrenal gland
Renal vein
Ureter
Large intestine
Bladder
Seminal vesicle
Cloaca
Cloacal opening

Figure 27-15. Excretory and reproductive systems of the male frog in ventral view. The seminal vesicle is poorly developed in the leopard frog, *Rana pipiens.*

Nervous System

The nervous system receives information from the body's internal and external environments, integrates this information, and responds by sending appropriate motor instructions to the muscles and glands of the body. The nervous systems of vertebrates are more complex than are any other animals'. The general morphology of the frog's nervous system is illustrated in Figure 27-17. Only major features of the frog's nervous system are considered here, as the nervous system of vertebrates is treated in considerable detail in Chapter 37.

The brain. The brain of the frog (Fig. 27-18) has two large **olfactory lobes** that are fused together, two large **cerebral hemispheres,** a **diencephalon,** two large **optic lobes,** a very small **cerebellum,** and a **medulla oblongata** that narrows into the spinal cord. The cranial nerves that leave the frog's brain are similar to those of higher vertebrates (see Chapter 37), except that all amphibians lack a **hypoglossal** (XII) **nerve.**

Figure 27-16. Excretory and reproductive systems of the female frog in ventral view. Left ovary is omitted. Note how much space is devoted to the ovary.

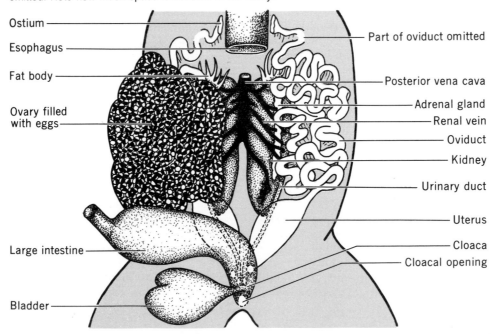

Ostium
Esophagus
Fat body
Ovary filled with eggs
Large intestine
Bladder

Part of oviduct omitted
Posterior vena cava
Adrenal gland
Renal vein
Oviduct
Kidney
Urinary duct
Uterus
Cloaca
Cloacal opening

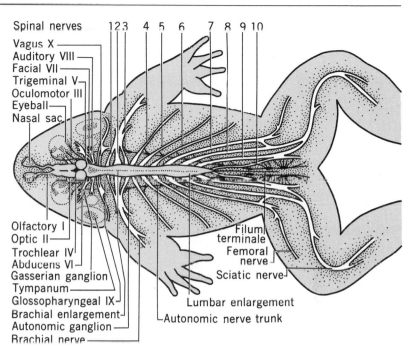

Spinal nerves   1 2 3   4   5   6   7   8   9   10

Vagus X
Auditory VIII
Facial VII
Trigeminal V
Oculomotor III
Eyeball
Nasal sac

Olfactory I
Optic II
Trochlear IV
Abducens VI
Gasserian ganglion
Tympanum
Glossopharyngeal IX
Brachial enlargement
Autonomic ganglion
Brachial nerve

Filum terminale
Femoral nerve
Sciatic nerve
Lumbar enlargement
Autonomic nerve trunk

Figure 27-17. Nervous system of the bullfrog. Note also the autonomic system.

The functions of the frog's brain have been partially localized by experiments in which parts of the brain are removed and the effects upon the animals' responses are observed. There is evidence that the cerebral hemispheres are involved in **associative memory.** When the diencephalon is removed with the cerebral hemispheres, the frog ceases spontaneous movement. When the optic lobes are removed, the spinal cord becomes more irritable, indicating that the optic lobes have an inhibiting influence on the reflex activity of the spinal cord. In humans, the cerebellum is a center of coordination, but experiments on the frog cerebellum have produced conflicting results. Many activities are still possible when everything but the medulla is removed: the frog breathes normally, snaps at and swallows food, leaps and swims regularly, and is able to right itself when turned on its back. Removal of the posterior region of the medulla results in the death of the frog. The brain as a whole controls the actions transmitted by the cranial nerves and the spinal cord.

The frog depends primarily on reflexes, as "conscious" control is limited. A brainless frog with only the medulla intact can survive for several weeks if food is stuffed down its esophagus and its trachea is kept clear.

Spinal Nerves. There are ten pairs of spinal nerves in the frog. Each arises by a **dorsal root** and a **ventral root** from the **gray matter** (Chapter 37) of the spinal cord. The two roots unite to form a **nerve** that passes laterally between the arches of adjacent vertebrae. The two largest pairs of nerves are the **brachials,** which innervate the shoulders and forelimbs and are each composed of a second spinal nerve and branches from the first and third pairs, and the **sciatics,** which are distributed to the hindlimbs and arise from branches received from the seventh, eighth, and ninth spinal nerves.

Autonomic Nervous System. The autonomic nervous system (Fig. 27-19) innervates the internal organs of the body: cardiac muscle, smooth muscles, and glandular tissues. It consists of two principal **nerve trunks** that begin at the cranium and extend posteriorly on either side of the vertebral column. Each trunk is pro-

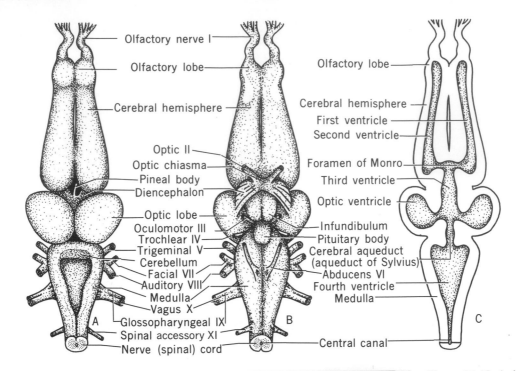

Olfactory nerve I
Olfactory lobe
Cerebral hemisphere
Optic II
Optic chiasma
Pineal body
Diencephalon
Optic lobe
Oculomotor III
Trochlear IV
Trigeminal V
Cerebellum
Facial VII
Auditory VIII
Medulla
Vagus X
Glossopharyngeal IX
Spinal accessory XI
Nerve (spinal) cord

Olfactory lobe
Cerebral hemisphere
First ventricle
Second ventricle
Foramen of Monro
Third ventricle
Optic ventricle
Infundibulum
Pituitary body
Cerebral aqueduct (aqueduct of Sylvius)
Abducens VI
Fourth ventricle
Medulla
Central canal

A   B   C

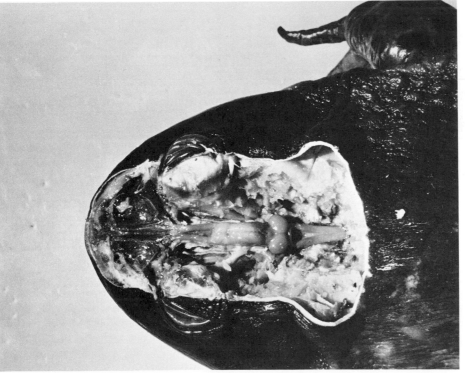

Figure 27-18. Left; photograph shows dorsal view of dissected frog brain. Line art shows three views of brain of the frog. (A) Dorsal view. (B) Ventral view. (C) Ventricles (cavities) in dorsal view.

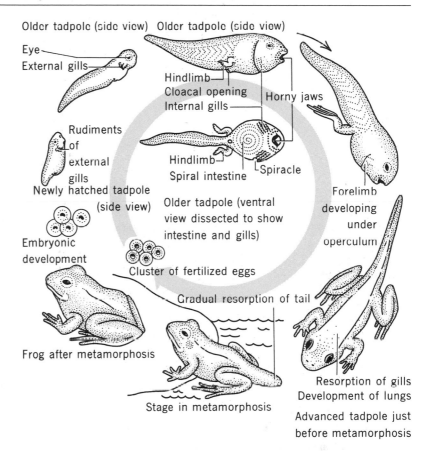

Older tadpole (side view) Older tadpole (side view)

Eye
External gills

Hindlimb
Cloacal opening
Internal gills
Horny jaws

Rudiments
of
external
gills
Newly hatched tadpole
(side view)

Hindlimb
Spiral intestine
Spiracle

Older tadpole (ventral
view dissected to show
intestine and gills)

Forelimb
developing
under
operculum

Embryonic
development

Cluster of fertilized eggs

Gradual resorption of tail

Frog after metamorphosis

Stage in metamorphosis

Resorption of gills
Development of lungs

Advanced tadpole just
before metamorphosis

Figure 27-19. Life cycle of the frog.

vided with ten **ganglia** at points where it unites with branches from the spinal nerves. The nerves of the autonomic system are distributed to internal organs and regulate many functions that are not under conscious or voluntary control, such as heartbeat, glandular secretions, and functions of the digestive, respiratory, urogenital, and reproductive systems.

Sense Organs

The most complex sense organs are the eyes, ears, and olfactory organs. There are many smaller structures on the surface of the tongue and floor of the mouth that probably function as receptors of taste. The skin is supplied with many sensory nerve endings that receive contact (tactile), chemical, temperature, and light stimuli.

Olfactory Organs. The **olfactory nerves** (Figs. 27-17 and 27-18) extend from the olfactory lobes of the forebrain to the nasal cavities, where they are distributed to the epithelial lining. The importance of the sense of smell in the life of the frog is not known.

Ears. There is no external ear in the frog. The **middle ear** is is a cavity that communicates with the mouth cavity through a **eustachian tube** and is closed externally by the **tympanic membrane, (eardrum)** visible behind each eye.

A rod called the **columella** extends across the cavity of the middle ear from the eardrum to the **inner ear.** The inner ear lies within the bony auditory capsule and is supplied by branches of the auditory nerve. Sound waves cause the eardrum to vibrate, and the vibrations are transmitted through the columella to the inner ear. The sensory ends of the auditory nerve are stimulated by the vibrations, and the nerve impulses that are carried to the brain give rise to the sensation of sound. The inner ears also serve as organs of **equilibrium.** Frogs from which they are removed cannot maintain an upright position.

Eyes. The eyes of the frog resemble those of humans in general structure and function but differ in certain details. The **eyeballs** lie in bony cavities (**orbits**) in the sides of the skull. They are rotated by six muscles and also pulled into the orbit. The **upper eyelid** does not move independently. The **lower eyelid** consists of the lower eyelid proper fused with the transparent third eyelid, or **nictitating membrane.** The **lens** is large and almost spherical. Its shape and position cannot be changed, so the eye can distinctly focus on objects at only a certain distance. Movements are noted much more often than is form. The amount of light that enters the eye can be regulated by contraction of the **pupil.** The **photoreceptors** in the **retina** are stimulated by light rays that pass through the pupil and lens. The nerve impulses that result are carried through the **optic nerve** to the brain, where they give rise to sensations of sight.

Endocrine System

Physiological processes of the frog, like those of other vertebrates, are regulated by **hormones,** chemical messengers that are produced by the ductless **endocrine glands.** These internal secretions pass directly into the blood and circulate to their specific target organs.

The **pituitary gland** at the base of the brain has three lobes. A **growth-stimulating hormone** is secreted by the **anterior lobe** in the larvae and the young. In the adult, the anterior lobe secretes a **gonad-stimulating (gonadotropic) hormone** that initiates the development of ova or spermatozoa. The **intermediate lobe** produces **intermedin,** a hormone that regulates the behavior of chromatophores in the skin. The **posterior lobe** regulates water intake by the skin.

Thyroxin, produced by the **thyroid gland,** regulates general metabolism throughout the body and is particularly important during **metamorphosis. In-**

sulin, produced by the **pancreas,** regulates sugar metabolism. **Epinephrine,** secreted by the **adrenal glands,** increases blood pressure, causes contraction of dark skin pigments, and has several other functions.

Reproductive System

In the frog, the sexes are separate (**dioecious**). Males can be distinguished from females externally by the greater thickness of the thumb.

Spermatozoa of the male arise in the **testes** (Fig. 27-15) and pass through the **vasa efferentia** into the kidneys, where they ultimately enter the ureter by a route that varies in different species. The posterior end of the urinary duct is dilated in some species to form a **seminal vesicle.** The spermatozoa pass from the urinary duct or seminal vesicle to the external environment through the cloacal opening.

Ova arise in the **ovaries** of the female (Fig. 27-16). During the breeding season, the eggs break through the ovarian walls into the body cavity (coelom). There they are moved anteriorly by the beating of cilia that line the peritoneum. Cilia at the entrance, or **ostium,** to the **oviduct** create currents that draw the eggs into the convoluted tube. They are then transported by cilia down the oviduct into the thin-walled distensible uterus. The glandular wall of the oviduct secretes gelatinous coats around the eggs.

Just in front of each reproductive organ is a yellowish glove-shaped **fat body** that serves as a reserve food supply that maintains the frog during its period of hibernation.

Chapter 39 contains a general discussion of vertebrate reproduction, embryology, and development.

Frogs lay their eggs in water during the early spring (Fig. 27-19). The male clasps the female firmly just behind her forelegs with his enlarged thumb pads. As eggs are extruded by the female they are fertilized by sperm that the male discharges over them. Each female lays several hundred or more eggs. The protective jelly that surrounds the eggs soon swells due to absorption of water.

Frog embryology is discussed in some detail in Chapter 39. The **tadpole** emerges from the egg and lives for a few days on the yolk in its digestive tract; then it feeds on algae and decayed vegetable matter.

The **external gills** form long branching gills. Water enters the mouth, passes through the gill slits, and leaves through an opening, the **spiracle,** on the left side of the body.

The hindlimbs appear before the forelimbs. The **tail** is gradually resorbed as the end of the larval period approaches. The gills are then resorbed, and the lungs develop to take their place. Finally, the typical adult form is acquired, and further growth involves increase in size and sexual maturity.

Other Anurans

Family **Hylidae,** and **tree frogs,** is a large group, perhaps containing up to 450 species. These **arboreal** amphibians have rough adhesive discs on their toes and fingers that usually enable them to climb trees. They are provided with large **vocal sacs** and have correspondingly loud calls. About two dozen species occur in the United States. The **gray tree frog,** *Hyla versicolor* (Fig. 27-20), is about 5 cm long. Other common tree frogs are the **spring peeper,** *(Hyla crucifer),* **the cricket frog** *(Acris gryllus),* and *Hyla andersonii* (Fig. 27-1).

The **true toads,** family **Bufonidae,** consists of about 200 species, most of which belong to the genus *Bufo* (Fig. 27-21). About 18 species of this genus have been been reported in the United States. The common

Figure 27-21. A common toad, *Bufo punctatus,* the red-spotted toad.

toad of the eastern United States, *Bufo americanus,* possesses a rough and warty skin, bright markings, and jewel-like eyes. Contrary to popular superstition, handling toads does not cause the appearance of warts upon the hands. During the day, toads generally remain concealed in dark, damp microhabitats, but at night they hop about feeding insects, worms, and snails that they capture with their sticky tongues.

Frogs and toads are often confused. In temperate regions, the main differences are the following:

1. Frogs have maxillary (upper jaw) teeth; toads lack such teeth.
2. Frogs lack a **Bidder's organ;** toads do have them. Bidder's organs are thought to be some type of potential ovary, but their function is not yet understood.
3. Frogs are almost invariably associated with aquatic habitats; toads are much more widely distributed. This is partly associated with differences in the integument: Frog skin is relatively thin, whereas toad skin is generally much thicker and warted.
4. Frogs are rather streamlined; toads are generally bulkier.
5. Frogs are relatively good jumpers; toads are poor jumpers.
6. Frogs have a relatively smooth area just above and behind the tympanum; toads' parotoid glands,

Figure 27-20. The tree frog, *Hyla regilla.*

which secrete a toxic mucus, form a bulge in that area.

7. Frogs usually lay their eggs in clusters; toads usually lay a string of eggs.

Order Caudata—Tailed Amphibians

Hellbenders

The family **Cryptobranchidae** contains two genera of giant salamanders. The American **hellbender,** *Cryptobranchus alleganiensis* (Fig. 27-1), is restricted to streams in the eastern United States; it attains lengths of from 45 to 70 cm. The **giant salamander,** *Andrias (Megalobatrachus) japonicus,* of Japan is the largest living amphibian, reaching a length of over 1.5 m.

Axolotls

The **tiger salamanders,** genus *Ambystoma,* range from New York to California and south to central Mexico. They attain lengths of 14 to 30 cm. In some parts of their geographic range, tiger salamanders fail to metamorphose, and yet they reproduce while still in larval state. Sexual reproduction by larval stages is a phenomenon called **neoteny.** A sexually mature larval tiger salamander is called an **axolotl.**

The aquatic axolotls were long considered a separate species because of the external gills that persisted in the adult. However, if an axolotl is fed even one or two meals of beef thyroid (whose hormone, **thyroxin,** plays a major role in normal metamorphosis), it loses its gills and becomes an air-breathing salamander. This is not considered to be a case of retarded evolution but, rather, a secondary specialization for survival in arid regions, where mudholes and a few streams offered a viable alternative to desiccation.

Salamanders and Newts

The salamanders and newts (Fig. 27-1 and 27-22) belong to the family **Salamandridae. The red-spotted newt,** *Notophthalmus viridescens,* is aquatic as a larva, but, when it reaches about 2.5 cm long, it loses its gills and migrates onto land. During the one or two years of its terrestrial life, this salamander is bright

Figure 27-22. Red-bellied newt, *Taricha rivularis,* in water.

coral red in color and is known as a **red eft.** It eventually returns to the water and changes to the adult coloration of yellowish green with black spots on the undersurface and a row of black-bordered crimson spots along both sides.

The cool, slimy skin of salamanders gave rise to the medieval belief that they could live in fire and not be injured. Such beliefs were probably initiated by the fact that a number of species live under the loose bark of dead trees. When such wood was placed on the fire, no doubt a number of these salamanders attempted to escape, thus suddenly appearing that they lived in and around fires. This magical attribute led to their inclusion in many alchemical formulas.

Although a number of salamanders produce a toxic substance in the skin, the skin of the **fire salamander** of Europe secretes a particularly poisonous substance from its skin. This salamander is black with bright yellow spots and therefore very conspicuous. Its striking colors seem to warn potential predators that it is not to be mistaken for a palatable morsel of food.

Mud Puppies

Mud puppies, genus *Necturus,* consists of strictly aquatic salamanders that inhabit the eastern United States. They are unusual because, instead of lungs, they have feathery, reddish gills extending from either side of the back of the head. They often slowly wave the gills back and forth in the water, thus enhancing diffusion of respiratory gases. A mud puppy is a carnivorous animal that may attain lengths of up to 0.5 m. Like an axolotl, the mud puppy is a permanently aquatic neotenic larva, but no experimental manipulation has ever induced metamorphosis into air-breathing adults.

Order Apoda—Legless Caecilians

The **Apoda** contains about 60 species of legless amphibians. These wormlike animals, called **caecilians,** inhabit tropical regions of the Americas, Africa, and Asia. Some burrow in the moist soil with their strong heads, whereas others are aquatic burrowers in the muddy substrata. Caecilians are nearly blind as their eyes are very small, degenerate, and at least partially concealed by skin and/or skull bones. A **sensory tentacle** can be protruded from between the eyes and the nose. These inconspicuous amphibians have many degenerate features, most notable of which is the complete absence of limbs and limb girdles; they are not closely related to the other modern amphibians.

Adaptive Modifications and Behaviors of Amphibians

Regeneration

The ability to regenerate lost body parts is remarkably well developed in many amphibians. For example, when a foot of a two-year-old axolotl was experimentally amputated, in 12 weeks a complete foot was regenerated in its place. Newts are able to regenerate both limbs and tail. Tailless amphibians are apparently unable to regenerate lost parts to any extent, except in earlier developmental stages.

As a general rule, younger tadpoles regenerate limbs or a tail more readily than do older individuals. There is a distinct advantage to regeneration, as amphibians may escape from predators by leaving behind a tail or a limb in exchange for their freedom.

Breeding Habits

Most amphibians are **oviparous,** and their eggs, as in the leopard frog, are generally fertilized by the male after extrusion from the female. In some tailed amphibians, however, the eggs are fertilized before they are laid. A few species bring forth their young alive, as in the **alpine salamander,** *Salamandra*.

Some curious breeding habits are exhibited by certain species. The male **midwife toad,** *Alytes obstetricans* (Fig. 27-1), of Europe carries the egg strings with him, wound around his hindlimbs; when the tadpoles are ready to emerge, he enters water and they escape.

The eggs of the **Surinam toad,** *Pipa pipa* (Fig. 27-1), are placed on the back of the female during copulation and held there by a sticky secretion until they are gradually enveloped by the skin. Within these epidermal pouches, the eggs develop and the tadpole stage is passed, after which the young toads escape as air-breathing animals.

The several species of **pouched,** or **marsupial, tree frogs,** *Gastrotheca*, of South America have a pouch with an opening in the posterior part of the back in which the eggs are placed and the young are reared. The female of another tree frog, *Ceratohyla*, carries her eggs in a depression on her back until they are almost ready for metamorphosis.

The **American bell toad** (Fig. 27-1), found in the Pacific Northwest, is fairly unique in that fertilization is internal, thus keeping the sperm from being washed away. The male grasps the female around the pelvis, and, by use of an external tail-like copulatory organ, the sperm are deposited in the cloaca of the female. Most salamanders and caecilians also have internal fertilization.

Hibernation and Estivation

Many amphibians bury themselves in the mud of pond bottoms during the autumn and remain there in a dormant condition until the following spring. During this period of **hibernation,** the vital processes are reduced. No air is taken into the lungs, and all necessary external respiration occurs through the skin. No food is eaten, but physiological activities are fueled by stored food reserves; the temperature of the amphibian remains only slightly above that of the immediate surroundings. The body temperature of fishes, amphibians, and reptiles generally conforms to that of their environment, a condition known as **poikilothermy.** Frogs cannot survive total freezing, as is sometimes reported, because death ensues if the heart is frozen.

Many amphibians seek moist places of concealment; in warmer climates they may pass the driest part of the year in a quiet, torpid condition called **estivation.**

Poison Glands

The poison glands of the leopard frog have already been mentioned. Certain toads, salamanders, and newts are also provided with poison glands. As a means of defense, the poison is very effective, because a predator that has encountered a poisonous amphibian will tend to avoid others of the same species. Some of the most poisonous species, for example, *Salamandra salamandra*, are brightly colored to facilitate identification. Dogs and cats that catch and bite the **marine toad,** *Bufo marinus*, often die from the toxic effects of their secretions (**bufotoxins**). Generally, the warty skin and the parotid glands located behind eye of toads is heavily laden with poisonous secretions. Although they are extremely irritating to a predator, the secretions of most species are not as potent as those found in the marine toad.

Relations of Amphibians to Humans

Amphibians are virtually all beneficial to humans. Many are so rare as to be of little value, but frogs and toads are of considerable importance. Frogs are used extensively for laboratory dissections, physiological experiments, human pregnancy tests, in pharmacology, and for fish bait. Frog legs are eagerly sought as food; more than 3 million pounds are eaten every year in the United States. Mud puppies are also edible. In Japan the giant salamander is considered a delicacy.

Many superstitious beliefs are held about amphibians, such as handling toads causes warts, salamanders are not injured by fire, a croaking frog predicts rain, and a toad has a jewel in its head. In China, toad skin is used as a medicine; its use may have some therapeutic value, as certain glands contain a digitalis-like secretion that increases blood pressure when injected into humans.

Frogs and toads are widely recognized as predators of injurious insects. Toads are of special value because they may live in gardens, where insects are most injurious. In France, gardeners even buy toads to help keep destructive insects under control. *Bufo marinus* has been introduced into the southern United States, especially where sugarcane is grown, to control insects.

Frog farming has been promoted for pleasure and profit, but it has generally proven to have been a great disappointment to those who engage in the enterprise. Many "farms" are simply favorable marshes in which natural reproduction is encouraged. Artificial rearing of frogs is not practical because it is difficult to find a satisfactory supply of food. Furthermore, unless frogs of different sizes are separated, the larger ones eat the smaller, and losses from parasites and disease increase with crowding.

Summary

There are three general types of amphibians: frogs and toads, salamanders, and caecilians.

The frog exhibits many features typical of all vertebrates and is used frequently for comparative study. Its loose-lying skin, which can change color, contains mucus and sometimes poison glands.

The frog's internal anatomy is like an abbreviated version of that of the higher vertebrates. The digestive system is relatively short; the respiratory system includes the skin as an accessory respiratory organ; the circulatory system incorporates a three-chambered heart; the reproductive systems of the male and female allow external fertilization; the skeletal system lacks ribs; the muscular system contains all three types of muscle tissue (smooth, striated, and cardiac); the nervous system is complex, as it is in all vertebrates, but reflexes play a predominant role in the frog's behavior. The frog is capable of audition (thought there is no external ear), vision, tactile sensation, olfaction, and, presumably, taste. The frog depends on hormones for many biological functions, including metamorphosis and color changes. Its life cycle is from egg to tadpole to mature frog.

All amphibians are confined to moist habitats. The semiterrestrial adults must usually return to fresh water to breed, and gill-breathing larvae generally metamorphose into lung-breathing adults. There are many exceptions to this generalized life history, as many interesting adaptations have evolved that further exploit terrestrial existence. But retention of the ancestral kidney, permeable integument, and eggs that must develop in a wet environment have limited the amphibians' invasion of land.

Review Questions

1. How can male and female frogs be distinguished by external examination?
2. How can you get a frog to eat a dead fly?
3. You are told that an organ has three lobes and has a function (among others) that is opposite to one characteristic of bone marrow. What is the organ?
4. What is the oxygen-combining respiratory pigment in frog's blood? Is it dissolved in the plasma?
5. How many chambers has a frog's heart? What effect does this have on the oxygen content of blood exiting the heart?
6. Why could Adam, of Adam and Eve fame, not have been a frog?
7. Name three differences between frogs and toads.
8. Why do you suppose caecilians are nearly blind? How do they compensate?
9. List some major differences between apodan amphibians and oligochaete annelids.
10. What hormone is in part responsible for maintaining amphibian water balance?
11. Which amphibians mentioned in this chapter are neotenous?
12. What is the most common genus of toads? Of frogs in the United States?
13. Name and describe the functions of the various glands found in the skin of amphibians.
14. Which of the following are true? Toads
    (a) are generally stronger than frogs.
    (b) subsist on fly larvae.
    (c) have a wider environmental range than frogs.
    (d) induce wart formation in *Homo sapiens*.

Selected References

Barbour, T. *Reptiles and Amphibians, Their Habits and Adaptations*. Boston: Houghton Mifflin, 1934.

Bishop, S.C. *Handbook of Salamanders. The Salamanders of the United States, of Canada, and Lower California*. Ithaca: Comstock, 1943.

Boolootian, R., and D. Heyneman. *An Illustrated Laboratory Text in Zoology*, 2nd ed. New York: Holt, Rinehart and Winston, 1969.

Cochran, D. N. *Living Amphibians of the World*. New York: Doubleday & Company, 1961.

Conant R. A Field Guide to Reptiles and Amphibians of Eastern and Central North America. Boston: Houghton Mifflin. 1975.

Gilbert, S. G. *Pictorial Anatomy of the Necturus*. Seattle: University of Washington Press, 1973.

Gilbert, S. G. Pictorial Anatomy of the Frog. Seattle: University of Washington Press, 1965.

Goin, C. C., and O. B. Goin. *Introduction to Herpetology*. San Francisco: W. H. Freeman & Company, 1962.

Hickman, C. P. *Integrated Principles of Zoology*, 4th ed. St. Louis: C. V. Mosby Co., 1970

Holmer, S. J. *The Biology of the Frog*. New York: Macmillan, 1927.

Hughes, G. M. (ed.) Respiration of Amphibious Vertebrates. New York: Academic Press, 1976.

Moore, J. A. (ed.). *Physiology of the Amphibia*. New York: Academic Press, 1964.

Noble, G. K. *Biology of the Amphibia*. New York: McGraw-Hill Book Company, 1931, reprinted 1955, Dover Publications.

Oliver, J. A. *The Natural History of North American Amphibians and Reptiles*. New York: D. Van Nostrand Company, 1955.

Romer, A. S. *The Vertebrate Body*. Philadelphia: W. B. Saunders Co., 1970.

Savage, R. M. *The Ecology and Life History of the Common Frog (Rana temporaria temporaria)*. London: Sir Isaac Pitman & Sons, Ltd., 1961.

Stebbins, R. C., *A Field Guide to Western Reptiles and Amphibians*. Boston: Houghton Mifflin, 1966.

Storer, T. I., R. L. Usinger, and J. W. Nybakken. *Elements of Zoology*. New York: McGraw-Hill Book Company, 1968.

Webster, D., and M. Webster. *Comparative Vertebrate Morphology*. New York: Academic Press, 1974.

Wright, A. H., and A. A. Wright. *Handbook of Frogs and Toads: Frogs and Toads of the United States and Canada*. Ithaca: Comstock, 1949.

Phylum Chordata, Subphylum Vertebrata, Class Reptilia—Turtles, Tuatara, Lizards and Snakes, Crocodiles and Alligators

Reptiles constitute one of the most interesting and generally least-known of the vertebrate classes (Fig. 28-1). The study of amphibians and reptiles is called **herpetology.** Reptiles are better adapted for terrestrial life than are amphibians. Some of the advances shown by reptiles are (1) a dry scaly skin, (2) limbs generally suited for rapid locomotion, (3) partial or nearly complete division of the ventricle, which reduces mixing of oxygenated and deoxygenated blood in the heart, (4) completely ossified skeletons, (5) a copulatory organ that permits internal fertilization, and (6) eggs with shells and protective embryonic membranes to prevent desiccation, thereby allowing development on land.

Reptiles are most abundant in the warmer climates of the world; fewer live in the colder parts of the temperate zone, and almost none are found in the Arctic or Antarctic regions. Between 6000 and 6500 species of living reptiles have been described and about 260 species, including several marine turtles that frequent the coast, are known to occur in this country. Reptiles occupy an important place in the vertebrate hierachy because their anatomy is intermediate between that of a typical amphibian and that of a typical bird or mammal.

Taxonomy and Evolution of Reptiles

The reptiles living today are but a fraction of the vast hordes that inhabited the earth's surface during prehistoric times. In fact, of the approximately 16 orders of reptiles recognized by herpetologists, only 4 contain living representatives. And one of these includes only a single endangered species, whereas another (the crocodilians) is diminishing in numbers. The four orders of living reptiles are as follows:

Order **Chelonia (Testudinata).** About 250 species of turtles, terrapins, and tortoises.

Order **Rhynchocephalia.** *Sphenodon punctatus*, the **tuatara,** a single species confined to New Zealand.

Order **Squamata.** About 6000 species of lizards and snakes.

Order **Crocodilia.** The crocodile, alligators, gavials, and caimans; about 25 species.

Figure 28-1. Some orders and families of Reptilia.

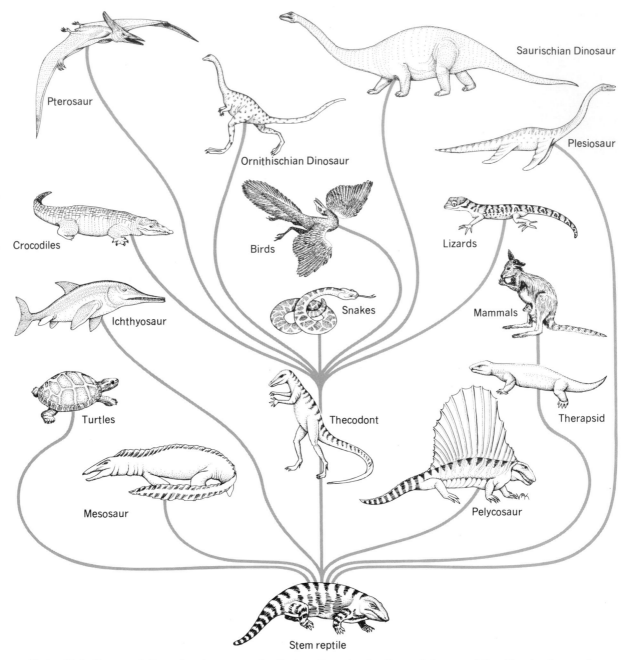

Pterosaur

Saurischian Dinosaur

Ornithischian Dinosaur

Plesiosaur

Crocodiles

Birds

Lizards

Ichthyosaur

Snakes

Mammals

Turtles

Thecodont

Therapsid

Mesosaur

Pelycosaur

Stem reptile

Figure 28-2. Representatives of major groups of extinct reptiles, showing the evolutionary relationships (no time scale intended; see also the classification appendix).

The living and extinct reptiles may be divided into five subclasses: Anapsida, Synapsida, Lepidosauria, Euryapsida, and Archosauria (Fig. 28-2 and classification appendix).

Because of the numerous extinct reptilian groups, all the subclasses and their evolutionary relationships will be surveyed before the more detailed discussion of representatives of the extant orders is presented.

Subclass Anapsida—Stem Reptiles and Turtles

The first reptiles evolved from **labyrinthodont** amphibians about 320 million years ago. Their waterproof skin and eggs allowed reptiles to exploit terrestrial habitats more fully than could amphibians, and for some 200 million years reptiles were the dominant land animals. The most primitive reptiles were the **cotylosaurs** (order **Cotylosauria**), lizardlike creatures of the subclass **Anapsida,** which also includes the turtles (order **Chelonia**). The turtles have a remarkably good fossil record, although their origin is somewhat vague.

Subclass Synapsida—Mammal-like Reptiles

The earliest divergence from the cotylosaur stem was the now extinct subclass **Synapsida,** which included the mesosaurs (order **Mesosauria**), pelycosaurs (order **Pelycosauria**), and the therapsids (order **Therapsida**). These were the so-called **mammal-like reptiles** because this line eventually led to the mammals. **Mesosaurs** were the first reptiles to return to the sea; their elongated snouts and tails gave them a crocodilelike appearance. The **pelycosaurs** were very successful land predators; the familiar *Dimetrodon* was characterized by extensions of the vertebrae that formed a sail-like structure along its back (Fig. 28-2). The **therapsids** were the dominant large animals after the decline of pelycosaurs about 250 million years ago, and they reigned until the rise of the dinosaurs, approximately 50 million years later. At about this time, small, shrewlike creatures evolved from the therapsids; these were the first mammals. The dinosaurs were more successful than the mammal-like reptiles and early mammals; the latter rose in number, diversity, and size only after the extinction of dinosaurs about 70 million years ago (see Fig. 28-3).

Subclass Lepidosauria—Snakes, Lizards, and Related Forms

This subclass consists of the lizards and snakes (order **Squamata**), the tuatara (order **Rynchocephalia**), and other extinct lizardlike forms. This subclass is the most highly represented group of reptiles living today and is considered in detail later in this chapter.

Subclass Euryapsida—Extinct Marine Reptiles

This extinct, primarily marine, subclass included the plesiosaurs and relatives (order **Sauropterygia**) and the ichthyosaurs (order **Ichthyopterygia**). Another common alternative taxonomy considers the latter group as a separate subclass. *Ichthyosaurus* was a fish-eating aquatic reptile. It was well adapted for life in the water, having evolved fins and a body very similar to dolphins and having been called the "whale" of its era. The remains of ichthyosaurs have been found in North America, Europe, Asia, Africa, and Australia. The **plesiosaurs'** fins also evolved from legs, but these widely distributed marine reptiles retained a distinct neck and probably swam more like turtles than like fishes or dolphins. Their body form is very close to that of many mythical sea creatures, and even today they are considered akin to the controversial "Loch Ness monster."

Subclass Archosauria—The Great Ruling Reptiles

The great ruling reptiles of the Age of Reptiles were the members of subclass **Archosauria.** The earliest members were the thecodonts (order **Thecodontia**), which were the first bipedal reptiles. These creatures commonly had armored plates, and most all were probably herbivorous. They were the ancestors to two important branches in evolution: (1) A return to quadrapedal locomotion eventually led to the extant order **Crocodilia.** (2) The first vertebrate attempts at flight characterized the familiar flying rep-

Brontosaurus

Tyrannosaurus

Stegosaurus

Trachodon

Triceratops

Ornithomimus

Figure 28-3. Representative dinosaurs (not drawn to scale).

tiles (order **Pterosauria**). *Pterosaurus* had forelimbs modified for flight. They resembled birds in certain skeletal characters, but differed from them in others. *Pteranodon* had a skull about 60 cm long and a wing spread of more than 8 m. Teeth were absent, and the tail was short. The largest pterosaur fossil, found recently in Texas, indicates that the reptile had a wingspan of over 15 m.

The most familiar archosaurs, the **dinosaurs,** probably originated from early thecodonts and evolved along two main lines (orders **Saurischia** and **Ornithischia**). The dinosaurs (Figs. 28-2 and 28-3) lived in various habitats, from swamps to uplands, and their fossils have been found on most continents.

Most were about the size of a modern rhinoceros or smaller, but some species measured over 26 m in length. Both herbivorous and carnivorous forms existed; the orhithischians were probably all herbivores. The **saurischians** are typified by the carnivorous *Tyrannosaurus* and by *Brontosaurus*, which was about 23 m long; remains of this giant herbivore have been found in Wyoming and Colorado. The Dinosaur National Monument in Colorado contains many dinosaur fossils. More advanced adaptive radiations within this order included a group of small, bipedal carnivores that had well-developed hindlimbs for running and elongated forelimbs. Recent studies suggest that these were the ancestors to modern birds, contrary to the previously well-accepted phylogeny (see Chapter 29).

The order **Ornithischia** was a diverse group of reptiles that also included both bipedal and quadrapedal forms. The well-known ornithischian *Stegosaurus* reached a length of approximately 8 to 9 m and was herbivorous. It possessed huge plates along its back (Fig. 28-3). *Protoceratops* was a small, hornless, herbivore about 2 m long; numerous fossils were discovered in the deserts of Mongolia, including immature forms.

A thorough examination of dinosaurs reveals a highly diverse (in size, anatomy, habitats, etc.) group of organisms rivaled in their variety only by the successors to their habitats—the mammals. The reasons for the decline and wholesale extinction (save the ancestors of the birds) of the dinosaurs about 70 million years ago is another dinosaur mystery that is being actively investigated by paleontologists. Of the many alternatives, no one explanation has satisfied a majority of investigators. The dinosaurs' evolutionary demise is somewhat redeemed if the recent evidence concerning the origin of birds is correct (Chapter 29).

The Origins of Homeothermy

Dinosaurs and other extinct reptiles have recently regained the interests of many paleontologists. An exciting ongoing controversy is the hypothesis that the therapsids, dinosaurs, and some thecodonts were warm blooded (**homeothermic** or **endothermic**), similar to birds and mammals. **Homeothermy** means that the body temperature is regulated to maintain a state of constancy (usually warmer than the environment), even during wide environmental temperature fluctuations. It is well accepted that the birds and mammals (or their immediate predecessors) developed homeothermy separately as they evolved from different cold-blooded (**poikilothermic** or **ectothermic**) reptilian ancestors. The question of homeothermy also involves another interesting hypothesis, namely, that dinosaurs did not represent an evolutionary dead end but, rather, that small, bipedal, warm-blooded saurischians were the immediate ancestors of all birds. This is further discussed in Chapter 29; the point here is that the accepted separate (**polyphyletic**) origins of homeothermy increase the acceptability of homeothermy in other groups, especially if a group is related

in structure and evolution to living homeotherms, as dinosaurs appear to be related to birds and therapsids to mammals.

What are the adaptive values and consequences of homeothermy? The major advantage is the constant physical readiness that it gives an animal because of the higher levels of metabolic activity that can be achieved and sustained for longer periods of time, as compared with the metabolism of poikilotherms. This is exhibited in the high energy requirements of the flying or the fast sprinting that characterizes the mammals and birds. The homeotherms' constant high metabolic rate is extremely wasteful of energy because most of the time they are inactive and must radiate the bulk of their metabolic energy as excess heat (hence the name "warm blooded"). This wastefulness is adaptive because it confers the ability to jump for an attack on a prey or to flee quickly from a predator at any time. A biochemical advantage is the constant temperature for all biochemical reactions; this allows optimization of all enzymic activities to one body temperature.

A "cold-blooded" animal is not always cold; it simply does not regulate its body temperature but allows it to fluctuate with environmental temperature changes. Hence, it is a more vulnerable prey during suboptimal temperature conditions. Because of this, many poikilotherms, such as lizards, may be seen basking in the sun after a cold night; they are much less active (and more vulnerable) until their internal temperature rises. An important benefit of a lower metabolic rate and higher energy efficiency is the relatively low quantity of food required for survival, which confers the ability to survive long periods without food. Thus, the physiological differences between homeothermy and poikilothermy has an important ecological implication. The number of homeothermic predators that can be sustained by a certain prey population is much less than the number of poikilothermic predators feeding upon a similar prey population; that is, the **predator–prey ratio** is much lower for homeotherms.

Living homeotherms exhibit many morphological adaptations, such as insulation (feathers on birds, hair on mammals) and the increased vascularization by the circulatory system, needed to transport higher amounts of oxygen to the tissues.

This understanding of homeothermy is necessary to

assess the recent evidence concerning the therapsids, dinosaurs, and thecodonts. Although there is an absence of complete support by all paleontologists, many results support the homeothermic hypothesis. Microscopic examination of fossilized bones ally the therapsids, dinosaurs, and primarily advanced thecodonts with the modern homeotherms rather than with other reptiles because of evidence of the former's higher vascularization. Specialized structures such as the parallel rows of plates on *Stegosaurus* (Fig. 28-3) and some of the structures of certain armored thecodonts have been interpreted as temperature-control mechanisms. The fossilized plates of *Stegosaurus* are bony, but spongy in internal appearance, suggesting that a large amount of blood could be pumped through the bones, dissipating excess heat. The amount of heat radiated from the animal could be regulated by its orientation to the direction of the wind; this would change the amount of plate surface area exposed to the wind and, hence, the amount of heat given off. Experiments have been performed to test the relative heat dissipation efficiencies of various geometrical shapes (e.g., cylinders, square rods, and variously shaped plates). The results indicated that a square plate with the corners cut off (forming an irregular octagon) was the most effective shape—this is very similar to the geometry of *Stegosaurus*'s plates! Anatomical analyses of many dinosaurs indicate highly active, fleet-footed creatures rather than the sedentary image of the well-known, larger representative such as *Brontosaurus*.

The origin of feathers in the active ornithischian ancestors to birds was probably also naturally selected for temperature regulation, in this case, for heat conservation. The large sizes of many dinosaurs also served to promote heat conservation, as it resulted in lower ratios of surface area to volume (Fig. 3-1), thus reducing the rate of heat dissipation. Even predator–prey ratios, controversial and difficult to obtain from fossils, have been interpreted as support for homeothermic dinosaurs and therapsids (data for thecodonts are scanty).

The lack of universal agreement among paleontologists concerning this issue may indicate that these reptiles represent an incomplete transition to modern homeotherms. That is, although the therapsids, dinosaurs, and certain thecodonts may have been homeothermic in the sense of constant temperature regulation, all might not have been truly endothermic (producing all their heat by internal metabolism).

Recent experiments on the largest living lizards, the **monitor lizards,** indicate that *any* large reptile (ectothermic or endothermic) that lived at the typically mild temperatures during the Age of Reptiles *must* have had a relatively constant internal temperature because of the low rate of heat exchange observed in large organisms. It does seem reasonable that the evolution of heat conservation (e.g., insulation or larger body sizes) and dissipation (e.g., cooling vanes) mechanisms needed for homeothermy would precede the biochemical changes that resulted in high heat production (endothermy). Thus the evolution of homeothermy and endothermy, as with most major evolutionary innovations, probably occurred in stages, and the dinosaurs, therapsids, and certain thecodonts were separate, parallel transitional stages to the modern endothermic conditions of birds and mammals.

Order Chelonia—Turtles, Tortoises, and Terrapins
Pseudemys—The Turtle

The turtle will be considered here as a representative reptile. The genus *Pseudemys* is a large species complex that includes most of those forms commonly used in anatomy and physiology labs. The turtle can live either in water or on land. Although slow moving on land, most can swim quite rapidly.

External Morphology

The turtle is distinguished from other reptiles by its **exoskeleton,** the **shell,** that protects the internal organs (Fig. 28-4). In some species, the head, limbs, and tail can be withdrawn into the shell.

The **neck** is long and flexible, and the **head** is dorsoventrally flattened. The **mouth** is large, but, instead of teeth, **horny plates,** which line the margin of the jaws, are used to crush food. The **external nares (nostrils)** are located together near the anterior tip of the snout. This adaptation allows the olfactory receptors to monitor the environment when the head is withdrawn into the shell; when underwater, the nostrils

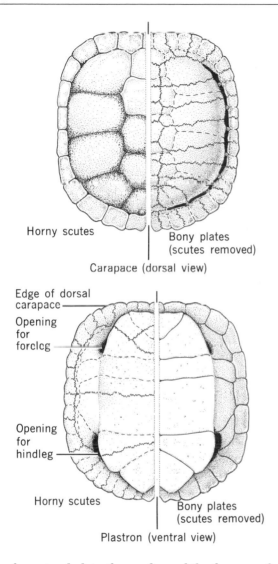

Horny scutes | Bony plates (scutes removed)

Carapace (dorsal view)

Edge of dorsal carapace

Opening for foreleg

Opening for hindleg

Horny scutes | Bony plates (scutes removed)

Plastron (ventral view)

Figure 28-4. Turtle shell showing the external horny scutes and the bony plates beneath. The living epidermis, which covers the bony plates, produces the scutes.

can be extended to the surface while the rest of the body remains submerged. The **eyes** on either side of the head are guarded by three **eyelids:** a short, thick, opaque **upper lid;** a long, thin **lower lid;** and a transparent **nictitating membrane** that can be drawn over the eyeball from the anterior corner of the eye. Just behind the angle of the jaw on either side is a thin **tympanic membrane.**

The **limbs** generally possess five **digits,** most of which are armed with large claws that are used in crawling, climbing, and digging. The skin is thin and smooth on the head, but it is thick, tough, scaly, and wrinkled over other exposed parts of the body.

Skeletal System—The Shell

Because the life of the turtle is influenced so strongly by its shell, the skeletal system will be considered primarily as an external feature.

The **shell** (Fig. 28-4) consists of a convex, dorsal portion—the **carapace**—and a flattened, ventral piece—the **plastron.** These two parts are formed by the union of numerous **bony plates.** Although expanded ribs and vertebrae form a large part of the carapace, **dermal bones** also play a role, especially in the plastron; as a result of these modifications, the **sternum** is

Figure 28-5. *Trionyx*, the soft-shelled turtle. Length of shell of adult is about 30 cm. Note leathery integument that is not divided into horny scutes.

missing in turtles. The dorsal carapace and ventral plastron are strongly bound together by bony **lateral bridges.** Both the bony carapace and plastron are usually covered by a number of symmetrically arranged horny plates, called **scutes (shields),** that are derived from the epidermis. The number and shape of the scutes vary according to the species. The bony plates do not correspond in either number or arrangement to the scutes above them.

Soft-shelled turtles (Fig. 28-5) have a leathery shell that is not divided into scutes and contains little bony substance.

Internal Morphology and Physiology

Digestive System

Turtles are **carnivorous, omnivorous,** or **herbivorous.** Animal prey consists of fishes, frogs, waterfowl, small mammals, and many types of invertebrates. The flexible neck enables the turtle to reach in all directions for food while at rest on the bottom. The jaws of large **snapping turtles,** *Chelydra serpentina,* are powerful enough to amputate a human finger.

The **digestive system** is simple (Fig. 28-6). A broad, soft **tongue** is attached to the floor of the **mouth cavity** but is not protrusible. At the base of the tongue is a longitudinal slit, the **glottis,** and a short distance posterior to the angles of the jaw are the openings of the **eustachian tubes.** The **pharynx** is thin walled and very distensible; it leads into the more slender and thicker-walled **esophagus,** which opens into the **stomach.** A **pyloric valve** at the posterior end of the stomach controls the passage of food into the first part of the **small intestine,** the **duodenum.** The **liver** discharges **bile** through the **bile duct** into the duodenum, as do several **pancreatic ducts** that lead from the **pancreas.** The rest of the small intestine functions primarily in the absorption of food. A valve separates the small intestine from the **colon** (large intestine). At the junction of both is a slight projection, the **cecum.** The terminal portion of the large intestine opens into the **cloaca.**

Circulatory System

The reptilian **heart** (Fig. 35-1), except in crocodilians, consists of two **atria** and a single **ventricle** that is

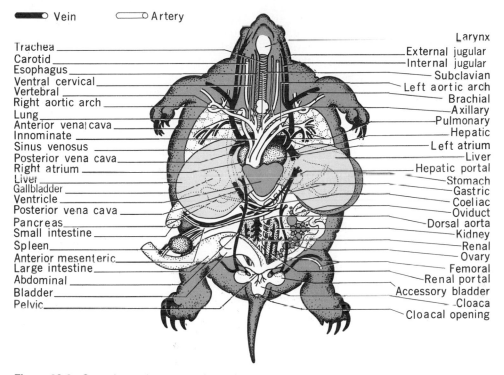

Vein   Artery

Trachea — Larynx
Carotid — External jugular
Esophagus — Internal jugular
Ventral cervical — Subclavian
Vertebral — Left aortic arch
Right aortic arch — Brachial
Lung — Axillary
Anterior vena cava — Pulmonary
Innominate — Hepatic
Sinus venosus — Left atrium
Posterior vena cava — Liver
Right atrium — Hepatic portal
Liver — Stomach
Gallbladder — Gastric
Ventricle — Coeliac
Posterior vena cava — Oviduct
Pancreas — Dorsal aorta
Small intestine — Kidney
Spleen — Renal
Anterior mesenteric — Ovary
Large intestine — Femoral
Abdominal — Renal portal
Bladder — Accessory bladder
Pelvic — Cloaca
Cloacal opening

Figure 28-6. Gross internal anatomy of a turtle.

partially divided into two chambers by an incomplete wall, or **septum.** In crocodiles and their allies, the **interventricular septum** is nearly complete, thereby essentially forming a four-chambered heart.

The arrangement of the atria, ventricle, and interventricular septum is such that most of the deoxygenated blood entering the right portion of the ventricle is pumped to the lungs, and most of the oxygenated blood is pumped into the **systemic arteries** from the left side of the ventricle. Deoxygenated blood from the body tissues is carried by the **posterior vena cava** and two **anterior vena cavae** into the **sinus venosus** and from there into the **right atrium** (Fig. 28-6). From there it is pumped into the right side of the ventricle. When the ventricle contracts, the deoxygenated blood is forced into the **pulmonary artery,** which sends a branch to each lung. Another vessel arises from the right side of the ventricle and meets the **left aortic arch,** which conveys blood to the viscera and the **dorsal aorta.**

Oxygenated blood returns from the lungs through the **pulmonary veins** and enters the **left atrium,** where it is pumped into the left side of the ventricle. This blood is pumped out through the **right aortic arch,** which merges with the left aortic arch to form the dorsal aorta. Because the septum dividing the ventricle is incomplete, the blood that enters the right aortic arch is a mixture of oxygenated blood from the left atrium and deoxygenated blood from the right atrium.

Certain species of turtles have a well-developed **renal portal system.** The **hepatic portal system,** which supplies the liver, is more highly developed than is that of the frog.

Respiratory System

Reptiles breathe primarily by means of **lungs** (Fig. 28-6), but in some aquatic turtles the **cloaca, pharynx,** and **integument** serve as auxiliary respiratory surfaces. Air enters the **mouth cavity** by way of the **nasal passages.** The **glottis** opens into the **larynx,** through which air passes into the **trachea.** The trachea divides, sending one **bronchus** to each lung. The lungs are more elaborate than are those of amphibians. The bronchi branch several times, and the lung cavity is broken up into many spaces (**alveoli**) that greatly increase the respiratory surface.

Unlike amphibians, which must swallow air, the lungs of reptiles are ventilated by muscular contractions that expand and compress the body cavity. The rigid shell of turtles prevents general expansion and compression of the body, and as a result turtles have evolved a unique method of breathing. Inspiration is accomplished by two **flank muscles.** When contracted, they cause an enlargement of the **pleuroperitoneal cavity,** and air is drawn into the lungs. Expiration is accomplished by a pair of **expiratory muscles** that enclose the viscera. Contraction of the expiratory muscles presses the viscera against the lungs, decreasing their volume and forcing air from the respiratory passages. Pulling in the legs and neck further decreases the size of the body cavity.

Some freshwater turtles pump water into and out of **accessory bladders** that connect to the cloaca. This method (analogous to the process in the sea cucumber and dragonfly nymph) may supplement respiration when the turtle is submerged. The highly vascularized pharynx is also elaborated into a respiratory surface in some aquatic turtles. Although important in amphibians, cutaneous respiration is of little consequence to most reptiles. However, soft-shelled turtles have been shown to take up to 70 percent of their oxygen through the leathery skin that covers their carapace and plastron.

Excretory System

Excretion is accomplished by two **metanephric kidneys** that are specialized to reabsorb water from the urine. Snakes, lizards, and desert turtles also convert ammonia into **uric acid,** a nontoxic nitrogenous compound that is insoluble in aqueous solutions. As water is reabsorbed, the urine becomes supersaturated and uric acid precipitates from solution. Dry crystals of uric acid are admirably suited for storage within the

Figure 28-7. Excretory and reproductive organs of a male turtle (ventral view).

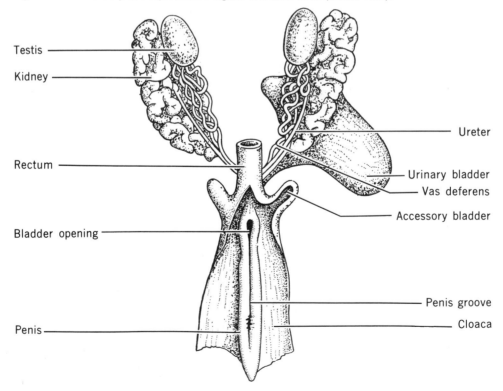

Testis

Kidney

Rectum

Bladder opening

Penis

Ureter

Urinary bladder

Vas deferens

Accessory bladder

Penis groove

Cloaca

confines of the egg and for elimination from adults with minimal water loss. The concentrated urine passes through the **ureters** into the **cloaca** (Fig. 28-7) and is stored in the urinary bladder before being voided through the **cloacal opening (anus).**

Nervous System and Sense Organs

The reptile **brain** is more highly developed than is that of amphibians. The **cerebral hemispheres** are larger and organized into a superficial gray layer and central white matter. The **cerebellum** is also larger, indicating a greater ability to coordinate movements. Reptiles have 12 pairs of **cranial nerves,** whereas amphibians have only 10.

The turtle **eye** is small. It has a round **pupil** and an **iris,** which is usually dark in terrestrial forms, but often colored in aquatic turtles. The sense of hearing is not well developed, but the turtle responds readily to vibrations detected through the skin, so it is easily frightened by noises. The sense of smell enables the turtle to distinguish between various kinds of food while submerged or on land. The skin over many parts of the body is very sensitive to touch.

Reproductive System

The sexes are separate. The male organs are a pair of **testes** and a pair of **vasa deferentia** through which **spermatozoa** pass into the **penis** attached to the ventral wall of the cloaca and normally lying in the tail. During mating, it is erected through the cloacal opening. Internal fertilization is the rule in reptiles, and this adaptation has freed them from the aquatic habitats of their amphibian ancestors. The development of a **copulatory organ,** a prerequisite for internal fertilization, is found in all reptiles, except the tuatara.

The female organs are a pair of **ovaries** and a pair of **oviducts** that open into the cloaca. Copulation is usually preceded by some display of courtship behavior by both sexes.

Turtles are **oviparous.** Their **eggs** are white, round or oval, and covered by a calcareous **shell** that protects the developing embryo from desiccation and physical damage. Turtle eggs are laid in holes dug by the female in soil or in decaying vegetation, where heat aids in incubation.

Other Turtles

Turtles live on land, in fresh water, or in the sea. The word **turtle** is often applied to semiaquatic species; **tortoise** mainly or entirely to terrestrial species; and **terrapin** to members of genus *Malaclemys*, which characteristically have angular rings on the carapace and are prized for human consumption (Fig. 28-8). Most land and freshwater turtles **hibernate** underground during the winter, and in warmer climates they **estivate** during the hotter months.

Some of the more interesting types of turtles are as follows: The **snapping turtle,** *Chelydra serpentina* (Fig. 28-9), is famous for its strong jaws and vicious bite. The **musk turtle,** *Sternotherus odoratus*, emits a disagreeable odor when molested or captured. The **painted turtle,** *Chrysemys picta* (Fig. 28-9), is brilliantly colored. The plastron of the **box turtle,** *Terrapene* (Fig. 28-10), is hinged transversely near its center so the shell can be completely closed when the animal is in danger. The **gopher tortoises** (*Gopherus*) live in burrows in dry sandy areas of the southern United States. Some of the **giant tortoises,** *Geochelone* (*Testudo*, Fig. 29-1), on the Galapagos Islands weigh over 230 kg and are probably more than 200 years old.

Sea turtles inhabit tropical and semitropical ocean waters and come on land only to lay their eggs in sandy beaches. The **green turtle,** *Chelonia mydas*, is a marine species that has been so extensively hunted for food that it may be in danger of extinction (Fig. 28-11). The green turtle received its name because its fat is green in color. The forelimbs are modified as flippers for swimming, and the hindlimbs are used for steering and as kickers. It weighs about 180 kg.

Figure 28-8. *Malaclemys,* the diamondback terrapin. It derives its common name from the markings on its shell. One of the most famous of all turtles as food for man. [Courtesy of Shedd Aquarium, Chicago.]

Figure 28-9. Common American turtles. The painted turtle is common in ponds. The snapping turtle is less protected by shell than are many turtles. It is said that a snapping turtle will snap as soon as hatched.

Figure 28-10. The western box turtle, *Terrapene.* The plastron is hinged and encloses the animal as though it were in a box. The turtle is known to live to the age of 123 years. [Courtesy of the American Museum of Natural History.]

Figure 28-11. The green turtle, *Chelonia.*

shelled turtles (family **Trionychidae,** Fig. 28-5) also have leathery shells without scutes.

Order Rhynchocephalia—The Tuatara

Sphenodon punctatus (Fig. 28-12) is the sole surviving species of these primitive reptiles, which are known mostly from the fossil record. The **tuatara,** as *Sphenodon* is commonly known, has a lizardlike body form but shares many skeletal characteristics with the oldest fossil reptiles. The front of the skull, for example, forms an overhanging beak that is set with teeth fused to the jawbone (rather than in sockets). *Sphenodon* is very similar to fossil reptiles dating from 100 million years ago.

The **leatherback turtle,** *Dermochelys coriacea* (Fig. 28-1), is the largest of all living turtles, sometimes weighing well over 680 kg. It has a leathery covering over the shell instead of horny scutes. **Soft-**

Figure 28-12. Tuatara, *Sphenodon punctatus,* at the entrance of its burrow. It has many characteristics of the early ancestral reptiles. A relic of a remote past, it is now found only on islands near New Zealand. Approximately 75 cm in length.

The **tuatara,** which is about 75 cm long, once roamed throughout most of New Zealand before the effects of western civilization reduced its populations to the brink of extinction. Its range is presently restricted to some small islands in the Bay of Plenty in New Zealand and in the Cook Strait. The tuatara is now protected, and its populations are increasing.

It lives in burrows, is nocturnal, and preys upon living animals. One of its most striking features, shared with many lizards, is the presence of a median, or **pineal,** eye in the roof of the cranium; this structure has a retina and other characteristics resembling a true eye in juveniles but is vestigial and scarcely visible in adults.

Order Squamata—Lizards and Snakes

The lizards and snakes are the most numerous and diverse living reptiles, comprising some 3000 species each. Snakes are essentially highly specialized lizards that have lost their limbs and elongated their bodies by a great multiplication of vertebrae and ribs. Also correlated with their elongated body form, snakes (and some lizards) have a reduction in size (or total loss) of the left lung.

Lizards

The lizards belong to the suborder **Sauria (Lacertilia).** Lizards usually have four well-developed limbs for running, clinging, climbing, or digging. Some (e.g., **glass snakes,** *Ophisaurus*) have no limbs or only vestiges of limbs. The tail is generally long and is easily detached in many lizards; a new tail, which does not possess vertebrae, is soon regenerated. The skin of the lizard is usually covered with small scales. Lizards range in length from a few centimeters to over 2 m. Some of the more familiar types are described below.

Geckos (family **Gekkonidae**) (Fig. 28-1) inhabit all the warmer parts of the globe; they are harmless and usually nocturnal. Many have specialized **lamellae** under the toes that enable them to climb over trees, rocks, walls, and ceilings with relative ease.

Several different kinds of lizards are called **chameleons,** but the 75 species of true chameleons (family **Chamaeleonidae**) are restricted to Africa, Madagas-

Figure 28-13. The American chameleon or green anole, *Anolis.* Anoles have an excellent facility to change their color (green to dark brown). Approximately 18 cm long.

car, Arabia, and India. Chameleons are noted for their ability to change colors rapidly. Many species of the New World chameleons, or **anoles** (*Anolis*, family **Iguanidae**), are common in the southeastern United States, the Antilles, Central America, and northern South America (Fig. 28-13).

Another New World lizard is the common **iguana** (*Iguana iguana*), which reaches a length of 2 m (Fig. 28-14). Iguanas inhabit tropical America and are consumed as human food there. The **horned toads** (Fig. 28-15) are not really toads but rather lizards. They live in arid regions of the western United States and in Mexico.

The **flying dragons** (*Draco*) are arboreal lizards in southern Asia whose sides are expanded into thin membranes supported by false ribs. These membranes are normally folded against the body, but when expanded they enable the lizard to glide from tree to tree.

Worm lizards (family **Amphisbaeridae**) are limbless, burrowing lizards that resemble worms in appearance. Only the Florida worm lizard, *Rhineura floridana*, occurs in the United States; other species inhabit mainly warmer areas of the world. Somewhat similar are the **glass snakes** (*Ophisaurus*, Fig. 28-1) in the United States and Mexico. These lizards have no limbs and move by lateral undulations of the body, as do the true snakes. They can be distinguished from true snakes by the presence of ear openings and movable eyelids. Their name is derived from the extreme brittleness of their long tail. Another type, called the **blindworm** (*Anguis*), inhabits Europe, western Asia, and Algeria. It looks much like a large, brightly colored worm and has, contrary to its name, well-developed eyes.

The largest of all modern lizards is the **dragon** (**monitor**) **lizard** (*Varanus*) of Komodo and some other small island in Indonesia. The natives of Komodo

Figure 28-14. *Iguana,* capable of changing color in response to ambient temperature, reaches a length of 2 m.

Figure 28-15. *Phrynosoma,* the "horned toad," is not a toad at all but, rather, a lizard that is common to the arid portions of the western and southeastern states. Natural size is approximately 18 cm.

claimed that dragons existed on their island, and in 1914 these "dragons" were identified as the largest living lizards. They reach lengths of 3 m and may weigh more than 115 kg. They are adept carnivores, capturing wild pigs and other prey, but they become quite tame in captivity.

Snakes

The true snakes belong to the suborder **Serpentes.** Snakes share many anatomical features with lizards but differ from them in four principal respects: (1) the right and left halves of the lower jaw are not firmly united; instead, they are connected by an elastic ligament; (2) there is no pectoral girdle, and generally there is a complete absence of limbs; (3) a urinary bladder is absent; and (4) the braincase is closed anteriorly.

Appendages are entirely absent except in a few species of pythons and boa constrictors that possess a pair of short spurlike projections on either side of the cloacal opening; these represent vestiges of hindlimbs (Fig. 28-16A).

Snakes are covered with **scales;** those on the head in particular are used in classification (Figs. 28-16B and 28-16C). The ventral surface anterior to the cloacal opening bears a single row of broad scales called **abdominal scutes,** to which the ends of the ribs are attached. The outer layer of the skin is shed several times each year.

The **eyelids** are fused over the **eyes,** but there is a transparent window that allows the snake to see. Just before the skin is shed, the **procorneal** portion overlying the eye becomes opaque, making the snake partially blind. Once the entire skin is shed, normal vision is restored.

The snake lacks an external ear opening. The **tongue** is a slender, deeply notched, protrusible organ that can be thrust out through grooves in the **jaws** even when the **mouth** is closed. The tongue serves as an auxiliary **olfactory organ,** transferring odorous particles from the external environment to the paired **Jacobson (vomeronasal) organs** on the roof of the mouth.

The **teeth** are sharp and curved inward, an adaptation that prevents prey from escaping by slipping forward once swallowing has commenced. In poisonous snakes, certain teeth are specialized for the injection of venom (Fig. 28-17).

Figure 28-16. (A) Vestigial hindlimb and girdle bones of the python. The skeletons of nearly all snakes are without limbs, but the pythons are among the exceptions. These remnants of hindlimbs suggest that the ancestors of snakes traveled on legs. (B) Scales on anterior end of the hognose snake or puff adder. (C) Scales on anterior end of the black snake.

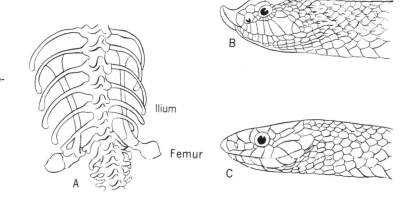

Ilium

Femur

A

B

C

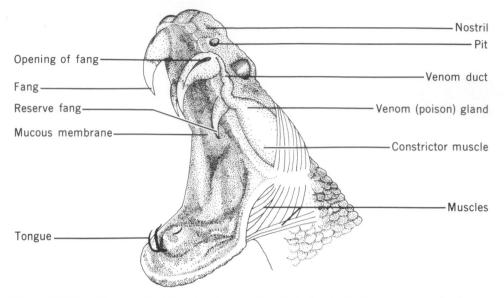

Figure 28-17. Rattlesnake *Crotalus*, drawing, showing the hollow teeth (fangs), venom gland, venom duct, and muscles used in forcing poison into the victim's flesh, as is done with a hypodermic needle. Note the pit between the nostril and eye, which is characteristic of pit vipers. Other teeth normally found have been deleted.

Figure 28-18. Boa constrictor feeding. As this picture shows, the snake captured the deer, which it suffocated by tightening its coils each time the prey exhaled. Eventually, the boa's prey can take in no more air. Arrow indicates the point on the deer to which it had been swallowed when the photographer started preparations for taking the picture. As a reaction to the photographer's presence, the snake began to disgorge its prey before the shutter clicked. This boa is native of tropical America. Boa's are among the longest snakes, growing to lengths of 3 to 6 m.

Snakes do not chew their food, but swallow it whole; animals much larger than the diameter of their own bodies can be swallowed (Fig. 28-18). Several structural adaptations make this possible: (1) The **lower jaw** articulates with the **skull** very loosely and is easily dislocated, allowing the **mouth** to expand to several times the diameter of the snake itself. (2) The lower jaw can also spread at the anterior midpoint, allowing for a moderate amount of lateral expansion. (3) The bones of the **palate** (roof of the mouth) are movable. (4) A noncollapsible extension of the **trachea** can be projected to the anterior end of the floor of the mouth, thus permitting the snake to breathe during prolonged swallowing. (5) The gut can be stretched sufficiently to accommodate such large meals.

Locomotion on land may be accomplished by several types of motion, but the two principal patterns are either lateral undulations of the body (most snakes) or looping of the abdominal scutes forward in alternate sections of the body (the **sidewinder,** *Crotalus cerastes*). Most snakes cannot move forward efficiently on a smooth surface. All snakes are able to swim.

The majority of snakes are **oviparous,** but some are **ovoviviparous,** as is the **garter snake** *(Thamnophis)* that brings forth its young alive (Fig. 28-19).

The tropics support a greater number of snakes than do the temperate zones. As in other groups of vertebrates, the snakes are found in almost every kind of habitat: Some species are marine, others live in fresh water, some are **arboreal** (climbing), and many are **fossorial** (burrowing).

Only five of the ten families of suborder Serpentes occur in North America. With a few exceptions, those described in the following paragraphs are found in the United States.

Two species of small **blind snakes** *(Leptotyphlops)* occur in the United States. They burrow long tunnels in the earth and feed on worms and insect larvae.

Pythons such as the **Indiana python,** *Python molurus* (Fig. 28-20), and the **boa constrictors,** *Constrictor* (Figs. 28-18 and 28-21), prey almost exclusively upon birds and mammals, which they squeeze to death in their coils. Death is caused by suffocation, not the breaking of bones; as the prey exhales, the coils tighten, until suffocation ensues. Neither pythons nor boas are venomous, and only a few are large enough to be dangerous to people. Of the approximately 60 species of boas and pythons, only 2 occur in the United States. Boa constrictors range up to 5 m in length. The longest members of the family **Boidae** are the **reticulated python,** *Python reticulatus,* of the Malay region that may reach 10 m and the **anaconda,** *Eunectes murinus,* of South America at 9 m.

The common garter snake, *Thamnophis sirtalis* (Fig. 28-19), of North America is the most abundant of our harmless snakes. It feeds largely on frogs, toads, fishes, and earthworms. The young are born alive, usually in August. The water snakes *(Nerodia)* are semiaquatic and are often mistaken for water moccasins. The **black snake,** *Coluber constrictor,* is a slen-

Figure 28-19. Garter snakes, *Thamnophis sirtalis,* are ovoviviparous. They are harmless to man and are frequently kept as pets. Average adult length is usually less than 1 m.

Figure 28-20. Indian python, *Python,* in its natural habitat. This is one of the world's largest snakes; it reaches a length of 8 m.

der, long-tailed snake that reaches a length of 2 m. West of the Mississippi it is replaced by a subspecies called the **blue racer** and another species, the **red racer,** in Texas. **King snakes** (*Lampropeltis,* Fig. 28-22) are of various sizes, but they are constrictors that have received their common name because they prey on other snakes. King snakes are immune to pit viper venom but not to coral snake venom. The scarlet king snake resembles the venomous coral snake in color. Another race of king snakes is the **milk snakes,** which derives their name from the unfounded superstition that they steal milk from cows. The **hog-nosed snakes** (*Heterodon*) are popularly known as **puff adders, spreading vipers,** or **blowsnakes.** They are nonvenomous, though they appear intimidating and also "play possum."

Figure 28-21. Boa constrictor (family Boidae).

Venomous Reptiles

Relatively few reptiles are poisonous. No turtles are venomous. Venomous reptiles that live in the United States include only 1 species of lizard and approximately 19 species of snakes; the only other poisonous lizard is a related species that inhabits Mexico and Central America.

Figure 28-22. The king snake is a constrictor. It is called "king" because it captures and kills other snakes, including poisonous species. Color is black with white or yellowish bands.

The **gila monster,** *Heloderma suspectum* (Fig. 28-23), inhabits the arid parts of Arizona, New Mexico, Utah, and Nevada. It is black and conspicuously spotted with pink or orange. A large specimen measures about 65 cm. The bite of the gila monster is fatal to small animals and dangerous to humans. The venom is as potent as that of some venomous snakes. Its method of injecting venom into the prey is less ef-

Figure 28-23. Head of gila monster, *Heloderma,* of the American Southwest, the only poisonous lizard in the United States. It is beautifully colored with black and orange patches; it has poison glands in the lower jaw and grooved teeth that carry venom into an animal that is bitten.

ficient, however, as it has to be ground into the prey by the nonspecialized teeth. Gila monsters are sometimes called **beaded lizards** because of their tuberculated skin and distinctive coloration.

The poisonous snakes that occur in the United States include 1 species of coral snake, the copperhead, the water moccasin (cottonmouth), and 15 species of rattlesnakes.

If you came across a snake that was ringed with red, black, and yellow, *with red rings bordered by yellow rings,* it is the venomous coral snake. Or, if the pupil is vertical and there is a pit (Fig. 28-17) between the eye and nostril on each side of the head (don't look too closely), the snake is a poisonous rattlesnake, water moccasin, or copperhead. All other snakes that live in the United States are harmless. Very few people in this country die as a result of a snakebite. In fact, more die of shock following bee stings. There are from 10 to 25 deaths from snakebite per year in the United States.

The **coral snake,** *Micrurus fulvius* (Fig. 28-24), of the southern United States is rather secretive in its habits and is rarely seen. The snake's docile disposition also has no doubt saved a number of people from potential bites. While the venom is very powerful, the coral snake's short, dull fangs often results in a less than fatal bite.

The **water moccasin,** *Agkistrodon piscivorus* (Fig. 28-25), lives in coastal swamps south of North Caro-

lina and in the Mississippi Valley from southern Illinois and Indiana southward. Its length averages about 1.25 m.

The **copperhead,** *Agkistrodon contortrix* (Fig.

Figure 28-24. The poisonous coral snake, *Micrurus,* lives in the southern United States and in tropical countries. It is the only representative of the cobra family in North America. It is about 1 m long.

Figure 28-25. Water moccasin (cottonmouth), *Agkistrodon piscivorus*. Its bite is occasionally fatal to man. Note the thick body and slender neck. It has a pit in front of the eye.

28-26), occurs from southern Massachusetts to northern Florida and west to Texas. An average specimen measures about 75 cm.

The **rattlesnakes** (Figs. 28-27 and 28-28) are easily distinguished by the **rattle** at the end of the tail in adults. The rattle consists of a number of horny, bell-shaped segments loosely held together. When disturbed, the rattlesnake rapidly vibrates the end of its tail, producing a loud and distinctive buzzing noise. This warns potential predators and keeps it from being stepped on. The **venom** is secreted by a pair of **glands** on each side of the head above the jaws (Fig. 28-17). These glands open through venom ducts into

Figure 28-26. The copperhead snake, *Agkistrodon contortrix*. Its common name comes from the fact that the top of the head is copper colored. Natural size about 1 m.

the two **fangs.** Each fang is pierced by a canal that opens near its pointed end, and functions very much like a very efficient hypodermic needle. The venom glands are surrounded by muscles that contract, squeezing the poison through the fangs and into the prey. When not in use, the fangs are folded back; this is not the case in coral snakes, cobras, and the sea snakes. Several smaller fangs lie just behind the functional pair; these reserve fangs replace those that are lost in struggles with prey or are normally shed. Rattlesnakes are most abundant in species and individuals in the deserts of the southwestern United States, but almost every region of this country is inhabited by one or more species. There are two genera of rattlesnakes: *Crotalus*, with 17 species of typical rattlesnake, and *Sistrurus*, with 2 species (**massasauga** and **pigmy rattlesnakes**).

The rattlesnake is one of the so-called **pit vipers** because it has a **pit** between the eye and nostril on each side of the head (Fig. 28-17). This depression houses the **pit organ,** which contains many blood vessels and nerve ending. Experiments have shown that this is a **heat-sensitive organ.** A rattlesnake can detect the movement of a moderately warm body at a few meters' distance; as a result it can hunt and kill rodents in complete darkness.

Rattlesnakes and other pit vipers usually strike from an S-shaped position of the body. Unless the venom is injected directly into a blood vessel, it usually travels slowly. The best first-aid measures in case of snakebite are to (1) Have the patient lie down and remain calm. (2) Apply a constriction band (tourniquet), handkerchief, a cord, or even a shoelace will do, but, whatever is used, it should be loose enough to force a finger under it, because only superficial circulation should be impeded. If you are unable to obtain medical treatment within 30 minutes, proceed as follows: About every 15 minutes release the tourni-

Figure 28-27. *Crotalus,* the rattlesnake. Rattlesnakes are dangerously poisonous; they range in size from about 80 cm to 125 cm.

quet for 1 minute (important—otherwise gangrene will develop); then retighten. (3) Make an incision with a sterilized sharp instrument parallel to the limb axis of the body (if the bite is on the trunk) directly over each puncture made by the fangs; each incision should be about 6 mm long and about 3 mm deep. (4) Apply suction to the wound by mouth (but not if your mouth has any cuts or open sores). (5) Move the tourniquet upward as the swelling advances. Seek a physician as soon as possible. A person suffering from snakebite should not run or get overheated, drink liquor, inject potassium permanganate into the wound, or cauterize the bite with heat, strong acids, or anything of a similar nature. The best and only true antidote is specific snake **antivenom.** A commercial polyvalent product* is effective against all poisonous snakes in this country, except that of the coral snakes.

Notable among the venomous snakes that do not occur in the United States are the sea snakes and the cobras. The **sea snakes** (family **Hydrophiidae**) are some 50 species of true sea serpents (Fig. 28-1). Most

*Wyeth, Inc., Box 8299, Philadelphia, Pa. 19101.

Figure 28-28. Rattlesnake radiograph showing the absence of limbs, limb girdles, or sternum; but the numerous vertebrae and ribs are much alike in structure. Two rattles are visible at the posterior (caudal) end of the body.

inhabit the Indian Ocean and the tropical western Pacific; one species occurs along the Pacific Coast of tropical America. Sea snakes attain lengths of 1.0–2.5 m or more. The tail (and sometimes the body) is laterally compressed, an adaptation for swimming. Their prey consists mainly of fish, which are quickly paralyzed by their venom. Laboratory tests have indicated that the venom of one species of sea snake is more potent than is that of cobras.

The **cobras** (Fig. 28-29) and their relatives constitute a family (**Elapidae**) of some 200 species. The coral snake is a member of this family, which is dis-

tributed worldwide except for the temperate regions. The **king cobra**, *Naja nafa*, of India, China, and the Malay Archipelago is easily aroused and hence is called the "vicious" cobra. When disturbed, it raises the anterior part of its body from the ground, spreads its hood with a hiss, and threatens to strike. In India, bare-legged natives are killed in large numbers by cobras and other venomous snakes. Some cobras are capable of spitting venom accurately for about 3 m and less accurately for greater distances. They aim for the eyes of the prey. When cobra venom strikes a person's eye, it usually causes only temporary blind-

Figure 28-29. The ringhals, a South African cobra, *Hemachatus haemachatus,* with its neck spread, ready to strike and inject venom; the venom can cause death in a few minutes. A few species are capable of spitting venom accurately up to a distance of approximately 3 m. [Courtesy of the American Museum of Natural History.]

ness (if it is washed out immediately), but a strong stinging sensation persists for some time.

Order Crocodilia—Crocodiles, Alligators, Gavials, and Caimans

Crocidilians are the most advanced reptiles living today, having a nearly complete four-chambered **heart** and other advanced characteristics. They are lizardlike in form, but their jaws are extended into a long snout, and the powerful tail is used for swimming. The **nostrils** are at the end of the snout, and the **eyes** protrude from the head, so that these animals can float near the surface with only the nostrils and eyes above the water.

The eyes of crocodilians are unusual: The **photopigments** receptive to light are different in the dorsal and ventral hemispheres of each retina. The dorsal retinal hemisphere, which looks down into the water, has a photopigment similar to that found in freshwater animals (**porphyropsin**), whereas the ventral retinal hemisphere has the pigment of terrestrial animals (**rhodopsin**).

The **skin** is thick and leathery and is covered with horny **epidermal scales.** There are dorsal and sometimes ventral dermal bony plates (**osteoderms**) that are somewhat like those in turtle shells. The **nostrils** and **ears** are provided with **valves** that are closed when the animal submerges.

The **limbs** are well developed. There are five **digits** on the forelimbs and four more-or-less webbed digits on the hindlimbs. The **tail** is a laterally compressed swimming organ. The **cloaca** is a longitudinal slit. Two pairs of **musk glands** are present, one in the throat and one in the cloaca.

About 21 species of living crocodilians are known. One is assigned to the family **Gavialidae,** and 20 to the **Crocodylidae.** The only New World crocodile, the **American crocodile,** *Crocodylus acutus* (Fig. 28-30), is an inhabitant of Florida, Mexico, and Central and South America. The African crocodile, *C. niloticus,* is one of the few man-eating species. It was held sacred by the ancient Egyptians, who preserved many specimens as mummies.

There are two species of the genus *Alligator:* The **American alligator,** *A. mississippiensis* (Fig. 28-31), inhabits the southeastern United States, and the other species is restricted to China. Alligators normally do not attack humans unless provoked.

The muscles of crocodilian jaws are specialized for closing rather than for opening. Whereas a person of medium strength would have no trouble holding a crocodilian's jaws closed, the powerful muscles that close the jaws are capable of snapping a human vertebral column.

Caimans are broad-snouted crocodilians found in tropical South America. The **true gavial,** *Gavialis gangeticus,* inhabits the major river systems of India and is characterized by its long, narrow snout, large numbers of teeth, and differences in skull bone structure as compared with other crocodilians.

Figure 28-30. The American crocodile, *Crocodylus,* with pointed snout, laterally compressed tail, webbed hindfeet, five toes in front and four behind, and claws on three inner digits. This is the largest known crocodile, reaching a length of over 7 m. [Courtesy of Science Software Systems, Inc.]

Figure 28-31. The American alligator, *Alligator.* Snout blunt, not pointed as in the crocodile. Natural size up to 5 m. [Courtesy of Science Software Systems, Inc.]

Relations of Reptiles to Humans

It may be generally stated that reptiles do very little damage by destroying animals and plants economically important to humans. They are, in fact, of considerable benefit, as they kill large numbers of insect pests and destructive rodents.

The turtles and tortoises rank first among reptiles as human food. Especially noteworthy are the green turtle, the diamondback terrapin, the snapping turtle, and the soft-shelled turtle. Certain lizards, such as the iguana of tropical America, are used locally as food. The flesh of the rattlesnake is said to have an agreeable flavor, and there is a growing market for canned rattlesnake meat, which tastes something like chicken.

The tanned skins of lizards, snakes, and crocodilians are used rather extensively for leather articles that combine beauty with durability. Populations of crocodilians in this country had decreased so rapidly from hunting that laws to protect the remaining populations were enacted. To date, this action represents one of the best successes of animal conservation, because the alligators' numbers have returned to levels high enough to allow regulated hunting in several Gulf Coast states.

Tortoise shell, especially the horny carapace of the hawksbill turtle and some others, is used for the manufacture of combs and ornaments of various kinds.

As noted, the poisonous snakes of the United States are of little danger to humans. In tropical countries, especially India, venomous snakes cause more deaths than do any other group of animals. The gila monster attacks people only when handled carelessly and rarely inflicts a fatal wound.

Summary

Representatives of four orders of reptiles exist today: Chelonia, Rhynchocephalia, Squamata, and Crocodilia.

The turtles, order Chelonia, are remnants of the most primitive fossil reptiles. The turtle's shell, which consists of a dorsal carapace and a ventral plastron, sets it apart from all other reptiles. The digestive system is simple; the circulatory system includes a heart with a single ventricle divided partially by a septum; the respiratory system depends upon special inspiratory and expiratory muscles because of the inflexible shell; the urogenital system incorporates a cloaca; the nervous system includes a brain in which gray matter and white matter are segregated.

Sphenodon punctatus, the tuatara, is the only living member of order Rhynchocephalia.

Order Squamata includes the lizards and snakes. Lizards are extremely mobile, poikilothermic creatures. Snakes characteristically lack appendages.

Crocodilians are lizardlike, but generally larger, and have elongated, powerful jaws.

Dinosaurs and other extinct reptiles reached their greatest success during the Mesozoic Era, the Age of Reptiles. These extremely large reptiles successfully radiated into terrestrial, aquatic, and aerial habitats. They may have been among the first homeotherms and are probably the ancestors of modern birds.

Review Questions

1. Name two advantages that reptiles have over amphibians in terms of adaptation for terrestrial life.
2. Reptilian anatomy is intermediate between _____ and _____ anatomies.
3. True or False: The turtle, like the frog has no ribs?
4. What structure separates the small intestine from the large intestine?
5. An increase in the size of the cerebellum indicates a probable enhancement of what type of body function?
6. Describe how a turtle manages to breathe.
7. What is the largest of all lizards?
8. True or False: All snakes have no appendages or vestiges of them?
9. Name two distinctive characteristics of snake jaws.
10. Describe how a snake ingests a large food item.
11. Name two venomous snakes other than the rattlesnake.
12. Name and describe the adaptive value of the photopigments in the crocodilian eye.
13. Contrast the advantages and disadvantages of homeothermy and poikilothermy; which category (and why) do the dinosaurs fit?

Selected References

Ashley, L. M. *Laboratory Anatomy of the Turtle.* Dubuque: W. C. Brown Company, 1955.

Bellairs, A. D. A. *Reptiles.* London: Hutchinson's University Library, 1957.

Bogert, C. M., and R. M. del Campo. "The gila monster and its allies," *American Museum of Natural History Bulletin,* **109**:1, 1956.

Buckley, E. E., and N. Porges. *Venoms.* Washington: American Association for the Advancement of Science, 1956.

Carr, A. *Handbook of Turtles.* Ithaca: Cornell University Press (Comstock), 1952.)

Colbert E. E. *The Dinosaur Book.* New York: McGraw-Hill Book Company, 1951.

Conant, R. *A Field Guide to Reptiles and Amphibians of Eastern and Central North America.* Boston: Houghton Mifflin Company, 1975.

Desmond, A. J. *The Hot-blooded Dinosaurs: A Revolution in Paleontology.* New York: Dial Press, 1976.

Ditmars, R. L. *Reptiles of the World.* New York: Doubleday & Company, Inc., 1928.

Johnson, W. H., L. E. Dolanney, E. C. Williams, and T. A. Cole. *Principles of Zoology,* 2nd ed. New York: Holt, Rinehart and Winston, 1977.

Klauber, L. M. *Rattlesnakes,* 2 vols. Berkeley: University of California Press, 2nd ed., 1972.

McIlhenny, E. A. *The Alligator's Life History.* Boston: Christopher, 1935.

Pope, C. H. *The Poisonous Snakes of the New World.* New York: New York Zoological Society, 1944.

Pope, C. H. *The Reptile World.* New York: Alfred A. Knopf, Inc., 1955.

Porter, K. R. *Herpetology.* Philadelphia: W. B. Saunders Company, 1972.

Romer, A. S. *Osteology of the Reptiles.* Chicago: University of Chicago Press, 1956.

Schmidt, K. P. *The Truth About Snake Stories.* Chicago: Natural History Museum, 1951.

Schmidt, K. P., and D. D. Davis. *Field Book of Snakes of the United States and Canada.* New York: G. P. Putnam's Sons, 1941.

Smith, H. M. *Handbook of Lizards of the United States and of Canada.* Ithaca: Comstock, 1946.

Stebbins, R. C. *A Field Guide to Western Reptiles and Amphibians.* Boston: Houghton Mifflin Company, 1966.

Wright, A. H., and A. A. Wright. *Handbook of Snakes of the United States and Canada,* 2 vols. Ithaca: Cornell University Press (Comstock), 1957.

Phylum Chordata, Subphylum Vertebrata, Class Aves—Birds

Birds are considered by many people to be the most interesting of all animals. This is largely due to their powers of flight, their beautiful and varied colors, their pleasant songs and call notes and other interesting behaviors, their migrations, and the many fascinating activities associated with their nests, eggs, and young. The study of birds is the science of **ornithology.**

Because birds have a reptilian origin, some have called them feathered reptiles. Birds have reptilian scales on their legs, and the earliest ones, which we know only from fossils, had reptilian teeth. Modern birds display many advances over their reptilian ancestors, most notably a superior nervous system and a **homeothermic** physiology. There are about 10,000 living species of birds, comprising about 27 orders (Fig. 29-1).

Taxonomy and Evolution of Birds

It is customary to divide the class **Aves** into two subclasses: **Archaeornithes,** the ancient birds, and **Neornithes,** all the living birds and fossil forms that are similar to modern birds.

The study of living birds is the interest of many professional zoologists and amateur ornithologists. However, the study of ancient birds, although of great interest, is a very restricted field. This is due in

part to the relative scarcity of avian fossils. The thin bones of birds do not preserve well, and the lack of hard teeth also contributes to their poor representation in the fossil record.

The earliest feathered creatures, fossilized *Archaeopteryx lithographica* (Fig. 29-2), have been described from five fairly complete skeletons that were found in the lithographic slates of Solenhofen, Bavaria. These sediments are dated at approximately 150 million years. The discovery of the first *Archaeopteryx* fossil in 1861 caused great excitement because this animal exhibited an unmistakable blend of reptilian and avian characteristics. As such it was hailed by proponents of the newly proposed theory of evolution as the "missing link" between reptiles and birds.

The discoveries of the two most recently identified *Archaeopteryx* fossils reflect a most interesting historical story concerning the origin of birds. The specimen was actually collected in 1855 but was labeled as a pterosaur. It was not until 1970 that J. Ostrom recognized the very faint impressions of feathers; further examination confirmed its true identity as *Archaeopteryx.* In 1973, F. S. Mayr announced the revelation of another misidentified fossil. For more than two decades, this specimen was thought to be the well-known saurischian dinosaur, *Compsognathus,* which existed along with *Archaeopteryx,* as shown by other more certain fossil specimens also found in the Solen-

Gruiformes (Coot)

Passeriformes (Barn swallow)

Apterygiformes (Kiwi)

Procellariiformes (Albatross)

Anseriformes (Mallard)

Piciformes
(Yellow-bellied
sapsucker)

Gaviiformes (Loon)

Charadriiformes (Great auk)

Falconiformes
(Hawk)

Struthioniformes (Ostrich)

Casuariiformes
(Cassowary)

Ciconiiformes (Flamingo)

Figure 29-1. Representatives of some major orders of class Aves.

Figure 29-2. Fosil remains of *Archae-opteryx* (ancient bird) showing claws on digits of forelimbs and long tail.

Restoration *Archaeopteryx,* about .5 m long, is oriented as in fossil and as it probably looked in its natural state. [Courtesy of the American Museum of Natural History.]

hofen limestones. Ironically, this is the genus that T. H. Huxley chose as the ancestor of *Archaeopteryx* in his 1868 defense of Darwin's theory of evolution, suggesting that *Archaeopteryx* was the intermediate link between dinosaurs and birds. Although others have supported this hypothesis since Huxley, early critics cited the diversity and adaptive flexibility of dinosaurs and explained the anatomical similarities between *Archaeopteryx* and particular dinosaurs as convergent evolution i.e., they had separate evolutionary origins but were physically similar because of similar modes of life; see Chapter 41). In 1926, Heilman's classic treatise, *The Origin of Birds*, supposedly established the ancestry of *Archaeopteryx* and birds from a hypothetical thecodont. His objections were based largely on the presence of the **furcula** (wishbone) in *Archaeopteryx* and birds; he argued that this bone was derived from paired **clavicles** (collar bones) and that no clavicles were observed in fossil dinosaurs.

This view was supported by most all evolutionists until 1973, when Ostrom's detailed and thorough analyses of *Archaeopteryx* and several small saurischians (*Compsognathus* and relatives) revealed many more similarities than could be ascribed to convergent evolution. Although some paleontologists still disagree, the likeness to the dinosaur examples certainly seems to outnumber those to any known thecodont fossil (Fig. 29-3). The incorrect rejection of dinosaur ancestry by Heilman and others was not so much a preference for a particular alternative predecessor but, rather, an overemphasis on one trait, the furcula. Ostrom cites reports of clavicles in three different saurischian dinosaurs and also notes the problems of preservability and correct identification of these fragile bones in fossils. Indeed, the furcula is

Figure 29-3. Comparisons of forelimb anatomy for the pigeon *(Columba), Archaeopteryx,* two saurischian dinosaurs *(Ornitholestes* and *Deinonychus),* and a thecodont *(Ornithosuchus).* All drawings are scaled to present all humeri the same; the bar scale at right of each limb represents 5 cm. Note the greater similarities between *Archaeopteryx* and the dinosaurs as compared with the thecodont skeleton.

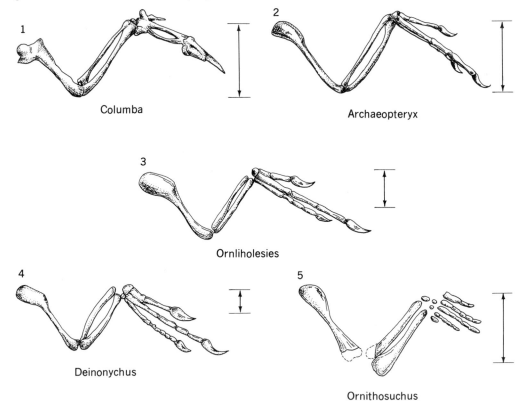

1
Columba

2
Archaeopteryx

3
Ornliholesies

4
Deinonychus

5
Ornithosuchus

present in only two of the five known samples of *Archaeopteryx*. The problem of avian origins is still considered controversial; the appearance of *Archaeopteryx* along with the diversification of many other small, bipedal, saurischian dinosaurs seems to be the most reasonable hypothesis for the present.

Archaeopteryx was about the size of a crow. It possessed teeth embedded in sockets, feathered forelimbs with three-clawed digits and unfused metacarpal bones, and a lizardlike tail with large feathers on both sides. The shape of their teeth suggests that they preyed upon insects and other small animals. The clawed digits were probably used for catching prey and maybe also for climbing, and its broad and rounded wings might have allowed it to swoop from tree to tree. Ostrom's studies reveal differences between fossil *Archaeopteryx* and modern avian skeletons, indicating that the former's wings could not be raised high enough to generate the thrust needed for sustained flight—*Archaeopteryx*, therefore, was probably only a glider at most. Some doubt its ability to fly at all, suggesting that feathers originated for the sole purpose of insulation and that the elongated forelimbs served as balancers during rapid pursuits of prey.

The descendants of *Archaeopteryx* and other ancient birds underwent a dramatic adaptive radiation during the Cretaceous period (Fig. 23-3) when both aquatic and terrestrial niches were invaded. *Hesperornis* (Fig. 29-4) was a loonlike diver that possessed teeth and atrophied, functionless wings. *Archaeopteryx* and *Hesperornis* are the only birds known to have teeth. All other known fossil birds and all modern birds lack teeth.

The major groups of birds evidently evolved rather rapidly. As a result, the taxonomic "tree" of class Aves can be best represented as a broad "bush" with many branches. This diverse class is conventionally divided into about 27 orders. Space does not permit a summary of each avian order, and so only a few of the major types of birds will be surveyed here, after a discussion of a representative bird, the common pigeon. Refer to the classification appendix for an overview of avian taxonomy and evolution.

Columba livia—The Common Pigeon

The common pigeon is derived from the rock dove, *Columba livia*, which originally ranged from Europe,

Figure 29-4. *Hesperornis,* a restoration, showing teeth, absence of wings, and laterally directed legs with lobed toes for swimming. Fossils of this bird have been found in the United States. [Courtesy of the Field Museum of Natural History.]

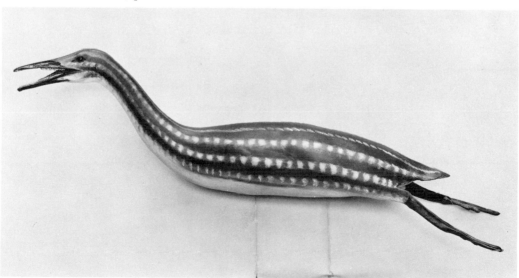

through the Mediterranean region, to central Asia and China. The pigeon is commonly used for studies of bird anatomy, not only because of its convenient size and availability, but also because it so well displays many adaptations for aerial life.

External Morphology

The body of the pigeon is streamlined, an adaptation that reduces air drag while flying. Three body regions may be recognized: head, neck, and trunk. The **head** is prolonged anteriorly into a pointed horny **bill** that covers the toothless **jaws.** At the upper base of the pigeon's bill is a patch of naked swollen skin, the **cere.** Opening between the bill and the cere are two oblique slitlike **nostrils.** The **eyes** are provided with upper and lower **lids** and a well-developed **nictitating** membrane. This third eyelid can be drawn across the eyeball from the anteriomedial corner. Posteroventrally to each eye is an **external** ear opening that leads to a tympanic cavity.

The **neck** is long and flexible. At the posterior end of the **trunk** is a projection that bears the tail feathers.

The two **wings** can be folded close to the body or extended during flight. The feet are covered with horny epidermal **scales,** and each of the four **digits** is provided with a horny **claw.**

Feathers

Feathers are peculiar to birds and distinguish them from all other animals. Feathers arise, like the scales of reptile, form **papillae.** A papilla consists of a projection of vascularized dermal tissue that grows out of an epidermal pit, called the **feather follicle.** A typical feather (Fig. 29-5) consists of a stiff **axial rod,** or **shaft.** The proximal portion of the shaft, the **quill** or **calamus,** is hollow and semitransparent, whereas the distal portion, the **rachis,** is angular and solid. The shaft bears two rows of branches, or **barbs,** which in turn support two rows of smaller, numerous **barbules.** The feathery **vane** is composed of a double series of barbs and barbules. The barbules on the side of the barb toward the tip of the feather bear hooklets, known as **barbicels,** that form bridges with ridges on the adjacent proximal barbules. The vane is thus

Figure 29-5. Left: Types of feathers. Right: Detail of a contour feather to show the curved edge, along which the hooklets slide to make the feather flexible.

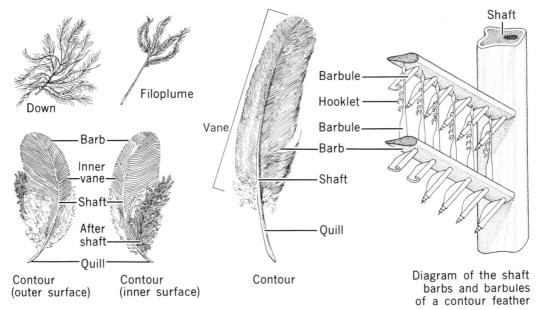

lightweight and pliable, but also extremely strong and resilient. At least once a year each feather is shed, and a new feather develops from the same papilla. There are three principal kinds of feathers:

1. **Contour feathers** are like the typical feather just described, with a stiff shaft and an interlocking vane. Contour feathers determine to a large degree the contour or outline of the body and include the wing (flight) and tail feathers.
2. **Down feathers** have a reduced shaft and no hooks on their barbules. As a result, the vane is not firm but rather a fluffy tuft. Natal down provides the first plumage of nestlings. In adult birds, especially waterfowl, down provides a layer of insulation beneath the contour feathers.
3. The **filoplumes** are hairlike feathers with a minute vane at the tip. These are the feathers singed off a chicken before it is sold for food. Filoplumes may have decorative or sensory functions.

Only certain regions of the pigeon's body bear contour feathers; such feather tracts are termed **pterylae**. The areas without contour feathers are called **apteria**. The feather tracts vary in different species of birds; those of the pigeon are shown in Figure 29-6. Down feathers are widely distributed in both the pterylae

and apteria. Filoplumes are always associated with contour feathers.

Birds usually shed, or molt, their old feathers during the late summer, before their fall migrations. Each replacement feather develops within the same follicle as the old feather. There may also be a partial or complete molt in the spring when the bird assumes a more colorful **breeding plumage.** The acquisition of breeding plumage may also result from wear or the breaking off of feather tips, thus exposing different colors beneath.

Internal Morphology and Physiology

Skeletal System

Adaptations for flight and bipedal locomotion highlight the principal differences between the skeletons of birds and reptiles. Most avian bones are very light, because many contain **air cavities.** The **forelimbs** and **pectoral girdle** of the pigeon are modified for flight, whereas the hindlimbs and **pelvic girdle** are adapted to bipedal locomotion.

The skeleton of the common fowl (Fig. 29-7) is similar to that of the pigeon in most respects but is larger and more easily studied. The **skull** is very light and strong, with most of the bones so completely fused that their boundaries can be distinguished only in a

Figure 29-6. Feather tracts of the pigeon reveal that feathers do not develop equally on all parts of the body as they do in some primitive birds such as the penguin.

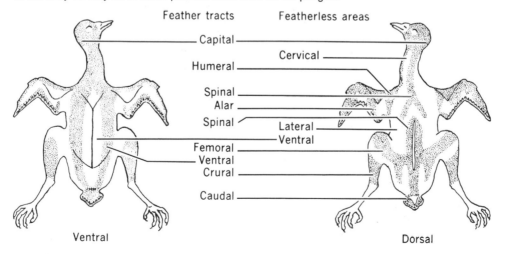

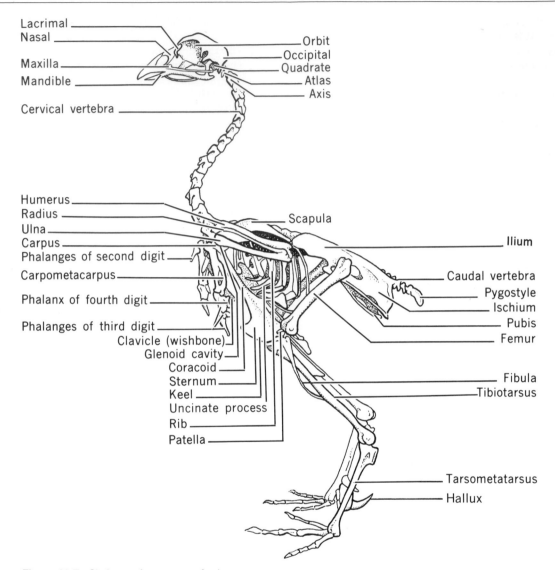

Lacrimal
Nasal
Maxilla
Mandible
Cervical vertebra

Orbit
Occipital
Quadrate
Atlas
Axis

Humerus
Radius
Ulna
Carpus
Phalanges of second digit
Carpometacarpus
Phalanx of fourth digit
Phalanges of third digit
Clavicle (wishbone)
Glenoid cavity
Coracoid
Sternum
Keel
Uncinate process
Rib
Patella

Scapula

Ilium

Caudal vertebra
Pygostyle
Ischium
Pubis
Femur

Fibula
Tibiotarsus

Tarsometatarsus
Hallux

Figure 29-7. Skeleton of a common fowl.

young bird. The **cranium** is rounded, the **orbits** are large, and no teeth are present in the **bill.** The skull only has a single **occipital condyle** that articulates with the first vertebra.

The **cervical vertebrae** are long and articulate freely with one another upon saddle-shaped surfaces. This makes the neck very flexible, enabling the bird to look in all directions. The **thoracic** and **lumbar ver-**

tebrae of the trunk are almost completely fused together into a rigid skeletal axis that supports the body while in flight.

There are two pairs of **cervical ribs** and four to five pairs of **thoracic ribs.** The second cervical and first four thoracic ribs each bear an **uncinate process** that arises from each rib's posterior margin and overlaps the succeeding rib, thus reinforcing the rib cage. The thoracic ribs articulate with the **breastbone (sternum)** ventrally. The sternum is united in front with the **coracoid** of the pectoral girdle and bears on its ventral surface a median ridge, the **keel,** to which the large pectoral muscles that move the wings attach.

There are four or five unfused **caudal vertebrae,** followed by a terminal **pygostyle** consisting of five or six fused vertebrae. The pygostyle supports the tail feathers, and the free caudal vertebrae allow movement of the tail.

The pectoral girdle consists of the **scapulae (shoulder blades), coracoids,** and **furcula** (wishbone). The scapulae are long, narrow, and bladelike; they lie above the ribs, one on each side of the vertebral column. The scapula and coracoid form the **glenoid fossa** in which the head of the humerus articulates.

The **forelimb,** or **wing,** of a bird is highly specialized and departs in many ways from the primitive vertebrate condition. For example, primitive vertebrates had five digits (numbered 1 to 5, beginning with the thumb), whereas a bird has only three digits (numbers 2, 3, and 4), and only the third digit is well developed. The upper arm, as in other vertebrates, contains a single bone, the **humerus.** The forearm possesses two bones, the **radius** and the **ulna.** The **wrist** contains two free **carpal bones;** the other carpal bones are fused with the three **metacarpals,** thus forming the solid **carpometacarpus,** which increases the rigidity of the wing. In addition to the carpometacarpus, the **hand** possesses the second digit, called the **alula,** consisting of two small **phalanges.** The alula supports a small tuft of feathers on the wing tip. The middle digit has three phalanges, and the fourth contains a single phalanx.

The main flight feathers, the **primaries,** are supported by digits 3 and 4, the **secondaries** by the ulna, and the **tertiaries** by the humerus.

The **pelvic girdle** consists of pairs of **ilia, ischia,** and **pubes,** as is true in nearly all tetrapod vertebrates. These bones are firmly fused together with the posterior part of the vertebral column, forming a rigid, stable unit known as the **synsacrum.**

The **hindlimbs** are used for bipedal locomotion. The thigh is concealed beneath the feathers; the **femur** is the short thick thighbone. Each lower leg contains a slender **fibula** and a long stout **tibiotarsus** bone of the drumstick, which consists of the **tibia** fused with the proximal row of **tarsal bones.** The ankle joint is between the tibiotarsus and the **tarsometatarsus;** the latter represents the fusion of the distal row of tarsal bones with the second, third, fourth, and fifth metatarsal. In addition to the tarsometatarsus, the foot also possesses four digits: the first digit **(hallux)** is directed posteriorly, and the other three are directed forward. Each digit bears a terminal **claw.** The number of phalanges in each digit is one more than the digit number; hence, digit 1 has two phalanges and digit 4 has five.

Muscular System

The muscles of the neck, tail, wings, and legs are especially well developed. The largest muscles are the **pectoralis majors** that produce the downward stroke of the wings. This pair of flight muscles comprise about one fifth of the body weight. Each pectoralis major originates along the sternal keel and inserts on the humerus. The **pectoralis minor** muscles, which underlie the majors, raise the wing. Together these muscles constitute what is popularly known as the bird's breast.

A perching reflex enables a bird to maintain itself upon a roost even while asleep. When the foot touches a perch, a pull is exerted on tendons that flex the foretoes and hallux (hindtoe) together thus holding the bird firmly to its perch.

Another interesting specialization of bird musculature is the large number of **cutaneous** muscles that allow a bird to ruffle and fluff the feathers. Some 12,000 such muscles occur in the skin of a Canadian goose.

Digestive System

Birds (and mammals) and **homeothermic** animals: They have relatively constant body temperatures that are independent of their external environment. The body temperatures of various birds generally range from 40–43°C (which is about 4–5°C higher than those of mammals). The body temperature of homeotherms (or **endotherms**) is derived from their own oxidative metabolism, and some 80–90 percent of an endotherm's food may go toward maintaining a stable body temperature (see Chapter 28).

Allied with their high metabolic rates, birds require large quantities of food, and digestion is rapid.

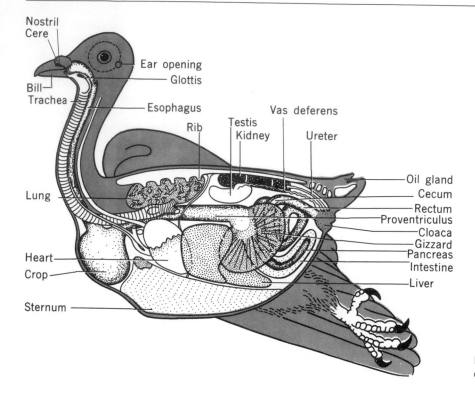

Nostril
Cere
Ear opening
Glottis
Bill
Trachea
Esophagus
Rib
Testis
Kidney
Vas deferens
Ureter
Oil gland
Cecum
Rectum
Proventriculus
Cloaca
Gizzard
Pancreas
Intestine
Liver
Lung
Heart
Crop
Sternum

Figure 29-8. The internal structure of a pigeon, *Columba*.

Pigeons feed principally upon vegetable matter such as seeds. The **mouth cavity** opens into the **esophagus** (Fig. 29-8), which enlarges into a **crop,** where food is stored and moistened. Pigeons are noted for their unique ability to produce "pigeon's milk." This creamy substance, similar in composition to mammalian milk, results from the degeneration of cells lining the crop. All pigeons and doves regurgitate this "milk" to feed their young. A short extension of the esophagus continues to the stomach, which consists of two parts: a small anterior **proventriculus** with thick glandular walls and a posterior muscular **gizzard** (**ventriculus**). The proventriculus secretes gastric juices, and the gizzard grinds food with the aid of grit sand grains and small pebbles that the bird habitually swallows. The slender, coiled **intestine** joins to the rectum at a point where two blind pouches, the **cecae,** are given off. The digestive tract terminates in the **cloaca,** which also receives the urinary and genital ducts. In young birds a thick glandular pouch of lymphatic tissue, the **bursa of Fabricus,** forms as a dorsal

outgrowth of the cloaca. The cloaca opens to the external environment through the **cloacal opening,** or **vent.**

The large, bilobed **liver** secretes **bile** into the small intestine through two **bile ducts.** The common pigeon has no gallbladder, although some birds and some other species of pigeons do have one. The **pancreas** pours digestive enzymes into the small intestine through three ducts.

Circulatory System

The relatively large **heart** (Fig. 29-9) is completely four chambered, having two muscular **ventricles** and two thin-walled **atria** (Fig. 35-1). This four-chambered heart of birds, with the complete separation of oxygenated from deoxygenated blood, is an advance over the three-chambered heart of their rep-

Figure 29-9. Circulatory system of a pigeon.

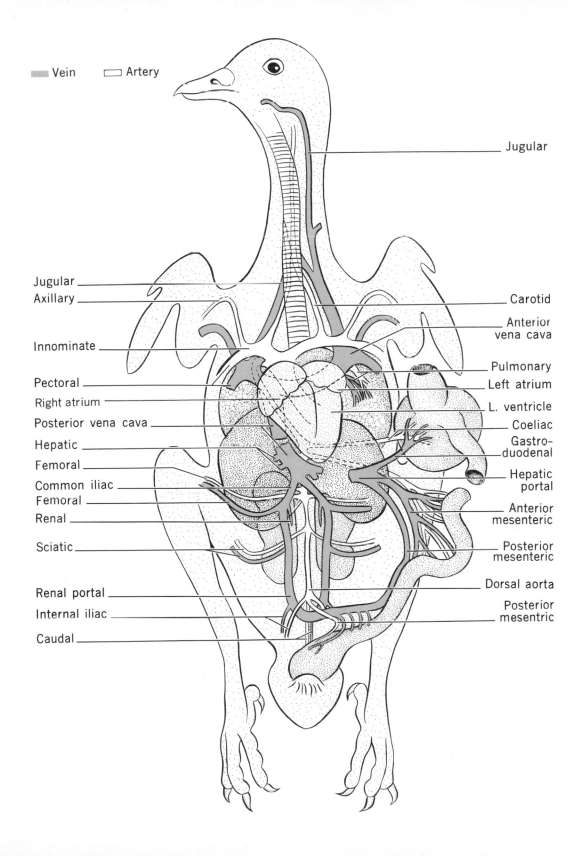

Vein   Artery

Jugular

Jugular
Axillary
Innominate
Pectoral
Right atrium
Posterior vena cava
Hepatic
Femoral
Common iliac
Femoral
Renal
Sciatic
Renal portal
Internal iliac
Caudal

Carotid
Anterior vena cava
Pulmonary
Left atrium
L. ventricle
Coeliac
Gastro-duodenal
Hepatic portal
Anterior mesenteric
Posterior mesenteric
Dorsal aorta
Posterior mesentric

tilian ancestors. The **right atrium** receives deoxygenated venous blood from the two **anterior** and one **posterior venae cavae.** The right atrium pumps this blood into the right ventricle, which then pumps it through the **pulmonary artery** to the lungs. Oxygenated blood returns from the lungs through four large **pulmonary veins** to the **left atrium.** Oxygenated blood is pumped from the left atrium into the **left ventricle,** which is then pumped through the **right aortic arch** into the **innominate arteries** and the **dorsal aorta** (Fig. 29-9). After completing the systemic circuit the now deoxygenated blood returns to the venal cavae. Venous blood from the posterior body is delivered directly to the heart, not through a renal portal system as in the lower vertebrates. The **jugular veins** of the pigeon are united by a cross vein;

this special adaptation ensures the blood flow when the neck is twisted and one of the jugular veins is blocked.

Circulation in birds is extremely rapid; the heart may beat several hundred times per minute when the bird is at rest and up to a thousand or more beats per minute when under stress.

Respiratory System

The two **lungs** in birds (Fig. 29-10) are assisted by a remarkable system of **air sacs.** During inspiration, relaxation of the thoracic and abdominal muscles allows elastic expansion of the thorax and abdomen; contraction of these muscles expels air during expiration. Air enters through the **external nostrils (nares)**

Figure 29-10. Diagram of the respiratory organs of a pigeon, showing the air sacs. Note that the hollow bones are connected to the respiratory system.

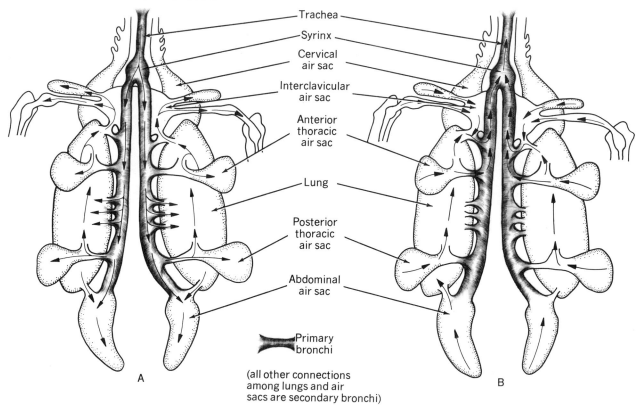

and enters the **pharynx** through the **internal nares.** Air passes through the slitlike **glottis** into the **trachea (windpipe),** which branches into two **primary bronchi;** the trachea and bronchi are held open by partially calcified cartilaginous rings. Here the respiratory system differs from that of other vertebrates: each primary bronchus leads directly into an **abdominal air sac,** which is a member of a system of large, thin-walled air sacs that extend throughout the body cavity and neck region *and* into the cavities of the larger bones (e.g., the humeri). Many **secondary bronchi** interconnect the primary bronchi, the lungs, and all air sacs, as detailed in Figure 29-10. As each secondary bronchus enters a lung, it breaks up into smaller tubes, called **bronchioles,** that ramify further into an intricate series of air passageways. This arrangement results in the unique respiratory cycle of birds. Inhalent air travels primarily into the posterior air sacs and lungs. Also during this phase, "old," stale air in the lungs enters the anterior sacs (Fig. 29-10A). During the expiratory phase the fresh air from the posterior sacs travels through the lungs, and the stale air leaves the anterior sacs via the primary bronchi (Fig. 29-10B). The net result is a fairly constant gas exchange in the lungs during both inspiration and expiration, obviously more efficient than that in all other vertebrate systems. The air sac system also helps to eliminate excess body heat by evaporative cooling; birds do not have sweat glands.

Birds have high respiratory rates. In fact, birds have the highest oxygen requirements of all animals due to their elevated metabolic rate. The breathing rate of the pigeon is 29 times per minute at rest, 180 while walking, and 450 in flight—a sharp contrast to the 14 to 20 times per minute in humans.

Where the lower end of the trachea divides into bronchi, it enlarges to form the vocal organ, or **syrinx,** a structure unique to birds. Extending up into the syrinx from the bifurcation of the trachea is a bony ridge to which is attached a short vibratory membrane, the **semilunar membrane.** This membrane vibrates like a reed when air is forcibly expelled from the lungs. Sound production also occurs through the aid of the **tympaniform membranes,** which form slit-like openings over the entrance to the bronchi. The different songs and calls of birds are made possible by variations in the tension of these vibratory membranes, which are controlled by delicate muscular actions.

Excretory System

The **kidneys** are a pair of dark brown, three-lobed bodies situated just below the synsacrum, as shown in Figure 29-9. Each kidney discharges urine through a duct, called the **ureter,** into the **cloaca.**

As do their reptilian ancestors, birds convert ammonia into **uric acid,** the whitish component of their droppings. Most of the water is reabsorbed from the urine, reducing it to a whitish paste.

Seabirds and a few terrestrial species also have a pair of **nasal,** or **supraorbital, glands** that help the kidneys to excrete excess salts from the body. These modified tear glands discharge concentrated salt solution through the nostrils. The nasal glands are so efficient that gulls, albatrosses, and other seabirds can subsist entirely on seawater with no ill effects.

Nervous System and Sense Organs

The **brain** of the pigeon (see Fig. 37-8) is very short and broad. The overall structure of the brain and nervous system is typically reptilian, but the **cerebrum, cerebellum,** and **optic lobes** are relatively large as compared with those of reptiles.

The **bill** and **tongue** serve as **tactile organs.** Tactile nerves are also present at the bases of the feathers, especially those of the wings and tail. Birds are usually unable to distinguish delicate odors, and on the whole their sense of smell is very poor. The sense of taste is also poorly developed, but it is nevertheless present, as is easily demonstrated by presenting a bird with a bad-tasting morsel of food.

Birds have an acute sense of hearing. The **cochlea,** the part of the inner ear that contains the auditory receptors, is more complex than is that of reptiles. The **eustachian tubes** open through a single aperture on the roof of the pharynx.

The **eyes** of birds are very large. The **visual acuity** (the ability to distinguish nearby objects) is eight times greater in some birds of prey than in humans. Birds also have a wide field of vision, as their eyes are placed laterally on the head. **Nocturnal** birds are adapted for night vision; an owl's ability to perceive objects in dim light is about ten times that of humans'. This remarkable ability results from the presence of a **tapetum lucidum,** a reflective layer situated behind the receptor cells of the retina. Birds also have extraordinary powers of eye **accommodation,** changing from farsighted to nearsighted vision in an instant, which helps them to fly rapidly among the branches of a tree or swoop down to the ground from a great height in the air.

Endocrine System

Birds have endocrine glands that are homologous with those of mammals: the **pituitary gland (hypophysis)** at the base of the brain, the **thyroid** in the neck, the **islets of Langerhans** in the **pancreas,** the **adrenals** on the ventral surface of the kidneys, and the **gonads.** The avian hormones most studied have been those associated with the gonads.

Reproductive System

Male birds have a pair of oval **testes** (Fig. 29-9). A duct called the **vas deferens** communicates each testis with the cloaca. The **sperm** pass through the vasa deferentia and are stored in the **seminal vesicles.** During copulation, the sperm are discharged into the male's cloaca and then transferred by contact into the cloaca of the female. There is no copulatory organ in most birds with the exception of male ducks, geese, swans, ostriches, and tinamous, which have a curved penis derived from the ventral wall of the cloaca.

The right **ovary** of the female bird usually disappears during development, and only the left ovary persists in the adult. The yolky **ova** are released from the ovary and enter the **oviduct.** During their passage through the upper end of the oviduct, **albumen,** commonly known as "egg white," is secreted. As the egg passes down the oviduct, first a thick albumen, the **chalaza,** is deposited over it. The egg rotates on its journey, and, during the process, the chalaza is twisted. Toward the lower end of the oviduct, a parchmentlike **shell membrane** is secreted around the egg. Finally the **shell** of porous calcium carbonate is added by a distended region of the oviduct ("uterus") shortly before deposition (Fig. 29-11).

Fertilization, if it occurs, takes place in the upper oviduct about 41 hours before the first egg of a clutch is laid.

Following an incubation period of about 14 days, the young pigeons have developed to such a stage that they can break through the shell and hatch. The nestlings are covered with fine down at first, but they soon acquire a covering of contour feathers. During early life, the nestling are fed pigeon's milk, which is ejected from the crop of the parent into the young bird's mouth.

It has been shown that experimental injection of pigeon with a **lactogenic hormone (prolactin)** will stimulate the secretion of pigeon's milk at any time of the year.

Figure 29-11. Diagram of the structure of a bird's egg.

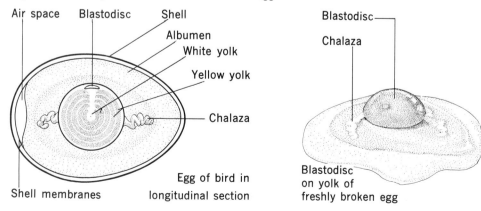

A Survey of Major Avian Orders

Wingless and Flightless Terrestrial Birds

Four orders of living birds fall into this category. These contain ostriches (**Struthioniformes**), rheas (**Rheiformes**), cassowaries and emus (**Casuariiformes**), and kiwis (**Dinornithiformes**). The flightless penguins are discussed with the aquatic birds. **Ostriches** (Fig. 29-1) are the largest living birds, attaining heights of more than 2.5 m and weights of more than 135 kg. Although flightless, they can run at speeds greater than 80 km/hr. The ostriches live in dry open regions of Africa and travel in groups. Contrary to popular superstition, they do not stick their heads in sand and think that they have hidden themselves. The nest is a hollow in the sand, and several females lay their eggs in a single nest. The males cover them with warm sand by day and uncover and incubate the eggs at night.

Rheas are ostrichlike birds that inhabit the pampas of South America. They are smaller than the ostriches, but their habits are quite similar. **Cassowaries** (Fig. 29-1) and **emus** are also ostrichlike birds. The emus (Fig. 29-12) are confined to Australia, and the cassowaries inhabit New Guinea and neighboring islands. **Kiwis** (Fig. 29-1) are wingless birds that live in New Zealand. They are about the size of a chicken, their wings are rudimentary, and they lack tail feathers. Kiwis bear a peculiar hairlike plumage throughout life and are credited with a keen sense of smell, which is unique among birds. Correlated with this is the fact that the nostrils are at the tip of the long, probing bill.

Aquatic Birds

Many birds are well adapted for spending a large part of their life in or near water. Perhaps the bird most conspicuously adapted for aquatic life is the **penguin** (order **Sphenisciformes**, Fig. 29-13). Its paddle-like forelimbs are modified for swimming underwater, its feet are webbed, water is easily shaken from its feathers, and an insulating layer of fat just beneath the skin serves to retain body heat. **Loons** (order **Gaviiformes**, Fig. 29-1) are large birds that swim and dive with great agility. The somewhat smaller **grebes** are also excellent swimmers and divers. **Albatrosses** (Fig. 29-1), **fumars**, and **petrals** (order **Procellariiformes**) possess exceptionally long and narrow wings that are adapted for soaring over the open ocean.

Figure 29-12. An adult emu with chicks that are strikingly striped. This characteristic disappears as the flightless birds become mature.

Figure 29-13. Emperor penguins are antarctic birds that are unable to fly.

Albatrosses rarely come onto land except to breed, and they are thought to sleep for short periods while gliding.

Pelicans (order **Pelecaniformes**) possess a huge membranous pouch between the mandibles of the lower jaw (Fig. 29-14). The pouch is used to hold small fish scooped up during feeding. This order also includes the **comorants** (Fig. 29-15). Among the common wading birds are the **flamingos** (Fig. 29-1), **herons** (Fig. 29-16), and **bitterns** (order **Ciconiiformes**). They possess long legs, broad wings, and short tails. The tropical flamingos are gregarious birds that congregate in the thousands of mud flats where they feed and build conical mud nests. Most flamingos have rosy-white plumage with scarlet wing feathers.

The ducklike birds (order **Anseriformes**, Fig. 29-1) are adapted for swimming, with short legs and fully webbed front toes. Their young are entirely covered with down, and they can swim and walk soon after hatching; that is, they are precocial. This order includes **swans, geese, river ducks, sea ducks,** and **mergansers.**

Figure 29-14. Pelican. Note the pouch that hangs from the lower bill; this serves as a scoop net for capturing and storing fish. [Courtesy of Science Software Systems, Inc.]

The marsh birds (order **Gruiformes**) are mostly of the wading type, with incompletely webbed front toes. This order includes the **rails, gallinules, coots, cranes,** and **limpkins.** The shorebirds (order **Cha-**

Figure 29-15. Cormorant, a fish-eating bird. [Courtesy of Science Software Systems, Inc.]

Figure 29-16. Great blue heron. Long neck, long legs, slender body, and a stilettolike bill are characteristics of this water-adapted species. [Courtesy of Science Software Systems, Inc.]

radriiformes) include a varied assemblage of **plovers, sandpipers, gulls, terns, jacanas, puffins,** and **auks** (Fig. 29-1) that frequent fresh or salt water. The jacanas are tropical shorebirds with very long toes and claws that enable them to walk over lily pads without sinking. The puffins and auks spend a large part of their life at sea; most are excellent swimmers and divers, but they are very awkward on land.

Birds of Prey

The falconlike birds (order **Falconiformes**) are active predators. These **diurnal** (active during the day) birds of prey generally possess powerful wings, a stout, hooked bill with a cere at the base (Fig. 29-8),

and strong toes armed with sharp, curved talons. The **New World vultures** (order **Cathartidae,** Fig. 29-17) have weak feet and live on carrion (dead animals). They are especially valuable as scavengers in warm countries, where they remove dead bodies before they become a health hazard. Some of the birds of prey that live in the United States and the **swallow-tailed kite** *Elanoides forficatus,* **osprey** (*Pandion haliaetus,* **bald eagle** (*Haliaetus leucocephalus,* (Fig. 29-18), **red-tailed hawk** (*Buteo jamaicensis*), Cooper's hawk (*Accipiter cooperii*), **kestrel** or **sparrow hawk** (*Falco sparverius,* Fig. 29-1, and **golden eagle** (*Aquila chrysaetos*).

Owls (order **Strigiformes**) are **nocturnal** (active at night) birds of prey. Owls possess large rounded heads, strong bills with the upper mandible curved downward, large eyes directed forward and surrounded by a radiating disc of feathers, strong legs, feet armed with sharp claws, and soft fluffy plumage that renders their flight almost noiseless. Owls prey upon mice, rats and other small mammals, insects,

Figure 29-17. The California condor is saprophagous; that is, it feeds on carrion (dead animals).

Figure 29-19. The screech owl is one of the smallest of the birds of prey. These owls show two color phases, some being grayish and the others reddish colored. They are well known because of their weird nocturnal cries. [Courtesy of the N.Y. Zoological Society.]

birds, and fish. The indigestible parts (e.g., bones) of their prey are regurgitated in the form of pellets. Most species of owls are beneficial to humans. Among the best-known North American owls are the **barn owl**, *Tyto alba*, **screech owl**, *Otus asio* (Fig. 29-19), **great horned owl**, *Bubo virginianus* (Fig. 29-20), **burrowing owl**, *Speotyto cunicularia*.

Perching Birds

More than half of all the known species of modern birds belong to the order **Passeriformes** (Fig. 29-1), the perching birds. Their feet are four toed and adapted for grasping (Fig. 29-24). The first toe, or hallux, is directed backward and is level with the three that are directed forward. The nonsinging perching birds (e.g., **flycatchers**) have a poorly developed syrinx. Most passerines, however, are noted for their melodious songs.

Other Well–Known Birds

The order **Cuculiformes** includes the cuckoos and the roadrunners. **Cuckoos** are mostly tropical birds. The majority do not build nests but lay their eggs in nests of other birds. The **roadrunners** *(Geococcyx)* inhabits deserts in the southwestern United States

Figure 29-18. Bald eagle. The head and tail are white and the beak hooked.

Figure 29-20. Great horned owl. Owls have a keen sense of hearing, which may help them in locating prey at night. The tufts of feathers around the large external ear openings are characteristic of this species.

where it lives among cacti, sagebrush, and mesquite. This long, slender bird is an excellent runner that preys upon lizards and snakes.

Swifts and hummingbirds (order **Apodiformes**) are small, extremely agile, and quick flyers. The **chimney swift,** *Chaetura pelagica,* formerly made its nest in hollow trees, but now it usually frequents chimneys (Fig. 29-29). When in the open air, it is always on the wing, catching insects or gathering twigs from dead tree branches for its nest. The twigs are glued together with saliva and firmly fastened to the inside of a chimney, forming a cup-shaped nest. Certain species of swifts inhabiting the East Indies construct nests from a salivary gland secretion, producing the edible birds nests relished by the Chinese for soup. The minute **hummingbirds** (Fig. 29-21) were appropriately called "glittering fragments of the rainbow" by Audubon. There are 688 known species and subspecies of hummingbirds, all confined to the New World.

Woodpeckers (order **Piciformes**) use their chisel-shaped bills to excavate holes in trees in search of insects that live beneath the bark. Most woodpeckers are of great benefit because of the insects that they

Figure 29-21. Annas Ruby-throated hummingbird.

destroy. But the yellow-bellied sapsucker, *Sphyrapicus varius* (Fig. 29-1), is harmful as it eats portions of trees and sucks sap, thus disfiguring and devitalizing fruit and ornamental trees and reducing the market value of lumber.

Recent Extinctions and Endangered Species

The **passenger pigeon** (*Ectopistes migratorius*, Fig. 29-22), **great auk** (*Pinguinu impennis*), **Labrador duck** (*Camptorhynchus labradorium*) **heath hen,** (*Tympanuchus cupido*), and various other birds have become extinct in recent times. Alexander Wilson in 1808 reported a flock of passenger pigeons in Kentucky that he estimated contained over 2 billion birds, and Audubon wrote of flocks that darkened the sky like the approach of a tornado. The last known passenger pigeon died of old age in The Cincinnati Zoo in September 1914. The enormous slaughter of this bird by hunters was undoubtedly a critical factor in its extinction. Every great food market from St. Louis to Boston received hundreds of barrels of pigeons until it was realized too late that this bird needed protection.

The great auk (Fig. 29-1) became extinct from overhunting in 1844. These penguinlike birds from the North Atlantic were destroyed for their feathers, and their eggs were used as food. All that remains today of the great auk are about 80 preserved specimens 75 eggs, and some 25 skeletons.

The North American **ivory-billed woodpecker,** *Campephilus principalis*, is in extreme danger of extinction or may already be extinct. In 1970, there were only 20 known individuals left.

Because of the destruction of feeding or breeding areas, the **American bald eagle,** *Haliaeetus leucocephalus*, the **whooping crane,** *Grus americana*, and the **California condor,** *Gymnogyps californianus*, are also among those in danger of extinction.

Adaptive Modifications and Behaviors of Birds

Different species of birds have become adapted to specific environments. Specialized adaptations are most easily observed in the wings, tails, feet, and bills.

Body Shape

Although there are a few flightless species, the great majority of birds can fly. One of the adaptations that makes flight possible is body form. An irregularly

Figure 29-22. The passenger pigeon, *Ectopistes migratorius*, became extinct after indiscriminate slaughter. Length: 40 cm.

shaped body meets significant wind resistance as it moves through the air. On the other hand, a body that is **fusiform** in shape, rounded, and tapered from the middle toward each end can pass with less resistance through either air or water. Birds generally have such spindle shapes, as do the bodies of most animals that actively fly or swim (Fig. 26-8).

Wings

The wings of most birds are used as organs of flight. Several different wing shapes characterize different types of flight. Some highly aerial birds (e.g., swallows, gulls, and albatrosses) have long pointed wings that enable them to soar in the air for long periods. On the other extreme, some birds (e.g., the **bobwhite** and **song sparrow**) possess short, rounded wings that enable them to take off quickly and fly rapidly for short distances. Many aquatic birds (e.g., penguins, auks, and murres) use their wings effectively for swimming and diving. Some flightless birds (e.g., ostriches, rheas, emus, and kiwis) possess only the vestiges of wings but have well-developed legs.

The ancestors of the modern birds probably used their forelimbs for climbing in addition to gliding and

Figure 29-23. The hoatzin of South America is unique in many ways. The young bird, as shown here, has claws on its wings and climbs about among the branches. It is regarded as a survivor of that ancient period when birds were distinctly reptilian in character. Note that avian and reptillian embryologies are similar.

flying. The extinct *Archaeopteryx* (Fig. 29-2) had three strong claws on its forelimbs. Among living birds, the young of the **hoatzin** (Fig. 29-23), a peculiar South American bird, are able to climb tree branches with the aid of two claws on each forelimb; small wing claws are found in numerous other species as well.

Wings may also serve as organs of offense and defense, as in large waterfowl, or for communication, as in the "drumming" of the ruffed grouse.

Tails

During flight, the tail acts as an aerial rudder; a long-tailed bird is thus able to fly in short curves or to follow an erratic course without difficulty. The tail is also used as an air brake. It is light and therefore easy to manipulate, and the tail feathers are firmly supported by the terminal bone of fused vertebrae, the **pygostyle** (Fig. 29-7). Movement of the tail is allowed by the freely articulating vertebrae just anterior to the pygostyle. While perching, the tail acts as a balancer. Birds that cling to the sides of trees or to other objects (e.g., woodpeckers and chimney swifts) brace themselves by means of their stiff tail feathers.

In many birds, the tail feathers of the male are more colorful than are those of the female. Notable among such sexually dimorphic birds are the **lyrebird**, **peacock**, and **turkey**.

Feet

The feet (Fig. 29-24) are used for locomotion, for obtaining food, and for building nests and as offensive and defensive weapons. Many ground birds have strong feet adapted for scratching, whereas perching birds have feet adapted for grasping branches. Most swimming birds have webbed feet, wading birds have long legs and long toes, and birds of prey possess very strong feet with long sharp **claws** (**talons**) for capturing prey.

Bills

The bills of birds (Fig. 29-24) function primarily to procure food, but they are also used to construct nests, to preen feathers, and to perform other duties. In **preening** the bill is used to extract a drop of oil from an **oil gland** at the base of the tail and then

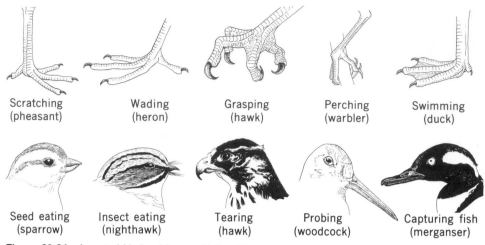

Figure 29-24. Assorted kinds of feet and bills of birds.

spread the oil through the feathers. The oil serves to keep the feathers in good condition and also prevents the horny covering of the bill from becoming brittle. Furthermore, experimental removal of the oil gland in some birds results in rickets, a disease marked by soft, deformed bones.

Seed-eating birds possess strong, thick bills for crushing seeds, whereas insect eaters generally have small, slender bills. Birds of prey are provided with strong curved beaks adapted for tearing flesh, ducks have flattened bills for straining food from water, and some birds have serrated bills for holding fish.

Coloration

Birds are among the most beautifully colored of all animals. Some colors are due to two basic pigments within the feathers; these are the **melanins,** pigment granules of brown, black, or yellow, and the **carotenoids,** which are either red or yellow. Green, blue, and iridescent markings as on some hummingbirds and others, are due to the peculiar surface and internal structure of their feathers. Absence of pigment results in partial or complete **albinism.**

The **juvenile plumage** of birds eventually gives way to the first **winter plumage.** This is usually worn throughout the first winter and is generally dull in color, often resembling the plumage of the adult fe-

male. Males and females frequently differ in color (**sexual chromatic dimorphism**), especially during the breeding season, when the male may have a brightly colored coat; the **cardinal** and **scarlet tanager** exhibit marked sexual chromatic dimorphism.

Color may often be of protective values to birds. The color patterns of many birds effectively camouflage them in their surroundings. A striking example of such **protective coloration** is the **ptarmigan,** which is white during winter when snow is on the ground, but a mottled brown during the other seasons.

Bird Songs

The songs of birds, as previously explained, are produced by the forceful exhalation of air through the syrinx. A familiarity with bird songs is very helpful to an ornithologist, as one generally hears a great many more birds than one is able to see. Songs should be distinguished from call notes. **Songs** are usually heard during the breeding season and are generally limited to the males. They function as a substitute for physical combat in the defense of the birds' territories and are important in establishing and maintaining the pair bonds and in synchronizing the reproductive activities of mating pairs.

Call notes, on the other hand, are produced throughout the year and are analogous to our conversations. Call notes serve to warn of danger, to communicate among parents and young, and to coordinate flocking, feeding, and migratory activites.

Bird Flight

As mentioned, the bodies of flying birds are shaped so that they offer minimal resistance to the air. Several adaptations result in a low center of gravity, which tends to prevent the body from turning over during flight. The wings are attached high up on the trunk, as are such light organs as the lungs, whereas the heavy flight muscles and digestive organs are positioned ventrally.

The speeds at which bird fly vary considerably. The **carrier pigeon** attains a maximum racing speed of about 96 km/hr. For ducks, 145 km/hr has been recorded, but their usual speed is 65–110 km/hr. The **peregrine falcon**, *Falco peregrinus*, has been clocked during a nose dive at 267–290 km/hr, and several **swifts** appear to have comparable speeds.

Bird Migration

The **Arctic** tern, *Sterna paradisea*, holds the record for long-distance migration. The extremes of its Arctic

nesting and Antarctic wintering ranges are 16,700 km apart. Because the routes taken are circuitous, these birds may fly 40,300 km each year. During the autumn, many birds gather in flocks and fly southward, returning the following spring. Certain species migrate east and west. Birds that breed in the far north may spend the winter in parts of the temperate zone. Some birds such as the **great horned owl**, *Bubo virginianus*, and the **bobwhite**, *Colinus virginianus*, do not migrate, whereas certain others move southward only when the weather becomes very severe.

Most birds migrate on clear nights at an altitude of about 1 km or more, but there are some daytime migrants. A flock of **snow geese**, *Chen hyperborea*, is shown in Figure 29-25. Each species has a more or less definite time of migration, and one can predict with some accuracy the date at which it will pass a given locality. As a rule the speed of migration is rather slow; for example, a daily rate of about 60 km is about the average for the **robin**, *Turdus migratorius*.

Many hypotheses have been advanced to account for the onset of bird migration, and factors such as temperature, amount of light per day (**photoperiod**), and condition of the local food supply have been proposed. Changes in the photoperiod seem to correlate most strongly with the onset of migration. There is

Figure 29-25. A migration of Canadian geese in formation.

still much to learn about how birds navigate along their migratory routes. Chapter 32 discusses migrations in more detail.

Nests, Eggs, and Nestlings

Some birds, such as **Canadian geese,** *Branta canadensis,* and eagles, usually mate for life, but the majority live together for only a single breeding season. The nesting period varies according to the species.

Many birds conceal their nests or construct them in relatively inaccessible places. Some species simply lay one or more eggs directly upon the ground (Fig. 29-26). A nest-building bird may construct a flimsy platform of twigs (**doves,** family **Columbidae**) or an intricate hanging basket (**orioles** and other **Icteridae**), but most species build distinctive nests that fall between these extremes.

A few birds do not build nests, or incubate their eggs, or take care of their offspring. This is true of the **European cuckoo,** *Cuculus canorus* (Fig. 29-27), and the **brown-headed cowbird,** *Molothrus ater.* Their

Figure 29-27. A fairy tern, like newly hatched ducks and pheasants are covered with down and leave the nest soon after hatching (precocial).

eggs are usually laid in the nests of other birds that are smaller than themselves. The young cuckoos or cowbirds are reared by their foster parents, and, because of the intruders' relatively larger size, they often starve or crowd out the rightful offspring.

Bird eggs vary in shape, size, color, and number per nest. The smallest eggs are those of certain hummingbirds, measuring about 1 cm in length. The largest bird eggs were those of the extinct **elephant bird,** *Aepyornis,* of Madagascar (Fig. 29-28), whose volume equaled that of 148 hen's eggs; the largest living eggs are those of the ostrich, which are about 25 times larger than are those of a hen. Eggs laid in dark places, such as those of woodpeckers, are generally white. Many eggs are colored or spotted, and those of one species are generally distinguishable from those of other species. The number of eggs laid in a **clutch** (setting) varies from 1 to 20 but averages about 4 to 6 for most passerine (perching) birds.

The average incubation period for the passerines is about 12 to 14 days. The eggs of the ostrich hatch in about 45 days. Those of the royal albatross, *Diomedia epomorpha,* have the longest period of incubation:

Figure 29-26. An egg of the fairy tern laid directly on a broken branch.

Figure 29-28. *Aepyornis,* an elephant bird. This is a restoration of the large, flightless, strange bird that once lived in Madagascar. [Courtesy of the Field Museum of Natural History.]

about 80 days. In some cases, only the female incubates the eggs. In others, the male and female work in shifts, and, in a few species of shorebirds and the ostrich, the male alone incubates the eggs.

Two general types of young birds are recognized: **Precocial** birds are covered with down and are able to run about soon after hatching (Fig. 29-29), whereas **altricial** birds are born blind and naked and must remain in the nest for some time before they are able to take care of themselves.

Relations of Birds to Humans

Birds are of great commercial value; they augment our food supply and furnish feathers for various pur-

poses. Before the wearing of wild bird feathers was prohibited by law, vast numbers of birds were killed for their colorful plumes. The **American egrets** (*Asmerodius albus,* Fig. 29-30) were almost exterminated in efforts to secure their distinctive nuptual plumes, called **aigrettes,** that resemble spun glass. In certain regions, the excrement **guano** of seabird, especially **cormorants** (Fig. 29-15), accumulates in great quantities and is a valuable source of fertilizer. The greatest quantities are on small islands near the coast of Peru, where guano may accumulate at annual rates of 2.75×10^3 kg/m².

One of the most beneficial services rendered by birds is their consumption of weeds and land insects that are not beneficial to humans. Practically all the insects devoured by birds are injurious to plants or animals. In Salt Lake City a monument has been

Figure 29-29. Right: Young chimney swifts are naked and blind at birth and are cared for in the nest for about three weeks (altricial). [Photo by Robert Knickmeyer from National Audubon Society.] Left: piping plovers, like the newly hatched ducks and pheasants, are covered with down and leave the nest soon after hatching (precocial). [Photo courtesy of Bertha Daubendiek.]

Figure 29-30. American egret, *Casmerodius albus,* in a cypress swamp in the southern United States. The long, plumelike feathers, known as aigrettes, are carried only during the breeding season. Length: About 1 m.

erected to the **California gull,** *Larus Californicus.* The crops of the Mormons in 1848 were threatened by a plague of locusts, "Mormon crickets," when flocks of gulls appeared, ate the locusts, and saved the crops. Other birds prey upon small mammals such as field mice, ground squirrels, and rabbits. Hawks, owls, and other birds of prey have been killed indiscriminately because of their alleged destruction of poultry and game birds. However, the majority of the species are chiefly beneficial.

Birds have also been domesticated as sources of meat, eggs, and feathers. The common hen was probably derived from the **red jungle fowl,** *Gallus gallus,* of northeastern and central India. The varieties of chickens that have been derived from this species are legion. The many varieties of domesticated pigeons (tumbler, fantail, pouter, etc.) are all descendants of the wild rock dove, *Columba livia.*

Geese are supposed to have been derived from the **graylag goose,** *Anser anser.* Most domesticated breeds of ducks have sprung from the **mallard,** *Anas platyrhynchos* (Fig. 29-1). The common **peacock,** *Pavo cristatus,* of the India and Ceylon has been domesticated since at least 1000 B.C. The **guinea fowl,** *Numida meleagris,* is a native of West Africa, and domesticated turkeys are descendants of bird domesticated in pre–Columbian Mexico.

Summary

The common pigeon is a standard representative of the class Aves. Its feathers are composed of a shaft and vane, with barbs bearing hooked and unhooked barbules. Three major types of feathers occur: contour feathers, down feathers, and filoplumes.

The avian skeleton is for the most part hollow and is modified for flight and bipedal locomotion. Most of the skull bones and other large portions of the axial skeleton are fused in the adult.

The pectoral muscles are responsible for avian flight. The digestive system includes a stomach that consists of a proventriculus and a muscular gizzard. Birds excrete very concentrated uric acid. Aves is the first class considered that have distinctly four-chambered hearts.

Birds are oviparous, and the shell is secreted as the egg passes through the oviduct.

Most birds are fusiform and winged and use their tails as rudders during flight. Their feet are variously adapted for swimming, wading, clawing, scratching, and grasping. The bills also show species-specific morphology that reflects adaption to their diets.

Bird migration is usually the result of a seasonally unsuitable habitat or other physical factors. Some migrations may involve flying as far as 40,300 km each year.

Some of the 27 orders of class Aves were summarized in the text.

Birds are generally extremely beneficial to humans, especially in terms of "pest" control.

Review Questions

1. Explain in some detail the morphology of a contour feather.
2. True or False: Feathers usually grow evenly over a bird's body, except over the legs, feet, bill, and eyes?
3. What is the structure to which the large pectorals are attached on the sternum?
4. Hallux is
   (a) a type of corn eaten by Old World flightless birds.
   (b) a rearward-facing digit of the foot.
   (c) a tree-climbing bird indigenous to South America.
5. What inorganic particles does the gizzard utilize in its function?
6. What is the main difference between a reptilian and an avian heart?
7. What muscles serve a function analogous to a diaphragm in birds?
8. Tail movement is possible because of the freely moveable vertebrae just anterior to the _____ .
9. A short, thin beak would indicate what form of nutrition?
10. What organ is responsible for bird songs?
11. Name two aquatic birds, and give a general description of each.
12. List some avian adaptations for endothermy.

Selected References

American Ornithologists' Union. *Check-list of North American Birds*. Lancaster, Pa: American Ornithologists' Union, 1957.

Audubon, J. J. *The Birds of America*. New York: The Macmillan Company, 1937.

Austin, O. L., Jr. *Birds of the World*. New York: Golden Press, 1961.

Baurvuel, P. *Birds of the World: Their Life and Habits*. New York: Oxford University Press, 1973.

Bent, A. C. *Life Histories of North American Birds*, 19 vols. Washington: U.S. National Museum, 1919–1953.

Bradley, O. C. *The Structure of the Fowl*. Philadelphia: J. B. Lippincott Company, 1950.

Cody, M. L. *Competition and Structure of Bird Communities*. Princeton: Princeton University Press, 1974.

Dorst, J. The *Migration of Birds*, translated from original French edition (1956) by C. D. Sherman. London: William Heinemann, 1962.

Dorst, J. *The Life of Birds*. New York: Columbia University Press, 1974.

Pearson, R. G. *The Avian Brain*. New York, Academic Press, Inc., 1972.

Jack, A. *Feathered Wings: A Study of the Flight of Birds*. London: Methuen & Company, Ltd., 1953.

Lincoln, F. C. *Migration of Birds*. New York: Doubleday & Company, Inc., 1952.

Marshall, A. J. (ed.). *Biology and Comparative Physiology of Birds*, 2 vols. New York: Academic Press, Inc., 1960–1961.

Parker, T. J., and W. A. Haswell. *A Textbook of Zoology*. New York: St. Martin's Press, 1962.

Peters, J. L. *Birds of the World*, 7 vols. Cambridge: Harvard University Press, 1931, 1934, 1937, 1940, 1945, 1948, 1951.

Peterson, R. T. *A Field Guide to the Birds*, 2nd ed. Boston: Houghton Mifflin Company, 1959.

Peterson, R. T. *A Field Guide to Western Birds*, 2nd ed. Boston: Houghton Mifflin Company, 1961.

Schorger, A. W. *The Passenger Pigeon*. Madison: University of Wisconsin Press, 1955.

Skutch, A. F. *Parent Birds and Their Young*. Austin: University of Texas Press, 1976.

Sturkie, P. D. *Avian Physiology*. Ithaca: Comstock Publishing Associates, 1954.

The Auk. Quarterly journal published since 1884 by the American Ornithologists' Union.

Tinbergen, N. *Social Behavior in Animals*. London: Methuen & Company, Ltd., 1953

Van Tyne, J., and A. J. Berger. *Fundamentals of Ornithology*. New York: John Wiley & Sons, 1959.

Wallace, G. J. *An Introduction to Ornithology*, 2nd ed. New York: Macmillan Publishing Co., Inc., 1963.

Welty, J. C. *The Life of Birds*. Philadelphia: W. B. Saunders Company, 1962.

Phylum Chordata, Subphylum Vertebrata, Class Mammalia—Mammals

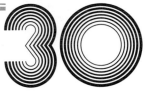

The name **mammalia** is applied to those animals that possess mammary glands that secrete milk for the nourishment of their young. Mammals also possess body hair at some time during their life history, and they are all homeothermic and endothermic. Parental care is generally highly developed in this class, as is evident in humans.

The majority of the mammals are **viviparous**, and their young are nourished before birth through an organ (**placenta**) that attaches the embryo to the uterus of the mother.

The 5000 or more species of living mammals are distributed in every major habitat, from the tropics to the polar regions, from the oceans to the driest deserts.

Taxonomy and Evolution of Mammals

The fossil record indicates that the mammals evolved from the mammal-like reptiles, which shared remote ancestry with the modern reptiles (see Chapter 28). Throughout the Age of Reptiles during the Mesozoic era, mammals apparently could not compete successfully against the dinosaurs. For over 100 million years the early mammals, derived from the therapsids, were generally mouse sized and were restricted to niches such as those occupied by modern shrews and moles. Many were probably nocturnal, another attribute that minimized their contact with the dominant reptiles. During the transitions from pelycosaurs to therapsids to early mammals, the distinctive characteristics of mammals evolved: endothermic physiology, with its associated homeothermic attributes such as insulating hair and increased circulation; efficient locomotion; vivipary and the nursing of young; and relatively high intelligence.

Upon the mass extinction of the dinosaurs about 70 million years ago, many potential niches on land, sea, and air were vacated, and an explosive radiation of mammals followed. The beginning of the Cenozoic era saw the evolution of many of the mammalian orders (Fig. 23-3 and the classification appendix). As a result, the mammalian pedigree (as with that of the birds and many other major adaptive radiations) resembles a broad bush more than a simple tree.

Class Mammalia (representative orders: Fig. 30-1) is customarily subdivided into two subclasses on the basis of reproductive physiology: subclass **Prototheria,** containing the **oviparous mammals,** and subclass **Theria,** containing the **viviparous mammals.** This division reflects the very early divergence of prototherians from the rest of the mammals. Subclass Theria subsequently diverged into two major branches, again based on differences in reproductive strategies: infraclass **Methatheria (pouched mammals)** and infraclass **Eutheria (placental mammals).**

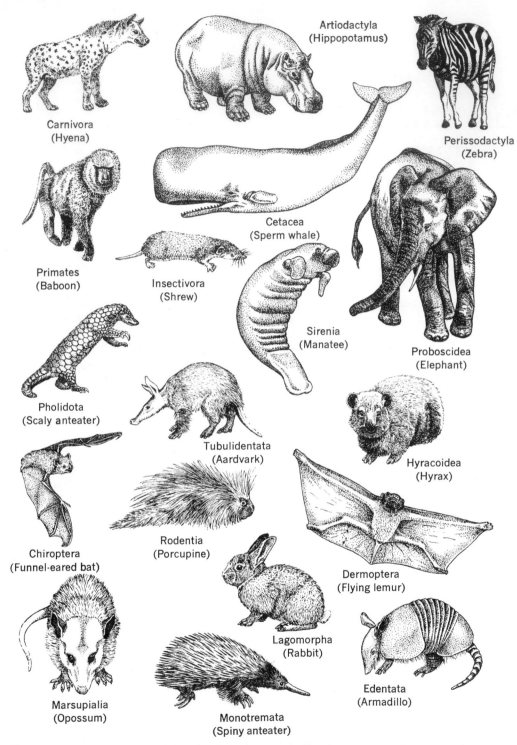

Figure 30-1. Representatives of the 18 orders of mammals. The figures are not drawn to scale.

Subclass Prototheria—Monotremes

These most primitive mammals survive today only in Australia, Tasmania, and New Guinea. There is only one prototherian order (**Monotremata**), and so these oviparous mammals are commonly called **monotremes.** Before hatching, the embryo lives on the yolk contained in the pliable egg. After hatching, the young are for a time nourished by milk from the mother's **mammary glands.** There are no teats, so the young lap up the milk as it oozes from the tubular glands. There are five living species of **spiny anteaters,** or **echidnas** (Fig. 30-2A), and one species of **duck-billed platypus,** *Ornithorhynchus anatinus* (Fig. 30-3B). The platypus is noted for its ducklike, horny bill. An aquatic carnivore, it feeds chiefly on

A

B

Figure 30-2. Monotremes, (A) Spiny anteater, *Echidna,* an egg-laying mammal. Like the duck-billed platypus, its egg shell is not limy but leathery. The egg is incubated in a pouch where the newly hatched young are kept for a time. The echidna has numerous hard sharp quills and an underfur of coarse hair. It possesses digging powers that enable it to disappear quickly. These animals are typical anteaters with long and slender bony tongues, adapted to catching ants and other insects. [Courtesy of H. C. Reynolds.] (B) Duck-billed platypus, *Ornithorhynchus anatinus.* About 75 cm long with thick fur and webbed feet. It possesses both reptilian and mammalian characteristics.

A

B

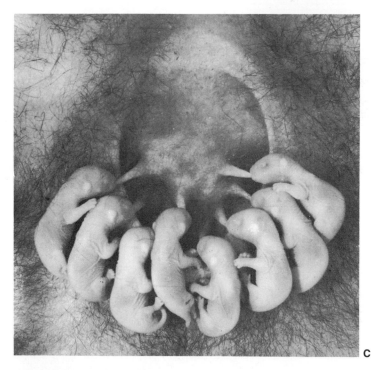

C

Figure 30-3. Marsupials. (A) A type of kangaroo, the wallaro, *Macropus robustus.* An adult female, about 1 m long; note the young in pouch. (B) Koala, the Australian "teddy bear." Female first carries young in pouch, then on her back. Adults are about 70 cm long. They live in trees and feed on the leaves of the giant gum (eucalyptus) tree of this region. [Courtesy of Science Software Systems, Inc.] (C) Young American opossums, *Didelphis virginiana,* attached to the mammary glands in the brood pouches of their mother. When born, the young opossum is strikingly underdeveloped and smaller than a honeybee. The opossum is considered the most primitive of America's mammals and is often referred to as a "living fossil." Except for its larger size, it is almost the same as its ancestor of millions of years ago.

freshwater snails and mussels. The female lays leathery eggs in an underground burrow and incubates them for about two weeks until the young hatch. The male platypus has a poisonous claw on its heel.

Subclass Theria, Infraclass Metatheria—Marsupials

Infraclass Metatheria also contains only one order (**Marsupialia**). Marsupials are most abundant in Australia and its neighboring islands, but a few survive in the Americas. The young are born in a premature condition, having no hair, eyes, or ears, but with good olfactory organs and well-developed front feet. For a while they live in a pouch on the mother's abdomen where they cling to the mammary teats. The young of the **American opossum**, *Didelphis virginiana* (Figs. 30-1 and 30-3C), remain in the mother's external pouch until they complete their natal development. Many species of marsupials, such as **kangaroos** (Fig. 30-3A) and **koalas** (Fig. 30-3B), are distributed over the Australian continent.

Subclass Theria, Infraclass Eutheria—Placentals

The earliest placentals were small, insect-feeding (**insectivorous**) mammals. Some of the earliest offshoots represented today are the edentates (order **Edentata**) in South America, bats (order **Chiroptera**), and the aboreal primates (order **Primates**). This was soon followed by a diversification characterized by

various modes of feeding, such as the primarily herbivorous rodents and rabbits (orders **Rodentia** and **Lagomorpha**), the carnivores (the primarily terres-

A

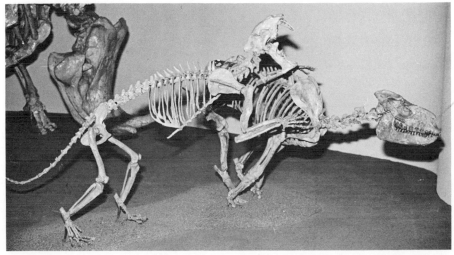

B

Figure 30-4. Some examples of North American mammals that become extinct during the Pleistocene epoch. (A) Giant ground sloth. (B) Saber-toothed tiger.

trial order **Carnivora** and aquatic order **Pinnipedia**), and the very diverse branch of herbivores that include and hooved mammals (orders **Perissodactyla** and **Artiodactyla**) and elephants (order **Proboscidea**). Habitat diversification, with the corresponding anatomical modifications, are also apparent during the adaptive radiations of mammals; this is especially evident in the whales and relatives (order **Cetacea**).

As with the other vertebrate classes, there were many early intermediates represented today only by their fossils. Besides these primitive forms, many of the relatively recent mammals, primarily the larger terrestrial types, are not present today because of a massive extinction that occurred within the last 20,000–30,000 years. This particularly affected the native populations of North and South America, resulting in the loss of horses, many camels, sabertoothed cats, and many others (Fig. 30-4). As with the extinction of dinosaurs, many hypotheses have been proposed, but no single cause is apparent. The **Great Ice Age** (the **Pleistocene epoch,** from about 55,000 to 10,000 years ago) was probably an important factor, but the survival of most groups through the major portion of the four or five series of extensive glaciation seems to require that additional influences were necessary to cause the sudden, massive extinctions. The rise of man as an efficient hunter was also probably influential in the extinction of some species, but probably not all of them, as some have suggested.

Thirteen of the seventeen placental orders are surveyed in the following paragraphs; all are described in the classification appendix.

Order Carnivora

Order Carnivora contains flesh-eating mammals such as wolves, cats, bears, weasels, and seals, to name a few. The teeth of carnivores (Fig. 30-23) are perhaps their most characteristic feature. The incisors (front teeth) are small and of little use; but the canines (eyeteeth) are very large and pointed, enabling these predators to capture, kill, and tear apart prey. The fourth premolar of the upper jaw and the first molar of the lower jaw bite on one another like a pair of scissors and are called shearing (or carnassial) teeth. The other molars are usually broad crushing teeth.

Felis domesticus—A Representative Mammal

The common house cat, *Felis domesticus*, is a carnivore that belongs to the same family as the lions, tigers, and smaller cats.

External Morphology

The domestic cat is a four-footed animal (**quadruped**), as are most mammals, reptiles, and amphibians. The cat possesses an external covering of **hair (fur).** The **mouth** is bounded by thin fleshy **lips.** Two narrow **nostrils** open at the anterior tip of the head. The large **eyes,** one on either side of the head, are protected by an **upper** and a **lower eyelid,** each bordered by fine **eyelashes,** and also by a white, hairless, third eyelid, or **nictitating membrane,** that may be drawn over the eyeball from its inner angle. Above and below the eyes and on the upper lip are long tactile hairs, the **whiskers (vibrissae).** There are two **external ears (pinnae)** evident behind the eyes.

The **trunk** may be separated into an anterior **thorax,** which is supported laterally by the ribs, and a posterior **abdomen.** Internally, a muscular **diaphragm** separates the **thoracic** and **abdominal cavities.** Beneath the base of the **tail** is the **anus,** just anterior to which is the **urogenital opening.** In males the **scrotum,** containing the **testes,** hangs beneath the anus. Both sexes generally have four pairs of small **nipples (teats)** on the ventral surface of the thorax and abdomen. In the female the ducts of the **mammary glands** open through the teats.

The **forelimbs** of the cat each possess five **toes** with fleshy **pads** and retractile **claws.** The **hindlimbs,** which are stouter and more muscular than the forelimbs, provide most of the power in locomotion. Each hindlimb bears only four toes, as the digit corresponding to the human great toe is absent. The cat walks on its toes, a posture that is termed **digitigrade.**

Internal Morphology and Physiology

Skeletal System

The skeleton of the cat (Fig. 30-5) contains bones that correspond closely to those in the human skele-

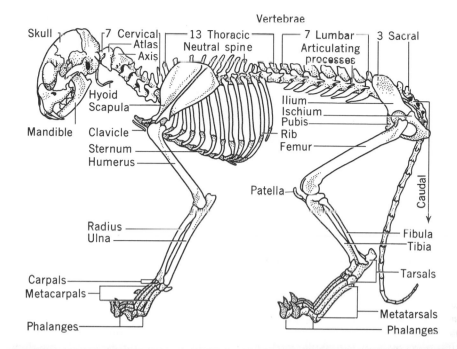

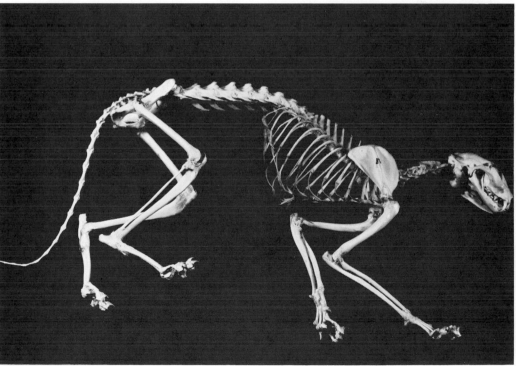

Figure 30-5. Skeleton of the cat. This represents a highly specialized skeleton. Note how much more highly developed the appendicular skeleton is in comparison with that of the fish in Figure 26-5.

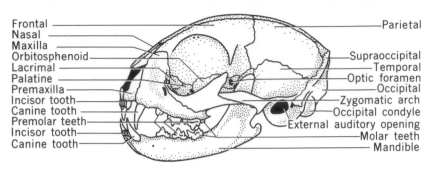

Frontal
Nasal
Maxilla
Orbitosphenoid
Lacrimal
Palatine
Premaxilla
Incisor tooth
Canine tooth
Premolar teeth
Incisor tooth
Canine tooth

Parietal
Supraoccipital
Temporal
Optic foramen
Occipital
Zygomatic arch
Occipital condyle
External auditory opening
Molar teeth
Mandible

Figure 30-5. Skull of the cat (lateral view). Some of the sutures are "dove-tailed."

ton. The cat skeleton consists principally of bone, but a small amount of cartilage is also present. As in other vertebrates, the skeleton includes both **cartilage bones** preformed in cartilage and **membrane bones** that arise from connective tissue (Chapter 3). A third type, called **sesamoid bones,** also occurs in the tendons of some limb muscles; the **kneecap (patella)** is such a bone.

The **axial skeleton** consists of the skull, vertebral column, ribs, and sternum. The **skull** is formed of both cartilage bones and membrane bones, with only a small amount of cartilage persisting in the adult. The individual bones are so immovably fused with one another that their boundaries are in many cases obliterated in the adult and can be distinguished only in an immature individual.

The **vertebral column,** as in other vertebrates, supports the body and protects the spinal cord. The vertebral column is flexible, as the bony **vertebrae** are separated (except in the sacrum) by **intervertebral discs** of fibrocartilage. There are 7 **cervical** vertebrae in the neck; 13 **thoracic** vertebrae articulate with the ribs that frame the chest; 7 **lumbar** vertebrae occur in the abdominal region; 3 fused **sacral** vertebrae support the pelvis; and 16 to 20 caudal vertebrae form the skeletal axis of the tail.

The **ribs** and **sternum** frame the thoracic cavity; they not only protect the heart and lungs but also play an important role in respiration. There are 13 pairs of ribs. The first 9 pairs articulate with the sternum, while the others terminate in the muscles of the abdominal wall. The long, laterally compressed sternum consists mostly of bone and is divided transversely into segments.

The **appendicular skeleton** contains the bones of the pectoral girdle, forelimbs, pelvic girdle, and hindlimbs. The **pectoral girdle** consists of two **scapulae (shoulder blades),** two knoblike **coracoid processes,** and two **clavicles (collar bones).** Each forelimb contains a **humerus, radius, ulna, 7 carpal bones, 5 metacarpals,** and 14 **phalanges.**

Each half of the **pelvic girdle** is called the **hipbone,** or **innominate** bone, and is made up of an **ilium, ischium,** and **pubis** fused together. The socketlike concavity in the innominate bone in which the head of the **femur (thighbone)** articulates is called the **acetabulum.** The **hindlimb** contains the **femur, tibia, fibula, 7 tarsals, 4** long **metatarsals** and a rudiment of the first (innermost) metatarsal, and 12 **phalanges.** The ankle joint of the cat lies between the lower leg bones (tibia and fibula) and the tarsal bones. The most important sesamoid bone of the hindlimb, the **patella (kneecap),** is situated anterior to the distal end of the femur.

Muscular System

Many of the muscles (Fig. 30-6) are more or less the same as in lower vertebrates, but mammals have more highly developed muscles on the head, neck, and limbs, and muscles on the vertebrae and ribs are somewhat reduced. A distinctive feature of mammalian musculature is the dome-shaped **diaphragm** that partitions the coelom into an anterior **thoracic cavity** containing the heart and lungs and a posterior **abdominal cavity** containing the other viscera. The **facial muscles** of the cat allow expression of emotional states to a small degree.

See Chapter 33 for additional information on the mammalian muscular system and its functions.

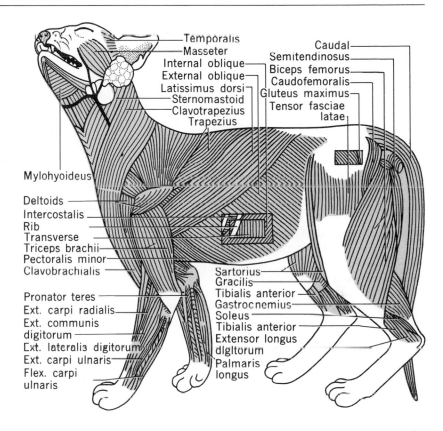

Figure 30-6. Superficial muscles of the cat; some of the abdominal muscles are cut to show the disposition of others that lie beneath.

Digestive System

The anterior roof of the **mouth cavity,** known as the **hard palate,** bears a series of transverse ridges that help hold food. Posterior to the hard palate and its bony foundation is a fleshy flap, the **soft palate,** which separates the mouth from the pharynx, the cavity formed by the meeting of the nasal and mouth cavities. Lateral to the posterior part of the soft palate are a pair of small reddish masses of lymphoid tissue, the **tonsils.** The two orifices of the **auditory (eustachian) tubes** are situated in the roof of the mouth in the region of the soft palate, as are the two openings of the **internal nares.** The **tongue** is attached to the floor of the mouth.

There are four pairs of **salivary glands:** (1) the **parotids,** below the ears; (2) the **infraorbitals,** below the eyes; (3) the **submaxillaries,** behind the lower jaws; and (4) the **sublinguals,** beneath the tongue. The sali-

vary glands pour watery, mucous secretions into the mouth cavity. The saliva moistens, lubricates, and partially digests the food.

On the floor of the **pharynx** is the **glottis,** the respiratory opening that leads into the **larynx,** or **voice box.** A bilobed cartilaginous flap, the **epiglottis,** covers the glottis while swallowing. Posterior to the glottis, the pharynx opens into the narrow, muscular **esophagus.** The esophagus penetrates the diaphragm and opens into the **stomach.**

The stomach connects, via the **pyloric valve,** to the first subdivision of the **small intestine,** the **duodenum.** This U-shaped tube also receives the **pancreatic duct** from the pancreas and the bile duct from the liver. The small intestine is several meters in length; the duodenum is followed by the **jejunum** and the terminal region, the **ileum.** At the junction of the ileum and **large intestines,** a short blind sac, the **cecum,** is given off. The cecum ends in a conelike sac, but there is no appendix as in humans. The large intestine terminates in the digestive waste storage portion, the **rectum,** which opens to the external environment at the **anus.**

Circulatory System

The red blood cells (**erythrocytes**) of the cat and other mammals are unlike those of the lower vertebrates. Instead of being relatively large and round, mammalian erythrocytes are small biconcave discs, lacking nuclei and other organelles.

The mammalian **heart** is completely four-chambered, as in birds. But the main artery, the **aorta,** arising from the **left ventricle,** forms the **left arch,** whereas in birds the right arch persists. The **right systemic arch** of the cat is represented by the **innominate artery,** which is the common trunk of the right **carotid** and **subclavian** arteries. A **hepatic portal system** is present, but there is no renal portal system. The elongate **spleen,** a dark reddish organ lying on the left side behind the stomach, produces **lympho-** cytes and removes old erythrocytes from the blood and serves as a storage organ for noncirculating erythrocytes.

Lymphatic System

The lymphatic system is extensive in the cat and other mammals. **Lymph vessels** collect excess intracellular fluid and filter it through the **lymph nodes** before emptying it into large veins in the neck region. A portion of the lymphatic system that collects nutriment (especially fats) from the small intestine are called **lacteals.**

Respiratory System

External respiration in the cat and in all other mammals occurs by means of the **lungs** (Figs. 30-7 and 30-8). The **glottis** opens into the **larynx** (**voice box**), from which a tube call the **trachea** (**windpipe**) arises. The larynx is supported by a number of carti-

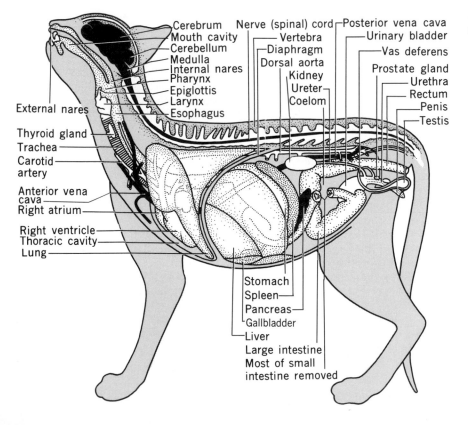

Figure 30-7. Gross internal anatomy of the cat.

Cerebrum
Mouth cavity
Cerebellum
Medulla
Internal nares
Pharynx
Epiglottis
Larynx
Esophagus

External nares

Thyroid gland
Trachea
Carotid artery
Anterior vena cava
Right atrium
Right ventricle
Thoracic cavity
Lung

Nerve (spinal) cord
Vertebra
Diaphragm
Dorsal aorta
Kidney
Ureter
Coelom

Posterior vena cava
Urinary bladder
Vas deferens
Prostate gland
Urethra
Rectum
Penis
Testis

Stomach
Spleen
Pancreas
Gallbladder
Liver
Large intestine
Most of small intestine removed

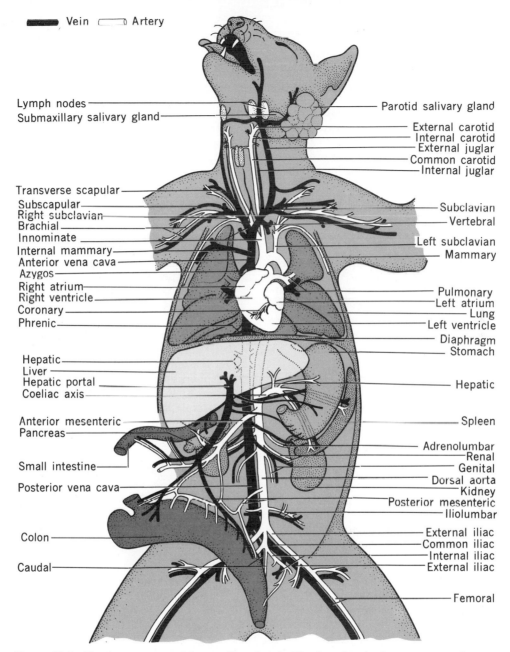

Vein ▬ Artery ▭

Lymph nodes
Submaxillary salivary gland
Parotid salivary gland
External carotid
Internal carotid
External juglar
Common carotid
Internal juglar

Transverse scapular
Subscapular
Right subclavian
Brachial
Innominate
Internal mammary
Anterior vena cava
Azygos
Right atrium
Right ventricle
Coronary
Phrenic

Subclavian
Vertebral
Left subclavian
Mammary

Pulmonary
Left atrium
Lung
Left ventricle
Diaphragm
Stomach

Hepatic
Liver
Hepatic portal
Coeliac axis

Hepatic

Anterior mesenteric
Pancreas

Spleen

Adrenolumbar
Renal
Genital
Dorsal aorta
Kidney
Posterior mesenteric
Iliolumbar

Small intestine

Posterior vena cava

Colon

External iliac
Common iliac
Internal iliac
External iliac

Caudal

Femoral

Figure 30-8. Circulatory system of the cat. How does it differ from the circulatory systems of vertebrates previously studied?

lages, and across its central cavity extend two elastic **vocal cords** that produce the cat's meows and cries. The trachea is held open by incomplete rings of cartilage. The trachea divides into two **bronchi;** one bronchus supplies air to each lung. The lungs are conical in shape and are suspended freely within the thoracic cavity by the bronchi. Each bronchus subdivides repeatedly within the lung into smaller and smaller branches, the **bronchioles,** that end in microscopic sacs, the **alveoli.** Each alveolus is surrounded by a capillary network of blood vessels, and this is the site of respiratory gas exchange between the external environment and the blood.

Air is drawn into the lungs as the thoracic cavity enlarges when the **external intercostal muscles** and the **diaphragm** contract. The intercostal muscles, attached to the outside proximal ends of the ribs, pull the ribs forward and separate them. The diaphragm is normally bowed forward, but it flattens during contraction, thus enlarging the thoracic cavity. This action creates a slight negative pressure around the lungs, and air rushes in (**inspiration**). Air leaves the lungs (**expiration**) because of the elastic recoil of the expanded lungs and by a reduction in the size of the thoracic cavity as the diaphragm and rib muscles relax. During labored breathing, contraction of internal intercostal and abdominal muscles participate, allowing larger volumes of air to be inhaled and exhaled with each breath.

Excretory System

Mammals evolved a new set of solutions to the problems of water balance and excretion. The basic **metanephric kidney** inherited from their reptilian ancestors was modified to accommodate the higher metabolic rate and increased circulation associated with homeothermic physiology. Higher blood pressure increases the filtration rate of blood plasma into the kidney tubules. This tendency toward water loss was balanced by other adaptations that helped to increase the reabsorption rate of water from the filtrate (see Chapter 34). Nitrogenous wastes are excreted in the **urine** of the mammalian kidney as urea, not uric acid, as in reptiles and birds.

The urine drains from the two kidneys through two slender tubes, the **ureters,** into a thin-walled expandable sac, the **urinary bladder.** At intervals the muscular walls of the bladder are voluntarily contracted, forcing the urine out through the **urethra.** In males the urethra passes through the penis (Fig. 30-6).

Nervous System

The cat's brain (Fig. 30-9), as do other mammals,' differs from those of lower vertebrates in the relatively larger size and complexity of the **cerebral hemispheres** and **cerebellum.** The surface area of the cerebral hemispheres is increased by slight depressions (**sulci**) that divide the surface into lobes (**convolutions**) not present in the pigeon. The entire surface of the cerebellum is likewise thrown into numerous folds.

Twelve pairs of **cranial nerves** leave the brain, and

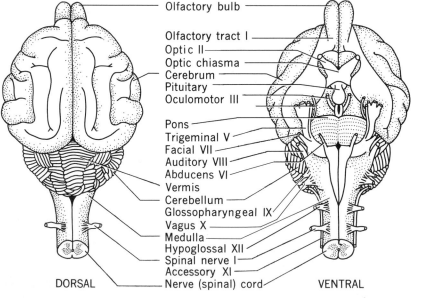

Figure 30-9. Brain of the cat.

Olfactory bulb
Olfactory tract I
Optic II
Optic chiasma
Cerebrum
Pituitary
Oculomotor III
Pons
Trigeminal V
Facial VII
Auditory VIII
Abducens VI
Vermis
Cerebellum
Glossopharyngeal IX
Vagus X
Medulla
Hypoglossal XII
Spinal nerve I
Accessory XI
Nerve (spinal) cord

DORSAL

VENTRAL

a pair of **spinal nerves** emerges from the spinal cord between each successive pair of vertebrae. An **autonomic** branch of the nervous system that controls involuntary visceral functions is also well developed.

Sense Organs

The locations and functions of the organs of smell, taste, sight, and hearing in the cat are very similar to those in humans. The cat's **pupil** varies noticeably in size and shape, depending on the amount of light striking the eye. Their eyes "shine" at night because of a layer of reflecting crystals behind the retina, the **tapetum lucidum,** which allows more light to be sensed by the retina.

The large **external ear (pinna)** serves to collect sound waves and funnel them to the eardrum, or tympanic membrane. The **middle** ear transmits the vibrations of the **tympanic membrane (eardrum)** to the auditory receptors of the inner ear by means of three **auditory bones (malleus, incus, stapes).** The middle ears of amphibians, reptiles, and birds have only one bone, the **columella.** The **cochlea,** which houses the auditory receptors, is spirally coiled and not simply curved as in the pigeon. The inner ear also houses the receptors for equilibrium, three **semicircular canals.**

As suggested by the large olfactory bulbs of the brain (Fig. 30-9), the sense of smell is well developed.

Reproductive System

The two **testes** of the male lie in an oval pouch of skin, the **scrotal sac (scrotum).** The **copulatory organ,** the **penis,** lies protected inside a fold of skin, the **prepuce,** except during urination and copulation. **Spermatozoa** pass from each testis through minute convoluted tubes called the **epididymis** into a **sperm duct,** or **vas deferens.** The two vasa deferentia pass into the abdominal cavity and open into the **urethra** (Fig. 30-7). During copulation the sperm are propelled by muscular contraction into the urethra and through the penis. Two kinds of accessory reproductive glands, the **prostate gland** and a pair of **bulbourethral (Cowper's) glands,** lie at the base of the urethra. Secretions from these glands are mixed with the spermatozoa, making the **semen** more fluid and neutralizing the acidity of the female vaginal tract.

The two **ovaries** of the female are ovoid organs with small, rounded projections on their surfaces; these are the outlines of the **ovarian follicles,** each of which contains a developing **ovum.** Each ovary lies in the dorsal abdominal coelom next to an **ostium,** the opening of an **oviduct.** Each oviduct of the cat continues posteriorly and widens into a thick-walled **uterus.** The two uteri unite medially to form the **body** of the uterus, from which the **vagina** extends to the **urogenital sinus (vestibule).** On the ventral wall of the vestibule lies a small rodlike body, the **clitoris,** which is homologous to the male's penis.

The **mammary glands** are modified sweat glands that secrete **milk** for the nourishment of the newborn young. Five groups of mammary glands are gathered around each **nipple,** which bears the numerous fine openings for their ducts.

The ova are fertilized and undergo cleavage as they pass through the oviducts, and they eventually enter the uteri, where further embryonic development occurs (Chapter 39). The **zygote** implants in the wall of the uterus, where it receives nourishment and disposes of wastes by way of the mother's blood circulation through a structure called the **placenta.** This is formed from an intimate union of the fetal membranes with the mucous membrane of the uterine wall. The interval between fertilization and birth, known as the **gestation period,** is 60 days in the cat.

Other Carnivores

Wolves, foxes, and other members of the dog family (**Canidae**) walk on their toes (digitigrade condition, Fig. 30-24). The **red fox,** *Vulpes fulva,* is the most common of all the foxes in America. The fur of the **Arctic fox,** *Alopex lagopus,* may become perfectly white in winter, helping it to approach on its prey unseen. The **gray** or **timber wolf,** *Canis lupus,* hunts in packs, a common social unit among mammals.

The cat family (**Felidae**) includes the **cat** (*Felis catus*), **puma** (*Felis concolor,* also called **mountain lion, cougar,** or **panther,** Fig. 30-10), **ocelot** (*Felis paradalis,* Fig. 30-10), **leopard** (*Leo pardus*), **lion** (*L. leo*), **tiger** (*L. tigris*), **lynx** (*Lynx canadensis*), and **cheetah** (*Acinonyx jubatus*). Felids in general are less social animals than the canids. When hunting, stealth and then a pounce on the prey play a large role in their success, whereas a long pursuit and lasting stamina play an integral part in canid hunting strategy.

Figure 30-10. Order Carnivora. (A) Puma, *Felis concolor,* a large predatory mammal; it kills calves, deer, and other large animals. (B) Ocelot, *Felis pardalis,* is an inhabitant of Texas and tropical America. (C) Grizzly bear, *Ursus horribilis.* (D) Timber wolf, *Canis occidentalis,* is related to the dog. (E) American otter, *Lutra canadensis,* has broad webbed feet and a strong muscular tail.

Bears (family **Ursidae**) are large animals with rudimentary tails (Fig. 30-10). Their **plantigrade** condition (Fig. 30-14) results in slower locomotion but allows grasping and tearing abilities suitable to these omnivores.

The **mustelids** (family **Mustelidae**) constitute a large family of small fur-bearing animals. About 50 species inhabit North America. The **otter,** *Lutra canadernsis,* is adapted for swimming, with its webbed feet and powerful tail (Fig. 30-10). The **mink,**

Mustela vison, is also fond of water and, unfortunately for the mink, is highly valued for its fur. The **weasels,** of which there are three species in North America, are the smallest mustelids. They are rapacious carnivores, usually killing their prey by biting it at the base of the skull. The **striped skunk,** *Mephitis mephitis* (Fig. 30-10), is notorious for the odorous secretion that it and other skunks can eject from a pair of scent glands that open into the anal canal.

Order Pinnipedia—Seals, Sea Lions, and Walruses

These aquatic carnivores share ancestry with the terrestrial carnivores, but they are now greatly modified for life in the water. Forelimbs or hindlimbs, or both, are fully webbed and function as swimming organs. The body has acquired a fishlike form suitable for progress through the water. They are chiefly marine, but a few inhabit fresh water or swim up rivers. The **northern fur seal,** *Callorhinus ursinus*, breeds on the Pribilof Islands in the Bering Sea and several other islands. In the fall, after the young are weaned, the females and juveniles migrate southward, some even to the California coast, a trip of approximately 4800 km. The bulls stay much closer to the breeding rookeries. The **California sea lion,** *Zalophus californianus* (Fig. 30-11), is often seen in captivity, especially in circuses. Only the females, which are one

Figure 30-11. Order Pinnipedia representatives. (A) Harbor seal, *Phoca vitulina.* (B) A dominant elephant seal bull, *Mirounga angustirostris.* (C) A population of California sea lions, *Zalophus californianus.*

third the size of the males, are trained to perform. Adult male **walruses,** *Odobensis rosmarus,* may weigh 1000 kg. The canine teeth of the upper jaw are very long and are used to dig up mollusks and crustaceans on muddy ocean bottoms and to climb up on blocks of ice in the Arctic seas that they inhabit.

Order Insectivora—Shrews and Moles

The insectivores are considered to be the most primitive placental mammals and probably resemble the early mammals in size and habits. Order **Insectivora** contains the **shrews, hedgehogs,** and **moles** (Fig. 30-12). They are entirely absent from Australia and most of South America. Insectivores are nocturnal and feed primarily on insects, which are seized by the projecting front teeth and cut into pieces by sharp-pointed cusps on the rear teeth. Most insectivores travel on the ground, but some live in burrows, a few are aquatic, and some live in trees. The shrews are of special interest because they are the smallest mammals (some weigh less than a dime) and have the highest metabolic rate of the class. The masked shrew, *sorex cinereus,* is reported to breathe 850 times per minute and have a pulse rate of about 800 beats per minute. As a result, shrews must consume

several times their body weight each day. Shrews are rarely seen but are quite common.

Order Chiroptera—Bats

Bats (Fig. 30-12) possess forelimbs that are modified for flight. The elongated forearm and four digits support a thin leathery membrane, along with the hindlimb and sometimes the tail. Because of their remarkable powers of locomotion, bats are very widely distributed, sometimes occurring on small islands that are devoid of other mammals. Most of the more than 800 species are small and nocturnal. During the day, most bats seek dark retreats such as caves, tree hollows, and houses, where they hang head downward, suspended by the claws of one or both feet. At night, bats fly about actively in search of food. Most feed on insects, some on fruit, and a few extract the blood of other mammals and birds. The little **brown bat,** *Myotis lucifugus,* is abundant in eastern North America.

The bat apparently is able to see well. Nevertheless, a completely blinded bat can fly about a room hung with criss-crossing silk threads without striking them. It guides itself by **echolocation,** emitting ultrasonic sound and monitoring the echoes that are bounced back from objects in its path. Bats can catch

Figure 30-12. Adaptations in mammals. (A) Order Insectivora, mole, *Scalopus aquaticus* (about 22 cm long); note forefeet adapted for digging, sensitive nose, and rudimentary eyes. (B) Order Chiroptera; the Pallid bat, *Antrozous pallidus.* (C) Order Edentata, giant anteater, *Myrmecophaga tridactyla* (about 2 m long); note long snout and sharp foreclaws for digging. (D) Order Edentata, nine-banded armadillo, *Dasypus novemcinctus* (5.5–7.0 kg); note bony shell, scanty hair, and long claws for digging. [Courtesy of New York Zoological Society.]

flying insects in pitch darkness while dodging all sorts of objects. Some moths are sensitive to the bats' echolocation signals and will automatically drop to the ground or initiate erratic flight patterns upon "hearing" them, thus increasing their chances of survival.

Order Edentata—Sloths, Armadillos, and Anteaters

These mammals either lack teeth or have only a few that are poorly developed. Edentata contains the sloths, anteaters, and armadillos. **Sloths** (family **Bradypodidae**) are leaf-eaters that inhabit the tropical forests of Central and South America, hanging from the underside of branches by means of two or three long curved claws. The **giant anteater**, *Myrmecophaga tridactyla* (family **Myrmecophagidae**, Fig. 30-12), has long front claws with which it tears open ant and termite mounds and a long snout and slender sticky tongue with which it captures its prey. **Armadillos** (family **Dasypodidae**, Fig. 30-12) are omnivores that possess an exoskeleton of bony scutes. When disturbed, some roll into a ball for protection. A number of species produce a single zygote that gives rise to a number of identical young, a phenomenon known as **polyembryology.**

Order Primates

Primates are of special interest because this order includes the human, *Homo sapiens*. Primates chiefly inhabit the warmer parts of the world. Most are arboreal climbers whose great toe and thumb are apposable to the other digits, which adapts the feet and hands for grasping. Most primates are social animals that live in groups. Fruits, seeds, insects eggs, and birds are their principal food. Only one offspring is usually born at a time, after which it receives a great amount of parental care. Eleven families of living primates are recognized; some interesting representatives are described here.

The **lemurs** (family **Lemuridae**) are small to moderate size and usually possess a long nonprehensile tail. They are mainly confined to Madagascar and neighboring islands. Their food is mostly plants and small animals. The **aye-aye**, *Daubentonia madagascariensis* (family **Daubentoniidae**), about the size of a cat, is one of the most remarkable lemurlike animals. Its toes are long and end in pointed claws. The aye-aye has a special finger longer than the others that is used to tap and locate grubs under the tree bark.

Tarsiers (family **Tarsiidae**) live in the Philippines and adjacent islands. They are about the size of rats and have large eyes, rounded pads at the ends of their toes, and an extra tarsal joint enabling them to make remarkable forward or backward jumps.

The typical South American (**New World**) monkey (family **Cebidae**, Fig. 30-13) is of small or medium size; the thumbs and great toes are apposable; all digits possess nails. The tail is usually long and prehensile, aiding in climbing; the space between the nostril openings is wide; and there is no vermiform appendix. The more common ones include the **squirrel monkeys** (*Saimiri*), **howlers** (*Alouatte*), **capuchins,** (*Cebus*), **woolly monkeys** (*Lagothrix*), and **spider monkeys** (*Ateles*).

The **marmosets** (family **Callithricidae**) are small primates that range from Central America to Brazil. They are about the size of small squirrels, their long tails are nonprehensile, and the body is covered with soft fur. Although other South American monkeys and the Old World monkeys, discussed in subsequent paragraphs, have nails on all their digits, the marmosets have a flat nail only on the great toe, and the other digits bear claws.

The **Old World monkeys** (family **Cercopithecidae**, Fig. 30-13) usually possess long tails that are never prehensile in the adults; their buttocks are provided with thick patches of callous skin on which they rest in a sitting posture. The nostrils are separated by a narrow space, and some have cheek pouches. Many species live in Asia and Africa, including the baboons, macaques, and langurs. The **rhesus monkey**, *Macaca mulatta* (Fig. 31-24), is commonly used as an experimental animal. It was in the rhesus monkey that the Rh blood factor was first identified (Chapter 35).

The **anthropoid apes** (family **Pongidae**) are the primates most closely related to humans. Like us, they have a vermiform appendix. A tail is absent; the forelimbs are longer than the legs. Locomotion is often bipedal; the feet tend to turn in when the ape is walking, with the knuckles helping to preserve equilib-

Figure 30-13. Primates. (A) Spider monkey, *Ateles ater,* a New World South American species; it has a prehensile tail (adapted for grasping) that functions as a fifth appendage and a great toe and thumb apposable to other digits. (B) *Gorilla gorilla;* it is the largest of the great apes. (C) Barbary ape, *Macaca sylvana,* an Old World monkey. (D) Chimpanzee, *Pan troglodytes;* it has large ears, long lips, and nails on fingers and toes. (E) Rhesus monkey, *Macaca mulatta* (formerly *M. rhesus*); the rhesus (Rh) factor, which is found in about 86 percent of the white population in the United States, was first found in this monkey. (F) Gibbon; the smallest of the apes and the only one that habitually walks erect like a man, it is arboreal, with very long arms, legs, hands, and feet. [Photos of chimpanzee, rhesus monkey, and gibbon courtesy of the N.Y. Zoological Society.]

E   F

rium. There are four genera in this family: *Hylobates*, the gibbons; *Simia* (formerly *Pongo*), the orangutans; *Gorilla*, the gorillas; and *Pan*, the chimpanzees.

Gibbons (Fig. 30-13) are aboreal; they have slender bodies and limbs. Most are herbivorous (except for eating bird eggs), reach a height of not over 1 m, and, when walking, are not assisted by the hands. More than a dozen species inhabit southeastern Asia and Indonesia.

The single species of **orangutan**, *Pongo pygmaeus*, is confined to Borneo and Sumatra. these herbivorous animals live principally in the treetops, where they construct a sort of nest which is used a few times before they move to other parts of the forest. They are about 1.4 mm in height, and, when walking, they use their knuckles as well as their feet. The brain of the orangutan is, of all animals, the most nearly like our own.

The **gorilla**, *Gorilla gorilla* (Fig. 30-13), inhabits the forests of western Africa. These primates are primarily terrestrial, and feed mainly on vegetation. They reach a height of 1.7 m, and the larger males may weigh about 225 kg. The gorilla walks on the soles of its feet, aided by the knuckles of the hands.

The **chimpanzee**, *Pan troglodytes* (Fig. 30-13), also lives in West Africa. It resembles a small gorilla, but it has shorter arms and a smoother, rounder skull. In many respects the chimpanzee is more nearly like us than is any other living mammal.

Family **Hominidae** contains a single living species, *Homo sapiens*. Humans differ from other primates in their brain size, which is about twice as large as that of the highest ape, and in their erect bipedal locomotion. The hairy covering is reduced, and the great toe is not apposable. The mental development of humans has enabled us to accommodate ourselves to almost every climate.

Order Rodentia—Rodents

In terms of numbers of species and individuals, the rodents are the most successful mammals. **Rodentia** includes the squirrels, beavers, rats, mice, pocket gophers, porcupines, chinchillas, and golden hamsters, among others. Their front teeth are specialized as efficient chisels and grow constantly. Many **squirrels** (family **Sciuridae**) are excellent climbers. **Chipmunks** usually live on the ground among rocks (Fig. 30-14). **Ground squirrels** are sometimes called "gophers" in part of their range. Ground squirrels should not be confused with **pocket gophers** (family **Geomyidae**), which are primarily subterranean in their habits. A mound marks the spot where the animal has deposited the earth removed during burrowing. Pocket gophers consume roots and the aboveground parts of plants near their burrow openings. When done at the surface, they plug their burrow

Figure 30-14. Order Rodentia. (A) Chipmunk, *Tamias*. A beautiful little rodent that lives in crevices in rocks, under logs, or in burrows in the ground. (B) Beaver, *Castor canadensis* (about 1 m long). It has chisel-like teeth, webbed hindfeet, and a flat scaly tail. It builds its home in forest ponds that it makes by building dams. (C) Chinchilla, *Chinchilla laniger*. These animals were introduced into the United States in 1923 by Chapman, who obtained 11 animals from Chile. Since then many chinchilla ranches have been established in the United States. Chinchilla fur has a downy softness, but it is not as serviceable as some other furs. [Courtesy of N.Y. Zoological Society.]

entrances with earth. **Prairie dogs** of our western plains are burrowing rodents that are very social animals. They often live in "towns" of several thousand individuals. **Woodchucks,** or **ground hogs** (*Marmota*),

also live in burrows. These chunky mammals become extremely fat by fall, as they prepare for winter hibernation. The **beaver,** *Castor canadensis* (family **Castoridae,** Fig. 30-14), is adapted for aquatic life, possessing webbed hindfeet and a broad, flat tail. Other adaptations include ear and nose valves that close under water. The beaver can also gnaw logs below the water surface.

The more common **mice** and **rats** (family Muridae) include the **house mouse,** *Mus musculis,* the **Norway rat,** *Rattus norvegicus,* and the **black,** or **roof, rat,** *Rattus rattus.* All have been introduced into this country from the Old World. They are responsible for destroying huge quantities of foodstuffs and also for the spreading of disease. The domesticated and colorful versions of the house mouse and Norway rat are extremely popular as research animals and pets.

Porcupines (family **Erethizontidae,** Fig. 30-1) possess sharp spines that normally lie flat on the back but can be elevated by integumentary muscles when they are disturbed. The **chinchilla,** *Chinchilla laniger* (family **Chinchillidae**), of South America is noted for its very soft fur (Fig. 30-14). While natural populations are endangered, thousands are raised in captivity for their pelts.

Order Lagomorpha

This order contains widely distributed herbivorous mammals that are most closely related to the rodents. The **hares** and **rabbits** (family **Leporidae**) are characterized by their long ears and short tails. Hares, such as the **snowshoe hare,** *Lepus americanus,* and **black-tailed hares,** *Lepus californicus* (erroneously called the jackrabbit, Fig. 30-15), do not construct nests and bear their young fully haired with eyes open. They are troublesome because of the crop damage that they cause in agricultural areas; this was a major problem when they were first introduced into Australia, which has no native forms. The **cottontail,** *Sylvilagus nuttalli,* and all other rabbits bear young that are blind and hairless and must be raised in fur-lined nests. Rabbits, as with certain rats, are commonly used in scientific research.

The **pikas** (family **Ochotonidae**) are common in the western Rockies; they differ from hares and rab-

Figure 30-15. The black tailed hare, *Lepus californicus*, Order Lagomorpha.

bits because of their short, rounded ears. None of the lagomorphs hibernate, and they are common prey to other mammals and birds.

Order Cetacea—Whales, Dolphins, and Porpoises

Order **Cetacea** contains the whales, dolphins, and porpoises. The **whales** include the largest living animals. In fact, the **blue whale**, *Balaenoptera musculus*, is the largest animal that has ever lived. It reaches lengths of 35 m and weights of over 100,000 kg. This giant mammal is remarkably well adapted to life in the ocean. It possesses a very large head with an elongated face and jaw bones; the nostrils are situated posteriorly and dorsally on the head. The forelimbs are modified as paddles (flippers), and there are no external traces of hindlimbs on its streamlined body. The tail is flattened horizontally into two propulsive flukes, the eyes are small, and there are no external ears. Only a few hairs near the mouth are present; beneath the skin lies a thick layer of fat, or blubber, which takes the place of the insulating hair of most mammals.

The blue whale is a member of a suborder of whales known as **whalebone (baleen) whales.** Other members include the **gray whale**, *Eschrichtius robustus* (Fig. 30-16); the **humpback whale**, *Megaptera novaengliae* (Fig. 30-16); the **bow-head right whale**, *Balaena mysticetus;* and others. The whalebone whales lack teeth. Instead, their upper jaw bears numerous horny plates of whalebone, or baleen. These baleen plates act as strainers: the whale takes in enormous mouthfuls of seawater and then forces it back through the sieve-shaped plates that trap small planktonic animals, which are termed krill. All baleen whales feed in the rich Arctic and Antarctic waters during the summer; they migrate to warmer latitudes in the winter when the mating and birth of their young occur.

The **sperm whale** reaches lengths of 23 m and is the largest toothed whale. Its enormous head contains a cavity filled with as much as 900 kg of fine oil (sperm oil). The sperm whale only has numerous, conical teeth on its lower jaw, but most members of the toothed whale suborder have teeth on the upper jaw as well. Other toothed whales and relatives include the **common,** or **harbor, porpoise**, *Phocaena phocaena;* the **common dolphin**, *Delphinus delphis;* the **narwhal**, *Monodon monocerus;* the **killer whale**, *Orca orca* (Fig. 30-1); and the **bottle-nosed dolphin**, *Tursiops truncatus*, which is commonly miscalled a porpoise. The toothed whales feed upon squid, fish, and other large aquatic animals. In the sperm whale, the nostrils form a single opening; blue whales have two nostrils. As warm moist air is forcefully exhaled through the nostril(s), it condenses in the cooler atmosphere, forming the "spout" of the whale. All whales have a distinct spout, and experienced whalers can identify a whale by its spout alone.

Several recordings of the communication sounds of the humpback whale are presently available commercially. The toothed whales do not make audible sounds. Instead, they utilize ultrasonic sounds, which function for prey detection.

The larger species of whales are threatened by extinction due to overhunting; in 1979, after many years of protesting, an international agency agreed on a law preventing the harvest of most whales in international waters. Whale hunting and the importation of whale products are now illegal in the United States.

Figure 30-16. Order Cefacea. Representative cefaceans. (A) California gray whale, *Eschrichtius gibbosus.* (B) Killer whale, *Orca rectipinna.* (C) Humpback whale, *Megaptera novaenglie.* (D) Dolphins of the genus, *Tursiops.*

Order Proboscidea—Elephants

Elephants are the largest land animals, weighing close to 7,000 kg. They are covered by a thick loose skin with a sparse coat of hair, and they have a long, muscular proboscis (trunk) with nasal openings at its tip. They are provided with tusks that are elongated incisors; canine teeth are lacking. Elephants also possess small eyes, small tails, and enormous ears, the latter serving as a heat-dissipation device. The skull is massive, due to thick bones containing air spaces. The grinding teeth (two at any time per jaw) are very large and possess complicated ridges for grinding the many forms of vegetation they consume.

The two living elephants are the **African elephant** (*Loxodonta africana,* Fig. 30-17A), and the **Indian,** or **Asiatic, elephant** (*Elephas maximus,* Fig. 30-17B). These two can be distinguished by their ear size: the African elephant's ears are larger than are those of its Asiatic counterpart. Also, the African elephant's trunk has two distal processes, whereas the Indian elephant's trunk has one. The Indian elephant is widely

Figure 30-17. Order Proboscoidea. (A) African elephant, *Loxodonta africana*. Note long prehensile trunk with tusks (incisor teeth) loose skin, large ears, and hollow in back. (B) Indian elephant, *Elephas maximus* (3 m high at the shoulder). Note smaller ears and arched back. The upper lip and nose of the elephant are drawn out into a trunk, a remarkable grasping organ.

used in circus performances and is used in Asia for hauling heavy loads, as oxen and other animals are used elsewhere.

Order Sirenia

Order **Sirenia** includes the **sea cows,** or **manatees.** Sirenians differ considerably in structure from cetace-

ans and seals. Their bones are heavy, enabling them to remain on the bottom; the lips are large and moveable and are used to gather the aquatic vegetation upon which they feed. Their teeth are broad and adapted for crushing. The forelimbs are flexible flippers, and the tail is more rounded and expanded into flukes as in whales. The **Florida manatee,** *Trichecus manatus* (Fig. 30-18), inhabits the coastal waters and sluggish rivers of the southern United States, eastern

Figure 30-18. Order Sirenia. Manatee, *Trichechus,* 4–5 m long. Note flipperlike forelimbs, absence of hindlimbs, and rounded tail with horizontal fin.

A

B

C

D

E

F

G

Figure 30-19. Order Perissodactyla representatives. (a) Tapir, *Tapirus bairdii.* Note long, prehensile nose. (b) Hippopotamus. (c) White rhinoceros, Order Artiodactyla representative. (d) Pronghorn antelope, *Antilocapra americana.* (e) African giraffe. (f) African antelopes. (g) South American llama.

Central America, and West Indies. They are sluggish, inoffensive animals. The sirenians probably account for some of the tales about mermaids.

Orders Perissodactyla and Artiodactyla—Hooved Mammals

Hooved mammals are divided into those with an odd number of toes (order **Perissodactyla**) and those with an even number of toes (order **Artiodactyla**).

Horses (Fig. 30-19), **tapirs,** and **rhinoceroses** have an odd number of hooved toes, the axis of symmetry passing through the third digit, which is the sole remaining one in the horse. No living species of this order is native to the United States, but many fossils of extinct species have been found in this country. The tapirs occur in Central and South America and in Sumatra, Java, and the Malay Peninsula. Unique to tapirs is their long, prehensile nose (Fig. 30-19).

The most successful of the hooved animals are the even-toed artiodactyls, the axis of symmetry passing between digits 3 and 4, which are the predominant supportive digits in most species. To this order belong the **peccaries** (family **Tayassuidae**), **pigs** (family **Suidae**), **hippopotames** (family **Hippotamidae**), **camels** and **llamas** (family **Camelidae**), pronghorn **antelopes** (family **Antelocapridae,** Fig. 30-19), **giraffe**

(family **Giraffidae**), **deer** (family **Cervidae,** Fig. 30-19), and the **hollow-horned ruminants** such as the many Asian and African **antelopes, oxen, cattle,** and **bison** (family **Bovidae**). The term ruminant has been given to mammals belonging to the camel, deer, giraffe, pronghorn antelope, and ox families, as they ruminate or "chew their cud." The food of these herbivores is swallowed without sufficient mastication but is later regurgitated in small quantities and thoroughly chewed. A typical ruminant (cow) possesses a stomach consisting of four chambers (Fig. 30-20). Populations of symbiotic bacteria digest the cellulose cell walls of the vegetable food and thereby assist their host.

Domesticated Mammals

The most common domesticated mammals are the dog, horse, ass, camel, cow, sheep, llama, goat, pig, rabbit, and cat. **Dogs** have been selectively bred for desired characteristics so that there are now more than 200 breeds. Although the dog *(Canis familiaris)* is treated as though it is a distinct species, most likely it is descended from a number of canids with which humans have had contact over thousands of years. The immediate ancestors of the horse *(Equus caballus)* are not known, but, like the dog, it is mostly likely derived from several species. The **ass** is descended from the African **wild ass,** *Equus asinus.* Males are the hybrid sterile offspring of a male ass and a female horse. They are valued for their endurance, sure footedness, and strength. The **Arabian,** or **dromedary, camel,** *Camelus dromedarius,* and the double-humped **Bactrian camel,** *Camelus bactri-*

Figure 30-20. A four-compartment stomach of a ruminant (cow). Arrows show the course of roughage such as hay. Grains such as corn go directly to the reticulum. This type of stomach permits the animal to consume large amounts of grass, after which it can rechew its food.

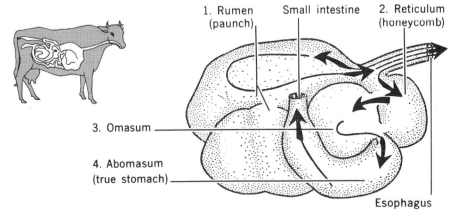

1. Rumen (paunch)   Small intestine   2. Reticulum (honeycomb)

3. Omasum

4. Abomasum (true stomach)

Esophagus

anus, have both been domesticated from wild stock; only the latter is still known in the wild state. The cattle of Europe and America were probably drived from the **auroch,** *Bos primigenius,* which inhabited the forests of Europe, northern Africa, and southwestern Asia until its extinction.

Sheep *(Ovis aries)* have probably arisen from one or more species of wild sheep of the genus *Ovis.* **Goats** *(Capra hircus)* have been domesticated since early times, and their wild relatives are abundant in many parts of western Asia. Our domesticated pigs are descended from the wild boar *(Sus scrofa)* of Europe and southeastern Asia. The remote ancestor of the **common house cat** *(Felis catus)* is unknown but is another animal suspected of multiple origins and subsequent interbreeding.

Adaptive Modifications and Behaviors of Mammals

Hair

A covering of hair distinguishes mammals from all other animals. Unlike feathers, which are modifications of horny scales, hairs are new structural elements of the skin (Fig. 30-21). Both, however, arise from sunken pits called **hair follicles** that penetrate the dermis. A major function of hair and feathers is to insulate the bodies of the homeothermic mammals and birds that they cover. As hairs are shed, new ones usually develop to take their place. Secretions from oil-producing **sebaceous glands** keep their hairs glossy.

Two main types of hairs are recognized: **guard hairs,** which are long and strong, and **underfur,** which is generally dense and fine, providing the bulk of insulation. The wooly underfur of some mammals (e.g., sheep) has a rough surface that causes it to cohere, giving it a felting quality. Certain of the stronger, specialized hairs such as spines may be moved by integumentary muscle fibers.

Claws, Nails, and Hooves

These are all horny modifications of the epidermis on the dorsal surface of the distal ends of the digits (fingers and toes). The foot rests partially or entirely upon the fleshy digital pads that are covered by **dermal papillae,** often arranged in concentric ridges such as those that produce human fingerprints.

Other horny epidermal thickenings are the **horn sheaths** of cattle and other ruminants, the **nasal horns** of the rhinoceros (composed of hair, actually), and the "whalebone" **(baleen)** of certain whales. **Dermal plates** of bone form the exoskeleton of armadillos.

Skin Glands

Mammals possess more integumentary glands than do reptiles or birds; there are for the most part sebaceous and sweat glands or modifications of them (Fig. 30-22). **Sebaceous glands,** which usually open into hair follicles, secrete a greasy substance called **sebum** that keeps the surface of the skin soft and the hair glossy. **Sweat glands** secrete a watery fluid containing a small amount of solutes, mainly sodium chloride and urea. The evaporation of sweat cools the skin

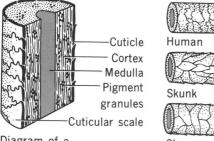
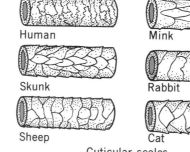

Diagram of a sectioned human hair — Cuticle, Cortex, Medulla, Pigment granules, Cuticular scale. Human, Mink, Skunk, Rabbit, Sheep, Cat. Cuticular scales

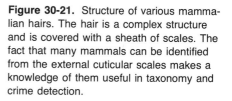

Figure 30-21. Structure of various mammalian hairs. The hair is a complex structure and is covered with a sheath of scales. The fact that many mammals can be identified from the external cuticular scales makes a knowledge of them useful in taxonomy and crime detection.

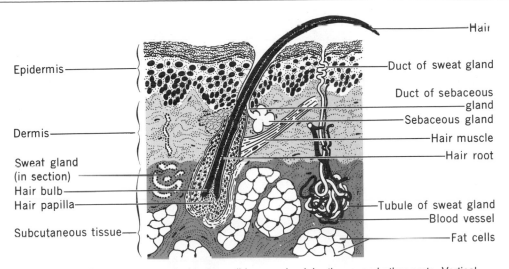

Epidermis

Dermis

Sweat gland
(in section)

Hair bulb

Hair papilla

Subcutaneous tissue

Hair

Duct of sweat gland

Duct of sebaceous
gland

Sebaceous gland

Hair muscle

Hair root

Tubule of sweat gland

Blood vessel

Fat cells

Figure 30-22. Skin of a mammal, showing cell layers, glandular tissue, and other parts. Vertical section.

and thereby helps to remove excess body heat. Sweat glands are much reduced in number in some mammals, notably carnivores; the panting of a dog utilizes evaporation of the tongue for thermoregulation. Highly specialized integumentary glands include **lacrimal glands,** which secrete tears that keep the eyeballs moist, the **scent glands** of many species, and **mammary glands.**

Teeth

The teeth of mammals indicate their respective food habits and are useful in classification. The adults of whalebone whales, monotremes (duck-billed platypus and spiny anteater), and many edentates (e.g., sloths and armadillos) lack teeth.

The embryological development of mammalian teeth is like that of other vertebrates (Fig. 30-23). The **enamel** is the hard, outer layer, the bonelike **dentine** constitutes the largest inner portion of the tooth, and the **cementum** usually replaces the enamel and covers the part of the tooth embedded in the jawbone. The central **pulp cavity** of the tooth contains nerves, blood vessels, and connective tissue. A tooth has an open pulp cavity during growth, but this generally becomes restricted with age.

The teeth of any fish, reptile, or amphibian generally have similar shapes; their dentition is said to be **homodont.** The dentition of mammals, on the other hand, is almost always differentiated into several kinds and is said to be **heterodont.** There are usually four kinds of teeth in each jaw: chisel-shaped cutting **incisors,** conical tearing **canines,** grinding **premolars,** and larger grinding **molars.**

In most mammals, the first set of teeth, known as the **milk (deciduous)** dentition, is pushed out by the **permanent teeth,** which must last throughout the life of the animal. The permanent teeth erupt as the jawbones grow large enough to accommodate them. The milk molars are replaced by the permanent premolars, but the permanent molars have no predecessors.

The shape of a mammal's teeth can be closely correlated with its diet. For example, sharp, conical teeth (porpoises) are adapted for capturing fish, whereas large canine teeth (lions) are suitable for capturing and killing terrestrial prey; large molars (cattle) are useful for grinding vegetation. Much about the lifestyles of extinct mammals can be determined from their fossilized teeth. The types of contemporary plant life can be reconstructed, and some conclusions can be drawn regarding the digestive organs of the animals. For example, animals with large premolars

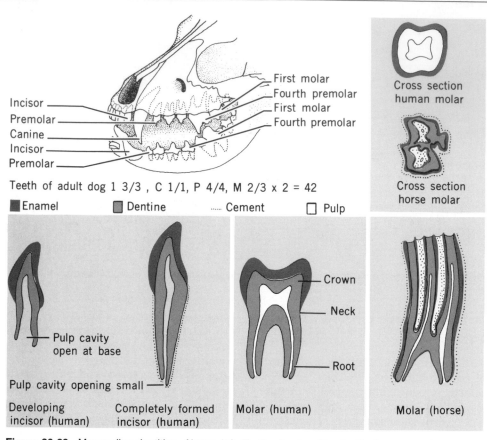

Incisor
Premolar
Canine
Incisor
Premolar

First molar
Fourth premolar
First molar
Fourth premolar

Cross section human molar

Cross section horse molar

Teeth of adult dog 1 3/3 , C 1/1, P 4/4, M 2/3 x 2 = 42

■ Enamel   ■ Dentine   ····· Cement   □ Pulp

Pulp cavity open at base

Pulp cavity opening small

Developing incisor (human)

Completely formed incisor (human)

Crown
Neck
Root

Molar (human)

Molar (horse)

Figure 30-23. Mammalian dentition. Above, left: Teeth of a dog. The teeth of a mammal are of a definite number and are collectively referred to as the dentition. Their number is expressed as a dental "formula" that includes those of the upper and lower jaws of one side. The dental formula of a dog is given and multiplied by 2 to show the total number of teeth. Above, right and below: the structure of mammalian teeth shown diagrammatically in cross- and longitudinal section.

and molars (e.g., horses) are obviously completely herbivorous, whereas animals with teeth that are reduced in size and modified for shearing large chunks of meat off their prey are distinctively carnivorous.

Foot Posture

The type of foot posture exhibited by humans is called **plantigrade;** the entire palm or sole rests on the ground, with neither the wrist nor the ankle raised above the ground (Fig. 30-24). Cats exhibit **digitigrade** posture, which entails walking upon their

digits, with the upper ends of the palms and soles raised above the ground. A third type of foot posture, called **unguligrade,** is characteristic of hooved mammals. These animals walk on modified nails or hooves that form over the tip of the distal phalange of one (horse) or more (cattle, pigs) digits.

Hibernation and Migration

During the winter, mammals either remain active (e.g., rabbits, deer) or hibernate. Woodchucks and ground squirrels, for example, hibernate in under-

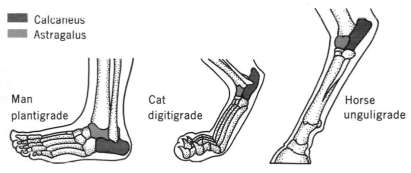

Figure 30-24. Feet of mammals (left hind-foot). Man, generalized, with five digits (toes); the entire sole rests on the ground (plantigrade). Cat, with four toes, heel raised, walks on digits (digitigrade). Horse, most specialized, with only one digit (third); walks on the hoof (homologous to toenails) that covers the end of its toe (unguligrade). How does a sprinter run to attain maximum speed on the track?

ground nests, and some bats hibernate in caves or buildings. During **hibernation** metabolic rates drop to a low level and body temperatures decrease, and the mammal falls into a profound torpor. Respiration almost ceases, the heartbeat is slowed, and no food is ingested. Fat deposits that were accumulated during the summer and autumn are slowly metabolized (especially brown adipose, see Chapter 3), and the animal awakens the following spring in a somewhat emaciated condition.

Relatively few mammals migrate to more temperate climates. Among those that do are the **fur seal** (*Callorhinus ursinus*), **caribou** (*Rangifer tarandus*), **bison** (*Bison bison*), **red bat** (*Lasiurus borealis*), and some whales. The female fur seals that breed on the Pribilof Islands in the Bering Sea spend the winter months making a circuit of about 9,600 km. For the caribou in Alaska and northern Canada, the distance between summer and winter ranges is often several hundred kilometers. Some bats (as is true for many birds), such as the red bat, leave their northern summer habitats when their insect food becomes scarce and migrate southward for the winter. Others move to caves and hibernate until spring arrives.

The **lemmings** (*Lemmus lemmus*) of Scandinavia are noted for their curious mass movements. Approximately every four years, these small herbivores reach rather high densities on the high plateaus that they inhabit. At this time they move into the valleys in vast hordes, swimming across lakes and streams, and eating crops along their paths. The survivors of this journey finally reach the ocean or some other large bodies of water, where they plunge in and drown. Many theories have been proposed to explain why the lemmings undertake such a doomed migration, but perhaps they simply attempt to swim across the ocean as

they have other bodies of water, in an attempt to find a new favorable habitat, and fail in the great undertaking.

Developmental Innovations

The zygotes of most mammals develop within the body of the mother (**viviparity**). Exceptions are the egg-laying monotremes, whose reptilian characteristics indicate the relatively primitive status of these mammals. The eggs of monotremes are incubated by the female and hatch in the same manner as do those of reptiles and birds. Among the viviparous mammals, two general developmental patterns are seen. The young of marsupials, such as the American opossum, *Didelphis virginiana* (Fig. 30-3C), are born in a very immature condition, after a gestation period of only 12–13 days. At birth, the young immediately climb by hand-over-hand movements through their mother's hair until they reach the pouch where they remain attached to a nipple for about 40 days.

The young of most mammals develop to a much more advanced stage before birth, because the embryo is nourished through a **placenta** within the uterus. (The placenta of the higher mammals arises as shown in Figure 39-10). The young embryo becomes connected to the uterine wall by means of its outer epithelial layer, called the **trophoblast.** This later becomes partially or wholly coated on its inner side by mesodermal tissues that eventually form a membrane, the **chorion.** This connection of the fetal chorion with the maternal uterine wall gives rise to the placenta, by means of which the fetus is nourished within the body of the mother. Note that the blood of the fetus and that of the mother do not normally mix, but the membrane of the **chorionic villi** that separates them is so thin that gases, nutrients, and nitrogenous wastes diffuse through it. Many variations in the development and the mode of connection of the placenta are exhibited among the different orders of mammals. For further details and comparisons to other vertebrates, refer to Chapter 39.

Summary

Mammals are vertebrate animals that possess milk-secreting mammary glands. They also characteristically have body hair and are homeothermic.

The cat may serve as a representative example of the class Mammalia. The skeleton is well constructed for a terrestrial existence and efficient locomotion. The muscular system includes a diaphragm for breathing; the digestive system is somewhat similar to our own; the circulatory system includes a four-chambered heart; a lymphatic system returns much of the tissue fluid to the blood; the excretory system includes a distensible urinary bladder and kidneys suited for the excretion of urea and the retention of water. Features of the brain are the large, complex cerebral hemispheres and the cerebellum; the sensory systems are also rather well-developed. The reproductive system can be divided into egg-laying, pouched, and placental types. Hair, claws, nails, hooves, skin glands, and heterodont dentition are all characteristic of mammals.

Review Questions

1. What structure divides the coelom of mammals into thoracic and abdominal cavities?
2. Name the major components of the axial skeleton.
3. Describe the female reproductive tract of the cat.
4. What is the name for the common trunk of the right carotid and subclavian arteries of the cat?
5. When a terrestrial mammal inspires, the diaphragm _____ (contracts or relaxes), which results in its moving _____ (up or down).
6. What is the reflective part of the cat's retina called?
7. What is the function of sebum?
8. If one were to have a dental carie (cavity), through which structures would the hole pass? Assume that it stopped in the pulp.
9. What is the proper term for a pouched mammal?
10. Because much of our knowledge of mammalian evolution is derived from fossilized teeth, why do you suppose that the relationship of the monotremes to other mammalian groups is unclear?
11. Generally describe the dental characteristics of a carnivore.
12. Which nonhuman primate has a brain that is most similar to our own?
13. What is one major feature differentiating humans from other primates?
14. How can one easily differentiate between African and Asian elephants?
15. How does the mammalian kidney differ from the other vertebrate kidneys studied in previous chapters?
16. Describe the development of the placenta.
17. Where does fertilization take place in the mammalian reproductive tract?
18. What is unique about the eating and digestive processes of a ruminant?

Selected References

Anderson, H. T. (ed.). *The Biology of Marine Mammals.* New York: Academic Press, Inc., 1969.

Bourliere, F. *Mammals of the World: Their Life and Habits.* New York: Knopf, 1955.

Burns, E. *The Sex Life of Wild Animals.* New York: Holt, Rinehart and Winston, 1953.

Burt, W. H., and R. P. Grossenheider. *A Field Guide to the Mammals.* Boston: Houghton Mifflin Company, 1952.

Calahane, V. H. *Mammals of North America.* New York: Macmillan Publishing Co., Inc., 1947.

Cockrum, E. L. *Manual of Mammalogy.* Minneapolis: Burgess Publishing Company, 1955.

Collins, H. H. *Complete Field Guide to American Wildlife.* New York: Harper and Row, 1959.

Davis, E. E., and F. B. Golley. *Principles of Mammalogy*. New York: Reinhold Publishing Corp., 1964.

Eadie, W. R. *Animal Control in Field, Farm, and Forest*. New York: Macmillan Publishing Co., Inc., 1954.

Glass, B. P. *A Key to the Skulls of North American Mammals*. Minneapolis: Burgess Publishing Company, 1951.

Gunderson, H. L. *Mammology*. New York: McGraw-Hill Book Company, 1976.

Hall, E. R., and K. R. Kelson. *The Mammals of North America,* 2 vols. New York: The Ronald Press Company, 1959.

Hamilton, W. J., Jr. *American Mammals: Their Lives, Habits and Economic Relations*. New York: McGraw-Hill Book Company, 1939.

Henderson, J., and E. L. Craig. *Economic Mammalogy*. Springfield, Ill.: Charles C Thomas, 1932.

Hill, W. C. O. *Evolutionary Biology of the Primates*. New York, Academic Press, Inc., 1972.

Leach, W. J. *Functional Anatomy of the Mammal*. New York: McGraw-Hill Book Company, 1946.

Miller, G. S., and R. Kellogg. *List of North American Recent Mammals*. Bull. 205, U.S. National Museum, Washington, 1955.

Palmer, R. S. *The Mammals Guide, North America North of Mexico*. New York: Doubleday & Company, 1954.

Righard, J., and H. S. Jennings. *Anatomy of the Cat*. New York: Holt, Rinehart and Winston, 1935.

Scheffer, V. B. *Seals, Sea Lions, and Walruses. A Review of the Pinnipedia*. Stanford: Stanford University Press, 1958.

Schmidt-Nielson, K. *Desert Animals*. New York: Oxford University Press, 1964.

Troughton, E. *Furred Animals of Australia*. New York: Charles Scribner's Sons, 1947.

Walker, E. P. *Mammals of the World*, 3rd ed. Baltimore: John Hopkins University Press, 1975.

Wolstenholme, G. E. W., and M. O'Conner (eds.). *The Lifespan of Animals*. The Ciba Foundation Colloqia on Ageing, Vol. 5. Boston: Little, Brown, and Company, 1959.

Young, J. Z. *The Life of Mammals*. 2d ed. New York: Oxford University Press, 1975.

Morphology, Physiology, and Continuity of Animals

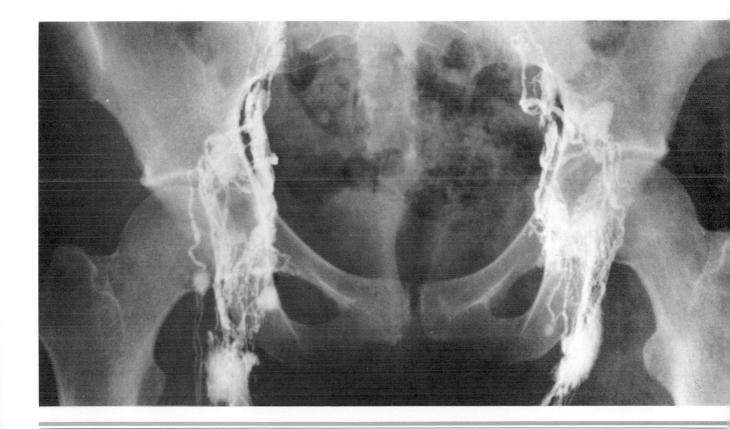

Animal Behavior

Throughout this text, great emphasis is placed on the anatomy and physiology of animals. Such knowledge is a prerequisite for understanding organisms as a whole, but our awareness would be incomplete if we ignored the expression of the functionally unified organism. The activities of an animal define its **behavior.**

Introduction

What is animal behavior, how it is studied, and what factors affect its expression?

Until several decades ago, the study of animal behavior was simply a catalog of anthropomorphic descriptions: animal behavior was interpreted purely in terms of human emotions. This approach proved rather unreasonable as it was incapable of analyzing behavior from the animal's standpoint. A mechanistic viewpoint developed in which animals were considered to be emotionless automatons reacting to their immediate environments. Both the anthropomorphic and mechanistic approaches failed to develop a scientifically sound basis for understanding and predicting animal behavior.

The foundations of modern animal behavioral theory were laid down by K. Lorenz and N. Tinbergen in the 1930s. These scientists attempted to analyze and understand the patterns of behavior that occur under natural conditions and adapt an animal to its environment. This approach to animal behavior, basically a field psychology of animals, is called **ethology.**

General Characteristics of Behavior

Behavior is dynamic. Animals respond to changes in their environments. For example, a flash of light (stimulus) may be followed by the blink of an eye (response).

The patterns of behavior may change over time. Postnatal development (development that occurs after birth) is usually accompanied by increasingly more specific actions. A newborn human's most frequent actions are crying and random arm, leg, and head movements. As the baby develops, its behaviors become more specific. As its repertoire of sound and movement increases, the behaviors become more individualized. The utterance "mama" elicits a different response from the utterance "waalupo," and the baby soon learns to modify its vocal behavior accordingly. Behavioral change is continous. Not only is an animal's environment constantly changing, but an individual's patterns of response may change with time.

Behavior may follow a periodic pattern. For example, some of an animal's actions may fluctuate with lengths of daylight (photoperiod). This usually takes the form of an uninterrupted cycle.

Much behavior is adaptational. For example, certain moths drop to the ground upon hearing the echolocation signals of predatory bats. We will consider orientation, social interaction, communication, and other behaviors that help animals to survive under the conditions of their environments.

Some Methods of Behavioral Study

Ethology is an interdisciplinary science that relies on contributions from the fields of anatomy, physiology, embryology, genetics, neurology, astronomy, ecology, physics, chemistry, sociology, psychology, and others. Many avenues of investigation are consequently open to the ethologist. Furthermore, the subjects of ethological investigations include all protozoans and metazoans.

Animal behavior may be studied in the field or in the laboratory. Controlled experiments in the laboratory yield concise data that complement field observations. Study in the wild has great relevance to ethology, for it ideally allows unadulterated observation of animals in their natural habitats.

Genetic Methods

The genetic approach attempts to assess the heredital components of behavior. Three major methods are used in genetic studies of behavior.

In **pleiotropic studies,** the behavioral as well as the morphological effects of single gene mutations are observed; dominant and recessive expressions are also evaluated, for example, as when the fruit fly *Drosophila melanogaster* receives two recessive genes for white eyes. In addition to the physical expression of white eyes, the white-eyed flies were found to copulate 25 percent less often than did red-eyed flies. Mating behavior is presumably modified because of this particular inheritance.

In **artificial selection studies,** animals with a particular behavioral trait may be bred with one another, or an animal with a particular trait is mated with an individual that lacks the trait. Geneticists may also observe the results of a cross in which the parents have different traits, the concerted expression of which is of interest. Such studies help assess the probability that a given trait is heritable. As Chapter 38 explains, many traits are polygenic—determined by the interactions of more than one pair of genes.

Cross-matings between individuals of different breeds or subspecies also help to determine the genetic component of particular behaviors. J. Scott and J. Fuller inbred dogs and concluded that, of all behav-

ioral variations between the breeds studied, 25 percent were directly attributable to genetic differences.

The genetic components of behavior can be surprisingly far reaching and persistent. S. Wecker worked with habitat selection in two subspecies of deermice: *Peromyscus maniculatus bairdi*, which, given a choice, prefers a grassland habitat, and *P. maniculatus gracilis*, which prefers to dwell in forests. Wecker maintained 20 generations of P. maniculatus bairdi in a forest habitat and then tested the twentieth generation for habitat preference; they preferred the grassland habitats of their ancestors.

Social Methods

Many animals interact socially with other members of their own species. An offspring may be isolated from its parent(s) to observe if any unusual behavioral patterns result. The role of parental care and teaching may then be objectively evaluated by statistical comparison of parent-reared and nonparent-reared progeny.

An interesting extension of this approach is to replace the parent with a **surrogate** (substitute) parent. This was done with rhesus monkeys in a series of significant experiments reported by Harlow in 1961. Infant rhesus monkeys were presented with cloth-covered and bare-wire surrogate "mothers," and their behaviors were observed through adulthood. In brief, the young monkeys preferred the cloth-covered substitutes to the wire versions, especially in stressful circumstances; they also showed several abnormal behaviors as adults, notably in the sexual realm.

Environmental Methods

Isolating an animal from its natural habitat is another method for delineating the effects of environment upon behavior.

Transplanting an animal into a foreign habitat may have minor, modest, or drastic (including fatal) effects. For example, when an animal is transferred from its natural habitat to a zoo, it may well die. During the last century, Carl Hagenbeck pioneered the introduction of ecological settings into zoos. This successfully reduced mortality rates, especially among newly captured animals. Even today there are still many zoos with animals simply caged; hence, by no means can their behaviors be considered normal or typical of wild animals.

The results of environmental studies also have application in conservation. Endangered animal species that are shown to be adaptable to alternative environments may be transferred to controlled game preserves with reasonable assurance of success.

A relatively recent innovation in field study is **biotelemetry,** in which animals may be tracked and monitored for several physiological variables (e.g., heartbeat, body temperature, wing beats) by attaching miniature electronic transmitters to their bodies. Biotelemetry allows physiological data to be gathered while an animal is in its natural setting.

The Sources and Mechanics of Behavior

As already mentioned, most behavior is very complex and dynamic. Following are some behavioral terminology discussed in their simplest context, before more detailed and thorough descriptions of these terms are considered.

Because all organisms are developed with instructions from their genes, all behavior may be considered to have a genetic component. In very simple organisms, much behavioral activity is totally **innate,** that is, genetically determined. In organisms with nervous systems, the development and final structure of this system controls many activities that come under the category of innate behavior. In these more complex organisms, this behavior, usually termed an **instinct** (discussed in a subsequent paragraph), is usually not passively read out like a tape recording but, rather, is **elicited** (triggered) by a particular signal, or **stimulus. A stimulus may be any perceptible (conscious or otherwise) variation in the external or internal environment of an organism. A response** is a change in behavior that occurs in reaction to the stimulus. Thus, although some simple behaviors may be considered largely innate, most are in the form of responses to internal, external, or a combination of stimuli. This simply reflects the fact that behavior, as with most all biological activities, is not absolutely

determined but is carefully regulated to respond to the constantly changing environment.

The simplest form of stimulus–response behavior is shown by the **taxes** (e.g., chemotaxes, Chapter 6) that allow many simple organisms to locate food or avoid harmful environments. Higher organisms also exhibit taxes (discussed later in the chapter). The simplest stimulus–response behavior in organisms with nervous systems is the **reflex.** Most often, a reflex involves one sense organ. Reflexive behavior is stereotyped: every time that an appropriate stimulus is presented to any member of a given species, an essentially identical response will be elicited. The reason for the stereotyped response is that most reflexes are built into the nervous system of the organism. The **receptors** and **effectors** (muscles and glands that respond to the stimulus) are coordinated through the central nervous system. A **spinal reflex** (Chapter 36) is the simplest of all: The neural impulse does not even have to reach the brain, as the integration between stimulus reception and motor response occurs within the spinal cord. However, most spinal reflexes include association neurons that inform the brain that a reflex has occurred. The magnitude of a reflexive response is directly proportional to the intensity of the stimulus, provided that the latter falls within the threshold interval of the receptor.

Reflexive behavior is most prevalent and important to the survival of animals with relatively simple body plans. Humans have relatively few reflexes; the knee jerk and sneeze are examples. But a decapitated frog may survive for weeks if fed through its esophagus. The fly has taste receptors on its tarsal pads ("feet"); if a fly lands on sugar, its proboscis (feeding tube) reflexively extends. In many invertebrates, copulation is reflexive.

During the development of organisms, environmental information is constantly bombarding it, so that each individual organism, even if it started with identical genetic information, is an individual different from all others. Organisms using nervous systems are constantly processing sensory information; some may be immediate stimuli for reflexive or instinctive behavior, whereas some information may cause long-term changes that influence behavior at some later time. This stored information about the environment is called **memory,** and the process of establishing this information in a manner that allows it to alter behavior for some useful purpose is called **learning.** Most behaviors discussed in the following paragraphs are probably a mixture of innate and learned information, and it is often impossible to decipher the exact contribution of each. This problem takes the form of the controversial "nature–nurture" question concerning complex human behaviors, such as intelligence. It should seem clear from your understanding of the complex system called the organism that it is a highly dynamic and regulated integration of genetic and environmental influences and that no complex behavior can be entirely innate or entirely controlled by external forces. Keep this in mind during the following discussions, especially when complex responses, instincts, memories, and other behaviors are considered.

Stimulus–Response

Most behaviors of animals with nervous systems may be viewed as stimulus–response relationships. The only exceptions are caused by random nerve firings ("mistakes" of the nervous system). A response may be quite uncomplicated, as in a simple reflex, or extremely complex, as in one's response to the death of a loved one. Responses, like stimuli, may be physiological or phychological. The majority of this chapter is devoted to descriptions of various responses.

Stimuli may vary greatly in duration. A pinprick is a stimulus lasting only milliseconds (though its effects may be felt for some time), whereas an illness (which acts as a stimulus to the homeostatic receptors of an organism) may last for years. The effectiveness of a stimulus in eliciting a response depends partly on its intensity. In the 1860s, Gustav Fechner introduced the study of **psychophysics,** which concerns itself with the physiological basis of behavior. A major concept in psychophysics is the importance of the **threshold.** A receptor will respond to a stimulus only if the stimulus is greater than a certain minimum intensity (threshold); otherwise the stimulus will not be perceived by the organism. The pressure receptors in our skin present a convenient example: If a weight of, say, 1 mg is placed on the skin, one may not be aware of it. However, a weight of 1 g will probably be felt. If so, somewhere between these weights at that locus

lies the threshold for pressure reception—the weight at which pressure may be felt but below which none can be perceived.

Note that both lower and upper thresholds may occur. For example, a normal human cannot hear a sound of 5 cycles per second (cps) or a sound of 30,000 cps. The normal range of human hearing is from 20 to 20,000 cps, and this interval is defined by thresholds of the auditory receptors.

Stimuli may be received from either the external or internal environments. External stimuli are perceived by the five senses (Chapter 36): optic (visual), olfactory (smell), auditory (hearing), tactile (touch), and gustatory (taste). These may elicit responses either singly or in complexes. For example, several simultaneous visual cues may elicit a response. Also, combinations of stimuli may all contribute to a sensation (e.g., the familiar interaction of taste and smell). Animal orientation may depend upon the stimuli of illumination, tactile contact with a substratum, chemical gradients, gravity, wind and water currents, and astronomical cues.

Internal stimuli may be either kinesthetic or psychogenic. **Kinesthetic stimuli** are perceived by receptors in the joints, muscles, ligaments, tendons, blood vessels, and various other organs that serve to inform the organism of its present state. One's sense of weight, position, balance, motion, and temperature are perceived, at least in part, kinesthetically. **Psychogenic** stimuli originate from mental processes and do not depend upon concrete physical stimuli. Any thought may be considered a stimulus. It may be triggered by external stimuli, however; for instance, a particular odor may "remind" one of a past experience (here, the thought is a response).

Motives are internal stimuli that may be psychological, physiological, or both. A **psychological motive** follows the principle of **hedonism**–that is, that an organism capable of purposeful acts will actively attempt to preserve or improve its present state. A **physiological motive** is based on those processes by which an animal unconsciously and automatically attempts to maintain a constant internal environment (homeostasis). An example of a complex psychophysiological motive is hunger, which is composed of a psychological component (food tastes good) and a physiological component (low levels of glucose utilization in the blood).

Learning, Conditioning, and Memory

Learning is a modification of behavior that due to experience results in a record of the learned experience (memory) in the brain.

A particular sensory image that is to be "memorized" must cause some physiological change within the brain; this is the **engram,** or **memory trace.** Memory is commonly divided into **short-term,** and **long-term memory.** A small amount of sensory information may be retained for a short period of time, and, if there is repetition of and concentration on the reception of this information, it will result in a long-term memory trace. The retrieval of a memory trace (transfer from the subconscious to the conscious condition) is called **recall.** This process is usually elicited by a particular stimulus; when the stimulus is external to the organism, the activation of a trace is called **recognition.** Recall may also be elicited by internal stimuli, that is, by other memory traces in the processes of **association.** Associative recall capacities increase with increasing brain size, culminating with its most advanced state in the human processes of logical thinking. The ability to recall, associate and form new memory traces is probably the basis for most of the human processes that are commonly called the mind.

Figure 31-1 presents a tentative model for human memory and information processing. This model probably has some relevance to the less complex memory activities of other animals.

The Physiological Basis for Memory

In the 1960s, many experiments were performed to find the chemical changes that constitute a memory trace. Especially popular were the experiments on planaria, in which certain worms were conditioned to do a particular task and then ground up and fed to unconditioned worms. The results showed that the unconditioned worms learned the same task faster (in fewer trials) than did worms not fed their conditioned comrades. Realizing that genetic information is stored in nucleic acids, some investigators demonstrated that RNA extracted from conditioned planaria and injected into untrained individuals also improved conditioning in the latter. These, and similar experiments in other animals, have received much criticism in recent years for lack of adequate controls and the

inability to reproduce certain results. There are clearly changes in the amount and types of RNA synthesized after learning experiences; this may simply be new messenger RNA's (see Chapter 38) and, hence, new proteins, needed for the neurons involved in memory, not information storage as in genes. There are still some chemical transfer experiments that have not been disproven, and the problem is still open to controversy.

Investigations that attempt to localize memory traces within certain regions of the brain also disagree with intracellular models of memory encoding. Results show that memory is dispersed over a large area of the cerebral cortex and probably involves lower brain centers as well. These results favor the encoding of memory as changes in the structure of the nervous system, that is , in the patterns and strengths of interneural connections. John Eccles found that neural pathways function better (electrical transmission is facilitated) when they have been used extensively in the organism's recent past. If a nerve is not used for a considerable period of time, then, like a muscle, it may atrophy (degenerate), at least partially. This could be the physical manifestation of forgetting. For example, each nerve in the cerebral cortex has one or several dendrites, and these are covered with thousands of protrusions called **spines.** When neural input to a cortical dendrite is reduced, over a period of time many of its spines completely degenerate.

It has been demonstrated that neural circuits, to remain intact, require neural stimulation above and beyond the intrinsic baseline firings that occur. For instance, in one experiment one eye of each member of a litter of newborn kittens was covered, and after a period of time they were tested for stereoscopic (binocular) vision, which requires the functional interplay of both eyes. These kittens were found to lack depth vision, even though the necessary neurons were present. If the masks were removed after a relatively short time the kittens eventually developed stereoscopic vision, but visual impairment was irreversible if the masks were worn too long.

Some scientists believe that learning is the result of newly manufactured cortical connections (synapses). If that is the case, it is possible that spine growth is essential to the new synaptic connections. It would then follow that forgetting, possibly due to disuse of particular synapses, would result from degeneration

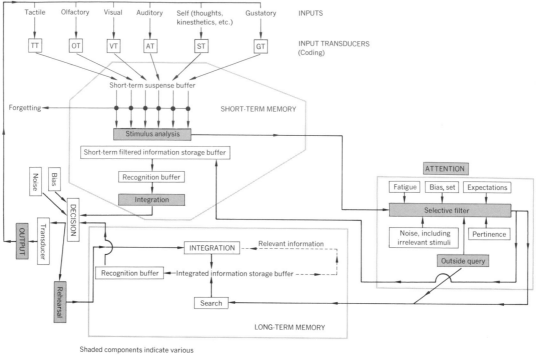

Tactile   Olfactory   Visual   Auditory   Self (thoughts,   Gustatory   INPUTS
kinesthetics, etc.)

TT   OT   VT   AT   ST   GT   INPUT TRANSDUCERS (Coding)

Short-term suspense buffer

Forgetting ←

SHORT-TERM MEMORY

Stimulus analysis

Short-term filtered information storage buffer

Recognition buffer

Integration

Noise   Bias

DECISION

OUTPUT   Transducer

Rehearsal

INTEGRATION   Relevant information

Recognition buffer → Integrated information storage buffer

Search

LONG-TERM MEMORY

ATTENTION

Fatigue   Bias, set   Expectations

Selective filter

Noise, including irrelevant stimuli   Pertinence

Outside query

Shaded components indicate various forms of rehearsal

Figure 31-1. Proposal for a human information processing system. The student is not expected to memorize this but merely to appreciate the complexity of human information processing. Stimuli enter the system through tactile, olfactory, visual, auditory, gustatory, and self inputs. Self is considered an input because of kinesthetics, volitional stimuli, and the like. Stimuli are coded in the Input Transducers and are sent to the Short-Term Suspense Buffer. If information is kept here too long, forgetting takes place. Information may be sent to Stimulus Analysis and acted on by Attention. Noise, Pertinence, Fatigue, Bias, Expectations, Set, and Outside Query all affect a Selective Filter (one based on more than mere physical attributes), constituting Attention. Information returning to Short-Term Memory enters the Short-Term Filtered Information Storage Buffer, where it may remain for a few seconds. Information then is acted upon by the Recognition Buffer and is Integrated so that a Decision, affected by both Bias and Noise, may be made.

After Decision, information may take one of two routes. One sends it to Rehearsal, strengthening it enough to enter Integration in the Long-Term Memory. The information is then held for an indefinite amount of time in the Integrated Information Storage Buffer, which is the main component of Long-Term Memory. Relevant Information may be run again through Integration and back to Integrated Information Storage Buffer, though it has been debated as to whether "thought" is possible at this level. Search is activated by Attention, especially the Expectations and Outside Query components. Here, the indexing of the Integrated Information Storage Buffer is used, and the Recognition Buffer collects appropriate information to be sent to Decision.

The other route out of Decision consists of a trip, first, through the Transducer, which changes the Decision impulses into physical and/or mental Output. The Output affects the outside world *as well as* one's self, as we can see, hear, smell, and so on our own acts. Note: Although there is a separate Rehearsal component, shaded components also contribute to rehearsal.

482

of spines, which would in turn unlink the synapses. A logical extension, of course, is that learning produces new spines and therefore new synapses.

Conditioning

No matter what the mechanism of learning, it is generally agreed that learning is adaptive. An organism's response to a given stimulus may be modified by the absence or presence of another stimulus normally associated with it. This form of learning is termed **conditioning**. Two major categories are recognized: **classical conditioning** and **instrumental (operant) conditioning.**

The Russian psychologist, Ivan Pavlov, in 1927 reported the results of a series of experiments that established classical conditioning as one of the most basic processes of learning. A brief description of his experiments will serve to explain the fundamental concepts of classical conditioning.

Pavlov placed meat powder in the mouths of hungry dogs, and the dogs responded by salivation, mastication, and swallowing. In classical conditioning terms, the meat powder was the **unconditioned stimulus(s) (UCS)** for the **unconditioned response(s) (UCR)** of salivation, mastication, and swallowing. A UCS is generally unlearned or "natural," as is the UCR that it elicits.

Pavlov then began ringing a buzzer just before he gave the dogs their food. The buzzer became the **conditioned stimulus (CS).** He observed that after a few training trials the buzzer alone would elicit salivation. Thus, salivation became the **conditioned response (CR)** to the CS, in other words, the UCR became the CR. The basic model of classical conditioning is

1. UCSUCR
2. CS + UCSUCR
3. CSCR (= old UCR)

Instrumental, or operant, conditioning does not involve a substitution of stimuli as in the case in classical conditioning. Instead, there is a trial-and-error type of learning: An animal makes many random responses until a desirable outcome is achieved. The attainment of a desirable situation (e.g., receipt of food, releases from pain, or any other satisfaction of a need) becomes **positive reinforcement** for the response that preceded it. The frequency of that re-

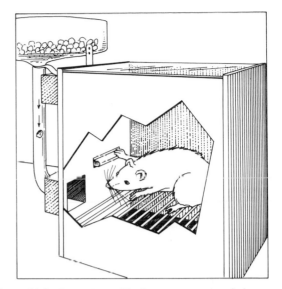

Figure 31-2. Operant conditioning cage, commonly known as the "Skinner box," after B. F. Skinner, originator of operant conditioning theory. The rat may press the bar to obtain a pellet of food, to turn on or off electrical shocks from the grid upon which it stands, or to respond to any number of different reinforcing stimuli that the experimenter may provide.

sponse increases as more positive reinforcement is experienced.

This was well demonstrated by B.F. Skinner's and E.L. Thorndike's experiments in animal conditioning. Skinner placed hungry rats into a box that contained a bar (Fig. 31-2). When the bar was pressed, a pellet of food was released into the "Skinner box." The hungry rats made random movements, and every once in a while they would press the bar. Gradually, the rats learned that pressing the bar would bring food. The response of bar pressing was positively reinforced by the delivery of food, which relieved their internal stimulus of hunger. Consequently, the rats "learned" to press the bar for food and would go straight for the bar if placed in the box at a later date. Thorndike's experiments were similar, except that it related to cats that learned to push a bar to escape from a confining box.

The following experiment suggests the power of operant conditioning. Hungry chickens were reinforced with food as they performed random activities in a barnyard. In a short time, the birds were consistently performing many individually bizarre activities, such as spinning around in circles, constantly walking to and from points in the barnyard, or pecking at a rock. It is assumed that the chickens happened to be performing these activities repeatedly when food was given to them, and so these behaviors were positively reinforced.

Negative reinforcement is defined as the removal of adverse or competitive stimulation. For example, just as a hungry rat will learn to press a bar for food pellets (here, food is a positive reinforcement), so too will rats learn to press a bar to stop an electrical current from crossing the floor of the cage (administering a shock is a negative reinforcement). In both cases, the bar-pressing response is reinforced by the attainment of a desirable outcome.

Punishment can also lead to behavior modification. If a rat is given an electrical shock whenever it stands on two feet, in a short time upright behavior will cease.

The physiological basis for conditioning is probably the establishment of a memory trace of the conditioning stimulus; recognition of this stimulus is linked by associative memory recall to a trace of the unconditioned stimulus, thus eliciting the same response (UCR) that would occur when the unconditioned stimulus is presented.

Habituation

Habituation essentially represents the extinction (or at least a decrease in frequency) of a response that has become nonadaptive. For example, a loud noise will cause a bird to take flight, but, if the bird hears the noise very often, soon it will not respond.

Insight

Insight is a modified type of trial-and-error learning in which an animal rapidly learns a particular (sometimes complex) response. Insight is also called **abridged learning.** Insight is most prevalent among the higher vertebrates and is usually not observed in lower vertebrates and invertebrates. It is exemplified in a situation in which food is visible to an animal but in which a barrier exists between the two. Most animals capable of recognizing the food would try indefinitely to break through the obstacle, but some dogs, cats, and primates will travel away from the food to circumvent the obstacle after only one failure at traversing the barrier.

Instinct

Instinct is an inherited, often complex, response or series of responses invoked by a particular stimulus or series of stimuli; the stimuli may be internal, external, or a combination of both. According to Breland and Breland, "Learned behavior drifts toward instinctive behavior." These experimenters observed that animals conditioned to perform a certain task tended to gradually cease that behavior in favor of a similar, instinctive behavior. Such a shift away from the conditioned response occurred even if it delayed or prevented positive reinforcement. What makes instinct so strong that it may overshadow direct hedonistic gratification?

The most basic characteristic of instinct is its innateness; that is, instinct is believed to be largely unlearned and neurally built into the organism so that such behaviors are stereotyped and hereditary. These qualities are exhibited by spider web patterns, which are specific to most arachnid groups. A spider will normally construct a web essentially identical to that of all members of its species, without ever having seen the web produced by others.

Lorenz and Tinbergen proposed a three-stage model of instinctive behavior. In the first stage, an animal experiences **appetitive behavior,** or restlessness, in preparation for the instinctive act. This tension gives rise to the second stage in which a **releasing stimulus** activates an innate releasing mechanism. For example, the male stickleback fish is known to attack other sticklebacks that have red bellies (the releasing stimulus). If an uncolored artificial stickleback is presented to a male fish, nothing will happen. But if a crude model with a red belly is displayed, an attack will occur. The attack represents the third stage of instinctive behavior: the final **consummatory act,** which relieves the tension.

A specific example of a releasing stimulus is found in ticks that climb foliage and there await the passing of a potential host. The odor of butyric acid (characteristic of mammals) will cause the tick to drop from its perch onto the host. And, if a rock coated with butyric acid is presented, the tick will attempt to bite the rock. Carbon dioxide elicits a similar instinctive response. Another example: Some artificial lures used by fishermen successfully represent releasing stimuli. The Lorenz–Tinbergen theory, reasonable as it sounds, still lacks firm experimental verification.

An interesting aspect of instinct is that the magnitude of response has no correlation with the stimulus magnitude. For example, it has been observed that a lion or tiger will flee at the sight of a frog, a foreign (and therefore "frightening") object.

Physiological Clocks

Almost all aspects of animal behavior are rhythmical, with activities occurring in cycles. Humans normally sleep at night and are active during the day. Reproduction is periodic in many animals. Several characteristic **cyclic periods** (lengths of time), such as the day, year, and lunar cycle, are encountered repeatedly in the study of zoology. A **circadian rhythm** has a period of about 24 hours. Circadian cycles seem to be coordinated with some period of the external environment (namely, solar days), as are the vast majority of biological cycles. Every animal studied so far exhibits some type of cycle corresponding to a geophysical frequency, and the biological rhythms rarely vary more than one hour on either side of the geophysical event.

Animals may be active only at certain times. **Nocturnal** animals, such as many owls, sleep during the day and are active exclusively at night. In the opposite case **diurnal** animals are active only during the daytime. **Crepuscular** animals, such as many bats, limit their activities to periods of dim light at dawn and dusk.

Two major hypotheses exist to explain the controlling factors of such biological clocks. The **exogenous hypothesis** states that physical factors outside the organism control the clocks. The physiological clock may remain intact even through such drastic changes as hibernation (as it does in the bat, **Myotis myotis**) or metamorphosis. The **endogenous hypothesis** maintains that the rhythms are based within the organisms but allows that external stimuli may affect (reset)

them. Many animals will continue their rhythms even if placed in steady-state environments (constant temperature, humidity, illumination, pressure, and chemical composition). For example, fiddler crabs, *Uca*, normally exhibit a circadian rhythm in their coloration. During daylight they are dark in color, while at night they are lighter. One might expect this chromatic change to be due to light impinging on the exoskeletal pigments, but the fiddler crab will go through circadian color changes in a steady-state environment as well as in its natural habitat. The clock may be reset however. If a fiddler crab is placed on ice for six hours, the whole cycle will be offset by six hours.

This resetting phenomenon is often realized by humans traveling by jet who find themselves suffering the effects of "jet lag." For example, a flight from California to Switzerland upsets one's physiological clock by approximately nine hours. Several experiments have been done on the resetting of physiological clocks. If resetting is allowed to occur autonomously, that is, if an organism is placed in a particular steady-state environment for several days or weeks and is allowed to do whatever it wishes, the activity rhythms may shift in roughly equal but significant increments (Fig. 31–3) in an attempt to accommodate the new conditions.

Attempts have been made to localize the clock, assuming that the endogenous hypothesis is correct. Experiments have shown, for example, that the nuclei of neurosecretory cells in the rat hypothalamus may exhibit diurnal fluctuations in volume, but no conclusive evidence exists that the central nervous system is the control center for the biological clocks.

Some Examples of Behavior

The behavior of protists such as the amoeba is based on the inherent responsiveness of the cytoplasm of a single cell. Extensive studies indicate that there are no fundamental differences between the behaviors of protists and the lower metazoans. The behavior of protists appears to be no more or less stereotyped than is that of simple metazoans; similar principles seem to govern both. Higher metazoans exhibit more complex behavior.

A

B

C

Figure 31-3. Opposite: Perching behavior in the house finch (bird) as a function of illumination. Each horizontal line represents one 24-hour period (successive lines represent successive periods), and the vertical marks represent perchings. (A) For this set of responses, light was turned on and off for the same period at the same time each day. We see that the finch only perched when the light was on. (B) Here, constant bright illumination was present. The finch perched erratically, and the longest "rest" period was about 3 hours. (C) In this situation, constant, very low, illumination was provided (0.1 foot-candle). This light level approximates the darkest environment that the bird ever normally experiences (complete lack of illumination is rarely, if ever experienced). We have evidence here for an internal circadian rhythm, for the finch receives no environmental time cues. Note that although this biological clock changes over time (the finch perches 45/60 min earlier each day), the period of activity remains relatively constant. [Courtesy of Science Software Systems, Inc.]

Kineses and Taxes

Animals exhibit various locomotor behaviors that orient them with respect to various aspects of their environments.

Kinesis is a general term for physical movement. Kineses are undirectional locomotor responses to particular environmental stimuli. An animal exhibiting kinesis is activated (or deactivated) by a particular environmental stimulus. For example, some organisms will move faster under unfavorable conditions but will slow down in a more favorable environment.

Taxes are directed locomotor responses, that is, the movement of an organism in a particular direction in response to an external stimulus. Many organisms move toward more favorable environments and, conversely, away from less favorable environments. For instance, fly maggots move away from illumination; worker honeybees move toward a light source.

Imprinting

This type of behavior is mostly limited to young animals. It is concerned with behavior elicited by particular visual, auditory, olfactory, or tactile stimuli. The classic example of imprinting was documented by O. Heinroth in 1910. He found that a chick will follow the first large object that it sees just after hatching and that it will do so completely ignoring anything else. If a human walks slowly by a newly hatched chick, the chick will follow the human even if it meets its mother along the way. A. Ramsay has observed that, under proper conditions, young fowl may imprint on inanimate objects such as boxes and footballs. These types of experiments were also done by K. Lorenz, who

termed such behavior **imprinting**. Lorenz successfully imprinted geese on humans.

Communication

Communicative behavior may be expressed through visual, tactile, auditory, or chemical cues, either alone or in combination.

An important example of **visual communication** is the hand language used by deaf humans. The vocabulary of this system is extensive. Other visual communications include the defense behavior of feigning death or disablement. The hognose snake, although relatively harmless, hisses and strikes when irritated (with its mouth closed, though): if all else fails, it will feign death.

Tactile communication is exemplified by the spawning behavior of the three-spined stickleback fish. When the male nudges the tail of the female, she releases her eggs. Another example is mutual grooming, in which two animals (notably primates, cats, and horses) clean or scratch each other. Such tactile behavior is thought to communicate trust and contentedness.

Auditory communication is perhaps the most common form of communication in higher animals. Its effectiveness is measured, as in all forms of communication, by the richness of the vocabulary. Vocabularies vary considerably among animal groups. One primate group is known to have nine distinct vocal signals, and those of many more species have been described. The vocabulary of humans seems to be the most extensive, but communication research with other primates, and with dolphins, is revealing examples of the auditory and vocal capabilities of nonhuman species.

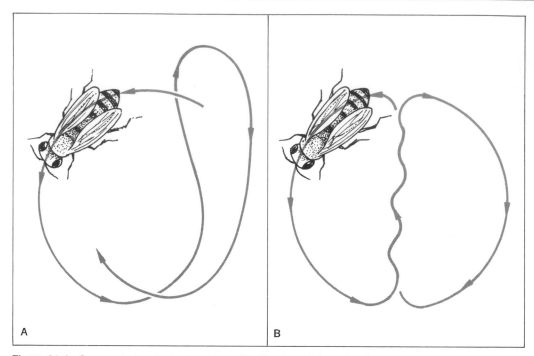

Figure 31-4. Communication in the honeybee. (A) The "round dance" indicates that food is nearby. (B) The "tail-wagging dance" indicates, via the wavy run the direction of the food source, and the time spent "wagging the tail" indicates the distance from the hive.

Many marine fishes produce sounds. These water-borne vibrations may be transmitted to the inner ear of other fishes through various means. The air bladders may emit sound vibrations because of muscular activity, or they may act as resonating chambers that amplify the sounds made by grinding teeth or beating the fins against the body wall.

The sea lion, *Zalophus californianus* (Fig. 31-6), uses different modes of auditory communication in different surroundings, and the males and females produce different sounds (though the sounds are understood by both sexes). The bark, belch, growl, high-pitched squeal, pup attraction call, and mother attraction call comprise the above-water vocabulary. Barks, bangs, clicks, buzzes, and growls are used underwater as is a sonar communication system.

The honeybee provides a well-studied example of a complex, primarily visual, communicative behavior. Karl von Frisch, who shared the 1973 Nobel prize for medicine and physiology with Lorenz and Tinberge,

studied in detail the "language" of honeybees during the 1920s. Using marked bees and observation hives, von Frisch discovered that the bees communicate with a dance, the movements of which indicate the location and abundance of a particular food supply (Fig. 31-4). The dance is performed on a vertical surface of a honeycomb; if the major direction of the dance is up, the food is located away from the hive toward the sun, and, if down, the food is away from the sun. The number of dance cycles performed per unit time indicated the distance of the food from the hive: the farther away the food, the lower the frequency of the dance cycles. Different dance forms are used to communicate different distances. The "round dance" indicates a distance of under about 85 m (the scout bee dances in a circle right then left several times), whereas the "tail wagging dance" is used to indicate farther distances. Some species change dance forms at increments of, for example, 3 m.

The information communicated by the dances is

very specific. The compound eye is very sensitive to light polarization, and hence the sun's position can be accurately observed. The behavior is so complex that the bees extrapolate the sun's position; even if the view of the sun is blocked, the dancers' orientation will shift in conformity with the sun's actual move- ments. In addition, different species use various "dia- lects" based on the stereotyped dance patterns.

Olfactory communication is best exemplified by the use of **pheromones.** These are chemicals that cer- tain animals release into the environment that influ- ence the behavior and/or development of other mem-

Figure 31-5. EEG (electroencephalograph) tracings. Each line on an EEG represents electrical changes recorded di- rectly over specified loci on the brain, though the recordings are made with electrodes placed on the integument of the head. Left: Normal EEG of a human in the waking state. Note the slow, large-amplitude waves (alpha waves) concern- ing which there is considerable current research. Right: Normal EEG of a sleeping human. Alpha waves are absent here; the general wavelengths are of small amplitude and high frequency. EEG recordings are useful in the diagnosis of brain pathology and have currently been used especially in the study of epilepsy (epileptic seizures seem to travel from one side of the brain to the other and are evidenced by erratic, very-high-amplitude EEG recordings). [Courtesy of Science Software Systems, Inc.]

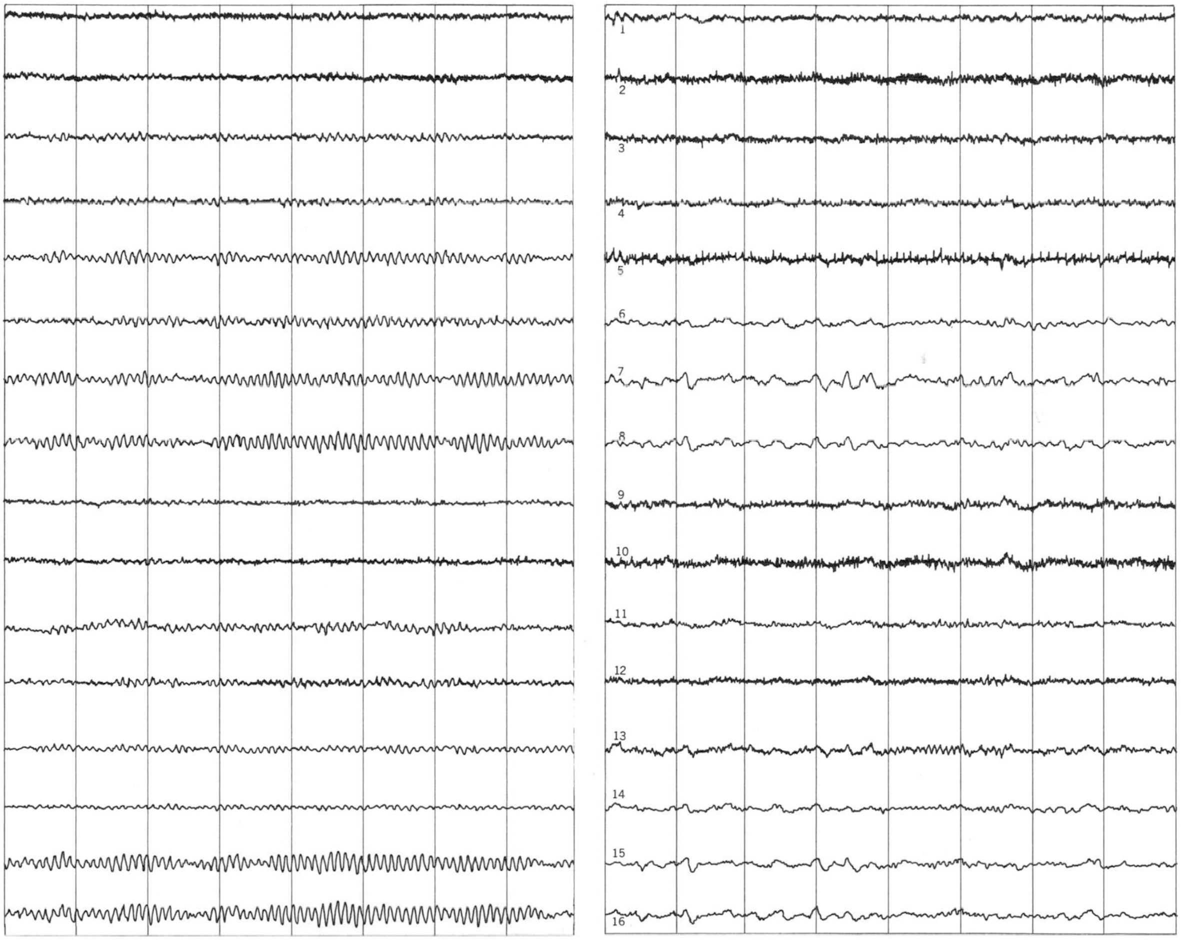

bers of the same species. Pheromones may function to attract mates. This form of communication may be long range; some moths can respond to pheromones released over 10 km away. The concentration of a pheromone may indicate the direction of its source. How might this work?

Chemical communication is also demonstrated by ants. They deposit pheromones on the substratum over which they walk. It has been found that separate "food" and "danger" pheromones are used by ants. Also, the symbiotic "guests" of many colonies, such as beetle larvae, may communicate chemically with the ants.

Sleep

An operational definition of sleep, among the vertebrates, is the deactivation of the reticular formation in the midbrain (Chapter 36). This part of the brain acts as a sensory filter that selects the incoming stimuli that are sent to the cerebral cortex. If the reticular formation is damaged, vertebrates seem to be in a constant state of sleep. Basically, sleep is a lack of attention—a psychological insensitivity to the environment.

Most sleep research has been done on humans.

Sleep is characterized in humans by deeper, slower breathing than when the human is awake and a slowing of most metabolic processes. Two stages of sleep are recognized. **Paradoxical sleep,** occurring four to six times a night, exhibits an **electroencephalogram** (**EEG**) trace similar to that when awake, and rapid eye movements (**REM**) occur (Fig. 31-5). Paradoxical sleep alternates with **slow-wave sleep** (**SWS**), characterized by large-amplitude, low-frequency EEG tracings.

Sleeping periods vary considerably in duration. Whereas the albatross, a bird that spends most of its life over the ocean, sleeps for extremely short periods during gliding flight, bears may enter a deep sleep that resembles hibernation for several months at a time.

Aggression

Aggressive behavior is widespread among animals. It may take the form of a physical attack or a warning display, such as the baring of teeth by a dog. Aggressive behavior may be communicated by bodily postures. Myriad stimuli may evoke aggression. Frustration, territoriality, sexuality, and social ascension may all stimulate aggressive behavior. The termination of

Figure 31-6. Territoriality in elephant seal. This is during the mating season, and we see two males vehemently establishing boundaries.

aggression may be either **physical** (for instance, the death of one of the combatants) or **symbolic** (a dog may bare its throat to another to indicate defeat).

The sea lion, *Zalophus californianus*, obtains and maintains territories on its breeding grounds by aggressive behavior. Figure 31-6 shows two elephant seals establishing territorial boundaries. Observers have noted the following sequence of events: When two territorial males see each other, they move rapidly toward each other, barking. The vibrissae (whiskers) are extended in this stereotyped sequence. When they are close, they fall on their chests and open their mouths, ceasing their barking behavior. After shaking their heads from side to side, they rear up as far as they can and stare at each other obliquely. This sequence may be repeated, until the boundaries of their respective territories are established. Such complex sequences of aggressive behavior are not uncommon in many other species.

Courtship Behavior

Courtship behaviors have been studied extensively. Most often they are cyclic or seasonal, varying with lunar, temperature, or estrous states. For example, the female rat is sexually receptive for a period of approximately 19 hours every 4 to 5 days.

General characteristics of courtship behavior, which often acts as releasing stimuli, include

1. species-specific amounts of mutual tactile activity (which may stimulate further hormone release)
2. a soliciting posture of one mate (usually, but not always, the female), which may facilitate intromission
3. special "calls"
4. protracted precopulatory periods
5. selective, as opposed to random, mate choice

At the beginning of precopulatory activity in the cockroach, *Nauphoeta cinerea*, the male strokes the female's antennae (a virgin female emits a special chemical stimulus that increases the probability of copulation). The female reciprocally strokes the male's antennae. A characteristic posture follows: the male turns his back to the female and holds his wings at a 90° angle for about 1 minute. This exposes special glands that secrete a substance that attracts the female. The female "mounts" the male, he seizes her genitalia, flops over, and copulation begins.

In addition to fights over mates and breeding territories, aggression may play a major role in courtship and precopulation behavior. In the yellow-hammer, *Emberiza citronella*, a bird of northern and central Europe, the first meeting between two potential mates frequently results in a fight. Gradually, the male and female become habituated to each other and leave the flock. For several weeks during the courtship period, the male performs mock attacks on the female. During early May, nesting and mating behaviors begin. The male exhibits two sexual displays, the "fluffed run" away from the female and the "raised-bill run" toward the female. The latter may lead to copulation. Many species seem to recognize sex by the reaction: if it fights, it's a male; if it's receptive, it's a female.

The jewel fish, *Hemichromis bimaculatus*, has a rather complex method for preparing a spawning site. One of the mates chooses a site and indicates the selection by quivering over it. The other mate indicates approval by following suit. Both then clean the spawning site with their mouths. Finally, the female skims over the stone, touching it with her belly and depositing eggs, and the male follows, depositing sperm.

A rather eccentric copulatory sequence is followed by the praying mantis. When a female mantis sees a male, she may decapitate him! The male physiology is such that decapitation removes neural inhibition from parts of the ventral nerve cord, allowing continuous adbominal and genital copulatory undulations.

Migration and Homing

Migration is a seasonal movement from one geographical locality to another. The best-documented migrations are those of birds, but some fish (salmon), reptiles (sea turtles), insects (monarch butterflies), and mammals (bats, whales, and many others) also migrate.

Most bird migrations are latitudinal, north–south routes. Another migration type is altitudinal; for example, elk and mountain goats migrate between highland and lowlands. Latitudinal migrations take varying lengths of time and may span considerable distances. A warbler may take up to 60 days to complete its summer migration from Central America to Canada. The Arctic tern's migrations cover a distance of 40,000 km each year! The California sea lions have an interesting migratory behavior. In the winter, the bulls (males) migrate as far north as British Columbia, whereas the females migrate south toward Mexico.

Migrations fall into two major categories. **Climatic migration** brings an animal from a zone of (at least potentially) unsuitable climate to another that will better support daily activity. One theory states that an unfavorable energy balance is the releasing stimulus for the onset of climatic migration. This theory stresses the primary importance of food resources in accounting for migration. Because of the importance of food in many cases, the term **alimentary migration** is sometimes used. The new habitat generally provides for more extensive hunting or foraging and often offers improved conditions for rearing progeny.

Reproductive (gametic) migration occurs at the beginning of a breeding season and is generally preceded by gonadal changes. Salmon serve as a good example of reproductive migration. The salmon lives in deep marine waters for three to four years. At spawning time, there is a reproductive migration to the headwaters of the stream where the salmon was born. The Pacific salmon die after spawning; of the Atlantic salmon, the species *Salmo salar* die, but others may return for three or four spawnings. The young salmon remain in their freshwater environment until salt-secreting osmoregulatory cells develop; then they migrate to the ocean.

Orientational Mechanisms of Migration

It has been proposed that salmon migration into freshwater streams is directed by olfaction. It has been demonstrated that each stream (and hence each estuary) has a characteristic "odor" due to its constituent solutes, to which salmon are sensitive.

Avian migratory activity is oriented by the sun, and nocturnal migratory birds can orient by the stars as well. Starlings (*Sturnus vulgaris*) confined during a migratory season tend to face the migratory direction if they are able to see the sun. If a stationary artificial sun is made visible to the birds, they shift their orientations approximately 15° per hour (which indicates a definite endogenous mechanism).

Additionally, migrating birds are capable of compensating for daily solar position changes by using the solar azimuth (Fig. 31-7). These migratory tendencies are evidenced by the directional hopping of caged birds in northerly or southerly orientations during the appropriate seasons. Pigeons, and probably other birds, have a most interesting orienting mechanism, an internal compass that orients them to the earth's magnetic field.

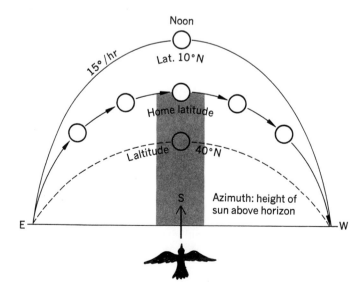

Figure 31-7. The solar azimuth is a navigation aid to bird migration. Shown here for the northern hemisphere. To get "home," if the sun is too high, the bird will fly away from it, whereas if it is too low, the bird will fly toward it. For the southern hemisphere, the reverse holds. Different "homes" will of course require different reactions to the azimuth.

Homing

The phenomenon of **homing** is intriguing. It consists basically of the return of an animal (usually a bird) after having been transported some distance away, frequently into unfamiliar territory. Some birds can even home after having been carried to a new location in a closed box. Perhaps the most familiar examples are **homing pigeons,** which have been selectively bred for over a century from domestic pigeon (*Columba livia*) stock. Homing pigeons have been known to return to their homes from a release point located as far as a 1600 km away.

Two major hypotheses have been proposed to explain homing behavior when the release location is unfamiliar to the bird. The first states that birds may extrapolate their position by using the sun's arc. This mechanism would be similar to that used in migration. The second hypothesis holds that homing may be accomplished by semirandom search. Some released birds are observed to fly in ever-increasing spirals, presumably searching for a familiar visual landmark.

Some dogs have been known to travel hundreds of kilometers to return to their lost owners. What mechanism might account for the occasional success of these homings?

Social Behavior

The main, apparent reasons for the existence of **social groups** (aggregations) in some animal species lie in environmental factors (temperature, illumination, presence of water, etc.). Allee has proposed three principles of animal social behavior:

1. **Dominant–subordinate hierarchies** occur in many animal societies.
2. The existence of **territorial rights** serve to control the acquisition of nutriment and mates.
3. Some forms of leadership exist in most animal societies.

T. Schjelderup-Ebbe, a Norwegian scientist, noted in 1922 that chicken societies are based on a **peck order.** The dominant hen may peck at all others without retaliation; the next in line may peck at all but the first; and so on. Rank in the linear peck-order hierarchy is maintained by such aggressive behavior. Leadership positions in animal societies is attained in different ways in almost every species. A pack of wolves is usually composed of one family, and the leader, a dominant male or female, heads the stable peck-order hierarchy. The pack is so structured that, if a female is pregnant, she stays in the den while the others hunt and bring her food.

The society of the European red deer, *Cervus elaphus,* is also somewhat complex. The herds are matriarchal (one female is the leader); once a male reaches sexual maturity, he leaves the main herd and joins a male herd. During the fall, the male herd breaks up and the individual bulls select and guard **harems** (groups of females). A male will keep the harem, copulating frequently, until challenged and beaten by another male.

Chimpanzee social behavior is extremely complex. There is much mutual grooming and tactile contact to indicate greeting and trust. A most interesting phenomenon of chimpanzee society is that of the "carnival." The animals get together and for several hours continuously beat trees, scream, dance, and jump on one another. Jane Goodall, who has contributed much to our understanding of chimpanzee societies, was permitted to participate in such carnivals, mostly because she learned the simple dance (come down heavily on one foot, lightly on the other, and repeat indefinitely).

Many fishes congregate in asocial schools. There is essentially no hierarchy to a school—the fishes react and move as a unit. This is made possible by their lateral line sense organs (which detect pressure waves) and eyes that allow each fish to follow what the rest of the school is doing. If a school is startled, it immediately closes ranks and equal fish–fish spacing occurs; it then flees as a unit.

The territoriality of many animals is well documented. Most of a bird's chirping, especially during the breeding season, is a statement of "This is my territory" (Fig. 31-8).

Most wild birds spend considerable time flying around the perimeter of their territories, warning others of the same species to keep out. Lions and tigers exhibit definite territoriality, though in captivity their defended territories extend only a few meters around them. The circus "lion tamer" is careful to keep an appropriate distance from the beasts. If the

Figure 31-8. Sea gull territoriality. A seemingly placid sea gull changes its temper when its territory is threatened.

trainer moves into the territorial space set by the animal, a rather ferocious, aggressive display or even an attack may ensue.

The Old World rabbit, *Oryctolagus cuniculus*, lives in a group, or **warren.** Each rabbit indicates its territory by means of pheromones secreted by anal and submandibular subcutaneous glands. The size of a rabbit's anal gland, whose secretions mark its territory, is of great importance to its position in the rabbit society hierarchy.

Other delineations of territorial boundaries include the familiar "posting" of a territory by a dog's scented urine, the claw scratching on trees of bears, and the urine marking by cats.

Tool Use

The use of tools is generally confined to animals with complex nervous systems. Only a few animals other than humans use tools. The Egyptian vulture, *Neophron percnopterus*, breaks the shells of ostrich eggs by dropping rocks on them. On the Galapagos Islands, some of Darwin's finches manipulate thorns or twigs with their beak to pry insects from crevices in tree bark. The wasp, *Ammophila*, uses a pebble to tamp the earth over its burrow, thereby sealing the

entrance. Chimpanzees insert a straw into a termite mound and then lick off the termites that attack it. Another form of tool use is the utilization of inanimate objects as shelters, such as the hermit crabs' use of empty snail shells. Man, of course, is the most obvious tool user; modern society is totally dependent on a vast collection of tools called **technology.**

Summary

Studies of animal behavior are currently based on comparing patterns of animal activity. Behavior is constantly changing, both linearly (developmentally) and in repeating patterns and cycles.

Behavior is studied in all environments, inside and outside the laboratory. Much laboratory study concerns itself with the genetic versus environmental components of behavior (nature versus nurture). The results of laboratory studies are combined with field observations to provide a unified view of an organism.

Most of behavior is reducible to stimulus–response sets. Stimuli may be extremely diverse, as may be the responses that they elicit. Stimuli may be generally divided into internal and external types.

Reflexes are not learned and do not require the conscious control of the organism. Reflexive behaviors become progressively more important as one descends the phylogenetic tree.

Memory is the storage of a prior image in the form of neuraltraces; memories are utilized and modified by recall, recognition, and association.

Learning is basically a change in behavior that is gained through experience, and it involves the formation of memory. Chemical theories and structural theories both attempt to explain the phenomenon of memory. Structural theories are generally favored at present.

Conditioning experiments have told us much about animal behavior. Classical conditioning pairs a novel stimulus with an established one, with the eventual result that the novel stimulus alone may elicit the established response. Operant conditioning depends on reinforcement of a response to increase its frequency.

Insight is an abridged form of learning that occurs only in animals with highly complex nervous systems.

Instinct is an often complex behavior that occurs

with no prior learning. Instinctive behavior seems to follow three stages: first, appetitive, in which the animal is in preparation for the instinctive act; second, a releasing stimulus, which sets a releasing mechanism into action; and third, the releasing mechanism, which leads to a final consummatory act.

The behavior of protists and simple metazoans are basically similar. Both groups, for the most part, exhibit stereotyped behaviors.

Kineses and taxes are movements. Kineses are undirected locomotor responses, whereas taxes are more complicated directed responses toward or away from particular factors in the environment.

Communication among animals may involve some or possibly all the senses. The richness of vocabulary determines the effectiveness of communication among a given group of animals.

Imprinting is an instinctive behavior by which the image of the first large object experienced by a new-born animal becomes the memory trace for the "parent." The recognition of this memory elicits behavior that keeps the animal near this object (in the wild, usually the natural parent).

Sleep is basically a lack of attention and a slowing of metabolic activity. Human sleep is characterized by slow breathing, slow metabolism, paradoxical sleep (which resembles the waking state in some respects and includes REM sleep), and slow-wave sleep.

Aggression is widespread among animals and may be physical or nonphysical. Some of the most complex behaviors are aggressive.

Courting behavior is part of the behavior repertoire of most higher animals. It may be characterized by mutual tactile activity, specialized postures, special calls, protracted time periods, and selected mate choice.

Migration is widespread among animals, usually resulting in a move to an environment more favorable for feeding or breeding. Migration in birds is often oriented by celestial navigation. Homing seems to use similar orientational methods.

Social behavior revolves around dominant–subordinate hierarchies, territorial rights, and leadership roles in a group.

Tool use is not widespread in the animal kingdom, but it prevails with man. Physiological clocks seem to operate in all organisms and are generally coordinated with environmental phenomena. Some scientists believe these clocks to be endogenous and subject to resetting in concordance with the environment.

Although behaviors are extremely diverse, major patterns are shared among almost all groups, and this indicates that there is an underlying unity to the behavior of animal life.

Review Questions

1. Early in the chapter, it was noted that purely mechanistic considerations are insufficient for understanding animal behavior. Yet it was shown later that conditioning, which is seemingly purely mechanistic, represents the most acceptable current explanation for behavior. How may this be resolved?
2. Of what use is biotelemetry?
3. You are studying an entire population of cats, and you choose two particular individuals (for instance, the only two with white around their eyes) and mate them. This is an example of
   (a) pleiotropy
   (b) artificial selection
   (c) cross-mating
   (d) (a) and (c)
   (e) (a) and (b)
   (f) (b) and (c)
4. What might the results of the study in Question 3 tell you?
5. The main difference between pleiotropic and artificial selection studies is that pleiotropic studies concentrate mostly on
   (a) protists
   (b) polygenic inheritance
   (c) phenotypic (bodily and or behavioral expression) as opposed to genotypic (chromosomal makeup) effects
   (d) pleiotropes
   (e) the results of nonrandom matings
   (f) small populations of animals
6. True or False: An unconscious human can still exhibit reflex activity?
7. Using operant conditioning techniques, how might you train a cat to walk backward?

8. Using classical conditioning techniques, how might you train a cat to walk backward?

9. Discuss the importance of memory, recall, recognition, and association to the higher thinking processes of humans.

10. Can something at once be both a stimulus and a response? Why or why not?

11. Gas pains are
    (a) psychogenic
    (b) caused by the Lorenz–Tinbergern syndrome
    (c) kinesthetic

12. Give an example, based on nonhuman animals, in which a response is not proportional to its stimulus.

13. It was mentioned that a lion or tiger may flee at the sight of a frog. What type of conditioning might account for this behavior?

14. What is a releasing mechanism? What are its limits?

15. Which is a directed movement, kinesis or taxis?

16. Do cnidarians sleep? Why or why not?

17. Name three general characteristics of courtship behavior.

18. What type of behavior usually involves only one sense organ?

19. An organism's most necessary type of behavior is
    (a) reproductive
    (b) homeostatic
    (c) muscular
    (d) individual
    (e) genetic

20. What are the three stages of instinctive behavior according to Lorenz and Tinbergen?

21. When birds chirp, what are they usually indicating?

22. True or False: The leader of a school of fish is the one in the front?

23. You and an ignorant friend are discussing physiological clocks. Your friend argues that all terrestrial animals are either nocturnal or diurnal. What examples do you use to convince your friend otherwise?

24. That same friend then tells you that a photoperiod is the time interval between light flashes of a firefly. Do you agree?

25. Might a sound of 42,000 cps be a potential stimulus for human behavior?

26. Give the rationale behind one theory of climatic migration.

27. What behavioral attribute did the Pilgrims and the Arctic tern have in common?

Selected References

Altman, J. *Organic Foundations of Animal Behavior.* New York: Holt, Rinehart and Winston, Inc., 1966.

Andrew, R. J. "The Aggressive and Courtship Behavior of Certain Emberizinae," *Behavior,* **10:**255–308, 1957.

Bastock, M. *Courtship: An Ethological Study:* Chicago: Aldine Publishing Co., 1967.

Beach, F. A. *Hormones and Behavior.* New York: Hoeber, 1947.

Bekleminshev, W. N. *Principles of Comparative Anatomy of Invertebrates.* Chicago: University of Chicago Press, 1969.

Breland, K., and M. Breland. "The Misbehavior of Organisms." In McGill, T. E. (ed.), *Readings in Animal Behavior.* New York: Holt, Rinehart and Winston, 1965.

Bunning, E. *The Physiological Clock.* New York: Academic Press Inc., 1964.

Communications Research Machines Books. *Biology Today.* Del Mar, Calif.: CRM Books, 1972.

Crook, J. H. *Social Behavior in Birds and Mammals.* New York: Academic Press, Inc., 1970.

Eccles, J. C. *The Neurophysiological Basis of Mind.* Oxford: Clarendon Press, 1952.

———— "The Physiology of Imagination," *Scientific American,* **199:**135–146, 1958.

Etkin, W. *Social Behavior from Fish to Man.* Chicago: University of Chicago Press, 1967.

Ewer, R. F. *Ethology of Mammals.* New York: Plenum Publishing Corporation, 1968.

Frings, H., and M. Frings. *Concepts of Zoology.* Toronto: Macmillan Co., 1970.

Funkenstern, D. H. "The Physiology of Fear and Anger," *Scientific American,* **192:**74, 1955.

Goldenson, R. M. *The Encyclopedia of Human Behavior.* Garden City, N.Y.: Doubleday & Co., Inc., 1970.

Gordon, M. S. *Animal Physiology,* 2nd ed. New York: Macmillan Publishing Co., Inc., 1972.

Hickman, C. P. *Integrated Principles of Zoology,* 4th ed. St. Louis: C. V. Mosby Co., 1970.

Johnsgard, P. A. *Animal Behavior.* Dubuque, Iowa: Wm. C. Brown Co., 1967.

Lorenz, K. "Der Kumpar in der Umwelt des Vogels," *J. Ornithol.,* **83:**137–213; 287–413, 1935.

—— "Comparative Study of Behavior." In Schiller, C. H. (ed.), *Instinctive behavior.* New York: International Universities Press, 1939, pp. 239–263.

—— "Morphology and Behavior Patterns in Closely Allied Species." In Schaffner, B. (ed.), *Transactions of the First Conference on Processes.* New York: Josiah Macy, Jr. Found. 1954, pp. 168–220.

Nelson, G. E., G. G. Robinson, and R. A. Boolootian. *Fundamental Concepts of Biology,* 3rd ed. New York: John Wiley & Sons, Inc., 1974.

Ramsay, A. O. "Familial Recognition in Domestic Birds," *Auk,* **68:**1–16, 1951.

Roeder, K. D. *Nerve Cells and Insect Behavior.* Cambridge, Mass.: Harvard University Press, 1963.

Roth, L. M., and E. R. Willis. "A Study of Cockroach Behavior," *Am. Midl. Nat.,* **47:**66–129, 1952.

—— "The Reproduction of Cockroaches," *Smiths. Misc. Coll.,* **122:**1–49, 1954.

Sanford, F. H. *Psychology: A Scientific Study of Man,* 2nd ed. Belmont, Calif.: Wadsworth Publishing Co., Inc., 1954.

Saunders, D. S. *Insect Clocks.* New York: Pergamon Press, Inc., 1976.

Scott, J. P. *Animal Behavior.* Chicago: University of Chicago Press, 1958.

Scott, J. P., and J. L. Fuller. *Genetics and the Social Behavior of the Dog.* New York: John Wiley & Sons, Inc., 1960.

Tavolga, W. *Principles of Animal Behavior.* New York: Harper & Row, 1969.

Tinbergen, N. "Social Releases and the Experimental Method Required for Their Study," *Wilson Bull.,* **60:**6–52, 1948.

—— *The Study of Instinct.* Oxford: Oxford University Press, 1951.

—— *Social Behavior in Animals.* New York: John Wiley & Sons, Inc., 1953.

Van der Kloot, W. G. *Behavior.* New York: Holt, Rinehart and Winston, Inc., 1968.

Von Frisch, K. "Dialects in the Language of the Bees," *Sci. Amer.* (reprint), August 1962.

Von Frisch, Karl *The Dance Language and Orientation of Bees.* Cambridge, Mass.: Harvard University Press, 1967.

Wilson, E. O. *The Insect Societies.* Cambridge, Mass.: Harvard University Press, 1971.

Skeletal Systems, Muscular Systems, and Locomotion

Skeletal systems perform several functions—they provide a framework that supports and protects the softer tissues of the body, and they provide a firm surface for muscle attachment. Perhaps the primary function of skeletal systems is to provide rigid structures that can be moved by muscular contraction. The result of these actions is often the movement of the organisms relative to its environment, that is, **locomotion.**

Skeletal Systems

Skeletons may be of three types: (1) **exoskeletons** that are formed on the outside of the body with muscles attached to their inner surfaces; (2) **endoskeletons** that are formed inside the body, surrounded by soft tissues and having muscles attached to their outer surfaces; and (3) **hydrostatic skeletons** that consist of one or more sacs of fluid whose shape may be changed by muscular contraction.

Protists

The body shape of "naked" protists is maintained by the **pellicle** (Fig. 8-3). Many protists secrete an external **shell;** for example, the shell of the amoeba *Arcella* is made of **chitin,** a nitrogenous polysaccharide. Others (e.g., *Diffugia*) construct a shell of tiny sand grains that are glued together in definite patterns. The calcium carbonate shells of untold billions of *Globigerina* make up the foraminiferan ooze that covers much of the ocean floor, along with the siliceous shells of radiolarians (Fig. 6-11). Still other protists secrete internal skeletons of calcium carbonate and silicon; notable among these are the freshwater heliozans.

Metazoan Exoskeletons

Exoskeletons are especially characteristic of the arthropods, but many other invertebrates secrete a covering that similarly serves to protect their soft body parts. The soft bodies of many cnidarians, such as that of the colonial hydroid *Obelia*, are supported and protected by a chitinous tube, the **perisarc.** Still other cnidarians secrete exoskeletons of calcium carbonate; the massive reefs secreted by coral polyps in shallow tropical marine waters attest to the durability of such exoskeletons.

Calcium carbonate is also used in the exoskeletons of many metazoans (e.g., mollusks, brachiopods, and barnacles).

The chitinous exoskeleton of an arthropod develops in a characteristic way. A brief discussion of insect molting follows.

As an insect grows, its soft body tissues gradually become too large for the confines of its rigid exoskeleton. Consequently insects and all other arthropods must periodically **molt** (a process known as **ecdysis**) and secrete a larger one in its place. Just before molting, the underlying epidermis (**hypodermis**) begins a rapid proliferation. The increased mitotic rate forces the epidermal tissue into folds under the exoskeleton. Meanwhile, an **epicuticle** of **cuticulin,** a lipoprotein, is secreted by the hypodermis. After sufficient epicuticle has been produced, a fluid is secreted between the epicuticle and the old exoskeleton. This fluid digests the inner layers of the old exoskeleton, freeing it for ecdysis.

Under the new epicuticle, an **exocuticle** is secreted. This layer contains the pigment granules that color the animal. The last layer to be secreted, the **endocuticle,** is the one closest to the soft internal organs and is made up of protein and chitin.

The final stage of molting involves the splitting of the old exoskeleton across the head and thorax. The animal slips out of its old exoskeleton; a rapid uptake of water by the body tissues expands the new exoskeleton before it is hardened by a tanning process.

The arthropod exoskeleton serves as protective armor and also provides attachment sites for muscles. The arthropod skeleton covers the entire body but is very thin and flexible at the joints, thus enabling the animal to move parts of its body and appendages (Fig. 20-2).

There are not many large animals with exoskeletons. The explanation for this phenomenon is analogous to that for the limit on cell size (Fig. 3-1). An exoskeleton can support only a certain mass. As the size of an organism increases, its surface area increases much more slowly than does its volume and its mass. Consequently, the physical properties of the exoskeleton limit the maximum size of the animal that it confines. The mass of the exoskeleton itself is also limited, because too great a thickness inhibits movement. Certain land crabs are the largest terrestrial animals with exoskeletons, attaining diameters of about 20 cm by virtue of having a relatively thick exoskeleton. Certain marine crabs may reach 2 m in diameter; this is possible because the aqueous medium buoys the body and so less stress is placed upon the exoskeleton. A further restraint on exoskeleton size is the expenditure of energy and materials that periodic molting requires. It is interesting to note in this regard that many arthropods eat their old exoskeleton after molting.

The bony fishes are covered by an exoskeleton of overlapping **scales** that allow lateral body movements. The skeletal elements of vertebrates are composed of living substances that grow with the animal, and so they do not have to be periodically shed and replaced.

What amounts to an exoskeleton is also present in certain terrestrial vertebrates, such as the bony **carapace** and **plastron** of the turtle shell, the **bony plates** of the armadillo, and the **epidermal scales** of the pangolin.

Metazoan Endoskeletons

The simplest metazoans, the poriferans (sponges), possess endoskeletons of calcium carbonate or siliceous spicules and/or flexible spongin fibers. Many simple metazoans employ a hydrostatic skeleton, adapting a fluid-filled pseudocoel, coelom, or hemocoel for that purpose. The hydrostatic skeleton is obviously not a solid structure; the skeletal properties of stiffness or resistance to pressure is simply a result of the incompressable nature of the volume of water that is trapped within the body cavity.

Endoskeletons have played decisive roles in the evolution of deuterostomes, and the internal skeletons of echinoderms and chordates are generally elaborate and highly specialized.

The Notochord

The embryos of all chordates have a supporting structure called the notochord (Gr. *noton*, back, + L. *chorda*, cord). The notochord is a semirigid rod (Fig. 32-1) that develops beneath the dorsal nerve

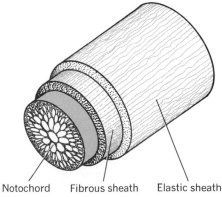

Notochord   Fibrous sheath   Elastic sheath

Figure 32-1. The notochord.

cord. Its rigidity stems from the turgid (fluid-filled) thick-walled cells of which it is constructed. Further stiffness is imparted by two concentric connective tissue sheaths that surround the notochord. The inner sheath is fibrous, whereas the outer sheath is less fibrous and more elastic.

The notochord serves various purposes including support, an axis around which the vertebral column develops, and an attachment site for certain muscles. The notochord persists as an adult structure only in the primitive cephalochordates *(Amphioxus)* and cyclostomes (e.g., lampreys). The notochord of adult cyclostomes becomes elaborated by the addition of calcareous neural arches in the trunk region. In all higher vertebrates the development of cartilaginous and osseous (bony) structures tend to replace the notochord.

Osseous Tissue: Development

The origin of osseous tissue may take one of two forms. The cranium (skull) and shoulder girdle arise by **intramembranous ossification,** which involves direct conversion of mesenchyme into bone. Most other bones arise from an intermediate connective tissue (**cartilaginous**) stage, known as **intracartilaginous ossification.** Mesenchyme gives rise to a cartilaginous model of the future bone, after which replacement of the cartilage by calcium salts eventually occurs. The mechanism of cartilage replacement is rather complex, so we will discuss only intramembranous ossification here. Keep in mind that intracartilaginous ossification is generally similar but more elaborate.

During embryonic development migratory mesenchymal cells aggregate and proliferate in the regions where bone is to be produced. The mesenchymal cells then differentiate into **osteoblasts,** cells that actively secrete the bone's protein matrix. The osteoblasts form **trabeculae,** which are scattered bars and plates of fibrous material. At this point, calcium salts are deposited on the protein matrix.

As the osteoblasts become surrounded by the matrix of the trabeculae, they are transformed into **osteocytes.** The osteocytes, housed in spaces known as **lacunae,** function to maintain the newly formed bone. The osteocytes are interconnected by minute processes through a network of channels, the **canaliculi,**

which reach to the capillaries that infiltrate and nourish the bone.

Meanwhile, the proliferating mesenchyme is still forming osteoblasts, which continue to add to the trabeculae. In time, the trabeculae fuse and form a continuous lattice, forming **spongy,** or **cancellous, bone tissue** (Fig. 3-3E). In spongy bone tissue, the areas between the trabeculae are filled with extremely vascular connective tissue. This is the **bone marrow,** responsible for the formation of blood cells and, in later life, for fat storage as well.

About the time a thin but tough membranous sheath, called the **periosteum,** forms over the bone. Beneath the periosteum, new osseous tissue continues to be produced. Eventually, instead of the scattered bars and plates, dense sheets of bone are laid down, resulting in compact bone that surrounds the cancellous bone.

Compact bone, the final stage of bone development (but not necessarily growth), contains a system of concentric canals (**Haversian systems,** Fig 3-3E) that communicate between the outer periosteum and the inner marrow. This network distributes nerves and blood vessels throughout the osseous tissue. Microscopic cross-sections reveal that bone tissue develops in concentric layers around each main Haversian canal from which may branch other radiating Haversian canals.

Osseous Tissue: Further Growth

Bone is a living tissue that is capable of growth and homeostatic response throughout the life of the organism. After a bone is formed, it may grow in length as well as in breadth. The maximum length to which a bone may grow is determined genetically, but actual growth may fall short of this maximum due to environmental influences. If a vertebrate lacks sufficient food supplies, its bones may not grow to their normal lengths. Dietary deficiency may also result in bone deformation, as in rickets, which is due to vitamin D deficiency.

At each end of a long bone there is an **epiphyseal plate,** or **epiphysis** (Fig. 32-2). The epiphyses are separated from the main shaft of the bone, the **diaphysis,** by thin cartilaginous zones of active bone growth. New cartilaginous tissue forms next to the epiphyses

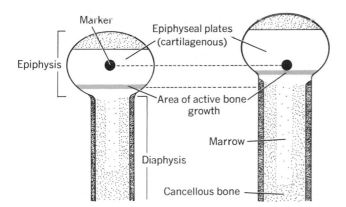

Figure 32-2. Diagram of bone growth. In the wake of the active bone growth area, the cartilage is transformed into bone. The "marker" cartilage (at left) is well within the epiphyseal plate area. Later (right), although it has remained in the same horizontal position, the bone growth area has moved up to it. A later view would show it turned into bone, whereas the cartilage area continues to produce more cartilage.

in the active zones, whereas older cartilage is converted to bone at either end of the diaphysis. As the individual ages, this growth declines and finally stops completely. When the epiphyseal cartilage becomes completely ossified (at about 20 years in humans), no further lengthwise growth is possible.

The skeletal system is influenced by several hormones. For instance, bone growth is under the influence of **growth hormone** secreted by the anterior pituitary gland.* Growth hormone speeds up protein synthesis in both the center and edge plates of extant bone. The mechanism for this process is still under debate. It is known that bone structure and growth are related to mechanical stress—the more stress, the more growth, up to a point.† For example, surgical removal of the tibia from the lower leg (Fig. 32-5) leaves the smaller fibula bearing more weight. The fibula responds to the mechanical stress by increasing significantly in thickness. Breadthwise growth is a

---

*The human disease **acromegaly** results from oversecretion of growth hormone. In acromegaly, the forehead, nose, hands, feet, and lower jaw become abnormally massive.

†This is roughly analogous to the necessity of frequent activity for maintaining neural circuit function or of exercise to prevent muscle atrophy. "If you don't use it, you lose it."

dynamic process that involves deposition of bone near the outer periosteum.

Major Divisions of Vertebrate Endoskeletons

The vertebrate endoskeleton may be subdivided into membrane bones and cartilage bone, depending upon how they develop. The bones of the skeletal system may also be differentiated according to where they develop. The vertebrate skeleton may be subdivided into visceral and somatic skeletons. The **visceral skeleton** occurs in the gill (pharyngeal) region and includes such structures as the jaws, the hyoid, and the ear ossicles. The remainder of the skeletal bones comprises the somatic skeleton.

The **somatic skeleton** is further divided into axial and appendicular skeletons. The **axial skeleton** includes the braincase, vertebrae, ribs, and sternum, which all form the central axis of the body. The **appendicular skeleton** includes the bones of the pectoral and pelvic girdles and their appendages.

Skeletons of Aquatic and Avian Vertebrates

The vertebrate skeleton provides support and protection to the body and furnishes a firm surface for muscle attachment. The characteristics of the skeleton differ between aquatic and terrestrial vertebrates. Because fishes are partially buoyed by the surrounding water, they neither need nor possess the relatively sturdy skeletons required by land animals. The appendages of fishes are paired pectoral and pelvic fins, and sometimes dorsal fins, that contain weak skeletal structures. These appendages serve primarily for balancing and steering rather than for locomotion. The skeletons of other types of aquatic vertebrates such as frogs, salamanders, turtles, alligators, penguins, whales, and seals also reflect various degrees of adaptation to the buoyant and fluid nature of water.

Avian bony structure is unique in its being hollow. The obvious result of such construction is that the mass of the skeleton is decreased, an obvious advantage for flight. In fact, if certain avian bones are broken, the bird may even breathe through the abnormal openings, because of the air sacs that communicate with the lungs (Chapter 30).

Primitive vertebrates tend to have a larger number of skull bones than do the more advanced forms. Some fishes have 180 skull bones, and amphibians and reptiles have 50–95 skull bones. Mammals may have 35 or fewer (humans have only 29). The fewer the bones, the stronger and more solid the braincase. In addition, fewer bones provide stronger points of attachment for head and facial muscles.

The Human Skeleton

In general the human skeleton resembles that of the frog in both structure and function (Fig. 24-14). The total number of bones in the average adult human skeleton is 206, consisting of

```
   28 skull
    1 hyoid
   26 vertebrae
   24 ribs
    1 sternum
    4 pectoral girdle
   60 upper limb
    2 pelvic girdle (excluding pubis)
   60 lower limb
  ─────────────────────────────────
  206 bones
```

The human **skull** includes the **cranium** and the **face** (Fig. 32-3). The cranium, or braincase, is composed of eight bones. The most anterior paired bone is the **frontal.** Posterior to the frontal on either side lies a **sphenoid.** Anterior to each sphenoid is an **ethmoid,** which contributes part of each eye socket. Medio-

lateral on each side lie the two **temporals** and directly above them are the two **parietals.** The single **occipital** bone forms the posterior portion of the skull.

There are 14 facial bones: Two **maxillae** form the upper jaw and bind together the upper facial bones. The two **nasal** bones form the hard bridge of the nose; the lower part is composed of cartilage. The two "cheekbones," called **zygomatics,** brace the face to the cranium. Two inferior **nasal conchae (inferior turbinate)** bones help to frame the inside of the nose. Two **lacrimal** bones contribute to the inner eye sockets, each perforated by a hole through which a tear duct passes. The two **palatines** frame the hard palate and floor of the nasal cavity. The single **vomer** forms the bony part of the **nasal septum.** The single **mandible,** or lower jaw bone, articulates with a depression in each temporal bone.

In each middle ear, there are three minute bones: the **malleus,** the **incus,** and the **stapes** (Fig. 37-14).

The horseshoe-shaped **hyoid** bone, which adds strength to the cartilaginous trachea, is suspended by ligaments between the tongue and larynx.

The **vertebral column** consists of 33 **vertebrae** in children, but in adults, 5 have become fused to form the **sacrum** (Fig. 32-4), and another 3 to 5 vertebrae fuse to form the **coccyx,** or vestigial tail. The sacrum and coccyx of females are less curved and more anteriorly pointed than are those of males; this is an adaptation for childbirth. The other vertebrae are the **lumbar** (5) in the lower back, the **thoracic** (12) in the thorax, and the **cervical** (7) in the neck. The first cervical vertebra, or **atlas,** articulates with the occipital bone of the skull in such a way that the head can nod up and down. The second cervical vertebra, or **axis,** has an upward peglike process around which the atlas rotates, allowing the head to turn from side to side. The vertebrae are semi-independent in their movements. Pads of fibrocartilage separate the vertebrae, and many ligaments bind them together. Each vertebra contains a dorsoventral cavity, or **neural canal,** which houses the spinal cord.

The twelve pairs of **ribs,** the **sternum** (breastbone), and the **costal cartilages** form the bony cage of the thorax (Fig. 33-7). The upper seven pairs of ("true") ribs are attached directly to the sternum by costal cartilages. The lower five pairs of ("false") ribs do not attach directly to the sternum: Ribs 8, 9, and 10 join the sternum indirectly; their costal cartilages fuse

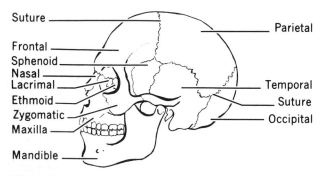

Figure 32-3. Human skull. Note the distinct sutures between the bones of the cranium; these sutures close and are nearly or completely obliterated in old age.

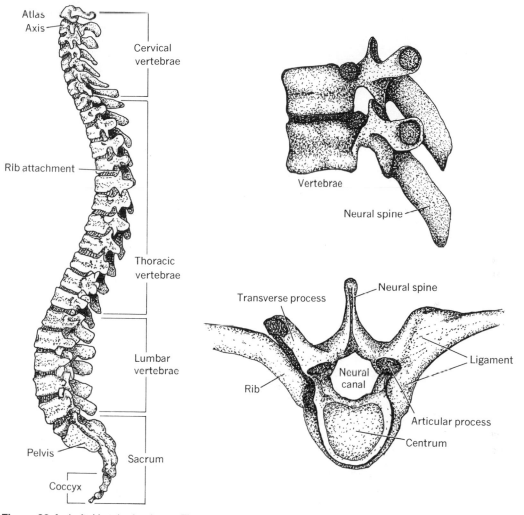

Figure 32-4. Left: Vertebral column. The sacrum and coccyx consist of fused vertebrae. Upper right: Lateral view of two thoracic vertebrae. Lower right: Superior view of vertebra, drawn to same scale as lateral view above.

with that of the seventh rib. The last two pairs are called **floating ribs** because they terminate in the muscles of the lateral abdominal wall.

The bones of the **appendicular skeleton** are those of the **pectoral** (shoulder) **girdle**, the **upper limbs**, the **pelvic girdle**, and **lower limbs** (Fig. 32-5). The pectoral girdle consists of two **clavicles** (collarbones) and two **scapulae** (shoulder blades). The bones in each arm are the **humerus**, running from the elbow to the

shoulder, the **ulna** (the larger bone of the forearm), and the **radius**, which helps the thumb to pivot. Each hand is composed of 3 **carpals**, 5 **metacarpals**, and 14 **phalanges.** Each side of the pelvic girdle is composed of three bones, the **ilium, ischium,** and **pubis,** that are fused into a single **innominate** bone (**hipbone**). Each leg is made up of the **femur** (longest bone in the body), **patella** (kneecap), **tibia, fibula,** 7 **tarsals,** 5 **metatarsals,** and 14 **phalanges.**

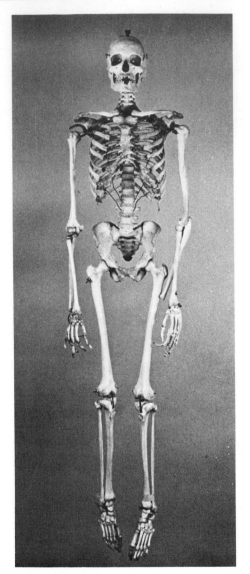

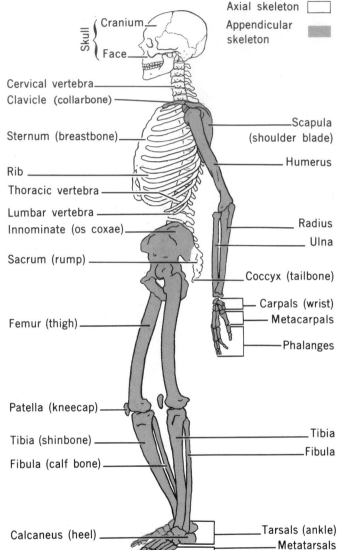

Figure 32-5. The human skeleton.

Joints

Any apposition of two bones, or a bone and carti-lage, constitutes a **joint (articulation).** Types of joints vary with their function and may range from a rigid junction to an articulation that is recognized: fibrous joints, cartilaginous joints, and synovial joints.

Fibrous joints are immovable. The bones are di-

rectly apposed and are held together by thin connec-tive tissue or by bony interdigitations known as **sutures,** with little connective tissue. The former type occurs between the distal ends of the tibia and fibula. Sutures occur exclusively in the skull, where they fuse the cranial bones together.

Cartilaginous joints are only slightly movable,

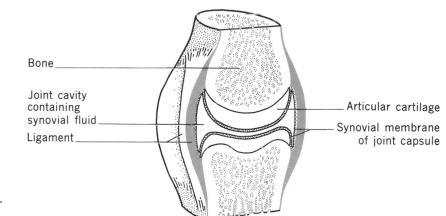

Figure 32-6. A diagram of a synovial joint surface (longitudinal section) showing some of its characteristics.

owing to the semirigid character of the cartilage sandwiched between the two bones. Cartilaginous joints occur between the ends (epiphyseal plates) of growing bone or in special areas such as where the two hipbones fuse anteriorly, known as the pubic symphysis.

Synovial joints (diarthroid joints) are freely movable in selected directions. Their flexibility is due, in part, to a lubricating **synovial fluid** secreted by a baglike synovial membrane that encloses the ends of the articulating bones (Fig. 32-6). Ligaments hold the bones together. The ends of the bones are covered with a cushioning layer of hyaline cartilage. Six categories of synovial joints are usually distinguished:

1. **Hinge joints,** such as in the elbow, move in only one plane.
2. **Pivot joints,** as between the radius and humerus of the lower arm, allow only rotation.
3. **Saddle joints,** such as between the carpal and metacarpal of the thumb, permit movement in two planes.
4. **Gliding joints,** such as between the vertebrae, are nearly flat and permit back-and-forth and side-to-side motion.
5. **Ovoid joints,** like those of the wrist, also permit side-to-side and back-and-forth motion, but here the articulating surfaces are not flat: an oval projection from one bone fits into a depression in the other.
6. **Ball-and-socket joints,** such as those of the shoulder and hip, allow multidirectional movement in three planes.

Muscular and Other Contractile Systems

Movement is characteristic of animal life, as heterotrophic life-styles often involve active searches for food or flight from predators and other harmful elements of their environments. These demands have fostered the evolution of complex and diverse muscular systems in the different animal groups. Except for amoeboid movements, all animal systems for locomotion have three things in common: (1) all involve contraction of protein elements such as actin and myosin; (2) contraction moves certain rigid skeletal parts (whether cilia, external shells, or internal bones); and (3) such contractions require energy, that is, the expenditure of ATP.

Muscle tissue is usually classified into three types: smooth, striated, and cardiac (Fig. 3-5).

Smooth Muscle

Smooth muscle is found in many invertebrates and in the visceral organs of vertebrates. Smooth muscle is composed of spindle-shaped cells, each with a single nucleus near its center. These muscles may be organized in three ways: they may be isolated, aggregated into small groups, or arranged in parallel layers of substantial thickness.

Smooth muscles contract more slowly than do other muscles and may take from 3 to 180 seconds to perform a single contraction–relaxation sequence. Contraction of smooth muscles changes the shape and size of the visceral organs that they surround. Smooth muscles are present in the walls of the digestive tract, blood vessels, and the bladder; around the openings of exocrine glands; in the skin (where they may perform a hair-raising function); in the trachea and bronchi; and in the reproductive ducts. Smooth muscles are not normally under voluntary control, and so they are sometimes referred to as involuntary muscles. It is believed that smooth muscles represent the most primitive type of muscle tissue.

Striated Muscle

Striated, or voluntary, muscle (Figs. 32-9 and 32-10) makes up about 40 percent of a person's body weight. Found in the chordates, it is also common in those invertebrates for which brief, rapid actions are necessary. Nearly all the muscles of arthropods are of the striated type. Each striated muscle cell is not derived from individual cells but, rather, from a number of cells that have joined together to form the multinucleated muscle fiber. Each fiber consists of a complex mass of muscle cytoplasm (**sarcoplasm**) with many nuclei embedded just below its surface membrane. When viewed with a microscope, each striated muscle fiber appears to be striped with alternate light and dark bands. This alternation of light and dark segments results from the exact alignment of many contractile fibrils (myofibrils) within the sarcoplasm. The endoplasmic reticulum of a muscle cell is a series of tubules and flattened saclike structures that surround the myofibrils; it is called the sarcoplasmic reticulum.

Skeletal muscles are the fastest to react (due to their rapid recovery rate) and have a contraction–relaxation period of about 0.1 sec.

Connective tissue binds the spindle-shaped fibers together, thus forming the muscle. The blood supply to striated muscle is rich—between 1.4 and 4.0×10^3 capillaries have been observed in a single cubic millimeter (mm^3) of striated muscle tissue. Each striated fiber is innervated by a branch of a motor nerve fiber.

Striated muscles are under voluntary control and are often called skeletal muscles because they form the musculature of our limbs and body wall. Striated muscle also occurs in the diaphragm, tongue, pharynx, larynx, upper third of the esophagus, and around the eyes.

Each skeletal muscle is attached by tendons to two points on the skeleton: the muscle's **origin** is the end attached to the more stationary part of the skeleton,

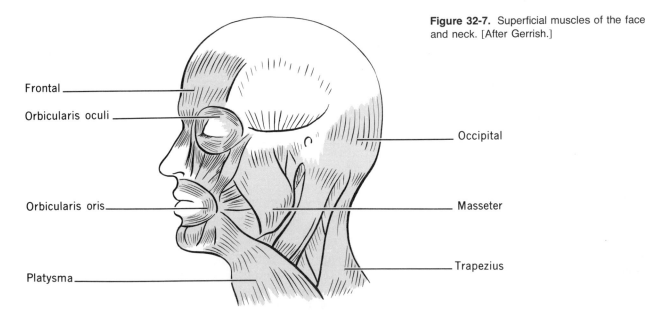

Figure 32-7. Superficial muscles of the face and neck. [After Gerrish.]

Frontal

Orbicularis oculi

Orbicularis oris

Platysma

Occipital

Masseter

Trapezius

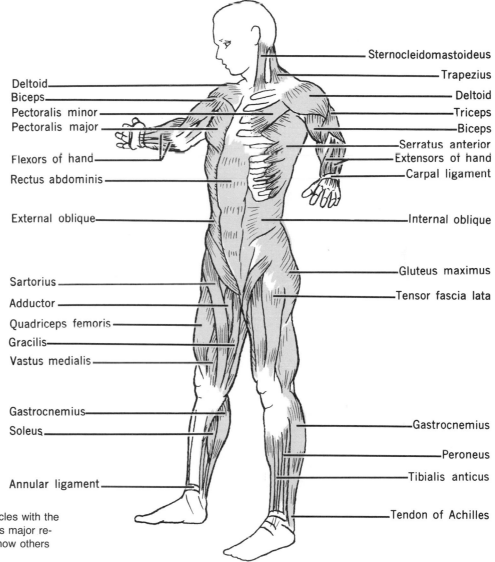

Figure 32-8. Superficial muscles with the external oblique and pectoralis major removed from the left side to show others that lie beneath.

whereas its **insertion** is attached to the bone or bones that are moved when the muscle contracts. Some of the skeletal muscles of the human body are shown in Figures 32-7 and 32-8. The origin, insertion, and function of some representative striated muscles are given in Table 32-1. In all, there are about 400 skeletal muscles in the human body.

Skeletal muscles are usually arranged in **antago-** **nistic pairs;** body parts are generally moved by pairs of muscles whose actions oppose one another. For example, the biceps' brachia, which flex the arm, are antagonistically paired with the triceps' brachia, which extend the arm. Antagonistic pairing is necessary because muscles act only by contracting (shortening). Once contracted, a muscle is lengthened only by relaxation and the pull of other muscles or gravity.

Table 32-1. Name, Origin, Insertion, and Functions of Some of the Skeletal Muscles of Man

Name of Muscle	Origin	Insertion	Function
Biceps	Scapula	Radius	Flexes* elbow and supinates† forearm and hand
Deltoid	Clavicle and scapula	Humerus	Abducts‡ the arm
Gastrocnemius	Femur	Calcaneus	Flexes leg and extends foot
Gluteus maximus	Ilium, sacrum	Femur	Extends, abducts and rotates femur
Gracilis	Pubis	Tibia	Adducts§ thigh and flexes leg
Pectoralis major	Clavicle, sternum, etc.	Humerus	Adducts and draws arm across chest and rotates it inward
Peroneus	Fibula	Metatarsals	Extends foot
Quadriceps femoris	Innominate (oscoxae), femur, etc.	Tibia	Extends leg and flexes thigh
Sartorius	Ilium	Tibia	Flexes leg on thigh and thigh on pelvis
Serratus anterior	Ribs and intercostals	Scapula	Carries scapula forward, assists trapezius and deltoid, etc.
Soleus	Fibia and tibia	Calcaneus	Extends foot
Tensor fascia lata	Ilium	Fascia lata	Abduction and rotation of thigh and tightening of fascia lata
Triceps	Scapula and humerus	Ulna	Extends forearm

* Flex means "to bend" (L. *flexus*, bent).
† Supinate means "to bring the palm upward" (L. *supinus*, bent back).
‡ Abduct means "to draw away from" (L. *ab*, from; *ducere*, draw).
§ Adduct means "to draw toward" (L. *ad*, toward).

Cardiac Muscle

Cardiac muscle tissue occurs only in the hearts of vertebrates and a few invertebrates. Cardiac muscle is composed of striated fibers, but the striations are less distinct than in skeletal muscle. The individual cardiac cells are all joined together by permeable structures, called **intercalated discs,** that lower the electrical resistance between cells. As a result, the walls of each heart chamber function as if they were a single muscle fiber. The contraction–relaxation period of cardiac muscle is about 1–5 sec, much slower than is that of skeletal muscle. Cardiac muscle differs from other muscles in that it does not require nervous stimulation to contract. A pacemaker (**sinoatrial node**) of very specialized muscle tissue in the wall of the right atrium produces excitatory signals that initiate contraction (Chapter 35).

Physiology of Muscle Contraction

The physiology of vertebrate striated muscle will be considered in some detail. Then the muscular systems of invertebrate animals will be briefly surveyed.

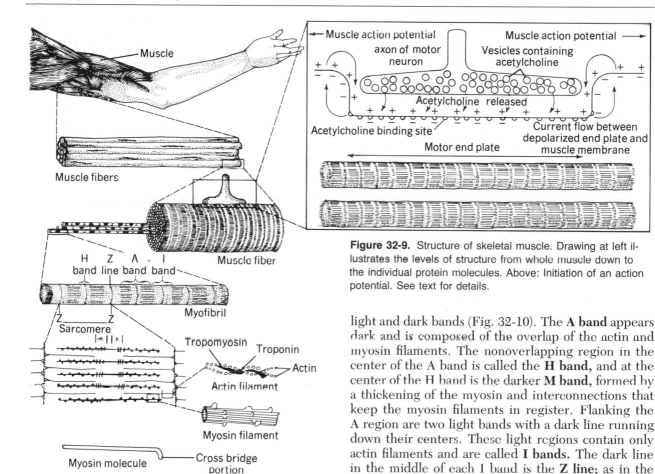

Figure 32-9. Structure of skeletal muscle. Drawing at left illustrates the levels of structure from whole muscle down to the individual protein molecules. Above: Initiation of an action potential. See text for details.

Studies with the electron microscope have helped to reveal the nature of muscle contraction in considerable detail. Recall that striated muscles are composed of many fibers (Fig. 32-9). Each fiber is about 100 μ in diameter and from a few mm to a few cm long. Embedded in the sarcoplasm are many longitudinally arranged myofibrils. The myofibrils are in turn, composed of two types of myofilaments. Thick myofilaments are formed by the association of many molecules of the protein **myosin,** whereas thin myofilaments are made by association of the proteins **actin, troponin,** and **tropomyosin.** These thick and thin myofilaments lie parallel to one another and interdigitate in a definite pattern.

Microscopically, a muscle appears as a series of light and dark bands (Fig. 32-10). The **A band** appears dark and is composed of the overlap of the actin and myosin filaments. The nonoverlapping region in the center of the A band is called the **H band,** and at the center of the H band is the darker **M band,** formed by a thickening of the myosin and interconnections that keep the myosin filaments in register. Flanking the A region are two light bands with a dark line running down their centers. These light regions contain only actin filaments and are called **I bands.** The dark line in the middle of each I band is the **Z line;** as in the M band; interconnections are present to keep, in this case, the actin filament in alignment. The distance between a pair of Z lines demarcates one muscle unit, or sarcomere. Note that the actin and myosin filaments do not completely overlap in the relaxed sarcomere.

Microscopic observation reveals that the bands shift during muscle contraction (Fig. 32-10B). The results of contraction are as follows:

1. The A band width remains constant.
2. The I and H bands become narrower.
3. The Z lines, which separate the sarcomeres, move toward one another.

These observations are explained by the **sliding filament model** of muscle contraction, which was pro-

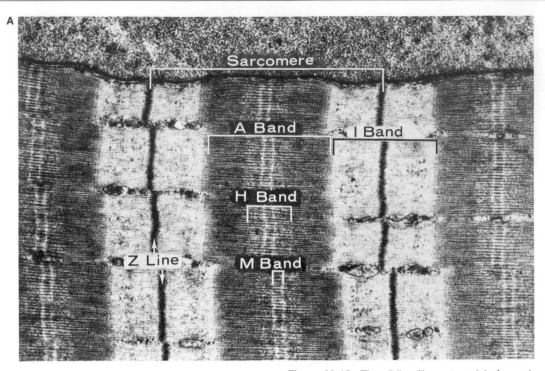

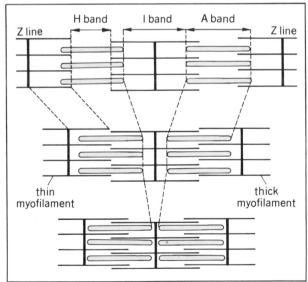

Figure 32-10. The sliding filament model of muscle contraction. (A) This is an electronomicrograph of a skeletal muscle. 3400 ×. See text for description. [Courtesy of W. Bloom and D. W. Fawcett, *A Textbook of Histology,* 9th ed. Philadelphia: W. B. Saunders, 1968.] (B) The changes in bonding patterns from resting (top) to maximum contraction (bottom).

posed independently by J. Hanson and H. E. Huxley in the 1950s. The sliding filament model recognizes that the interdigitating pattern of the myofilaments

within each sarcomere forms the structural basis for muscle contraction. According to this theory the myofilaments in each sarcomere do not shrink but rather slide over each other, thus shortening the muscle fiber as a whole. As a sarcomere contracts, the actin myofilaments approach each other by sliding over the myosin filaments. This increased actin–myosin interdigitation causes the widths of both the H bands and the I bands to decrease. Note that the width of the A band does not change because the myofilaments themselves do not contract. The sliding brings the Z lines closer together; that is, the length of the sarcomere decreases. The simultaneous contraction of all the sarcomeres in a muscle fiber causes the fiber as a whole to contract.

The sliding of myofilaments past one another requires energy, and in living systems this energy is provided by adenosine triphosphate (ATP). A corollary to the sliding filament model states that **cross bridges** (portions of the myosin protein molecules that project away from the main axis of the filament, Fig. 32-9) act as a ratchetlike apparatus by forming temporary attachments to the actin filaments. In the process of forming and breaking these bonds, the bridges "pull" on the actin filament, probably not unlike the action of rowing a boat.

Skeletal muscles do not normally contract without our willing them do so.* The central nervous system sends instructions to the muscle fibers via **motor (efferent) nerve fibers,** which branch repeatedly as they approach their target muscle. Each branch ends in a single muscle fiber on a flattened area called the **motor end plate.**

When a nerve impulse reaches the tip of the nerve axon, it causes a **transmitter substance** to be released, which diffuses across the **synaptic cleft** that separates it from the motor end plate. **Acetylcholine** is the transmitter substance released at the motor end plates of vertebrate striated muscle. When sufficient acetylcholine is released, the motor end plate potential reaches a **threshold intensity,** and an **electrical action potential** is initiated in the cell membrane. Once initiated, the action potential travels as a wave of depolarization along the entire length of the muscle fiber and into the internal portions of the fiber through a system of **transverse (T) tubules** that is continuous with the outer plasma membrane and contacts the sarcoplasmic reticulum.

As an action potential sweeps along the membranes, it causes relatively large amounts of calcium ions (Ca^{++}) to be released from the sarcoplasmic reticulum (from regions called the **lateral sacs**) into the sarcoplasm surrounding the myofilaments. Calcium ions bring about muscle contraction in the following manner.

The release of CA^{++} into the sarcoplasm removes the inhibitory influences of certain proteins of the thin myofilament that prevent a resting muscle from contracting. These proteins, called tropomyosin and troponin, cover active sites on the actin, thus interfering with the formation of cross bridges between the actin and myosin filaments. Although all the molecu-

lar details are not resolved, it seems that the uncovering of the actin's active sites allows a myosin–ATP complex to bind to these sites. The ATP is cleaved into ADP and phosphate, and the energy released somehow facilitates the movement of the attached myosin cross bridges relative to the actin filaments, thus resulting in contraction of the sarcomere. Only a few cross bridges are connected at any one moment; cycling of many individual bridges results in a smooth sliding motion of the filaments relative to each other. When the stimulus from the nerve fiber ceases, the calcium leaves and the cross bridges are again inhibited.

Enough ATP is available in resting vertebrate muscle to sustain maximum contraction for only 1 sec. Muscle tissue must rapidly and continually replenish its ATP supply.

Most of the ATP expended during short periods of muscle contraction is formed by the rephosphorylation of ADP to ATP; the phosphate is donated by the higher-energy intermediates **creatine phosphate** (vertebrates and some invertebrates) and **arginine phosphate** (most invertebrates). The creatine phosphate, whose concentration is 4–8 times that of ATP, serves as a short-term energy reservoir that supplies energy (via the phosphate) very quickly to ADP, forming more ATP needed for contraction. During sustained contraction, the creatine phosphate supply is also depleted; extra ATP and the ATP needed to recharge the creatine back to creatine phosphate are obtained from the breakdown of **glycogen** to glucose and then **glycolytic** degradation of glucose to **lactic** acid (Chapter 2). If even more contraction is called for by continual stimulus from the motor neurons, the rate of glycolysis will not be fast enough to supply all the ATP required, and the muscle reaches **fatigue.** The glucose and glycogen used are resynthesized using ATP produced by **cellular (aerobic) respiration.** Thus the immediate energy source for muscular contraction is from *anaerobic* glycolysis, and it is later replenished by *aerobic* respiration. During vigorous exercise, therefore, the energy used must be replaced by the energy-producing process that requires oxygen (aerobic, or oxidative, respiration, Chapter 2), so one incurs an **oxygen debt.** This is easily seen in the heavy breathing that follows vigorous activity. This aerobic ATP synthesis is greatly enhanced by the presence of the protein **myoglobin,** which stores oxygen for re-

---

*A notable exception is reflex postural adjustment.

lease when needed. This oxygen is donated by hemoglobin from the blood.

When a muscle contracts, it converts stored potential chemical energy into mechanical energy and heat. The efficiency of changing chemical energy into mechanical energy in an animal is only about 25 to 40 percent. Most of the energy is released in the form of heat, and about four fifths of all our body heat is derived from muscle contraction. Exercise stimulates blood circulation and enhances the size, strength, and tone of muscles. The increase of muscle size is not an increase in the number of muscle cells but, rather, an increase in the number of myofibrils per cell.

How are the contractions of individual motor units amplified and coordinated into complex muscular movements? Each motor neuron innervates many muscle fibers, which all contract simultaneously when stimulated past a particular threshold. This so-called **all-or-none** behavior characterizes the aggregate of fibers, called the **motor unit.** The speed and tension of contraction within a motor unit may be controlled by the frequency of stimulation by its motor neuron. The response of a motor unit to a single nervous impulse is shown in Figure 32-11. Note the lag, the latent period, before contraction is observed, and another delay, the relaxation period, after the impulse has ceased. Although the all-or-none characteristic yields a constant tension during each twitch, if another impulse arrives from the motor neuron before the relaxation period is over, the motor unit contracts again and tension is increased. This modulation of the time between stimuli (**temporal summation**) occurs with increasing frequencies of stimulation until the maximum tension of the motor unit is reached (**tetanus** Fig. 32-11); each muscle fiber maximally contracts to about one third of its resting length.

The amount of tension is also controlled by the number of motor units that are simultaneously stimu-

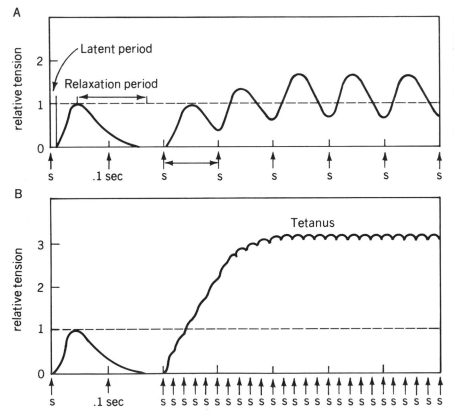

Figure 32-11. Isometric contraction. The single curve in each graph, starting shortly after a single stimulus at time 0, shows the contraction response. In the right portion of (A), the frequency of stimulation is 10 per sec; in (B), 50 per sec. Note how the responding tensions add when time between stimuli shorten (temporal summation) until a limit is reached (tetanus).

lated within the muscle; this is called **spatial summation.** The frequency and number of motor units stimulated is under the control of the **central nervous system** (Chapter 37).

During the movement of body parts, the entire muscle shortens by what is termed isotonic contraction. The more common type of contraction is **isometric contraction,** which produces tension against relatively immovable parts of the body, producing no movement (e.g., the maintenance of body posture).

Locomotion

Locomotion in protists and multicellular animals invariably involves the contraction of protein elements; this contraction moves skeletal structures and requires ATP as an energy source.

Protozoans

Many protists contain **actin** and **myosin myofilaments** that initiate body contraction. These myofilaments are similar in structure and mechanism to those in muscle cells, but they are not as regularly structured and are not surrounded by their own membrane system (i.e., they are **organelles** rather than cells). The amoeba moves by extending pseudopodia. This process involves the interconversion of plasmasol and plasmagel. It is interesting to note that actin or proteins very similar to actin have been identified in all most all protists examines thus far, including many algae.

Mastigophorans and ciliophorans move by **flagella** and **cilia,** respectively, Flagella and cilia are both composed of longitudinally arranged **microtubules** that are arrayed in a characteristic 9 + 2 pattern (Fig. 2-21B). Current theory holds that contraction of the organelle is brought about by a sliding of the microtubules past one another, an activity that requires the dephosphorylation of ATP.

Metazoans

The larvae of poriferans are ciliated and free swimming, but the adults are sessile and permanently fixed to the substratum. The osculum and certain pores can be closed, however, by the contractile cells, called myocytes, that ring them.

The **epitheliomuscular cells** of the cnidarians are generally considered to be the most primitive type of muscle cells. They are arranged to form what may be considered a muscular system. The epitheliomuscular cells in the epidermis are arranged in a network of **longitudinal muscle fibers,** and those in the gastrodermis form **circular muscle fibers.** Contractions of the longitudinal fibers shortens the entire body or bends the body and tentacles. On the other hand, contractions of the circular fibers extend the body or tentacles and are also responsible for the peristaltic waves that push food into the gastrovascular cavity.

Three sets of muscle fibers are present in planarians and many other flatworms: an outer **circular layer,** an inner **longitudinal layer,** and a series of **dorsoventral fibers** (Fig. 13-1). These muscles, as with those in all the more complex animals, are mesodermal in origin.

The annelid body wall is very muscular, containing an outer **circular layer** and an inner **longitudinal layer** (Fig. 17-3). The individual contraction of muscles in each segment allows for efficient control of locomotion, revealing an important advantage of metamerism. Special muscles are present for moving the setae and, in polychaetes, for moving the parapodia as well. Leeches are provided with muscular suction discs that are used for attachment and locomotion.

The circular and longitudinal muscle layers that occur in annelids are divided in arthropods into individual muscles that extend from one body segment or appendage to another. These muscles are fastened to the inside of the exoskeleton and span flexible regions of the cuticle that serve as joints. In the crayfish, sudden contractions of the powerful **flexor abdominal muscles** bend the abdomen forward, propelling the animal backward through the water. Crayfish are also able to walk in any direction by means of muscles within their legs. In the grasshopper and most other insects, wings also serve as locomotor organs.

Insect flight is quite different from that of birds or bats. There is no musculature in the wings of insects. Each wing articulates with the dorsal thoracic plate (**tergum**) at a movable hinge. Contractions of primary **flight muscles** do not move the wings directly but rather indirectly by moving the tergum up and down.

In mollusks, locomotion is generally due to contraction of muscles in the foot. These contractions vary the shape of the hemocoel that acts as a hydrostatic skeleton. The shell of bivalves is closed by powerful **adductor muscles** that span the valves.

All echinoderms possess a **water vascular system,** consisting of radially arranged canals that bear rows of **ampullae** and **tube feet.** Muscular contractions squeeze the ampullae and force fluid into the tube feet, thereby initiating movement. Several echinoderm groups also possess a movable endoskeleton to which muscles are attached.

The larvae of the urochordates swim by means of special muscles in the tail, and some adult urochordates swim by forcing water from their atrial cavity. Cephalochordates swim by lateral undulations of their body wall, brought about by the coordinated contractions of metamerically arranged muscle bands.

Summary

Skeletal structures provide support, protection, and attachment sites for muscles. Exoskeletons cover the outside of an organism, whereas endoskeletons are located within the body tissues. Hydrostatic skeletons involve sacs of fluid within the body whose shapes can be changed by muscular contraction.

Most deuterostomes have an endoskeleton, whereas protostomes, notably the arthropods, are more likely to have exoskeletons.

The human skeleton contains approximately 206 bones, which comprise the visceral, axial, and appendicular skeletons.

Joints are divided into three major categories: fibrous, cartilaginous, and synovial. Synovial joints may be of the hinge, ball-and-socket, pivot, saddle, gliding, or ovoid type.

The vertebrate muscular system contains three major muscle tissues: smooth, striated (skeletal), and cardiac. Smooth and cardiac muscle is under involuntary control and is found in visceral organs. Striated muscle is under voluntary control, and its fibers are made up of contractile units called sarcomeres. Muscle contraction is accomplished by the sliding of actin and myosin filaments within each sarcomere. Muscles are generally directed to contract by nervous stimulation, and they use the energy stored in ATP.

Review Questions

1. How many bones are there in the average adult human skeleton?
2. The term "ambulacral" has to do with
   (a) myosin filaments
   (b) interfilament bridges
   (c) sarcolemma
   (d) tube feet
   (e) the occipital bone
   (f) marine mammals
3. What is a suture?
4. What is the first structure secreted during molting?
5. The prefixes sacro- and myo- usually refer to something having to do with _____ ?
6. What is the intermediate substance in intracartilaginous ossification?
7. Muscle contraction depends on energy derived from what molecule?
8. Why are some ribs referred to as "floating ribs"?
9. Which band(s) becomes narrower during muscle contraction, and why?
10. What is the difference between a metatarsal bone and a metacarpal bone?
11. What type of myofilament is attached to the Z line?
12. Growth hormone is produced by which endocrine gland?
13. What types of muscle are multinucleated?
14. What are epiphyses?
15. Darwin reportedly smoked a pipe made from an albatross's wing bone. What characteristic of avian skeletons makes this possible?
16. Name a crustacean that has an exoskeleton heavily impregnated with calcium carbonate.
17. Which of the three muscle types contracts the fastest?
18. The term "synovial" applies to
    (a) siphoning
    (b) lubrication
    (c) pubis
    (d) the H and I bands

19. What is the difference, other than size, between the vertebral column of a human child and an adult?
20. Is it reasonable to subdivide skeletal systems into two categories: exoskeletons and appendicular skeletons? Justify your answer.

Selected References

Arcy, L. B. *Human Histology Textbook in Outline Form*, 2nd ed. Philadelphia: W. B. Saunders Company, 1963.

Balinsky, B. I. *An Introduction to Embryology*, 3rd ed. Philadelphia: W. B. Saunders Co., 1970.

Barkam, J. M., and W. L. Thomas. *Anatomical Kinesiology*. London: The Macmillan Company, 1969.

Beklemishev, W. N. *Principles of Comparative Anatomy of Invertebrates*. Chicago: University of Chicago Press, 1969.

Best, C. H., and N. B. Taylor. *The Human Body: Its Anatomy and Physiology*, 4th ed. New York: Holt, Rinehart and Winston, Inc., 1963.

Bourne, G. H. (ed.). *The Biochemistry and Physiology of Bone*. New York: Academic Press, Inc., 1971–1972.

Bourne, G. H. *The Structure and Function of Muscle*. 2nd ed. New York: Academic Press, Inc., 1972–1973.

Carlson, A. J., and V. Johnson. *The Machinery of the Body*, 4th ed. Chicago: The University of Chicago Press, 1953.

Crouch, J. E., and J. R. McClintic. *Human Anatomy and Physiology*. New York: John Wiley & Sons, 1971.

Davson, H., and M. G. Eggleton. *Starling and Evans, Principles of Physiology*, 13th ed. Philadelphia: Lea & Febiger, 1962.

Esch, D., and M. Lepley. *Musculoskeletal Function: An Anatomy and Kinesiology Laboratory Manual*. Minneapolis: University of Minnesota Press, 1974.

Gray, J. *How Animals Move*. London: Cambridge University Press, 1953.

Hancox, N. M. *Biology of Bone*. Cambridge (Eng.) University Press, 1972.

Huddart. H. *The Comparative Structure and Function of Muscle*. New York: Pergamon Press, Inc., 1975.

Huxley, A. F., and R. Niedergerke. "Structural Changes in Muscle During Contraction," *Nature*, **173**:971–973, 1954.

Huxley, H. E. "The Contraction of Muscle," *Sci. Amer.* **199**:66–82, 1958.

Jahn, T. L. "A Possible Mechanism for the Effect of Electrical Potentials on Apatite Formation in Bone," *Clin. Orthopaedics*, **56**:261.273. 1698.

———, and J. J. Votta. "Locomotion of Protozoa," *Ann. Rev. Fluid Mechanics*, **4**:93–115, 1972.

Kahn, F. *Man in Structure and Function*. New York: Alfred A. Knopf, Inc., 1953.

Langley, L. L., and E. Cheraskin. *The Physiology of Man*, 2nd ed. New York: McGraw-Hill Book Company, 1958.

Montagna, W. *Comparative Anatomy*. New York: John Wiley & Sons, 1959.

Morton, D. J. *Human Locomotion and Body Form*. Baltimore: Williams & Wilkins Company, 1952.

Scheer, B. T. *Animal Physiology*. New York: John Wiley & Sons, Inc., 1963.

Stiles, K. A. *Handbook of Histology*. New York: McGraw-Hill Book Company, 1956.

Toldt, C. *An Atlas of Human Anatomy*. New York: Macmillan Publishing Co., Inc., 1928. Torrey, T. W. *Morphogenesis of the Vertebrates*. 4th ed. New York: John Wiley & Sons, 1965.

Usherwood, P. N. R. (ed.). *Insect Muscle*. New York: Academic Press, Inc., 1975.

Vander, A. J., J. H. Sherman, and D. S. Luciano. *Human Physiology: The Mechanisms of Body Function*. New York: McGraw-Hill Book Co., 1970.

Villee, C. A. *Biology*. Philadelphia: W. B. Saunders Co., 1967.

Weisz, P. B. *The Science of Biology*, 4th ed. New York: McGraw-Hill Book Co., 1971.

Yapp, W. B. *An Introduction to Animal Physiology*, 2nd ed. Fair Lawn, N.J.: Oxford University Press, 1960.

Digestive Systems

Our survey of the various animal phyla has emphasized the diverse ways in which animals capture and digest food and eliminate metabolic wastes. This chapter will present a general overview of digestion and excretion, with emphasis on human anatomy and physiology.

Digestion and **excretion** are complementary aspects of **nutrition.** Nutrition may be defined as the sum of the processes by which an animal utilizes food substances. Animal nutrition generally involves five steps: (1) the **ingestion** of food, (2) the **digestion** of food into its basic nutrient components, (3) the **absorption** of these products of digestion into the blood or other circulating fluid, (4) the **transport** of nutrients throughout the body by the circulatory system, and (5) the **assimilation** of nutrients into the body's cells.

Metabolism is the sum of all the biochemical processes that occur within the individual cells of an organism (Chapter 2).

Once inside the body's cells, the nutrient products of digestion may be metabolized. Metabolism is divided into two components. (1) Nutrients may be **anabolized** (constructive metabolism) and incorporated into living protoplasm or stored for further use, or (2) they may be immediately **catabolized** (destructive metabolism) to release the energy stored in their chemical bonds. Catabolism breaks down organic molecules into their simpler inorganic components —some of these are toxic, whereas others are merely useless.

Nutrition and Ingestion
Food

All substances taken into the body that are used to produce protoplasm and energy are foods. Because many of the chemicals in this section are essential building blocks of all animals, the reader should review the relevant discussions in Chapter 2. The principal foods of animals are organic compounds (carbohydrates, fats, proteins, nucleic acids, and vitamins) preformed by other organisms, plus water, inorganic salts, and oxygen. The ultimate energy source for all living organisms is the sun. Plants utilize radiant energy via the process of photosynthesis and synthesize glucose from water and carbon dioxide. Unlike plants, animals cannot synthesize their food out of inorganic substances but must feed on the tissues of other living organisms.

Now let's consider some general categories of food substances, starting with the simpler inorganic components.

Oxygen

Oxygen enters cells from the medium in which the cells live (water in aquatic species, and tissue fluid or blood in higher animals). The oxygen is used in the production of metabolic energy by oxidative catabolism.

Water

Water is ingested in greater amounts than are all other substances combined and makes up about 60 to 90 percent of cytoplasm by weight. It is also the medium in which digestion occurs, nutrients are absorbed, and metabolic wastes are excreted. In fact, there is hardly a physiological process in which water is not of fundamental importance.

Mineral Salts

Inorganic salts make up only about 1 percent by weight of the cytoplasm and body fluids. Nevertheless, such minerals are crucial to many physiological processes. For example, calcium ions are necessary for muscle contraction; calcium salts are also a main component of bones and teeth. The principal mineral salts contained in cytoplasm are as follows.

Calcium is a constituent of all cytoplasm and body fluids, but 99 percent of the calcium in our bodies is contained in bones and teeth. Milk and leafy vegetables are the best sources of calcium. Milk is also the best source of **phosphorus,** which is necessary for normal bone development. **Iron** functions principally in hemoglobin, the protein within erythrocytes that transports oxygen and carbon dioxide throughout the body. Good dietary sources of iron include beef liver, egg yolk, whole grains, fruits, and green vegetables. **Iodine** is incorporated by the thyroid gland into the hormone **thyroxin** (and others), which consists of about 65 percent iodine by weight. Insufficient iodine in the diet results in enlargement of the thyroid gland, a condition known as **goiter.** Iodine may be supplied by iodized salt, certain fishes, lobsters, oysters, milk, leafy vegetables, and fruits from soils rich in iodine. Other minerals, such as **copper, magnesium, manganese, zinc,** and **selenium,** are also necessary; their primary functions are usually as cofactors for enzymes (Chapter 2).

Carbohydrates

Carbohydrates are important as fuels in the metabolic oxidation reactions that supply energy for all our metabolic functions (Chapter 2). Three general types of carbohydrates are recognized on the basis of molecule size: **monosaccharides** (simple sugars), **disaccharides** (double sugars), and **polysaccharides** (complex carbohydrates of three or more sugar subunits). **Glucose** is a simple sugar that occurs in certain fruits, notably grapes. Glucose is also the most important carbohydrate in our bodies. **Fructose** is another simple sugar that occurs in fruits. Disaccharides are hydrolyzed during digestion into two simple sugars. **Sucrose,** or cane sugar, is a disaccharide found in vegetables, fruits, and many plant juices. **Lactose** occurs in the milk of all mammals. **Maltose** is a product of starch digestion and occurs in germinating cereals, malts, and malt products.

Polysaccharides are large molecules that are polymers composed of many simple sugar subunits. **Starch** is a polymer of many simple sugars. We obtain plant starch especially from grains, tubers, and roots. The solid matter of cereal grains consists of 50 to 75 percent starch. Starches are digested into simple sugars, which are then absorbed into the blood stream. Another polysaccharide that we ingest in large quantities is **cellulose,** a component of plant cell walls. The cellulose molecule resists digestion as it passes through our alimentary canal because we, as is true for most animals, cannot synthesize the enzyme cellulase.* However, by its bulk it aids the peristaltic muscular contractions that push materials through the digestive tract.

Once absorbed into the bloodstream, the simple sugars may enter the body cells for immediate use, either to be anabolized into protoplasm or catabolized to produce energy. Or the sugars may be stored as **glycogen** (animal starch) in the liver and muscles until needed.

Lipids

Lipids are integral components of cell membranes and are largely responsible for their semipermeability (Chapter 2). Excess carbohydrates may also be converted into **fats** and stored as adipose tissue. Mobilization of these fat reserves requires more time than is

---

* In ruminants, such as cattle, cellulose is digested by symbiotic protozoans that populate the cattle's multichambered stomachs; the by-products as well as some symbionts are then digested and assimilated by the mammalian hosts.

the case with glycogen. The fats that we use are derived from plants and animals and are commonly ingested with meat, lard, butter, olive oil, and cod-liver oil.

Besides the true fats, lipids also include the **sterols** and the **phospholipids.** The best-known sterol is **cholesterol,** which is excreted in bile and is widely distributed in the body. The best-known phospholipid is **lecithin,** which is abundant in egg yolk; it contains nitrogen and phophorus.

Proteins

Proteins are extremely complex molecules that may be composed of hundreds or thousands of amino acid subunits. Proteins always contain nitrogen in addition to carbon, hydrogen, and oxygen; often, sulfur, phosphorus, and iron are also found. Dietary sources of protein are animal tissues, such as meat, milk, fish, and albumin, and certain plant sources, especially beans, nuts, and seeds.

The **amino acids** that result from protein digestion are much more important than are carbohydrates or fats in the synthesis of protoplasm. Proteins play structural roles as components of all membranes, muscles, bone, skin, hair, nails, and other structures. Proteins also have crucial metabolic functions as enzymes that foster the myriad chemical reactions that take place within a cell. Amino acids may also be oxidized to produce energy, but this is a secondary function. The first step (**deamination**) in amino acid catabolism removes the nitrogen atom, forming **ammonia,** a toxic metabolic waste that must be excreted from the body. Very few amino acids are stored in the body and accordingly there exists a group called the essential amino acids. However, when carbohydrates and fats are used up, the proteins in the cytoplasm itself are digested.

Nucleic Acids

Nucleic acids are broken down first into **mononucleotides;** they may be further hydrolyzed into their component sugars, nitrogenous bases, and phosphate. These small organic molecules are utilized as intermediates in many metabolic pathways.

Vitamins

Vitamins are specific organic molecules that are required in very small amounts in the diet. Vitamins are essential to the growth and health of all living systems, as is demonstrated by the diseases that result with vitamin deficiencies (Fig. 33-1). Plants are able to synthesize the vitamins that they require from inorganic molecules and sunlight. Animals can synthesize some of the vitamins that they need, but others must be supplied in the diet. Humans require approximately a dozen vitamins that must be supplied in the diet.

Vitamins were originally designated by capital letters, but, as their chemical structures became known, they have been given chemical names. To these are sometimes added a third group of functional names. Vitamins A, D, E, and K are oil soluble; the others are soluble in water. All the water-soluble vitamins are the very important coenzymes necessary for enzymic catalysis in many biochemical reactions (Chapter 2).

Vitamin A: Antixerophthalmic Vitamin

The precursor of this vitamin occurs in plants as the accessory photopigment **carotene,** which gives some fruits and vegetables their yellow color. Night blindness, the inability to see in dim light, has been known for centuries. This condition results from insufficient amounts of a photochemical substance in the eye called rhodopsin (visual purple). To regenerate visual purple, vitamin A is required. Lack of this vitamin also disturbs the secreting powers of mucous membranes; for example, if the lacrimal glands do not keep the eye moist, it leads to a condition called **xerophthalmia,** "dry eye." Also, resistance to certain infections decreases, and growth is retarded. Few individuals show the more pronounced effects of vitamin A deficiency, such as xerophthalmia, but a large number do suffer from some degree of night blindness. The chief sources of vitamin A are such vegetables as spinach, asparagus, carrots, and sweet potatoes and such animal products as butter, cream, eggs, liver, and fish-liver oils, especially halibut-liver oil.

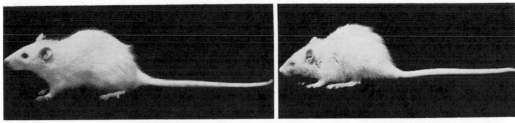

A. Normal   B. Deficient

Vitamin A

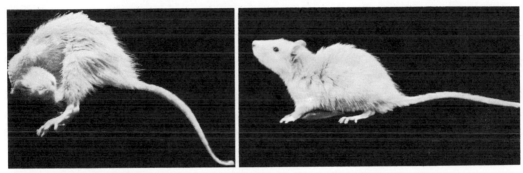

A. Deficient   B. After treatment

Vitamin B$_1$

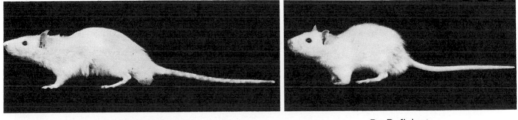

A. Normal   B. Deficient

Vitamin D

Figure 33-1. Effects of vitamin deficiency in rats. With vitamin A and two rats from the same litter: (A) weighed 123 grams and had bright eyes and sleek fur; (B) rat had no vitamin A and weighed 56 grams and had infected eyes and rough fur. Vitamin B$_1$: (A) rat 24 weeks' old did not have enough vitamin B$_1$; (B) same rat 24 hours after receiving food rich in vitamin B$_1$. Vitamin D: (A) normal rat; (B) rat that did not receive vitamin D and, hence, is suffering from rickets. [Courtesy of U.S. Bureau of Human Nutrition and Home Economics.]

Vitamin B$_1$: Thiamine, the Antineuritic Vitamin

Several important vitamins belong to the vitamin B complex. Many of the B vitamins function as coenzymes in metabolic reactions. Vitamin B$_1$ was the first member of the B complex to be differentiated. Its presence prevents a disease known as **beriberi,** which is characterized by loss of appetite and degenerative changes in the nervous system. Thiamine is abundant in brewer's yeast and in the outer husk of cereal grains. Processed cereals such as white rice and white flour that have been polished or milled are deficient in vitamin B_1. Beriberi is prevalent among Oriental peoples that live on a diet consisting largely of polished rice. Dietary sources of thiamine also include peas, beans, nuts, and liver.

Vitamin B_2: Riboflavin

Deficiency of riboflavin in our diet results in soreness at the corner of the mouth (**cheilosis**) and inflammation of the cornea of the eye. Vitamin B_2 is contained in milk, green and leafy vegetables, egg white, liver, meat, and yeast.

Vitamin B_3: Pantothenic Acid

Pantothenic acid contributes to the formation of **coenzyme A,** which participates in the metabolism of carbohydrates, lipids, and proteins. Yeast, cane molasses, meat, egg yolks, milk, and liver are some of the many dietary sources of this vitamin. Vitamin B_3 is so widely distributed in our foods that no deficiency disease is known for humans. However, dietary deficiencies cause dermatitis in chicks, decreased adrenal cortex function in rats, and diarrhea and nerve degeneration in swine.

Vitamin B_6: Pyridoxine

Although no deficiency disease for humans has been indentified, vitamin B_6 has been shown to be essential for our health. Pyridoxine plays an important role in amino acid metabolism within cells. Yeast, whole-grain cereals, milk, and liver are good dietary sources of B_6.

Vitamin B_{12}: Cyanocobalamin

Vitamin B_{12} is necessary for the formation of red blood cells. A deficiency in humans causes **pernicious anemia.** Vitamin B_{12} is also essential for the growth of young animals. Sources of this vitamin are milk, liver, kidney, and lean meat. Vitamin B_{12} is also notable because it is the only known physiologically essential molecule that contains an atom of cobalt.

Vitamin C: Ascorbic Acid

The principal role of vitamin C is prevention of **scurvy,** a disease caused by the breakdown of capillaries, with symptoms of loosened teeth, bleeding gums, and fragile bones. Ascorbic acid was isolated in pure form in 1932. It is present in citrus fruits and tomatoes. Animals other than primates and guinea pigs can synthesize vitamin C and thus do not have to rely upon dietary sources.

Vitamin D: Calciferol, the Antirachitic Vitamin

Vitamin D is also essential for the formation of strong bones and teeth. Vitamin D is synthesized within the human body through the action of ultraviolet radiation from the sun upon an organic compound (**ergosterol**) in the skin. Deficiency of vitamin D causes **rickets,** a softening of the bones which leads to deformities, especially among young children. Urban populations have limited exposure to the sun, especially during the winter months, and not many foods contain vitamin D. For these reasons, milk and flour are often fortified with this vitamin.

Vitamin E: Tocopherol

This vitamin is necessary for normal reproduction in some animals such as rats, mice, and poultry. However, experiments have also shown that the reproductive capacity of goats, sheep, and rabbits is not impaired when on a diet deficient in vitamin E. These results warn against making broad generalizations from experiments on just one type of animal. This vitamin is widely distributed in the foods we eat, such as green leaves and vegetable fats, that there appears to be no reason for concern about it. Vitamin E is an antioxidant that preserves easily oxidizable vitamins and fatty acids in foods or in the body.

Vitamin H: Biotin

This vitamin is necessary in the food of birds but is not a dietary requirement of humans because it is supplied by symbiotic bacteria in our large intestine. Deficiency symptoms are diarrhea, dermatitis, and nervous disorders. Dietary sources for biotin are liver, kidney, and yeast.

Vitamin K: The Antihemorrhagic Vitamin

Deficiency of vitamin K delays the clotting time of the blood and results in excessive bleeding. It is necessary for the formation of **prothrombin,** one of the precursors of blood clots. Our chief sources of vitamin K are green leafy vegetables and certain bacteria that inhabit our large intestine. It appears that only under extraordinary circumstances do humans suffer from a deficiency of this vitamin. Bile salts are necessary for the absorption of vitamin K in the intestine.

Vitamin M: Folic Acid

Deficiency produces **anemia** and **sprue** (a tropical disease) in humans. Vitamin M is contained in green leaves, soy beans, yeast, and egg yolk.

Niacin, the Antipellagric Vitamin

Niacin is used to make certain coenzymes necessary for cellular functions. This vitamin constituent of the B complex prevents **pellagra,** which is characterized by roughened skin on the hands, arms, feet, face and neck, a sore mouth, pink tongue, diarrhea, and nervous disturbances. A concentrated source of niacin (**nicotinic acid**) is brewer's yeast, and it is also present in wheat germ, brown rice, lean meat, milk, green vegetables, peas, nuts, and beans.

Nutritional Modes

Plants and certain protists (e.g., *Euglena*) that photosynthesize the organic compounds that they require from simple inorganic precursors are called **autotrophic.** Most animals feed on preformed organic matter, that is on plant and animal tissues. Nutrition involving the ingestion of such preformed organic matter is said to be **heterotrophic.** Some parasites and a few free-living organisms absorb dissolved organic substances through the surface of their bodies; this type of nutrition is termed **saprozoic.**

Three principal categories of holozoic animal nutrition may be differentiated: herbivores, carnivores, and omnivores. **Herbivorous** animals feed on vegeta-

tion and are adapted to this kind of diet. For example, leaf-eating insects possess chewing mouthparts, seed-eating birds often have short, thick bills for breaking hard seed coats; rabbits and rodents are provided with sharp incisor teeth suitable for gnawing and biting. **Ruminants** such as cattle have enormous stomachs that serve as fermentation chambers in which protozoan symbionts digest the cellulose cell walls of their host's food. The cheek teeth in all mammalian herbivores are modified to provide extensive grinding surfaces for finely masticating the food before it continues through the digestive tract.

Carnivorous animals feed on animals. Many of the animals that we have studied are carnivores: paramecians eat other protists; coral polyps ingest planktonic animals; and spiders, lampreys, and owls (to list a few) all subsist on animal tissues. **Insectivorous** animals, a subset of the carnivores, consume primarily insects. Frogs, certain turtles, a few rodents, certain birds, and other animals are insectivores. Carnivorous mammals have relatively large canine teeth for seizing their prey and ripping meat into shreds.

Omnivorous animals feed on both animal and vegetable tissues. Some protists eat both microscopic plants and animals, as do many filter-feeding metazoans, such as mussels and clams. Earthworms feed largely on decaying vegetation but they also ingest animal food. And, of course, we humans are generally omnivorous.

Plants and certain protists (e.g. *Euglena*) that photosynthesize the organic compounds that they require from simple inorganic procursors are collectively called **autotrophic.** There are within this group, however, organisms (e.g. bacteria) capable of deriving primary energy from inorganic compounds instead of sunlight. These organisms are said to be **chemotrophic.** Organisms capable of deriving primary energy from sunlight using carbon dioxide, water, and nitrogen compounds are said to be **photosynthetic** or **holophytic.**

Organisms whose nutrition involves the ingestion of such preformed organic matter are said to be **heterotrophic.** Two principal groups of heterotrophic nutrition can be identified: **Holozoic** nutrition involves the ingestion of whole food; *Ameba by* phagocytosis, *Paramecium* through the cytostome and man through the mouth. **Saprozoic** nutrition is whereby organisms that can absorb its nutrients by diffusion (no cytostome or mouth) through the cell membrane (e.g. Opalina, Sporozoa, and tapeworms).

Summary of Nutritional Types

Autotrophic

1. Self nourishing organisms that can use inorganic substances to form inorganic compounds.
2. **Chemotrophic** Bacteria-primary energy is derived from inorganic compounds instead of sunlite.
3. **Phototrophic** or **Holophytic** Primary energy is derived from the sunlight using CO_2, H_2O and N compounds.

Heterotrophic organisms are dependent on outside food sources and require complex organic compounds of nitrogen and carbon.

1. **Holozoic** ingest whole food.
2. *Ameba* by phagocytosis.
3. *Paramecium* through cytostome man through mouth.
4. **Saprozoic** organism that can absorb its nutrients by diffusion (no cytostome or mouth) through the cell membrane, for example, *Opalina, Sporozoa* and tapeworm.

Food Capture

Most animals that are capable of locomotion move about in search of food. Those that are sessile either wait for food to come to them or possess some method of bringing it within reach. For example, the sessile oyster draws suspended organic materials into its shell via undulipodia-generated water currents. Sponges and many other aquatic animals are also filter feeders.

Protists may capture food by means of flagella, pseudopodia, or cilia. Tentacles armed with nematocysts are characteristic of cnidarians. The nematocysts are used for paralyzing and entangling prey. Starfishes are able to pry open bivalve shells by a long, steady pull with their tube feet. Some spiders build elaborate webs that snare flying insects. The frog captures prey with its long sticky tongue. Many carnivores depend on speed for overtaking their prey. Birds of prey rely on the speed of flight as well as on

sharp claws (talons) and hooked beaks. Claws and teeth are a familiar and effective means of mammalian offense and defense. Human civilizations, by their domestication of animals, have in great measure eliminated the necessity of hunting.

Food Selection

The mouthparts of most animals are supplied with sense organs that help them to distinguish different kinds of food and to avoid poisons. Such chemical sense extends even into the protist realm: amoebas seem to prefer the small flagellate *Chilomonas*. Many herbivorous insects and their larvae feed exclusively on one or a few kinds of plants and will starve if these specific foods are not available. The diets of most vertebrates are likewise more or less limited. This has led to such descriptive common names as "fish hawk" (osprey), "carrion crow" (black vulture), "duck hawk," "herring gull," "kingfisher," "flycatcher," "anteater," "sea cow," and "fruit bat."

The character of the food depends largely on the habitats of the animal—be it marine, freshwater, or terrestrial—which in turn are influenced by climatic factors. Another factor is the animal's period of activity—be it **diurnal, nocturnal,** or **crepuscular** (active at dawn or dusk). The size of the animal, its method and speed of locomotion and capture, its sense organs, the structure of its mouthparts (e.g., teeth, Fig. 33-2), and the structure of its digestive system all play significant roles in food selection. These factors all contribute in general to the character of an animal's food, but it is important to realize that they simultaneously represent successful adaptations to the physical and biological environment in which the species evolved (Chapter 41).

Ingestion of Food

In many animals, the structures used in obtaining food are also employed for its ingestion. Examples include the pseudopodia of amoebas, the tongue of the frog, and the muscular pharynxes of planarians, earthworms, and certain sucking insects. Other animals employ special methods of ingestion. In sponges, food particles drawn into the water canals by means of long flagella are phagocytosed by the collar cells. In the crayfish and in many insects, the food is

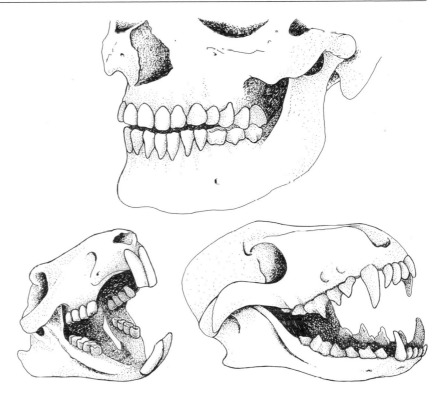

Figure 33-2. Variations in teeth—adaptations to different foods. Top: Dentition in man. Lower left: Dentition in a herbivore; note molar teeth for grinding. Lower right: Dentition of a carnivore; note sharp teeth for tearing and biting. Man, an omnivore, shares dental characteristics with both herbivores and carnivores.

held by specific mouthparts while it is crushed or cut into small pieces by the mandibles. A starfish does not directly ingest the large bivalves and other prey that it attacks; rather, it everts its stomach and digests the soft tissues of the prey outside its body . The lamprey attaches to fishes with its suckerlike mouth and rasps away the flesh with its horny tongue; then it feeds on blood. Most fishes do not chew their food but hold it with their teeth and swallow it at once. Many other vertebrates, such as snakes and birds, also swallow their food whole. Mammals, for the most part, chew their food before swallowing, and their heterodont teeth are modified for this purpose in specific ways that reflect their diets (Fig. 33-2).

Digestion

Digestion is the process of breaking down food material into minute soluble components so that they can be absorbed into the animal's body. Certain simple compounds such as water, oxygen, mineral salts, and monosaccharides (such as glucose) can be absorbed in the same form in which they are ingested. Most foods, however, consist of complex carbohydrates, fats, and proteins that as such cannot pass through the membranes that line the digestive system.

Digestive Enzymes

The carbohydrates must be broken down into their component simple sugars; the fats into glycerol and fatty acids; the nucleic acids into sugars, nitrogenous bases, and phosphate; and the proteins into amino acids. These catabolisms are all accomplished through the activities of digestive enzymes. Enzymes are complex molecules produced by living cells. Their characteristics are discussed in Chapter 2. Enzymes are secreted into various parts of the digestive tract, bringing about hydrolysis of different substances in different organs, as summarized below.

Carbohydrates

Saliva contains two enzymes, **ptyalin** and **maltase,** that initiate carbohydrate digestion in the mouth. Ptyalin (salivary amylase) hydrolyzes starch into the disaccharide maltose. Maltase converts maltose into glucose. You may demonstrate this by placing a soda cracker in your mouth and letting it remain there for a few minutes. You will soon taste sweetness, which indicates that the starch is being digested into the simple sugar, glucose.

Because of the short time in the mouth and low acidity of the stomach (pH of 1–3), 90–95 percent of the carbohydrates are undigested until they reach the small intestine. There, **amylase** from the pancreas hydrolyzes starch into maltose. Enzymes secreted by the intestinal wall complete the digestion of carbohydrates: maltase hydrolyzes maltose into glucose; **sucrase** digests **sucrose** into glucose and fructose; and **lactase** hydrolyzes **lactose** into glucose and galactose. Glucose, fructose, and galactose are all simple sugars that can be absorbed into the blood.

Fats

No digestion of fat occurs in the mouth. In the stomach very little occurs, although a gastric **lipase** may begin the digestion of emulsified fats such as cream.

Most fat is digested in the small intestine. **Bile salts,** produced in the liver and stored in the gallbladder, break up (emulsify) fat globules into smaller droplets, thereby increasing the surface area exposed to enzymic attack. The enzyme **lipase,** from the pancreas, splits fats into their component glycerol and fatty acid subunits, which can be absorbed.

Proteins

No protein digestion occurs in the mouth. In the stomach, **hydrochloric acid** transforms the inactive enzyme **pepsinogen** into **pepsin,** which hydrolyzes proteins into polypeptides.

In the small intestine the enzyme **trypsin** from the pancreas also breaks up proteins and partially digested proteins into polypeptides. These are then digested into their amino acid subunits by **chymotrypsin** and **carboxypeptidase** from the pancreas and by various peptidases from the intestinal walls. The amino acids thus are absorbed by the intestinal cells.

Nucleic Acids

A variety of **ribonucleases** and **deoxyribonucleases,** which digest RNA and DNA, respectively, are synthesized and secreted by the pancreas. These enzymes break the long nucleic acid polymers into water-soluble mononucleotides (Chapter 2) that can be absorbed by the small intestines.

In most animals, certain mechanical processes contribute to digestion, such as mastication (the chewing of food), peristaltic movement (wavelike muscular contractions that push food through the digestive canal), and various grinding or mixing activities within the digestive tract.

Two general types of digestion can be recognized in animals, **intracellular** and **extracellular.** In protists and poriferans, food particles are digested inside **food vacuoles (phagosomes)** within cells. Cnidarians such as hydra initially digest food extracellularly in the gastrovascular cavity; then digestion is completed intracellularly, in phagosomes within the gastrodermal cells. Most metazoans have a tubular gut with two openings, a mouth and an anus, and practice extracellular digestion within the digestive tract.

We will now concern ourselves with what actually happens to food during the digestive process. The remainder of this discussion will emphasize human digestion. Note that a great variety of systems and methods of digestion have evolved in the animal kingdom; many of these were surveyed in previous chapters.

The Human Digestive System

The digestive tracts of the earthworm, crayfish, grasshopper, mussel, and vertebrate are all constructed on the same general plan. The bodies of most complex metazoans are essentially double tubes, as is clearly seen in the earthworm (Fig. 17-3). The outer tube is the body wall, and the inner tube is the digestive canal. Between the two tubes is a body cavity (a coelom or hemocoel). All structures that make up the digestive canal, together with the accessory organs that aid in digestion, constitute the digestive system. The process of digestion is, in general, quite similar throughout the animal kingdom. More is known about digestion in humans than in any other animal. The human digestive system resembles that of other vertebrates more closely than it does that of invertebrates, but even in such animals as the earthworm, the digestive organs are given similar names, such as the mouth, pharynx, esophagus, stomach, intestine, anus, and digestive glands. This is because their functions are somewhat analogous to those of vertebrates.

The parts of the human digestive tract are as follows (Fig. 33-3):

1. **Mouth cavity,** containing **teeth, tongue,** and openings of ducts from **salivary glands.**
2. **Pharynx,** or throat cavity, shaped like an inverted cone.
3. **Esophagus,** a muscular tube about 23 cm long.
4. **Stomach,** a very muscular, saclike dilation of the digestive canal.
5. **Small intestine,** a muscular tube about 7 m long and from 2.5 to 3.8 cm in diameter, divided into three regions: **duodenum,** about 30 cm long; **jejunum,** about 2.5 m long; and **ileum,** about 4.25 m long.
6. **Large intestine,** a muscular tube about 1.5 m long and 6.25 cm in diameter; may be divided into **cecum,** a large pouch with a **vermiform** (worm-shaped) **appendix** about 7.5 cm long; **colon,** the

Figure 33-3. The digestive system in man. Not all the coils of the ileum (small intestine) are included.

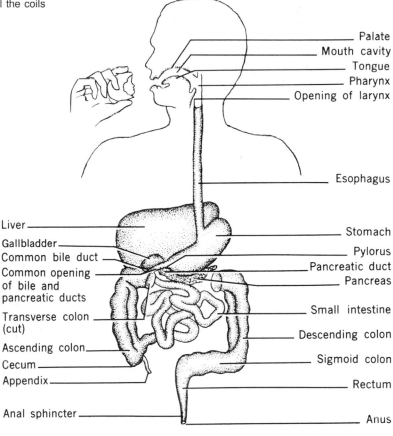

main part of the large intestine; **rectum,** about 12.5 cm long; and **anal canal** about 3.8 cm long.

The Sequence of Events for Digestion

The mouth cavity receives the food. Here it is **masticated** (chewed) by the teeth, aided by the tongue, which helps to position the food between the jaws. The tongue also houses the **gustatory receptors (taste buds)** and assists in the swallowing of food. The mouth cavity is lined by a mucous membrane that contains minute **oral glands** that secrete lubricating mucus. This mixes with the more abundant secretions from the three pairs of **salivary glands:** the **sublingual glands** lie under the tongue, the **submandibular glands** are farther back and below the mandible (lower jawbone), and the **parotid glands** are at the back of the throat. The salivary glands secrete saliva, which contains two enzymes, ptyalin (salivary amylase) and maltase, that are both involved in the digestion of carbohydrates. Approximately 1 liter of saliva is produced each day. Secretion of saliva is an involuntary act stimulated by the taste, sight, smell, or even thought of food; some saliva is constantly secreted even with no outside stimulus.

A portion of the mixture of saliva and well-chewed food, collectively called **chyme,** is periodically formed into a lump (**bolus**) by the tongue and jaws. The bolus is passed by the tongue back to the pharynx, which dilates and allows it to pass into the tubular esophagus. A cartilaginous flap called the epiglottis closes over the larynx as food is swallowed, thus preventing food from entering the respiratory tract. Rhythmical wavelike contractions (**peristalsis**) of the smooth (involuntary) muscles in the esophagus push the bolus toward the stomach without the aid of gravity (proven by successful ingestion by astronauts in weightless conditions).

The esophagus penetrates the diaphragm and opens into the stomach through a ring of smooth muscle called the **cardiac sphincter.** The wall of the stomach is very muscular and is thrown into temporary folds (**rugae**) that allow for expansion. The anterior rounded end of the stomach above the cardiac sphincter is called the **fundus,** the central region is called the **body,** and the smaller end, which communicates with the small intestine, is called the **pyloris.** The

chyme is usually churned in the stomach for about three or four hours and is acted upon by digestive juices secreted by glands in the stomach wall.

Mucus glands in the cardiac and pyloric regions secrete mucus, which protects the stomach itself from digestion. **Fundic glands** in the fundus region secrete hydrochloric acid and pepsinogen, an inactive precursor of pepsin. Hydrochloric acid breaks up connective tissues and cell membranes. The acidity also converts pepsinogen into pepsin. **Rennin,** which coagulates milk proteins, is also secreted in the fundus region. In addition, certain cells in the mucous membrane that lines the lumen of the stomach secrete the hormone **gastrin,** which, when stimulated by the presence of food in the stomach, stimulates further production of the gastric juices.

When the chyme is in the appropriate semiliquid state, the pyloric sphincter leading to the duodenum, the first part of the small intestine, opens. The chyme is acted upon by bile from the liver and pancreatic juice from the pancreas; both usually enter the duodenum through a common orifice. Simple tubular glands in the mucous membrane throughout the small intestine and more complex **duodenal glands** in the submucous membrane of the duodenum also secrete a digestive fluid that contains enzymes. Certain cells in the mucous membrane of the duodenum are stimulated to secrete the hormone **secretin** when acidic chyme from the stomach enters the intestine. Secretin is carried by the blood to the pancreas, where it stimulates the secretion of bicarbonate by the liver and the pancreas.

The pancreas is an important accessory organ of the digestive system. Pancreatic juice is distinctly alkaline (pH of 8–9) and serves to neutralize the acidic chyme. The pancreas also secretes several enzymes: Trypsinogen, an inactive enzyme, is converted in the presence of **enterokinase** from the intestinal wall into trypsin. Chymotrypsinogen secreted by the pancreas is converted into active chymotrypsin through the action of trypsin. Both trypsin and chymotrypsin complete protein digestion in the small intestine. Pancreatic amylase converts partly digested starches into simpler sugars (primarily maltose), and lipase splits fat molecules into glycerol and fatty acids.

The liver, another accessory organ to the digestive system, secretes bile, a liquid that may be stored in the gallbladder until needed in the duodenum. Bile contains bile salts that aid in the digestion of fat by bringing about physical division (**emulsification**) of fat droplets into smaller particles. This increases the surface area of the substrate upon which lipase acts. Some other functions of the liver include regulating the concentration of each kind of amino acid in the blood; storing glycogen, vitamins, iron, and copper; metabolizing fats producing **heparin** (a blood anticoagulant); and detoxifying numerous chemicals that are harmful to the body.

Peristaltic actions transport the chyme through the small intestine, but circular folds tend to retard the movement and promote mixing, thereby enhancing both digestion and absorption. Minute fingerlike projections called **villi** occur throughout the small intestine. The membranes of the villi are likewise organized into fingerlike **microvilli**; together, the villi and microvilli vastly increase its internal surface area (Fig. 33-4). The villi are highly vascularized, and it's through their thin walls that nutrients are absorbed into the blood capillaries and lymph vessels (lacteals).

An **iliocecal valve** allows material to pass from the small intestine into the large intestine (colon) and prevents backflow. The small pouch at the beginning of the large intestine is the cecum. Attached to the end of the cecum is a narrow tube about 7 cm long, the vermiform appendix. Material is carried through the colon by peristalsis. Digestion is continued here by digestive fluids that enter with the chyme, and absorption also takes place. While most of the water in the chyme is absorbed in the small intestine, this process continues as the chyme passes through the colon. Eventually the contents of the posterior end of the colon are in a semisolid condition and are referred to as **feces**. This fecal material consists largely of cellulose and other indigestible substances as well as bacteria and excretions of the colon itself such as excess calcium and iron. The feces are temporarily stored in the rectum until they are eliminated (defecated) through the anus.

Absorption

Water, minerals, salts, and digested nutrients must be absorbed from the digestive tract into the circulatory system, which distributes them then to the tissues of the body. Very little absorption takes place in

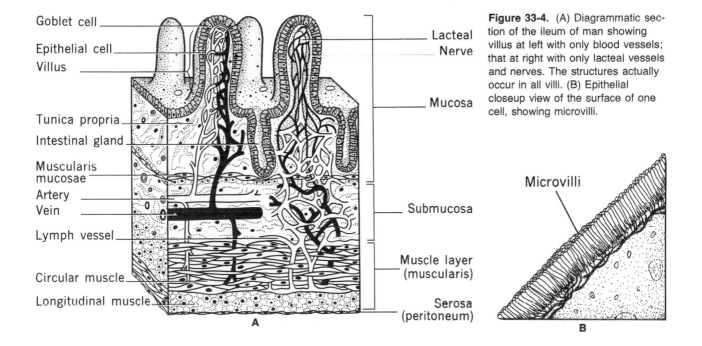

Figure 33-4. (A) Diagrammatic section of the ileum of man showing villus at left with only blood vessels; that at right with only lacteal vessels and nerves. The structures actually occur in all villi. (B) Epithelial closeup view of the surface of one cell, showing microvilli.

the stomach, although alcohol and some drugs are readily absorbed across its walls. Most of the water is absorbed by **osmosis** from the small intestine, but the large intestine also plays a role in water absorption. Almost all food nutrients are absorbed across the walls of the small intestine. The millions of villi and microvilli that project into the lumen of the small intestine greatly increase the absorptive surface area.

Diffusion and osmosis contribute to absorption, but **active transport** is more important (Chapter 2). Glycerol and inorganic salts are absorbed by passive diffusion through the small intestine, and water is absorbed by osmosis if the fluids of the small intestine are hypotonic with respect to the blood. Other nutrients (except fats) are actively transported into the mucosa that lines the small intestine and then into the bloodstream. The fatty acids and glycerol that result from fat digestion are coalesced into microscopic aggregates called **micelles** in the lumen of the small intestine. The micelles enter the intestinal mucosa, where triglycerides are resynthesized. These tiny fat droplets then enter blind-ending lymph vessels (**lacteals**) that are found in each villus. By traveling through the lymphatic system, the fats enter the bloodstream via the thoracic duct of the lymphatic system. Absorption of the fat-soluble vitamins A, D, and E is enhanced by the presence of bile salts. Without bile, a deficiency of these vitamins may occur. The bacteria that inhabit the colon are important sources of vitamins, especially when dietary intake is very low; these are absorbed through the walls of the colon.

Summary

The principal foods of animals are organic compounds synthesized by other organisms (including many vitamins), water, inorganic salts, and oxygen. Mobile animals usually search for food, whereas sessile animals utilize means such as filter feeding to capture food.

The digestive systems of animals are rather diverse, but they are generally compartmentalized into various organs. In humans these include the mouth, pharynx, esophagus, stomach, small intestine, and large intestine. Food moves through the gut due to peristalsis. In the mouth and stomach, enzymes act on the food, but little is actually absorbed. The small intestine is responsible for most nutrient digestion and absorption, whereas the large intestine absorbs bile salts, some vitamins, and much of the remaining water. The pancreas and liver function as accessory digestive organs: pancreatic enzymes and the liver's bile facilitate digestion in the small intestine.

Digestion, a catabolic process, involves the changing of food material into an absorbable state. Digestive enzymes play an important role in this process, both when it occurs within cells (intracellular digestion) and outside them (extracellular digestion). Water can be absorbed without first being digested. Mineral salts of calcium, phosphorus, iron, iodine, copper, magnesium, and manganese are important ions for animal nutrition. carbohydrates are digested by the enzymes ptyalin and maltase, whereas fats and lipids are emulsified by bile and digested by lipases. Proteins are broken down to their constituent amino acids by pepsin, trypsin, chymotrypsin, peptidase and hydrochloric acid. Nucleases digest nucleic acids into component parts.

Vitamins' major actions are those of coenzymes.

Review Questions

1. What makes a human capable of swallowing upside-down?
2. How does your body make use of cellulose?
3. Almost all the body's calcium is contained in the _____ and _____ .
4. What element is of great importance in the formation of thyroxin?
5. If you swallowed a glass marble, would it be considered food? Why or why not?
6. A tapeworm is
   (a) holozoic
   (b) saprozoic
   (c) holophytic
7. The presence of large, flat teeth would indicate what mode of nutrition?
8. What factors contribute to an animal's particular diet?

9. Is the "esophagus" of the earthworm analogous to the human "esophagus"? Are they homologous?
10. What is a cecum? Do humans have any?
11. Name the three pairs of salivary glands, and describe their locations.
12. What does amylase do?
13. What function do villi and microvilli serve?
14. Is a bolus made of fecal material?
15. What is the process called whereby fats are physically but not chemically broken up? What is the ultimate source of all biological energy?

Selected References

Adolph, E. F. (ed.). *The Development of Homeostasis*. New York: Academic Press, Inc., 1960.

Bullock, R. H. (ed.). *Physiological Triggers*. Washington: American Physiological Society, 1957.

CRM Books. *Biology Today*. Del Mar, Calif: Communications Research Machines, 1972.

Frings, H., and M. Frings. *Concepts of Zoology*. Toronto: Macmillan Co., 1970.

Gaffon, H. (ed.). *Research in Photosynthesis*. New York: Interscience Publishers, 1957.

Gerard, R. W. (ed.). *Food for Life*. Chicago: University of Chicago Press, 1952.

Griffin, D. R. *Animal Structure and Function*. New York: Holt, Rinehart and Winston, 1962.

Guyton, A. C. *Textbook of Medical Physiology*. Philadelphia: W.B. Saunders Company, 1956.

Jacobson, E. D. and L. L. Shanbour (eds.). *Gastrointestinal Physiology*. Baltimore: University Park Press, 1974.

Kimber, D. C., C. E. Gray, C. E. Stackpole, and L. C. Leavell. *Textbook of Anatomy and Physiology*. New York: Macmillan Publishing Co., Inc., 1956.

Ponder, E. "The Red Blood Cell," *Scientific American*, **196**:95, 1957.

Potts, S. T. W., and G. Parry. *Osmotic and Ionic Regulation in Animals*. New York: Macmillan Publishing Co., Inc., 1964.

Prosser, C. L., and F. A. Brown, Jr. *Comparative Animal Physiology*. 2nd ed. Philadelphia: W.B. Saunders Co., 1961.

Schmidt-Nielsen, K. *Animal Physiology*. 2nd ed. Englewood Cliffs: Prentice-Hall, Inc., 1964.

Scholander, P.F. "The master switch of life," *Scientific American*, **209**:92, 1963.

Smith, H. W. "The kidney," *Scientific American*, **188**:40, 1953.

Smith, H. W. *The Kidney*. New York: Oxford University Press, 1951.

Wiggers, C. J. "The Heart," *Scientific American*, 1957.

Excretory Systems

Cellular metabolic processes produce potentially toxic waste products that must be removed, or excreted, from the body. Carbon dioxide is a metabolic waste product that will be considered in our discussion of respiration (Chapter 36). Excretory systems are generally concerned with the elimination of (1) nitrogenous compounds (e.g., ammonia, guanine, urea, uric acid; Fig. 34-1) that result from amino acid catabolism, (2) excess inorganic salts such as sodium chloride, and (3) excess water. The control of the internal concentrations of salts and other solutes and of water is called **osmoregulation,** and it is the major (and sometimes sole) function of excretory systems. Wastes usually diffuse from the cells into the circulatory system from which they are collected by the excretory system and are eliminated from the body. Several general types of excretory systems will be surveyed before considering the human kidney in some detail.

Protistan Excretory Mechanisms

Protistan water-expulsion vesicles were once thought to eliminate soluble wastes as well as water. It has now been demonstrated, however, that the liquid in the expulsion vesicles is in fact almost pure H_2O. Hence, water-expulsion vesicles are no longer considered to excrete nitrogenous wastes and salts, despite their well-documented role in osmoregulation.

The plasma membrane is the primary site of excretion in protists. Nitrogenous waste in the form of **ammonia** diffuses easily through the cell membrane into the external environment. Actually, the cell membrane is the primary unit of most metazoan excretory systems as well. Metabolic wastes produced within the cells of multicellular animals first diffuse or are transported through the cell membrane into the interstitial fluid, and from there the wastes diffuse into the circulatory and excretory systems.

Simple Metazoan Excretion

In poriferans, cnidarians, and many other animals, diffusion continues to play an important role in eliminating wastes. In addition, wandering amoeboid cells called **amoebocytes** may transport waste products. The amoebocytes come alongside other cells, allow the excretions to diffuse into them, and then transport the waste out of the organism. Amoebocytes also phagocytose particulate wastes in some invertebrate animals.

Flame and Renette Cells

Platyhelminths, nemerteans, and aschelminths often have excretory units known as **protonephridia.** They consist of blind-ending excretory tubules that originate from specialized cells known as **flame cells** (also called **flame bulbs,** Fig. 34-2A). A portion of these flame cells contain tufts of lengthy cilia that appear to "flicker" when viewed under the microscope, hence their name. The degree to which flame cells actively participate in excretion is uncertain, as ammonia can diffuse directly across the soft body wall of these simple worms. Some authors have ascribed ex-

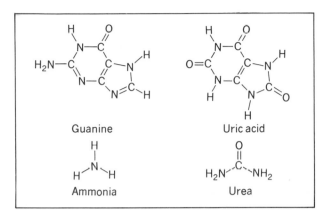

Figure 34-1. Molecular structures of the most common nitrogenous wastes of animals.

clusively osmoregulatory functions to flame cells. Another hypothesis states that, by stirring up the fluids in the tubes, more of the fluid is exposed to the walls, thus enhancing diffusion. In this way flame cells would increase the efficiency of the excretory tubule but not alter its actual function.

Nematodes have excretory tubes known as **renette cells** that vary considerably in morphology. Located in the pseudocoel, they communicate via short excre-tory ducts with the external environment. Some nem-atodes have rather complex systems of longitudinal canals as part of the system. Again, the function of renette cells appears to be primarily osmoregulation, as ammonia diffuses freely through the body wall and digestive tract.

Nephridia

Most annelids and many other coelomate animals have more complex excretory organs called **meta-nephridia,** or simply **nephridia** (Fig. 34-2B). A nephridium has three basic parts. A funnel-shaped opening into the body cavity called the **nephrostome** is lined with cilia that draw fluid into an unbranched but highly coiled tubule. Between the coiled tubule and nephridiopore is a bladder. The distal end of the tubule opens into the external environment through a **nephridiopore.**

While the arrangement is simple in annelids (Fig. 17-2), the "kidney" of mollusks is actually an ex-tremely folded group of nephridia. Also, the **green glands** that open at the base of the antennae of crusta-ceans are modified nephridia. Unlike the other excre-tory plans thus far discussed, the nephridia are the first to concentrate nitrogenous wastes as well as to function in osmoregulation.

Figure 34-2. Excretory apparatus of some invertebrates. (A) Flame cells. (B) Nephridia. (C) Malpighian tubules.

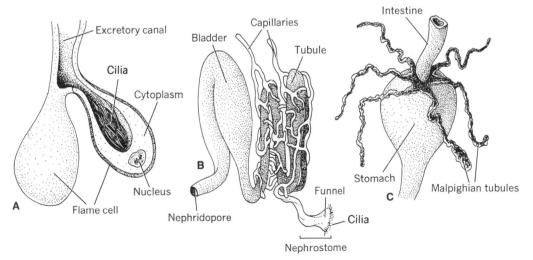

Malpighian Tubules

This type of excretory organ occurs in terrestrial arthropods (Fig. 34-2C). **Malpighian tubules** originate at the junction of the stomach and intestine, and their blind ends move about in the hemocoel. The tubules gather soluble waste from the blood, concentrate it, and channel the remainder into the intestine, where it undergoes further concentration before elimination from the body.

Recall that insects and other terrestrial animals conserve water by converting toxic ammonia into nontoxic nitrogenous compounds (e.g., **guanine,** or **uric acid**), that can be eliminated with little water loss.

The Vertebrate Excretory System

The principal excretory organs of vertebrates are the **kidneys** (Fig. 34-3). The kidneys excrete **urine,** which is carried by two **ureters** to the **bladder,** where it is temporarily stored prior to elimination from the body through the **urethra.** The human kidneys are a pair of reddish-brown, bean-shaped organs, about 12

cm long, that lie against the dorsal abdominal wall just above the waist. Each kidney is supplied with blood from a **renal artery** that branches from the abdominal aorta. **Renal veins** drain blood from the kidneys into the inferior venae cavae.

A translucent **renal capsule** of epithelial cells encases each kidney. Blood vessels and the ureter enter the kidney through a median notch (**hilium**). A smooth **cortex** underlies the renal capsule. Below the cortex is an inner **medulla,** which is divided into approximately a dozen striated wedges, called **renal pyramids.** These lead to a large central cavity, called the **renal pelvis,** that is continuous with the ureter and serves as a funnel, gathering the urine from the renal pyramids.

The Nephron: The Functional Unit of the Vertebrate Kidney

Each kidney is composed of approximately 1 million microscopic units called **nephrons,** which function to form urine. Each nephron is a blind tube with walls of epithelial cells one cell in thickness. The tube begins at a blind cup called **Bowman's capsule,** which

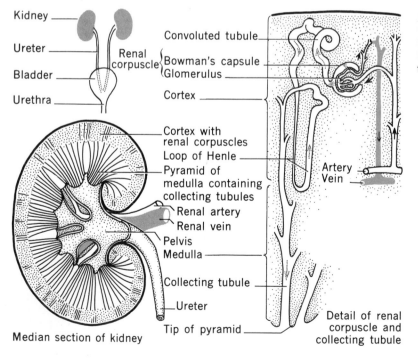

Figure 34-3. Upper left: Positions of kidneys, ureters, bladder, and urethra in human. Lower left: Cross-section of the human kidney. Right: Nephron, functional unit of the kidney. See text for description of function.

opens into a coiled **proximal convoluted tubule.** The next region is an ascending and descending loop, called the **loop of Henle,** which leads to a coiled **distal convoluted tubule.** The distal convoluted tubules from all the nephrons communicate with collecting ducts in the renal pyramids that eventually empty into the renal pelvis. Bowman's capsules and the proximal and distal convoluted tubules are located in the renal cortex, whereas the loops of Henle and the collecting ducts are found in the medulla.

Blood enters the kidney via the renal artery, which branches into a multiplicity of arterioles known as **afferent arterioles.** Each afferent arteriole branches into a compact group of capillaries called a **glomerulus.** Each glomerulus is surrounded by a Bowman's capsule and thus is in intimate association with the tubular elements of the nephron. The glomerular capillaries recombine to form another arteriole (**efferent arteriole**) that soon branches into another capillary network, called the **peritubular capillaries,** that are closely associated with the remaining portions of the nephron's tubules. The peritubular capillaries recombine to form **venules** that join with others to eventually form the renal vein.

Each glomerulus and Bowman's capsule together form a **renal corpuscle,** and it is here that the process of filtration occurs. As blood is confined to the small volume of the glomerulus, a hydrostatic pressure gradient across the walls of the capillary and capsule drives blood plasma into the capsule. Water, salts, glucose, urea, and other low-molecular-weight components enter Bowman's capsule, but blood cells and plasma proteins do not. The fluid in the capsule is now called the **glomerular filtrate.** The blood cells and other substances left in the blood leave the glomerulus through the efferent arteriole.

About 20 percent of the plasma that enters the kidney is filtered into Bowman's capsules. The entire plasma volume (3 liters) is filtered by the kidneys some 60 times a day. This amounts to about 180 liters a day—approximately 45 gallons! Obviously, this enormous volume of filtrate is not all excreted. As the filtrate flows from Bowman's capsule through the tubular portions of the nephron, its composition is altered so that valuable solutes are recovered into the bloodstream and the ionic and water content of the blood are regulated within narrow limits.

As the glomerular filtrate passes through the proximal convoluted tubule, loop of Henle, and distal convoluted tubule, its composition is changed by two processes: tubular reabsorption and tubular secretion. **Tubular reabsorption** involves the recovery of water and valuable solutes (e.g., glucose, inorganic ions, and amino acids) into the surrounding peritubular capillaries. The proximal and distal convoluted tubules are formed of simple squamous or columnar epithelium, and they reabsorb virtually all glucose, amino acids, and other substances that are not waste. The reabsorption process involves active transport— as a matter of fact, the cells of the convoluted tubules expend more energy (that is, use more ATP) per unit mass than does cardiac muscle!

Tubular secretion involves the transport of materials from the peritubular capillary into the tubule. The most important of these secreted substances are hydrogen ions, potassium ions, and ammonia. Others include foreign substances taken for medicinal purposes, such as penicillin.

In addition to removing metabolic wastes, the kidneys also regulate the concentrations of many useful solutes in the blood plasma, chiefly water, inorganic ions, and minerals. The concentrations of these solutes in the blood plasma also determine their concentrations in the interstitial fluid that bathes the body's cells, because blood plasma, minus the plasma proteins, permeates quite freely across the capillary walls. A constant interstitial fluid composition is necessary for the proper functioning and viability of the body's cells. The maintenance of such a constant internal environment is called homeostasis, and the kidneys play a critical role in this process.

Homeostasis is the maintenance of a steady-state condition, and its regulation involves a system of feedback controls. That is, a change in the internal environment acts as a stimulus that results in a corrective response. The integration systems of the body— the nervous and endocrine systems—recognize the stimulus and respond in a manner that readjusts the internal environment.

For example, if the water concentration of the plasma is lowered, receptors in the hypothalamus of the brain are stimulated. The **hypothalamus** (Chapter 37) then directs the increased secretion of **antidiuretic hormone (ADH)** from the **posterior pituitary gland** into the bloodstream. At the kidneys, ADH increases the tubular reabsorption of water. Conversely, if excess water is ingested, blood volume (and hence pressure) increases. This stimulates the hypothalamus to direct the posterior pituitary gland to secrete less ADH. The kidney tubules respond by decreasing the tubular reabsorption of water, and the excess water is excreted in the urine.

As the glomerular filtrate nears the end of its journey through the nephron, most materials necessary for bodily function have been reabsorbed, leaving metabolic wastes (especially urea) and excess water and mineral ions. The resulting urine is about 70 times more concentrated than is the original glomerular filtrate. The urine travels from the distal convoluted tubules into the collecting tubules and renal pelvis, then through the ureters and into the urinary bladder. A normal adult human excretes 1.2–1.5 liters of urine daily. Contents of urine are presented in Table 34-1.

The kidneys are so efficient that people can frequently survive with only one functional kidney. Nevertheless, the kidney is rather delicate and is subject to several serious diseases. In **diabetes** (Gr., to go through), ADH secretion or **insulin** (a pancreatic hormone) secretion is upset, and this affects the resorption properties of the nephrons. In **diabetes insipidus,** too little ADH is secreted, and tremendous

volumes of urine are secreted, leading to extreme thirst, potential dehydration, and ion imbalance. In **diabetes mellitus,** insulin is secreted in insufficient quantities, cellular uptake of glucose decreases, and the concentrations of glucose in the blood and urine increase beyond the normal limits. It can be controlled by careful dietary monitoring of glucose intake and by daily injections of insulin.

Kidney "stones" commonly consist of solid clusters of uric acid and/or calcium phosphate and oxalate. When the concentration of these relatively insoluble molecules in the urine becomes too great, they precipitate into solid masses. Kidney stones may clog ureters, leading to severe renal problems.

Nephritis, as the name implies, is an inflammation of the nephrons. It may be caused by a bacterium that allows the glomerulus to let through larger molecules than it would normally filter. Sometimes even whole blood cells may be excreted. Nephritic individuals also lose essential proteins from their blood, and the resultant change in blood osmolarity results in edema (a watery swelling) that is especially manifested in the appendages.

Other Vertebrate Excretory Routes

A considerable quantity of water is excreted through the lungs of vertebrates by **evaporation.** **Sweat glands** also excrete water but are of little importance in the elimination of other excretory material. The sweat glands are very valuable, however, in the regulation of body temperature. The **liver** excretes certain substances, such as the decomposition products of hemoglobin. These are carried in the bile to the duodenum and then out of the body in feces. The feces are not chiefly excretory products of metabolism, as most of their contents have never been part of the body's internal environment, consisting largely of undigested food materials plus large populations of bacteria.

Summary

The types of excretory structures used varies throughout the protistan and animal kingdoms. In protists, the plasma membrane is generally the organ-

Table 34-1. Human Urine Contents (approximate values)

H$_2$O	96.0%
Organic materials	2.5
Urea (50% of solids in urine)	
Uric acid	
Urochrome (gives urine its color)	
Inorganic materials	1.5
Sodium chloride	
Potassium	
Calcium	
Magnesium	
Ammonia	
Ammonium sulfate	
Phosphate	
Bicarbonate	
	100.0% = 1200–1500 ml/day

Total solids: 60 g
        pH: 4.8–8.0

elle of excretion. Simple metazoans generally excrete wastes across their body walls by means of diffusion. Flame cells seem to assist in osmoregulation in platyhelminths, nemertineans, and aschelminths. Nephridia are excretory organs present in annelids, mollusks, and some other more complex invertebrates. Malpighian tubules occur in terrestrial arthropods; they excrete concentrated wastes into the digestive tracts of these animals, where they are dehydrated even further for maximum water conservation.

The mammalian kidney's basic morphology consists of a surrounding renal capsule, an external cortex, an inner medulla, and a pelvis. The functional units of the kidney, the nephrons, are made of a Bowman's capsule, glomerulus, proximal convoluted tubule, loop of Henle, distal convoluted tubule, and collecting tube. Surrounding these structures is a dense system of capillaries that exchange substances with the nephron. The results of glomerular filtration and tubular secretion include the removal of metabolic wastes and toxins from the blood, homeostatic regulation of water and ions in the blood, and the concentration of urea.

Review Questions

1. Summarize the more primitive excretory systems; at which stage did the removal of nitrogenous wastes become important?
2. List the materials involved in tubular reabsorption.
3. What hormone is most responsible for water reabsorption in the kidney?
4. What type of excretory structure most resembles a funnel?
5. What is the structure residing within the Bowman's capsule, and what is it made up of?

Selected References

Adolph, E. F. (ed.). *The Development of Homeostasis.* New York: Academic Press, Inc., 1960.

Bullock, R. H. (ed.). *Physiological Triggers.* Washington: American Physiological Society, 1957.

Campbell, J. W. (ed.). *Comparative Biochemistry of Nitrogen Metabolism.* New York: Academic Press, Inc., 1970.

CRM Books. *Biology Today.* Del Mar, Calif: Communications Research Machines, 1972

DeetJen, P., J. W. Boylan and K. Kramer. *Physiology of the Kidney and of Water Balance.* New York: Springer Verlag, 1975.

Frings, H., and M. Frings. *Concepts of Zoology.* Toronto: Macmillan Co., 1970

Gaffon, H. (ed.). *Research in Photosynthesis.* New York: Interscience Publishers, 1957

Gerard, R. W. (ed.). *Food for Life.* Chicago: University of Chicago Press, 1952.

Griffin, D. R. *Animal Structure and Function.* New York: Holt, Rinehart and Winston, Inc., 1962.

Guyton, A. C. *Textbook of Medical Physiology.* Philadelphia: W. B. Saunders Company, 1956.

Kimber, D. C., C. E. Gray, C. E. Stackpole and L. C. Leavell. *Textbook of Anatomy and Physiology.* New York: Macmillan Publishing Co., Inc., 1956.

Moffat, D. B. *The Mammalian Kidney.* New York: Cambridge University Press, 1975.

Ponder, E. "The Red Blood Cell," *Scientific American,* **196**:95, 1957.

Potts, S. T. W., and G. Parry. *Osmotic and Ionic Regulation in Animals.* New York: Macmillan Publishing Co., Inc., 1964.

Prosser, C. L., and F. A. Brown, Jr. *Comparative Animal Physiology,* 2nd ed. Philadelphia: W. B. Saunders Co., 1961.

Schmidt-Nielsen, K. *Animal Physiology,* 2nd ed. Englewood Cliffs: The Master Switch of Life," *Scientific American,* **209**:92, 1963.

Smith H. W. "The Kidney," *Scientific American,* **188**:40, 1953.

Smith, H. W. "The Kidney," *Scientific American,* **188**:40, 1953.

Smith, H. W. *The Kidneys.* New York: Oxford University Press, 1951.

Wiggers, C. J. "The Heart," *Scientific American,* **196**:74, 1957.

Circulatory and Lymphatic Systems

The principal function of the circulatory system is the transport of materials within the body. Oxygen diffuses into the blood through capillaries in the lungs. Nutrients, water, vitamins, and minerals are absorbed into capillaries in the intestinal wall and are then transported in the blood throughout the body. Waste products of metabolism are transported away from the cells to elimination sites. Carbon dioxide is taken to the lungs, where it is eliminated during exhalation; nitrogenous wastes go to the kidneys for elimination.

The circulatory and lymphatic systems are intimately connected in the body's defense against injury and infection. Another major shared function is the maintenance of a constant internal environment (**homeostasis**) for the body's cells.

The Circulatory System
Circulatory Systems in Invertebrates

In protists, intracellular transportation is accomplished by the streaming of the cytoplasm that carries the nutrients supplied by the food vacuoles. In poriferans, digested food is passed intercellularly or is carried from place to place by **amoebocytes.** In hydras and other cnidarians, digestion is completed within the gastrodermal cells that distribute some of the food to neighboring cells, and these may in turn pass on a portion of the nutriment to other cells. The gastro-

dermal cells are also responsible for sending nutriment through the mesoglea to the epidermal cells. A similar method of distribution occurs in the multi-branched intestine of the nonparasitic platyhelminths. The parasitic cestodes rely upon their relatively permeable cuticle and on diffusion for nutrient uptake and waste elimination.

In the earthworm, we encounter a complicated system of closed vessels, the circulatory system, which carries digested food, oxygen, and other substances to all part of the body. A similar circulatory system is present in most types of higher animals and is referred to as a **closed circulatory system.**

The **blood** is not always confined to a system of vessels, however. Arthropods and most mollusks have **open circulatory systems.** In the arthropods, the body cavity, or **hemocoel,** is filled with blood that bathes the tissues. A dorsal heart pumps blood, called hemolymph, but the majority of hemocoel blood movements are initiated by muscular respiratory, locomotory, and peristaltic motions. In mollusks, the blood does not bathe the organs; instead, blood flows through vessels to the tissue spaces (**sinuses**) within the various organs before it is recollected and returned, after passing through the gills, to the heart.

The Evolution of the Vertebrate Heart

The vertebrate heart has evolved from a relatively simple two-chambered structure in cold-blooded, water-inhabiting fishes to a complex four-chambered structure in the active, warm-blooded, terrestrial-inhabiting birds and mammals (Fig. 35-1).

The fish heart consists of a single thin-walled **atrium** and a single muscular **ventricle.** Blood enters the atrium from the body. When the atrium contracts, blood passes into the ventricle; it is prevented by **valves** from returning to the atrium. The ventricle pumps the blood into arteries leading to the gills, where it is oxygenated and carried directly to the body tissues before again returning to the atrium.

In the amphibians is seen the beginning of a new trend in the evolution of circulatory systems: the separation of circulation to the respiratory organs from that of the rest of body. In early amphibians the use of lungs and subsequent advances into terrestrial habitats probably necessitated new adaptations for circulation and resulted in the **three-chambered heart.** Note that, although the blood leaving the ventricle is a mixture of oxygenated and deoxygenated blood, the blood returning from the lungs and body is pumped separately by the left and right atria, respectively.

Reptiles also have two atria, but the ventricle is partly divided into two chambers. Nonoxygenated venous blood (from the body) entering the right atrium is thus kept more or less separated from the oxygenated blood (from the lungs) that flows into the left atrium and "left" ventricle from the lungs. When the "right" ventricle contracts, the nonoxygenated

Figure 35-1. Diagrams showing the comparative structure and evolution of the heart among different types of vertebrates. Valves are omitted. Arrows indicate the direction in which the blood flows.

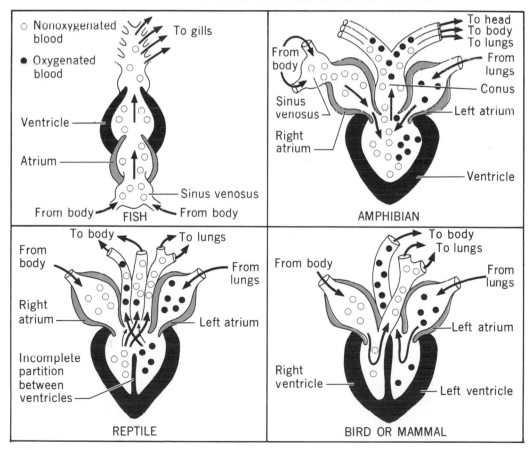

blood is forced through the pulmonary arteries to the lungs and also through the left aortic arch into the dorsal aorta. At the same time, contraction of the "left" ventricle sends oxygenated blood through the right aortic arch, which merges into the dorsal aorta. Thus the dorsal aorta contains a mixture of both non-oxygenated and oxygenated blood.

In birds and mammals, the ventricle is completely separated into two chambers, forming a **four-chambered,** double-pump heart; thus the nonoxygenated and oxygenated blood are kept entirely separate, hence providing more efficient pulmonary circulation. In birds, it is the right aortic arch that flows to the dorsal aorta; in mammals, it is the left aortic arch.

The Mammalian Heart

The body of the mammalian heart consists of cardiac muscle and is referred to as the **myocardium.** Surrounding the outside of the myocardium is an epithelial covering, the **epicardium.** The lining of the heart chambers is known as **endocardium.** The heart itself is enclosed in a fibrous sac called the **pericardium.** The adult human heart is about the size of a clenched fist and lies above the diaphragm in a **pericardial cavity** (formed by the pericardial sac) anterior to the lungs.

The **ventricles** in all vertebrates are more muscular than are the atria, and their contractions are more powerful. The **artria** receive blood into the heart and only have to pump it into the ventricles, which pump the blood to the lungs and body.

The atria communicate with their respective ventricles through **atrioventricular valves** (Fig. 35-2). The **tricuspid valve** between the right atrium and right ventricle is so named because of its three flaps. The **bicuspid valve** separates the left atrium and left ventricle. These valves allow blood to move unidirectionally from the atrium into the ventricle, thus preventing backflow. As a ventricle contracts, blood pressure closes the atrioventricular valve, preventing blood from reentering the atrium. Stringlike elastic cords, called chordae tendinae, which extend between the atrioventricular valves and the heart wall, help to keep the valves shut when the ventricles contract.

Semilunar valves occur between the ventricles and their communicating blood vessels (**pulmonary arteries and aorta**). The action of the semilunar valves is analogous to that of the atrioventricular valves in preventing backflow. Sometimes the semilunar valves are known more specifically as the **aortic valves** (junction of left ventricle and aorta) and the **pulmonary valves** (junction of right ventricle and pulmonary artery).

The Heartbeat

Cardiac muscle is unique in that it does not require stimulation from the nervous system to contract (myo-

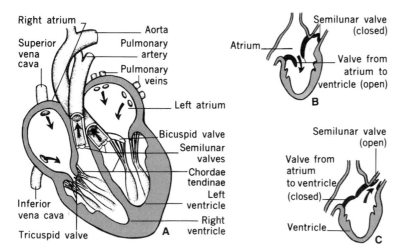

Figure 35-2. Diagram illustrating chambers of the heart and the action of the valves in man. (A) Internal structure of heart and direction of blood flow through it. (B) Blood is flowing from atrium to ventricle, and the semilunar valve is closed. (C) The valve between the atrium and ventricle is closed, and blood is forced past the semilunar valve into the artery. Note that, because of the upright position of the body of man, the anterior vena cava is called the superior vena cava and the posterior vena cava is termed the inferior vena cava. The tricuspid and bicuspid valves are also known as atrioventricular valves.

genic). Near the tricuspid valve is an area of special ized myocardial cells called the **sinoatrial node (SA node).** This "pacemaker" produces rhythmical electrical signals that lead to the contraction of the heart. The activity of the SA node is self-generated (intrinsic), but the rhythm may be modified by nervous stimulation, hormones, and many other factors.

Cardiac muscle is composed of **striated fibers** fused together at the **intercalated discs** to form a unified mass of tissue (Chapter 3). As a result, the walls of the atria function as if they were a single muscle fiber. Upon stimulation from the SA node, a wave of contraction spreads simultaneously outward through both atria.

In less than 0.1 sec, the excitatory impulse from the SA node reaches another group of specialized cardiac cells, the **atrioventricular node (AV node),** located at the base of the right atrium near the septum that separates the ventricles. The AV node pauses for 0.1 sec before sending its signal to the ventricles, thereby allowing them sufficient time to fill. The AV impulse passes through the insulative septum via the **atrioventricular (AV) bundle,** after which it spreads throughout the ventricles by means of the **Purkinje fibers** that ramify throughout the ventricular myocardium.

Due to the force exerted by the heart during contraction and the fact that the blood is contained in vessels, it is always under some pressure. **Blood pressure,** like barometric pressure, is measured in millimeters of mercury (Hg). Blood pressure in part controls and is controlled by the heartbeat, along with vasodilation, vasoconstriction, and other factors. When the ventricles contract, the heart is said to be in **systole.** The average systolic blood pressure is 125 mm Hg. During the refractory (waiting) period between ventricular contraction and atrial contraction, the heart is in **diastole,** and blood pressure normally drops to about 75 mm Hg. The average human blood pressure is about 91 mm Hg; this value is calculated by biasing the diastolic contribution, as diastole has a longer duration than systole.

Chemical control of the heart rate is mostly hormonal. For instance, epinephrine (adrenaline), which directs certain blood vessels to dilate and others to constrict, is released in response to fright. The medulla (Chapter 37) in the brainstem acts as an integra tion center for nervous stimuli controlling the heart rate. Its **vasomotor center** controls the tonus (amount of contraction) of the arteries. When vasomotor center activity is high, the volume of the circulatory system decreases because the vessels constrict somewhat; hence blood pressure is increased. If the vasomotor center is inhibited, arterioles relax, the volume of the circulatory system is increased, and blood may pool in the capillaries; in some cases, fainting may result.

Medullar nervous control of the heart rate stems from dual innervation by the **autonomic nervous system** (Chapter 37). **Sympathetic** stimulation from the autonomic nervous system increases the frequency of heartbeat, whereas **parasympathetic** stimulation decreases its rate. Receptors located on the aorta and right atrium monitor blood pressure. When, for instance, blood enters the aorta, the increased pressure stimulates the pressure receptors, and they send impulses to the medulla in the brainstem. If blood pressure is too high, the medulla directs the heart, through its parasympathetic fibers to the SA node, to slow down (and also dilates the capillaries to lower pressure). This is known as a negative feedback system and is another example of a homeostatic mechanism in action. Low blood pressure may have an opposite effect—the pressure receptors would not inhibit the vasomotor center, and so normal tonus would be reinstated.

The **electrocardiogram (EKG, ECG)** is an important diagnostic tool that traces the polarizations and depolarizations of the atria and ventricles during their cycles of contraction and relaxation (Fig. 35-3). An EKG tracing has three parts: a **P section,** a **QRS complex,** and a **T section.** Atrial depolarization shows up as the P section. The QRS complex, which appears approximately 0.1 to 0.2 sec after the P section, represents the combined ventricular depolarization and atrial repolarization. The T section indicates ventricular repolarization. Various cardiac anomalies may be diagnosed with the aid of such EKG tracings.

Heart sounds may be diagnosed using a **stethoscope.** There is a basic sound that is approximated by "lubb-dup." The "lubb" sound is made by closure of the atrioventricular valves plus simultaneous vibrations of the chordae tendinae. The subsequent "dup" sound stems from the arterial walls.

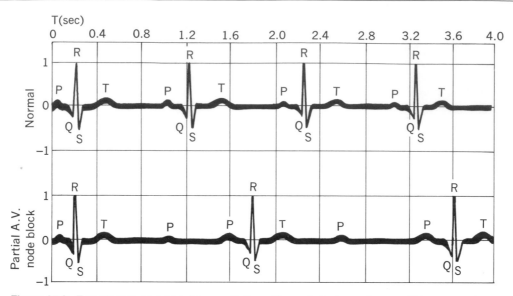

Figure 35-3. Top: Normal electrocardiogram. Bottom: Aberrant electrocardiogram. This represents a partial block of the AV node. Note that the QRS and T wave forms are absent after every other P; hence, although the atria depolarize, the ventricles do not always get the message.

Blood Vessels

The heart pumps blood through a closed circuit of blood vessels. **Arteries** deliver blood from the heart to the body tissues. The arteries branch into smaller **arterioles** as they near their target organs. The arterioles in turn branch into microscopic **capillaries** that penetrate the tissues of the target organ. All exchange of materials between the blood and the body's cells occurs across the thin (one cell in thickness) capillary walls. Blood leaves the body tissues through small veins, called **venules,** that are formed by the merger of the capillaries. The venules merge to form **veins** that return blood to the heart.

Arteries and veins differ in the amounts of muscle and elastic tissue that comprise their walls (Fig. 35-4). Arterial walls are thick and muscular, with elastic layers. Arteries are consequently able to expand and contract, thereby exerting continuous pressure that forces blood through the rest of its circuit. Venous walls are thinner, with less muscle and elastic tissue, so blood is under less pressure.

When an artery is cut, blood spurts from it, indicating that it is under considerable pressure. Blood from a cut vein flows out continuously, indicating lower pressure. Blood pressure depends in part on the force of the contracting ventricles of the heart, on the elasticity of the walls of the blood vessels, and on the resistance to flow through the vessels. Chronic high blood pressure may result from a reduction in the elasticity and interior diameter of the small arteries. This condition is caused by increasing growth of arterial connective tissue, as part of the aging process, and by deposits of cholesterol and is called arteriosclerosis, or "hardening of the arteries."

If a finger is placed on the **radial artery,** found on the thumb side of the wrist, or on other arteries, a periodic distension will be felt corresponding to the beat of the heart. This **pulse** is due to the alternating dilation and elastic recoil of the artery as blood is pumped through it by the contractions of the heart. The general character of the heart's action can be monitored by feeling the pulse.

The capillaries are minute tubes with a very thin wall (Fig. 35-4), consisting of a single layer of simple squamous epithelial cells. It has been estimated that, if all the capillaries in the human body were placed end to end, they would form a tube 10^5 km long. Because each capillary is only about 1 mm long, their total number is almost inconceivable. The wall sur-

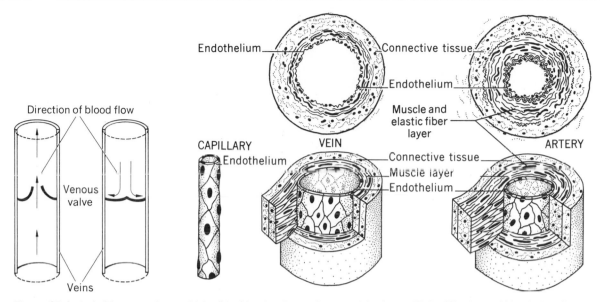

Figure 35-4. Left: Venous valves, which allow blood to flow only toward the heart. Right: Histology of blood vessels. The muscle layer of an artery is thicker than the muscle layer of a vein of the same size. A vein usually has a lumen of greater diameter than its corresponding artery.

face of these capillaries is about 820 m², or 1.5 acres, and no cell is more than 0.13 mm from a capillary. Capillaries are so small that red blood cells have to squeeze through them in single file. At the beginning of each capillary, there is a **precapillary sphincter,** a circular muscle that can constrict and regulate local blood flow. Capillary walls are permeable to water and to small solute components of plasma, but not to larger particles, such as blood cells and plasma proteins. Exchange of materials does not occur directly between the capillaries and body cells, however. They are separated by an **interstitial fluid** that is formed from the fluid components of plasma that pass through the capillary walls. Dense capillary beds are found throughout the body; in fact, usually one can find no more than four consecutive cells without a capillary next to at least one of them.

Capillaries carry out the principal functions of the circulatory system, such as absorbing oxygen and giving up carbon dioxide in the lungs, delivering metabolic wastes to the kidneys, absorbing nutrients from the digestive tract, taking up hormones from ductless glands, and exchanging metabolites and wastes with the interstitial fluid that bathes the body's cells. All

other functions of the circulatory system are dependent upon providing an adequate blood flow to the tissues via the capillaries.

The rate of circulation is so rapid that it almost defies imagination. The entire blood supply circulates from the heart, through the body, and back to the heart again in about 20 to 40 sec. This means that the 5.3 liters of blood in an average-sized man make some 3000 to 4000 circuits through the body every day. This accounts for the rapid distribution of substances that gain entrance into the bloodstream. The velocity of blood flow varies in different arteries and veins, depending largely on the diameter of the vessel. In the large arteries, blood moves along rapidly; in the smaller arteries and capillaries, it moves more slowly. The rate of movement in the veins is less than it is in the arteries.

One-way **valves** within veins (Fig. 35-4) prevent the backflow of blood and aid in returning it to the heart. Most of the blood pressure within the veins arises from skeletal muscle contraction—when a muscle near a vein contracts, it may squeeze the vessel, decreasing its volume and forcing blood toward the heart.

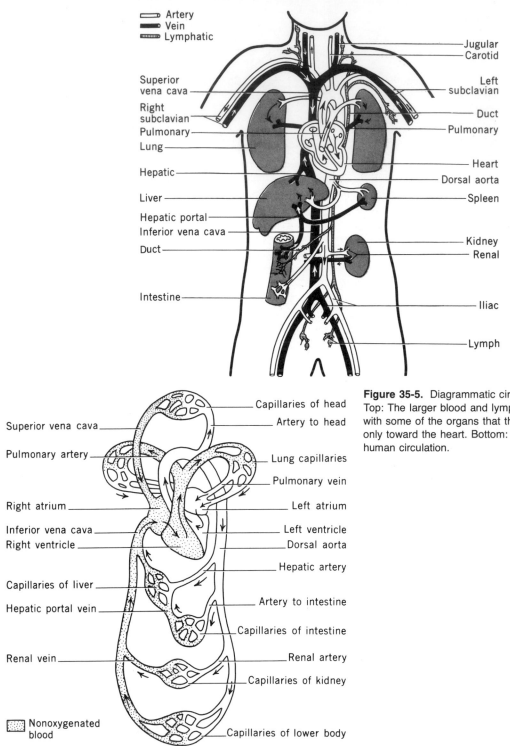

Artery
Vein
Lymphatic

Jugular
Carotid
Superior vena cava
Left subclavian
Right subclavian
Duct
Pulmonary
Pulmonary
Lung
Hepatic
Heart
Dorsal aorta
Liver
Spleen
Hepatic portal
Inferior vena cava
Duct
Kidney
Renal
Intestine
Iliac
Lymph

Capillaries of head
Superior vena cava
Artery to head
Pulmonary artery
Lung capillaries
Pulmonary vein
Right atrium
Left atrium
Inferior vena cava
Left ventricle
Right ventricle
Dorsal aorta
Hepatic artery
Capillaries of liver
Hepatic portal vein
Artery to intestine
Capillaries of intestine
Renal vein
Renal artery
Capillaries of kidney

Nonoxygenated blood
Capillaries of lower body

Figure 35-5. Diagrammatic circulatory system in man. Top: The larger blood and lymphatic vessels together with some of the organs that they serve. Lymph flows only toward the heart. Bottom: General scheme for human circulation.

General Layout of the Mammalian Circulatory System

Blood is circulated throughout the body via the pulmonary and systemic systems. The **pulmonary system** uses the pumping action of the right side of the heart to circulate blood from the heart to the lungs and back to the heart. Reduced hemoglobin from the body tissues enters the **right atrium,** which pumps it into the **right ventricle.** The right ventricle pumps the deoxygenated blood into the **pulmonary artery,** which divides into two branches that enter the right and left lungs. The arteries in turn subdivide into arterioles, which subdivide into capillaries. Capillaries surround the microscopic air sacs (**alveoli**) within the lungs. It is across these capillary walls that carbon dioxide diffuses into the lungs and oxygen diffuses into the blood. The capillaries leave the alveoli and merge into venules, which join to form veins. The **pulmonary veins** carry oxygenated blood to the **left atrium** of the heart, where it enters the systemic system.

The **systemic system** uses the pumping action of the left side of the heart to circulate oxygenated blood from the heart to all the tissues of the body and back to the heart. Oxygenated blood enters the left atrium and is pumped into the **left ventricle,** which pumps it into the aorta. The **aorta** is the main trunk of the arterial system and branches into smaller arteries that supply all the organs. A summary of the systemic circuit is illustrated in Figure 35-5; see also Figure 30-8 for more details.

The venous circulation is generally analogous to the arterial system, and the names of most veins are the same as their corresponding arteries (for instance, iliac artery and iliac vein). The **anterior** (**superior**) and **posterior** (**inferior**) **venae cavae** deposit the systemic venous flow into the right atrium. Note, in Figure 35-5, that pulmonary veins are the only ones to carry oxygenated blood.

The Lymphatic System

Capillaries are under some pressure, and hence significantly more fluid leaks out than is reabsorbed. Only about 40 percent of the interstitial fluid is reabsorbed into the capillaries before they merge to form venules. The remaining 60 percent of the interstitial fluid returns to the blood by a different path; it is collected by a separate division of the circulatory system called the lymphatic system.

The lymphatic system is composed of a series of blind-end vessels with thin walls and many valves. **Lymph nodes** (or glands) are located at intervals along their course. The fluid that the vessels transport back to the bloodstream is called **lymph.** The lymphatic system has no central pump as does the heart, so the transport of lymph through the lymphatic vessels is effected by muscular activity throughout the system. Lymph from the right portion of the body enters the blood stream at the **right subclavian vein** (Fig. 35-6); lymph from the rest of the body drains into the **left subclavian vein.**

The lymphatic system performs other functions as well. The digested fat from the small intestine enters the general circulation via the **lacteals** of the lymphatic system. Also of great importance are the lymph nodes, which produce **antibodies** (see the discussion of antibodies (the immune system) later in this chapter) that destroy bacteria and other foreign cells and viruses that invade the body. Lymph nodes strain particulate matter from the lymph. The **lymph nodes, spleen, thymus,** and **tonsils** all produce **lymphocytes,** a type of white blood cell, discussed in the next section.

The Composition of Blood and Lymph

Human blood and lymph will be described in this section. The oxygenated blood in arteries and arterioles appears bright red, whereas the deoxygenated blood in veins and venules has a dark red color. Blood makes up about 7 percent of the body weight. Its volume consists of about 55 percent liquid plasma and about 45 percent cellular components. Plasma is that part of the blood that is liquid and noncellular, that suspends the corpuscles, and in addition that contains dissolved salts and proteins. The cellular components of blood are

1. **erythrocytes** (red blood cells, RBCs), which contain **hemoglobin** molecules that transport oxygen and carbon dioxide
2. **leukocytes** (white blood cells, WBCs), which are

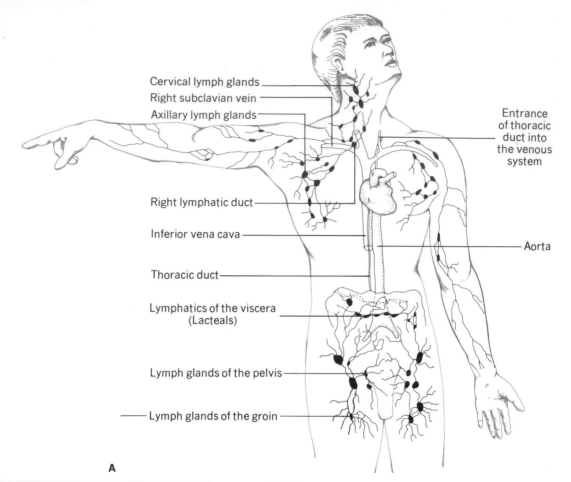

Cervical lymph glands

Right subclavian vein

Axillary lymph glands

Entrance of thoracic duct into the venous system

Right lymphatic duct

Inferior vena cava

Aorta

Thoracic duct

Lymphatics of the viscera (Lacteals)

Lymph glands of the pelvis

Lymph glands of the groin

A

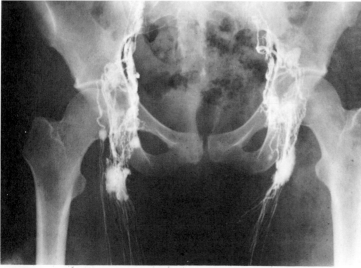

B

Figure 35-6. (A) Some major lymph vessels, glands, and nodes in humans. (B) An X ray of the lymph glands and vessels of the groin area. Lymph vessels are distributed throughout the body in an extensive yet delicate meshwork.

mobile cells that serve defensive functions and are subdivided as follows:

   a. neutrophils
   b. eosinophils
   c. basophils
   d. monocytes
   e. lymphocytes

3. platelets, which are cell fragments that initiate blood clotting

Lymph contains no erythrocytes but many leukocytes; the cellular concentration is much lower than in blood, but the liquid portion of lymph is very similar to blood plasma in composition.

Human erythrocytes are enucleated (lacking a nucleus) discs approximately 8.0 μ in diameter and biconcave in cross-section (Fig. 35-7). RBCs are the principal cellular components of blood, with a density of about 5×10^6 per mm³ of blood. Each erythrocyte is capable of living for about four months in the human circulatory system, and the formation of new erythrocytes takes place continuously in the **bone marrow.** The total surface area of all the red blood cells in an average adult is approximately 3,000 m². The erythrocyte is sufficiently elastic so that it can squeeze through capillaries smaller than its diameter. Each erythrocyte contains about 200 to 300 million molecules of **hemoglobin** in its cytoplasm. Hemoglobin consists of four protein (**globin**) subunits, and each has a nonprotein, iron-containing pigment molecule (**heme**). Hemoglobin is a respiratory protein that readily bonds with oxygen in the lungs and readily releases oxygen at the tissues (Chapter 36).

There are about 6 to 10×10^3 leukocytes (white blood cells) per cubic millimeter of human blood. Leukocytes are subdivided according to their morphologies, functions, and origins. **Granular leukocytes,** which originate from cells in the bone marrow, contain granules in their cytoplasms and have irregular nuclei. Granular leukocytes include neutrophils, eosinophils, and basophils.

Guided by chemotaxis, **neutrophils** are the first leukocytes to arrive at the site of bacterial infection, where they **phagocytose** (ingest) bacteria. The granules are actually **lysosomes** (Chapter 2) that digest the phagocytized material. Neutrophils are rather mobile and can penetrate capillary walls (diapidesis) and move into the interstitial fluid. Neutrophils are formed in bone marrow and make up about 65 to 75 percent of the blood's leukocytes. **Eosinophils** are

Figure 35-7. Blood platelets and corpuscles of man.

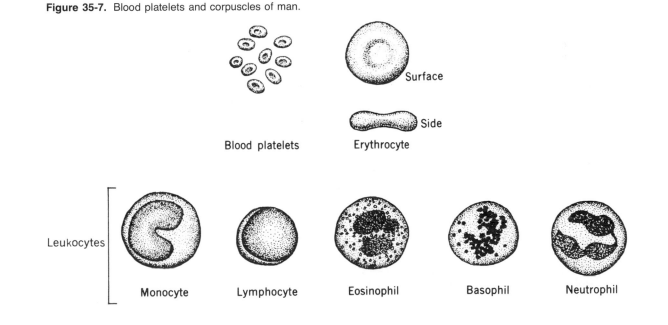

Blood platelets

Surface

Side

Erythrocyte

Leukocytes

Monocyte   Lymphocyte   Eosinophil   Basophil   Neutrophil

nonphagocytic leukocytes that contain one or two nuclei. **Basophils** have several nuclei. Their function is unspecified, but they may be involved in allergic reactions. Basophils have also been observed to proliferate during periods of disease. Nongranular leukocytes are formed in the spleen and lymphoid tissue. Nongranular leukocytes include monocytes and lymphocytes. **Monocytes** are large, nongranular, uninucleated cells that, like granular neutrophils, phagocytose bacteria and debris. Dead monocytes and neutrophils are the main constituents of pus. Lymphocytes do not phagocytose the invader but are involved with the production of antibodies (see the following discussion of the immune system).

Blood **platelets** are minute discoidal cell fragments that are produced in the bone marrow. Their density varies but averages about 2.7×10^4 per mm^3 of blood. Platelets initiate blood clotting.

Blood Clotting

Hemostasis (blood coagulation) maintains the integrity of the circulatory system in response to hemorrhage. The formation of a blood clot involves the interaction of cellular and plasma components of the blood. When a blood vessel is ruptured, the rough edges of the vessel break open the blood platelets, releasing the enzyme **thrombokinase (thromboplastin).** Thrombokinase plus calcium ions and other factors (including vitamin K) convert the plasma protein **prothrombin** into **thrombin.** Once thrombin is formed, it can react with **fibrinogen** (another plasma protein) to form insoluble **fibrin.** Fibers of fibrin form a lacework mesh that ensnares erythrocytes and platelets, forming a barrier that prevents blood loss from the wound. It also provides a framework upon which connective tissue can repair the ruptured vessel. Briefly summarized, the steps in clot formation are as follows:

Thrombokinase + prothrombin + calcium $\longrightarrow$
Thrombin

Thrombin + fibrinogen $\longrightarrow$ Fibrin

Fibrin + erythrocytes $\longrightarrow$ Clot

Certain individuals lack one or more of these factors, and as a result their blood clots so slowly that a minor wound produces extensive bleeding. For example, the lack of so-called factor VIII results in the genetically transmitted disease known as **hemophilia;** the hereditary aspects are discussed in Chapter 38.

If blood is allowed to clot, the remaining plasma is known as **serum. Heparin** is an anticoagulant that circulates in the blood to prevent accidental blood clotting. Heparin is administered clinically to reduce embolisms (blood clots within a blood vessel), as in the disease **phlebitis** (inflammation of the inner walls of veins) or in strokes (embolisms that clot vessels in the brain). Heparin is also used to prevent clotting in blood taken for medical tests or for transfusions.

The Immune System

A most important cooperative function between the circulatory and lymphatic systems is the protection of the body against disease; this is called the immune system. The **immune system** is best described as a kind of chemosensory system that has the ability to recognize foreign proteins, viruses, bacteria, and other cells. This recognition takes place at the molecular level in a manner entirely analogous to the recognition of a particular substrate by an enzyme. The class of proteins involved is the **immunoglobulins,** some of which are soluble proteins called **antibodies.** These are synthesized and excreted by certain leukocytes (**B cells**) originating from the bone marrow, and they attack foreign substances. This process is termed the **humoral immune response.**

Other immunoglobulins are attached to the cell membranes of certain leukocytes; when a foreign substance interacts with immunoglobulins in the membranes of B cells, it triggers the production and release of antibodies that specifically attack that foreign substance. There are also immunoglobulins acting as chemical sensors on the surface of the phagocytic leukocytes discussed earlier; this allows the ingestion of only foreign substances during the process called the **cellular immune response.** Other immunoglobulins do not reside in leukocytes at all, such as those of the mucous membranes of the lungs, that are involved in **allergic reactions** (e.g., the recognition of pollen grains in hay fever).

The most amazing property of the immune system is its ability to react against almost any foreign cell.

Two identical regions (**antigen-binding sites**) of an antibody or other immunoglobulin have the ability to recognize a specific molecular configuration (**antigenic determinant**) on the foreign substance (**antigen**). Thus a particular cellular antigen has thousands of possible antigenic determinants. The antigen binding sites of immunoglobulins must be different to recognize different antigenic determinants; this is accomplished by allowing certain amino acids to vary on each of an antibody's polypeptide chains (**heavy** and **light chains**, Fig. 35-8). These variable (V) regions form the antigen-binding sites of an antibody or any other immunoglobulin. The rest of an immunoglobulin's amino acids are the same (C, or constant regions) on comparable chains for a particular type of immunoglobulin. The C region of the heavy chains

Figure 35-8. The structure of an immunoglobulin, an antibody molecule. Note that this protein is actually composed of four polypeptide chains, two identical heavy chains, and two identical light chains. Each type of light or heavy chain has a region that is constant among a particular type of antibody and another that is relatively variable.

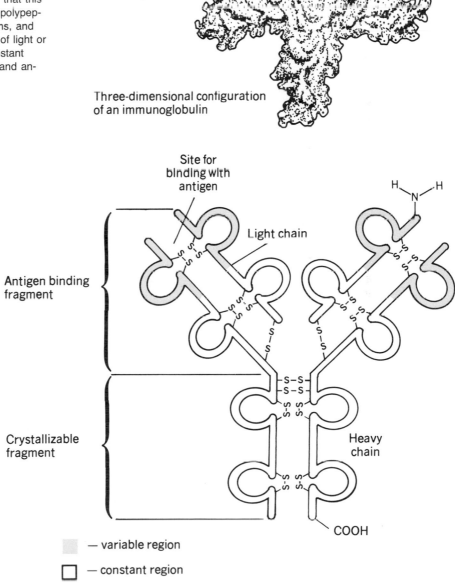

Three-dimensional configuration of an immunoglobulin

Site for binding with antigen

Light chain

Antigen binding fragment

Crystallizable fragment

Heavy chain

COOH

— variable region

— constant region

determines whether the immunoglobin will attach to the membrane or will float around as an antibody. Because all proteins are made from DNA (genes), this poses an interesting genetic problem: How does the cell select the proper V and C regions needed and make the correct immunoglobin? And how can there be so many possible V regions? This is accomplished by complicated genetic regulation processes, discussed in Chapter 38.

The diversity and regulation of immunoglobulin genes are only part of the complexities of the immune system. The body is constantly producing B cells; each eventually becomes **committed** to a unique antigen-binding capacity (i.e., it uses only one pair of V regions on all its immunoglobulins). Most of these committed cells reside in the lymph nodes, where they are exposed to possible antigens as the lymph is continually filtered. The conversion of a particular B cell into a cell that starts proliferating and producing large quantities of antibodies (a **plasma cell**) requires the participation of a morphologically similar cell (**T cell**) that originates in the **thymus gland** of the lymph system. The T cells also have immunoglobulins on their cell surfaces, but they do not produce antibodies.

The production of large quantities of antibodies after the first exposure to an antigen takes a few days and is called the **primary immune response.** If the same (or very similar) antigen is encountered again, larger amounts of antibodies are produced in a much shorter time; this is the **secondary immune response** that is responsible for immunity against certain diseases. During the primary response, certain B cells recognize the antigen but do not produce antibodies. Instead, these **memory cells** somehow remember this first encounter and initiate a secondary response when it meets the same antigen at a later time.

This is also the basis for vaccination; the body is exposed to a disease-causing agent that has been killed but can still cause a primary immune response. When the live cells or viruses invade the body, the quicker and stronger secondary immune response can attack before the disease takes hold.

Antibodies act against foreign substances in two ways. Because there are two identical antigen-binding sites on each antibody (Fig. 35-8), they can link antigens, such as foreign cells, together. The joining of many antigens and antibodies into an insoluble network is called **agglutination;** an example of this for red blood cells is discussed in the following section. The agglutination of foreign cells increases the probability of being found by the wandering phagocytic leukocytes during the cellular immune response. Antibodies may directly destroy a foreign cell by triggering the lysis of the cell with the help of a complex group of proteins called **complement.**

By means of a complicated process that occurs near the time of birth, the immune system is trained not to react against its own cells. Malfunctions in this process result in the often fatal **autoimmune diseases,** in which the immune system attacks the cells of the body. Most types of cancer are malfunctions in a certain cell type that prevents proper regulation of cell growth and other cellular activities. Because they are not foreign cells, but aberrant body cells, the immune system often does not react. There are usually some changes in the surface proteins of a cancer cell that may elicit an immune response; a major strategy of cancer research is to stimulate the body's immune system to increase the probability of recognition and destruction of cancer cells.

The A.B.O. Blood Groups— A Simple Antigen-Antibody System

The many surface molecules of erythrocytes, as with all other cells, can elicit the production of antibodies that cause the clumping (**agglutination**) of such cells. This phenomenon was observed after certain blood transfusions from one person (donor) to another (recipient) that were harmful instead of helpful. In 1900 Landsteiner discovered that such agglutinations were primarily due to an antigen–antibody system now known as the A-B-O blood groups. There are other erythrocyte antigen systems, but their effects are not as drastic.

When blood from a donor belonging to the same blood group as the patient (recipient) is transfused, unfavorable reactions do not occur. Incompatibility of bloods results when the plasma from person contains an antibody that reacts with the complementary antigens on the erythrocytes of another causing the cells to clump. Landsteiner discovered that there were two possible **agglutinogens** (antigens), designated **A** and

Table 35-1. The A-B-O Blood Group System

Blood Group	Antigens on Erythrocytes	Antibodies in Blood Serum
A	A	Anti–A
B	B	Anti–B
AB	AB	None
O	None	Anti–A and Anti–B

B, on human erythrocytes and that the serum may contain two kinds of **agglutinins** (antibodies), called **Anti-A** and **Anti-B.** A diploid organism has two copies of each gene (Chapter 38); this makes four possible types of A-B-O blood groups: A, B, AB, and O. Anti-A cross-links erythrocytes that have antigen A, forming clumps of cells. Likewise, Anti-B causes clumping of red blood corpuscles containing antigen B. Obviously, Anti-A must be absent from the blood of a person who has the A antigen, and Anti-B must be absent from the blood of an individual who possesses the B blood antigen. Some individuals lack both antibodies and are said to have type AB blood. Individuals with type O blood lack both antigens A and B. These relationships are summarized in Table 35-1.

Because the donor's blood is diluted in the recipient's bloodstream during transfusions, the effects of any antibodies in the donor's blood are usually insignificant. The primary concern is whether antibodies in the recipient's blood will react with antigens in the donor's erythrocytes. Thus it is possible, for example, to give a type O blood transfusion to a type A person, but the reverse—giving type A to a type O person— would cause severe agglutination and possible death. Because a type O person lacks A and B antigens, this person is known as a **universal donor.** Likewise, the lack of both Anti-A and Anti-B allow an AB-type person to be a **universal recipient.**

Many other blood groups have been discovered in recent years, some of which must be taken into consideration in making blood transfusions, thus complicating the work of the blood specialist or hematologist.

The blood of anthropoid apes is assignable to one of the four human groups; for example, types A and O are found in chimpanzees. This is evidence of a close biochemical relationship between the apes and humans. Blood groups are also known in rabbits, dogs, and cattle, but none of these is identical with the human types.

Rh Factor

The Rh blood factor is another important blood antigen found in humans and monkeys. About 86 percent of the white population of the United States have this antigen and are designated Rh-positive. The remaining 14 percent lack this antigen (Rh-negative). Since the discovery of the Rh factor, it has been established that there are no less than six different antigens involved in this antigen system, but of these only the Rh antigen is usually of serious clinical significance.

Unlike the A-B-O blood groups, no regularly produced antibody accompanies the Rh antigen. However, the Rh antigen will cause formation of antibodies if the blood of an Rh-positive person is transfused into an Rh-negative person. If another transfusion of Rh-positive blood is subsequently made to this Rh-negative person, then the antibodies will react with the Rh antigens, causing a serious agglutination, which may result in death.

Another danger of Rh incompatibility may occur when an Rh-negative woman bears an Rh-positive child (whose father must be Rh-positive). Rh-positive erythrocytes may pass from the bloodstream of the fetus through small breaks in the placenta and into the maternal circulation, where they stimulate production of antibodies. More likely, fetal blood cells enter the maternal system when the placenta ruptures during birth. During a subsequent pregnancy with an Rh-positive fetus, the maternal antibodies may get into the bloodstream of the fetus and cause agglutination of the fetal erythrocytes. This gives rise to a condition called **erythroblastosis fetalis,** which results in varying degrees of injury to the fetus, depending on the concentration of antibodies present. The effects range from a very mild anemia to more severe anemic conditions or various structural and nervous abnormalities, which sometimes cause mental deficiency, miscarriage, or stillbirth.

There are many other blood groups such as the P, M-N-S, Kell, Lewis, Lutheran, Kidd, and Duffy systems. This is an active field of research, and new blood group antigens are frequently found. Because the presence of the antigens of one blood group system is independent of another, a person may have any one of many different blood groupings. Considering all these blood group systems, the number of different blood-type combinations is possibly well up in the millions, which makes them useful in autopsy and in determining the parentage of a child. The inheritance of some of the blood groups will be discussed in Chapter 38.

Anthropologists have used blood groups to study the relationships and migrations of races. By a special technique it is possible to determine the blood groups of some ancient bones and mummies.

Summary

The circulatory system transports materials within the body, plays important roles in defense against injury and infection, and generally helps maintain a constant internal environment (homeostasis) for the body's cells.

Key events in the evolution of circulatory systems include the transition from open to closed circulatory systems in invertebrates and the evolution from the two-chambered heart of fishes to the four-chambered heart of birds and mammals.

The mammalian heart has two ventricles and two atria. Between the ventricles and atria are the tricuspid and bicuspid valves. The ventricles, which pump blood out of the heart, communicate with either the pulmonary artery or the aorta through semilunar valves. Heartbeat is governed by the sinoatrial and atrioventricular nodes. Blood pressure is dependent on heartbeat and vasodilation or vasoconstriction states. Heartbeat and blood pressure are interrelated through the negative feedback system, incorporating the aortic depressor nerve and the medulla, with its cardiac inhibitor center.

One can tell much about cardiac function by using either EKG tracings or the stethoscope.

Blood flows through the arteries because of the pressure created by the beating heart, whereas the flow in veins depends primarily on skeletal muscle contraction. Blood flow in veins is directed by internal valves. The arterial and venous systems are roughly parallel in their morphologies.

Human blood contains liquid plasma plus erythrocytes, five types of leukocytes, and platelets. The erythrocytes are most important in oxygen–carbon dioxide exchange, whereas the leukocytes serve for bodily protection. Platelets are necessary for blood clotting.

The lymphatic system picks up excess interstitial fluid that has leaked from capillaries and returns it to the blood stream. Lymphocytes are also active in bodily defense, via the antibody-producing immune system.

Antibody–antigen interactions of an individual are extremely complex and are important for the recognition of foreign materials.

The A-B-O blood types (and others) are concerned with characteristic antigen and antibody complements. Agglutination will result if blood of one type, containing antigen X, mixes with blood that contains the antibodies Anti-X. Rh antibodies are normally not present in humans, but if Rh^+ blood (containing Rh antigens) is transfused into an Rh^- person, the Rh^- individual will form Anti-Rh antibodies.

Review Questions

1. What is the name of the cavity in insects that contains the blood?
2. What type of blood cells contain hemoglobin? What does hemoglobin do?
3. What specific type of blood cell comprises about 70 percent of human white blood cells?
4. Place the following in the correct order:
   (a) Thrombin + fibrinogen $\longrightarrow$ Fibrin
   (b) Thrombokinase + prothrombin + calcium $\longrightarrow$ _____
   (c) Fibrin + erythrocytes $\longrightarrow$ Clot
   (d) Platelets rupture
5. Describe the major functions of the circulatory system.
6. An antibody could conceivably act as an antigen. How?
7. Diffusion requires that the respiratory surfaces be _____ .

8. Describe the action and function of the semilunar valves.
9. What is systole?
10. The chordae tendineae are attached to which valves? What is their function?
11. To which persons may an Rh⁻, type B blood group male donate blood without fear of agglutination (in the recipient)?

Selected References

Crouch, J. E., and J. R. McClintic. *Human Anatomy and Physiology*. New York: John Wiley & Sons, Inc., 1971.

Elves, M. W. *The Lymphocytes*. 2nd. ed., Chicago: Year Book Medical Publishers, 1972.

Good, R. F., and D. W. Fisher (eds.). *Immunology*. Stanford: Sinauer Assoc., Inc., 1971.

Griffin, D. R. *Animal Structure and Function*. New York: Holt, Rinehart and Winston, 1962.

Guyton, A. C. *Textbook of Medical Physiology*. Philadelphia: W.B. Saunders Company, 1956.

Kimber, D. D., C. E. Gray, C. E. Stackpole, and L. C. Leavell. *Textbook of Anatomy and Physiology*. New York: Macmillan Publishing Co., Inc., 1956.

King, B. G., and M. J. Showers. *Human Anatomy and Physiology*. Philadelphia: W.B. Saunders Company, 1969.

Ponder, E. "The Red Blood Cell," *Scientific American*, **196**:95, 1957.

Prosser, C. L., and F. A. Brown, Jr. *Comparative Animal Physiology*, 2nd ed. Philadelphia: W. B. Saunders Company, 1961.

Roeder, K. D. (ed.). *Insect Physiology*. New York: John Wiley & Sons, Inc., 1953.

Schmidt-Nielson, K. *Animal Physiology*. 2nd ed. Englewood Cliffs: Prentice-Hall, Inc., 1964.

Vander, A. J., J. H. Sherman, and D. S. Luciano. *Human Physiology: The Mechanisms of Body Function*. New York: McGraw-Hill, 1970.

Vroman, L. *Blood*. New York: Natural History Press, 1967.

Whitteridge, G. *William Harvey and the Circulation of the Blood*. New York: American Elsevier Publishing Co., Inc., 1971.

Wiggers, C. J. "The Heart," *Scientific American*, **196**:74, 1957.

Respiratory Systems

Aerobic respiration is concerned with the exchange of oxygen and carbon dioxide in the body. **Oxygen** is the terminal **oxidizer (electron acceptor)** in the series of reactions required for the metabolic oxidation of organic nutrients. This oxidation occurs within every living cell of an animal's body, and it produces the energy that drives all life functions. Waste products of oxidation include **water** and **carbon dioxide.** The actual use of oxygen by the cells is called **internal** (or **cellular**) **respiration,** and its details are essentially the same throughout the animal kingdom (Fig. 2-11). **External respiration** involves the transport of oxygen and carbon dioxide between the external environment and the body's cells. The particulars of external respiration differ among the various groups of animals, but generally two steps are involved: (1) the exchange of respiratory gases between the external environment and the internal circulating fluids, through the body wall, gills, tracheae, or lungs and (2) the exchange of these respiratory gases between the circulating fluid and the body's cells.

The Evolution of Respiratory Systems

External respiration in the protists occurs principally by the diffusion of oxygen, dissolved in water, through the plasma membrane and into the cytoplasm, with carbon dioxide diffusing in the opposite direction. Oxygen has a higher concentration in the water than in the cytoplasm, and the opposite concentration gradient exists for carbon dioxide; this is what causes the diffusion patterns noted across the permea-

ble cell membrane. Diffusion over a concentration gradient is the basis for external respiration in all the metazoans as well.

In poriferans, cnidarians, platyhelminths, annelids, and small and simple aquatic animals, respiratory gas exchange occurs by diffusion across the general body wall.

In more complex aquatic animals with circulatory systems, respiration often takes place with the aid of **gills,** specialized projections of the body that have thin walls and are highly vascularized. For example, the crayfish possesses a set of gills in branchial chambers on each side of its thorax (Fig. 20-4B). As water flows through the branchial chamber, oxygen diffuses through the cells of the **gill filaments** and into the blood. Similarly, carbon dioxide diffuses out of the blood through the gill filaments and into the chamber water. External respiration is carried on in an analogous fashion by the gills of fishes. Certain specialized lungfishes acquire oxygen and give off carbon dioxide by means of modified lunglike **air bladders** as well as by gills.

Diffusion requires that the respiratory surfaces be vascularized and moist, and this poses potential problems for terrestrial animals. For example, earthworms are restricted to moist soils because respiratory gas exchange occurs by diffusion across their moist permeable skin, which is underlaid with numerous capillaries.

Insects and most other terrestrial arthropods possess respiratory systems consisting of networks of air-filled tubes, the **tracheae,** that ramify throughout the body from openings (**spiracles**) in the impermeable exoskeleton (Figs. 21-2 and 21-7). The terminal branches of the finest tracheae, the **tracheoles,** are filled with liquid, and gases diffuse through this liquid into and out of the tissue cells that are relatively inactive, however, the tracheolar fluid leaves the terminal tracheoles when the cells become active—thus there is an efficient exchange of gases. Because the tracheae branch throughout the body tissues, the insect circulatory system is of little value in oxygen transport.

Lungs, found in all vertebrates except the fishes, are specialized organs for breathing air that develop embryologically as diverticula of the gut. The essential feature of lungs is that only a thin, moist epithelial lining separates oxygen-containing air from blood in capillaries. Oxygen from the lung cavities diffuses across the epithelial cells and from there into the capillaries; meanwhile carbon dioxide diffuses in the opposite direction.

Amphibians, such as the frog, respire by means of lungs and also through their moist body surface, both while in the water and while in the air. In fact, the skin is the only means of respiration during hibernation. One large family of salamanders, the plethodontids, are lungless and depend solely on cutaneous respiration. The lungs of the frog are simple sacs, whereas those of higher vertebrates are more complex.

The lungs of reptiles are more complex than are those of amphibians, attesting to their more terrestrial life-style and waterproof skin. Certain aquatic turtles possess **cloacal sacs** that act as auxiliary respiratory organs; these are analogous to the **rectal gills** of dragonfly naiads and the **gill trees** of sea cucumbers (Fig. 22-8). In birds the lungs are supplemented by **air sacs** (Fig. 29-10), which results in their unique and very efficient respiratory process (Chapter 29). The hollow bones of birds may also serve a minor respiratory function. Air capillaries replace the alveoli, in birds.

The Mammalian Respiratory System

Our respiratory system (Figs. 3-9 and 36-1) is similar to that of other mammals. Air passes through the **nose** and **mouth** and into the lungs by way of the pharynx, larynx, trachea, bronchi, and bronchioles (Fig. 3-9). Incoming air is warmed, moistened, and filtered in the nose. The **glottis,** a slitlike opening, separates the **pharynx** from the **larynx.** A leaf-shaped lid of car-

tilage, called the **epiglottis,** closes over the glottis during swallowing. The larynx, the "Adam's apple," is a triangular organ that in humans has nine pieces of cartilage in its walls to prevent it from collapsing.

Within the larynx are the **vocal cords.** Air forced out of the lungs may vibrate these cords, and the column of air is modified to produce sounds. This may be compared with an organ pipe in which a reed and a column of air are set into vibration by a blast of air. Certain notes can be played on the organ that are remarkably similar to those in the human voice. The cavities of the pharynx, mouth, and nose act as resonators. The amplitude of the vibrations and the volume and force of the air current determine the loudness and intensity, and the length, tightness, and vibrational frequency of the cords determine the pitch of the voice. In women and children, the vocal cords are usually short, and the voice is high pitched. In men, the cords are usually longer (brought about by the male sex hormone), and the voice is correspondingly lower.

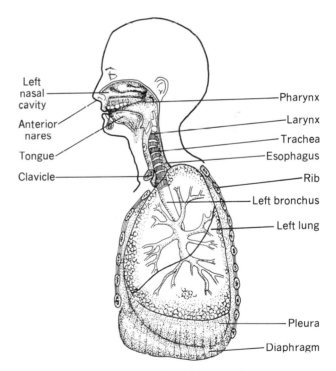

Figure 36-1. Major organs of the human respiratory system.

Esophageal speech results from sounds formed by vibrations of air in the esophagus and lips and tongue. Air is swallowed and then, by stomach control, expelled through the esophagus and into the mouth. Esophageal speech is used by people who have had their larynxes removed. The trachea is modified in such instances to a breathing hole in the neck.

The **trachea** is a tube about 10 cm long, supported with C-shaped rings of cartilage in its walls. The trachea branches into two **bronchi,** one to each lung, and these branch within the lungs into many smaller **bronchioles.** Each bronchiole terminates in an elongated saccule, the **alveolar duct,** that bears many **alveoli** on its surface (Fig. 3-9). The number of alveoli in the human lungs has been estimated to be approximately 750 million, exposing a surface area of over 60 m², or more than 50 times the skin's surface area.

The **lungs** are membranous sacs, each lying in a cavity lined with a serous membrane, the **pleural sac,** lessens friction between it and the constantly moving lung by secreting a small amount of lubricating serous fluid. If the pleural sac becomes inflamed, **pleurisy** is the result. The total volume of air that can be contained in the lungs averages 3.5–4.0 liters and is referred to as one's **vital capacity** (the maximum air that can be inhaled and exhaled). Under resting conditions, about 0.5 liter enters and leaves the lungs, the so-called **tidal volume.** During maximum inspiration, 2.5–3.5 liters of **inspiratory reserve volume** is available; exhalation may also be forced past the average, but about 1.0 liter of **residual volume** remains at maximum exhalation.

Breathing and the Respiratory Function of Blood

Air must be periodically brought into and expelled from the lungs—if air were merely allowed to stay in the alveoli, it would soon come to equilibrium with the gases in the blood; then no further diffusional exchange would occur, oxygen would not be transported to cells, and death would result. Note that the lungs, which reside in the thoracic cavity, have no musculature. Consequently, they remain passive during the body's respiratory movements. The lungs are entirely dependent on muscular movements of the diaphragm and intercostal muscles.

During **inspiration** (taking in air), the **external intercostal muscles** attached to the ribs contract, bringing the rib cage slightly up and slightly out and, in so doing, generally increasing the volume of the thoracic cavity. This expansion is greatly augmented by contraction of the **diaphragm,** the muscular sheet dividing the thoracic and abdominal cavities. The **stomach muscles** may also play a role in increasing or decreasing the size of the thoracic cavity. These efforts increase the volume of the thoracic cavity and thereby decrease its internal pressure. Because atmospheric pressure is now higher, air moves through the respiratory passageways and into the lungs to regain equilibrium.

Expiration is the opposite of inspiration. Now the **internal intercostal** muscles on the insides of the ribs contract, while the others relax. The diaphragm relaxes during expiration, arching up to its "resting" position. These events decrease the volume of the thoracic cavity, making the pressure in the lungs greater than atmospheric pressure, so that air flows outward. An awake adult normally breathes 11 to 14 times per minute.

It should be noted that these thoracic pressure changes also play a significant role in circulation. The **inferior (posterior) vena cava** passes through the thoracic cavity, and, when pressure is increased, blood is pushed through this great vein.

One of the controlling factors in respiration is resistance of the air passages to air flow. In the disease **asthma,** the smooth muscles in the smallest bronchioles go into spasm, making their diameters smaller and drastically increasing resistance to air flow. Thus, breathing becomes much more difficult.

As already mentioned, the process of diffusion is responsible for the exchange of oxygen and carbon dioxide across the walls of the alveoli. The changes in concentration of these molecules are not great, as shown in Table 36-1.

Most of the oxygen that diffuses into the blood enters into a loose chemical combination with the **heme** portions of hemoglobin, producing **oxyhemoglobin,** which gives oxygenated blood its bright red color. **Hemoglobin** itself is a tetramer, made of four polypeptide chains: two alpha (α) chains and two beta (β) chains. Each protein subunit has a covalently-bound heme group. In the disease **sickle cell anemia** (which,

Table 36-1. Concentration Changes of Atmospheric Gases During Respiration

	Inspired Air (%)	Expired Air (%)	Net Difference (%)
Oxygen	21.00%	16.02%	Gain: 5.80%
Carbon dioxide	0.05	3.60	Loss: 3.55
Nitrogen	78.30	75.00	Gain: 3.30%

in the U.S. population, affects blacks more often than it does whites), the β chain is different from the normal condition (actually, only one amino is different because of a mutation in the gene for the protein; see Chapter 38). In a person homozygous for the sickle cell trait, both β chains of the hemoglobin molecule are aberrant. Most of the homozygous individual's erythrocytes are actually sickle shaped. This results in a decreased surface area and capacity for holding oxygen, as well as in some interference with smooth blood flow; crippling or death occurs in childhood or early adulthood if treatment is unavailable. In the heterozygote, half the β chains are normal and half are mutant; the random association of α and β chains therefore yields both mutant and normal hemoglobin molecules, and latter greatly improving one's ability to survive. The heterozygote's erythrocytes generally appear discoidal, as do normal erythrocytes, unless there is a relatively small amount of oxygen around, in which case they sickle also.

Oxyhemoglobin in the systemic capillaries gives up its oxygen, after which it diffuses through the interstitial fluid and into the body cells to take part in metabolic respiration. The carbon dioxide resulting from metabolic respiration diffuses across its concentration gradient from the cells and into the blood, where it is less concentrated. A small portion (about 10 percent) of the carbon dioxide is dissolved in plasma. Approximately 20 percent combines with hemoglobin to form **carboaminohemoglobin,** and the remaining 70 percent is transported through the bloodstream in the form of **bicarbonates.** In the lungs, carbon dioxide is freed from these bicarbonates, the hemoglobin, and the plasma and diffuses into the air in the lung cavities, where the gas is exhaled into the atmosphere.

Control of Respiratory Rate

The respiratory rate is under both chemical and neural control. Chemical control depends primarily upon the concentration of carbon dioxide in the blood, which affects the hydrogen ion (pH) and oxygen concentrations. Although the exact mechanism is a very controversial subject, it is believed that control is affected by changes in the pH of the blood, according to the following equation:

$$CO_2 + H_2O \rightleftharpoons HCO_3^- + H^+$$

Deviations in the concentration of CO_2, and hence pH, cause corresponding changes in the pH of the respiratory control centers of the brain. This results in the appropriate compensating stimuli, that is, an increased respiratory rate when CO_2 is high and a lower rate when CO_2 is low. Another control mechanism operates when the concentration of dissolved oxygen decreases. **Chemoreceptors** in the **aorta** and the **carotid artery** are stimulated by a high concentration of CO_2 or a low concentration of O_2; these receptors in turn stimulate the respiratory center in the **medulla** to increase the respiratory rate.

Carbon monoxide (CO) **poisoning** occurs because hemoglobin bonds more easily and strongly to CO than to O_2, and, if CO is abundant in the air (e.g., for auto exhausts), the blood will carry little or no O_2 to the cells. Because of the very low effective levels of CO, the dissolved O_2 and CO_2 levels and respiratory control mechanisms are not affected. Thus fainting, followed by death, occurs without increased ventilation.

H. Houdini, the famous escape artist, used to have himself locked in a trunk, and the trunk would be cast into water. He would have to escape before suffocating. One of the things he did just before being locked into the trunk was to **hyperventilate** (breathe very fast) using pure oxygen. This vastly increased the oxygen content of his blood and allowed him to hold his breath much longer than he would have normally been able. This removed much CO_2, which is the main chemical stimulus of breathing.

Neural control of breathing is primarily involuntary but is under some voluntary control. The medulla can be stimulated by motor instructions from the cerebral cortex. Nevertheless, we can cease breathing until the CO_2 concentration is so great that this chemical stimulus overpowers our voluntary effort. The motor nerves from the respiratory center go to the intercostals, diaphragm, larynx, and abdominal muscles, all of which participate in respiratory movements to a greater or lesser extent.

Summary

The respiratory system is mostly concerned with obtaining oxygen and ridding the body of carbon dioxide. The respiratory process is diffusional. All animal cells require oxygen and obtain it in diverse ways: for example, through the integument, through holes in the body (for instance, spiracles), through gills, and through lungs. Air is brought into and out of the lungs because of thoracic pressure changes initiated by certain muscular contractions (of intercostals, extracostals, and diaphragm). Air is brought into and out of the lungs because of thoracic pressure changes initiated by certain muscular contractions (of intercostals, extracostals, and diaphragm). Feedback systems also control respiratory rate. The medulla, hypothalamus, cerebral cortex, and various peripheral receptors are involved in this regulation.

Review Questions

1. What is the earthworm's equivalent of the human lung?
2. What is the function of the pleura?
3. What is the major function of the respiratory system?
4. The internal intercostal muscles contract during
   (a) inspiration
   (b) expiration
5. Approximately what percentage gain in oxygen is realized during one respiratory cycle?
6. What good is a yawn? Some people say it "gets the blood moving." How could it possibly do that?
7. The part of the brain involved in control of both blood pressure and respiratory rate is the _____ .

Selected References

Crouch, J. E., and J. R. McClintic. *Human Anatomy and Physiology*. New York: John Wiley & Sons, Inc., 1971.

Dejours, P. *Principles of Comparative Respiratory Physiology*. New York: American Elsevier Publishing Co., Inc., 1977.

Good, R. F., and D. W. Fisher (eds.). *Immunology*. Stanford: Sinaver Assoc., Inc., 1971.

Griffin, D. R. *Animal Structure and Function*. New York: Holt, Rinehart and Winston, Inc., 1962.

Guyton, A. C. *Textbook of Medical Physiology*. Philadelphia: W.B. Saunders Company, 1956.

Kimber, D. D., C. E. Gray, C. E. Stackpole, and L. C. Leavell. *Textbook of Anatomy and Physiology*. New York: Macmillan Publishing Co., Inc., 1956.

King, B. G., and M. J. Showers. *Human Anatomy and Physiology*. Philadelphia: W. B. Saunders Company, 1969.

Ponder, E. "The Red Blood Cell," *Scientific American*, **196**:95, 1957.

Prosser, C. L., and F. A. Brown, Jr. *Comparative Animal Physiology*, 2nd ed. Philadelphia: W. B. Saunders Company, 1961.

Roeder, K. D. (ed.). *Insect Physiology*. New York: John Wiley & Sons, Inc., 1953.

Schmidt-Nielson K. *Animal Physiology*, 2nd ed. Englewood Cliffs: Prentice-Hall, Inc., 1964.

Vander, A. J., J. H. Sherman, and D. S. Luciano. *Human Physiology: The Mechanisms of Body Function*. New York: McGraw-Hill Book Co., 1970.

Wiggers, C. J. "The Heart," *Scientific American*, **196**:74, 1957.

Nervous Systems, the Senses, and Endocrine Systems

Coordinated activities and behavior in animals is accomplished largely by nervous systems and sense organs and by means of hormones secreted by endocrine glands.

The Evolution of Nervous Systems

Irritability and **conductivity** are fundamental characteristics of cytoplasm. Amoebas and other unicellular protists respond to changes in the environment much as do animals that possess sense organs and nervous systems.

The simplest organisms to exhibit definite nervous systems are the cnidarians, represented by the hydra (Fig. 11-2). In this type of nervous system, **sensory (receptors)** cells for reception of stimuli are in direct contact with fibers from **motor neurons (conductors)** that send out processes to the contractile fibrils in epitheliomuscular cells (**effectors**). Together, the receptors are conductors constitute a **nerve net,** or **plexus.** The conduction of nerve impulses in this simple system in omnidirectional, that is, in both directions along the nerve fibers.

Platyhelminths have a slightly more complex nervous system. For example, the planaria has two **longitudinal nerve cords** connected by **transverse nerves,** exhibits **cephalization,** has **ganglia,** and has anterior **sense organs** resembling eyes and auricles for chemoreception and tactile reception. (Fig. 13-1).

The annelids have ganglia and **reflex arcs** that allow them to react to certain stimuli automatically (Fig. 17-6). Arthropods, in addition, have relatively complex sensory systems (Fig. 21-8), including such structures as the **antennae, eyes** (Fig. 20-5), and **chemoreceptors** on various appendages.

The chordates have **dorsal nervous systems** as opposed to most other animals with ventral nervous systems. *Amphioxus* is a good example of the primitive chordate plan (Fig. 23-6).

It is evident that the appearance of unidirectional nerve cells, cephalization, and the specialization of receptors all represent advances in the phylogenetic development of nervous systems.

The Vertebrate Nervous System

The nervous system of vertebrates is vastly more complex than is that of other animals, and the nervous system of mammals is considered to be the most advanced of all. We will use the human nervous system as our primary example.

The most basic division of the vertebrate nervous system is between the **central nervous system (CNS)** and the **peripheral nervous system (PNS).** The CNS is composed of the **brain** and **spinal cord** and is concerned with the interpretation and integration of incoming data. The PNS is comprised of all nerves outside the CNS: Some of these peripheral nerves convey sensory information to the CNS; others send motor instructions from the CNS to the muscles and glands of the body. The PNS is incapable of interpreting stimuli and formulating appropriate responses; these are the roles of the CNS. The PNS is recognized as having two subdivisions: the autonomic nervous system and the somatic nervous system.

Voluntary functions such as most skeletal muscle contractions are controlled by the **somatic nervous system.** The **autonomic nervous system** regulates involuntary functions such as digestion, thermoregulation, and other visceral functions.

The autonomic nervous system is in turn subdivided into sympathetic and parasympathetic divisions, with antagonistic actions. The **sympathetic nervous system** plays an activating role by helping the organism to react to stressful stimuli from the external environment. The sympathetic system prepares the animal for "fight or flight" by simultaneously increasing heart rate and blood glucose levels, dilating blood vessels that supply the skeletal muscles, constricting the blood supply to the digestive tract, and generally arousing the organism. The motor neurons of the sympathetic nervous system leave the CNS from the thoracic and anterior portion of the spinal cord and then from ganglia near the cord before innervating their target organs. Sympathetic activity is reinforced by secretion into the bloodstream of the hormone epinephrine (adrenaline) by the adrenal glands situated atop the kidneys. A related chemical, norepinephrine is probably the actual transmitter molecule for sympathetic synapses.

The **parasympathetic nervous system** is concerned primarily with internal homeostasis, which involves the optimatization of digestion, blood flow, body heat, and other visceral functions. Its actions are generally opposite to those of the sympathetic system—decrease of heart rate and blood glucose levels, return to normal diameter of blood vessels, and so on. The motor neurons of this system leave the spinal cord rostrally (toward the head end) and caudally (toward the tail end), and parasympathetic ganglia occur near their target organs.

The Neuron

The basic functional units of nervous systems are nerve cells called **neurons** (Fig. 3-6). All neurons share the following characteristics:

1. Neurons are single cells that are capable of reacting to a stimulus and transmitting the resulting excitation to another neuron, muscle, or gland.
2. The nucleus is within the **cell body** (soma). A variable number of processes or fibers, called dendrites and axons, extend from the cell body.
3. **Dendrites** are processes that conduct impulses toward the cell body. A neuron has several or many dendrites.
4. The **axon** is the cytoplasmic fiber that conducts nerve impulses away from the cell body. A neuron has only one axon, although the axon commonly branches near the tip.
5. Electrical impulses are normally propagated in only one direction.
6. All functional neurons communicate with others via the **synapse** at which the axon of one neuron abuts a dendrite or the cell body of another.
7. Generally, neurons in the brain and spinal cord cannot be regenerated, so damage to the CNS is permanent.

Four major types of neurons are recognized: bipolar, sensory, interneurons, and motor neurons are distinguished by their morphologies and functions within the nervous system. A **bipolar neuron** has a more or less centrally located soma, and its axon may or may not have an insulating covering of myelin. The dendrites of a bipolar neuron have a threadlike, spiny appearance (Fig. 37-1). Bipolar neurons occur in various parts of the nervous system. The cell body of a sensory bipolar neuron projects from the main line of the neuron in a "T" configuration. **Sensory neurons** conduct impulses from receptors to the CNS and are usually myelinated. **Interneurons,** or connective neurons, occur exclusively within the CNS; they are usually nonmyelinated and have profuse dendritic connections with other neurons in the CNS (Fig. 37-1). **Motor neurons** have radiating dendrites and long, usually myelinated, axons. Motor neurons carry impulses from the CNS to the muscles and glands of the

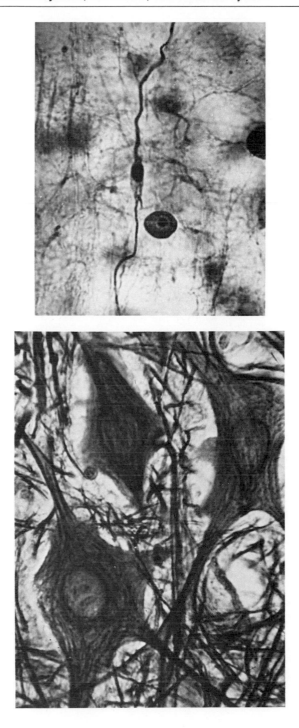

Figure 37-1. Top: Bipolar neuron. Bottom: Multipolar neurons (connective or interneurons). Note multiple connections. 200 ×.

body. The dendrites and axons of both sensory and motor neurons may extend for hundreds of centimeters from the cell bodies to reach their targets.

Bundles of nerve fibers held together by connective tissue are called **nerves** when they occur outside the brain and spinal cord and nerve tracts within the CNS. Similarly, a mass of nerve cell bodies outside the CNS is called a **ganglion,** and within the CNS a **nucleus** (not to be confused with the nucleus within a cell).

Myelination

Many nerve axons are covered by a segmented sheet of white insulative, fatty substance called **myelin.** This covering consists of the membranes of **Schwann cells** that wrap around the axon in concentric layers (Figs. 3-6 and 37-2). Between adjacent Schwann cells are gaps without myelination called **nodes of Ranvier.** Such myelinated axons conduct nerve impulses from 10 to 20 times faster than nonmyelinated fibers. This is because myelin acts as an electrical insulator, and so the impulse jumps from node to node along the axon.

The Mechanism of Nervous Function

The common language of the nervous system consists of **action potentials,** which are transient localized changes in electrical potential that are transmitted along a neuron and from one neuron to another. To understand the flow of an electrical impulse along a neuron, it is first necessary to consider the neuron at rest (Fig. 37-3). The neuron is normally surrounded by a blanket of positively charged sodium ions (Na^+).

Active transport by a sodium pump within the neuron cell membrane maintains a higher concentration of Na^+ outside than inside the neuron. Conversely, a potassium pump maintains a higher concentration of potassium ions (K^+) inside the membrane.

Chlorine ions (Cl^-) and negatively charged proteins are also more concentrated within the neuron. These three ions are so distributed that the neuron is posi-

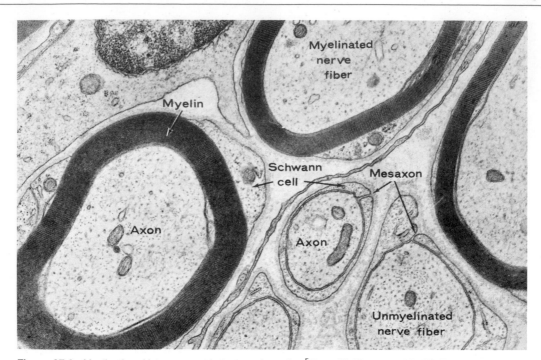

Figure 37-2. Myelination. Note concentric layers of myelin. [From W. Bloom and D. W. Fawcett, *A Text-book of Histology.* Philadelphia: W. B. Saunders, 1968.]

tively charged outside and negatively charged inside. This separation of charge causes the **resting,** or **membrane, potential** of approximately -70 millivolts (mv). In this resting state, the permeability of the membrane for Na^+ is almost zero, and, hence, although the Cl^- and Na^+ ions attract each other, they are kept apart. A membrane in this state is said to be electrically polarized because the charges are separated.

When the neuron is activated, a rapid succession of changes occurs (Fig. 37-3). The membrane becomes locally very permeable to Na^+ ions, and these ions diffuse rapidly into the neuron. The membrane also becomes locally permeable to potassium ions (K^+), but K^+ moves out more slowly than Na^+ enters. This **depolarizes** the membrane from its resting potential, because the inside (momentarily) becomes more positive than the outside. If the initial stimulus passes a certain level, the **threshold potential** of -60 mv, the membrane potential is changed from -70 mv to a

value of about $+30$ mv. At this point the Na^+ pump and K^+ pump locally **repolarize** the membrane to its resting potential. Because of the threshold effect, the neuron either fires or it does not. This is known as the **all-or-none law** of neuronal conduction.

The local chemical changes just described serve to depolarize the next part of the membrane by creating an electrical current (which may be defined as the movement of charges). The current affects the membrane just ahead of the depolarization by increasing that region's permeability to Na^+ and K^+ ions. Thus, the wave of depolarization and subsequent repolarization—an **action potential**—sweeps along the neuron (Fig. 37-3). The passage of an action potential is followed by a waiting, or a **refractory, period,** during which time the neuron repolarizes and cannot respond to another stimulus.

Recall that the presence of myelination serves to increase the velocity of nerve impulse propagation along some axons. Because the myelin in thick and

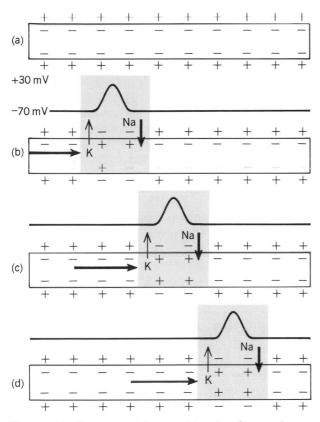

Figure 37-3. Sequence of the neural impulse. See text for description.

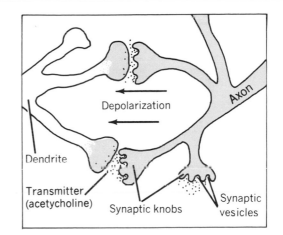

Figure 37-4. The synapse. Here we see acetylcholine being released from the synaptic vesicles across the synaptic cleft.

nonconductive, the gaps at the nodes of Ranvier provide the only loci for ionic movements across the neuron's membrane. And so permeability changes need not occur at every point along the axon. The action potential can "jump" from gap to gap (the nodes of Ranvier) in the myelin. This results in substantially faster conduction time.

The Synapse

Neurons never occur in isolation but always "connect" with other neurons, receptors, or effectors (muscles and glands). Adjacent neurons do not actually touch but are separated by a very small gap (about 200 Å) called a **synaptic cleft.** Most nerve impulses are chemically transmitted across such synapses.

Electrical transmission is extremely rare in the vertebrates but it does occur in many invertebrates. In these animals the current flow from the axon is sufficient to induce the adjacent dendrite(s) to depolarize to threshold.

The tips of an axon are enlarged (**synaptic knobs**) and contain small sacs of chemicals (**synaptic vesicles,** Figs. 32-9 and 37-4). When a wave of depolarization reaches an axon's synaptic knob, many synaptic vesicles release **transmitter chemicals** into the synaptic cleft, affecting the permeability of the adjacent dendrite's membrane in one of two ways. At **excitatory synapses** (Fig. 37-4), release of **acetylcholine** or **norepinephrine** induces the adjacent membrane to depolarize. Conversely, **inhibitory synapses** make it more difficult for the adjacent membrane to be depolarized to threshold.

Let's consider the release of a common transmitter substance, acetylcholine. Molecules diffuse across the synaptic cleft, contact the membrane of the dendrite, and make it locally permeable to Na^+. The diffusion of ions across their concentration gradients depolarizes the membrane. If the membrane is depolarized to its threshold potential, an action potential is initiated and propagated along the entire length of the neuron. If acetylcholine were allowed to accumulate in the synaptic cleft, repolarization would be impossible, and the neural pathway would be blocked. However,

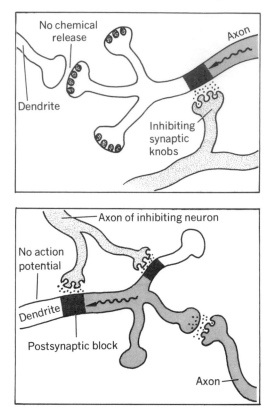

Figure 37-5. Top: Presynaptic inhibition. Bottom: Postsynaptic inhibition. See text for description.

an omnipresent enzyme called **cholinesterase** hydrolyzes acetylcholine rapidly, thereby allowing repolarization of the dendritic membrane in preparation for other incoming impulses.

Synaptic inhibition has the reverse effect of synaptic excitation and may occur in two ways (Fig. 37-5). **Presynaptic inhibition** occurs when an inhibitory axon synapses on an excitatory axon. Firing of the inhibitory axon reduces the amplitude of an action potential traversing the excitatory axon, causing less transmitter substance to be released across the excitatory synapse. **Postsynaptic inhibition** involves the synapse of an inhibitory axon upon a dendrite, just beyond an excitatory synapse. In this case, the inhibitory axon releases a transmitter substance that hyperpolarizes the dendrite, making it more stable by further decreasing its permeability to Na$^+$.

Summation Effects

Neurons usually have multiple connections. In fact, the vertebrate dendrites of motor neurons often synapse with hundreds or even thousands of other neurons. When the motor neuron fires, each of the muscle fibers in its motor unit contracts (Chapter 32). Yet how does the motor neuron integrate the potentially hundreds of excitatory and inhibitory signals that it may receive from other neurons?

This integration is brought about by the same two types of summation discussed in Chapter 32 for skeletal muscle control. If one axon fires repeatedly in rapid succession, a sufficient volume of acetylcholine may be released to depolarize the adjacent dendrite to threshold; this is **temporal summation.** Or, if several excitatory axons fire simultaneously, the combined volumes of acetylcholine released may depolarize the dendrite to threshold; this is **spatial summation.**

Outline of the Vertebrate Nervous System

The Spinal Cord

The spinal cord represents the part of the CNS extending from the base of the medulla (at the foramen magnum of the skull) down to just below the ribs. The adult spinal cord is approximately 45 cm in length and about as thick as your little finger. The body of the cord is composed exclusively of neurons (Fig. 37-6).

The butterfly-shaped core of **gray matter** is composed of nonmyelinated neurons. The **horns** of gray matter contain mostly interneurons (that synapse on other neurons) and motor neurons (that leave the spinal cord to their target organs.) Surrounding the central core is **white matter** composed of myelinated axons, grouped into nerve tracts that transmit impulses along the spinal cord.

Thirty-one pairs of **spinal nerves** emerge from the spinal cord and pass laterally through the spinal column between the vertebrae. Each spinal nerve has two roots. The **ventral root** contains motor fibers that leave the CNS and innervate the muscles and glands of the body. The **dorsal root** contains sensory fibers that carry information from peripheral receptors to the CNS.

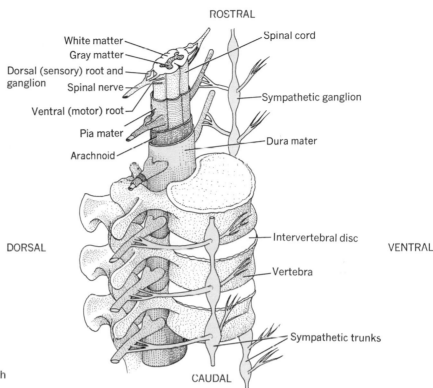

Figure 37-6. The human spinal cord with associated structures.

The spinal cord and brain are protected by bones (vertebrae and cranium) and by fibrous and elastic membranes called **meninges.** The inner **pia mater,** middle **arachnoid,** outer and **dura mater** successively enclose the CNS. **Cerebrospinal fluid,** which is similar to blood plasma but contains less protein and cells, cushions and nourishes the delicate CNS within the central canal surrounding the spinal column. Within the brain there is a system of channels, or **ventricles,** that are continuous with one another and with the central canal of the spinal cord.

The Simple Reflex Arc

The spinal cord cannot interpret incoming sensory impulses, but it can direct some effectors without involving the brain. A basic example of this function is a **spinal reflex** (Fig. 37-7A), which generally involves the rapid withdrawal of a body part away from an area of injurious stimulation. For example, assume that your finger is burned by a match. Sensory neurons in the finger are stimulated, and they send action potentials through the dorsal root into the spinal cord. In vertebrates, the sensory axon synapses with an interneuron, which in turn synapses either **ipsilaterally** (on the same side) or **contralaterally** (on the other side) with one or more motor neurons. The motor neurons exit through the ventral root(s) and stimulate the appropriate muscles to withdraw your hand immediately from the flame. In more primitive organisms, a reflex arc may consist of simply two cells, one sensory neuron that synapses directly to one motor neuron.

Note that no cerebral control has been exerted, although impulses along other interneurons may ascend the nerve cord to inform the brain. Depending on the level at which the reflex occurs, one may be "conscious" of the pain before, during, or after the reflexive muscle contraction has occurred. Note also that the spinal reflex lacks fine motor control: not only

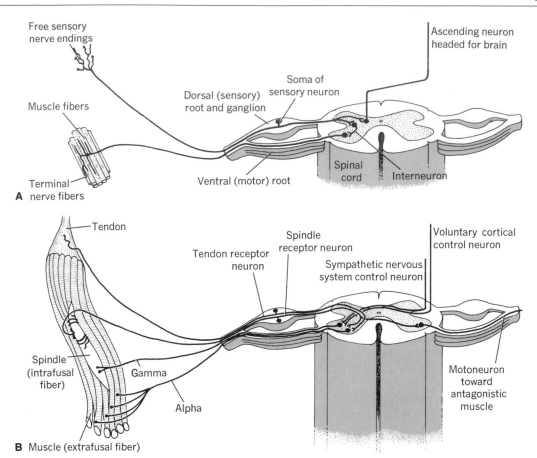

Figure 37-7. (A) Spinal reflex arc (see text for description). (B) Muscle control mechanism (see text for description).

the burned finger but also the hand and arm suddenly jerk away from the flame.

Neural Control of Tonus

At any one moment, a number of muscle cells are contracting; thus the total muscle is always in a state of partial contraction, a condition called **tonus.** If muscles were completely relaxed, they would **atrophy** (degenerate), just as unused neural synapses degenerate. Much of the control of tonus occurs in the spinal cord.

Special **stretch receptors** are imbedded in modified skeletal muscle cells called spindle fibers (Fig. 37-7B). The axons of the receptor neurons enter the

spinal cord and synapse on two sets of the so-called **alpha motor neurons.** The alpha motor neurons leave the spinal cord and synapse with the same muscle group that the spindle fiber monitors.

As a muscle is stretched, the spindle stretch receptors are stimulated. They send action potentials to the neurons, which stimulate the muscle to contract, thereby reinstating tonus. What prevents a muscle from contracting too far? Special **tendon receptors** also synapse on the motor neurons, but their influence is inhibitory. If a muscle becomes too short, the tendon is stretched to the point at which receptors become stimulated and send inhibitory impulses to the motor neuron. There are degrees of contraction between tendon receptor activation and spindle fiber activation in which neither of these homeostatic mechanisms is stimulated: this range defines the acceptable state of tonus for the muscle.

Reflexive spinal control of muscle tonus can, of course, be overridden by the brain. Voluntary control of muscle contraction originates in the cerebral cortex of the brain.

Brains of Vertebrates

Differences in the brains of the various vertebrate classes are probably more striking than are those of any other organ, as Figure 37-8 indicates. Note especially the progressive increase in size and importance of the cerebellum (the center of unconscious control of skeletal muscles) and the cerebrum (the seat of vol-

untary and higher mental functions). Conversely, the olfactory (smell) lobes, optic (vision) lobes, and medulla constitute progressively smaller percentages of the total brain mass.

In mammals the **olfactory lobes** are well developed but are nearly hidden by the very large **cerebral hemispheres.** The cerebellum is large, and the medulla is comparatively short.

During development, the brain differentiates into three lobes: the **forebrain, midbrain,** and **hindbrain.** The forebrain consists of the cerebrum and diencephalon. In humans, the **cerebrum** is the most complex part of the brain (Fig. 37-9). The outer surface of the cerebrum, called the **cerebral cortex,** consists of gray matter. There are an estimated 10 billion nerve cells in the cerebral cortex of a human brain. Each cerebral

Figure 37-8. Diagrams showing the dorsal surface of the brain of six kinds of vertebrates. They illustrate differences in the degree of development of the different parts. Note especially the progressive increase in size of the cerebellum and cerebrum. (Not all same scale.)

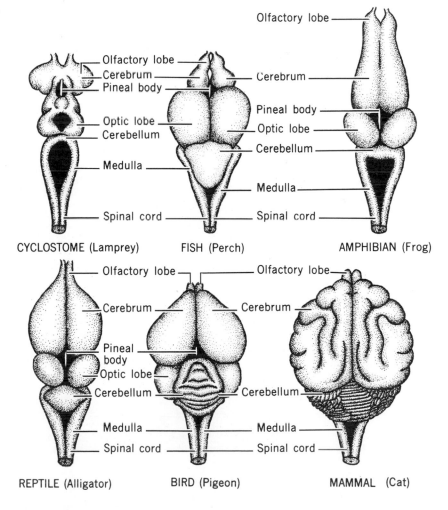

CYCLOSTOME (Lamprey) FISH (Perch) AMPHIBIAN (Frog)

REPTILE (Alligator) BIRD (Pigeon) MAMMAL (Cat)

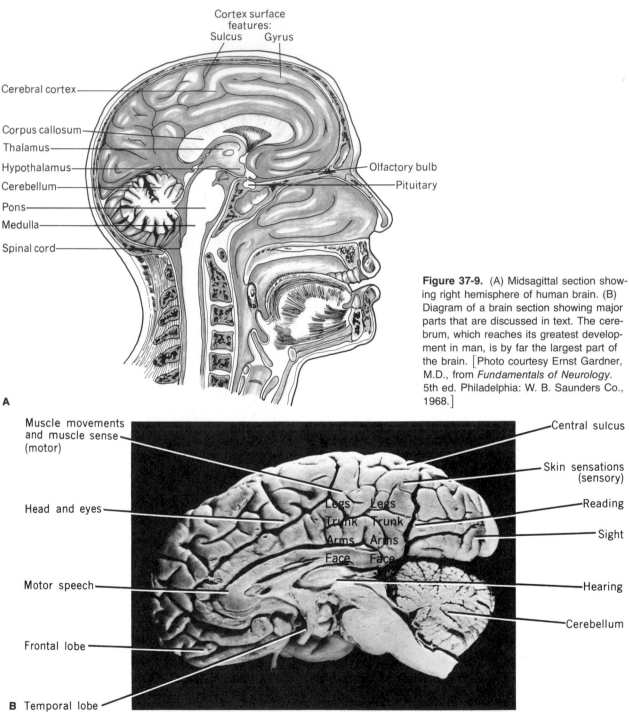

Cortex surface features:
Sulcus   Gyrus

Cerebral cortex

Corpus callosum

Thalamus

Hypothalamus

Cerebellum

Pons

Medulla

Spinal cord

Olfactory bulb

Pituitary

Figure 37-9. (A) Midsagittal section showing right hemisphere of human brain. (B) Diagram of a brain section showing major parts that are discussed in text. The cerebrum, which reaches its greatest development in man, is by far the largest part of the brain. [Photo courtesy Ernst Gardner, M.D., from *Fundamentals of Neurology.* 5th ed. Philadelphia: W. B. Saunders Co., 1968.]

A

Muscle movements and muscle sense (motor)

Central sulcus

Skin sensations (sensory)

Head and eyes

Legs   Legs
Trunk   Trunk
Arms   Arms
Face   Face

Reading

Sight

Motor speech

Hearing

Cerebellum

Frontal lobe

B Temporal lobe

hemisphere is divided by prominent fissures into the **frontal, temporal, parietal,** and **occipital lobes.** The entire cortex is extensively wrinkled by infoldings of the gray matter upon itself, forming grooves (**sulci**) and ridges (**gyri**).

In general, the cerebral cortex is responsible for our consciousness of sensations, our ability to initiate voluntary muscular movements, and our capacity to integrate sensation and movement with higher mental functions of the cortex, such as memory, emotion, personality, and intelligence. Different regions of the cortex control different functions and parts of the body. For example, the integration of sensory information from receptors in the thumb is localized in the cortex approximately 10 cm above the middle ear in an adult. The two cerebral hemispheres are joined by a large mass of nerve fibers, which is called the **corpus callosum.**

Below the cerebral cortex is the **diencephalon,** which includes two important neural centers, notably the thalamus and hypothalamus. The **thalamus** is the integration and relay center for sensory information (other than olfaction). It attributes general feelings of pleasantness or unpleasantness to sensations and then relays the information to the appropriate regions of the cerebral cortex. The **hypothalamus** (situated below the thalamus) controls internal homeostasis and is a major relay center between the cerebrum and the autonomic nervous system. The **hypophysis (pituitary gland)** extends downward from the hypothalamus and lies just above the roof of the mouth in higher vertebrates. The hypophysis releases at least nine different hormones of great physiological significance (discussed in the endocrine system sections.)

The **olfactory bulbs** (Figs. 37-9 and 37-10) develop from the forebrain area of the embryo and integrate sensory impulses from the olfactory (smell) receptors.

The midbrain is much more developed in lower animals: it contains reflex centers and houses sensory and motor fibers on their way to and from the forebrain.

The hindbrain includes the cerebellum, pons, and medulla oblongata. The **cerebellum** is the second largest part of the brain. It lies below the posterior portion of the cerebrum and is covered by a wrinkled cortex of gray matter. The cerebellum is almost exclusively concerned with movement. It coordinates muscle groups so that voluntary movements, equilibrium, and postural adjustments are smoothly executed. The **pons** serves mainly as a communication bridge between the two halves of the cerebellum and between the cerebrum and the cerebellum, synchronizing the

Figure 37-10. Human brain, ventral surface, showing the 12 pairs of cranial nerves.

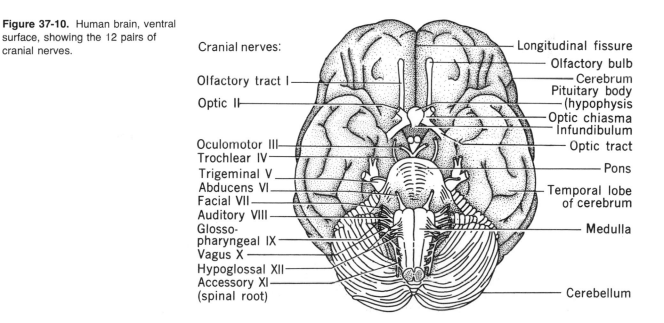

Cranial nerves:

Olfactory tract I

Optic II

Oculomotor III
Trochlear IV
Trigeminal V
Abducens VI
Facial VII
Auditory VIII
Glosso-
pharyngeal IX
Vagus X
Hypoglossal XII
Accessory XI
(spinal root)

Longitudinal fissure
Olfactory bulb
Cerebrum
Pituitary body
(hypophysis
Optic chiasma
Infundibulum
Optic tract
Pons
Temporal lobe
of cerebrum
Medulla
Cerebellum

muscular activity induced by each. Other activities of the midbrain include the processing of visual and auditory inpulses and the maintenance of muscle tone and posture. The **medulla oblongata,** or simply the medulla, contains sensory and motor tracts that communicate the higher brain centers with the spinal cord. Most of these axons cross (**decussate**) from one side to the other within the medulla. As a result, each side of the brain receives sensory information from, and sends motor instructions to, the opposite side of the body. The medulla also houses vital reflex centers that control heart action, breathing (involuntary), the diameter of blood vessels, swallowing, and vomiting.

The **reticular formation** spans an area within the hindbrain and acts as a major sensory filter. This group of neurons is responsible for attention and general activation of the organism. The reticular formation is capable of selecting important stimuli and ignoring others, so that, for example, the sound of your name may awaken you from slumber, whereas the sound of someone else's may not. Dyrby syvjd smf kpjm rfestdf. Your ability to ignore the previous "sentence" involved your reticular formation.

Twelve pairs of **cranial nerves** (Fig. 37-10) emerge from the ventral surface of the brain. These nerves carry sensory and motor information from and to the head and upper torso. Starting anteriorly, the cranial nerves are customarily numbered using Roman numerals. Their distributions and functions are summarized in Table 37-1.

Table 37-1. The Number, Name, Origin, Distribution, and Function of the Cranial Nerves of Vertebrates.

Number	Name	Origin	Distribution	Function
0	Terminal	Forebrain*	Lining of nose	Probably sensory
I	Olfactory	Olfactory lobe	Lining of nasal cavities	Sensory
II	Optic	Diencephalon	Retina of eye	Sensory
III	Oculomotor	Ventral side of midbrain	Four muscles of eye	Motor
IV	Trochlear	Dorsal side of midbrain	Superior oblique muscle of eye	Motor
V	Trigeminal	Side of medulla	Skin of face, mouth, and tongue and muscles of jaws	Sensory and motor, mostly sensory
VI	Abducens	Ventral side of medulla	External rectus muscle of eye	Motor
VII	Facial	Side of medulla	Chiefly to muscles of face	Motor and sensory, mostly motor
VIII	Auditory	Side of medulla	Inner ear	Sensory
IX	Glossopharyngeal	Side of medulla	Muscles and membranes of pharynx, and tongue	Sensory and motor
X	Vagus	Side of medulla	Larynx, lungs, heart, esophagus, stomach, and intestines	Sensory and motor
XI	Spinal accessory	Side of medulla	Chiefly muscles of shoulder	Sensory and motor
XII	Hypoglossal	Ventral side of medulla	Muscles of tongue and neck	Motor

*According to one authority it originates from both the hindbrain and diencephalon in most species.

The Senses

The survival of all animals depends on their abilities to react to changes in their environments. Receptor cells receive information about environmental fluctuations and transmit it to the CNS, which then formulates the appropriate responses that are carried out by the muscular and endocrine systems. Receptors are specialized nerve cells that convert different types of environmental energy, such as light or sound, into action potentials, the common language of the nervous system. Each type of receptor cell is most sensitive to a specific type of stimulus. Furthermore, the frequency of the neural impulses generated is in most cases directly proportional to the duration and intensity of the stimulus, and in this way quantitative information is relayed to the CNS. Sense organs are specialized combinations of receptor cells with accessory cells that protect the fragile receptors and present the stimulus in an advantageous manner.

Vision

Vision is the ability to perceive position, shape, and sometimes color by receiving wavelengths of the electromagnetic spectrum from about 4000–7000 Å. The sense of vision has evolved many times in the animal kingdom in a variety of forms. Yet however varied the visual organs appear, their basic biochemistry is similar: A photosensitive pigment (often **rhodopsin**) ab-

Figure 37-11. Left: General structure of the eye of man, showing the mechanism of sight. Right: Changes in the shape of the lens to focus on distant and near objects. The lens is normally stretched by the zonules of Zinn, which are attached to the ciliary bodies. For closeup vision, the ciliary muscles, which act like a sphincter, contract. This decreases the size of the eye; therefore, much of the tension on the zonules of Zinn is released. The somewhat elastic lens rounds out, focusing for shorter distances.

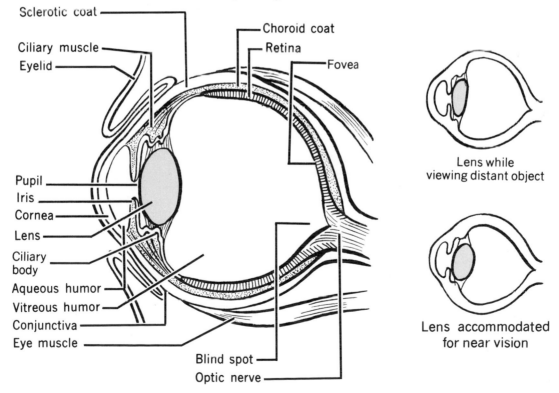

Sclerotic coat
Ciliary muscle
Eyelid
Choroid coat
Retina
Fovea
Pupil
Iris
Cornea
Lens
Ciliary body
Aqueous humor
Vitreous humor
Conjunctiva
Eye muscle
Blind spot
Optic nerve

Lens while viewing distant object

Lens accommodated for near vision

sorbs light energy and undergoes a change in molecular shape that initiates a nerve impulse; the photopigment subsequently regenerates. The various types of visual sense organs (eyes) invariably refine the image that impinges on the photopigments so that as much information as possible is extracted from incoming light and, hence, from the environment.

Imagine for a moment that the photosensitive pigment is a regenerating photographic film. Primitive light sensitivity by isolated cells containing visual pigments (for example, in the lobster tail) is akin to naked film; it detects light, but no image is formed. The planaria, with its semicircular ring of pigment cells, is capable of detecting the direction of incoming light, as would a simple box camera. The vertebrate eye would then be analogous to a very sophisticated camera, one capable of detecting light intensity, direction, shape and sometimes wavelength (color), thereby furnishing the brain with enough information to form images.

The human eyeball (Fig. 37-11), which is rather typical of vertebrate eyes, is composed of three tissue layers. The outer **sclera** is the "white of the eye." This bag of firm fibrous connective tissue surrounds the posterior five sixths of the eyeball. A transparent anterior window, the **cornea,** admits light and, by its curvature, helps to focus the light waves. Beneath the sclera lies the **choroid,** a vascular and heavily pigmented layer that nourishes the receptors and darkens the cavity of the eye, preventing reflection. A biconvex **crystalline lens** is suspended under tension from the choroid behind the cornea by a suspensory ligament. Contraction of smooth muscle fibers relaxes the suspensory ligament, allowing the lens to assume a more spherical shape. The colored part of the eye as seen through the cornea is the **iris.** This circular disc has a central aperture, the **pupil.** Muscle contraction regulates the diameter of the pupil and, hence, the amount of light that enters the eye. The color of the iris is determined by the quantity of pigment cells and has no known physiological importance as long as the iris is sufficiently opaque to block light transmission. The chamber between the cornea and the lens is filled with a clear and watery solution called **aqueous humor.**

The **retina** (Fig. 37-12) is the innermost layer and contains the receptor cells. It is held in place by a transparent semigelatinous substance, the **vitreous**

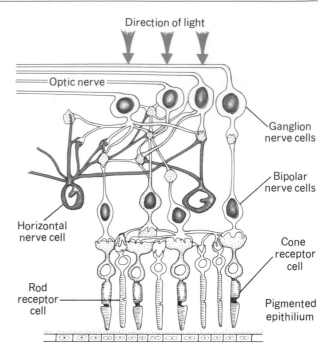

Figure 37-12. Cross-section through the human retina. Note that the actual light receptor cells (cones and rods) are the *bottom* layer of cells, so that incoming light must pass through the other cells.

humor, that fills the large chamber behind the lens. That explains why a blow on the head can cause a detached retina. It can be reattached by laser surgery, during which the retina is "welded" to the choroid coat.

The human retina contains two types of photoreceptor neurons: rods and cones. **Rods** contain **rhodopsin** and are specialized for vision in dim light; rods cannot detect color but they can interpret light intensity as various shades of gray. Cones contain three types of **iodopsin,** and these photopigments are stimulated by the wavelengths that we visualize as red, green, and blue. There are approximately 120 million rods and 6 million cones in a human retina. The cones are densely concentrated in a small depression, the **fovea,** at the exact center of the retina. This is the region of greatest **visual activity,** the ability to distinguish one object in space from a nearby object.

Vision results from the following three events: (1) An image is formed on the retina, (2) the radiant energy of this image is converted into electrical action potentials by the receptor cells of the retina, and (3) these impulses are processed and transmitted to the visual centers in the cerebral cortex.

Light reflected from nearby objects is focused on the retina by three muscular adjustments; a rounding of the lens bends the divergent light waves reflected from nearby objects sufficiently to focus them on the retina. Constriction of the pupil directs light through the central, most perfect part, of lens, and convergence of each line of sight medially achieves single binocular vision for viewing near objects.

As early as 1500 B.C., ox liver was eaten to treat night blindness (**nyctalopia**) in Egypt. Liver is high in vitamin A, and in 1931 Kathren Tansley conducted experiments that demonstrated that vitamin A is involved in vision. Extensive research has demonstrated that rhodopsin is composed of a protein (**opsin**) associated with the chromophore (the light-absorbing molecule, called **retinal** is the aldehyde of vitamin A.) Retinal can occur in two possible molecular shapes (isomers). It happens that the isomer that will bond to opsin is rather unstable. When radiant energy (mostly at approximately 5000 Å) is absorbed by a retinal molecule, it springs into another conformation and splits from the opsin. This initiates a neural discharge involving a complicated series of biochemical reactions. The rhodopsin is subsequently regenerated by rearrangement of the retinal and recombination with opsin.

Cones may contain one of three similar iodopsin pigments that differ in the wavelengths of light that they absorb. One type absorbs mainly in the blue range of the spectrum (4000 Å), another the green (5250 Å), and a third the red (7000 Å). Any light source will activate the three types of cone cells to different degrees. Nervous integration of the action potentials resulting from their selective activation produces the sensation of viewing different colors.

Rods have a lower threshold than cones and, hence, predominate in night vision. The nerve impulse travels from a rod across the synapse to a **bipolar cell;** hundreds to thousands of bipolar cells connect to a **ganglion nerve cell** (whose axon contributes to the optic nerve). In addition, there are **horizontal nerve cells** (interneurons) that multiply these impulses in the case of rods: these multiple connections increase sensitivity and are necessary for dim-light

vision. In most cones, the impulse proceeds directly from a cone cell to a bipolar cell and then to a single ganglion cell of the optic nerve without much cross-communication. This is an advantage for visual acuity because no scrambling of the image occurs. The axons of the ganglion cells converge to form the optic nerve

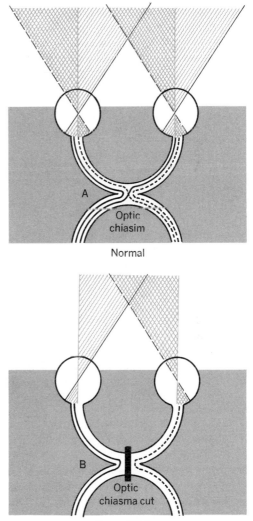

Figure 37-13. "Tunnel vision" results if the optic chiasm is cut because the fibers from the medial retinae, which receive light from the periphery of the visual field, cannot provide the brain with information.

and leave the eye as a unit. This results in the **blind spot,** as this point on the retina contains no receptor cells. Hundreds of thousands of receptors may be stimulated whenever an image falls on the retina. The resulting nerve impulses travel through the optic nerve. At the **optic chiasma** (Fig. 37-10), fibers from the nasal side of each retina cross over and continue along the opposite optic tract. Hence, a cut through the chiasma, as illustrated in Figure 37-13, would cause "tunnel vision."

As a person grows older, the lens become stiffer and accommodation for close-up vision is impaired. The lens also becomes yellowed, thus making it more difficult to see blue. This has been suggested as an explanation for the aging Michelangelo's rather heavy use of intense blue pigments in the Sistine Chapel.

Myopia (nearsightedness) is a condition in which the eyeball is too long, and so the light waves from distant objects are focused on a point in front of the retina, producing blurred and indistinct vision. **Hyperopia** (farsightedness) occurs if the eyeball is too short, causing the image to be focused behind the retina.

The aqueous and vitreous humors are secreted by the **ciliary body** and produce the **intraocular pressure** that maintains the shape of the eyeball. As new humors are produced, old ones leave the eye through the **canal of Schlemm** that circles the cornea. If this passage is blocked, increasing intraocular pressure (**glaucoma**) may close blood vessels, and the retina may die. In addition, the increased internal pressure may cause hydration of the cornea and/or lens, rendering them opaque, a condition known as **cataract.**

Astigmatism is a condition in which irregularities of the cornea or lens curvature cause light waves to be focused unevenly on the retinal surface.

Audition

Audition may be defined as the reception and interpretation of vibrations that impinge on an organism from its external environment. Fishes and larval amphibians have receptors that are stimulated by water-borne vibrations. The **lateral line system** extends along each side of bony fishes. This organ is composed of aggregations of **sensory cells, sensory hairs,** and **cupulae** that protrude from a lateral canal. When the cupulae are disturbed by pressure waves in the water,

the sensory hairs are stretched and nerve impulses are initiated in the sensory cells. Specialized receptors resembling hair cells appear on the snouts of cartilaginous fishes. These **ampullae of Lorenzini** respond to temperature, pressure, salinity (salt content), and electrical field changes. The salamander receives vibrations from the ground with its legs; the vibrations are transmitted through physical (not electrical) bone conduction up to the ear bones in the head.

The mammalian ear (Fig. 37-14) is specialized to be stimulated by air-borne vibrations. Sound waves, consisting of cycles of compressed and rarefied air, are funneled by the trumpet-shaped **outer ear,** or **pinna,** and channeled through the external **auditory canal** to the **eardrum,** or **tympanic membrane.** This membrane is an elastic sheet of tissue that separates the external auditory canal from the middle ear cavity. The eardrum bulges inward when sound waves strike it and then quickly springs back to its resting state. The distance that it is displaced is almost exactly proportional to the forces and frequencies of the sound wave.

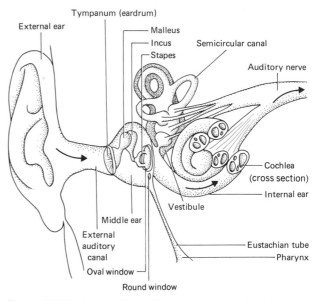

Figure 37-14. A dissection of the human ear to show the parts that have to do with receiving sound and the part that is concerned with balance.

The **middle ear** is a small air-filled cavity within the temporal bone. The tympanic membrane separates it from the external ear, and a thin bony wall separates it from the fluid-filled **inner ear.** This wall has two small membrane-covered openings, the **oval window** and the **round window.** Three tiny movable bones (**ossicles**) span the middle ear cavity, connecting the tympanic membrane with the oval window. The **malleus** (**hammer**) transmits the tympanic membrane's vibrations through the **incus** (**anvil**) to the **stapes** (**stirrup**) and from there to the oval window. The ear bones function to transform air movements from the outer ear into fluid movements within the inner ear. The anterior wall of the middle ear has a small opening into the **eustachian tube,** which communicates with the nasopharynx, allowing air pressure within the middle ear cavity to be equalized with atmospheric pressure.

The **cochlea** (Gr. *kochlias*, snail) is a coiled structure within the inner ear that contains the receptors for hearing. A thin **basilar membrane** divides the cochlea into two fluid-filled passageways. The receptor hair cells are housed in an **organ of Corti** on the basilar membrane. The rocking motion of the stapes against the oval window creates fluid pressure waves that displace the basilar membrane. This stretches the **hair cells** and initiates an action potential in the auditory nerve. Different regions of the basilar membrane are maximally displaced by different frequencies, and in this way sound waves of different frequencies (pitch) are sorted out along the length of the basilar membrane. The pressure waves are eventually dissipated by bulging out the round window. If the round window did not exist, the oval window could not be deflected as the liquid in the cochlea is essentially incompressible.

Equilibrium

The inner ear of vertebrates also house the receptors for equilibrium. These monitor any movement of the head and also the position of the head with respect to gravity.

The gravity sense organ is located in two membranous sacs, the **utriculus** and the **sacculus.** These fluid-filled sacs contain patches of specialized nerve cells with stiff, hairlike processes. Atop these hairs in each sac is an **otolith** ("ear stone") of calcium carbon-ate. When the head changes position, the otoliths move accordingly, and the movement of the hairs upon which the otoliths contact initiates action potentials in the receptor cells.

The acceleration sense organs are located in three mutually perpendicular **semicircular canals** (Fig. 37-14) that communicate with the utriculus. A swelling called an **ampulla** occurs at the end of each canal. Receptors with long hairlike processes in the ampullae are stimulated by fluid movements as the head turns in the three planes of space.

Chemoreception—Taste and Smell

Some receptors (Fig. 37-15) are specialized to be stimulated by chemicals. Such **chemoreceptors** are involved in the senses of **gustation** (**taste**) and **olfaction** (**smell**), by which we obtain information about chemicals in the external environment. The gustatory and olfactory senses act primarily to detect harmful substances, to stimulate appetite, and to initiate the secretion of digestive enzymes.

The organs of gustation are called **taste buds** (Fig. 37-15). These barrel-shaped groups of epithelial cells lie within the mucous membrane on the upper surface and sides of the tongue. Each taste bud communicates with the fluids of the mouth by a taste pore. The receptor cells are modified epithelial cells that have short hairlike processes that extend through the taste pore. The other ends of these receptor cells are surrounded by dendrites of the **gustatory nerve.** We begin life with about 10,000 taste buds, but this number decreases significantly with age.

Only water-soluble and lipid-soluble chemicals in solution can stimulate the taste receptors. Substances that can be tasted produce a change in the receptor cell that initiates a nerve impulse that is transmitted to the gustatory nerve.

The variety of the taste sensations that we experience are actually combinations of only four primary modes of sensation that are recognized: **sweet, sour, salty,** and **bitter.** Sweet and salty substances are detected on the front of the tongue, sour on the sides, and bitter on the back. The sensation of taste depends on the information supplied to the CNS from the differential activation of many taste cells, which is compared and integrated within the cerebral cortex.

The 10 to 20 million receptors for our olfactory

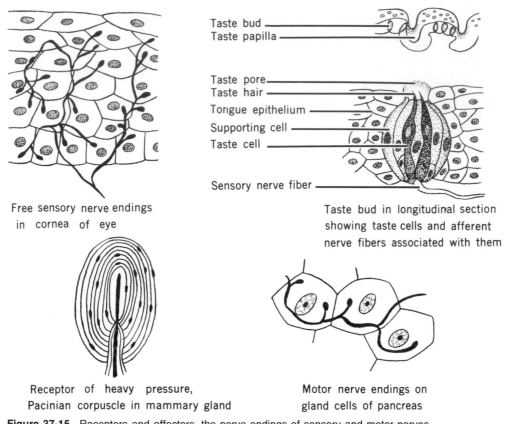

Free sensory nerve endings
in cornea of eye

Taste bud
Taste papilla

Taste pore
Taste hair
Tongue epithelium
Supporting cell
Taste cell

Sensory nerve fiber

Taste bud in longitudinal section
showing taste cells and afferent
nerve fibers associated with them

Receptor of heavy pressure,
Pacinian corpuscle in mammary gland

Motor nerve endings on
gland cells of pancreas

Figure 37-15. Receptors and effectors, the nerve endings of sensory and motor nerves.

sense lie concentrated in a small patch of mucous membrane in the upper part of the nasal cavity. Only about 2 percent of inhaled air normally passes over the olfactory receptors, and so smell is enhanced by **sniffing,** which forcefully draws air into the upper nasal passages.

The olfactory receptor cells are specialized neurons whose dendrites are modified into many hairlike filaments that enormously increase the receptive surface area. The axons of the olfactory receptors go directly to the **olfactory bulbs** in the forebrain.

Our sensation of smell results from the following conditions: A substance must release molecules that diffuse into the air. These volatile molecules must be brought into contact with the olfactory mucosa, where they must dissolve in the moist mucus that covers the receptors. At the receptor cell, the molecule depolar-izes the neuron's membrane, which initiates an action potential. The sensation of smell quickly diminishes when the stimulus is repeated; the olfactory receptors become easily adapted to the stimulus. This accounts for our ability to become accustomed to unpleasant odors.

Tens of thousands of different odor qualities can be perceived and differentiated with practice. Many attempts have been made to classify odors into primary categories (e.g., ambrosial, burnt, floral, putrid, etc.), but no underlying chemical, electrophysiological, or behavioral correlations have yet been found.

Touch

The senses of touch inform us of the numerous forces that impinge upon the exterior of our body.

This awareness is derived from receptors that are located in the inner, living layer of the skin, which is called the dermis. Specialized receptors mediate the five basic touch sensations: tactile, pressure, cold, warm, and pain.

Meissner's corpuscles are the receptors of light touch, whereas heavier pressures stimulate **Pacinian corpuscles** (Fig. 37-15). Meissner's corpuscles lie closer to the skin surface than do Pacinian corpuscles. Both types of **mechanoreceptors** are stimulated when their distal membranes are bent or displaced; the momentary change in morphology changes the permeability to selected ions, and action potentials are generated in proportion to the degree of stretch or displacement. Sensory nerves associated with body hair are also stimulated by any side displacement of the hair shaft. Specific receptors for cold and warmth, **thermoreceptors,** are also located in the skin.

Pain is a sensation that is not limited to any one receptor or type of stimulus. However, there are specialized receptors for pain located in the skin and internal organs. These receptors are branching unmyelinated nerve endings that ramify through the tissues. Any stimulus that damages the tissues stimulates these receptors. Thus pain serves as a warning that injury is occurring.

Endocrine Systems

The nervous system is intimately allied with another bodily control system, the **endocrine system.** Whereas the nervous system is very effective in quickly regulating activities of specific muscles and internal organs, the endocrine system usually acts more slowly and in a more general fashion. The endocrine system is composed of specialized tissues (**endocrine glands**) that secrete chemical messengers called **hormones.** Most multicellular glands (e.g., the liver) have ducts (e.g., common bile duct) that deliver their secretions to a particular point (e.g., the duodenum). Glands with ducts are called **exocrine glands.** The endocrine glands, however, are ductless; their hormones are secreted directly into the blood or other body fluids. As a result, these glands are highly vascularized. A hormone is a chemical coordinator that moves through the bloodstream to another part of the body, where it is effective in very small concentrations in regulating the growth or activity of its target cells. A hormone may be either excitatory or inhibitory in its influence on the activities of various tissues and the general behavior of an animal.

Hormones of Invertebrates

There is good evidence of hormones in cnidarians, platyhelminths, arthropods, mollusks, and echinoderms. The study of invertebrate hormones is a recent and growing field of investigation. The arthropods have been the most extensively studied, especially with regard to the roles of hormones in metamorphosis and molting. Endocrine glands also function in growth and regeneration of annelids, chromatophore functioning in crustaceans and cephalopods, and mating and reproductive activities in many groups. The types, locations, and functions of endocrine glands vary greatly among the different phyla.

Hormones of Vertebrates

Many investigations of vertebrate endocrine function have been conducted by observing the effects of either removing an endocrine gland or injecting hormones. Additional information has been obtained from animals with diseased endocrine glands. Other experiments have involved transplanting endocrine glands from one animal to another. Several hormones such as thyroxin have been synthesized by chemists. The hormones that are now prescribed clinically are either commercially synthesized or are isolated from the tissues of other mammals. Insulin, which allows diabetics to live normal lives, is extracted from beef pancreas.

The Human Endocrine System

The following discussion deals with the human endocrine system, but understand that the hormones of most vertebrates have similar actions. The most important endocrine glands are the thyroid, parathyroids, adrenals, pituitary, islets of Langerhans, and gonads (Fig. 37-16). Hormones are also produced by the hypothalamus, the walls of the intestine, and stomach.

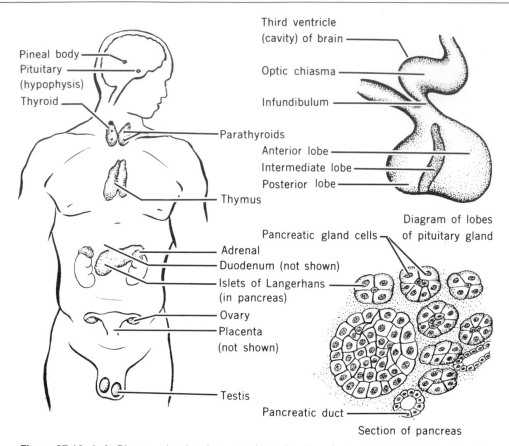

Figure 37-16. Left: Diagram showing the approximate location of some of the endocrine glands in man. Although the pineal body and thymus are included, they are not known definitely to be organs of internal secretion. Right: Section of a pituitary gland showing lobes and microscopic structure of the pancreas.

Thyroid

The thyroid gland is shaped like a bow tie and lies anterior to the trachea, just below the larynx. The thyroid secretes a hormone called **thyroxin** that regulates the overall metabolic activity of the body's cells. Thyroxin contains about 65 percent iodine by weight.

When insufficient iodine is present in the diet, the thyroid may become much enlarged, forming a **goiter.** When the thyroid fails to function properly in early life, children may become dwarfed and mentally deficient, a condition known as **cretinism.** If the thyroid of an adult atrophies or is removed, **myxedema** occurs, which results in dry, puffy skin, lethargy, slowed heartbeat, and a feeling of being constantly cold. Overactivity of the thyroid, **hyperthyroidism** or **Grave's disease,** is diagnosed by an array of symptoms that result from hyperactivity of the body's cells.

Parathyroids

The parathyroids are four small glands on the posterior surface of the thyroid. They secrete a hormone called **parathormone** that stimulates osteoclastic activity that releases calcium and phosphorous from the bones. Calcitonism inhibits osteoclastic activity that removes calcium from the blood. A deficiency of parathormone results in low blood calcium and violent twitching of the muscles, a condition known as **tetany.** An excess, as in parathyroid tumors, may cause excessive withdrawal of calcium from the bones.

Adrenals

The adrenals are two small glands, each perched on top of a kidney. Each adrenal gland consists of an outer **cortex** and an inner **medulla.** The adrenal cortex secretes about 50 known hormones that are collectively called **steroids.** One of the best known of these steroids is **cortisol,** which has anti-inflammatory properties that are used in the clinical treatment of some types of arthritis and allergies.

Injury to the adrenal cortex results in **Addison's disease,** which is characterized by anemia, low blood pressure, intestinal disturbances, and sometimes deep bronzing of the skin; this may terminate in death if not treated. Removal of the adrenal cortex results in death.

The adrenal medulla secretes the hormone **epinephrine (adrenaline),** which mimics the "fight or flight" action of the sympathetic division of the autonomic nervous system. Under emotional excitement, such as fear or anger, the sympathetic nervous system activates the adrenal medulla, causing epinephrine to be secreted. This results in an acceleration of the heartbeat and in blood vessel constriction, which raise blood pressure; in addition, blood sugar is elevated by the breakdown of glycogen in the liver. Epinephrine is clinically used in treating asthma because of its relaxing effect upon the bronchioles of the lungs.

Pituitary, or Hypophysis

The pituitary gland is about the size of a pea and is located at the base of the brain. It consists of an anterior lobe and a posterior lobe (Fig. 36-14). The **posterior lobe** stores and releases two hormones, collectively called **pituitrin,** that are actually secreted by neurons in the adjacent hypothalamus. Injection of pituitrin into the circulatory system raises blood pressure, decreases the amount of urine formed, and causes contraction of smooth muscles, especially those of the uterus and mammary glands.

The **anterior pituitary lobe** secretes six hormones that possess important regulatory activities. **Prolactin** stimulates breast development and milk production; **growth hormone** affects the development of several organs and tissues; **TSH (thyroid-stimulating hormone)** stimulates the adrenal cortex to secrete cortisol; **FSH (follicle-stimulating hormone)** stimulates the ovaries to release ova; and **LH (lutenizing hormone)** stimulates corpus luteum formation following ovulation. FSH and LH are also involved in both females and males in the control of sex hormone levels.

Excessive secretion of growth hormone in the young stimulates development of the long bones, resulting in **giantism.** When this occurs in adults, the bones, especially those in the limbs and face, become thickened, a condition called **acromegaly.**

Because the pituitary gland influences the activities of other endocrine glands, it is often referred to as the "master gland."

Ovaries

Several hormones, collectively called estrogens, are secreted by the ovary. Estrogens are secreted in the **ovarian (Graafian) follicles;** they contribute to the control of the menstrual cycle, sexual behavior, and the development of accessory genital organs and secondary sex characteristics.

Progesterone and estrogen are secreted by the **corpus luteum,** a body that forms from the area in which the ovum was released (Chapter 39). Both hormones regulate the menstrual cycle and play a part in preparing for pregnancy; they are also secreted by the **placenta** during pregnancy.

Testes

The interstitial cells of each testis secrete **testosterone.** This hormone influences sexual behavior and controls the development of male accessory genital organs and secondary sex characteristics, such as a deep voice and relatively heavy body hair. Testosterone and related substances are also produced in females by the adrenals; they influence sexual behavior in females as they do in males.

Pancreas

The pancreas is basically an exocrine gland, situated behind the stomach, that contains a million or more islands of endocrine tissue known as the **islets of Langerhans** (Fig. 36-14). These glands secrete the hormones insulin and glucagon. **Insulin** increases the uptake of glucose from the blood and its storage as glycogen in muscle and other cells. Failure of insulin secretion causes **diabetes mellitus,** characterized by high levels of glucose in the blood and sugar loss in the urine. **Glucagon** enhances the breakdown of glycogen by the liver and the release of glucose into the bloodstream. The balanced activities of these two hormones help to maintain the blood glucose concentration at a fairly constant level. Insulin can cause hypoglycemia or hyperglycemia.

Gastric Mucosa

When proteinaceous food is present, the mucous membrane located at the pyloric end of the stomach secretes a hormone called **gastrin,** which is carried in the blood to the gastric glands and stimulates the secretion of gastric juice.

Intestinal Mucosa

Parts of the mucous membrane of the intestine produce a hormone called **secretin,** which was the first hormone discovered (1902). Secretin is carried in the blood to the pancreas, where it stimulates the immediate secretion of pancreatic juices rich in bicarbonates that neutralize stomach acids. Secretin also increases the flow of bile. A second intestinal hormone, **cholecystokinin,** stimulates the emptying of the gallbladder. A third hormone, pancreozymin, stimulates the secretion of the enzyme portion of pancreatic juice.

Summary

The cnidarians are the simplest animals that possess a true nervous system, with definite receptors, conductors, and effectors. Ascending the phylogenetic tree, successively more refined nervous systems have evolved with integrating ganglia, specialized receptors, reflexes, and cephalization.

The vertebrate nervous system consists of central and peripheral divisions. The CNS integrates sensory information and formulates motor instructions for the muscles and glands of the body. The PNS conducts impulses to and from the CNS. Another way of dividing the nervous system is into somatic and autonomic branches; the former is concerned mostly with voluntary skeletal movements and the latter with involuntary visceral functions.

The autonomic nervous system is subdivided into the sympathetic division, which has mostly an excitatory preparative function, and the parasympathetic division, which has homeostatic functions that are almost diametrically opposed to sympathetic activation.

Neurons, the basic functional units of nervous systems, have three major parts: cell body (soma), dendrites, and axon. The cell body contains the nucleus; the dendrites conduct impulses toward the cell body; and the single axon conducts impulses away from the cell body. Neurons interconnect at chemical synapses. Many nerve axons are myelinated, and this increases the velocity of nerve impulse propagation.

The common language of the nervous system consists of action potentials, which are transient localized changes in electrical potential that are conducted along a neuron and from one neuron to another neuron, muscle, or gland.

The synapse is the point at which neurons come together. Release of acetylcholine or other transmitter substances from the axon of one neuron induces membrane permeability changes in the adjacent dendrite, thus initiating impulse transmission. Synaptic inhibition may also occur, either presynaptically or postsynaptically.

Neurons have characteristic thresholds, below which they cannot respond to stimulation and above which they respond all or none, by initiating an action potential. Temporal and spatial summation can facilitate depolarization of an adjacent neuron to its threshold level.

The spinal cord has a butterfly-shaped core of nonmyelinated gray matter and a coating of myelinated white matter. Motor fibers leave the cord ventrally; sensory fibers enter dorsally. The reflex arc is an example of noncortical control of movement.

The brain is generally divided into the forebrain, midbrain, and hindbrain. The multilobed cerebral cortex is responsible for thought and sensation localization and interpretation. The cerebellum oversees most of the involuntary control of locomotion. All impulses entering or leaving the brain pass through the pons and medulla, and within the medulla the ascending and descending tracts decussate. The reticular formation filters incoming sensory information and is responsible for general activation of the organism. The brainstem also contains the vital reflex center.

A sensation may be defined as the discrimination of incoming stimuli from external or internal receptors. Vision depends on molecular changes in photosensitive pigments. Light traveling into a vertebrate eye first encounters the fluid over the eye, the cornea and aqueous humor, then travels through the pupil, lens, vitreous humor, and hits the retina. Rod and cone cells in the retina contain the photosensitive pigments. Rods are sensitive to dim light, and cones to colors and daylight illumination.

Audition depends on vibrations of the surrounding medium. The vibrations are ultimately transferred to hairlike sensors that, when bent (stimulated), send action potentials to the brain. The complex auditory sense organs of vertebrates presents the stimulus so as to obtain an optimal response from the hair cells and maximize the amount of information received. The membranous labyrinth that houses the equilibrium sense organs is also part of the auditory system.

Chemoreception involves the sense organs of gustation (taste) and olfaction (smell). Receptors for touch in the dermis and internal tissues respond to touch, pressure, cold, warmth, and pain.

Hormones are chemical communicators secreted into the bloodstream by ductless endocrine glands. Minute concentrations of these chemicals affect specific target organs, either as inhibitors or as excitors. The principal endocrine glands are the thyroid, parathyroid, adrenals, pituitary, islets of Langerhans hypothalamus, and gonads. The functions of their major hormones are summarized in this chapter.

Review Questions

1. An omnidirectional nervous system implies a _____ (high, low level) of nervous system development.
2. A nervous impulse depends mostly on a change in the neural membrane's permeability to which ions? During depolarization, these ions diffuse in which directions?
3. What is the function of the canal of Schlemm?
4. The chemical responsible for most synaptic transmission is
   (a) adrenaline
   (b) acetylcholinesterase
   (c) acetylcholine
   (d) epinephrine
5. What type of inhibition causes hyperpolarization of a dendrite?
6. Is the ulnar nerve, running along the ulna in the forearm, part of the CNA or the PNS?
7. What part of the nervous system exerts its influence on the visceral organs?
8. The more the myelination, the _____ the nerve impulse.
   (a) stronger
   (b) weaker
   (c) faster
   (d) slower
9. Account for the difference between a resting potential and an action potential.
10. Where are our olfactory receptors located?
11. Where does the "blind spot" occur, and how do you account for its insensitivity to light?
12. What fills the brain's ventricles?
13. Describe the events at a synapse during transmission of a nervous impulse.
14. What are the three bones of the human ear, and what part of the inner ear does the third attach to?
15. What are the islets of Langerhans?
16. Otoliths play what role in sensation?
17. What portion of the brain links the mental and somatic functions of the organism?
18. Coordination of skeletal muscles is mostly under the control of what portion of the brain?
19. Iodopsin has to do with
   (a) goiter
   (b) the hypophysis
   (c) cones
   (d) olfaction
   (e) the sympathetic nervous system
20. Diagram the nerve tracts in the optic chiasma.
21. Describe the general morphology of the semicircular canals.

22. What effect does the sympathetic nervous system have on the adrenal glands?
23. Diagram a simple spinal reflex arc. Indicate the directions of the nervous impulses.
24. What purpose do you suppose that saliva serves in gustation?
25. The organ of Corti occurs in what sense organ? Describe its function.

Selected References

Anthony, C. P., and N. J. Kolthoff. *Textbook of Anatomy and Physiology*, 8th ed. St. Louis: C. V. Mosby Co., 1971.

Barrington, E. J. W. *An Introduction to General and Comparative Endocrinology*, 2nd ed. Oxford: Clarendon Press, 1975.

Beck, L. H., and W. R. Miles, "Some Theoretical and Experimental Relationships Between Infra-red Absorption and Olfaction," *Science*, **511**, p. 547, Nov. 28, 1947.

Brady, R. O. (ed.). *The Basic Neurosciences*. New York: Raven Press, 1975.

Burdette, W. L. (ed.). *Invertebrate Endocrinology and Hormonal Heterophylly*. New York: Springer-Verlag, 1974.

Case, J. *Sensory Mechanisms*. New York: Macmillan Publishing Co., Inc., 1966.

Communications Research Machines. *Biology Today*. Del Mar, Calif. CRM Books, 1972.

Constantinides, P. C., and N. Carey. "The Alarm Reaction," *Scientific American*, **180**:20, 1949.

Crouch, J. E., and J. R. McClintic. *Human Anatomy and Physiology*. New York: John Wiley & Sons, Inc., 1971.

Funkenstern, D. H. "The Physiology of Fear and Anger," *Scientific American*, **192**:74, 1955.

Gardner, E. *Fundamentals of Neurology*, 5th ed. Philadelphia: W. B. Saunders Co., 1968.

Goldenson, R. M. *The Encyclopedia of Human Behavior*. Garden City, N.Y.: Doubleday & Co., 1970.

Guyton, A. C. *Structure and Function of the Nervous System*. Philadelphia: W. B. Saunders Co., 1972.

Henning, H. *Der Geruch*. Leipzig: Barth, 1924.

Huxley, H. E. "The Contraction of Muscle," *Scientific American*, **199**:66, 1958.

King, B. G., and M. J. Showers. *Human Anatomy and Physiology*. Philadelphia: W. B. Saunders Co., 1969.

Laird, D. A. "How the Consumer Estimates Quality by Subconscious Sensory Impression," *J. App. Psych.*, **16**:241–246, 1932.

Lentz, T. L. Primitive Nervous Systems. New Haven: Yale University Press, 1968.

Nelson, G. E., G. G. Robinson, and R. A. Boolootian. *Fundamental Concepts of Biology*, 3rd ed. New York: John Wiley & Sons, Inc. 1974.

Parker, G. H. *Animal Colour Changes and Their Neurohumours*. London: Cambridge University Press, 1948.

Romer, A. S. *The Vertebrate Body*, 2nd ed. Philadelphia: W. B. Saunders Co., 1960.

Sarnat, H. B., and M. G. Netsky. *Evolution of the Nervous System*. New York: Oxford University Press, 1974.

Storer, T. I., and R. L. Usinger. *General Zoology*, 4th ed. New York: McGraw-Hill Book Co., 1965.

Stover, T. I., R. L. Usinger, and J. W. Nybakken. *Elements of Zoology*. New York: McGraw-Hill Book Co., 1968.

Tower, D. B. *The Nervous System*. New York: Raven Press, 1975.

Van Bergeijk, W. A., J. R. Pierce, and E. E. David, Jr., *Waves and the Ear*. Garden City, N.Y.: Doubleday Anchor, 1960.

Genetics and Heredity

So far we have studied the anatomy and physiology of animal cells, tissues, organs, and, finally, organisms. We have been exposed to the overwhelming diversity in animal form and behavior. Our attention has been focused on the physiological systems that, when summed, form the organism. We are now ready to review the genetics underlying the similarities and differences of living creatures.

Chromosomes and DNA

We have come a long way since believing that the similarities and differences between members of a given species are determined by "blood lines." In 1869, the German physiologist Friedrich Miescher found that a major component of chromosomes is **deoxyribonucleic acid (DNA).** At that time, the structure and function of DNA were not known, but further research led to some intriguing results. It was discovered that, for a given organism, the amount of DNA in almost every cell is constant. Eventually, it was found that our distinguishing characteristics, or traits, are determined by DNA, not by the versatile proteins, as was suspected earlier. In 1944, Avery, Macleod, and McCarty showed that one type of pneumonia-causing bacterium could be transformed into another type by treating a growing culture of the first with purified DNA extracted from the second. Then in 1952 Hershey and Chase demonstrated that a bacteriophage (a virus that parasitizes bacteria) infects *Escherichia coli,* a common intestinal bacterium, by injecting only its DNA into the host cell. When infected, the bacterium produces complete bacteriophages, with DNA, protein, and other structural components, until it burst.

In 1953, J. D. Watson and F. H. C. Crick proposed a "double helix" structure for the DNA molecule, which led to a rapid understanding of its basic function. DNA, as the key molecule in heredity, must contain the information for its own reproduction (replication) and for the biosynthesis of the other components of the cell and organism as well.

A **chromosome** consists of an extremely long molecule of supercoiled DNA that is complexed with RNA and proteins (Chapter 2). The twisted DNA molecule would look something like a ladder if it were uncoiled (Fig. 2-5). The "sides" of the ladder are composed of alternating **deoxyribose sugars** and **phosphate** groups. Attached directly to each sugar is one of four **nitrogenous bases.** Each phosphate–sugar–base subunit is called a **nucleotide.** The four nitrogenous bases in DNA are **adenine (A), thymine (T), guanine (G),** and **cytosine (C).** These molecules are lined up in pairs to form the hundreds of thousands of "rungs" of the DNA ladder. The chemical structure of the bases allows only certain combinations to match, that is, to form hydrogen bonds (Chapter 2 and Fig. 2-8). Thus, adenine can bond only with thymine and cytosine only with guanine. This pair specificity is the basis for the genetic code, which directs biosynthesis within the cell (discussed in the next session).

The double helical structure of the Watson–Crick model suggested that new DNA molecules form directly, with each strand acting as a template for a new "daughter" strand. Each new DNA molecule contains

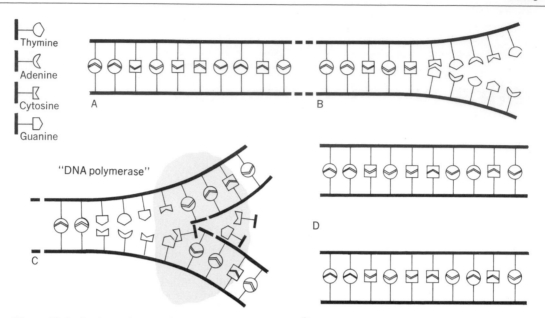

Thymine

Adenine

Cytosine

Guanine

"DNA polymerase"

A

B

C

D

Figure 38-1. Basic mechanics of DNA replication. (A) Diagrammatic section of the DNA chain, uncoiled for clarity. (B) As the DNA chain "unzips," deoxyribonucleotide triphosphates (containing A, T, C, and G) from the surrounding medium link up with their counterparts. (C) As the "unzipping" process continues, we see that two daughter chains, identical to the mother chain, are being formed, with the aid of a large complex of enzymes that unwind the double helix and catalyze the addition of new nucleotides. Here they are shown schematically as "DNA polymerase." (D) Completed daughter chains. Compare with part A. See text for more detailed description.

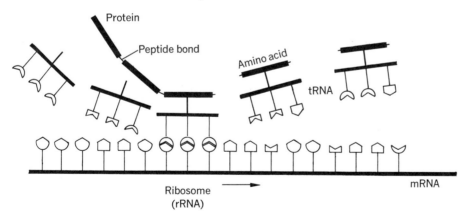

Protein

Peptide bond

Amino acid

tRNA

Ribosome
(rRNA)

mRNA

Figure 38-1. Protein synthesis. Note that each tRNA is specific for a given amino acid, and that as the protein is formed, the tRNAs return to the cytoplasm to repeat their work.

one of the parental strands and its newly synthesized complement. Figure 38-1A represents a small portion of an intact DNA double helix; as the chain starts to "unzip," the relatively weak hydrogen bonds between the A-T and C-G nucleotide pairs are broken. The uncoiling and separation of the double helix is accom-plished by several enzymes working together. Two stretches of single-stranded DNA are now exposed, and unattached nucleotides within the nucleoplasm now hydrogen bond to the nucleotides on each single strand, again aided by an enzyme (**DNA polymerase**). Note that a free nucleotide can bond only with its complement on the DNA strands: A with T, T with A, C with G, and G with C. Replication is complete when two equivalent DNA molecules are formed.

The sequence of replication assures that the integrity of parental DNA chains is preserved; an equivalent double strand is passed on to each daughter cell during mitosis. This form of duplication was called semi-conservative because each new DNA molecule has one strand derived from the parent bonded to a newly synthesized complementary strand.

The requirements for DNA synthesis were experimentally documented by Kornberg and his associates in 1956. They synthesized DNA *in vitro** by mixing a small amount of native DNA with unattached nucleotides of adenine, cytosine, guanine, and thymine plus an enzyme that they had isolated and called DNA polymerase. The new DNA had the same composition of bases as did the initial DNA that provided the template for synthesis. It is highly probable that *in vivo* **DNA synthesis is generally similar to this** *in vitro* synthesis, although today it is know that the enzyme that Kornberg isolated was not the primary DNA polymerase but, rather, a repair enzyme. The polymerization reaction yields the same result, using either polymerase.

Further support for the general Watson–Crick model was provided in 1963 by Cairns, who demonstrated new growth in the chromosome of *E. coli*. Since those early years in the field called molecular biology, the advances in our understanding DNA has been tremendous. During the 1970s, a biological revolution, equivalent to the Watson–Crick discoveries, started and is still continuing. With the discovery of hundreds of enzymes that can cut, polymerize, or link together pieces of DNA *in vitro* with extreme specificity, the techniques of recombinant DNA have resulted. These techniques allow different genes or other pieces of DNA from any organism (including humans) to be isolated from their source organism, grown in bacterial cells, and studied in the laboratory.

DNA, RNA, and the Genetic Code

Why is it so important that DNA maintain its sequence of nucleotides during replication?

*In vitro means inside a test tube; *in vivo* means inside a living organism.

Recall that the frameworks of all cells and organelles are made of different proteins. Proteins are also important as metabolic enzymes. Proteins are generally long polymers of hundreds to thousands of amino acids arranged in a linear series. The specific sequence of amino acid subunits determines the particular characteristics of each protein. The synthesis of a protein is entirely dependent upon the sequence of bases (nucleotides) in the DNA molecule. The major function of the DNA (which comprise the genes) and almost all the various RNA molecules (made from some of these genes) is to arrange amino acids during **protein synthesis** in a sequence that forms a precisely determined protein.

It is obvious that only four bases (A, T, C, and G) cannot individually code for the 20 amino acids used to build proteins. It has been experimentally determined, however, that sequences of three bases form the genetic code (Table 38-1) and that each triplet of bases (**codon**) codes for a specific amino acid or acts as a terminator, signifying the end of the protein.

RNA Synthesis

Chromosomal DNA occurs only in the **nucleoplasm** (in eucaryotes), yet protein synthesis occurs in the **cytoplasm**. **Ribonucleic acid (RNA)** acts as an intermediary in protein synthesis. There are three types of RNA: **messenger RNA (mRNA), transfer RNA (tRNA),** and **ribosomal RNA (rRNA).** All three types are very similar to DNA, with the following notable exceptions:

1. The sugar in DNA is deoxyribose, whereas the sugar in RNA is **ribose** (which has one more oxygen that does deoxyribose).
2. In DNA, adenine complements thymine, whereas, in RNA, adenine pairs up with uridine (U).
3. DNA replicates itself, whereas RNA is usually formed by synthesis from a DNA template (except in certain viruses).
4. Different enzymes catalyze DNA and RNA syntheses.

All types of RNA are synthesized using one strand of an "unzipped" DNA molecule as a template. This synthesis proceeds just as in DNA replication (Fig. 38-1), but only one strand is copied. Not the entire

Table 38-1. The Genetic Code

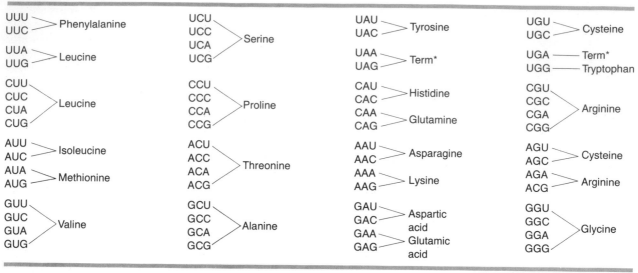

* Terminating codon.

DNA molecule, only the region necessary, of course, is copied at one time; this ranges from only about 100 bases for tRNAs to tens of thousands for the mRNA for large genes. Even though there is only one strand, almost all RNA molecules fold back on themselves to form short stretches of double helix. RNA synthesis is called **transcription** because a complementary copy of the DNA's nucleotide sequence results. When the transcription is complete, the RNA leaves the DNA template as a unit, is modified by a special set of enzymes (see the later discussion on regulation), and moves into the cytoplasm. Synthesis of tRNA and rRNA also involves transcription from DNA. The three sizes of rRNA molecules (120, 2000, and 4000 bases each) are assembled with many different proteins subunits that together migrate into the cytoplasm, where they finish assembling into **ribosomes.** The tRNA molecules spontaneously fold into their proper three-dimensional shape, but they too are modified by special enzymes.

Protein Synthesis: RNA Translation

Extensive research has deciphered the genetic code, revealing the base triplets that code for each amino acid (Table 38-1); these triplets are read from the mRNA. Note the **redundancy** in a code; several different codons may code for the same amino acid. For example, the simple amino acid glycine (Fig. 2-9) may be coded by the following codons: GGU, GGC, GGA, and GGG.

Protein synthesis occurs in the cytoplasm in several complicated steps (Fig. 38-2). Basically, each tRNA delivers an amino acid to the mRNA, whose sequence of codons orders the sequence of amino acids during protein synthesis. Because of the large sizes of these macromolecules, the individual molecules must be brought together in the proper orientation by the ribosome, which is acting analogously to an enzyme (Chapter 2) in bringing together the RNAs; actually many enzymes are part of the ribosome.

Once a tRNA molecule has been synthesized in the nucleus, it moves into the cytoplasm and bonds to an amino acid. There is a different tRNA for each codon of the genetic code, and each has an exposed triplet of nucleotide bases, the **anticodon,** that can potentially bond to a complementary codon on an mRNA molecule a ribosome.

Recalling the redundancy of the genetic code, there are 61 different RNAs (64 possible triplets

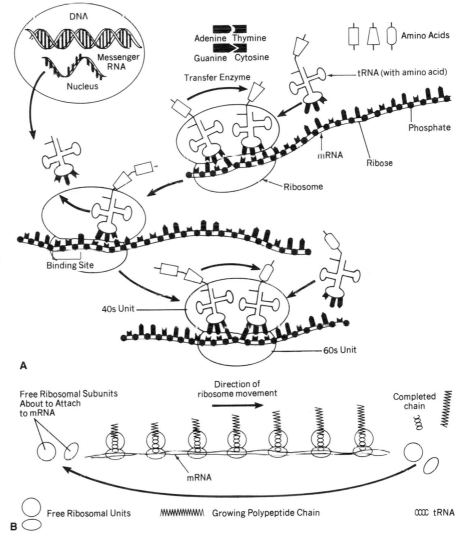

Figure 38-2. Protein synthesis (translation). (A) Schematically, the steps in joining amino acids together by the matching of tRNA–amino acid complexes to the mRNA, according to the genetic code. Note that, after the first amino acid is transferred to the second amino acid (still attached to its tRNA), the first tRNA leaves the ribosome. The ribosome then moves exactly three bases to the next codon, and the process is repeated as shown. (B) As the ribosome moves down the mRNA, the polypeptide chain grows; more than one ribosome may be translating at the same time on one mRNA—the polyribosome. At the end of translation, the ribosome falls apart into two subunits (called 60s and 40s, denoting their relative sizes); the subunits and tRNAs are recycled to make more proteins. The rRNAs are complexed with many ribosomal proteins, which together make up the ribosomal subunits.

minus the three **terminators**). The proper tRNA is joined to the appropriate amino acid by an important set of enzymes (aminoacyl-tRNA synthetase). There is one enzyme for each amino acid; for example, the enzyme for glycine will bind only to glycine and any of the four tRNAs that have the anticodons specifying glycine. It is important to note that these enzymes are the actual translators of the genetic code and that if they make a mistake, the wrong amino acid will be joined to the tRNA and will result in a mutation in the protein.

An mRNA molecule attaches at one end to an rRNA molecule in the ribosome; this ensures that the ribosome is positioned at the right place on the mRNA. Then a tRNA molecule with the appropriate anticodon for the first amino acid attaches to the proper codon farther down on the mRNA. The codon–anticodon triplets coincide, and the first amino acid is in place (Fig. 38-2). Then a second tRNA, complementary to the next codon on the mRNA, falls into place, delivering the second amino acid. A peptide bond forms between the adjacent amino acids, and the first two amino acids of the protein chain are complete. Because the first amino acid left its tRNA to bind to the second amino acid, the "empty" first tRNA leaves the ribosome. The ribosome then moves along the mRNA molecule exactly three bases, so the second tRNA is in the same position on the ribosome that the first tRNA previously occupied. The third tRNA amino acid is delivered and another peptide bond formed. This occurs over and over as the protein chain grows. Nearly all codons code for amino acids, but some—the three **terminating codons** (Table 38-1)—code for the cessation of synthesis. A specific protein recognizes the terminators and causes the separation of the new protein, ribosome, mRNA, and last tRNA.

The Regulation of Gene Expression

The conversion of genetic information within the nucleus of a zygote into an extremely complex and dynamic organism is one of the most amazing and least understood processes of life. The past two decades of research in molecular biology and genetics have revealed much about these **regulatory mechanisms,** especially in the simpler bacterial cells. Here we see a complicated interplay of regulatory proteins and nucleic acids, allowing genes to be turned off and on only when appropriate to the needs of the cell. For example, how does a bacterial cell produce the enzyme, or the set of enzymes, needed to metabolize a particular nutrient only when this nutrient is present in the environment? A particular protein molecule called a **repressor** is able to bind to DNA at a particular nucleotide sequence next to the genes for the enzymes in question and to prvent the RNA polymerase from transcribing the genes into mRNA (genes are turned off). If the nutrient appears, it is capable of binding to its repressor, thus changing its shape. The nutrient is acting as an **inducer,** and the **inducer–repressor complex** is not able to bind to DNA, thus allowing the mRNA and enzymes to be synthesized. The region of DNA that includes the genes for a set of enzymes and the adjacent regulatory nucleotide sequences (some that bind a repressor, others that bind

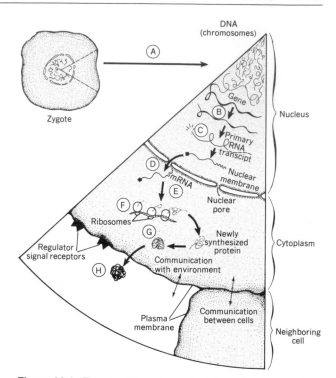

Figure 38-3. The regulation of genetic information processing in an animal cell. In the fertilized zygote (A) the DNA of a cell may undergo rearrangements that affect the kind of cell it will become. Throughout development and during normal activities of the differentiated cell, the bordering cells and environment communicate with a cell by means of molecules that interact with special receptors on the membranes or that pass directly through membranes. Regulatory signals come in a variety of forms, from simple salts, to hormones, to special proteins that interact with genes in the nucleus. (B) A specific gene is chosen to be transcribed; it is first made as a long "primary transcript." (C) This transcript is processed during a complicated series of steps that cut away excess RNA and chemically modify and add other constituents. (D) The transport of mRNA out of the nucleus is also a process under control. (E) Even in the cytoplasm, the translation of mRNA into proteins does not always occur immediately, but further processing or regulatory signal is often needed. (F) The rate of translation (*e.g.,* many ribosomes on a mRNA polyribosome) may be regulated by a number of different mechanisms. (G) The newly synthesized protein is often not immediately active, but must be chemically modified (*e.g.,* the addition of coenzymes or cofactors) or combined with other polypeptide chains to form multisubunit proteins (*e.g.,* hemoglobin, DNA, or RNA polymerases). (H) Many proteins are excreted from the cell, a process also regulated (see Chapter 2).

RNA polymerase) is called the **operon.**

We understand the details of regulation, such as in this example, in relatively few bacterial genes of only a few bacteria; the variety and intricacies of these regulatory systems are astounding. Recent investigations of gene regulation in animal cells reveals even greater complexity. The importance of regulation may be illustrated by the fact that the nucleotide sequence of DNA for chimpanzees is over 96 percent identical to that of humans; the differences are primarily in the regulation of the expression of genes during development.

The general features of genetic information processing and the many levels of regulation in a typical animal cell are shown in Figure 38-3. A recent and particularly surprising discovery is that, in all the eucaryotic genes examined in detail to date, they are broken up into nucleotide sequences that are separated by nucleotides that do not code for amino acids. As discussed earlier, the mRNA must have a continuous series of triplet nucleotide codons to be properly translated into a protein. The mRNAs isolated from eucaryote cells do not contain these noncoding spacer sequences; the primary RNA molecule that is transcribed from the gene does contain spacers, but they are cut out by certain enzymes during **RNA processing** (Fig. 38-3). This process explains some of the regulatory problems of immunoglobulin genes (see Chapter 35): The different V and C regions of the protein are coded for by many sections of DNA organized in separate clusters along the DNA of a chromosome. A long primary Rna is transcribed, and, during RNa processing, the correct combination of pieces are put together to yield an mRNA with a single continuous V–C series of nucleotides. Also important is the reorganization of genes that occurs during the differentiation of a B cell, which is the process that helps "choose" the type of immunoglobulin that a particular cell will make. Researchers working on immunoglobulin genes believe that similar DNA recombination events may be important in the regulation of a number of other genes during development (Fig. 38-3).

Recent trends in evolution theory stress the importance of mutations (discussed in later sections) in regulatory processes, for they are probably the major causes of the evolution of new traits. For example, the human–chimpanzee differences may simply be the result of regulatory mutations that alter the structures and quantities of cells and tissues and, hence, the morphology of the organism.

Genes and Chromosomes— Some Genetic Terminology

Somatic cells contain two sets of **homologous chromosomes.** Each chromosome of a homologous pair contains genes for similar types of proteins, and the linear positioning (**loci**) of the genes is extremely similar on each homologous chromosome. Alternative forms of genes at the same locus on homologous chromosomes are called **alleles.** Each allele may encode for a slightly or markedly different protein, or for no protein, as a result of mutations (discussed in the next section). An individual is **homozygous** for a trait when both the alleles that control the trait are identical. An individual is **heterozygous** for a trait if the two genes at corresponding loci on homologous chromosomes are different. An organism is invariably homozygous for many pairs of genes and heterozygous for many others; the percentage of heterozygous genes in an organism indicates the amount of genetic diversity available.

The two different alleles that form a heterozygous pair may be of two types, dominants and recessives. An allele is **dominant** when the gene is expressed in the heterozygote, that is, when a protein is made that determines the development of the trait. The other allele that is not expressed, or produces an ineffective protein, is said to be **recessive.** For example, in the fruit fly, *Drosophila*, the gene for red eyes is dominant and that for white eyes is recessive. The gene for white eyes is expressed only when it is homozygous. The result of dominance and recessiveness in allelic genes is that individuals do not express all the genes that they inherited from their parents. It is important to note that most "**traits**" (such as height) are the result of the actions of hundreds or thousands of proteins and, therefore, genes. Often, however, a particular enzyme is crucial to the development the trait and is thus a primary determinant of the final form of the trait.

When we say that we inherit certain traits, we mean that it is the developmental potentialities (the genes) that are inherited, never the developed traits. We do not inherit our father's or mother's nose, but we may inherit the genes that code for proteins used to construct theirs. The genotype is the particular set of alleles that are present in every somatic cell of an organism. In sexually reproducing organisms, half the genotype is inherited from each parent—one set of homologous chromosomes in each gamete. The **genotype** includes the sum of all the dominant genes, whose traits are expressed in the individual, and the recessive genes, whose expression is masked in the heterozygous condition. The **phenotype** of an individual is the total result of the expressed traits. For example, a person's genotype may include both a dominant allele for black hair and a recessive allele for blond hair, but the phenotype will be black hair.

In **somatic (autosomal) cells** (all cells in an organism that are not sex cells), **mitotic cell division** yields two genetically identical cells (Fig. 2-26). Each daughter cell contains the same number of chromosomes and the same genes as the parent somatic cell. **Meiosis,** the division of **sex cells** to produce **gametes,** differs in four major respects from mitosis:

1. Whereas mitosis yields two daughter cells, meiosis results in four cells.
2. The products of mitosis have genetic complements identical to that of the mother cell, but the gametes produced by meiosis contain only half the original chromosomal complement.
3. In mitosis there is only one division, but in meiosis there are two: the first is a **reduction division** in which the number of chromosomes is reduced by one half and the second is an equational division.
4. During the initial condensation of chromosomes, the homologous chromosomes are brought together before lining up at the **metaphase plate.** The side-by-side juxtaposition of homologous chromosomes allows **crossovers** (the exchange of parts of chromosomes) to occur; this does not happen in mitosis.

Meiosis of sex cells to produce spermatozoa and ova will be described more fully in the next chapter. Of major importance to heredity is the second reduction division of meiosis. The **diploid,** or **2n,** number of chromosomes in somatic cells is reduced by one half:

each gamete is haploid. When two haploid gametes fuse during **fertilization,** the diploid chromosome number is reinstated.

Alteration of Genes: Mutations

A **mutation** is a change in the sequence of nucleotides in chromosomal DNA or a change in the arrangement of genes that alters its expression. Mutations are normally random events that occur at very low (but measurable) frequencies. The possible results of genetic mutation are various. For one, no change at all may result due to codon redundancy (e.g., UUU to UUC—both code for phenylalanine). Or protein synthesis may be prevented or incomplete; a particular protein or set of proteins may be produced continually, even if unnecessary; or a different protein than originally coded may be synthesized.

The long-term results, as far as the individual organism is concerned, may be minimal, may cause disease, may be fatal, or (infrequently) may be beneficial. If a different protein is coded, the new protein may either be usable or not. Premature synethsis termination may have far-reaching effects. If a particular protein is not formed, a necessary cellular constituent may not be made, resulting in either impaired function or death. For example, a biochemical pathway leading to the anabolism or catabolism of a particular substance may be blocked (usually because a necessary enzyme is not synthesized). Mutations in control regions of DNA will not alter the amino acids of a protein but may have drastic consequences by altering the regulation of genes. In fact, it is now believed that regulatory mutations are probably the major source of evolutionary change, as most changes during evolution do not need new proteins but, rather, are changes in the organization and regulation of existing proteins.

Mutations that occur within the sex cells (gametes) or their precursors are passed on to future generations, and this is the source of the genetic variability upon which natural selection operates.

Missense Mutations

A **missense** mutations involves alteration of a single base pair on the DNA molecule. Here we'll concern ourselves with only the template strand of the DNA, although the mutation affects the complementary strand as well. Let us assume that part of the DNA had the following bases on the template strand*:

$$\cdots \; CG, \; \textbf{TTT}, \; C \; \cdots$$

The mRNA transcribed from this template would read:

$$\cdots \; GC, \; \textbf{AAA}, \; G \; \cdots$$

and the codon in bold type would code for lysine (see Table 38-1). A missense mutation might alter the sequence so that the DNA would read:

$$\cdots \; CG, \; \textbf{CTT}, \; C \; \cdots \qquad \text{(alteration: T to C)}$$

The mRNA would then read:

$$\cdots \; GC, \; \textbf{GAA}, \; G \; \cdots$$

and a different amino acid, glutamic acid, would be coded. Glutamic acid would replace lysine during synthesis of the protein coded by this gene. A missense mutation is the cause of abnormal hemoglobin in sickle cell anemia (Chapter 37).

Nonsense Mutations

A **nonsense mutation** is similar to a missense mutation, but it results in the production of a terminating codon. Both missense and nonsense mutations involve the substitution of one base pair in DNA for another, but a nonsense mutation results in premature termination of protein synthesis.

Example:

Original DNA	. . .	GC, **ATA**, CC . . .
Original mRNA (code: tyrosine)	. . .	CG, **UAU**, GG . . .
Mutated DNA (change: A to T)	. . .	GC, **ATT**, CC . . .
Resulting mRNA	. . .	CG, **UAA**, GG . . .

(UAA is a terminating codon)

* The normally continuous sequence of nucleotides is shown as groups of triplets in order to reveal the genetic code.

Deletion Mutations

Deletion mutations involve the omission of one or more nucleotides from a gene. If the number of bases deleted is a multiple of three, the result is synthesis of a shorter protein that lacks the amino acids that were coded by the missing codon(s). Deletions that are not multiples of three are frameshift mutations:

Original
DNA   · · · AAA, GTT, CTT, TGA, TCT   · · ·
Mutated
DNA   · · · AAA, TGA, TCT   · · ·

(GTTCTT have been deleted, and the protein will lack glutamine and glutamic acid.)

Insertion Mutations

An **insertion mutation** involves the addition of one or more base pairs to a gene and may change amino acids in a manner analogous to deletions. This may add an extra amino acid if three base pairs are inserted, or it will result in "frameshift" if some nonmultiple of three base pairs is inserted. Insertion of a termination codon will result in premature termination of RNA translation.

Frameshift Mutations

Frameshift mutations involve deletions or insertions of one, two, four, five, or other nonmultiples of three base pairs. The result is gibberish: the mRNA codes for no utilizable protein past the mutation site. It is called frameshift because the "frame" of three base pairs shifts at the point of deletion or insertion, resulting in a new set of codons that bind sequentially to different tRNAs, thereby upsetting the amino acid sequence.

For example, suppose that the original mRNA coded for

$$UUA, \; GUU, \; ACU, \; UAC, \; UUA, \; U \; \cdots$$

If a deletion of the sixth base (A) in the DNA sequence occured, the boldface U would be deleted, and the frame would shift, resulting in

$$UUA, \; GUA, \; CUU, \; ACU, \; UAU \; \cdots$$

Originally, leucine, valine, threonine, tyrosine, leucine, . . . , were coded. The frameshift mutation yielded leucine, valine, leucine, threonine, tyrosine, . . . , and so on.

Metabolic Pathway Blocks

The result of gene mutation may be blockage of a metabolic pathway, implying that the synthesis of a functioning protein (enzyme) has been prevented because of detrimental amino acid changes, insertions, or deletions. The first indication of such blocks was recognized by Garrod, an English physician, in 1902. Labeling it an "inborne error in metabolism,"

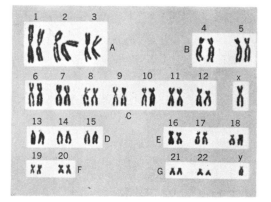

Figure 38-4. Normal human karyotype. Is this a male or a female?

Garrod found that the disease **alkaptonuria** was due to a genetic defect. In this disease, **homogentisase,** an enzyme necessary for the breakdown of **homogentisic acid** (a part of the biochemical pathway for the use of phenylalanine), is not synthesized. The acid consequently accumulates in the blood, is excreted in the urine, and turns the urine black upon contact with air (it becomes oxidized at this point, transformed into alkapton).

Albinism is another result of genetic mutation. Tyrosine is normally transformed into **dopamine,** and the reaction is catalyzed by the enzyme **tyrosinase.** Dopamine is further converted into **melanin,** a skin pigment that albinos lack. The error here, caused by genetic mutation, is that tyrosinase is not made by the body; hence, dopamine is not made, and the skin cells synthesize no pigment.

Some other diseases with genetic etiologies are listed in Table 38-2.

Chromosomal Abnormalities

Thus far, we have seen the mutations that affect individual nucleotides within genes. If the position of a gene on a chromosome (locus) is altered, the expression of a trait may be affected. Thus mistakes that occur during the segregation of crossing-over of chromosomes are also considered mutations.

A **karyotype** is a photographic representation of all the chromosomes in a somatic cell of an organism. A normal human karyotype is shown in Figure 38-4. The

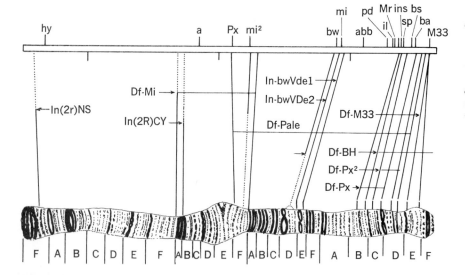

Figure 38-5. (A) Stained *Drosophila melanogaster* chromosome. Note the obvious bands. Upon close inspection, chromosome puffs may be seen. 384 X. (B) The chromosome (or linkage) map of a section of a *Drosophila melanogaster* chromosome. The chromosome section is drawn as it appears with appropriate staining methods.

22 pairs of human autosomes and 1 pair of sex chromosomes are differentiated morphologically into seven groups, labeled A through G. The groupings are made on the basis of relative size and centromere position. Karyotypes are useful tools in detecting chromosomal abnormalities. Further refinements include the **fluorescent karyotype** (which indicates certain characteristic **bands** on the chromosomes) and the **trypsin-stained karyotype** (which highlights the bands even better); these methods allow positive identification of each chromosome type. Figure 38-5 shows the chromosome bands of *Drosophila melanogaster*.

Figure 38-6. Karyotype of Down's syndrome (trisomy 21). Note the three number 21 chromosomes, making the total chromosomes number equal to 47.

Trisomies

A major type of chromosomal mutation is a **trisomy,** in which three instead of two chromosomes of a given type are present in each somatic cell. One relatively common trisomy is **trisomy 21,** otherwise known as **Down's syndrome** or "mongolian idiocy" (Fig. 38-6), which involves inheritance of an extra chromosome 21. Down's individuals thus have 47 chromosomes instead of the normal 46. They are mentally retarded and have a characteristic appearance.

Trisomy E people have three number 18 chromosomes (which appear in the E group, Fig. 38-7). Sixty

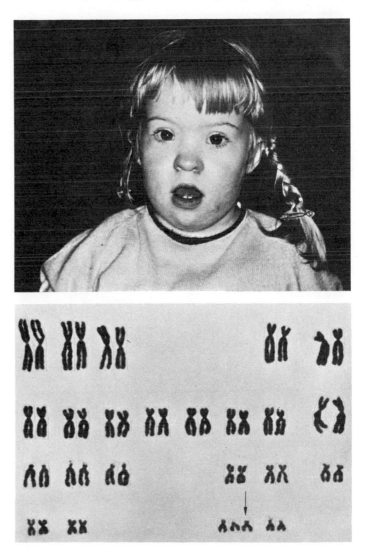

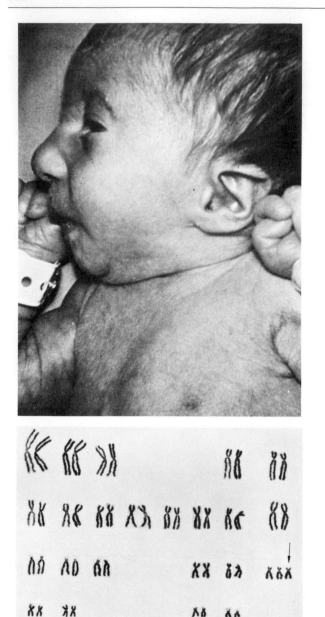

Figure 38-7. Trisomy E (trisomy 18) syndrome. Note large occipital lobe and extra number 18 chromosome.

the third. Trisomy E people are subject to heart defects. Of the 1 in 3000 live births of these individuals, approximately two thirds are female.

Trisomy D occurs in approximately 1 in 5000 live births. Characteristics include **cleft palate, polydactyly** (extra fingers), cardiac abnormalities (about 80 percent exhibit this trait), and premature death (70 percent die when they are approximately five to six months old.

Trisomies in many chromosome groups, such as A, B, and C, result in spontaneous abortion; consequently, living persons are not observed with such mutations. The most commonly accepted explanation

to 70 percent of these individuals die before the age of three months. They have enlarged occipital lobes, small jaws, and frequently the second finger overrides

Table 38-2. Inheritance modes of some human traits

Trait	Domi-nant	Reces-sive	Several Genes
Structural			
Brown eyes			●
Blue or grey eyes			●
Premature greyness of hair	●		
Brachydactyly	●		
Skin color			●
Albinism		●	
Polydactyly	● R-P		
Split hand "lobster claw"	●		
Extra teeth	●		
Physiological			
A-B-O blood groups			● M-A
Ability to taste PTC (phenyl thiocarbamide)	●		
Psychological			
Huntington's chorea	● R-P		

Table 38-2. (continued)

Trait	Domi-nant	Reces-sive	Several Genes
*Diseases**			
Alkaptonuria		●	
Red-green color blindness		● S-L	
Deaf-mutism		●	●
Nearsightedness (myopia)	●	●	
Absence of iris (anirida)	●		
Bleeding (hemophilia)		● S-L	
Cancer of eye (retinoblastoma)	● R-P		

Table of some human traits and their usual mode of inheritance. Key to the symbols. ● = inheritance of trait; R-P = reduced penetrance; S-L = sex-linked; M-A = several alternative genes; if more than one mode of inheritance is given for a trait, it means that in some families it is inherited in one way and in others in another.

*This is an artificial classification, because these diseases actually fall into structural or physiological categories.

for trisomies is that the chromosomes did not separate during the first meiotic division, a condition called **nondisjunction,** so that one of the gametes contained both homologous chromosomes instead of a single one, and another gamete received no chromosome at all.

Translocations

Translocations are the result of the **crossing-over** (Fig. 38-8) of homologous chromosomes during meiosis; the exchange of chromosome pieces is usually equal and results only in the exchange of alleles. Sometimes, the crossover is not equal, so that one chromosome receives more DNA and the other loses some. This, in fact, is the primary mechanism for producing the insertions and deletions that have already been described. Obviously, this type of mutation can cause dramatic effects on the expression of genes, and it is probably also required for the evolution of new genes.

Translocation sometimes occurs in conjunction with trisomy. For example, in Down's syndrome it is frequently observed that the number of chromosomes is 46 (normal) but that the individual still exhibits trisomy 21 traits. A close examination of the karyotype in Figure 38-9 will reveal an unusual-looking chromosome number 15 in the D group. What has happened? An extra (third) chromosome 21 has become translocated on the upper portion of one chromosome number 15. Chromosome-band staining techniques are useful tools in revealing such abnormalities.

Balanced carriers may appear karyotypically abnormal, having perhaps one less chromosome than the norm, but they have the normal complement of genes. The explanation lies in a translocation. Although there may be only 45 chromosomes, one will be somewhat larger than its homolog, and it may be deduced that the larger one contains the "missing" chromosomal material.

Another error that occurs during crossing-over is **inversion,** in which the orientation of a transferred piece of chromosome is inverted. Again, upsetting

Figure 38-8. Crossover in the pachynema stage. (A) The two synapsed homologous chromosomes. (B) The tetrad, composed of the two chromatids of each of the two synapsed chromosomes. (C) Crossover is occurring between adjacent nonidentical chromatids. (D) After the bivalent splits, there are two sets of chromatids that are genetically different from those in part A. Thus, the gametes will not have identical genetic complements.

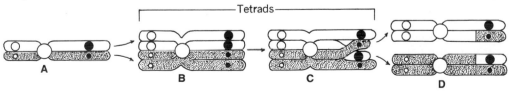

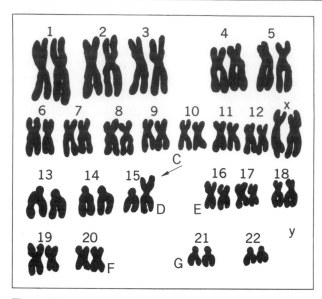

Figure 38-9. Representation of a translocated 21 chromosome in Down's syndrome.

the usual order of genes or the chromosome can alter the regulation and expression of genes.

Ultraviolet light, cosmic rays, high temperature, X-rays, formaldehyde, mustard gas, hydrogen peroxide, and many synthetic chemicals are all capable of causing chromosomal mutations.

Lethal Genes

Many phenomena are difficult to detect in genetic studies due to **lethal genes,** which for various reasons cause death of the embryo. In many organisms such early death results in the mother's reabsorbing the embryo. Thus, it appears to an observer that only a certain number of offspring are born (hatched, etc.), but actually more than that number were conceived. The presence of a lethal gene is more obvious if the gene is **sex linked.** If the lethal gene is on an X chromosome, half the males will die during early development as the male has only one X chromosome.

The action of a lethal gene usually involves the interruption of a vital biochemical pathway. For instance, the synthesis of a necessary enzyme may be blocked by many of the types of mutations already discussed.

There also occur sublethal genes that cause the death of the carrier between the time of birth and the attainment of reproductive age. One sublethal gene is known to cause **infantile amaurotic idiocy,** in which degeneration of CNS neurons results in death at age two to three years. The allele responsible for the trait is recessive. **Delayed lethal genes** exhibit their effects after the attainment of reproductive age.

Genetics and Heredity

The study of **genetics** deals with the origin of similarities and differences between parents and offspring. It is concerned with the nature of these similarities and differences, their sources, and how they develop. Briefly stated, geneticists investigate the transmission of genes from generation to generation (**heredity**) and look at the way in which the developmental potentialities of genes become expressed.

Heredity refers to all traits of an organism, but variations in anatomical and physiological characteristics that are passed from parents to offspring are particularly important. Thus we are not ordinarily concerned that both parents and children have two eyes but, rather, about the color and other characteristics of their eyes. We may find that the eye color of a child differs from its parents. Heredity, therefore, involves differences as well as likenesses between parents and offspring.

Representation of the Genotype and Crosses

Geneticists have devised a simple method to represent genotypes. A particular trait that is controlled by one gene is usually assigned a letter of the alphabet. For example, peas (after drying) may have either round or wrinkled seed coats. Experimental observations have shown these traits to be controlled by one gene locus, with the round allele dominant over the wrinkled allele. In our shorthand notation, we might assign this gene locus the letter "R."

Dominant alleles are customarily written in uppercase letters and recessive alleles in lowercase letters. Therefore, a pea plant homozygous dominant at that particular locus would have the genotype RR; a homozygous recessive would be symbolized as rr; and a heterozygote would be Rr. Note that both RR and Rr individuals will exhibit the round-seeded phenotype, whereas rr individuals will have wrinkled seeds.

When two individuals are **crossed** (animals mated, plants pollinated, etc.), it is represented as follows:

$$RR \times Rr$$

Here one parent is homozygous dominant for that particular gene, and the other is heterozygous. The first in a series of crosses is labeled P, referring to the parental generation. Each successive generation is called a **filial generation** and is indicated as F_1, F_2, etc.

Mendelian Genetics

Gregor Mendel (1822–1884) laid the foundations of modern genetics by his controlled breeding of pea plants in an isolated monestary garden in Austria. Mendel is noted for his careful experimentation, varying one parameter at a time, and for his meticulous recording and mathematical analysis of data.

One of his famous experiments concerned pea seed coats. Mendel true bred several successive generations to the point at which no phenotypic changes were observed when any generation was **selfed** (cross-fertilized with others from the same generation). He ended up with two different strains, one with round and another with wrinkled seed coats. The repeated breeding of like phenotypes eventually leads to a high probability of **homozygosity,** so we will assume that the round and wrinkled strains were homozygous at the seed-coat locus (although this was unknown to the pioneering Mendel).

Mendel then crossed these two strains, round-seeded plants with wrinkle-seeded plants:

P: RR $\times$ rr
F_1:   All Rr

He observed that all the offspring had rounded seed coats. Note that Mendel could only observe the phenotypes. His great discovery was that genotypes can be inferred from phenotypes. But to do this he needed more data. Mendel produced another generation by cross-fertilizing the F_1 individuals:

F_1:   Rr $\times$ Rr
F_2:   RR, Rr, Rr, rr

Mendel observed that roughly three quarters of the F_2 generation had rounded seed coats. And from these phenotypic ratios he inferred the genotypes shown. The phenotypic ratio of the F_2 generation in this **monohybrid cross** (in which only one trait is varied) is 3 to 1 (3:1), three rounded to one wrinkled. In most crosses, the genotypic ratio is not the same as the phenotypic ratio. In this example, the genotypic ratio was 1:2:1, that is, one RR to two Rr to one rr. These characteristic ratios (3:1 for the phenotype and 1:2:1 for the genotype) are observed rather consistently in monohybrid crosses.

Through several series of such experiments, Mendel formulated two laws, the law of segregation and the law of independent assortment. In modern terms, the **law of segregation** states that the members of a homologous pair of genes are separated during maturation of reproductive cells, so that each gamete contains one of the alleles. Thus when two gametes unite at fertilization, the two genes for each trait (one from each parent) are brought together in the offspring. Research in later years that uncovered the mechanism of reduction division during meiosis substantiated Mendel's law of segregation.

The **law of independent assortment** states, in modern terms, that the segregation of each pair of alleles to the gametes is entirely independent of the distribution of any other pair. This second law holds only if the different pairs of genes are not on the same homologous chromosomes; this condition is not always met (see the section on Linkage). Mendel was either very lucky to have examined traits governed by genes on nonhomologous chromosomes, or, more likely, he chose those traits (after conducting many preliminary experiments) because they gave more interpretable results.

With Mendel's laws, it is easy to understand how the F_1 genotypes in the example given came about. Let's assume that the P male plant was homozygous dominant (RR) for rounded seed coat. This male parent segregated its genes for a rounded seed coat so that each haploid gamete (pollen grain) received a dominant R gene. Similarly, the homozygous recessive (rr) female parent's gametes (ova) could only contain r genes. When an R pollen fertilized an r ovum, it

	♀ Female	
	p(0.5) R	p(0.5) r
♂ Male p(0.5) R	RR	Rr
p(0.5) r	Rr	rr

Figure 38-10. Punnett square. Here we are crossing two Rr individuals. The four squares in the diagram represent possible zygotes from this cross. The probability for each square is computed by multiplying the marginal values. For example, the probability for a RR individual is 0.5(0.5) = 0.25, whereas that for a Rr individual is twice that (as there are two Rr squares). The term *p*(0.5) represents a probability of 0.5.

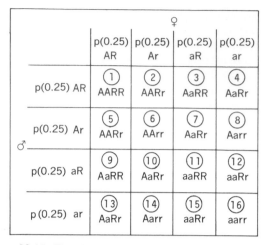

Figure 38-11. The dihybrid cross. This is a "four-by-four" Punnett square, and zygote probabilities are computed just as in Figure 38-10.

was inescapable that the zygote would be heterozygous (Rr).

Now let's consider the F_2 generation. The F_1 heterozygotes could produce two types of gametes—one type would contain R and the other type would contain r. Furthermore, these two types of gametes should be produced in approximately equal numbers. To figure out the probability of obtaining any one type of genotype (or phenotype) in a genetic cross, one may use a **Punnett square**. In this device each possible gamete of one parent is listed across the top, and the gametes of the other parent are listed along the side. Figure 38-10 shows a Punnett square for this monohybrid cross-fertilization. Each square represents a possible zygote, and each, in this simple example, represents a unit of equal probability. Therefore, the probability of obtaining heterozygous (Rr) offspring is two out of four, or 0.50; and, for either RR or rr, it is one in four, or 0.25.

The Dihybrid Cross

A **dihybrid cross** is a cross in which two traits are considered simultaneously. For example, consider the human traits of albinism (lack of integumentary pigmentation) and the ability to roll the tongue. Pigmentation (A) is dominant over albinism (a), and rolling ability (R) is dominant over nonrolling (r). Let us

assume that both parents are both heterozygous (AaRr), that is, that both phenotypically exhibit the dominant traits, but both are carriers of the recessive alleles that are not expressed. Mendel's laws tell us that each of their gametes has an equal probability of having either the dominant or the recessive allele for each trait. Figure 38-11 shows the Punnett square for this dihybrid cross. Note that there is a 25 percent probability of each of the following combinations of alleles in the haploid gametes: AR, Ar, aR, ar. A "four-by-four" Punnett square is necessary in a dihybrid cross because parents heterozygous for both traits can form four types of gametes.

The Punnett square can further be used to compute the numerical probabilities of each genotype and phenotype by multiplying the probabilities of each parental contribution. In this example, each box has a one-sixteenth probability of occurring. (The boxes are usually not numbered, but numbers are included here to help reference this discussion.)

First, how many different phenotypes result from this standard dihybrid cross? Recall that, when one allele exhibits complete dominance, a heterozygote will exhibit only the dominant phenotype. And so the following genotypes all exhibit the dominant phenotype pigmented roller: AARR, AaRR, AaRr, AARr. Similarly, the genotypes for the phenotype albino roller are aaRR and aaRr. Pigmented nonrollers are AArr and Aarr; and the only albino nonroller is aarr. Four different phenotypes result from a standard dihybrid cross. What is the phenotypic ratio? Pigmented rollers appear in boxes 1, 2, 3, 4, 5, 7, 9, 10, and 13, or nine in all. Albino rollers appear in boxes 11, 12, and 15, for a total of three. Pigmented nonrollers show up in boxes 6, 8, and 14—three again. The double homozygous recessive albino nonroller appears only once, in box 16. The phenotypic ratio of a standard dihybrid cross is 9 : 3 : 3 : 1.

Compute the genotypic ratio for yourself.

Non-Mendelian Genetics

Linkage, Crossover, and Chromosome Maps

In the fruit fly, *Drosophila melanogaster*, two of the many genes identified so far are C, the gene for wing shape, and B, the gene for body color. Both represent cases of simple dominance: C codes for normal

wings, and its recessive (c) allele codes for curved wing; B codes for gray body, and b codes for black body.

Assume that a heterozygous individual, BbCc, mates with a homozygous recessive individual, bbcc. According to Mendelian theory, the heterozygote BbCc can produce four kinds of gametes: BC, Bc, bC, and bc. The homozygous recessive can produce only bc gametes. The Punnett square in Figure 38-12 lists the genotypes of the possible zygotes. The four corresponding phenotypes, in order, are gray body, normal wing; gray body, curved wing; black body, normal wing; and black body, curved wing. We would expect, on the average, to observe equal numbers of these four phenotypes.

When this experiment is carried out, however, only genotypes BbCc and bbcc (and their corresponding phenotypes, from which the genotypes are inferred) are found. One explanation might be that the combinations Bbcc and bbCc are lethal, but that cannot be because both these types have been observed in other crosses. What explanation can account for these results?

It was mentioned earlier that genes are located in a linear array on the chromosomes and that homologous chromosomes contain the same genes (although often in different allelic forms). Chromosomes are essentially linear, and the genes along the DNA strand occur in a definite sequential order. If we postulate that the loci of alleles B and C occur on the same chromosome, the unexpected results described may be easily understood. In Figure 38-13, we see that the

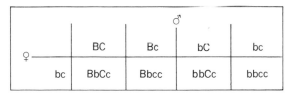

	♂			
♀	BC	Bc	bC	bc
bc	BbCc	Bbcc	bbCc	bbcc

Figure 38-12. Possible zygotes of the *BbCc x bbcc* mating. The bc trait is only given once because it represents the only type of gamete produced by the bbcc individual.

Figure 38-13. Possible gametes of the heterozygote and the homozygous recessive as described in the text.

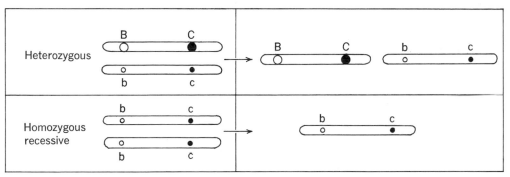

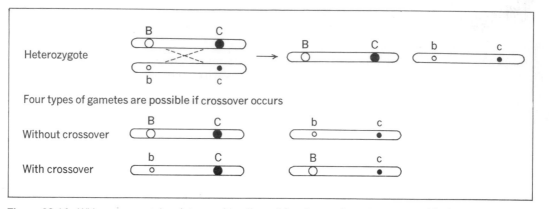

Figure 38-14. With crossover taken into consideration, all four types of zygotes, as noted in the text, may occur.

double heterozygote can only produce two types of gametes upon anaphase chromosomal splitting, whereas the homozygote can only produce one type. Thus there are only two combinations of gametes possible to form zygotes: BbCc and bbcc. The genes B and C are said to be **linked** because they are located on the same chromosome.

Reality is still more complicated, however, as such a cross rarely produces such neat results. Instead, all four genotypes are generally observed, but their proportions are not as Mendel would have predicted. The vast majority of offspring are generally BbCc and bbcc, as predicted, but some turn out to be Bbcc and bbCc. This is explained by the phenomenon of **crossover,** as mentioned in the discussion of chromosome mutations. Quite frequently during meiosis, two chromatids from homologous chromosomes cross over each other, and sections of the chromatid are actually exchanged (Fig. 38-14). A crossover leads to the production of two new gamete types that occur in addition to the normal types, for not every chromatid pair will cross over. These four gamete types, in combination with the homozygous recessive's single gamete type, will produce the four observed kinds of offspring. Because crossover occurs relatively infrequently, greater proportions of BbCc and bbcc zygotes are observed.

Another aspect of crossover is **crossover frequency.** Taking the linear order of gene loci into consideration, Morgan and Sturtevant postulated that genes close

together on a chromosome would experience crossovers between them less frequently than would genes that are farther apart. And so the percentage crossover, as monitored by the incidence of "unexpected" genotypes in the progeny, is directly related to the distance between the marker genes on chromosome. Relative gene positions with respect to the centromere can then be determined by statistical analysis, and a chromosome map can be constructed (Fig. 38-5).

Incomplete Dominance and Codominance

In all Mendel's pea experiments, dominance was essentially complete, so that there was no appreciable difference between the phenotypes of homozygous dominant and heterozygous individuals. For example, RR peas could not be directly distinguished from Rr peas. However, there are many genes whose expressions give **partial,** or **incomplete, dominance,** in which the heterozygote is phenotypically different from either homozygote.

A good illustration of incomplete dominance is provided by the blue Andalusian fowl. A cross between black homozygous and white-splashed homozygous parents produces "blue" heterozygous offspring, with very fine alternating black and white stripes (Fig. 38-15). When blue Andalusians (F_1) are interbred, the phenotypic ratio of the offspring (F_2) is one black to two blues to one white. Before basic genetics was

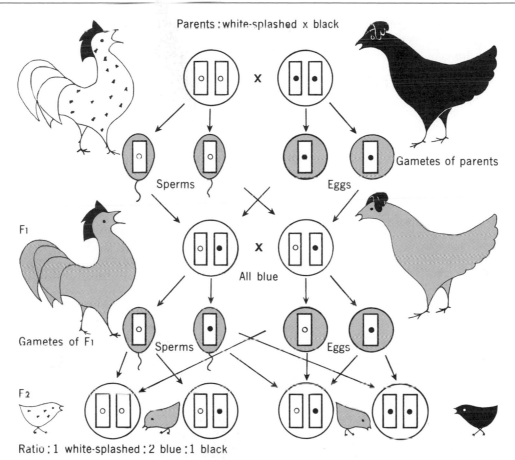

Figure 38-15. Incomplete dominance. Cross between a white splashed Andulasian fowl and a black Andulasian fowl. The hybrid (heterozygote) is called a blue Andulasian.

understood, poultry breeders had tried in vain to obtain pure blue Andalusian fowls by interbreeding blue parents. This is obviously impossible because only the hetcrozygote expresses the blue plumage.

Incomplete dominance seems to contradict Mendel's assumption that no blending of traits occurs. Note, however, that the genotypic ratio of the F_2 generation agrees with Mendel's prediction of $1:2:1$. Only the phenotypic expression $1:2:1$ is at variance with Mendel's. When the contribution of each allele to a heterozygous trait is approximately equal, they are called **codominant.** This is exemplified by the M–N blood group system in humans. It has been found that two types of genes, a^M and a^N, control M–N blood type. When two homozygous a^M people mate, their children are of the M type. Likcwise, two a^N parents produce all N children. But, if two heterozygotes mate ($a^M a^N \times a^M a^N$), 25 percent of the children are type M, 25 percent are type N, and 50 percent are type MN. Note that the heterozygotc expresses both alleles; onc allele does not preclude expression of the other. Molecularly, incomplete dominance and codominance are identical, as they both involve the simultaneous production of two different proteins (one from each allele). Codominance is seen as a more distinct combination of traits resulting from the action of both proteins, whereas, in incomplete dominance, the activities of each protein contribute to the appearance of blending in the phenotypic characteristic.

Incomplete Penetrance

Implicit in Mendel's assumptions is that dominance is always expressed when it occurs, but this is not always the case. For example, the trait of polydactyly (extra fingers) in humans is dominant over normalcy. Studies of **kindreds** (**pedigrees** or **family trees**) show that, although the polydactyly allele is dominant, it does not always get expressed—many carriers of the dominant gene still have a normal number of fingers. Incomplete penetrance is very important to our understanding of the interrelationship of genes and the environment, for it is the environment that allows the phenotypic expression of a gene. Keep in mind that the inheritance of a gene is the inheritance of a potentiality, not of a trait itself.

Penetrance refers to the frequency with which a gene is expressed in a detectable way. If every individual possessing a dominant gene develops the trait, the gene is said to have 100 percent, or **complete,** penetrance. But, if possession of a dominant gene does not always result in development of the trait, the gene is said to have reduced, or **incomplete, penetrance.**

All Mendel's experiments involved genes with 100 percent penetrance, and hence a $3:1$ phenotypic ratio was observed in monohybrid crosses. This ratio is not realized if the penetrance is reduced. Incomplete penetrance of a dominant gene is sometimes called **irregular dominance** because the gene sometimes becomes expressed in the heterozygous condition, and sometimes it does not. The causes of reduced penetrance is most often an indication of the effects of other genes or environmental influences on the trait. That is, although one gene may be the dominant factor, the contributions of various alleles of other genes, or the environmental conditions that affect the expression of genes, modify the final result.

Striking effects on the penetrance of some genes are produced by differences in environmental temperature. On Siamese cats (Fig. 38-16), the dark fur pigment is produced only on the relatively cool extremities of the body. The color of Himalayan rabbits corresponds to that of Siamese cats; the fur is lightly colored except on the feet, nose, ears, and tail, where

Figure 38-16. Siamese cats. Pigment develops on the extremities as a result of temperatures below that of the rest of the body.

it is black. Experiments have shown that expression of the gene responsible for black fur pigment depends upon the relative temperatures of different body areas where the fur develops.

Polygenic Inheritance

Polygenic inheritance involves systems of genes, each of which makes a small contribution to a particular trait. Human body stature and intelligence both involve polygenic inheritance, although here again the phenotypic expression of the genes may be modified by the environment (e.g., diet). A good example of polygenic inheritance is body weight in breeds of rabbits. If a Flemish giant rabbit weighing 12 kg is mated with a mature Polish rabbit weighing 2 kg, the weights of their offspring when mature will vary in a range that is intermediate between the extremes of the parents. The F_2 generation shows even more variation, and the gradations approach continuity. It is now understood that many gene loci code for body weight and that each allele at these loci contributes an increment or decrement of approximately equal magnitude to the trait.

Multiple Alleles

The inheritance of human A-B-O blood groups (Chapter 35) is determined by three alleles of one gene locus. One allele causes development of the A antigen, another development of the B antigen, and the third results in the absence of both antigens. Because each individual must have two genes at this locus (on homologous chromosomes), six different genotypes are possible. Designating the three alleles by the antigens that they produce as A, B, and O (for none), group A will have the genotype AA or AO; group B, the genotype BB or BO; group Ab, the genotype AB; and group O, the genotype OO. Allele A and allele B are codominant to each other, but either is dominant to the O allele.

The inheritance of the Rh blood groups is more complex and may be explained in two ways. One school of thought holds that three closely linked (i.e., with loci very close to each other on the same chromosome) pairs of genes are involved, whereas the other explanation is based on an allelic series of at least eight (and probably more) genes. We will concern ourselves only with the one locus that is most commonly involved in clinical problems. Two alleles seem to control development of the Rh antigen. The dominant allele, R, causes development of the Rh antigen, whereas the recessive allele, r, does not. Thus persons of genotype RR or Rr will have the antigen and will be Rh-positive, but homozygous recessives (rr) will lack the antigen and be classified as Rh-negative.

In matings between Rh-positive and Rh-negative individuals, two patterns of inheritance have been observed. If the Rh-positive parent is homozygous

Figure 38-17. Diagram of the Rh mechanism in pregnancy. This illustration shows what may take place in pregnancy when an Rh⁺ man mates with an Rh⁻ woman. The Rh⁺ erythrocytes of the fetus cause the production of Rh antibodies in the mother. The Rh antibodies from the mother may pass through the placenta and cause destruction of the red blood corpuscles of the fetus.

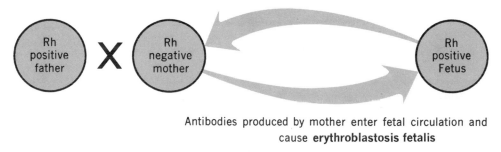

Rh positive erythrocytes of fetus enter maternal circulation and stimulate production of Rh antibodies

Rh positive father **X** Rh negative mother → Rh positive Fetus

Antibodies produced by mother enter fetal circulation and cause **erythroblastosis fetalis**

(RR), then all the children will be Rh-positive, because they will all be heterozygous. If, however, the Rh-positive parent is heterozygous (Rr), then one half of the offspring are expected to be Rh-positive (Rr) and the other half Rh-negative (rr). This is because the chances are equal that either an R or r allele from the Rh-positive (Rr) parent will combine with an r allele from the Rh-negative (rr) parent.

The first and usually the second child born of parents with incompatible Rh blood groups usually escape harm, but by the time the third child is conceived there may be sufficient antibodies in the mother's blood to produce serious effects. The mechanism for Rh incompatibility (**erythroblastosis fetalis**) is illustrated in Figure 38-17.

Note that, if the mother has ever received a transfusion of Rh-positive blood, this would serve as a stimulus for the formation of anti-Rh antibodies. In such a case, even the first child might suffer the effects of Rh incompatibility. The transfer of antibodies and red blood cells between fetal and maternal circulation occurs only in the small percentage of cases in which the placenta is torn or defective in other ways. Consequently, the mating of 28 Rh-positive males and Rh-negative females will show any signs of the effects of Rh incompatibility about 1 out of 28.

Inheritance of Sex

Each human somatic cell has 46 chromosomes (or 23 homologous pairs). Of these, two particular chromosomes are the determinants of sex: females normally have two **X chromosomes,** whereas males have one X chromosome and one **Y chromosome.** A female can produce only ova that contain an X chromosome, but a male can produce two types of sperm: some with an X chromosome and an equal number with a Y chromosome. As a result, there is a 50 percent probability of the zygote's receiving an X chromosome from the father, a 50 percent probability of receiving a Y chromosome from the father, and a 100 percent probability of receiving an X chromosome from the mother. Therefore, an average of 50 percent of all children are girls and 50 percent are boys.

The sex chromosomes form a homologous pair, but sometimes they may not segregate during meiosis, as discussed earlier for autosomes. Such a **nondisjunction** may result in an ovum that contains two X chromosomes or in a spermatid that contains an X and a Y chromosome, or in a gamete with neither sex chromosome. This may result in certain **syndromes** (groups of symptoms that usually occur as a unit). **Kleinfelter's syndrome** appears when a zygote gets two X chromosomes and one Y chromosome. The XXY individual is phenotypically a male but is sterile and may show signs of secondary female characteristics (such as breast development); mental retardation may be included among the symptoms. **Turner's syndrome** afflicts individuals with only one X chromosome and no Y chromosome (XO). Phenotypically a female, the ovaries are either lacking or vestigial; additional symptoms may include an abnormally short stature, a webbed neck, and mental retardation. Other human sex chromosome abnormalities (for example, XXX, XXXX, etc.) are also known.

A quick and reliable method of determining the number of X chromosomes in a human cell is to observe **Barr bodies** (Fig. 38-18). Only one of a female's X chromosomes is active at any one time. The inactive X chromosome shows up in the normal cell, under the microscope, as a darkened area in the karyoplasm. This darkened area is the Barr body. A normal female will have one Barr body per cell.

Sex-Linked Inheritance

The X chromosome carries genes for many traits in addition to those influencing sex. For example, in cases of color blindness (Fig. 38-19) involving an inability to distinguish red from green, a color-blind fa-

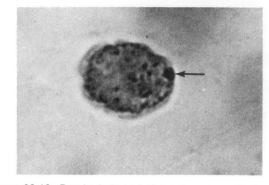

Figure 38-18. Barr body (arrow), the darkened area that represents an inactivated X chromosome.

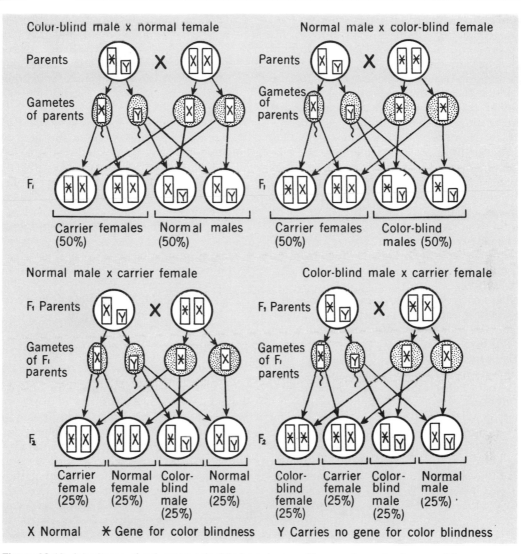

Color-blind male x normal female

Parents

Gametes of parents

F₁

Carrier females (50%)   Normal males (50%)

Normal male x color-blind female

Parents

Gametes of parents

F₁

Carrier females (50%)   Color-blind males (50%)

Normal male x carrier female

F₁ Parents

Gametes of F₁ parents

F₁

Carrier female (25%)   Normal female (25%)   Color-blind male (25%)   Normal male (25%)

Color-blind male x carrier female

F₁ Parents

Gametes of F₁ parents

F₂

Color-blind female (25%)   Carrier female (25%)   Color-blind male (25%)   Normal male (25%)

X Normal   ✳ Gene for color blindness   Y Carries no gene for color blindness

Figure 38-19. Inheritance of red–green color blindness in man. Diagram shows the four possible combinations of color blindess, excluding homozygous crosses, and the ways in which the defect is inherited from these combinations. The female, designated as a carrier, is one heterozygous for the gene for color blindness but is normal in color vision. Marriages of brothers and sisters in the human species are so very rare that usually no F_2 is shown in texts; however, it is designated here to illustrate what commonly occurs in other species.

ther and a normal mother will produce no color-blind children. The XY zygotes develop into normal males possessing the normal X chromosome from the mother and one Y chromosome, which carries no gene for color blindness. Similarly the XX zygotes develop into normal females, as only one X chromosome (from the father) bears the recessive gene for color blindness. In an F_2 generation, however, half the grandsons and half the granddaughters will be free from this defective allele. The other half of the females would carry the gene for color blindness in heterozygous condition. The other half of the males would be color blind, having the defective allele in their single X chromosome.

Of particular interest to human society are the numerous alleles that produce defects such as hemophilia, blindness, and deafness. These alleles are segregated during the reduction division of gametogenesis and are distributed to offspring just as normal genes are. The sex-linked recessive allele for one type of **hemophilia** is located on the X chromosome; males containing one X chromosome bearing this allele and females with two such X chromosomes are hemophilic. Females with only one allele for hemophilia do not exhibit the condition, but serve as carriers.

There are many sex-linked traits. Genes for over 30 sex-linked traits are known for humans, and about 150 sex-linked genes have been found in the X chromosome of the fruit fly.

Another kind of sex-linked inheritance is **holandric inheritance,** where the gene appears on the Y chromosome; thus the trait will be expressed in all males and never expressed or carried by females. This phenomenon is rarely observed, and so far the only human trait (other than sexuality) with even a minimal probability of being substantiated as holandric is hairy ears in males.

If a recessive allele for a defect is located on an autosome (somatic chromosome), two alleles must be present for the defect to be expressed, in either males or females. Usually it is impossible to determine for sure if an individual carries only one autosomal recessive gene for a defect, but pedigree charts allow genetic counselors to assign probabilities for prospective parents.

Population Genetics

The mathematically based discipline of **population genetics** investigates the genetic variations within and among specific groups called **populations**. Populations are commonly defined geographically—a set of boundaries is demarcated, and all members of a given species within that region form a population. Geneticists define a population much more rigorously. The following criteria define an ideal population:

1. The group must be capable of interbreeding. That is, all the members of a population belong to the same species (Chapter 4). Note that a species often consists of many populations that are geographically isolated from one another. All members of a species can at least potentially interbreed, however.

2. The group must share a common gene pool. This is essentially a refinement of the first criterion. A **gene pool** may be defined as the collection of all alleles of every gene present in the members of the population. This implies that the alleles of each member of a population must have the potential to be combined during fertilization with any others in the population to produce an offspring possessing a novel combination of genes. Note for now that different populations of a species may have quite different gene pools—the types and frequencies of alleles may vary significantly. As we shall see, this is a criterion for the evolution of new species.

Populations at Genetic Equilibrium

Population geneticists investigate changes in the gene pool of a population over time, because this is a way of monitoring the process of **evolution.** But first we must consider a population at **genetic equilibrium.** Recall that equilibrium implies no net change: any changes occurring in one direction must be counterbalanced by equal changes in the opposite direction. Actually, populations at genetic equilibrium for alleles at all loci are never encountered in nature. This is, however, a useful device for formulating genetic theory. A number of unrealistic assumptions apply:

1. Either no mutation occurs or the mutation rates are equivalent in both directions: Frequency A $\longrightarrow$ a = Frequency a $\longrightarrow$ A.
2. Each adult member is capable of reproducing.
3. Mating is random (panmixia).
4. Zygote formation (fertilization) is random.
5. Every zygote is equally viable.
6. Members of each sex enjoy equal fertility.
7. Members of each sex have equivalent mortality (death) rates.

8. Generations do not overlap (the P and F_1, etc., cannot simultaneously reproduce).
9. Segregation of alleles is Mendelian (nondisjunction and the like do not occur).
10. The population approximates infinite size.

Allelic Frequency

To monitor changes in a gene pool, one must be able to calculate the frequencies of particular alleles at selected gene loci. Allelic frequency is simply expressed as:

$$\frac{\text{Total no. of a particular allele in the population}}{\text{Total no. of genes at that locus in the population}}$$

For a simple example, assume that allele A is dominant to its counterpart a and that, out of a population of 100 organisms, 50 are homozygous dominant (AA), 20 are heterozygous (Aa), and 30 are homozygous recessive (aa). What is the frequency of allele A?

We know that 100 of the alleles for this trait occur in the homozygous dominants:

$$2 \times 50 = 100$$

In the heterozygotes, half the genes are A (20); the homozygous recessives contain no A alleles (0):

$$100 + 20 + 0 = 120$$

The allelic frequency of A in this population, then, is $120/200 = 0.60$, or 60 percent.

By convention, the allelic frequency of a dominant allele is a assigned the letter p, and the frequency of its recessive allele is given as q. If these are the only two alleles at that locus in the population, then

$$p + q = 1$$

In a hypothetical population at genetic equilibrium, the allelic frequencies would remain unchanged from generation to generation. Let us see how this might work: Assume that there is a P generation in which 70 percent are HH and 30 percent are hh (with no heterozygotes in the population at this

Figure 38-20. Probability computation using the Punnett square. See text.

point). It is easy to compute that $p = 0.7$ and $q = 0.3$. Let us further assume that there are equal numbers of males and females in this P generation. What will be the allele frequencies of the F_1 generation?

Based on the assumptions, 70 percent of the sperm and ova contain allele H, and 30 percent of the gametes contain h (Fig. 38-20). If panmixia prevails, then simple multiplication gives the frequencies in the diploid zygotes. We can now compute the total frequency of each allele in the F_1 population. For H;

$$(0.49 \text{ HH}) + (0.5)(0.21 + 0.21 \text{ Hh})$$
$$+ (0)(0.09 \text{ hh}) = 0.7 = p$$

For h;

$$(0)(0.49 \text{ HH}) + (0.5)(0.21 + 0.21 \text{ Hh})$$
$$+ (0.09 \text{ hh}) = 0.3 = q$$
$$p + q = 0.7 + 0.3 = 1.0$$

The allele frequencies have remained the same even though the genotype frequencies have changed. The genotype frequencies of this particular F_1 generation are:

HH	0.49	$= p^2$
Hh	0.42 = 0.21 + 0.21	$= 2pq$
hh	0.09	$= 2$
	1.00	1.00

It should be apparent that the genotype frequency of

HH is p^2 (0.7 × 0.7); of hh, it is q^2; and of Hh, it is $2pq$. The 2 in $2pq$ comes about because there are two ways to obtain a heterozygote, namely, Hh and hH (as shown in the Punnett square in Fig. 38-20). At genetic equilibrium, $p^2 + 2pq + q^2 = 1$. This is known as the **Hardy–Weinberg law,** devised by the English mathematician G. Hardy and the German physician W. Weinberg in 1908. The Hardy–Weinberg law is very useful in genetic studies. If one assumes that the population is at genetic equilibrium, then, by knowing the frequency of one dichotomous allele, the genotypic frequencies in the population can be calculated.

Natural Populations Are Not at Genetic Equilibrium

Although a consideration of equilibrium populations is useful in understanding the basics of population genetics theory, it has limited relevance to natural populations. The ten prerequisites for an equilibrium model cannot all be met. For example, members of different sexes have different mortality rates, all zygotes are not equally viable, and so on. But the most important factors that contribute to the dynamic nature of a population's gene pool are those that foster new combinations of alleles and the selective forces that act on their phenotypic expressions.

Genetic Drift

One aspect of natural populations is **genetic drift,** or chance changes in allele frequencies that result from random samples. Genetic drift becomes more prevalent as population size decreases. Two phenomena associated with genetic drift are the founder and bottleneck effects. In the **founder effect,** either a very small population with a limited gene pool colonizes a new habitat or a new gene (or group of genes) is introduced into a small population. Either case results in much higher allele frequencies for certain genes in the "founder" population than in the species at large, simply because there are not enough founder individuals to carry all possible alleles, so many are lost. Recessive traits are therefore more apt to be phenotypically expressed (homozygous). In the **bottleneck effect** there is a rapid reduction in the number of individuals in the general population, and this also results

in a relatively small gene pool. The bottleneck effect faces the great whales and other species that have been killed to near extinction.

Both these effects contribute to a reduction in genetic diversity, which characterizes genetic drift; the frequency of p or q may become fixed at zero (the lost allele) or unity (the allele retained).

Selection Effects: Assortative Mating

The criterion of panmixia (random mating) is also seldom realized. When mates are not randomly chosen, it is called **assortative mating.** This results in changes in the genotypic and phenotypic frequencies but not necessarily in the allele (gene) frequencies.

Let's assume that a population has the following characteristics:

$$P: \quad \tfrac{1}{2}\,AA,\ \tfrac{1}{2}\,aa$$
$$F_1: \quad All\ Aa$$
$$F_2: \quad \tfrac{1}{4}\,AA,\ \tfrac{1}{2}\,Aa,\ \tfrac{1}{4}\,aa$$

Note that the frequencies of p and q are 0.5 throughout. Now, let's assume an assortative mating in which only similar genotypes mate: homozygous dominants mate only with homozygous dominants, heterozygotes mate only with heterozygotes, and homozygous recessives mate only with homozygous recessives (with respect to a single gene, of course). What result will this have on the F_3 generation?

Because the AA individuals only mate with AA individuals, they will produce only AA offspring. Similar considerations hold for the aa genotypes. The Aa individuals, however, produce a 1:2:1 ratio of AA to Aa to aa. Because there were originally $\tfrac{1}{2}$ Aa individuals, the new ratios of the three genotypes produced by Aa x Aa crosses are:

$$\tfrac{1}{2}\left(\tfrac{1}{4}\right) = \tfrac{1}{8}\,AA$$
$$\tfrac{1}{2}\left(\tfrac{2}{4}\right) = \tfrac{2}{8}\,Aa$$
$$\tfrac{1}{2}\left(\tfrac{1}{4}\right) = \tfrac{1}{8}\,aa$$

The total F_3 genotypic ratios are:

$\tfrac{2}{8} + \tfrac{1}{8} = \tfrac{3}{8}\,AA$ Original $\tfrac{1}{4}$ from AA plus the contribution from Aa crosses

$\tfrac{2}{8}\,Aa$ from Aa crosses

$\tfrac{2}{8} + \tfrac{1}{8} = \tfrac{3}{8}\,AA$ Same as AA reasoning

The genotypic (and phenotypic) frequencies have clearly changed, with relatively fewer heterozygotes in the F_3 generation. Have the allele frequencies also changed?

For A: $\frac{3}{8} + \frac{1}{2}\left(\frac{2}{8}\right) = \frac{1}{2}$

For a: $\frac{3}{8} + \frac{1}{2}\left(\frac{2}{8}\right) = \frac{1}{2}$

Selection Effects: Directional Selection

The role of natural selection (Chapter 41) in changing gene frequencies is well documented in the case of a British moth, *Bistan betularia*, that may appear in two phenotypic forms, dark and light colored. Years ago the light phenotype was predominant, presumably because the dark moths were more easily detected by avian predators as they rested on light lichen-covered tree trunks. The industrial era brought environmental pollution that killed the lichens and deposited dark soot on the tree trunks. This change in the environment affected pre–predator relations considerably, for now the light-colored moths stood out, and over a period of time the dark phenotypes became more numerous. These changes in phenotypic frequencies reflect underlying shifts in the frequencies of the genes that control body pigmentation. The reversal of the allelic frequencies was a result of directional selection in *B. betularia*. The genetic changes enhanced survival of the species in a changing environment.

Selection Effects: Selection Against Homozygous Recessives

Eugenics is the science concerned with the application of genetic principles to the improvement of the human species. The study of human genetics is difficult for several reasons: (1) the impracticality of experimental breeding, (2) the small number of offspring per family, and (3) the relatively long time between generations. Despite these restrictions, significant achievements are being made in this field, as evidenced by the establishment of many centers that provide practical genetic counseling to prospective parents.

Physicians are realizing more and more that genetics has an important bearing on clinical problems, and relatively rapid progress is being made in accumulating data that may prove helpful in the prevention and diagnosis of disease.

Figure 38-21. Genotypic ratios may be derived directly from probabilities. Here, we have a ratio of 4 GG to 4 Gg to 1gg.

However, certain genetic manipulations, proposed by some to be extremely effective, have been shown to be of little value.* We will give an example of a hypothetical population, the Hwrembleton insects of eastern Eldorado. In this culture, the majority of insects have antennae that are green, due to possession of a dominant autosomal gene G. This is a case of simple dominance, so heterozygotes (Gg) also display the green-antennaed phenotype. Homozygous recessives, however, have purple antennae and are not recognized by the green-antennaed majority. The green-antennaed insects begin to kill off all the purple antennaed individuals to drastically reduce the occurrence of purple antennae. What effect does this have on the gene pool?

For simplicity, let us assume that the genotypic population is described as $\frac{1}{4}$ GG $+ \frac{1}{2}$ Gg $+ \frac{1}{4}$ gg. If the homozygous recessives are kept from breeding, the effective gene pool of the P generation is $\frac{1}{3}$ GG and $\frac{2}{3}$ Gg. What is the makeup of the F_1, assuming panmixia among green-antennaed individuals? The allele frequency of g (from Gg only, as no gg individuals are breeding) is $\frac{1}{2}\left(\frac{2}{3}\right) = \frac{1}{3}$.

According to Figure 38-21, we have a new genotypic ratio of 4GG: 4Gg: 1gg.

The F_1 parents are $\frac{1}{2}$ GG and $\frac{1}{2}$ Gg as homozygous recessives are still not allowed to breed:

---

*Genocide (the organized extermination of a racial, cultural, or political group) is not only barbaric but is ineffective in its aims because it does practically nothing to change allele frequencies (see Chapter 41).

♂ \ ♀	3/4 G	1/4 g
3/4 G	9/16 GG	3/16 Gg
1/4 g	3/16 Gg	1/16 gg

Figure 38-22. The inevitable return to 9:3:3:1.

$$\text{Allele frequency of G:} \quad \tfrac{1}{2} + (\tfrac{1}{2} \times \tfrac{1}{2}) = \tfrac{3}{4}\ G$$
$$\text{Allele frequency of g:} \quad \tfrac{1}{2} \times \tfrac{1}{2} = \tfrac{1}{4}\ g$$

When these gametes are crossed (Fig. 38-22), the results are a 9:6:1 genotypic ratio. In review, the population has gone from 25 percent purple antennae to 11 percent and to 6 percent by the F_2 generation. But note that, as long as there are heterozygotes that "hide" the g allele, some gg genotypes will keep appearing. As successive generations are selected against, smaller and smaller percentage decrements in homozygous recessives are realized. The human generation time of about 25 years would make eugenic selection against homozygous recessives even less effective.

Selection Effects: Consanguinity

Consanguinity refers to the mating of genetically related individuals ("blood relatives"). The major nonmoral objection to consanguinous matings is rooted in genetic theory—consanguinous matings increase the incidence of homozygous recessives. This is usually seen as disadvantageous, as many diseases and defects are coded for by recessive genes; the majority of the population is "normal" for these traits by virtue of the masking effect of dominant genes in heterozygotes.

A simple example will suffice, but first some background in reading kindreds (pedigrees) is necessary. A **kindred** is a diagram that indicates mating relationships in a succinct, at-a-glance manner. Males are indicated as squares and females as circles. An individual affected by a defective trait is represented by a

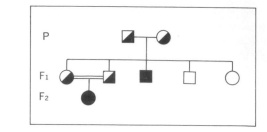

Figure 38-23. Kindred with consanguinity.

darkened symbol. A horizontal line running between two symbols represents a mating, and a double horizontal line represents a link to the offspring, who are connected by T lines, as shown in Figure 38-23.

In the example used, the third child, a male, displays the defective trait. Both parents must have been carriers (heterozygotes) if the trait is caused by a recessive allele on an autosome. Carriers are sometimes indicated as ◍ or ◐ . The first two children mated and produced a female who displayed the trait. Because this child was homozygous recessive, the parents must have each been heterozygous for the defective allele.

Now, refer to Figure 38-24. Note that individuals III-1 and III-2 are first cousins and, thus, consanguineously related. Given that III-1 is Aa, where a is a rare autosomal recessive, what is the probability that III-2 also carries that allele? The implication here is that, in outcrossings (nonconsanguinous matings), the unrelated (foreign) mate will most likely not carry the rare gene.

If III-1 has a, she must have inherited it from either II-2 or II-3, with equal probability of .50. The female II-3, if she carries the a in heterozygous condition, received it from either I-1 or I-2, again with an equal

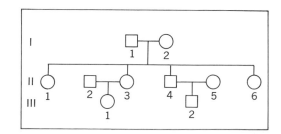

Figure 38-24. Individuals III-1 and III-2 are first cousins; hence they are consanguinously related.

probability of .50 of transmitting a, so the probability that II-4 carries allele a is $\frac{1}{2}$ $(\frac{1}{2})$ = $\frac{1}{4}$. Now, the probability that III-2 receives allele a from II-4 is .50, given that II-4 may have the gene, as calculated. That yields a $\frac{1}{4}$ $(\frac{1}{2})$ = $\frac{1}{8}$ probability that the two cousins III-1 and III-2 carry the same recessive allele a.

Remember, we specified that allele a was rare. A conservative estimate of its rarity would be that its allele frequency was $\frac{1}{100}$ (0.01). According to the Hardy–Weinberg law ($p^2 + 2pq + q^2 = 1$), the probability that an unrelated mate carries allele a as a heterozygote is $2pq$. Given the allele frequency of a as $q = 0.01$, then $p + q = 1$, and $p = 0.99$. Therefore, $2pq = 2$ (0.01) (0.99) = $0.0198 \cong 0.02$.

This is the probability of anyone in the population's being heterozygous for these alleles. Therefore, if we know for sure that one partner is heterozygous, the probability of the other's being heterozygous is 0.01. If we compare that to the $\frac{1}{8}$ = 0.125 probability of heterozygosity in the first cousins, the abrupt increase in recessive gene expression in consanguinous relationships becomes obvious.

Some Methods that Advance Genetics and Heredity Studies

Twin Studies

Twins occur about once in every 85 human births. Twins are of particular genetic interest. There are two types: fraternal and identical. **Fraternal twins** develop from two separate zygotes and therefore differ both genotypically and phenotypically, as do the other progeny from the same parents. **Identical**

Figure 38-25. Diagram to show the (relative) components of heredity and environment in the development of traits. The black symbol lettered H illustrates a hypothetical trait, having about equal genetic and environmental components. The other symbols represent different traits (phenotypes); the length of the solid and dotted lines shows the relative importance of the genetic and environmental components. No trait can occur at either extreme, because all traits possessed by an organism are influenced by both genetic and environmental factors. In many cases, one cannot measure the effects of one or the other, but in this diagram traits have been selected for which the relative influences of environment and heredity are known. A symbolizes the phenotype albinism (absence of pigment), a trait that develops because of the presence of an allelic pair of recessive genes. Here, the environment provides only the minimal essentials. C_1 represents a cancer of the eye, retinoblastoma, a trait that is largely influenced by a single dominant gene. The environment, however, must play a part in its development. C_2 symbolizes another malignancy, breast cancer, a trait for which there is statistical evidence of genetic factors, but there are also several important environmental factors known to be operating in the development of this trait. W represents the trait "writing," a behavior characteristic for which most of mankind possesses the necessary physical equipment. However, such a behavioral trait also requires training, an environmental component lacking in many cultures. An additional aspect of hereditary and environmental interactions not illustrated by the diagram is the fact that the same phenotype may result from both a "weaker" and a "stronger" genetic component if the "weaker" is augmented by a "stronger" environmental component; in other words, the same genetic component may result in different phenotypes if different environments are provided. The best evidence for this may be found in the differences observed in "identical" twins in different environments.

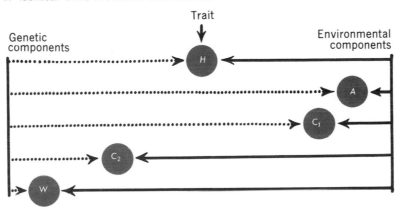

twins, on the other hand, arise from a single zygote that splits into two during the earliest mitotic division. Identical twins, therefore, have the same genotypic constitution. They are always of the same sex, and their phenotypic characteristics, both physically and mentally, are remarkably similar.

Triplets, quadruplets, and quintuplets may include mixtures of fraternals and identicals.

Because of the fortuitously (for the geneticist) identical chromosomal makeup of identical twins, such environmental factors as dietary deficiency can be studied with proper controls. If one identical twin is kept in a constant "normal" environment and another is exposed to the variable in question, all differences may be assumed to be environmental in origin (Fig. 38-25).

The Test Cross

A test cross is a genetic manipulation in which a known homozygous recessive (for the trait in question) is mated with an individual of unknown genotype. If the unknown is homozygous dominant, all offspring will be heterozygous for the trait. If the unknown is heterozygous, half the offspring will be expected to exhibit the recessive trait.

Amniocentesis

Many genetic disorders such as trisomes and translocations may be determined before birth by a relatively new procedure called **amniocentesis:** The karyotype of a fetus is determined by examining fetal cells obtained by the withdrawal of amniotic fluid (using a large syringe) through the mother's abdomen. The sex of the embryo is easily determined by this method.

Cell Fusion

A recent technique called **cell fusion** has proved useful in uncovering linkage groups in human chromosomes. In this method, mouse and human cells are made to fuse into a single hybrid cell by infecting them with a special inactivated virus. In these hybrid cells, the DNAs from both the mouse and the human chromosomes remain active in synthesizing RNA and protein. However, as the hybrid cells undergo mitosis, some of the chromosomes (and, of course, the genes that they contain) are eliminated from the dividing cells. Investigators can determine which enzyme activities no longer occur in these cells and correlate this information with the chromosomes that have been lost. For example, the genes controlling the synthesis of lactate dehydrogenase (an enzyme required for lactic acid production) and peptidase (a protein digestive enzyme) are lost simultaneously; therefore, they are almost certainly linked on the same chromosome.

Summary

The DNA molecule is the carrier of genetic information. The sequences of DNA nucleotides specify which proteins will be synthesized by the organism. RNA molecules, transcribed from DNA, translate the genetic code into amino acid sequences (protein) within the cytoplasm. Genes are transcribed only when needed; this, and the many other steps involved in the expression of genetic information, are carefully regulated.

Missense, nonsense, deletion, frameshift, and insertion mutations all involve an alteration of base pairs in DNA so that different, or no amino acids result. This alteration may result in no effect, the synthesis of a different protein, or premature termination of protein synthesis.

Chromosomal abnormalities also influence heredity. Trisomies, in which certain chromosomes appear more than twice in each cell, may result in disease or death. Translocations, where parts or all of certain chromosomes may be transferred or exchanged with other chromosomes, may or may not have an effect on the organism.

Meiosis occurs in the sex cells and results in the formation of haploid gametes, either sperm or ova (in animals), which contain only one half the normal complement of genes. During the reduction division of meiosis, the pairs of homologous chromosomes split, one going to each gamete.

Genetics is the study of the transmission of genes and their phenotypic expressions. Pioneering work in genetics was done by Mendel, who described simple dominant–recessive relationships from his experiments with pea plants.

Genetic crosses may be diagrammed in Punnett squares that afford immediate estimates (probabilities) of expected genotypic and phenotypic outcomes.

Mendel's laws have been shown to be incomplete by the subsequent discoveries of incomplete dominance, codominance, incomplete penetrance, polygenic inheritance, sex inheritance, and other phenomena.

Much can be learned through the study of sex inheritance. In humans, Kleinfelter's syndrome and Turner's syndrome are sex chromosome abnormalities that arise from meiotic nondisjunction. The discovery and study of sex-linked diseases such as hemophilia give us more insight into the mechanisms of inheritance.

The mathematically based study of population genetics has great utility in evolutionary studies. Ideal populations at equilibrium provide the basis for the study of nonequilibrium populations. Populations are not at genetic equilibrium because of one or more of the following disruptive tendencies: genetic drift (in which very few alleles form the gene pool), assortative mating (in which mating is not random), directional selection (in which natural selection operates and those most fit to the environment survive to produce the most offspring), consanguinity (in which close relatives mate, thus increasing the frequency of homozygous recessives), lethal genes (that often affect an organism's capacity to reproduce), and linkage and crossover (both of which increase the random rearrangement, or recombination, of genes).

Twin studies, test crosses, amniocentesis, cell fusion, and recombinant DNA techniques are currently used to extend our knowledge of genetics.

Review Questions

1. True or False: Chromosomes are made exclusively of DNA.
2. Why wouldn't you find uracil in DNA? Be specific in your answer.
3. What is the difference between a gene mutation and a chromosome mutation?
4. Does a cell with a diploid number of chromosomes have twice the number of chromosomes? How about twice the number of genes?
5. What role does the nucleolus play in protein synthesis?
6. Does RNA contain the same sequence of nucleotides as its DNA template?
7. Matching (more than one entry in column II may apply):

I	II
1. Translation	A. Haploid
2. Missense	B. G
3. XXY	C. Flat feet
4. Always made	D. DNA to RNA
5. Phenotype	E. RNA to protein
6. Genetic drift	F. U
7. No antibodies normally present	G. Mutation
8. Dihybrid cross	H. Rh
9. Zygote	I. Strong winds
10. Transcription	J. Turner
11. A	K. Diploid
12. Erythroblastosis fetalis	L. $9:3:3:1$
13. C	M. Kleinfelter
14. Libdenoch	N. GAUUAU to GAAUAU
	O. T
	P. Repressor
	Q. Operon
	R. $1:2:1$
	S. GAAUGU to CAAUGA
	T. p to 1, q to 0
	U. Nonsense

8. Given a trait with $p = 0.75$ and $q = 0.25$, what percentage of the F_1 will be homozygous recessive?
9. True or False: Pedigrees are nearly perfect canine specimens?
10. Can a human with a sublethal gene attain the age of 85? 45? $7\frac{1}{2}$? birth?
11. Three number 21 chromosomes indicates what abnormality?
12. Balanced carriers are the result of
   (a) equal strength in both arms
   (b) trisomies
   (c) alkaptonuria
   (d) translocations
   (e) the test cross

13. Incomplete penetrance occurs when
    (a) ribosomal RNA fails to pierce the nuclear membrane
    (b) a dominant gene is not always expressed
    (c) there is selection against the heterozygote
14. What are two phenomena mentioned in this chapter in which a heterozygote is phenotypically distinct from either type of homozygote?
15. A somatic cell will normally contain (number) alleles for a particular gene locus?
16. Name three prerequisites for a population to reach genetic equilibrium.
17. Does directional selection change genotypic frequencies? phenotypic frequencies?
18. Explain how a gene mutation may disrupt a biochemical pathway.
19. If two genes are known to be tightly linked, the probability that crossover will occur between them is (greater, lesser) than the probability of crossover if their loci are far apart on the same chromosome.

Selected References

Balinsky, B. I. *An Introduction to Embryology*, 3rd ed. Philadelphia: W.B. Saunders Company, 1970.

Beadle, G. W. "The Genes of Men and Molds," *Scientific American*, **179**:30, 1948.

Benzer, S. "The Fine Structure of the Gene," *Scientific American*, **206**:70, 1962.

Bodner, W. F. and L. L. Cavalli:-Sforza. *Genetics, Evolution, and Man*. San Francisco: W. H. Freeman, 1976.

Chai, C. K. *Genetic Evolution*. Chicago: University of Chicago Press, 1976.

Crow, J. F. *Effects of Radiation and Fallout*. Public Affairs Pamphlet No. 256, Public Affairs Committee, New York, 1957.

Dobzhansky, T. *Evolution, Genetic, and Man*. New York: John Wiley & Sons, 1955.

Dunn, L. C., and T. Dobzhansky. *Heredity, Race and Society*, rev. ed. New York: Mentor Books, New American Library, 1952.

Gates, R. R. *Human Genetics*. New York: Macmillan Publishing Co., Inc., 1946.

King, R. C. *Genetics*. New York: Oxford University Press, 1962.

Knight, C. A., and D. Fraser. "The mutation of viruses," *Scientific American*, **193**:74, 1955.

Lerner, I. M. *Population Genetics and Animal Improvement*. New York: Cambridge University Press, 1950.

Levine, R. P. *Genetics*. New York: Holt, Rinehart and Winston, Inc., 1962.

Moore, T. A. *Heredity and Development*. New York: Oxford University Press, 1963.

Müller, H. J., C. C. Little, and L. H. Snyder. *Genetics, Medicine and Man*. Ithaca: Cornell University Press, 1947.

Osborn, F. *Preface to Eugenics*. New York: Harper & Row, 1951.

Penrose, L. S. *The Biology of Mental Defect*. New York: Grune & Stratton, 1949.

Peters, J. A. (ed.). *Classic Papers in Genetics*. Englewood Cliffs: Prentice-Hall, Inc., 1959.

Reed, S. C. *Counseling in Medical Genetics*. Philadelphia: W.B. Saunders Company, 1955.

Robinson, R. Gene Mapping in Laboratory Mammals. New York: Plenum Publishing Corporation, 1971–72.

Sinnot, E. W., L. C. Dunn, and T. Dobzhansky. *Principles of Genetics*. New York: McGraw-Hill Company, 1958.

Snyder, L. H., and P. R. David. *The Principles of Heredity*, 5th ed. Boston: D. C. Heath, 1957.

Stern, C. *Principles of Human Genetics*, 2nd ed. San Francisco: W.H. Freeman & Company, 1960.

Stubbe, H. *History of Genetics, from Prehistoric Times to the Rediscovery of Mendel's Laws*. Cambridge, MIT Press, 1972.

Waddington, C. H. *The Strategy of the Genes*. New York: Macmillan Publishing Co., Inc., 1958.

Wagner, R. P., and H. K. Mitchell. *Genetics and Metabolism*. New York: John Wiley & Sons, 1955.

Watson, J. D. *Molecular Biology of the Gene*, 3rd ed. Menlo Park, Calif.: W. A. Benjamin, 1976.

White, M. J. D. *Animal Cytology and Evolution*, 2nd ed. New York: Cambridge University Press, 1954.

Reproduction and Embryology

Reproduction is the production of new individuals from one or two parents. **Sexual reproduction** involves the union of haploid gametes, which increases the genetic diversity of the offspring. Fertilization produces a diploid zygote that contains the full gene complement for the new individual. **Embryological development** involves a sequential differentiation and specialization of cells derived from the zygote, culminating with the hatching or birth of an independent organism.

Reproduction

Asexual Reproduction

Asexual reproduction is considered more primitive than is sexual reproduction because it does not require meiosis and haploid gametes. Four major types of asexual reproduction are displayed by a wide variety of protozoans and animals: binary fission, multiple fission and sporulation, fragmentation, and budding (Fig. 39-1)

Binary fission in protists is a simple mitotic division in which the nucleus divides and each of the two daughter cells receives half the cytoplasm of the original cell (Fig. 6-10). In metazoans, binary fission of individual cells is followed by a splitting of the animal into two equal parts, each of which assumes an independent existence. Binary fission is a common mode of reproduction in protists and some simple metazoans (e.g., sea anemones).

Multiple fission differs from binary fission in that the nucleus undergoes several mitotic divisions before cytokinesis occurs; the cytoplasm then cleaves into several new cells, each containing one nucleus. Multiple fission is exemplified by some amoebas and sporozoans (e.g., *Plasmodium*, which causes malaria). **Sporulation** is (in zoology) essentially the same process whereby many nuclear divisions and multiple fission produce many new cells within a cyst. These cells may then combine in pairs, forming **zygotes** that secrete protective coverings. This method of reproduction occurs mostly in parasitic sporozoans, and the dispersal usually involves ingestion of the cyst by a suitable host (see Chapter 7).

Fragmentation is the breaking off of body parts that subsequently regenerate into complete organisms. Some poriferans, cnidarians, platyhelminths, nemerteans, annelids (e.g., some earthworms), and echinoderms may reproduce by fragmentation.

Budding involves an unequal division of the organism's body in which the smaller individual is the "bud." Budding may be either internal or external. **Internal budding** occurs in poriferans when several cells within the organism's body become encapsulated by dense coverings, forming **gemmules.** The parent's body disintegrates during winter, and each gemmule eventually forms a new sponge the following spring. **External budding** involves the growth of a new individual from the parent's body wall. The bud may either remain attached (as in the colonial cnidarian *Obelia*) or separate, becoming an independent individual (as in *Hydra*).

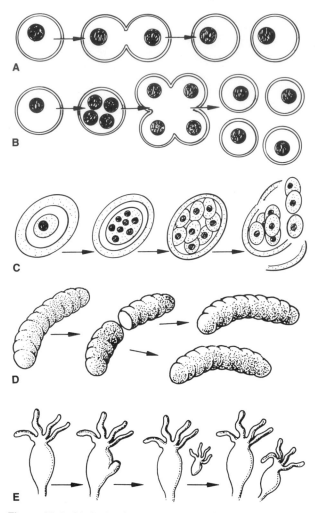

Figure 39-1. Methods of asexual reproduction (diagrammatic). (A) Binary fission. (B) Multiple fission. (C) Sporulation. (D) Fragmentation. (E) Budding. Note that sporulation is essentially the same as multiple fission except that it occurs within a cyst.

The Development of Sexuality

The **Volvocales,** an order of flagellated plantlike protists, can be arranged to illustrate different grades of organization, from loose associations of cells to complex colonies approaching a metazoan organization. This series also serves to illustrate one pathway by which sexuality evolved (Fig. 39-2).

Spondylomorum forms colonies of 16 cells that are practically independent. Each cell reproduces by binary fission to form a new colony. No gametes are known.

Chlamydomonas is a noncolonial type that may reproduce asexually by simple fission into two, four, or eight daughter cells. This flagellate also produces gametes of equal size and morphology (**isogametes**) that fuse together in pairs; the resulting zygote undergoes fission, producing several individuals. The fusion of gametes during fertilization results in new combinations of genes. The expressions of these novel recombinations varies the phenotypes found within the population, which enhances the survival of the species in the face of changing selective pressures.

Pandorina forms colonies of 4 to 32 cells embedded in a gelatinous matrix, each cell independent of the others. New colonies may be produced asexually by division of each cell in the mother colony into a new daughter colony. Or sexual reproduction may occur through the formation of gametes of unequal size. The larger "female" gametes fuse with the smaller "male" gametes.

Eudorina forms colonies of 32 cells. Each cell may produce a new colony asexually by fission. Sexual reproduction also occurs, with male and female gametes forming in different colonies. At times, the cells of certain colonies enlarge and develop into ova. In male colonies, each cell divides to form 16 or 32 small spermatozoa. These fertilize the large ova in a female colony to form zygotes.

Volvox globator represents the most complex stage in this series. The thousands (one species has 40,000) of cells in a colony are united by cytoplasmic strands; physiological continuity is thus established between the cells, a condition not found in the colonies previously described. If a single cell is removed and isolated from its neighbors, it will assume a teardrop shape and swim about by means of its two flagella, but after a time the isolated cell dies. Most cells in the colony contain an eyespot, chloroplasts, water-expulsion vesicles, and two flagella; these are called **somatic cells.** The production of daughter colonies is accomplished by specialized reproductive cells.

A colony may have as many as 50 reproductive cells, which are larger than somatic cells and lack flagella. Some of these enlarge to form **ova;** others divide longitudinally into mobile, spindle-shaped **sperm-**

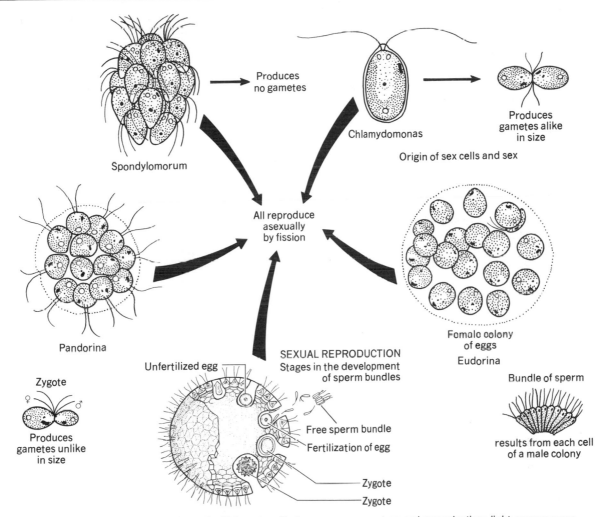

Figure 39-2. Evolution of sex in the order Volvocales. Dark arrows represent sexual reproduction; light arrows represent asexual reproduction.

atozoa. At fertilization a single sperm fuses with an egg, forming a zygote that secretes a surrounding wall. Eventually the zygote breaks out of the wall and produces a new colony by binary fission. The young colonies are retained within the parent colony until the somatic cells of the latter die and disintegrate.

Reviewing this series, note that the development of sexual reproduction is accompanied by a trend from a single aggregation of cells to a large and highly com-

plex colony composed of cells that show considerable specialization and division of labor. In *Volvox*, true somatic cells, which constitute the bulk of the individual colony and are unable to reproduce to form new colonies, are encountered. In the other genera just described, every cell has the capacity to reproduce the whole colony. *Volvox* also contains true germ cells—that is, cells that are specialized for reproduction. Furthermore, a clear case of natural death occurs in the somatic cells when they fall to the bottom of the pond and disintegrate. The situation in *Volvox* is similar to that in most metazoans in which the body consists of somatic cells and germ cells. The latter produce male or female gametes, which normally must fuse (fertilization) to produce a new organism. The germ cells maintain the continuation of the species by producing new individuals, whereas the somatic cells perish when the organism dies.

Sexual Reproduction in Metazoans

Sexual reproduction usually involves the fusion of a male gamete, or spermatozoa (sperm), with a female gamete, or ovum (egg), in a process called fertilization. Many protozoans and simple metazoans are **monoecious (hermaphroditic)**, with both male and female gametes produced by the same individual. The vast majority of metazoans are **dioecious,** with male and female gametes produced by separate individuals.

Monoecious organisms sometimes have limited locomotive capabilities, or they infrequently encounter others of the same species; hence, in some cases hermaphroditism may be an adaptation to a reproductively isolated existence. The majority of hermaphrodites **cross-fertilize;** for example, during copulation each earthworm receives sperm from the other. **Self-fertilization** is rare in the animal world (it occurs in tapeworms and flukes). As discussed in Chapter 38, this reduces the incidence of homozygous recessive gene combinations, whose expressions are most often deleterious.

A combination of self-fertilization and cross-fertilization occurs in some hydras; the sperm are released into the surrounding water and are carried at random by water currents to other hydras, sometimes even back to the releaser.

Another interesting variation of sexual reproduction is **sperm storage.** Here, the female receives sperm, but she may use only some of them immediately and store the rest, often in a specialized **seminal receptacle,** for future fertilizations. For example, the queen honeybee mates only once in her lifetime, yet she may lay a million eggs over several years.

The development of viable individuals from unfertilized eggs, as in some rotifers, nematodes, arthropods, and others, is called **parthenogenesis.** Males have never been observed in some species of rotifers and aphids. Ova that develop parthenogenetically are usually diploid, and the young are genetically identical to the mother. Some aphids and rotifers reproduce parthenogenetically throughout the summer months and then sexually as winter approaches. The sexual recombination that results from fertilization increases the genotypic (and phenotypic) diversity of the population, which enhances survival of the species when the zygotes hatch in the unpredictable spring environment.

Artificial parthenogenesis may be experimentally induced in some nonparthenogenic organisms (as in the frog) by various manipulations of the egg membrane such as pinpricks, dilute acid solution, warmth, and other stimuli.

Most aquatic invertebrates release their gametes into the surrounding water, where fertilization occurs. This is **external fertilization,** in which the gametes unite outside the bodies of the parents. Most fishes and amphibians also employ external fertilization. For example, when frogs mate, the male mounts the female dorsally and stimulates her to release eggs into the water; the male simultaneously releases sperm over the egg pile. Internal fertilization generally characterizes terrestrial animals. The transmission of sperm from male to female during copulation has evolved several times in the animal kingdom, often as an adaptation to protect the gametes from a dry external environment.

Animal embryo development takes place in one of three general ways: **Oviparous** animals lay eggs that hatch outside the body of the mother—for example, birds. **Ovoviviparous** animals produce eggs that hatch within the mother's body, but the embryos are nourished by yolk, not by the mother's bloodstream—for example, certain insects, many sharks, and many reptiles. **Viviparous** animals develop embryos within the body of the mother that are nourished through close contact with her bloodstream—for example, placental mammals. The following discussions refer primarily to mammalian reproductive systems.

The Formation of Gametes—Meiosis

Somatic cells and the primary germ cells that give rise to the male and female gametes are diploid, having inherited half their chromosomes from the father and half from the mother. Each parent contributes one autosome of each type plus an X or Y chromosome; that is, each parent provides half the genes necessary to determine each trait. The genes are distributed on specific chromosomes, and two chromosomes carrying complementary sets of genes are called **homologous chromosomes.** For example, in human somatic cells there are 46 separate chromosomes, or 23 pairs of homologous chromosomes. During meiosis the diploid chromosomal complement is reduced by one half, so that each (haploid) gamete receives one of each homologous chromosome pair.

Spermatogenesis

The production of **spermatozoa (spermatogenesis)** occurs in the male gonads (**testes**) (Fig. 39-4). The process of spermatogenesis involves two **meiotic divisions:** a **primary spermatocyte** divides to form two **secondary spermatocytes,** each of which undergoes a reduction division to produce two haploid **spermatids.** The spermatids then undergo extensive morphological changes that convert them into **spermatozoa.**

The first, or **leptonema,** stage of spermatid formation looks just like the prophase of mitosis. Next, in the **zygonema** stage, the homologous chromosomes pair up in a process called **synapsis,** in which first at selected points near the centromere and then all along their lengths the homologous chromosomes become applied to each other. The synapsed homologs are called **bivalents.** The **pachynema** stage follows, in which the synapsed homologs twist around each other and become indistinguishable. After a while, each synapsed chromosome splits into two chromatids. The unit of four chromatids is called a **tetrad.** It is during the pachynema stage that **crossover** occurs: sometimes, two chromatids of homologous chromosomes in the tetrad cross over each other, and regions of the chromatids containing corresponding loci are actually exchanged. Crossover increases genetic variability by randomly producing new combinations of genes.

The subsequent **diplonema** stage is characterized by a distinct split between the chromosomes of each bivalent. The chromatids may also be seen, and connection sites called **chiasmata** mark the point at which crossover has occurred.

The next stage of spermatogenesis, called **dia-kinesis,** contains the first meiotic division. The recently split bivalents have produced two sets of chromatids. As these sets are pulled apart, the chiasmata break and two chromatids from each tetrad are pulled toward each pole of the cell. Each of the two secondary spermatocytes that results from this division has only half the somatic number of chromosomes. This has been a reduction division as it transforms the original diploid cell into two haploid cells.

After a very short **interphase,** a second meiotic division occurs in each of the secondary spermatocytes. This is similar to mitosis, with the chromatids' splitting at the centromeres. This second meiotic division is sometimes described as **equatorial division,** because the separation of sister chromatids results in an equal (but still haploid) number of chromosomes in the nuclei of the spermatids. The spermatids then metamorphose, losing most of their cytoplasm, into spermatozoa.

Oogenesis

Up to the diakinesis stage, ovum formation is rather similar to spermatid formation, with the notable exception that the primary oocyte grows considerably before the reduction division. The mass increase results from the accumulation of nutritive substances in the cytoplasm. From this point on, oocyte formation is quite unlike spermatid formation due to the formation of **polar bodies.**

At the beginning of diakinesis, the nuclear membrane of the primary oocyte disintegrates, and the synapsed chromosomes move to the periphery. The bivalents separate during this reduction division and each daughter cell receives half the somatic number of chromosomes. The subsequent cytokinesis is unequal. One set of chromosomes pushes out into a very small cytoplasmic bulge, which pinches off and becomes the **first polar body.** Most of the cytoplasm in the primary oocyte has now become incorporated into the secondary oocyte, which contains one homolog of each chromosome type, each consisting of two chromatids.

The second equatorial division distributes the chromatids. Again cytokinesis is unequal. A bulge is produced and pinches off, becoming the **second polar body.** The polar bodies die, leaving a single functional haploid ovum.

The Male Reproductive System

Within the vertebrate testes lie the **seminiferous tubules** in which spermatogenesis occurs. A cross-section of a seminiferous tubule shows a concentric arrangement of developing sperm: closest to the inner tubule wall are the spermatogonia, which by mitosis give rise to primary spermatocytes; and toward the central **lumen** secondary spermatocytes, spermatids, and maturing spermatozoa are progressively encountered. Associated with the tubules are **ertoli cells**, which presumably nourish the developing sperm.

A mature spermatozoan is composed of three general areas: the head, middle piece, and tail (Fig. 39-3). The head is partly covered by an acrosomal cap, which contains enzymes that help the sperm penetrate the egg membranes during fertilization. The

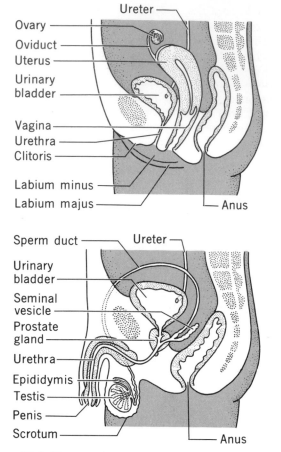

Figure 39-4. Diagram of a median section of the human male and female reproductive organs showing their relation to the urinary bladder and urethra.

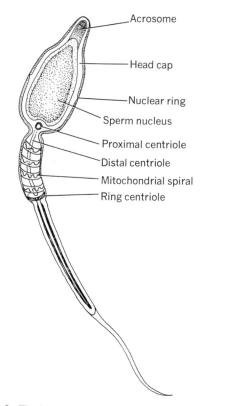

Figure 39-3. The human sperm (mature), diagrammatic. See text for detailed description.

main portion of the head contains the haploid nucleus. The middle piece houses many mitochondria and is separated from the head by two mutually perpendicular centrioles. The tail, a flagellum, obtains energy from the mitochondria.

The seminiferous tubules unite to form the **epididymis** (Fig. 39-4), a highly coiled tube about 6 m in length. Here mature sperm may be stored for several months.

The epididymis leads into the **sperm duct (vas deferens)**, a muscular tube that passes into the abdominal cavity where it joins the duct of the **seminal vesi-**

cle to form the **ejaculatory duct.** These relatively short ducts pass through the **prostate gland** and open into the **urethra.** The prostate gland is about the size of a chestnut, and on either side of it is a pea-sized **bulbourethral (Cowper's) gland.** Secretions from the seminal vesicles, prostate gland, and bulbourethral glands are added to the sperm, thereby forming the **seminal fluid (semen).** These secretions act as a medium of transport, provide nutrients, allow activation of the sperm, and neutralize acidity encountered in the vagina. Smooth muscle contractions propel semen through the ejaculatory ducts into the urethra and then out of the body through the **copulatory organ,** or **penis.**

The Female Reproductive System

In the female, there are two **ovaries** that produce ova and female sex hormones, two **oviducts** (or **Fallopian tubes**) about 10 cm long, a **uterus,** a **vagina,** and **external genetalia** (Fig. 39-4). The ovaries are almond shaped and weigh from 2.0 to 3.5 grams. At birth they contain about 70,000 developing egg cells, but most of these later disappear. Mature ova are discharged into the coelomic cavity and are swept by ciliary currents into an oviduct. Cilia and probably peristaltic muscular contractions transport the ovum to the uterus. Here, if the ovum has been fertilized, the zygote remains during embryonic development.

The Estrous and Menstrual Cycles

In mammals, females **ovulate** (release ova or an ovum from the ovaries) at regular intervals; the cycle of hormonally induced changes in the wall of the uterus during this "period" is called the **estrous cycle.** The characteristics of estrous cycles vary somewhat among different mammals. We will consider the human estrous cycle.

The follicular phase of the hyman estrous cycle begins with gradual maturation of one or more **ovarian (Graafian)** follicles within the ovaries (Fig. 39-5). The follicles are stimulated to mature by the presence of **follicle stimulating hormone (FSH),** which is secreted into the bloodstream by the anterior pituitary gland. FSH also stimulates **nurse cells** in the follicles to sup-

ply nutrients to the single primary oocyte within each follicle. As the follicles mature, they secrete **estrogen,** a female sex hormone. The name estrogen applies to several closely related female sex hormones that have similar functions. Estrogen stimulates the uterus to **hypertrophy,** that is, to grow by cell enlargement. When the estrogen concentration in the blood reaches a certain level, the anterior pituitary is stimulated to release a burst of **lutenizing hormone (LH).** LH almost immediately induces the mature Graafian follicle to rupture, allowing the ovum to escape (ovulation). The ovum is drawn into the funnel-shaped ciliated opening of a Fallopian tube and is slowly conveyed to the uterus. Fertilization usually occurs somewhere within the Fallopian tube.

As soon as the follicle releases the ovum, it undergoes drastic changes. It changes color, becoming a **corpus luteum** (yellow body) and begins prolific secretion of another hormone, **progesterone,** in addition to estrogen. This is the beginning of the **luteal phase** of the estrous cycle. Progesterone takes over where estrogen left off—it stimulates proliferation of cells in the uterine wall. If the ovum is fertilized, the corpus luteum remains intact, continuously secreting its hormones, and the zygote becomes implanted in the wall of the uterus. If the ovum is not fertilized, the corpus luteum stops secreting its hormones. In response to the cessation of progesterone secretion in particular, the membrane lining the uterus (the **endometrium**) begins to break down. In certain primates, the endometrium is suddenly sloughed off, along with varying amounts of blood, and is lost through the vaginal opening in a process called **menstruation.** For this reason, the estrous cycle in humans, anthropoid apes, and Old World monkeys is commonly referred to as the **menstrual cycle.** The menstrual cycle in mature human females has a period of about 28 days.

The blood estrogen level drops when the corpus luteum stops its secretions, and this decrease stimulates (via the hypothalamus) FSH secretion by the anterior pituitary. As noted, this in turn stimulates the maturation of new follicles. The follicles secrete estrogen, inhibiting further FSH secretion and stopping the menstrual flow so that the uterus can be prepared for the next ovulatory period.

Birth control pills are usually composed of synthetic estrogen and progesterone. The supplementary

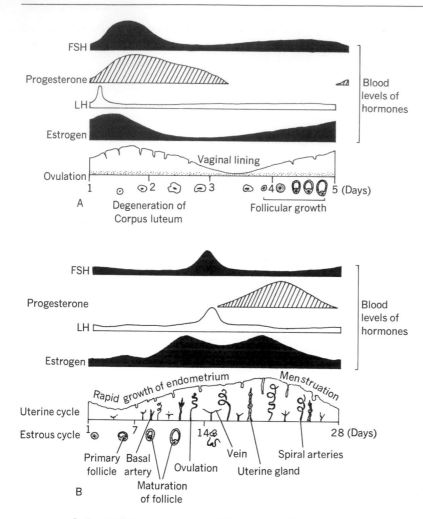

Figure 39-5. Human estrous and menstrual cycles. Top: Estrous cycle. Bottom: Menstrual cycle. [After R. A. Boolootian, *Human Reproduction,* slide sequence. New York: John Wiley, 1971.]

estrogen halts FSH secretion, and hence ovulation does not occur. A birth control pill is generally ingested each day during the first three weeks of the menstrual cycle, followed by seven days of placebos that contain no estrogen. The "pill," in effect, creates a false menstrual cycle in which ovulation, and therefore possible conception, is suppressed.

The estrous cycles of nonprimate mammals do not include menstruation. A brief estrous (or "heat") period coincides with ovulation; at that time, and at no other, will the female copulate with a male. The estrous cycle of rats, hamsters, and mice lasts about four days. No menstruation occurs, and, instead, the endometrium is partly reabsorbed if no fertilization occurs. The estrous cycles of many mammals are much longer, sometimes six months (cattle) or one year (deer). Female rabbits do not ovulate regularly —their ovulations are initiated by copulation.

Fertilization

Usually, if copulation has occurred at the proper time, several million sperm approach the ovum as it passes through the Fallopian tube, but it is extremely rare for more than one to penetrate the ovum. The mechanism of sperm penetration is primarily dependent upon the **acrosomal cap,** which contains lysosomal enzymes that digest the ovum's outer membrane. Penetration by a sperm stimulates the production of a protective membrane (**fertilization membrane**) that keeps other sperm out.

The nucleus of the ovum is known as the **pronucleus.** Upon entering the ovum, the sperm's head separates from its tail to form the **male pronucleus.** The haploid male pronucleus fuses with the haploid female pronucleus to form the diploid nucleus of the **zygote.** This event restores the diploid number of chromosomes and completes the process of fertilization.

Embryology—The Study of Development

Early Developmental Stages of Metazoans

Cleavage and Morula Formation

Cleavage is the repeated mitotic division of the zygote into a mass of small cells. If some of the cells separate in an early stage of cleavage, identical twins, triplets, and so on may result.

After mitotic division of its diploid nucleus, **cyto-**

Figure 39-6. Types of cleavage patterns and the resulting blastulae and gastrulae. (A) Archenteron (gastrocoel). (B) Blastocoel (segmentation cavity).

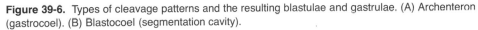

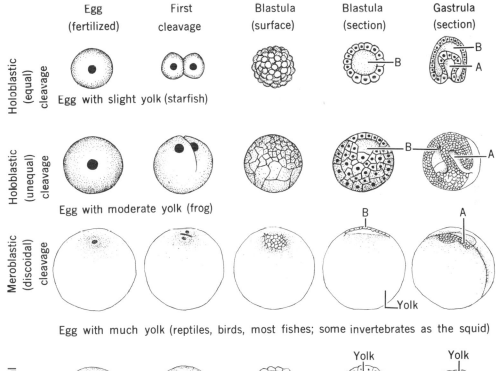

Egg with slight yolk (starfish)

Egg with moderate yolk (frog)

Egg with much yolk (reptiles, birds, most fishes; some invertebrates as the squid)

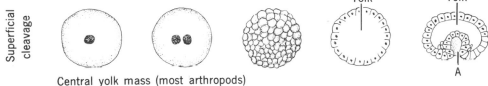

Central yolk mass (most arthropods)

kinesis divides the zygote into two cells, or **blastomeres,** that remain attached to one another. The blastomeres in turn undergo a sequential series of mitotic divisions. Because the ovum has not implanted onto the uterine wall, little growth occurs, and the mitotic cleavage divisions result in progressively smaller and smaller cells. The resemblance of the group of blastomeres to a mulberry suggested the term **morula** (L. mulberry), which is sometimes used to describe the zygote during the early cleavage stages. Cleavage characterizes the early development of all metazoans, but several distinct types of cleavage patterns are recognizable (Fig. 39-6).

If the ovum contains relatively little yolk, the entire zygote divides into first 2, then 4, 8, and so on blastomeres. If these daughter cells are approximately equal in size, the pattern is called **equal holoblastic cleavage.** Such a cleavage pattern is characteristic of starfishes and *Amphioxus.* If cleavage is complete but the daughter cells are of unequal size because of the difficulty of cleavage through areas of high yolk content, the process is known as **unequal holoblastic cleavage;** such a cleavage pattern is illustrated by the frog morula after the first two or three cleavage divisions.

If the ovum contains a considerable amount of yolk, the first cleavage furrow does not completely bisect the zygote, and only restricted portions of the cytoplasm undergo cleavage. If cell division is restricted to a small cap or disc on one side of the fertilized egg, as in birds, the process is called **meroblastic,** or **discoidal, cleavage.** In insects, superficial cleavage is restricted to a layer of cytoplasm surrounding a central mass of yolk.

The factor determining whether a zygote will cleave holoblastically, meroblastically, or superficially is the yolk concentration in the egg (**lecithality**). Although there are exceptions (notably in humans), generally the higher the yolk concentration, the more the cleavage is restricted to a particular region of the zygote.

Oligolecithal eggs contain relatively little yolk, and this condition is conducive to holoblastic cleavage, as in *Amphioxus.* In insects and many other arthropods, the yolk is distributed for the most part in the center of the egg. Such **centrolecithal** eggs lead to superficial cleavage.

Telolecithal eggs have high yolk concentrations, and the yolk is usually confined to one pole of the egg. Consequently, meroblastic or discoidal cleavage results, as in avian eggs. Telolecithal eggs are generally divided into animal and vegetal poles. The **animal pole** usually contains most of the cytoplasm and is more active mitotically, whereas the **vegetal pole** contains mostly yolk, and mitosis is slowed.

Cleavage may be further characterized as either determinate or indeterminate. In **determinate cleavage,** each blastomere is destined to become a particular portion of the embryo. In **indeterminate cleavage,** early blastomeres may develop into any part of the embryo; they are equipotent. Different parts of the embryo are determined later in its **ontogeny (life history).** Protostomes are usually determinate, whereas deuterostomes are indeterminant.

Early Cleavage Patterns

In the beginning stages of embryogenesis, the blastomeres of different animal groups orient themselves in characteristic spatial patterns (Fig. 39-7). In **radial cleavage,** successive tiers of equal-sized blastomeres lie directly above and below one another, and their fate is indeterminate. Radial cleavage characterizes almost all deuterostomes: vertebrates and echinoderms. In almost all protostomes, such as annelids, nemertineans, and most mollusks, **spiral cleavage** results from a shift of the upper tier of blastomeres. This forms a latticelike arrangement in which upper blastomeres lie directly over the junctions of lower blastomeres. In spiral cleavage, blastomeres in the upper tier are usually smaller (**micromeres**) than are those in lower tier (**macromeres**). The fate of the blastomeres in this case is determinate. **Bilateral cleavage,** which occurs in ctenophores and tunicates, arises when two sizes of blastomeres are confined to separate areas in the morula, resulting in an apparent plane of bisymmetry.

Blastula

As cleavage proceeds, a cavity forms in the center of the murula (Fig. 39-6) and enlarges until the embryo resembles a hollow ball of cells a single layer thick. At this stage the embryo is called a **blastula.** The cavity is called the **blastocoel,** and the cellular layer is the **blastoderm.**

RADIAL CLEAVAGE   SPIRAL CLEAVAGE

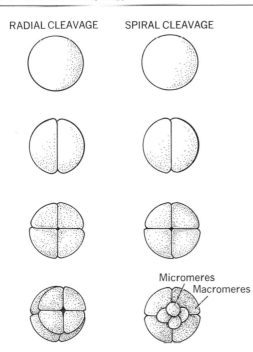

Micromeres
Macromeres

Figure 39-7. Left: Radial cleavage. Note that the cleavage furrows line up in successive tiers. Right: Spiral cleavage. Note that the cleavage furrows are at 90° angles to each other and that, in this and many cases, one tier (macromeres) has larger cells than the other tier (micromeres).

The blastocoel may not be concentric within the blastula (e.g., the frog; Fig. 39-6), or it may be lacking (e.g., most arthropods). Also, the blastula wall in some cases is more than one cell thick.

Gastrula

Gastrulation is a series of cell migrations that reorient the blastomeres into positions that define the prospective tissues of the adult body plan. The cells on one side of the blastula may begin to invaginate (infold) into the interior (Fig. 39-6) through a temporary opening called the **blastopore.** The blastocoel is gradually obliterated during invagination, whereas a new cavity, called the **gastrocoel (archenteron),** becomes bounded by the invaginated cells. The gastrocoel represents the primitive **gut cavity.** The embryo is called a **gastrula.**

Many variations of gastrulation occur in the animal kingdom. For example, invagination does not occur in hydras, but certain cells of the blastula divide and fill up the blastocoel. These and other variations in developmental patterns are used by the systematist to interpret evolutionary relationships.

Germ Layers

The gastrulas of the simplest metazoans, cnidarians, and ctenophores consist of two germ layers; an outer **ectoderm** and an inner **endoderm.** These phyla are said to be **diploblastic.** However, in most metazoans a third germ layer, the **mesoderm,** arises as a result of gastrulation. Thus most metazoans are **triploblastic.** The origin of mesoderm varies in different groups, originating either by multiplication of a few special blastomeres (which may be recognized in early cleavage stages) or from pouches arising from the walls of the archenteron. All the tissues and organs of the body are differentiated from these primary germ layers. The body parts that develop from the germ layers are listed in Figure 39-8.

Induction: The Spemann—Mangold Experiments

In the 1920s, an ingenious and illuminating series of experiments was performed by Hans Spemann and his pupil Hilde Mangold. They investigated the factors that determined the fate of a given piece of tissue in an embryo.

In the early gastrula, these experimenters found that, if a small piece of prospective ectoderm (that would have become epidermis if left alone) was transplanted to the **neural crest** region of another early gastrula, the transplanted cells differentiated into nervous tissue instead of epidermis. Several similar experiments were performed on early gastrulas, and in every case the transplanted cells, no matter what their origin in the original gastrula, differentiated in accordance with their environment.

Spemann and Mangold found that similar transplants with late gastrulas produced the reverse results: The transplanted tissue differentiated according to its original environment independently of the environment into which it was transplanted. Some change had occurred during gastrulation that determined the fate of the cells.

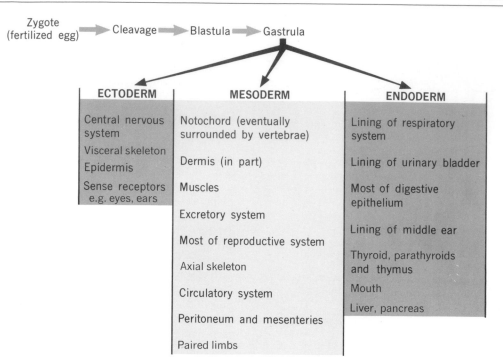

Figure 39-8. Simplified diagram of the embryonic differentiation in a vertebrate.

Mangold investigated this phenomenon further. In one of her most fascinating experiments, she transplanted part of the **dorsal lip** of the blastopore to the **ventral lip** (see Fig. 39-9), and soon a set of **rudimentary organs** was produced. The new organs were induced to develop by the presence of the cells taken from the dorsal lip of the blastopore. **Induction** occurs when one embryonic tissue directs the differentiation of another. Note in Figure 39-9 that the cells of the dorsal lip normally invaginate late in gastrulation to form the roof of the archenteron. Further experimentation has shown that, after invagination, the inducing capacity of the dorsal lip cells is effective on mesoderm and endoderm. Spemann, in 1938, called this special part of the gastrula the **primary organizer.** He found that, even before cleavage, its position is discernible as the gray crescent (Fig. 39-9).

With further experimentation, it was found that, after a particular portion of gastrula tissue had been induced to differentiate in a particular way, it in turn was frequently able to induce nearby tissues; this is known as **secondary induction.**

The sequences of embryonic induction have been outlined by many subsequent investigators, but discovery of the causative agents of induction, which are believed to be gene-mediated proteins, remains one of the most active pursuits of developmental biology.

Gross Comparative Embryology

Different animal groups display vastly different patterns of embryological development. This section serves as an overview of basic trends.

Multicellular animals share several common embryological characteristics. First, some form of cell division occurs before gastrulation. Second, the embryos of related animals go through a series of developmental stages that, initially, look very similar. For example, all vertebrates (including humans) have tails and gill clefts during some period of their early life history. As development proceeds, dissimilarities appear, as the embryos differentiate into the body plans that characterized their phyla, class, and so on.

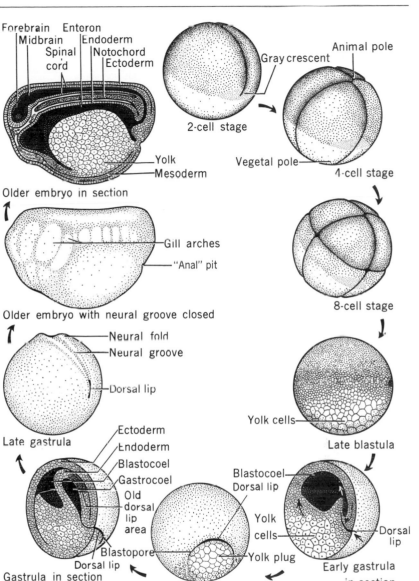

Figure 39-9. Embryology of the frog. Note the position and movements of the dorsal lip of the blastopore.

Usually these developmental stages roughly mirror phylogentic development and so can be used to trace evolutionary relationships (Chapter 41).

The major division of protostome and deuterostome lines of metazoan evolution is made on the basis of cleavage patterns, embryonic origin of the mouth (Gr. *stoma,* mouth), and coelom mesoderm development. In protostomes (annelids, mollusks, arthropods, and others), cleavage is determinate, the mouth is derived from the blastopore, and the coelom is formed from mesodermal rudiments. In deuterostomes (echinoderms, chordates, and others), cleavage is indeterminate, the mouth is not derived from the blastopore, and the coelom and mesoderm are derived from pouches in the gut region.

Table 39-1 lists some basic embryological trends in

Table 39-1. Gross Comparative Embryology—Some Basic Embryological Trends in Animalia*

Structure	Mammals Viviparous	Reptiles and Birds Oviparous	Fishes Oviparous	Amphibians Oviparous
Yolk	Lacking	Telolecithal or oligolecithal; absorbed during development	Large amounts, nonsegmented	Oligolecithal
Cleavage	Complete	Discoidal	Meroblastic	Complete; uneven
Yolk sac	Hypoblast spreads under inner surface of trophoblast, encircling cytoplasm	Area opaca (periphery of blastodisk) spreads down. Endoderm adheres to surface	Blastodisk spreads over yolk, endodermal portion does not go with it	Yolk is within gastrula
Body-yolk considerations	Via body folds under and around embryo.	Same as mammals	Same as mammals	Yolk is within gastrula, in early stages fills in entire gastrula except archenteron and blastocoel remnant
Amnion	Body folds fuse—later envelops embryo	Develops with chorion. Fuses around embryo, produces amniotic cavity. Function; prevent desiccation and shocks		
Chorion	Outer surface of chorion is continuous with trophoblast	Develops with allantois, also called serosa		
Allantois	Used exclusively for nutrition and O_2–CO_2 diffusion.	Function; urinary bladder, CO_2–O_2 diffusion. Between amnion, yolk sac and chorion. Spreads evenly under chorion		
Primitive/Streak/Hensen's node/Neurulation	Present: shorter than in birds; ephemeral	Present; ephemeral	Neural plates formed but no tube formed	Neural tube eventually formed

*The information in this table is highly simplified and generalized, and should only be used as a general guide. Note that entries are lacking under fishes and amphibians for amnion, chorion, and allantois. These organisms are anamniotes, without these structures. Fishes and amphibians develop in aqueous environments, and diffusional processes can readily occur without the aid of such structures.

the major vertebrate groups. Note that numerous variations exist in all categories. This table concentrates only on the placental mammals. Reptiles and birds are considered together because of the striking similarity of their embryological developments—evidence of a close evolutionary relationship.

The blastodisc of fishes is somewhat different. It lies directly over the blastocoel. Under the blastocoel is the **periblast,** which functions only in the metabolism of yolk, not directly participating in the formation of the embryo itself. Yolk sac formation in fishes is also accomplished by the blastodisc's spreading down over and enclosing the yolk. In the amphibians, generally, the yolk is more or less passively enclosed during gastrulation, and so there is no yolk sac per se.

In reptiles and birds, the blastodisc at the animal pole may be visually differentiated into two areas. Looking down on the blastodisc, its opaque periphery that lies directly over the yolk is called the **area opaca.** The central **area pellucida** is much less opaque as it overlies the blastocoel. The area opaca and area pellucida make up the **epiblast.** Below the blastocoel is a thin layer of cells called the **hypoblast,** and under that resides the **subgerminal cavity.** The subgerminal cavity separates the hypoblast from the yolk to form the yolk sac, and, as it does this, endodermal material adheres to the yolk surface.

In the primates, the blastodisc is formed of three major layers: the epiblast, the hypoblast, and a new layer, the **trophoblast.** The hypoblast spreads downward, under the trophoblast and around the cytoplasm, to form the yolk sac (it is called a yolk sac even though there is no yolk in it). The inner cell mass is mostly responsible for the formation of the embryo itself.

In all groups mentioned except the amphibians, the body separates from the yolk by an infolding of germ layers, usually under the embryo, so that the embryo is left sitting atop the yolk sac. The infolding results in the formation of the **yolk stalk** or, in the case of higher vertebrates that are viviparous, the **umbilicus.** In amphibians, the yolk, as mentioned, is taken into the blastula.

The infolding process is responsible for the formation of the extraembryonic structures in reptiles, birds, and mammals. These structures—the **amnion, chorion,** and **allantois**—serve for nutrition, waste excretion and storage, and sometimes protection from shocks (Fig. 39-10). Amphibians and fishes generally lack such membranes, as diffusion of nutrients and wastes occurs directly between the embryo and the surrounding aquatic environment.

The amnion and chorion develop from the germ layers differently in mammals than in reptiles and birds. In reptiles and birds, gaseous diffusion occurs from the vascularized allantois through the permeable egg shell. But the mammalian embryo, so securely enclosed within its mother, cannot obtain gases as easily. The highly vascularized allantois supplies the placenta with blood vessels. Carbon dioxide and nitrogenous wastes (urea) diffuse into the maternal bloodstream, and oxygen and nutrients diffuse into the fetal circulation.

In mammals, the amniotic cavity is developed by cavitation (a type of splitting), either between the inner cell mass and the trophoblast or within the inner cell mass itself. The chorion is developed from the trophoblast, which, upon contact with the maternal uterine wall, develops villi (maximizing surface area) that interdigitate with maternal tissues. The combination of maternal uterine tissues and chorion is known as the **placenta.** Diffusion between the maternal and fetal circulations here is the means by which nutrient and waste transport in placental mammals occurs.

The mammalian chorion and amnion are therefore not developed together. This is not the case with the reptiles and birds. In these terrestrial oviparous or ovoviviparous animals, the distal surfaces (facing away from the embryo) of the body folds become the amnion, whereas the proximal surfaces (facing the embryo) form the chorion. The chorion of reptiles and birds is sometimes referred to as the **serosa,** to differentiate it from the mammalian chorion. The allantois in reptiles and birds serves as a urinary bladder, collecting nitrogenous wastes in the form of nontoxic uric acid crystals. The necessity of this is obvious. Terrestrial eggs are provided with a waterproof shell as protection against dehydration, but this barrier prevents the elimination of metabolic wastes. And so they must be rendered nontoxic and stored within the egg. In higher mammals, the placenta takes care of waste disposal, and so this function of the allantois is unnecessary.

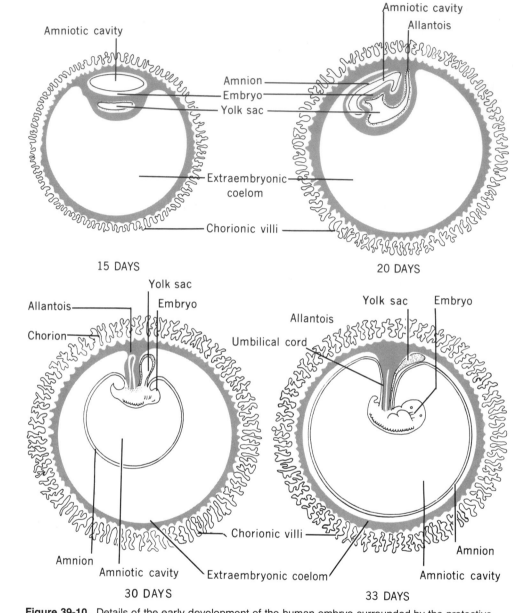

Figure 39-10. Details of the early development of the human embryo surrounded by the protective embryonic membranes. Note the rootlike villi that grow into the wall of the uterus. The placenta is a composite organ; that is, it is made of part of the uterus and part of the extraembryonic membranes.

Development of the Frog Embryo

The frog embryo provides a good example of one type of embryological development (Fig. 39-9). Cleavage is holoblastic and unequal. Due to an abundant supply of yolk, which nourishes the developing embryo, micromeres in the animal hemisphere are differentiated from macromeres in the vegetal hemisphere. Cleavage eventually produces a blastula with a cavity, the blastocoel, near the center. Gastrulation is modified due to the large amount of yolk present. The micromeres gradually grow over the vegetal pole (epiboly) until only a circular area of the yolk, the yolk plug, is visible. The gastrula now has two germ layers: an outer ectoderm and an inner endoderm. A third layer, the mesoderm, soon appears between the other two. The mesoderm itself splits into two layers: an inner **splanchnic mesoderm,** which forms the supporting tissue and musculature of the digestive canal, and an outer **parietal mesoderm,** which forms the connective tissue, muscles, and peritoneum of the body wall. The cavity surrounded by these two mesodermal layers is the coelom.

Soon after gastrulation, two **neural ridges** appear as dorsal thickenings of the ectoderm. The neural ridges fuse to form a hollow **neural tube,** which eventually will develop into the brain and spinal cord of the embryo. The neural tube lies along the **middorsal line,** and the embryo now lengthens along this axis. The region where the yolk plug was situated lies at the posterior end and eventually becomes the anus. The invagination that becomes the anal opening appears beneath the elongating tail at the posterior end. On either side near the anterior end, two gill arches appear, and in front of each of these a depression arises that unites with its complement and migrates to the ventral surface, becoming the ventral sucker. An invagination soon appears anterior to the ventral sucker; this oval pit develops into the mouth. On either side above the mouth, a thickening of ectoderm represents the beginning of an eye, and just above the gills appear invaginations that form the inner ears. The markings of the muscle segments show through the skin along the sides of the body and tail.

Development of the Human Embryo

Figures 39-10 and 39-11 illustrate the development of the human embryo.

At approximately 30 hours after fertilization, the first cleavage division occurs while the human zygote is still within the Fallopian tube. Some time between 40 and 60 hours, the second cleavage takes place, and the third occurs at about 72 hours. Approximately two days after fertilization, the zygote leaves the oviduct and enters the uterus. It is now a morula composed of 32 cells. Soon thereafter the **blastocyst** (mammalian blastula) and blastocoel are formed. Within the blastocoel, eccentrically, lies the inner cell mass that is the **primordium** of the embryo proper. Opposite the inner cell mass, the trophoblast begins to proliferate, and at approximately day 6 the complete blastocyst is attached to the uterine wall. Then the mother–embryo connection is extended as the trophoblast invades the uterine wall and actually dissolves part of it. This dissolution results in a number of spaces, or lacunae, that fill with blood and diffusionally communicate with the maternal circulation.

At this point, day 8, the embryo is fully imbedded in the uterine wall. The inner cell mass splits, forming the amniotic cavity.

By day 18, much differentiation has occurred. A **primitive streak** that gives rise to mesoderm has developed on the dorsal surface of the blastodisc, in the region where the notochord develops. At the anterior end, **Hensen's node** represents the area of the head process. The mesoderm splits, and the cavity formed is the coelom.

By day 20, the nervous system has begun to develop. The mesoderm at the head process induces the overlying ectoderm to form the neural tube. Mesoderm is also divided into blocks of tissue, called **somites,** that are destined to become muscle and bone tissue.

At week 3, the embryo is between 2 and 4 mm in length. The chorionic villi are present. Leg and arm buds appear. Between the amnion and the chorion, within the yolk sac, blood islands appear, and a network of blood vessels is forming. By the fourth week, the circulatory system is functioning as is part of the nervous system.

At week 5, the head of the embryo is relatively large, and the ears are delineated. The embryo is about 22 mm long by week 6, and the limbs are evident, positioned across the abdomen. The hand and foot rudiments are visible, and some digitation is occurring. By week 7, the hands and feet are clearly differentiated. The tail is almost gone. At week 8, the embryo appears humanoid and is about 40 mm long. The organ rudiments continue development, and bone ossification begins.

At week 10, the intestines, which previously projected into the umbilicus, are drawn into the body. The spinal cord is definitive, and its internal structure is well on the way to completion. During week 12, the sex of the embryo is visually obvious. The notochord

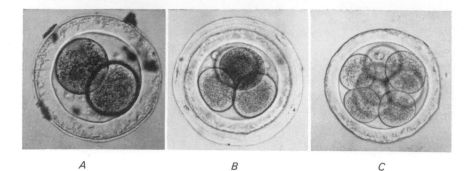

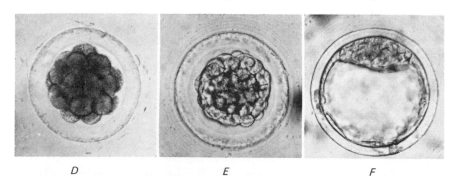

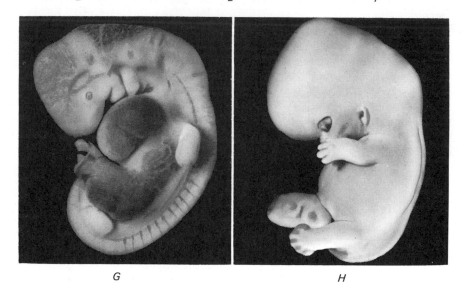

Figure 39-11. Photographs illustrating the development of the mammalian egg. A–F are states in the development of the rabbit's egg. (A) Two-cell stage, 24 hours after fertilization; the surrounding membrane is the zona pellucida in which are embedded several sperms; the polar body lies near the upper end of the cleavage furrow. (B) Four-cell stage (29 hours). (C) Eight-cell stage (32 hours). (D) Morula stage (55 hours). (E) Trophoblast stage ($17\frac{1}{2}$ hours), showing the differentiation of cells into the outer trophoblast layer and the inner cell mass from which the embryo arises. (F) Blastocyst stage (90 hours), showing the segmentation cavity, trophoblast, embryonic disc (above), and zona pellucida with part of the albuminous coat around it. (G) Human embryo, one month old (6.7 mm), showing arm and leg buds, growing at caudal end, umbilical cord, heart, branchial bars, olfactory fit, and eye. (H) Human embryo, six weeks old (19 mm), showing developing hands and feet, elbow and knee, and nose, eye, and ear. The cranial vault appears too large for the body, due to the fact that it is a hollow vesicle. [Photos A–F courtesy of P. W. Gregory; G–H courtesy of Carnegie Institute of Washington, Department of Embryology.]

begins degenerating, and the ossification of some bones is almost complete. At week 16, the embryo, or fetus, may move about to a limited extent within the uterus. It has some hair on the head, and the head grows more slowly relative to the body. The eyes, nose, and ears appear (externally) complete.

Between weeks 20 and 40, the embryo progressively approaches a more human form, and the proportions of the body become established. In the male, the testes descend into the scrotum; and in the female, the reproductive tract appears. In earlier stages most of the blood was formed in the liver, but now the bone marrow is gradually taking over this function. The nose and ears ossify, though the baby is deaf at birth.

At birth, all organ systems are functional. The sutures of the skull are not completely ossified, and an opening, the **fontanel,** in the skull has yet to close. The brain is covered by a thick membrane at the fontanel. In addition, there is a hole called the **foramen ovale** in the interatrial septum of the heart that closes shortly after birth.

Molting and Metamorphosis

Most insects, crustaceans, and amphibians are born in an immature **larval,** or **nymphal,** stages. For example, the **tadpole** is the larval stage of the adult frog. The process by which an animal is transformed from a larva into a sexually mature adult is called **metamorphosis.** A few amphibians retain the larval body form but do acquire reproductive capacities—this is called **neoteny.**

Molting (see Chapter 33) involves the shedding of a rigid exoskeleton and the secretion of a new and larger one to accommodate growth. The mechanism of molting is known to be neural, in that the brain controls the release of certain hormones. At the onset of an insect molt, neurosecretory cells in the brain secrete a hormone whose sole function is to activate the **prothoracic gland,** which is located in the head or thoracic region.

The prothoracic gland in turn secretes the hormone **ecdysone,** which initiates the epidermal and other changes that define the molt. But the type of molt, be it a **larval molt** (leading to the next larval stage), a **pupal molt** (resulting in a pupa or advanced larval stage), or an **imaginal molt** (from pupa to adult), is not decided by the ecdysone secretion. The paired **corpora allata** control the type of molt by secreting juvenile hormone, which inhibits metamorphosis. During each nonimaginal molt, the corpora allata are active, producing **juvenile hormone** that inhibits a full imaginal molt. The one time that no juvenile hormone is produced, the molt encompasses a metamorphosis to the adult body form.

Amphibian metamorphosis is also hormonally controlled. The tadpole is continuously growing, and various organs become functional during its short aquatic existence. Once it reaches a particular stage of development, the anterior lobe of the pituitary hypophysis produces **thyrotropic hormone.** Thyrotropic hormone activates the already differentiated thyroid gland, which in turn produces **thyroxin,** a hormone utilizing iodine. Thyroxin directly affects the tissues, causing either degeneration or proliferation of certain structures. In frogs, the following major changes occur, with each successive change requiring higher concentrations of thyroxin:

1. The hindlegs are formed.
2. The intestine shortens.
3. The forelegs, developing under the skin, break free.
4. The tail is resorbed.
5. The fin folds are resorbed.
6. The gills are resorbed.

Regeneration

The capacity of an organism to rebuild lost or damaged body parts is referred to as **regeneration.** The general trend is that the more complex the organism, the lower its capacity for regeneration. Whereas we humans can at most heal a large wound or reossify a broken bone, newts may regenerate entire limbs, an earthworm may regenerate a new head, and a bisected planariam may regenerate into two individuals. Various identifiable stages occur in regeneration, as the following discussion of regeneration in the salamander indicates.

Assume that a limb has been lost. First, and rather rapidly, the epidermis around the break grows over the wound. Soon the epidermis begins to bulge out due to the proliferation of underlying cells. The bulge of undifferentiated cells is called the **regeneration**

blastema. As it grows, it flattens distally in a dorso-ventral plane, and gradually the blastema cells differentiate into rudiments of digits, muscle, and so on. In time, the regenerated limb is indistinguishable from the original.

It has been demonstrated that, in most regeneration studies, cells near a damaged area **dedifferentiate,** that is, they become "general cells." Soon after dedifferentiation, these same cells proliferate and gradually differentiate into tissues essentially indistinguishable from the lost or damaged tissues.

Summary

Asexual reproduction, which does not require haploid gametes, may occur as binary fission, multiple fission or sporulation, fragmentation, or budding.

In sexual reproduction, haploid gametes fuse during fertilization to form a zygote. Individuals that produce both sperm (male gametes) and ova (female gametes) are monoecious (hermaphroditic), whereas those that produce only one type of gamete are dioecious. Hermaphrodites are usually self-sterile and hence must cross-fertilize to produce new individuals.

Parthenogenesis is the production of viable individuals from unfertilized eggs. For some organisms such as the aphid this is a normal or seasonal mode of reproduction; for a few other such as the frog, an ovum may be experimentally induced to develop parthenogenetically.

Oviparous females lay eggs that hatch in the external environment; viviparous females retain and nourish their embryos within their bodies; and ovoviviparous females retain but do not nourish their embryos within their bodies.

Spermatogenesis takes place in the testes where one diploid spermatocyte undergoes meiosis to produce four haploid spermatozoa. The sperm are composed of a head, with its acrosome and chromosomes, a middle piece (containing mitochondria), and a tail (a flagellum).

Oogenesis is the meiotic production of one haploid ovum from a diploid oocyte. The mammalian ovum is released from the follicle and passes down one of the Fallopian tubes. If it is fertilized during this passage,

it becomes implanted in the wall of the uterus. Ovulation is hormonally controlled and is regulated by the pituitary gland and corpus luteum secretions. Primate mammals have a menstrual cycle that culminates, if no fertilization has occurred, in the discharge of portions of the uterine endometrium.

Sperm penetration into an ovum depends on lysosomal enzymes released from the acrosome atop the sperm.

Cleavage, or mitotic division of the zygote, may follow several patterns that characterize different animal groups. Cleavage produces a hollow ball of cells, called a blastula, that reorganizes by cellular migration (gastrulation) to form a gastrula. Ectoderm, mesoderm, and endoderm are the three primary germ layers formed during the gastrulation of most metazoans. Each differentiates into specific tissues and organs in the embryo. This may be followed easily in the study of frog embryology.

Spemann and Mangold conducted an important set of experiments that outlined the general sequence of determination in tissue differentiation. Generally, groups of cells migrate in one way or another to particular loci within the developing embryo. Various groups of cells affect one another (induction), causing differentiation. As development proceeds, undetermined cells become less common.

The comparative embryologies of animals are roughly similar in the earlier stages, but they gradually diverge, generally tracing the phylogeny of the different groups. In the case of humans, it takes about 15 to 20 weeks before the embryo (fetus) is easily recognizable as human.

Review Questions

1. What effect does sexual reproduction have on the genetic composition of the offspring?
2. If your parents had really told you about the birds and the bees, what information would they have had to have changed from their original lecture?
3. In which type of asexual reproduction do nuclear divisions precede the first cytoplasmic division?
4. Are "fertilization" and "copulation" synonymous? Why or why not?
5. Humans are

(a) monoecious
(b) dioecious
(c) hermaphroditic
(d) (a) and (c)

6. Because hermaphroditic organisms are usually _____ , most must cross-fertilize.

7. Distinguish between ovoviviparity and viviparity.

8. Must fertilization always precede cleavage?

9. What is the difference between a menstrual and an estrous cycle?

10. Insect molting is under the control of what hormones?

11. The middle piece of a sperm contains a great number of mitochondria. What essential function do they serve?

12. When a blastula, in the eight-cell stage, has one layer of blastomeres directly over another layer, it is an example of
(a) bilateral cleavage
(b) Müllerian cleavage
(c) spiral cleavage
(d) radial cleavage

13. Draw a concise diagram indicating the most important hormonal changes during the human menstrual cycle. Indicate the sources of the hormones.

14. If you were presented with an ovum that had a high yolk concentration at one pole, in what portion of the egg would you expect the embryo to develop?

15. In amphibian (and other) embryology, invagination of part of the blastula leads to the
(a) blastocoel stage
(b) gastrula stage
(c) diploblastic germ layers
(d) production of thyroxin

16. What parallelism is there between the Spemann—Mangold experiments and the story of King Midas?

17. In frog embryology, the mesoderm splits into layers. Between these layers is the _____ _____ . One of the layers, the _____ , mesoderm forms the musculature of the digestive canal.

18. Cleavage in the human zygote is _____

_____ , which is somewhat unexpected because the ovum lacks yolk.

19. Which extraembryonic membrane functions as a "Urinary bladder" in oviparous vertebrates.

20. The placenta is composed of
(a) tissues exclusively from the embryo
(b) tissues exclusively from the mother
(c) tissues from the mother and the embryo
(d) chitin, though there is no exoskeleton in many cases

21. What causes an insect molt to be an imaginal molt?

22. What is the result of an imaginal molt?

23. What might an insect "molt control" pill consist of, if its purpose were to keep the insect from experiencing an imaginal molt? If this organism never did go through an imaginal molt, but it nevertheless became sexually mature, this would be an example of _____ .

Selected References

Arey, L. B. *Developmental Anatomy*. Philadelphia: W. B. Saunders Co., 1954.

Balinsky, B. I. *An Introduction to Embryology*, 3rd ed. Philadelphia: W. B. Saunders Co., 1970.

Barth, L. G. *Embryology*. New York: Dreyden Press, 1953.

Brachet, J. *Chemical Embryology*. New York: Interscience Publishers, 1950.

Crouch, J. E., and J. R. McClintic. *Human Anatomy and Physiology*. New York: John Wiley & Sons, 1971.

Ebert, J. D., and I. M. Sussex. *Interacting Systems in Development*. 2nd ed. New York: Holt, Rinehart, and Winston, Inc., 1970.

Fulton, C., and A. O. Klein (eds.). *Explorations in Developmental Biology*. Cambridge, Mass.: Harvard University Press, 1976.

Hamilton, W. J., J. D. Boyd, and H. W. Mossman. *Human Embryology*, 3rd ed. Baltimore: Williams & Wilkins Company, 1962.

Hickman, C. P. *Integrated Principles of Zoology*, 3rd ed. St. Louis: C. V. Mosby Co., 1970.

McKenzie, J. *An Introduction to Developmental Biology*. New York: John Wiley & Sons, 1976.

Nelson, G. E., G. G. Robinson, and R. A. Boolootian. *Fundamental Concepts of Biology*, 2nd ed. New York: John Wiley & Sons, 1970.

Patten, B. M. *Foundations of Embryology*, 2nd ed. New York: McGraw-Hill Book Company, 1963.

Perry, J. S. *The Ovarian Cycle of Mammals*. Edinburgh: Oliver and Boyd, 1971.

Prosser, C. L. (ed.), D. W. Bishop, F. A. Brown, Jr., T. L. Jahn, and V. J. Wulff. *Comparative Animal Physiology*. Philadelphia: W. B. Saunders Co., 1950.

Romer, A. S. *The Vertebrate Body*. Philadelphia: W. B. Saunders Co., 1953.

Rugh, R. *Experimental Embryology: Techniques and Procedures*, 3rd ed. Minneapolis: Burgess Publishing Company, 1962.

Shumway, W., and F. B. Adamstone. *Introduction to Vertebrate Embryology*. New York: John Wiley & Sons, 1954.

Spemann, H. *Embryonic Development and Induction*. New Haven: Yale University Press, 1938.

Torrey, T. W. *Morphogenesis of the Vertebrates*. New York: John Wiley & Sons, 1962.

Villee, C. A. *Biology*. Philadelphia: W. B. Saunders Co., 1967.

Waddington, C. H. *Principles of Embryology*. New York: The Macmillan Company, 1956.

Willier, B. H., P. A. Weiss, and V. Hamburger (eds.). *Analysis of Development*. Philadelphia: W. B. Saunders Company, 1955.

Witschi, E. *Development of Vertebrates*. Philadelphia: W. B. Saunders Company, 1956.

Species Dynamics

Zoogeography and Ecology

Zoogeography is the study of the geographic distribution of animals. **Ecology** investigates the interrelationships between organisms and their environments.

The Earth's Zoogeographical Regions

The science of taxonomy was formulated in the eighteenth century, and extensive voyages of exploration resulted in the identification and description of animals and plants from all parts of the earth.* Trade had fostered an awareness since antiquity that different parts of the world harbored "foreign" types of animals. The accumulation of information generated as a result of visiting these new lands allowed this notion to be quantified for the first time. In 1857 Sclater named six **avifaunal regions,** and shortly after 1876 Wallace delineated six **faunal regions** for all animal life, which did not differ radically from Sclater's scheme. With minor revisions, these regions are still recognized today: **Palearctic, Nearctic, Neotropical, Ethiopian, Oriental,** and **Australian** (Fig. 40-1). Each of these faunal provinces was found to support a more or less distinct assemblage of animals.

There are, of course, many areas of overlap, and some of them are puzzling. For example, why are

*That is, from the terrestrial areas that cover only about 29 percent of the planet Earth. It is interesting to note that analogous voyages are being made today, but with the expressed purpose of investigating the ocean and its life.

marsupial mammals so abundant in Australia and South America—half a world apart? And why do certain animals such as side-neck turtles and lungfishes live in the southern but not the northern continents?

Fossil discoveries added to the puzzle of animal distribution. The discovery that giant mammoths had once roamed North America, for example, indicated that the geographical distributions of animals had changed over time.

Paleogeographic Theories

Three paleogeographical theories have attempted to explain the distributions of animals through geological time.

Theory of Uniformitarianism

In 1833 Charles Lyell, the founder of modern geology, proposed the theory of **uniformitarianism.** This states that the continents have generally maintained their relative positions throughout geological time. Zoogeography may be explained by assuming that narrow bridges of dry land sometimes connected the continents.

This concept has lost much of its appeal in recent years, largely due to new evidence that favors the theory of continental drift (discussed in a subsequent section). However, all the continents contain rock strata

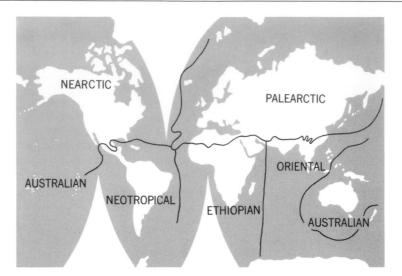

Figure 40-1. The paleogeographical regions of the world. The Palearctic is of temperate climate. It contains 28 mammalian families (most of wide distribution); endemic (exclusive) families are the Spalacidae and Selviniidae, both types of rodents, as well as the hedge sparrow; most families of birds are represented with wide distribution; there are few reptiles and many urodeles (types of amphibia). The Nearctic is also of temperate climate. It contains 24 families of mammals; endemic families include three of rodents, one of musk-ratlike animals, and the pronghorns, which are deerlike; reptilian organisms urodeles are abundant; many Palearctic and Neotropical families are absent. The Neotropical climate is mostly tropical. There are 32 mammalian families, 16 of which are marsupials, 2 families of monkeys, 5 of bats, and 11 types of rodents; half the avian families is present and endemic; many reptiles and almost all anurans (tailless amphibians) are represented; aquarian species are dominated by charlinsfishes, gymnotids, and catfishes. The Ethiopian has a varied climate, from tropical to desert (arid). It contains 38 mammalian families of which 4 are endemic and 2 are primates (tree shrews and tarsiers), with abundant lizards, snakes (especially poisonous species), and turtles and many anurans but few urodeles; fish include mostly carp and catfish. The Australian region has tropical, arid, and temperate climates. There are 9 mammalian families, most of which are marsupial; 8 mammalian, 10 avian, and 2 reptilian families are endemic; the bird population is varied, few amphibians exist, and no urodeles are present (why might one expect that?); there are few vertebrate families and a conspicuous absence of placental carnivores.

that date back at least 1 billion years. In addition, fossil deposits on the present continents show that these areas have been alternately covered with shallow seas or exposed during the past 600 million years. Clearly, the continents themselves have existed for as long as we can read the geological record. It is their relative positionings that has been in dispute.

Land Bridges

The land bridge theory was an offshoot of the uniformitarianism theory and suggested that substantial areas of dry land once connected the continents. One

such hypothetical continental bridge was South Atlantis, presumed to occur between Africa and South America; this was originally postulated to explain the

Figure 40-2. Continental drift. (A) A schematic reconstruction of Pangaea, showing Laurasia and Gondwanaland, the first two areas that broke apart. (B) The present arrangement of the continents, showing the borders of the six major tectonic plates: 1, American; 2, Eurasian; 3, African; 4, Australian; 5, Pacific; and 6, Antarctic. At these borders, movement of the crust, such as earthquakes and volcanic activity is most noticeable. These major plates are frequently broken into smaller plates, not shown in this figure.

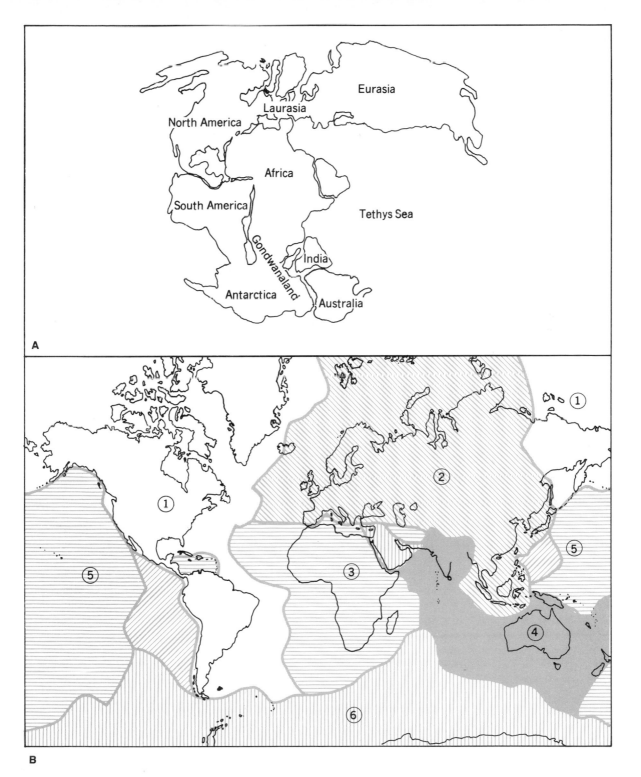

A

B

many faunal similarities between these two continents.

The land bridge theory was popular during the nineteenth century, but both it and the uniformitarianism theory have been largely replaced by the continental drift theory.

Continental Drift

The terms **continental drift, sea floor spreading,** and **plate tectonics** have become the watchwords of modern geology. Each emphasizes a different aspect of a revolutionary new theory of the earth's history. Prior to this century, the earth's crust was considered to be a relatively stable layer, and the geographic distribution of the continents and ocean basins was considered to have remained the same throughout geological history. But several puzzling phenomena seemed to be in variance with this view. Most intriguing of all is the apparent "fit" of the eastern coast of South America with the western coast of Africa. The complementary outlines of these distant coastlines led Taylor and Wegener to independently propose at the turn of this century that South America and Africa were once united but have subsequently drifted apart.

Many lines of evidence have since indicated that this is indeed the case. Major geological strata coincide when maps of the two continents are abutted along the edges of their continental shelves. And fossil deposits older than Cretaceous times (135 million years ago) are very similar, indicating that there was once free migration between the continents. But strata and fossils younger than Cretaceous times are dissimilar, which suggests that South America and Africa separated about then.

More evidence was discovered on other continents. There are parallels in the Late Paleozoic rocks in India, Australia, South Africa, and South America. Early and Late Paleozoic rocks in the northern Appalachians and northwestern Europe also show correspondences. By piecing together these similarities, one can reconstruct a unified landmass as it might have existed before the present continents drifted apart. The original land mass is called **Pangaea** (Fig. 40-2A). By about 100 million years ago, this broke into the two large areas: **Laurasia** to the north (pres-ent-day North America, Europe, and Asia) and **Gondwanaland** to the south (South America, Africa, India, Australia, and Antarctica).

The most conclusive evidence for continental drift has come from explorations of the ocean floor during the past few decades. The history of the drift is written in the rocks of the ocean floor, and sensitive new techniques have been devised to decipher these tracks.

The earth's crust has been found to consist of at least six large **blocks,** or **tectonic plates** (Fig. 40-2B). These plates are generally larger than individual continents and include both continental areas and parts of the ocean floor. Their perimeters do not correspond to the classical continent–ocean boundaries. Instead, their margins are regions in which crust is either being formed or destroyed. These are geologically active regions that are noted for earthquakes, volcanoes, and ongoing mountain building. The crustal blocks themselves are apparently rigid.

The lines of evidence, notably recent discoveries in submarine geology, are beyond the scope of this text. Suffice to say that the dynamic nature of continental drift is forcing major reassessments in paleontology (the study of fossils) and zoogeography, as it has in geology.

Zoogeographers generally deal with the worldwide distributions of animals. The interrelationships and distribution of animals and plants to particular surroundings (**habitats**) within a geographical area is the study of ecology.

Ecology

Every living organism is continually influenced by, and continually influences, its environment. Ecology investigates the interrelationships between organisms and their environments.

One approach to ecology is to study environmental factors, both physical and biological. **Physical (abiotic) factors** include light, temperature, chemicals, water, and substratum composition. **Biological (biotic) factors** are interactions between members of the same or different species. This approach often involves studying a particular **ecosystem** (e.g., a pond, a forest).

Another approach is to study the organisms themselves. An ecologist may consider a particular plant or animal population (or species) and try to discover the various environmental conditions that limit its geograpical range and population size.

It is significant that modern ecology is much less concerned with organisms as individuals than as groups. The life experiences of any one individual are so variable and subject to chance that the individual may be atypical of others of its kind. Ecologists have come to realize that it is the interaction of environmental factors with relatively stable populations that is important to the survival and evolution of a species. And so it is populations, not particular individuals, with which modern ecology is chiefly concerned.

An ecologist may attempt to study the physical environment and the populations that it contains in terms of each other. This approach stresses the constant interaction between the two and is called the **ecosystem approach.** Odum defines an ecosystem as "any entity or natural unit that includes living and nonliving parts follows circular paths."

The Physical Environment

If the presence or absence of a particular environmental factor in a certain minimal quantity is necessary for the survival of a population, it is called a limiting factor. Important abiotic limiting factors include light, temperature, wind, water currents, fire, soil texture, pH (hydrogen ion concentration, acidity or alkalinity) of water and soil, presence or absence of certain inorganic salts, and concentrations of oxygen, carbon dioxide, and other gases.

Light

Light is an important physical factor. We have considered how light influences the behavior of animals such as the amoeba, euglena, paramecium, hydra, earthworm, crayfish, insect, and frog. Furthermore, without light, life on earth would be impossible.

Radiation from the sun (solar radiation) supplies both light and heat. The sun emits a wide spectrum of energy, but certain wavelengths are most important ecologically. These are the wavelengths of visible light, around 400 to 700 nm (nanometers), and the shorter ultraviolet and longer infrared wavelengths.

The greatest ecological importance of solar radiation is in supplying energy for photosynthesis. Green plants convert solar radiation into chemical energy that is stored in the bonds of organic molecules. Thus the tissues of green plants ultimately supply energy for all animal life. On land, sunlight arrives relatively unhampered by the atmosphere, but solar radiation is quickly absorbed by the surface waters of the ocean. As a result, the photosynthetic producers in the ocean are generally restricted to the upper 100 m.

By means of **photoreceptors,** many animals are able to orientate and navigate through their environment. Sensory organelles that detect light are present in some protists, such as *Euglena* (Fig. 5-2), and complex light receptors and sense organs occur in most metazoans. We have encountered lensless eyes in planarians (Fig. 13-1), simple eyes in sandworms, compound eyes in crayfish and insects, eyes analogous to those of vertebrates in squid, and various types of homologous eyes in fishes, frogs, turtles, birds, and humans.

Illumination has a profound effect upon the animals that live in an area. For example, the length of daylight (**photoperiod**) appears to be a factor that stimulates the migration of birds and the breeding cycles of many animals. In both fresh and marine waters, many of the planktonic animals rise to the surface at night but migrate downward at sunrise. Many **nocturnal** animals possess adaptations such as very large eyes and pupils and a reflecting tapetum lucidum behind the retina. Examples of nocturnal terrestrial animals that possess large eyes are the tarsiers, night monkeys, and owls. Another adaptation of many nocturnal (and burrowing) animals is tactile-sensitive bristles on the face, as in cats and shrews. Many fishes that live in the dark ocean depths have large eyes and **bioluminescent organs** that aid in species recognition and food capture. Among the animals that produce light by means of biochemical reactions within luminescent organs is the dinoflagellate *Noctiluca*, some jelly fish and ctenophores, certain polychaetes, fireflies, glowworms, and many deep-sea fishes, squids, and shrimps.

Ultraviolet light may be both beneficial and harmful to animals. Ultraviolet light provides energy for the synthesis of vitamin D in the skin, hair, or feathers of many vertebrates. However, exposure to too much ultraviolet radiation may lead to skin cancer. Skin pigmentation influences ultraviolet sensitivity. Fair-skinned people produce more vitamin D per unit area of exposed skin than do dark-pigmented individuals. An interesting correlation can be made between the degree of skin pigmentation of different human populations and the latitudes in which their ancestors evolved. Dark-skinned peoples are most abundant in tropical regions, and fair-skinned peoples are found in more temperate latitudes. Tanning is an adaptation that regulates exposure to sunlight by proliferating pigment granules in skin cells. It is a homeostatic response that screens out excessive and potentially harmful ultraviolet radiation.

Temperature

Variations in temperature are generally less local than are variations in light, especially in aquatic habitats. Most animals, even the amoeba and paramecium, react to changes in temperature, and each species has a temperature range in which physiological functions are optimal. For example, the optimal temperature range of the paramecium is 24–28°C. An increase in temperature generally speeds up metabolism within the cells of the body and accelerates the activities of many animals. However, activity usually ceases before 45°C is reached because some proteins begin to denature above that temperature. Houseflies, for example, begin to move at about 6°C, carry on normal activity at about 17°C, become increasingly active up to 28°C, cease activity at about 45°C, and die at about 47°C.

Temperature has variable effects on the presence or absence of animals in different habitats. Certain species will not reproduce until a favorable temperature is reached. Eggs develop more slowly at low temperatures and will fail to develop if the temperature is low or too high. In some species the adults die as winter approaches after laying resistant eggs that can withstand the cold weather; other species escape the cold by **hibernation** or **migration.** Certain protozoans and mosquito larvae can live in hot springs at a temperature (50°C) that would soon kill other animals. In general, life is most abundant at temperatures of 10°C to 45°C, the **biokinetic zone.**

Animals handle temperature variations in various ways. **Homeotherms** are animals that maintain a constant body temperature by producing heat by metabolic oxidations (muscle contractions) and losing excess heat by evaporative cooling (e.g., sweating). Mammals and birds are homeotherms. Most other animals are **poikilotherms** and must obtain heat directly from their environment. As a result their body temperatures and hence metabolic rates vary with the temperature of their surroundings.

Some animals are **heterothermic:** they can switch from homeothermy to poikilothermy and thereby temporarily avoid the high energy expenditures of maintaining a constant body temperature. One manifestation of heterothermy is the ability to **hibernate.** Some mammals such as bats and woodchucks can lower their basal metabolic rates during a cold season. **Estivation** is a similar process, but it is in response to heat instead of cold. Estivation is characteristic of some birds and ground squirrels that live in hot, arid zones.

Water

Seawater accounts for about 97.2 percent of the earth's water supply; approximately 2.15 percent of the earth's water is frozen in ice caps and glaciers, and .001 percent exists as water vapor in the atmosphere. Surface waters in inland seas, freshwater lakes, streams, and rivers contain only 0.017 percent of the earth's water.

The waters of the earth are all components of a dynamic cycle called the **hydrological cycle,** familiar manifestations of which include rainfall, flowing waters, ocean currents, and clouds. Less obvious are the advance and retreat of glaciers and the percolation of subsurface ground waters. Superimposed on this grand physical cycling is the cycling of water through living systems. We continually ingest water with food and drink and eliminate it via urine, sweat, feces, and exhalations.

The abundance and availability of liquid water fostered the origin and evolution of life on earth. Indeed, life as we know it would be impossible without water, because the biochemistry of living systems occurs in a aqueous medium. Our bodies contain two-thirds water by weight.

Life evolved in the ocean, and many animal phyla

(e.g., echinoderms) and classes are still restricted to this ancestral habitat. Only a few animal groups have successfully invaded dry terrestrial habitats.

Animals that live in dry regions usually possess adaptations that prevent water loss. The horned lizard, by means of a waterproof integument, can survive for about four months in a drying chamber, and the camel, by storing water in its **reticulum** (second stomach), can live for a week on only dry food or for a month or more on green food. Pronghorned antelopes, jackrabbits, and certain ground squirrels can obtain all the water they require from green food.

Complete drying kills most animals, but members of some species can withstand considerable **desiccation** (dehydration). Small organisms such as some protists, rotifers, and minute crustaceans survive droughts by **encysting** or by laying eggs with heavy shells. Tardigrades (water bears) are famous for their ability to withstand desiccation for several years. Lungfishes and certain bats and birds estivate during dry hot weather.

Too much moisture may be a hazard to terrestrial animals. Earthworms, for example, are flooded from their burrows by heavy rains. Terrestrial animals are adapted to breathe air, and submergence underwater usually leads to rapid drowning.

Gases

The importance of oxygen (O_2) and carbon dioxide (CO_2) for photosynthesis and respiration in plants and for respiration in animals has been stressed. Air usually contains approximately 210 cc (cubic centimeters) of oxygen per liter. Fishes must work harder for oxygen than do terrestrial animals because saturated fresh water contains only approximately 7.2 cc of oxygen per liter, whereas seawater has a maximum of only 5.8 cc of oxygen per liter.

Aquatic habitats exchange gases with the atmosphere by diffusion only at their surface interphase. As a result, a concentration gradient with depth is often observed, especially in large bodies of water. Below the **euphotic zone** (where photosynthesis occurs), the oxygen concentration generally decreases as the carbon dioxide concentration increases—both the result of plant and animal respiration and decomposition. In certain waters, stagnation or decreased circulation may result in an **anaerobic** (without oxygen) zone near the bottom and may be nearly devoid of life.

Other Chemical Considerations

Green plants, which synthesize the organic molecules that feed all animals, depend on a number of mineral salts, notably those of nitrogen and phosphorous. In addition, potassium (K), calcium (Ca), sulfur (S), and magnesium (Mg) and are needed by all organisms.

Seawater contains approximately 40 elements, the most abundant of which are chloride (Cl^-) and sodium (Na^+) ions. Minor ionic constituents of seawater include calcium, magnesium, potassium, carbonate, sulfate, and bromide.

Biogeochemical Cycles

The chemical elements that compose all organisms come from the environment and eventually return to the environment after death of the organism. The constant interchange of these elements between living organisms and the physical environment results in natural **biogeochemical cycles,** of which three of the most important for life are considered here.

Much has been learned about the flow of minerals through the biosphere by radioisotope tagging. A particular chemical is made radioactive (unstable) by incorporating extra neutrons into the nuclei, and its presence can then be tracked by a Geiger counter or other detector.

Carbon Cycle

Carbon atoms form the framework of all the organic molecules that characterize life. Carbon atoms are extracted from their reservoirs in the atmosphere and ocean by green plants during photosynthesis. The energy of sunlight is transformed into chemical energy during this process and is packaged in the chemical bonds of organic molecules. Organic molecules serve as sources of energy and raw materials for the **heterotrophic** animals and decomposers of the biosphere. **Decomposers,** primarily bacteria and fungi, play a major role in recycling organic carbon back into the environment as carbon dioxide.

Two minor leaks are present in this cycle. Incomplete decomposition of plant materials has produced deposits of carbon-rich fossil fuels. The carbon atoms in these materials are being returned to the atmosphere at an accelerating rate by our use of these fuels (Chapter 42). Some animals incorporate carbon atoms into these skeletons, and these may become deposited in marine sediments. These are recycled in geological time through the processes of sedimentation, uplift, and erosion.

Nitrogen Cycle

Nitrogen constitutes about 78.0 percent by volume of the earth's atmosphere and about 3.5 percent by weight of living organisms. Nitrogen is an integral component of amino acids and nucleotides. It may seem surprising that elemental nitrogen is not available to plants and animals. For example, we inhale nitrogen gas with every breath, but we exhale it back to the atmosphere unused. Animals must ingest nitrogenous organic compounds that are ultimately derived from plant proteins. Plants must synthesize nitrogenous organic compounds from inorganic nitrogen salts that they absorb from the environment. These nitrates and nitrites enter the soil and water through the fixation of atmospheric nitrogen by microorganisms, notably certain free-living bacteria and blue-green algae, and symbiotic bacteria that are associated with the root nodules of certain plants. Nitrogen salts also enter the environment through the decomposition of dead plants and animals and their excretory products. The cycle is completed when these nitrogen salts are absorbed into plants. The availability of usable nitrogen is the critical limiting factor in the productivity of most ecosystems.

Phosphorous Cycle

Phosphorous is notable in that there is no atmospheric component in its cycle. Combined with oxygen atoms, phosphorus passes from soil and water through living systems as phosphate, in soluble and insoluble compounds.

Phosphate atoms are important components of ATP and nucleic acids, as well as vertebrate bones and teeth. Because DNA and ATP are characteristic constituents of all cells, animals ingest phosphate with virtually every bite. Plants must absorb phosphorus from the physical environment. Most of the inorganic phosphorus in soils is insoluble and hence unavailable to plants. Animal excretions and the decomposed remains of plants and animals liberate phosphate into the soil, but most of this is quickly absorbed by soil minerals. These phosphates are indirectly liberated by the activities of soil microorganisms and are then readily absorbed into the root hairs of plants.

The ocean eventually receives phosphates from land runoff. Phosphorus is a limiting factor of plant growth, and consequently most marine productivity occurs near the margins of continents. Much of the phosphate that enters the ocean precipitates, forming insoluble mineral salts that are incorporated into the bottom sediments.

Biomes

The distributions of animals and plants throughout the world vary according to regional climatic conditions such as temperature and rainfall. Areas with similar climates (**biomes**) tend to support organisms of the same general types. The principal terrestrial biomes are outlined here (Fig. 40-3).

Deserts

Deserts (Fig. 40-4) cover approximately 20 percent of terrestrial habitats and are found on all the continents. Deserts are characterized by extremely low annual rainfall, usually less than 6 inches per year. The desert's lack of humidity results in extreme temperature variations: days are very hot and nights cold, with the temperature varying as much as 50°C on a daily basis.

Desert plants are known as **xerophytes** because they are resistant to desiccation. These plants store water very effectively (a cactus plant may expand twofold in diameter after a cloudburst) and use the water efficiently. Small leathery leaves or sharp spines reduce evaporative water loss and foil herbivores and their seeds germinate rapidly in the presence of moisture.

Perhaps the most representative desert animals are the reptiles: lizards, snakes, and tortoises. Arthropod

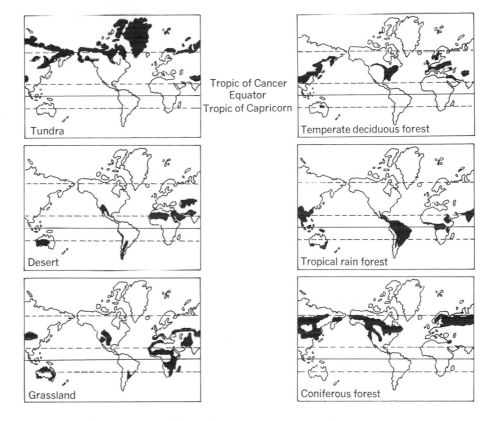

Tropic of Cancer
Equator
Tropic of Capricorn

Tundra

Temperate deciduous forest

Desert

Tropical rain forest

Grassland

Coniferous forest

Figure 40-3. Distribution of the major biomes.

Figure 40-4. Desert biome in Arizona.

Figure 40-5. Tundra biome in the Arctic.

life in the desert is also abundant; representatives include scorpions, millipedes, and ants. Common desert birds include the roadrunner and the shrike, both of which eat small reptiles, mammals, and insects. Small mammals are generally nocturnal, avoiding high daytime temperatures (which would necessitate water loss by evaporative cooling) by remaining inside their underground burrows during the day. The kangaroo rat is so well adapted to arid conditions that it obtains all its water needs from food alone.

Tundra

The **tundra** biome (Fig. 40-5) borders the Arctic Ocean in North America, Europe, and Asia. Arctic tundra is characterized by almost sunless winters and extremely short summers of constant illumination. During the 60-day growing season, the uppermost 15 to 20 cm of soil defrosts. The tundra is generally tree-less, with grasses and lichens composing the dominant vegetation. Precipitation is relatively low, and winds are strong and constant.

Most of the mammals that inhabit the tundra (e.g., the brown lemming) weigh less than 100 grams. Large mammals include the caribou and musk ox. Poikilotherms (cold-blooded animals) are generally excluded from the tundra, but swarms of insects arise during the brief summer. Waterfowl flock to the melted arctic bogs and ponds for breeding pruposes.

Aquatic biomes will be discussed in the next section.

Grasslands

Grassland biomes (Fig. 40-6) are characterized by a complete lack of trees. The predominance of grasses results from the low seasonal rain distribution. Precipitation generally varies from 25 to 75 cm a year,

Figure 40-6. Grassland biome in Wyoming (the Pawnee National Grassland).

which is too low to support forest plants and too high for desert plants.

Grasslands are divided into the **temperate** and **tropical grassland** biomes. Grasslands are widespread over the globe and generally occur in the interior of continents. The grasslands receive regional names: in South America, it is the **pampas;** in South Africa, the **veld;** in Russia, the **steppe;** and in western North America, the **prairie.** For the last, many people distinguish between the **short-grass prairie** found in western plains and the **tall-grass prairie** of the eastern plains. The **savanna** is a tropical grassland on which trees are interspersed and is found in parts of Australia, Africa, and South America.

Grasslands generally support populations of large herbivorous mammals, such as the antelope and the bison. Many small mammals (e.g., ground squirrels and prairie dogs) make their homes in underground burrows. Grasslands are also the homes of many in-sects, such as grasshoppers. A common grassland bird is the bobwhite quail, which nests on the ground and eats seeds and insects.

Chaparrals

The **chaparral** biome results from extremely long, hot, and dry summers and relatively mild, moist winters. The predominant vegetation is broad-leaved evergreens, such as pinion and juniper, that can tolerate low-moisture conditions. Chaparrals occur mostly along some Mediterranean coasts and in California and Mexico, Chile, and southern Australia. In fact, many of the herbs used in French cooking can also be found in California's chaparral.

The mule deer is one of the few large vertebrates represented in the chaparral biome. Common small vertebrates include quail, chipmunks, brush rabbits, lizards, and snakes.

Coniferous Forests

Coniferous forests (Fig. 40-7), also called **taiga**, or **boreal**, forests, usually support a dense population of evergreen conifers that do not completely defoliate during the winter months. The coniferous forests form a belt across northern North America and Eurasia. Rainfall is frequently low, although it may vary from 75 to 375 cm per year.

So dense is the coniferous forest that little light filters through to the forest floor. Moist temperate coniferous forests that range along the western coast of North America have thicker understories of shrubs and herbs. Coniferous forests are subject to marked seasonal changes in temperature and may be snow covered during the winter months. Snow limits the availability and accessibility of food; many animals consequently migrate or hibernate during the winter months.

Animal life is somewhat scarce in the coniferous forest, because of the lack of an understory and the winter snow that limit the supply of food. Small seed-eating rodents such as squirrels are now uncommon, however. Large mammals generally forage in clearings (elk), in and around bodies of water (moose), or on other animals (wolf).

Deciduous Forests

The temperate **deciduous forest** biome (Fig. 40-8) is characterized by relatively rapid seasonal changes, a flora of broad-leaved trees that drop their leaves in the fall, and an abundance of small mammals. The climate is generally moderate, but summers may be quite warm and winters cold. Precipitation averages from 75 to 150 cm per year and is evenly distributed throughout the seasons. This biome occurs mostly in

Figure 40-7. Coniferous forest in Boulder, Colorado.

Figure 40-8. Deciduous forest biome.

the northern hemisphere and forms a belt south of the coniferous forest biome.

Most of the insects, birds, and other animals are adapted to an **aboreal** (tree-dwelling) mode of life. Characteristic mammals are the squirrel and the raccoon as well as the deer and the bear.

Tropical Rain Forests

Tropical rain forests (Fig. 40-9) occur in equatorial regions where the temperature is usually high. They are varied in type, but all experience very high precipitation, over 200 cm per year. The flora is restricted almost exclusively to trees, which are frequently so dense that very little direct light reaches the ground.

Tropical rain forests are known for their arboreal animals (e.g., monkeys), many of which are nocturnal. Many animal species occur, but there are relatively few individuals of each species. This unique combination of few individuals of each species but many species makes the tropical forest one of the most diverse biomes.

Aquatic Environments

Freshwater Environments

Freshwater habitats occur on all substantial land masses throughout the world. **Lakes** and **ponds** form where water fills preexisting depressions in the land's surface. These may have originated in several ways, including glacial erosion, earthquakes and volcanic activity, and beaver dams. The depth of these bodies of water ranges from a few meters to approximately 1500 m. Flowing water, in the form of **rivers, streams, creeks** and **brooks,** contains the result of precipitation that was not absorbed and retained by the soil. Flowing water follows paths of least resistance down gravitational grades. Rivers may run thousands of miles across continents before emptying into the ocean.

Lakes and ponds exhibit various kinds of stratification that affect both plant and animal life. The most important of these are temperature and light penetration, both of which experience considerable seasonal variability in temperate latitudes. Running water, although more thoroughly oxygenated than standing water, is by virtue of its movement more limited in the life that it can support. Of the few algae that live in running water, the bottom-dwelling filamentous types, are well adapted to hold their position in the current. Many of the consumers, such as insect larvae and crustaceans, live in crevices and under stones on the bottom.

Many of the animals that we have studied live in freshwater communities; these include many protists, rotifers, roundworms, hydras, planarians, annelids, clams and snails, crayfish, insect larvae, and fishes.

Marine Environments

The **oceans** cover 71 percent of the earth's surface and have an average depth of about 4000 m. Ocean

Figure 40-9. Tropical rain forest biome.

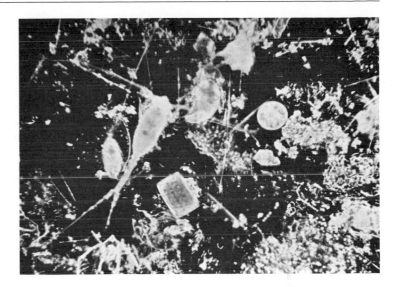

Figure 40-10. Plankton. Note the many diverse forms in this small sample of water.

water is, of course, "salt water," with an average salinity of 3.5 percent.

Perhaps the most familiar marine habitat is the **intertidal zone,** which is alternately exposed and inundated by the tides. The relatively few animals and plants that can survive here are adapted to withstand periodic desiccation and temperature extremes. Barnacles and mussels are common intertidal animals.

The **subtidal zone** below the lowest low-tide level contains a greater diversity of life. Seaweeds are abundant, as are starfishes, sponges, sea urchins, sea anemones, and fishes.

Most marine algae do not live attached to the bottom, however. Microscopic **phytoplankton** (Fig. 40-10), notably diatoms and dinoflagellates, that drift in surface currents are the most important primary producers in the ocean. These minute plants are consumed by microscopic animals (**zooplankton**), such as copepods and euphausids, which are in turn eaten by ctenophores, arrow worms, and many larval forms.

Higher levels of marine food webs are generally filled by larger and more mobile consumers, such as fishes. **Pelagic** (living in open water) animals are generally most abundant in the productive euphotic zone near the surface.

Bottom-dwelling organisms comprise the **benthos.** Familiar benthic animals include clams, polychaetes, starfishes, crabs, and flatfishes. Many are adapted for filter feeding or scavenging.

Shallow tropical ocean waters are characterized by coral reef communities (Chapter 11).

Terrestrial Environments

Terrestrial biomes are highly variable and can be contrasted with aquatic environments on the basis of several distinct challenges that they pose to plants and animals.

On land, water is often a limiting factor. Adaptations to conserve water include specialized excretory organs and the detoxification of nitrogenous wastes. Air also allows relatively rapid and extreme variations in temperature. Air further lacks the supportive capacity of water (buoyancy), and thus land animals need strong skeletal structures and special means of locomotion.

Terrestrial animals may live on the surface of the ground, under the surface, or largely in the air. Most terrestrial animals, such as protists, insects and their larvae, nematodes, and moles, live within about 15 cm of the surface. Others, such as rodents (gophers and prairie dogs), earthworms, and ants, may burrow much deeper. Adaptations for a burrowing type of life have already been mentioned.

The character of the **soil** and the amount of water that it contains largely determines the types of animals that live in subterranean communities; for example, earthworms need moisture and organic debris.

The principal kinds of soil are clay and sand. Representatives of the comparatively few species of animals that live in pure clay soils are insect larvae and isopods. **Clay soil** rich in humus is usually well populated, as it contains an abundance of food and is easily penetrated by burrowing. Earthworms actually eat their way through humus. Burrows in **sandy soil** will cave in unless the walls are treated; this plus poor water retention means that fewer species live in this type of soil. Soils that are too rocky are unfit for burrowing, but many animals find the crevices between them to be a satisfactory hiding place.

Islands

Islands are of two types, continental and oceanic. **Continental islands** are, as the name suggests, pieces of a continent that have been separated by water.

They are generally not too far offshore and resemble the adjacent continent in geographical features and in faunal complement. The British Isles, Formosa, Japan, Borneo, Java, Ceylon, Trinidad, and much of the West Indies are representative continental islands.

Oceanic islands are usually islands of volcanic origin and were never connected to continental land masses. The flora and fauna on these islands are varied because of the winds and water currents that largely determine the types of colonizing organisms that arrive. Oceanic islands exhibit more unique, isolated life forms than do the continental islands. Generally, the vegetation and fauna are sparse. Mammals, amphibians, and freshwater fishes are not frequent inhabitants due to the difficulties of passage. Bats, rats, and reptiles may be present, however.

Representative oceanic islands are the Hawaiian

Figure 40-11. A freshwater pond ecosystem. An ecosystem is a natural unit that includes living and nonliving parts, interacting to produce a stable system in which the exchange of materials between the living and nonliving follows a circular path. The ecosystem is the largest functional unit in ecology because it includes both the living (biotic) and nonliving (abiotic) environment.

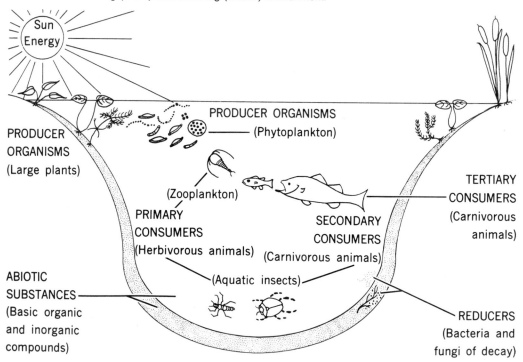

Islands, the Galapagos Islands, and Krakatoa. Krakatoa is extremely interesting because in 1883 it exploded, killing everything. The sequence of its subsequent repopulation has been extensively studied. Because of adaptations of certain animals and plants to high altitudes, many mountain tops are **ecological islands,** as they are just as isolated as if they were surrounded by water.

The Ecosystem

As mentioned previously, an ecosystem is composed of biotic (living) and abiotic (nonliving) components (Fig. 40-11). The term **community** describes the biotic components (populations) and their interactions. An ecosystem (e.g., a pond) contains many particular types of **habitats** (e.g., the surface water, the surfaces of submerged plants, etc.). Populations of organisms are generally confined to particular habi-

tats and have particular roles (**niches**) in the ecosystem (e.g., primary producer, carnivore, herbivore, decomposer).

Abiotic components of an ecosystem include all the nonliving material. The air (or dissolved gases), water, soil, and minerals are all abiotic components.

The biotic components can be divided into several categories, called **trophic levels** (Fig. 40-12), that define the flow of energy through a community. The entry level is that of the **producers**—photosynthetic plants that utilize the sun's energy directly. These **autotrophs** produce the chemical energy in organic molecules that provide support for all animals in the community.

The **heterotrophs**, which require preformed organic molecules, are called **consumers.** The **primary consumers,** or **herbivores,** consume the tissues of the producers directly. **Secondary consumers,** or **carnivores,** consume herbivores and so on. The final trophic level is that of the **decomposers**—fungi, bac-

Figure 40-12. Extremely simplified food web in a forest community. Some of the possible food relationships have been omitted. Arrows point from animal or plant eaten to the animal that eats it.

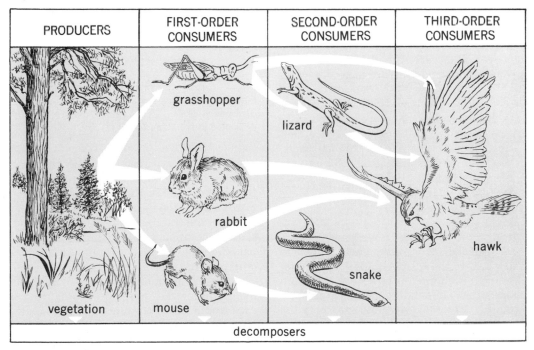

PRODUCERS	FIRST-ORDER CONSUMERS	SECOND-ORDER CONSUMERS	THIRD-ORDER CONSUMERS

grasshopper

lizard

rabbit

snake

hawk

vegetation   mouse

decomposers

teria, and some protists that obtain energy from dead plant and animal tissues.

The heterotrophs release simple substances into the environment as by-products of their metabolism. The producers can use these inorganic molecules, and hence there is essentially a continuous cycling of materials through the ecosystem, but not of energy. Energy that is expended in maintenance, growth, and reproduction is irretrievably lost from the community.

Energy flows in one direction through a community, and, at each trophic level, considerable energy is dissipated into the environment as heat. The flow of energy through a community is very difficult to quantify, but several independent measurements have found an efficiency of energy transfer between trophic levels of about 10 percent. Another way of putting this is to say that each successive level must eat approximately ten times the biomass that the preceding level had to eat to gain the same amount of energy (Fig. 40-13).

The Food Web

The scheme just described was a simple food chain, in which energy is transferred linearly from one trophic level to the next. Nature is not so simple and clear cut, however. If it were, the elimination of any "link" in the food chain would disrupt the entire community, and that rarely happens. Instead, complex **food webs** exist. A food web is established because animals generally have alternate sources of nourishment. Intertidal starfishes, for example, may feed not only on mussels but also on barnacles, chitons, snails, and dead fishes. This adds stability to the community because the disruptive effects of elimination or reduction of a population are dampened.

Ecological Pyramids

One way to appreciate the trophic interactions in a food web is to visualize the relationships using a pyramid model. Several such models may be constructed for a given community, depending on which parameters are employed. A **biomass pyramid** measures the mass of living tissue at each trophic level. Its base represents the total mass of the producers, and each successive tier stands for the mass of consumers (pri-

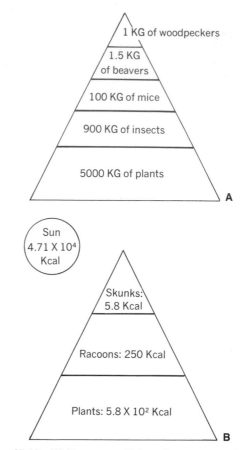

Figure 40-13. (A) Mass pyramid for a forest community. (B) Energy pyramid for a forest community. Values are approximate in both diagrams.

mary, secondary, etc.) that the producers support (Fig. 40-13). The biomass pyramid says nothing about the flow of energy through a community, however. For example, the biomass pyramids for marine plankton communities are often inverted: the relatively short generation time of the heavily grazed phytoplankton allows them to support a larger **standing crop** of herbivore biomass.

An **energy pyramid** (Fig. 40-13) can be constructed by computing the energy (in kilocalories) contained in each trophic level per unit of time. Such energy models always assume a pyramid shape, with the producers forming a broad base. This emphasizes the one-way flow of energy through biological communities.

Principles Associated with an Ecosystem

Symbiosis

When two or more dissimilar organisms form partnerships, they are in a **symbiotic** ("living together") relationship. Three categories of symbiosis are recognized: mutualism, commensalism, and parasitism.

In **mutualism,** both (or all) organisms concerned benefit from the relationship (Fig. 40-14A), and neither (none) can usually survive without the other(s). Mutualism may occur between plants, between plants and animals, or between animals. For example, honeybees obtain nectar from certain flowers and in the process may transfer pollen from one flower to another. Termites ingest wood but cannot digest it directly because they lack the enzyme cellulase—the insect relies on symbiotic protozoans in its gut to digest the wood (cellulose); then the nutrients are digested by the termite.

An interesting example of mutualism is found in the family of African honey guides (Indicatoridae). This is an example of nonobligatory mutualism, or **proto-cooperation.** Friedmann has studied these birds extensively. The African honey guide eats bees wax, but it hasn't the power to break into the hive. It circumvents this problem by making a very obvious call, a "churring," that attracts the attention of a foraging animal, such as a human or a honey badger, *Mellivora capensis.* It then flies in a very conspicuous manner, leading its "symbiont" to the bees' nest. When the nest has been broken into (and the honey removed), the honey guide moves in and takes its fill of wax.

Figure 40-14. (A) Mutualism. Here, an insect pollinates a flower "in return" for a bit of nectar.

A

Though Friedmann believes this behavior to be instinctive, other scientists tend to attribute it to operant conditioning.

Commensalism is a relationship in which one species benefits and the other is not affected. Usually, a larger host will accommodate a smaller guest. The guest may have become structurally adapted to this life-style. In most cases, the dependence is for shelter, locomotion, feeding, or a combination of these (Fig. 40-14B).

Shelter commensalism is exhibited, for example, between the Portuguese man-of-war and its guest, the *Nomeus* fish, that resides within the tentacles, safe from predators. A commensal relationship exists between sharks and remoras (sucker fish). The remoras are transported from meal to meal by their unwitting host. Many bacteria inhabit the digestive tracts of metazoans, obtaining nutriment with no apparent adverse effect on the host.

Parasitism is a symbiotic relationship in which one organism benefits at the expense of another. More detailed discussions of parasitism may be found in Chapters 8, 9, and 13.

The Community

A **community** is essentially a group of interacting populations within an ecosystem. Hundreds of different types of communities have been described and classified by ecologists, but many show common fea-

Figure 40-14. (B) Commensalism. Here a fish benefits from the shelter provided by a starfish; the starfish neither benefits nor is harmed by the fish.

B

tures. Communities may be named for some conspicuous characteristic such as plant cover (e.g., beech–maple, or deciduous, forest) or physical feature (tropical rain forest). Ecosystems with similar physical environments are likely to support similar communities of plants and animals, even though their species compositions are rarely identical. For example, temperate freshwater ponds are never exactly alike, but their abiotic environments pose similar selective pressures, and we expect to find the same general kinds of plants and animals living in them. Communities vary temporally as well as spatially. Pond animals may be permanent or temporary residents, or merely brief visitors. Among the permanent residents are protists and some crustaceans, a variety of snails, and other small animals. Frogs, toads, and some salamanders may be temporary residents, and water birds such as ducks and geese visit the pond to feed on the animals and plants there.

Ecological Succession

Every community undergoes a history of progressive changes in its species composition; such changes constitute **ecological succession** (Fig. 40-15). A new habitat, for example, a pond resulting from the overflow of a river, may contain certain types of pioneering plants and animals. Bass and sunfishes, introduced with the river water, may thrive at first. But soon the sides of the pond become overgrown with vegetation and the clean bottom becomes covered with deposits. The bass and sunfishes disappear because such an environment is not suited to them; however, catfishes may persist. Gradually, deposits accumulate and the pond becomes a swamp; mud minnows may replace the catfishes. Finally, the swamp is reclaimed by terrestrial plants, until the aquatic organisms are entirely eliminated. An ecological succession of terrestrial communities may then occur at that site, with grasses, shrubs, and trees eventually succeeding one another. The final and most stable stage in the series of succession is a **climax community.**

Ecological succession can be observed in an exposed jar of pond water containing a source of organic matter. For example, if a few pieces of rice are placed in a beaker of pond water, countless bacteria will cloud the water in a day or two. Soon afterward, minute flagellated protists will become more numerous as they feed on the bacteria and the products of bacterial decomposition. Next, the number of carnivorous ciliates will increase; however, these predators will eventually die unless green plants (algae) continually renew the organic matter in the jar. If the populations of plants and animals become adjusted to one another, a comparatively stable community may be established. Such a balanced aquarium is essentially a climax community.

Disruption of any habitat invites colonization by opportunistic species and temporarily sets back the development of a climax community. For example, an avalanche can completely change the face of a mountainous habitat within seconds by knocking down trees and completely covering its path with snow. Volcanoes, floods, plowed fields, logging operations, and fires also change the environment and disrupt succession.

Fire succession in coniferous forest and chaparral biomes involves a cyclic series of changes that involve periodic conflagrations. Such fires are natural events that occur frequently as ground litter accumulates. Generally, the dominant plants are adapted to resist periodic small fires. But, when people keep such an area free of fires, ground litter accumulates to the point that when a fire does eventually take hold it becomes a destructive inferno that may kill all the vegetation.

Many organisms depend on fire for their existence. The condor has a great wingspan (about 3 m) and requires wide clearings for take off and landing. The dense foliage that results from fire prevention limits its mobile range. In addition, certain plants depend on fire for germination of their seeds. Some coniferous trees rely on fire to melt the resin that seals in their cones, thereby allowing the seeds to escape.

The moose provides another example of fire dependence. Moose consume different plants at different times of the year. During the winter they forage on willow, alder, and aspen; these deciduous trees are preclimax stages in coniferous forest biomes. In the absence of fire, these trees cannot become established, and the moose population dwindles. This was observed to happen at the National Moose Preserve on the Kenai Peninsula in Alaska. At one time an "unauthorized" fire occurred, and the moose population subsequently increased dramatically. Ecologists real

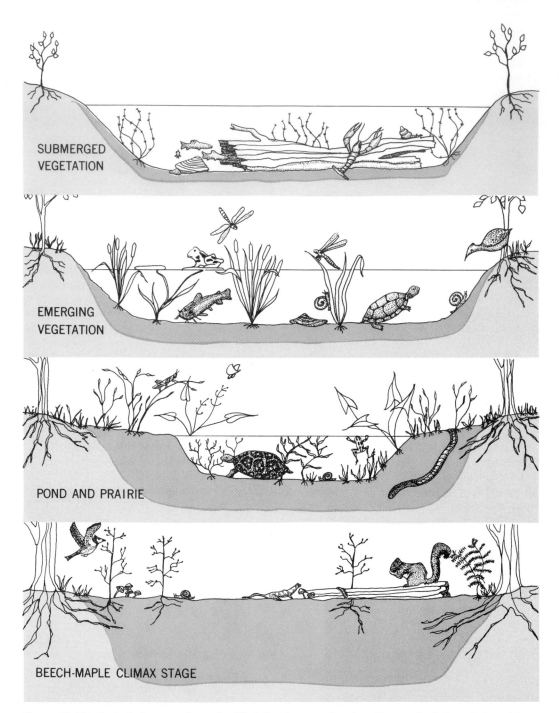

Figure 40-15. Ecological successions, simplified. Pond succession, from the practically bare bottom (pioneer) stage, which is gradually replaced by sequential stages, each more mature than the preceding one, until a climax is reached at which point the community becomes relatively stable. The stages shown here are based on studies in the Middle West. The succession in ponds with a different climate will not be the same in detail, and the climax will be different.

ized that fire was necessary for the maintenance of the moose population and began to allow controlled fires periodically.

Ecological Dominance

In all communities, there are producers, consumers, and decomposers, but certain organisms at each level will be more important than others because of their size and/or numbers. These **dominant species** largely control the energy flow and have much to do with determining the community's character. For example, in a beech–maple community, these tall trees cast shade that limits the growth of shorter plants to certain types. In turn, the available habitats and plant foods limit the types of animals that can survive there.

Community Stratification

Communities tend to stratify or arrange themselves into vertical layers. **Stratification** increases the number of habitats, and this increases the biotic diversity of an area by reducing interspecific competition for food and space. In aquatic communities, stratification is usually due to physical factors, such as light, temperature, and oxygen content of the water. As a result, the species that live near the bottom of a deep lake will not be the same as those near the surface. Stratification in terrestrial communities usually results from the varying heights of vegetation, such as grasses, shrubs, and trees.

Stratification often exhibits seasonal and daily changes. In the ocean, certain animals that live below the euphotic zone during the daytime migrate up to the surface to feed at night. Freshwater lakes often experience seasonal changes in the thermal layers, and certain animal species will follow these shifting strata of temperature.

Community Periodicity

Communities tend to exhibit rhythms or cycles. These involve recurring changes in the activities or movements of organisms. Certain animal species may be active only during the day (diurnal), whereas others are active only at night (nocturnal). In aquatic communities, many zooplankton move toward the surface at night and return to deeper waters during the day. Such cycles with 24-hour periods are called circadian rhythms.

Communities also exhibit seasonal rhythms, especially in biomes that experience pronounced climatic changes during the course of a year. Periodicity in the activities and sizes of plant and animal populations results. Lunar rhythms are well known in marine environments (for instance, spawning in polychaete palolo worms of the Pacific and the grunions, a West Coast fish).

Population Growth

When a species first colonizes a new habitat, its population size increases rapidly if it can find ample food and living space. Fig. 40-16 represents the growth rate of such a hypothetical population. The curve is described as a **sigmoid,** or S-shaped **curve.**

This initial period of rapid growth tapers off as the population reaches an **equilibrium** with environmental factors that tend to limit its size; the equilibrium population size is called the **carrying capacity (K).** Equilibrium is established when the birth rate approximately equals the mortality rate.

Many factors may limit the size of a population. The amount of food may become limited, either through overharvesting to the point that recovery is slowed or through competition from more efficient grazers or hunters. Living space may also decrease below optimum. Increased mortality is likely to result from predators and parasites. As a population grows, potential predators and parasites are presented with more targets, and their own populations tend to increase accordingly.

Populations tend to ensure their own continuation by producing excess offspring because many individuals die before reproductive age. When a species invades an unoccupied habitat, the local population increases rapidly until the particular carrying capacity is reached. If changing conditions alter the carrying capacity, the population size will correspondingly change. For example, the populations of many small freshwater metazoans (e.g., *Daphnia*) are greatly reduced during the cold winter months but proliferate during the productive summers.

It has been demonstrated, both in laboratory and field studies, that, when carrying capacity is raised,

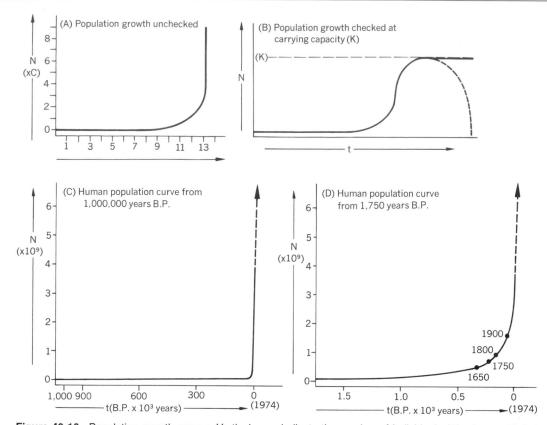

Figure 40-16. Population growth curves. Vertical axes indicate the number of individuals (*N*) when multiplied by the number indicated. Horizontal axes indicate time. (A) All populations grow exponentially if unchecked (solid line). (B) Because there is a finite amount of food and space, populations are eventually always checked (solid line) at some carrying capacity (*K*). The dashed curve represents a decline in population due to some population trauma. (C) Human population curve from over 1 million years B.P. (figures based on the work of Deevey, 1960). (D) Human population curve within the last 1700 years (to show clearly the recent rapid increases). At current rates of increase, within 100 years, the world population will require far more food than can possibly be produced on earth. Dashed lines in (C) and (D) represent projected populations at present rate of increase (which is actually a low figure because *r* is increasing due to the younger age of parents).

the population size increases at a characteristic rate. The rate depends on several factors: the original population size, the birth rate, and the death rate. The rate of increase is simply the number of individuals multiplied by the births minus the deaths.

This relationship can be expressed as an equation that is applicable to any population: $N = N_0 e^{rt}$. Here, N_0 is the original number of individuals; r is the growth rate $(b - d)$; b is the number of births per individual per unit time and d is the number of deaths per individual per unit time; t the amount of time; and e is a mathematical constant (the base of natural logarithms), which is equal to about 2.7. This equation allows us to calculate the population size (N) for any time, given the original number, the birth and death rates of individuals, and the elapsed time.

We might predict what the world human population will be in the year 2005. Based on census figures from 1969,

N_0 = 3.5 billion
b = 34 births/1000 people/year = 0.034
d = 15 deaths/1000 people/year = 0.015
r = 0.034 − 0.015 = 0.019 (This figure increases as the average age in the population becomes younger.)
t = 2005 − 1969 = 36 years

If we insert these data into the population equation, we note that

$$N = N_0 e^{rt}$$
$$N = (3.5 \times 10^9)\, e^{(0.019)(36)}$$
$$N = (3.5 \times 10^9)\, e^{0.684} \text{ where } (e^{0.684} \text{ is}$$
about equal to 2)
$$N = 3.5 \times 10^9 \times 2$$
$$N = 7 \times 10^9, \text{ or 7 billion}$$

From this calculation we can see that, if the carrying capacity does not intervene and if birth rate is not controlled, in 36 years the world population will have doubled (this is a conservative figure). Because the size and resources of the earth are finite, this kind of growth cannot continue for very long. Without such limitations, the human population would outweigh the visible universe in about 5000 years.

The rate and direction of change in population size depends on the differential between birth and death rates. When carrying capacity is reached and population size stabilizes, then $r = 0$, which is equivalent to saying that $b = d$. The larger b is, the larger d will be. If the human population continues to reproduce at a high rate, the mortality rate will increase (dramatically, when the carrying capacity is reached). High death rates and accompanying human suffering are clearly undesirable, and the only way to avoid them is through birth control.

Summary

The major faunal regions are considered to have resulted from the isolation of continental blocks during geological time. The theory of continental drift has been receiving convincing empirical support, especially from observations of the deep-sea floor.

The major terrestrial biomes include grasslands, deserts, coniferous forests, deciduous forests, chaparral, tropical rain forests, and tundra.

Ecology is the study of the interrelationships between organisms and their environments. The environment includes both abiotic (light, temperature, etc.) and biotic (presence of prey, predators, other members of one's species, etc.) factors.

Freshwater environments result from precipitation. Marine environments are more saline and form a continuous ecosystem that covers 71 percent of the planet. Terrestrial habitats are more variable than aquatic ones, especially with regard to moisture and temperature.

Islands are categorized as being either continental or oceanic. Continental islands mimic their mother continents, whereas oceanic islands are much more diverse, owing to their distance from the mainland and independent origin. Mountain tops are often ecologically equivalent to islands because of extreme isolation.

Symbiosis is any partnership between dissimilar organisms. It may benefit all concerned (mutualism) or only one symbiont without harming the other (commensalism); or it may benefit one at the expense of the other (parasitism).

The biotic components of an ecosystem are arranged in trophic levels: photosynthetic (primary) producers, herbivorous or omnivorous consumers, carnivorous consumers, and decomposers define the community.

Complex trophic relationships are shown in the food web. Bioenergetics may be represented through the use of pyramid models that estimate the relative amounts of mass, energy, and numbers in each trophic level of a community. Energy is a one-way flow that depends ultimately on the sun.

Ecological succession leads from a pioneer community through a number of stages until an equilibrium stage (climax) is reached. Ecological dominance defines the relative importance of a species in a community.

Population growth ideally follows a sigmoid curve. Population size is eventually regulated by lack of food and/or living space, predation, and parasitism; it fluctuates around the carrying capacity of the habitat.

Review Questions

1. Describe the six major crustal blocks by correlating them with current political boundaries.
2. Describe the major lines of evidence that support the theory of continental drift.
3. Do plants utilize all wavelengths of the electromagnetic spectrum during photosynthesis? Explain.
4. What is the range of temperatures to which life is adapted on earth?
5. In a very hot climate, one would be likely to find animals capable of
   (a) hibernation
   (b) estivation
   (c) migration
6. Describe the pathway of one of the biogeochemical cycles described in this chapter.
7. What basic difference is there between the pathway of energy through the biosphere and those of carbon, nitrogen, and phosphorus.
8. Which biome has the shortest growing period?
9. What is the euphotic zone, and why is it important in aquatic ecosystems?
10. Would you be more likely to find a monkey on an oceanic or on a continental island? Why?
11. Would you expect there to be community stratification in the ocean? Give your reason.
12. Given a value of one for the energy in a primary consumer, how much of that energy would a tertiary consumer obtain?
13. Cows ingest vegetable material, but they cannot digest it. The digestion of cellulose is performed by protozoans in the cows' stomachs, which are specialized to provide suitable environments for the microorganisms. What type of symbiotic relationship is this?

Selected References

Ager, D. V. *Principles of Paleoecology*. New York: McGraw-Hill Book Company, 1963.
Allee, W. C., A. E. Emerson, P. Park, O. Park, and K. P. Schmidt. *Animal Ecology*. Philadelphia: W.B. Saunders Company, 1949.
Bates, M. *Man in Nature*, 2nd ed. Englewood Cliffs, N.J.: Prentice-Hall, 1964.
Benton, A. H. and W. E. Werner, Jr. *Field Biology and Ecology*, 3d. ed. New York: McGraw-Hill Book Co., 1974.
Buschbaum, R., and M. Buschbaum. *Basic Ecology*. Pittsburgh: Boxwood Press, 1957.
Buzzati-Traverso, A. A. (ed.). *Perspectives in Marine Biology*. Berkeley: University of California Press, 1958.
Cold Spring Harbor Symposia on Quantitative Biology, Vol. 22. Population Studies: "Animal Ecology and Demography," Cold Spring Harbor: The Biological Laboratory, 1957.
Collier, B. D. *Dynamic Ecology*. Englewood Cliffs, N.J.: Prentice-Hall, Inc., 1973.
Cott, H. B. *Adaptive Coloration in Animals*. London: Sidgwick & Jackson, 1941.
Cushing, D. H. *Marine Ecology and Fisheries*. New York: Cambridge University Press, 1975.
Fiennes, R. N. *Ecology and Earth History*. New York: St. Martin's Press, 1976.
George, W. *Animal Geography*. London: Heinemann Educational Books, Ltd., 1962.
Hardy, A. C. *The Open Sea. Its Natural History: The World of Plankton*. Boston: Houghton Mifflin Company, 1956.
Hedgpeth, J. W. (ed.) *Treatise on Marine Ecology and Paleoecology*, Vol. 1, "Ecology"; Vol. 2, "Paleoecology." New York: The Geological Society of America, 1957.
Heirtzler, J. R. "Sea-floor spreading," *Sci. Amer.*, **219a:**60-70, 1968.
Hesse, R., W. C. Allee, and K. P. Schmidt. *Ecological Animal Geography*. New York: John Wiley & Sons, 1937.
Hubbs, C. L. *Zoogeography*. Washington: American Association for the Advancement of Science, 1958.
Kennedy, C. R. *Ecological Aspects of Parasitology*. Amsterdam: New Holland Pub. Co., 1976.
Kevan, D. K. M. *Soil Zoology*. London: Butterworth's Scientific Publications, 1955.
Klopfer, P. H. *Behavioral Aspects of Ecology*. Englewood Cliffs: Prentice-Hall, Inc., 1962.
MacArthur, R. M. *Geographical Ecology*. New York: Harper & Row, 1972.

Neill, W. T. *Biogeography: The Distribution of Animals and Plants.* Boston: D. C. Heath & Co., 1964.

Odum, E. P. *Fundamentals of Ecology,* 3rd ed. Philadelphia: W. B. Saunders Company, 1971.

Ricklefs, R. E. *Ecology,* 2nd ed. Portland, Oregon: Chiron Press, 1979.

Simpson, G. G., A. Roe, and R. C. Lewontin. *Quantitiative Zoology,* rev. ed. New York: Harcourt, Brace and Company, 1960.

Simpson, G. G. *Evolution and Geography.* Eugene, Oregon: State System of Higher Education, 1953.

Sondheimer, E. and J. B. Simeon (eds.): *Chemical Ecology.* New York: Academic Press, Inc., 1970.

Wagner, R. H. *Environment and Man.* New York: W. W. Norton & Co., Inc., 1971.

Wiens, H. J. *Atoll Environment and Ecology.* New Haven: Yale University Press, 1962.

Evolution

History of Evolutionary Thought

Where did the first animals and plants come from? And why do zebras have stripes, and elephants their trunks? Or, for that matter, how did living things in general come to look and act the way that they do? These are questions that have been the subject of speculation throughout human history. The most frequently proposed answers fall roughly into two categories: special creation and evolution.

The idea of **special creation** of life, that living organisms were created in their present forms at the beginning of time by a supernatural force, has been shared by many myths and religions of the past and present. Special creation was not an unreasonable belief, for it was rooted in the everyday experiences of life. Cows gives rise only to cows, dogs to dogs, and humans to humans. Although slight variations in offspring enable individuals to be distinguished, one never finds a cow or a dog in a cat litter. Because one sees no marked change in individuals of a species from generation to generation in the lifetime of human observers, or even after many human generations, how can one reasonably expect one species to change into another species?

However, as science uncovers the principles and patterns of nature, many commonly held beliefs have been discarded. One such belief was that the world was stationary, for, if the world were spinning as some astronomers thought, wouldn't everyone be thrown off into the sky?

Myths from other cultures contained evolutionary ideas, such as the possibility that one form of organism could change into another. For example in some places (Southeast Asia, ancient Persia, Polynesia, and South America), it was thought that humans arose from animals or plants. Such ideas sprang from observations that we are in many ways similar to other living things and, hence, have a common heritage with them.

A more systematic approach to the question appears to have begun in the sixth century B.C. by the Greek philosopher Thales. His method was to seek natural causes for phenomena rather than to attribute them to the whims of anthromorphic (humanlike) gods. By studying the Aegean Sea and by observing that all living things contain water, Thales concluded that animals arose in and from the ocean. Anaximander believed that life arose from moist elements that evaporated and that all animals, including humans, evolved from fishes. Aristotle attempted to classify organisms anatomically and noted that, with increased complexity, there was a corresponding progression of complexity in their embryological development.

Before the seventeenth century, little was known about biology or the fossil record, and special creation was a scientifically reasonable hypothesis. Advances in the young science of geology helped to prepare the scientific community for evolutionary theory. Nineteenth-century geologists were grappling with the age of the earth. A popular idea, proposed by Archbishop Ussher, was that the earth had been created at 9:00 A.M. on October 12, 4004 B.C. But geologists found it difficult to correlate the short time span of this biblical estimate with the great numbers and thicknesses of the rock formations. So geologists attempted to measure the age of the earth indirectly. One way was to calculate the amount of salts carried annually by rivers into the ocean and then to compare that estimate with the total salinity of seawater; this method gave an estimate of 50 million years. An age of 100 million years was obtained by comparing the annual rate of sediment deposition with the thickness of sedimentary rocks exposed on land. Each of these methods assumed constant rates of increase, however, and we now realize that, due to recycling, both yielded gross underestimates. From radioactive dating of certain minerals in lunar material and meteorites, it is now estimated that the earth was formed about 4.5 billion years ago. The oldest dated rocks on earth are about 3.8 billion years old. They nevertheless served to greatly expand conceptual horizons at that time.

Evolutionary ideas began to be proposed in the mid-eighteenth century. Buffon (1707–1788) was a French naturalist who viewed living organisms as being descendants of common ancestors and not the products of independent creation. Buffon used the empirical approach; that is, he made judgments based upon direct observation. Buffon based his support of an evolutionary theory on the observation that terrestrial vertebrates, whether adapted for running, burrowing, swimming, or flying, all share the same general body plan. He also defined a species as a group of organisms that can interbreed.

Erasmus Darwin (1731–1802), the grandfather of Charles Darwin, expressed in poetry and prose a belief in organic evolution, hypothesizing that changes result from the inheritance of environmentally related modifications.

Lamarck (1744–1829), a student of Buffon, was convinced that evolution best accounted for known biological phenomena and proposed a mechanism (something that Buffon did not do) by which it might operate—the **inheritance of acquired characters.** According to the Lamarckian hypothesis of evolution, traits acquired during the lifetime of an animal are passed on to its offspring. For example, the giraffe would have acquired its long neck as a result of its habit of foraging, generation after generation, on the leaves of trees. **Adaptation** was viewed as the cumulative inherited effects of responses to the environment. Many investigators have attempted without success to obtain experimental evidence of inheritance of acquired traits. Modern genetics has shown the Lamarckian mechanism of evolution to be incor-

rect—a case of putting the "cart before the horse."

During the nineteenth century awareness among scientists of the continuity of life continued to grow. In 1838, the German biologists Schleiden and Schwann proposed the theory that all living organisms are composed of **cells.** The implications of this cell theory were expanded in 1865 by Virchow, who proposed that new cells arise only from preexisting cells. These studies were important in the refutation of **spontaneous generation,** that individual organisms arise from inorganic matter. Once it was realized that living organisms can only come from other living organisms, the consistency of this idea with an evolutionary continuity among all creatures could be appreciated.

The Theory of Natural Selection

It might seem remarkable that two men, Charles Darwin (1809–1882) and Alfred Wallace (1823–1913), independently, and from studies in different parts of the world, arrived at the same conclusions to account for the natural history of all life forms. However, keep in mind that information about the physical and biological world had been accumulating, and it remained for someone familiar with this knowledge to integrate and interpret it, although its actual synthesis produced the most electrifying and dramatic new concept in modern history.

Darwin and Wallace were of greatly different backgrounds and dispositions. Darwin was from a wealthy family of physicians and didn't need to worry about money. He was always eager to please his demanding father and scrupulously avoided conflict. Wallace, on the other hand, came from a middle-class family.

As a boy, Darwin liked to collect stones, insects, and plants. Later he turned to collecting rats and hunting game. His father, anxious over Charles's lack of interest in school, sent him to medical school. Charles detested the gore of the business and was later sent to study for the clergy. Upon graduation, he received an invitation to sail on the H.M.S. *Beagle* as an unpaid naturalist. But his father offered great resistance; he wanted Charles to settle down in a career. Yet in 1831 Charles sailed on the five-year journey that would take him around the world (Fig. 41-1) as the *Beagle* went about its objective of preparing navigation maps for the British navy. During the voy-

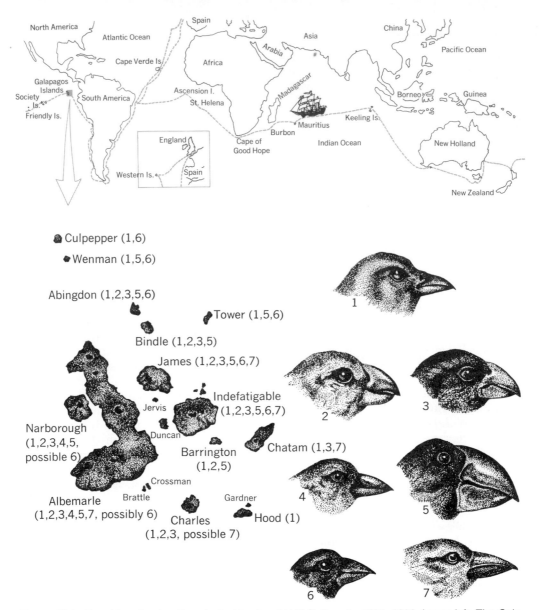

Figure 41-1. Top: Map showing the principal tracks of H.M.S. *Beagle,* 1831–1836. Lower left: The Galapagos Islands. Numbers indicate which of the finches (at right) inhabit any particular island. Lower right: Seven of the thirteen species of Darwin's finches showing beak modifications for feeding. 1, an insect-eating warbler finch, *Certhidea olivacea;* 2, the large insectivorous tree finch, *Camarhynchus psittacula* (it sometimes eats plants); 3, the vegetarian tree finch, *Camarhynchus crassirostris;* 4, the mangrove finch, *Camarhynchus heliobates,* eat insects and sometimes plants; 5, the large ground finch, *Geospiza magnirostris,* eats primarily plants and sometimes insects; 6, the sharp-beaked ground finch, *G. difficilis,* eats plants and some insects and is believed to resemble closely the ancestral South American stock that invaded the Galapagos and evolved into the existing species; 7, the tool-using, or woodpecker, finch, *Camarhynchus pallidus,* is unique in that it selects and uses cactus spines and twigs to extract insects from trees. Although it primarily eats insects, it occasionally eats plants.

age, he read Lyell's newly printed *Principles of Geology*. This book greatly influenced both Darwin and Wallace.

Lyell had rejected the prevailing biblical notions of geology—that the earth was created at about 4,000 B.C. and that a supernaturally induced flood caused all geological phenomena. He gathered together much of the known geological observations and argued that they could be best understood in terms of continuous, predictable, natural causes. This theory of **uniformitarianism** (Chapter 40) holds that the forces that shape nature have not changed over millions of years and that the earth's present geology is considered to have been gradually shaped by the action of rain, wind, volcanoes, and earthquakes rather than by a few enormous cataclysms. He further argued that the extinction of species is caused by geological changes.

Using Lyell's book, Darwin was able to interpret the cliffs and other geological phenomena that he encountered in terms of physical forces that were still acting in the present. Darwin witnessed an earthquake and measured several of its effects: Some coastal areas of South America had risen a full meter; in other locations, dying mussels were attached to rock 3 m above the water! The face of the earth was slowly and constantly changing. What effects might this have on the life in an area?

During the *Beagle's* voyage, Darwin kept detailed records on the nature and distribution of local plants, animals, and fossils; he also collected many specimens. His observations undermined his belief in special creation. If species were static, one would expect living species to be exactly like their fossil ancestors, but in most cases they differed greatly.

The distribution of organisms that Darwin encountered in the Galapagos Islands seemed unreasonable if one accepted special creation. Why would there be different species of finches, mockingbirds, and tortoises on such a closely grouped series of islands? Darwin collected many specimens of these birds, especially the finches (Fig. 41-1). Back in England, he carefully examined them and was struck by the fact that closely related but distinct species of finches, with a perfect gradation of structure, inhabited a neighboring continent. If one mainland species had settled on an island and then diversified into existing forms, the distribution of these finches could best be

understood. From such observations, Darwin began to doubt the hypothesis that species were immutable. But then how did species change?

Darwin knew that animal breeders were able to produce new breeds by selecting and mating individuals that possessed certain desired traits most fully. But, in nature, what acts as the selecting force, what serves as the breeder? The answer came from Thomas Malthus's "An Essay on the Principle of Population," which also influenced Wallace. Malthus stated that unchecked human populations tend to increase geometrically (for example, 2, 4, 8, 16, 32, 64, . . ., 2^n) but that food supplies could only increase arithmetically (such as 2, 4, 6, 8, 10, 12, . . ., $2n$). Human populations will consequently reach a limit of subsistence in which their size will be checked by famine, war, and disease. Darwin realized that resources placed similar pressures on all organisms and that, in the struggle for these resources, organisms with any advantage will more often succeed in reproducing. Variations that give an organism any advantage in a changing environment and that are passed on to the offspring will over a period of time give rise to a population quite different from that of its ancestors. In this way, the environment acts as the breeder, selecting organisms best suited for survival and producing a species that is better adapted to its environment.

Observations:

1. All living organisms within a species vary, and some of these traits are inherited.
2. Organisms reproduce in a way that potentially results in geometric growth of their populations.
3. Because more organisms are produced than survive, there is a struggle for existence.

From this Darwin deduced that, in the struggle, the fittest (those with the most favorable variations in that environment) survive to produce the most offspring, which in turn survive to pass on the favorable traits to their progeny.

Darwin realized this in 1838 but would not consider presenting his theory until he was certain that it was correct and had substantiated every point with voluminous evidence. It was, in fact, this drive for accumulating an enormous compilation of data that separated his works, and his influence, from those who went before. Others suggested, but Darwin marshalled the evidence that overwhelmed the massive inertia of the tradition his ideas challenged. In 1844, he wrote an essay on evolution in preparation for his future tomes on the subject that were to contain the vast stores of supporting evidence he was gathering. Something happened in 1858 to hurry Darwin along—he received a letter from Wallace containing a theory of evolution identical to his own. Wallace had also traveled extensively around the world and observed the same phenomena of variation and competition of species that Darwin had noted. Darwin was prepared to stop all his work and give full credit to Wallace, but, at the insistence of his friends (Lyell among them), Darwin presented his paper jointly with Wallace's before the Linnean Society, a scientific society that is still in existence today.

In 1859, Darwin's *On the Origin of Species by Means of Natural Selection* was published, a work he considered an "essay" or preliminary outline! Its bombshell effect echoed and reechoed throughout the western world. Although some readers welcomed the book and were swayed by the weight of its evidence and arguments, many balked at any system that diminished their anthropocentric image of nature. As with other revolutionary ideas, evolution proved to be deeply disturbing and frightening. It was popularly rejected much as Copernicus's theory that the earth orbits the sun was rejected, because it implied that our planet was not the center of the universe. Darwin's evolutionary theory seemed to imply that it wasn't even our planet—that humans were simply here, for no particular reason, and were "nothing but" animals. These philosophical implications have no bearing on the validity of the scientific theory, of course, but they have been a focus of intense opposition on religious grounds for many years. Eventually the overwhelming evidence forced widespread acceptance of the evolution concept, although sporadic resurgence of opposition on fundamental religious (or antiscience) grounds still occurs.

Objections to evolutionary theory on a scientific rather than on a philosophical basis also developed. For example, how did variations arise in sufficient numbers to account for evolution? Darwin had no way of knowing how the individual variations that he had observed in nature came about. Indeed, not until six years after the *Origin of Species* was published did Mendel formulate his laws of genetic inheritance

(Chapter 38), and his work was then overlooked for another 30 years.

Modern Synthetic Evolutionary Theory

Darwin's work provides a framework for the **synthetic theory of evolution,** so called because it blends new evidences from other biological sciences. During this century scientists have discovered that two types of molecules, nucleic acids and proteins, characterize all living organisms, and the major pathways by which they interact have been elucidated. Geneticists study **heredity** and variation between organisms. Mathematicians have also contributed to modern evolutionary thought; the statistical treatment of **population genetics** has proved to be a powerful tool in understanding evolution. **Comparative anatomy** and **embryology** have also given major insights into the evolutionary process.

Evolution is genetically defined as any change in allelic frequency. Genetic variation results primarily from mutations of genes and chromosomes but more significantly from new combinations of genes produced by meiotic crossovers and random recombinations at fertilization and mutations that affect the expression and regulations of genetic information (Chapter 38). These sources of genetic variation do not significantly alter allele frequencies, but they do result in the expression of different phenotypes. Each individual in a sexually reproducing population is unique. The variations in phenotypes (anatomical, physiological, and behavioral) are acted upon by selective pressures, and differential survival and reproductive rates of individuals results. This changes the gene frequencies, because the alleles of the survivors become incorporated into the gene pool of the next generation.

In the chapter on genetics, gene and chromosome mutations were discussed in some detail. This chapter is less concerned with the detailed mechanisms of mutations than with the fact that they occur. It was noted earlier that, in the absence of other factors, allelic frequencies of sexual organisms remain essentially unchanged in successive generations, and quantitative examples were presented by using the **Hardy–Weinberg formula.** Using this simplified model, we can analyze the amount of influence exerted by selection factors that are encountered in a natural environment.

The rate at which mutations occurs has been determined to be roughly 1 in every 10,000 meiotic events. Considering that, in one human male ejaculation there are perhaps 200–400 million sperm, the frequency at which mutations enter the gene pool is quite low. Assuming that organisms are adapted to their natural environments, it is easy to see that mutations are usually disadvantageous to an organism. The greater the effect of the mutation, the more likely it is to be harmful. However, as we shall note, many alleles are frequently retained despite their harmful effects. It is also possible for neutral and harmful genes to become adaptive when environmental conditions change; these genes give the gene pool versatility. Mutations furnish the new genetic material upon which other evolutionary influences can act. Now let us consider these other evolutionary influences that change allele frequencies.

Mutation Pressure

Mutations alter allele frequencies, thereby exerting **mutation pressure.** For example, a mutation from allele A to a would by definition change the frequencies of A and a. But, unless these mutations are selected for or against, their effect by themselves on gene frequency in a population's gene pool are probably negligible.

Meiotic Drive

During meiosis, a male heterozygous at a particular locus (Aa) will normally produce equal proportions of sperm containing the A and a alleles. However, if the sperm from this male were to contain only the a allele (with little or no A sperm produced), the frequency of a would increase in the gene pool. Certain alleles, called segregation distorters (SD), do result in all sperm containing only the SD allele. In *Drosophila melanogaster* the SD gene is located on the centromere of one chromosome. Its influence is exerted during meiotic synapsis (pairing of homologous chromosomes). One may think of SD genes as alleles that penetrate themselves by eliminating competition. Though uncommon in natural populations, SD genes

do occur. However, the extent of their role in producing evolutionary change is not known.

Gene Flow

Gene frequencies can be greatly changed by **immigration,** if one **deme** (interbreeding population) breeds with immigrants from other demes (of the same species). The changes in gene frequency will be related to the proportion of breeding immigrants and to the extent to which their allelic frequencies differ from those of the host deme. The larger these differences, the greater the changes in gene frequency.

However, when there are restrictions to **gene flow** (e.g., geographical barriers), genetic differences may become too large (as when the gene pools have diverged to the point of speciation); later immigrants may not be able to mate, or mating may result in sterile offspring.

Genetic Drift

Genetic drift is any change in gene frequency that results from chance and not from selection pressures. It is a statistical phenomenon that depends on the size of the gene pool and is only important in small, isolated populations (Chapter 38).

You can simulate this phenomenon by flipping coins. After many trials, one expects heads to have turned up as often as tails. Expressed another way, one expects the alternatives to occur with equal frequency. However, with fewer trials, the variance from equal frequency increases. That is why after four trials one would not be surprised to have flipped all heads, but, after four hundred, one would be flabbergasted!

In a population, let allele A represent heads and allele **a** represent tails. If many trials correspond to many individuals (a large gene pool), we can see at once that the variance in such a large population is negligible and that the allelic frequencies of A or a will not be changed by chance alone. However, in a small population, the variance may be large and an allele may even be lost by chance (as when all heads were tossed—the tails were lost).

Hence, in small populations that are not subject to frequent immigration, genetic drift can be an effective agent of evolutionary change. This type of evolutionary change is random, so there is no way to predict what direction it might take. Harmful mutations may even be retained and spread in the population in this way.

Natural Selection

The most important way in which allelic frequencies are changed is when organisms with particular genotypes are prevented from being fully represented in the next generation. Organisms that are more successful in reproduction cause their alleles to be more prevalent in the population, and hence they will alter the allelic frequencies of the gene pool. The process that determines which organisms will reproduce and which will not is referred to as **natural selection.** Selection for certain phenotypes will alter the gene pool, because phenotypes are expressions of underlying genotypes.

Selection may result from many circumstances. Diseases and predators may reduce certain phenotypes. Climatic and geological changes could have the same effect as could competition among phenotypes for food and mates. These are just a few examples. When one considers that these circumstances operate on the phenotype at every stage of life (through reproductive age), the complexity and pervasiveness of selective pressures become apparent.

The environment thus indirectly changes genotypes through the selection of phenotypes. Accumulation of small genetic changes by continuous natural selection over many generations results in "adaptation" of individuals to the particular environment (adaptation is perhaps an unfortunate choice of words as it implies that changes are due to a conscious effort on the part of the organism, which clearly is not the case). The course of evolution appears to be guided by the adaptation of individuals within a population to the environment.

Let us consider a normal distribution curve for some trait, say, nose length, in a hypothetical population to illustrate some ways in which selection operates (Fig. 41-2). Most organisms will have noses of an average length that we will arbitrarily call 5 cm. Fewer individuals will have extremes of 1 cm and 9 cm. If for some reason the extreme phenotypes are selected against, **stabilizing selection** will cause fewer individuals with extremes of nose length to become

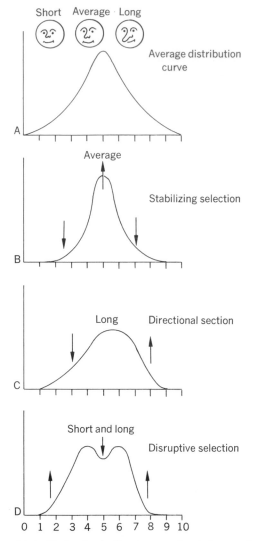

Figure 41-2. Kinds of selection (arrows directed upward depicted selection for and directed downward selection against the nose length indicated). (A) Before selection, the average distribution curve of nose length in a hypothetical population. (B) After stabilizing selection (most individuals with average noses). (C) After directional selection (most individuals with long noses). (D) After disruptive selection (most individuals with either long or short noses, fewer with average nose length).

parents, in favor of those with the average length. If the long nose is seen as attractive and is selected for, and the short nose is selected against, **directional selection** will produce a population tending toward longer noses. If both short and long noses are selected for and average noses are selected against, **disruptive selection** will result in a splitting of the curve between the extremes.

Stabilizing selection may act to sustain a heterozygous condition. Consider the genetics of sickle cell anemia: the "normal" hemoglobin allele A and the allele for the sickle cell trait a. The heterozygote Aa is more resistant to malaria than is the homozygote AA. Hence, heterozygotes have a selective advantage in malarious regions. The homozygous recessive aa is usually lethal before the breeding age and is thus strongly selected against. Despite this, in malarial regions the a allele is retained in the gene pool in high frequencies as heterozygote Aa is stabilized. Stabilization is actually at some equilibrium point that is determined by the amount of selection against AA by malaria. As the incidence of malaria approaches zero, the equlibrium is shifted and the selection is directional, against the a allele. This is, in fact, precisely what has developed with Africans in malarial areas, as compared with American blacks in the United States, where there is no malaria.

In regions where malaria does not occur, only the homozygous recessive aa is selected against, and so selection is directional. However, the change in allelic frequency is slow, as shown by the fact that the sickle cell trait aa is still found in the black U.S. population at twice the level of its occurrence among whites. If the frequency of a was one in 10,000 or 10^{-4}), it would take at least 9,900 generations to reduce the frequency to 10^{-8}.

Disruptive selection can give rise to new species, especially when the diverging phenotypes mate nonrandomly, such as only with others like themselves. This latter case is important because it prevents gene flow even though there is the potential for interbreeding; eventually, mutations that prevent successful interbreeding may occur and supply the reproductive isolation needed for speciation (discussed in a subsequent selection).

Any and all of these influences may interact to produce new genotypes and hence result in evolutionary change. The interactions are complex, and mathematical models are often devised to understand the actual process of evolution.

Adaptation

An adaptation is any characteristic of an organism that improves its chance of survival and reproduction. That animals are adapted to the environments in which they live is obvious to anyone who has considered their structure, physiology, and habits. Adaptations result from the winnowing effects of natural selection: as the fittest phenotypes survive, their underlying genotypes are propagated in the gene pool of the population.

An adaptation may be anatomical or physiological, or it may involve inherent patterns of behavior. We have noted, for example, how the basic body plan of insects has been variously modified: the wings are adapted for flight; the legs for running, swimming, or other purposes; the mouthparts for biting or sucking; and the digestive tract for digesting solid or liquid food. In birds the wings and tail are adapted for different types of flight; the feet for perching, wading, or swimming; and the bill for capturing insects, crushing seeds, or tearing flesh. Each species exhibits a constellation of adaptations that allow it to survive and reproduce in its particular habitat. Adaptation to the physical and biological environment is a dynamic process, and the gene pool of a population responds to any fluctuations of selective pressures.

Adaptive Radiation, Divergent Evolution, and Convergent Evolution

Because of competition for food and living space, there is a tendency for each species to specialize in a particular role or niche in a community. This tendency is dramatically illustrated when a species or group of species colonizes a new habitat; examples include the invasion of land in the geologic past and the colonization of oceanic islands. The successful invasion of an unoccupied habitat often leads to an explosive **adaptive radiation** that populates the area with new species over a relatively short time.

Figure 41-3 illustrates how, from primitive mammalian stock, different types have adapted to a variety of habitats (**divergent evolution**). The impetus for this adaptive radiation was the extinction of the dinosaurs at the close of the Mesozoic era, which opened many niches in terrestrial and aquatic habitats. The great flexibility of the mammalian body plan has allowed

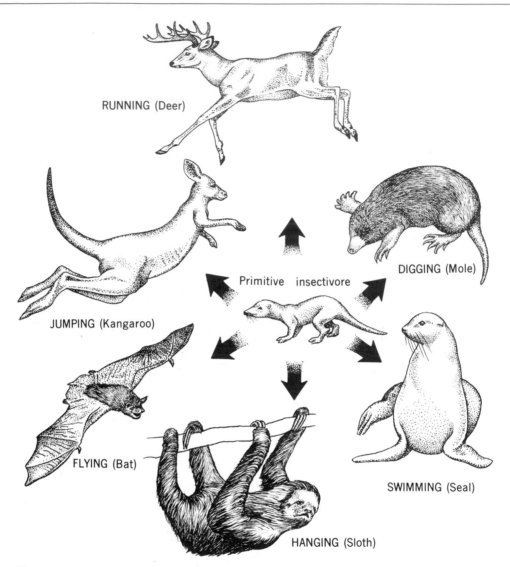

Figure 41-3. Divergent evolution of homologous appendage. The various mammals have evolved from a common ancestory, the five-toed land mammal shown in the center of the diagram. Note how the structure of the limbs has been modified (specialized) to adapt them to a wide variety of environments.

members of this class to compete successfully with other types of animals both on and under the ground, in trees, in the air, and under water.

Darwin's finches on the Galapagos Islands present another clear example of adaptive radiation (see Fig. 41-1). Fourteen species of finches inhabit the islands,

and these are thought to have all evolved from a single pair of birds that made the 600-mile flight from the mainland approximately 1 million years ago. The islands, previously uninhabited by birds, provided unoccupied niches into which the immigrants could enter, breed, and prosper. As the offspring dispersed

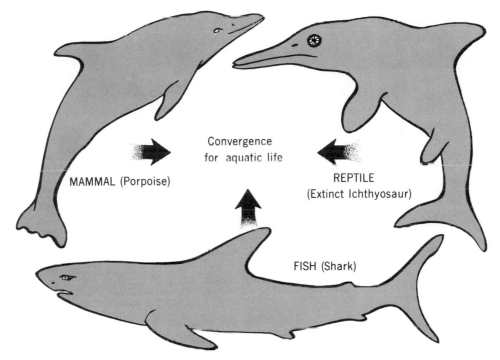

Figure 41-4. Adaptive convergence or convergent evolution of analogous structures. Although the fish, reptile, and mammal shown are not closely related, they have a marked superficial similarity because they are all adapted to living in the same environment.

over the islands, they encountered somewhat different climates, ground conditions, insect types, and so on. Different adaptations to particular niches eventually led to speciation. The most obvious of these adaptations are differences in beak shape and feeding behavior.

Convergent evolution results in the acquisition of some similar characteristics analogous structures among groups of organisms that are closely related. The evolution of such analogous traits usually represents similar adaptations to the demands of similar environments. For example, in Figure 41-4, it can be seen that the streamlined fusiform bodies of a fish (shark), an extinct aquatic reptile (ichthyosaur), and an aquatic mammal (porpoise) adapt them all for rapid locomotion underwater by minimizing frictional drag.

Convergent evolution thus produces the opposite effects of divergent evolution. In either case, however, the units of speciation are populations of individuals that seek to maximize survival in particular environments. Now let us take a closer look at the process of speciation.

Speciation

In the discussion of population genetics, the concept of gene pools and their permutations through time (for example, through genetic drift or natural selection) were outlined. Speciation involves the genetic divergence of populations of the same species to the point at which cross-mating between individuals of the two populations no longer produces fertile offspring. This divergence of genotypes is usually expressed by a divergence of phenotypes as well.

Speciation usually results when populations become geographically or spatially isolated from one another. Without interbreeding, their gene pools may then diverge to the point of speciation as each population independently adapts to a different environment by novel means.

Of course, the limiting effect of a physical barrier is a direct function of the organism's ability to traverse it; for example, an earthworm may not be able to cross a river, but a bird can do it with ease. Some examples of physical barriers are bodies of water (oceans, rivers, large lakes, streams), mountain ranges, deserts, forests, and, in some cases, concrete and asphalt (in cities and highways).

Ecological barriers, those due to climate, vegetation, or biological considerations, are also important. An organism cannot survive in a habitat in which it cannot find or successfully compete for food or in which it may freeze or dehydrate. Climate can also have more subtle effects: Some low-lying islands are populated by few, if any, winged insects because strong winds blow them off.

The ocean is by no means a homogeneous habitat either. Marine organisms are, for the most part, limited to rather specific regions by temperature, salinity, and depth. Water has a high heat capacity, and so marine and freshwater organisms experience less severe temperature variations than do terrestrial organisms. Nevertheless, there are definite thermal layers in the ocean. As one goes deeper, the water becomes colder. Furthermore, water currents tend to enhance the thermal heterogeneity of surface waters.

An interesting case concerns the fish populations along the Atlantic and Pacific coasts of South America. The species are different, presumably because the polar waters at the Cape of Good Hope provide a thermal barrier that prevents the migration of temperate species. Another example occurs on the Atlantic Coast of the United States. The warm Gulf Stream flows in a northerly direction, and so several tropical fish species can survive as far north as Cape Hatteras, North Carolina.

Another important difference between marine and terrestrial environments is that primary production in the ocean is confined to a shallow photic zone at the surface. Thus the availability of food generally decreases with depth. Many marine animals have planktonic larval stages that forage in the productive surface waters. These larvae tend to disperse the species in the prevailing water currents.

Reproductive patterns, such as the time of estrous

or receptivity for mating, often varies among species. Even in the absence of a spatial barrier, two populations may have diverged because they mate at different times of the year. Their gene pools have now become effectively isolated by a temporal barrier.

Evidence in Support of Organic Evolution

The compelling evidence for organic evolution has been derived from many scientific disciplines. The principles of evolution are based on a number of different types of evidence, some of which are presented briefly in the following paragraphs.

Comparative Anatomy

The study of comparative anatomy focuses on the similarities and dissimilarities in body plans (Fig. 41-5) and offers much evidence in favor of organic evolution. For example, the flipper of a seal, the wing of a bird, the leg of a horse, and the human arm are all similar in the type and arrangement of bones and have evidently all evolved from the same type of ancestral appendage—that is, they are homologous.

Vestigial organs, which are especially evident among vertebrates, furnish striking evidence of changes from ancestral conditions. The human eye, for example, has a vestigial nictitating membrane, and the modern horse possesses vestigial "splints" in place of what in its ancestors were functional metacarpals and metatarsals. The functions of many organs have changed during the course of evolution. For example, the salivary glands in certain snakes have become modified into poison glands.

Comparative Embryology

The **biogenetic law** was formulated in the 1860s by the German biologist E. Haeckel, who observed that early stages of vertebrate embryos showed remarkable similarities (Fig. 41-6). He concluded that animals summarized (recapitulated) in their developmental (ontogenetic) stages the phylogenetic history of the species, or that "ontogeny recapitulates phylogeny."

A more accurate statement is that the embryo of an animal may resemble the embryonic stages (but not

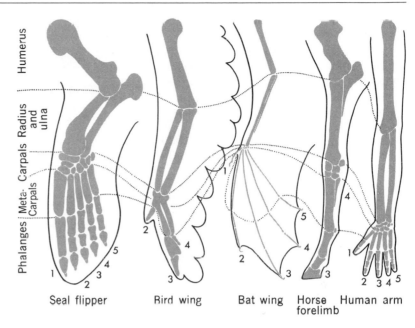

Figure 41-5. Homology and adaptation in bones of the forelimbs of vertebrates. The limbs are homologous, being of the same fundamental structure, based on common descent, but in each kind of animal they are adapted for special functions by modifications of an ancestral appendage.

the adult forms) of other animals on the phyletic scale (Fig. 41-6). This modification of the original statement of the biogenetic law is necessary because it is evident that evolution has also taken place in embryonic and larval stages. For example, certain stages may be omitted, whereas others may be added. Such additions, of course, have no ancestral significance. Despite such criticism of the biogenetic law, there can be no doubt that a study of the developmental stages of an animal gives us important clues to its ancestry. As discussed briefly in Chapters 38 and 39, mutations

in gene regulation that alter developmental processes are among the most important mechanisms for generating changes in adult traits.

Comparative Biochemistry

Within recent years the biochemical characteristics of animals have been found to furnish convincing evidence in favor of organic evolution. For example, comparative studies have been made of the amino acid sequences of hemoglobins isolated from various

Figure 41-6. Gill slits in embryos of animals belonging to three different classes of vertebrates. We find gill slits in adult fish, but they are present only in embryonic birds and mammals.

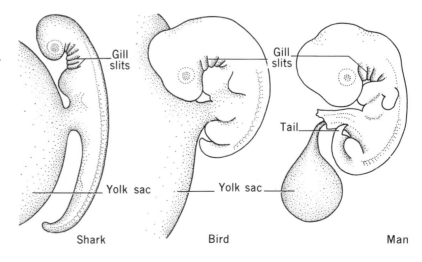

organisms. This shows that the sequences of closely related species are more nearly alike than are those of distantly related species. The sequences (and, hence, the molecular structure) of species belonging to the same genus resemble each other more closely than do those of species belonging to different genera. The degrees of relationship of families, orders, and so on to each other can also be measured in this way. For example, among carnivores the seals and bears have been found to be more closely related to each other than they are to dogs.

Additional evidence of relationships is provided by the degree of similarity between many other proteins of various animals. Blood tests indicate that humans are closest genetically to the great apes; next in order are the Old World monkeys, the New World monkeys, and the lemurs. Blood protein similarities also indicate that dogs, cats, and bears are closely related, whereas sheep, goats, cows, antelopes, and deer form another closely related group. Sea lions and other seals are more closely related to the carnivores than to other mammals, which agrees with anatomical and other correlations. With recent advances in molecular biology, techniques have been developed to sequence RNA and DNA molecules directly; comparisons of these data are also being used to derive phylogenies.

Hormones show similar reactions in different animals. If beef or sheep thyroid is fed to frog tadpoles from which the thyroid gland has been removed, they metamorphose into normal frogs as if their own thyroid glands had produced thyroxin. The hormones used to treat human endocrine deficiencies are generally obtained from other vertebrates.

Another method of comparison is the distribution of various enzymes and, thus, the metabolic pathways in which they participate. For example, trypsin, which degrades proteins, occurs in many animals from protists to humans, and amylase, which acts on starches, is present from sponges to humans. It seems reasonable to suppose that similarities in biochemistry occur in such widely different animals because they have evolved from common ancestors. The close correlations among biochemical, embryological, and anatomical lines of evidence is one of the most convincing demonstrations that evolution has in fact occurred (as distinct from the theory of natural selection to account for these facts.)

Gradations in Form

One hundred years before organic evolution was generally recognized, taxonomists had classified animals according to increasing degrees of similarity into phyla, classes, orders, families, genera, and species. For the most part, the successively more similar groups into which Linnaeus and others placed the various members of the animal kingdom correspond to degrees of genetic relationship.

This natural taxonomic scheme enabled biologists to arrange animal phyla according to degrees of complexity, beginning with the sponges and proceeding to the more complex groups. In other words, the early taxonomists had erected a phyletic tree even before the evolutionary relationships of its various branches were recognized.

Geographical Distribution

According to students of zoogeography (Chapter 40), each species arose in a definite area, known as its **center of origin,** and, from this area, the species tended to disperse to its present distribution. Many facts of present-day distribution can only be explained by assuming that organic evolution has taken place. For example, continental islands possess faunas that are closely related to those of nearby continents, whereas the communities on oceanic islands appear to have arisen by chance introduction of colonizing species and their subsequent evolutionary divergence. Isolation plays a critical role in evolution because it is an essential prelude to the accumulation of genetic variants upon which natural selection can act to yield a new species. Many animal populations on islands have diverged to the point of speciation as a result of isolation from their ancestral gene pools on the mainland.

Genetics

The universality of the genetic mechanisms underlying information processing and reproduction certainly implies an evolutionary unity among all organisms. The presence of DNA and RNA in all cells, the similarities of cell structures, and the processes that they undergo leave little doubt of relationship. The genetic code is identical for all organisms of the same species. Models constructed on the basis of evolution in **population genetics** (Chapter 38) are used to explain field observations.

Fossil Animals: An Historic Record of Evolution

From before Darwin's time to the present, the fossilized remains of ancient animals and plants have been regarded as strong evidence for evolution.

Fossils are usually petrifaction—that is, body parts that have been replaced by mineral matter. The hard parts of the animals, such as bones, shells, and teeth, may be preserved intact. In addition, animals may be preserved in ice, amber (primarily insects), natural asphalt, and tar or oil-bearing soil (mammoths). **Casts** of animals resulting from the dissolution of the original parts and the filling up of space with mineral matter also give some idea of external features.

Which animals become fossilized and which of these are eventually discovered are largely matters of chance. Animals with hard body parts became fossilized most often, especially those that lived near seasonal floodplains. Furthermore, fossils are usually broken by the forces of nature and are often fragmentary when they are found. Nevertheless, from fossil evidence paleontologists have been able to judge the approximate time of origin of the different animal groups with the geological periods of the earth's history (Fig. 23-3).

Such data show that the various invertebrates appeared before the vertebrates, because their remains occur in older layers of rock. The order of appearance is much the same as that of their consideration in this text. The fishlike ostracoderms were the first vertebrates to leave fossils, and these were followed by other fishes, amphibians, reptiles, birds and mammals in the order that would be expected from a study of their comparative anatomy.

Supportive evidence derived from paleontology also results from investigating the lines of descent of single groups, such as horses, camels, or elephants.

Evolution of the Horse

A classic example of paleontological evidence for organic evolution is the horse (Fig. 41-7). The horses now living in America are descendants of domesti-

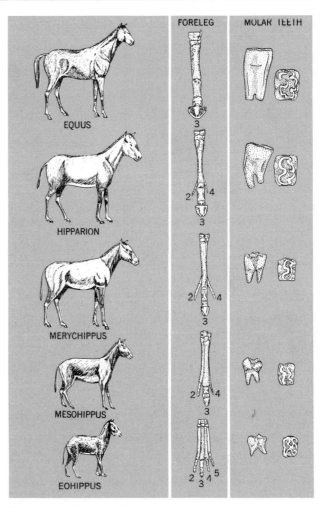

Figure 41-7. Diagram illustrates the evolution of the horse. Digits or rudiments of digits (splints) are designated by numbers 1–5.

cated animals that were introduced by settlers from Europe. However, in prehistoric times the ancestors of the modern horse were native here, and some of the finest fossil remains of these early horses have been found in America.

The evolution of the horse has been traced back through many distinct stages, extending from the Eocene epoch about 60 million years ago. A brief description of five of these stages will serve to illustrate the principal changes that took place during this evolution.

Horse evolution is highlighted by adaptations to life on the open plains, where food consisted of dry grasses. The feet of the horse gradually lost the side toes, and only the middle toe and rudiments (**splints**) of the second and fourth digits* remain in modern horses. The body gradually increased in size, and the limbs became longer, enabling the animal to run more rapidly; this change was correlated with an elongation of the head and neck for browsing. The front teeth were modified as chisel-like cropping structures. The back teeth evolved from simple molars into effective grinding organs with wrinkled ridges of hard enamel supported by layers of dentine and cement. During later periods the molars elongated and thus became adapted for grinding dry grasses, which increased the rate of tooth wear more rapidly than did softer vegetation.

Eohippus (dawn horse). This earliest-known member of the horse family was only 27 cm high at the shoulder and about the size of a fox terrier. *Eohippus* lived in North America and Europe during the early Eocene epoch and was a browsing forest dweller. Each of its forefeet had four complete toes but no trace of the first (thumb), and the hindfeet had three complete toes plus rudiments (splints) of the first and the fifth. The simple teeth indicate that it browsed on soft vegetation.

Mesohippus (intermediate horse). This horse lived during the Oligocene and reached the size of a sheep. Its forefeet each possessed three complete toes but no splint. All three toes touched the ground, but the middle toe was larger and bore most of the body weight.

Merychippus (ruminating horse). This horse lived in the Miocene but became extinct in the Pliocene. It was a transitional form in the shift from a browsing to a grazing diet. The milk teeth were short crowned and uncemented, like those of the primitive horse, but the permanent teeth were long crowned and fully cemented grinders, suited to the harsh vegetation of the plains. Its forefeet and hindfeet each possessed three toes.

*The digits are numbered 1 to 5, beginning on the thumb side.

Pliohippus (Pliocene horse). This horse, from the Upper Miocene and Pliocene, was the first one-toed horse. Both forefeet and hindfeet were single-toed, with the second and fourth toes represented by splints. The crowns of the upper molars were similar to those of the modern horse, but they did not possess as complex a pattern of surface ridges. *Pliohippus* had a shoulder height of some 100 cm, about the size of a modern pony.

Equus (horse). The modern horses of the Pleistocene and recent epochs have entirely lost the first and fifth digits, and the second and fourth digits are represented by splints. The third toe alone supports the weight of the body. The crowns of the molar teeth are much elongated and bear complex enameled ridges well adapted for grinding dry, harsh vegetation. The lengthened skull accommodates a larger and more complex brain. The modern horse is about 1.5 m tall, considerably larger than any of its ancestors. The evolution of the horse has resulted in an intelligent, long-legged, swift-running animal that is suited to live and feed on open grasslands.

At the present time, true wild representatives of the genus *Equus* occur only in Asia (the Asiatic wild ass, *E. hemionus*, and Przewalsky's horse, *E. przewalskii*) and in Africa (the African wild ass, *E. asinus*, and the zebras, *E. greyi* and *E. burchelli*). The mustangs and bronos of our western plains and South America are descendants of domesticated horses (*E. caballus*) brought over from Europe.

The evolutionary sequences of the elephant, camel, dog, and many other mammals have also been carefully reconstructed by paleontologists. These likewise provide well-documented evidence of the existence and course of evolution.

Artificial Selection

Many populations of animals (and plants) have been maintained under domestication for centuries by human breeders. During this time various species have undergone artificial selection to produce some desirable traits. In certain cases, we know that various breeds have all arisen from a single wild species. For example, the fantail, pouter, tumbler, and other types of pigeons may all have evolved from the rock dove, *Columba livia*. Horses, cattle, sheep, dogs, and fowls have undergone similar changes. If it were not for the fact that nearly all the breeds are capable of successfully interbreeding, they would be considered distinct species.

Origin of Life
Spontaneous Generation

Although many religious accounts of the origin of life state that the kinds of animals and plants were immutably fixed at the time of creation, it was widely believed until the eighteenth century that living organisms could spring spontaneously from nonliving matter. This theory is called **spontaneous generation,** or **abiogenesis.** For example, insects were believed to originate from dew, frogs and toads from the muddy bottoms of ponds under the influence of the sun, butterflies from cheese, and so on. Even today, some people believe that mosquitoes are generated by stagnant water and that horse hairs that fall into water change into living nematomorph worms.

The classical experiment in 1668 of the Italian, F. Redi, clearly refuted this theory that animals could arise from nonliving sources. It was widely believed at the time that maggots sprung spontaneously from rotten meat. Redi placed meat in wide-mouthed flasks; some flasks were left open, some were covered with gauze, and others with paper (Fig. 41-8). The meat decayed in all vessels. Flies entered the open vessels and laid eggs that hatched into maggots. No larvae developed on the meat in the covered vessels, however. Some maggots did develop on top of the gauze though, and, from this, Redi concluded that maggots result from eggs laid by flies and are not a direct by-product of putrifaction.

Old beliefs die hard, however, and the discovery of the microscope added a new dimension to the problem. Microorganisms appeared to be everywhere, and their ubiquitous distribution was generally believed to have resulted from spontaneous generation. It was not until 200 years later, in 1864, that Louis Pasteur demonstrated that microorganisms never arise from sterile matter. Pasteur boiled broth in flasks that were equipped with long swan necks that allowed air to enter only through the long tube. He reasoned that any airborne microorganisms would be trapped in the S-shaped bend of the neck. Sure enough, the broth usually remained sterile, unless he broke off the neck so that air-borne microorganisms could fall into the broth.

Biogenesis

The theory of spontaneous generation, abiogensis, thus gave way to that of biogenesis, which maintains that all life arises from preexisting life. If living animals do not arise from nonliving matter, it is natural to inquire how the world became populated, as both geologists and astronomers tell us that at one time life could not have existed on earth. The doctrine of special creation is sufficiently refuted to the satisfaction of most biologists by the evidence of organic evolution. Consequently, life must have either originated on the earth from nonliving matter or have been brought to the earth from some other part of the universe. The latter idea, known as the **cosmozoic theory,** or **pangenesis,** seems highly improbable, but, even if it were true, it would not explain the origin of life, only how life reached the earth.

The fossil record has a curious discontinuity. About 600 million years ago, metazoan fossils became abundant in sedimentary rocks. Most of the modern phyla are recognizable in such early Paleozoic rocks, and virtually all appear within the next 300 million years. Why metazoan fossils older than 600 million years are scarce is uncertain, but the reason can be correlated with a general lack of hard skeletons that enhance fossilization.

The sudden appearance of complex metazoan fossils at the onset of the Paleozoic era suggests that life must have evolved during the earlier three quarters of the earth's history. Precambrian paleontologists have recently examined ancient rocks using microscopes and modern methods of biochemical analysis. Their discoveries have pushed back the timetable of evolution.

These studies have revealed organic **microspheres, colonies,** and **filaments** that are morphologically extremely similar to modern **bacteria, blue-green algae** (including many filamentous forms), and, in the later part of the Precambrian, **eucaryotic algae** and **fungi.** Most of these fossilized structures are found in **chert,**

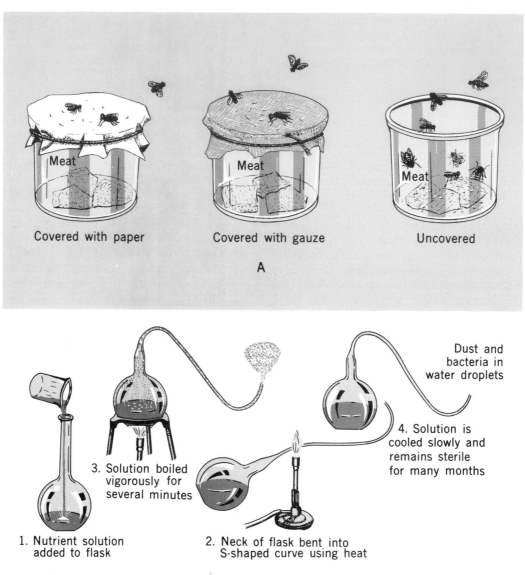

Figure 41-8. Spontaneous generation. Diagrams illustrate methods of disproving this theory when applied to existing organisms. (A) The three flasks illustrate Redi's method of experimenting with blow flies. (B) Pasteur's flask as used in his experiments. This peculiar glass flask was half filled with water that contained large numbers of microscopic organisms. By boiling at intermittent intervals, all the living organisms were killed, and the entire contents of the flask was made sterile. At the same time, escaping steam condensed, and water filled the lowest part of the bent tube, so that a complete barrier was formed there. The water in the bent tube captured all organisms so that air from the outside that finally reached the sterile mixture at the bottom of the flask was sterile. The result was that, within the flask proper, no life developed, whereas, in the water barrier in the bent tube, a host of minute organisms developed.

a quartz mineral that has the fortuitous property of very fine crystals. This allows the very delicate cells to be trapped within the chert matrix without being destroyed by growing crystals, which is what happens with most other minerals. Generally, only the outer cell walls retain their original size and shape. Delicate internal organelles and nuclei are usually altered beyond recognition, making the identification of eucaryotic cells a difficult task. The original biochemical composition of these organisms has been degraded and repolymerized into a complex, but sturdy, structure during the very long time since they died; this is exactly the same process that occurs when dead organic matter is converted into petroleum and coal. Because of this, it is very difficult to try to draw conclusions about the biochemistry of the original organisms; only their general size and shape remain for sure. The oldest organic microspheres have been found in cherts in South Africa that are over 3.2 billion years old. Interesting speculations about these simple spheres include the hypothesis that all may not have been living cells but, rather, organic structures that might have been precursors to the first bacteria.

Another important type of fossil is the preserved **stromatollite,** which is a multilayered macroscopic structure formed by the growth of certain bacteria and algae. They grow in a thin layer of cells or filaments on a surface such as mud; thus a larger surface area may be exposed to nutrients or, in the case of photosynthetic bacteria and algae, light. After a layer of sediment covers them, many cells move up and grow into a new layer. This may happen over and over, and, if the stromatollite is infiltrated by minerals, their layered structure may be fossilized. Living stromatollites exist only in a few areas of the world today because of competition by other organisms, but they existed in large numbers during most of the Precambrian. An exciting new find is a fossilized stromatollite from Australia that has been dated at about 3.6 billion years!

Such ancient evidence of life gives us a new perspective on evolution. it seems that life originated fairly early on the earth and was quite successful for over three fourths of its history in the form of bacteria, algae, and fungi. The more difficult evolutionary transition seems to have been the origin of multicellular organisms at around 600 million years ago; the rate

of evolution increased after that time to produce the multitude of organisms of kingdoms Plantae and Animalia, the latter being the main concern of zoology.

Physiochemical Theory

Our concepts of time and evolution have changed radically since the scientific revolution of the last century, and new vistas continue to unfold. Scientists no longer consider evolution only in terms of living organisms but, rather, now include the prebiological evolution from inorganic compounds to biochemicals to the macromolecules necessary for living systems.

The father of the **physiochemical theory** is A. I. Oparin, the Russian biochemist who, in the 1920's, first formally described the evolution from the inorganic world to living cells. Oparin's experimental work on this subject concerned the aggregation of macromolecules into spheres (**coacervates**) that could demonstrate biochemical activity if certain enzymes were trapped inside. Although these were made from biologically produced materials, they were important model systems that demonstrated the necessity of membrane-enclosed systems.

A large amount of interest and experimentation followed Oparin's pioneering efforts. It was soon realized that, to form organic compounds, no oxygen could be present in the early atmosphere. Investigators such as H. Urey, who were interested in the origin of the earth, concluded that the early atmosphere was indeed reducing (no O_2) and was probably composed of N_2, CH_4 (methane), H_2O, CO_2, CO, and NH_3 (ammonia), much like the atmosphere observed on Jupiter today. A student of Urey, S. Miller, performed a now classic experiment in the early 1950s. A mixture of NH_3, CH_4, CO_2, and H_2O gases was sealed in an apparatus that had electrodes to produce an electric discharge (spark), simulating lightning as an energy source. Of the products of the gaseous reactions that were collected in a water-filled trap, Miller identified many amino acids, simple hydrocarbons, and a variety of other organic compounds. There have been many arguments concerning the exact composition of the early atmosphere, but experiments since Miller's have shown that the important factor is the lack of O_2. As long as there is some NH_3, almost all mixtures of H_2O, CH_4, CO, and CO_2 produce a variety of organic compounds. The scientific literature now contains thousands of experiments demonstrating the production of various important biochemicals under a large variety of conditions and using various energy sources (ultraviolet light, electric discharge, heat, etc.). There are also many accounts of polymerization experiments that produce polysaccharides, proteinlike polymers, lipids, polymeric hydrocarbons, and simple nucleic acids. It must be realized that the point of these experiments is not to discover the particular series of reactions that led to the first living systems but, rather, to demonstrate the large variety of possible reactions and products that could have been produced in the hypothetical early earth environment.

At the time of the origin of life, the earth's surface was probably heterogenous, as it is today, producing local environments that do not represent the average conditions. The key to understanding the origin of life is to realize the variety of environments and chemical reactions within each. Recent theoretical advances have greatly aided our understanding of the problem of evolving from a collection of organic chemicals and polymers to the first living cells. I. Prigogine won the 1978 Nobel Prize in chemistry for his theories on **irreversible thermodynamics** and **self-organization** in chemical reactions. His detailed mathematical analyses of the thermodynamics of irreversible, dynamic systems are very important to the understanding of life. Complex living systems are at a **steady state** that sustains its complex organization at the expense of energy from the environment. This is seen in the need for food in all animals. Similarly, in a complex chemical environment, such as that described for the early earth, there were available energy sources to drive a variety of complex reactions. Prigogine's work describes how such systems led to self-organization into even more complex systems.

The concept of the **cycle** is very important. In a complicated series of chemical reactions, if the product of a particular reaction enhances the production of a precursor to itself, thus forming a cycle (e.g., the Krebs cycle, Fig. 2-11), it is called **autocatalytic** (Fig. 41-9). Autocatalytic cycles will grow at the expense of other materials; larger cycles, whose components are themselves cycles, may arise. This situation is called the **hypercycle** (Fig. 41-9) by M. Eigen, who has mathematically analyzed such systems (published in the 1970's). Eigen's application of such theoretical

considerations to specific models for the origin of genetic systems has yielded very interesting results. His hypotheses even include predictions for the origin of the genetic code; the most probable codes for the first amino acids agree with the most common amino acids produced by the Miller spark-discharge experiment! This logically follows the self-organization principle; the most abundant amino acids were utilized to make the first proteins by linking these reactions (by autocatalytic cycles) to the most probable codons available, thus inventing a primitive genetic code that has retained early characteristics even until today.

A common criticism of this "random" origin of life scenario is that the probability of arriving at a complex system, or even the amino acid sequence of any particular protein, by random reactions is extremely low. This argument is wrong because it assumes a particular goal; the explanation just given shows that there is no particular set of reactions or products as a goal but that any set of reactions that becomes autocatalytic will "grow" and predominate over other reactions. This is simply natural selection at the molecular level. These early complex systems only continued if they can (1) extract energy from the other chemicals in the environment (metabolism), (2) sustain and increase the intermediates of the autocatalytic cycles, (3) develop informational molecules that code for the production of specific catalysts in a controlled manner (genetics and regulation), and (4) replicate this information (reproduction). This line of reasoning does not predict the origin of a particular living system, but it does help to explain how complex systems arose. Changes in these reactions in the form of mistakes (mutation) or changes in the environment continually occurred, as they did throughout the history of biological evolution. The changes that supported the system, such as a new catalyst or improvements in regulating the system, would allow it to continue and grow. Thus, natural selection occurred until the remarkably complex yet highly regulated entity, the first bacterial cell, marked the beginning of what we recognize as life (Fig. 41-9).

Current Evolution

One seemingly puzzling aspect of evolution is that, if it is indeed an ongoing process, why can't we ob-

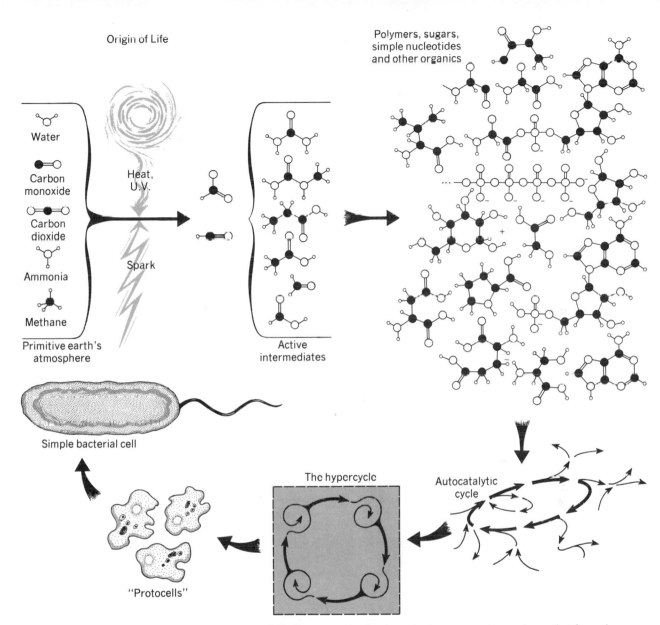

Figure 41-9. The origin of life. Progressing clockwise, some simple atmospheric compounds are shown that form simple organic molecules when energy (electric discharge, ultraviolet light, heat, etc.) is added. These simple molecules react with each other to form an enormous variety of organic chemicals, including various polymers. In these reactions, some cyclic connections are made, yielding autocatalysis. Larger networks of cycles form, called hypercycles. The individual cycles within a hypercycle may be very different; some may form energy-storing compounds (*e.g.,* ATP), others may be by simple replication of nucleic acids, and so on. When products from the individual cycles influence (catalyze) the others, as shown, the hypercycle arises. The dashed box indicates that some form of membranous enclosure was probably important at this stage to help prevent the required intermediates from drifting apart. This originally might have been simply spaces within clays or other inorganic structures. Eventually, the plasma membrane was synthesized as part of the complex chemical network, and the precursors (protocells) to the first bacteria developed.

serve its occurrence? Actually we can, but observation of current evolution is limited to its manifestation in rapidly reproducing organisms. This is because, in organisms that reproduce more slowly, speciation may take about 1 million years. The rapidly reproducing organisms are usually quite small, and evolutionary changes do not seem very spectacular. However, adaptation to selective pressures in these organisms occurs through mutations and changes in allelic frequencies, and such changes have been observed to follow the similar patterns in larger organisms.

A good example is the resistance that often arises when insects and microorganisms are subjected to poisons. Experiments have been conducted to determine whether this increased resistance is due to exposure or to evolution (mutations that quickly spread in the gene pool due to severe directional selection). One technique used to distinguish between resistance due to exposure and that due to changes in allelic frequencies is the replica plating method. A bacterial culture known to be susceptible to **streptomycin** (an antibiotic that kills some forms of bacteria) is placed on an agar substrate and is allowed to grow. After about 18 hours of incubation, each cell will produce millions of new cells (a colony) that are observable as a small dot on the plate; one plate may have thousands of individual colonies. A piece of sterile cloth is then pressed on the surface of this agar and many of the bacteria will stick to it. When the cloth is lifted, its surface will retain the locations of the original colonies from the surface of the agar plate. The cloth is then pressed onto two agar plates that have streptomycin mixed with the agar.

After another 18-hour incubation period, the streptomycin-containing agar plates are observed, and a few small colonies of streptomycin-resistant bacteria will appear on these plates. The cells in these colonies are descendants of the cells placed on these plates by the replica plating cloth. Most of the cells transferred by the cloth were killed by the streptomycin, but these few colonies represent descendants of cells that are resistant to streptomycin.

Have these colonies become resistant because of exposure to the antibiotic or has their resistance been due to a mutation that occurred before they were transferred? This question can be answered by comparing the two streptomycin-containing replica plates made from the same cloth. On both plates, the streptomycin-resistant colonies are found in identical positions. If the resistance to streptomycin was caused by exposure to streptomycin, one would expect the streptomycin-sensitive colonies to be independently affected on each plate, thus showing different locations of streptomycin-resistant colonies on the two plates. However, if the resistance is due to a mutation that occurred before they were transferred, and selected for after they were transferred, one would expect to find streptomycin-resistant colonies at identical positions on the two replica plates made from the same cloth. This is what one finds. Therefore, selec-

Figure 41-10. The Ancon (short-legged) mutation in sheep (ewe in center, ram at right) compared with a normal ewe at left. This is the earliest recorded mutant among domestic animals.

tive pressure in the environment (containing strepto-mycin) enhanced the survival of the bacteria with the allele conferring resistance to streptomycin.

This exact phenomenon is occurring naturally and is the reason why many viral and bacterial diseases are so troublesome. If antibiotics or vaccinations are widely used, the majority of bacteria or viruses may be killed, but this reduces the natural competition with related naturally occurring mutant forms, allow-ing them to flourish. This is why many organisms that cause venereal diseases that previously would be treated with penicillin are insensitive to this antibiotic today. A similar problem in agriculture occurs when mutant insects appear that are resistant to previously lethal pesticides. Unfortunately, this has prompted the production of thousands of new pesticides each year in an evolutionary game that obviously can't be won and only causes dangerous side effects to other organisms, including humans (Chapter 42).

Other signs of evolution are seen everywhere. An insect-eating Galapagos finch has begun to exploit a new food resource, blood from the rump of another bird (where it formerly picked ticks). Moths in indus-trial areas of England have changed color due to allele frequency changes in response to shifts in selective pressures.

Short-legged mutant sheep appeared in a flock in Massachusetts in 1791 and from these were devel-oped the Ancon breed of sheep. This breed was con-sidered valuable because these sheep could not jump over the low stone walls of New England. The breed became extinct about 90 years ago, but some 50 years later a short-legged lamb appeared in the flock of a Norwegian farmer. From this, a new strain of Ancon sheep has been bred (Fig. 41-10).

Summary

The basic theory of evolution by natural selection was formulated independently by Darwin and Wal-lace about a hundred years ago. They observed that a wide range of variations exist among individuals in a population and that populations introduced into an unoccupied habitat multiply by geometric progres-sion; yet population sizes in nature remain remarka-bly constant over long periods of time. From this they concluded that not all embryos become adults and not all adults reproduce as many offspring. In this strug-gle for existence, a process of natural selection is op-erative that results in survival of the fittest. Only those individuals that are best adapted to their envi-ronment reproduce, and their offspring presumably inherit the favorable variations responsible for their survival.

The modern theory of evolution has expanded Dar-win's theory by incorporating the knowledge from other fields of biology, especially genetics, embryol-ogy, and anatomy. Evolution is now seen as any change in allelic frequency or the regulation of the expression of alleles. Such changes can result from mutations, meiotic drive, gene flow, and genetic drift, upon which natural selection may act.

Speciation is the evolutionary divergence in the gene pools of isolated populations that leads to the origin of new species. Adaptations are hereditable changes in the morphology, physiology, and behavior that enhance an organism's survival and reproduction in its particular environment. Adaptive radiation is the evolutionary explosion of one group to fill a num-ber of different niches. Divergent evolution is specia-tion followed by subsequent evolution to different forms. Adaptive convergence results in a superficial similarity of different groups that reflects similar ad-aptations to similar environmental demands. The process of speciation requires spatial or temporal iso-lation of populations of a species until their gene pools have diverged sufficiently to prevent interbreeding.

The course of evolution appears to be a combina-tion of directional adaptation (following on the heels of changing physical and biological demands) and changes that are essentially random, as a result of mutations, gene flow, and genetic drift.

Recent experiments demonstrating synthesis of organic molecules in a primitive earth environment, coupled with an understanding of natural selection on complex chemical reaction systems, has extended evolution theory into the prebiotic realm as an expla-nation for the origin of life on earth.

Review Questions

1. How did the theory of uniformitarianism from geology influence the formulation of the theory of organic evolution?

2. List four sources of genetic variation.
3. What is the major source of changes in allelic frequency?
4. What is wrong with the word "adaptation." Can you suggest a better word?
5. Explain what conditions could change stabilizing selection into directional selection. Design an example.
6. Darwin's finches probably diverged from a single South American species that invaded the Galapagos Islands. Describe factors that probably enhanced their evolution.
7. Assume a population in which all sexually mature individuals practice the rhythm method of birth control. What might the menstrual characteristics of the female offspring be 20 generations from this time?
8. Why doesn't genetic drift operate in a large population?
9. Describe in some detail one line of evidence that supports the theory of organic evolution.
10. How can selection against phenotypes possibly influence the frequency of genes?
11. Explain why cyclic networks of chemical reactions are important to the problem of the origin of life.

Selected References

Antinsen, C. B. *The Molecular Basis of Evolution.* New York: John Wiley & Sons, Inc., 1964.

Brown, J. L. *The Evolution of Behavior.* New York: W. W. Norton & Company, Inc., 1975.

Chai, Chen Kang. *Genetic Evolution.* Chicago: University of Chicago Press, 1976.

Chiselin, M. T. *The Economy of Nature and the Evolution of Sex.* Berkeley: University of California Press, 1974.

Colbert E. H. *Evolution of the Vertebrates.* New York: John Wiley & Sons, 1955.

Creed, R. *Ecological Genetics and Evolution.* Oxford: Blackwell Scientific Publications, 1971.

Darwin, C. *On the Origin of the Species by Means of Natural Selection, or the Preservation of Favored Races in the Struggle for Life.* New York: Appleton-Century-Crofts, 1875.

DeBeer, Sir Gavin. *Atlas of Evolution.* London: Thomas Nelson and Sons Ltd., 1964.

Dobzhansky, T. *Evolution, Genetics and Man.* New York: John Wiley & Sons, 1955.

Dobzhansky, T., F. J. Ayala, G. L. Stebbins, and J. W. Valentine. *Evolution.* San Francisco: Freeman and Co., 1973.

Dodson, E. O. and P. Dodson (2nd ed.) *Evolution: Process and Product.* New York: D. Van Nostvard, 1976.

Downdeswell, W. H. *The Mechanism of Evolution.* New York: Harper and Row, Publishers, 1958.

Ehrlich, P. R., R. W. Holm and Dr. Pamell *The Process of Evolution.* New York: McGraw-Hill Book Company, 1974.

Grant, V. *Organismic Evolution.* San Francisco: W. H. Freeman & Company, Publishers, 1977.

Hayes, W. *The Genetics of Bacteria and Their Viruses.* New York: John Wiley & Sons, Inc., 1964.

Lack, D. *Darwin's Finches.* New York: Harper and Row, Publishers, 1961.

Leakey, L. S. B. *Adam's Ancestors,* 4th ed. New York: Harper and Row, Publishers, 1960.

LeGros, C. W. E. *The Fossil Evidence for Human Evolution.* Chicago: University of Chicago Press, 1955.

Mayr, E. *Animal Species and Evolution.* Cambridge: Harvard University Press, 1963.

Muller, H. J. *The Modern Concept of Nature: Essays on Theoretical Biology and Evolution.* Albany: State University of New York Press, 1973.

Ross, H. H. *A Synthesis of Evolutionary Theory.* Englewood Cliffs: Prentice-Hall, Inc., 1962.

Savage, J. M. *Evolution.* New York: Holt, Rinehart and Winston, Inc., 1963.

Sheppard, P. M. *Natural Selection and Heredity.* New York: Harper and Row, Publishers, 1960.

Simpson, G. G. *Horses.* New York: Oxford University Press, 1951.

Simpson, G. G. *The Major Features of Evolution.* New York: Columbia University Press, 1953.

Simpson, G. G. *The Meaning of Evolution.* New York: The Yale University Press, 1951.

Strickberger, M. W. *Genetics.* New York: Macmillan Publishing Co., Inc., 1976.

Stebbins, G. L. *Process of Organic Evolution,* 2nd ed. Englewood Cliffs, N.J.: Prentice-Hall, Inc., 1971.

Homo sapiens

Human Origins

It is remarkable that evolution produced an animal aware of its own evolution. How is it that this one species developed such great advantages that it may now control all others? Archeologists, paleoanthropologists, geologists, and others have attempted to recreate our past to understand our own evolution. Much of the information about early people emerges from the study of fossil remains, especially teeth and jaws that have fossilized better than other body parts. Fossil remains and other fragments of evidence (such as tools, campsites, and various other artifacts found with the fossils) have produced the following picture of our past.

The first primate fossils are known from about 70 million years ago. These **prosimians** were small nocturnal mammals that lived in trees and fed on insects. The lemurs are the largest of the family of five that constitute present-day prosimians and probably evolved from this early primate stock some 55 million years ago. Mountain building about 15 million years ago created a rain shadow that produced the open forests and savannas of eastern Africa. Our distant ancestors were able to adapt successfully to these new habitats.

Ramapithecus

The first fossil of a supposed hominid ancestor was discovered in 1932 in Pakistan. Named **Ramapithecus**, this animal was judged to be an early hominid on the basis of its short face and reduced canine teeth.

This fossil dates from about 14 million years ago. Other fossils, mostly jawbones, of *Ramapithecus* have been discovered in Europe, Asia, and Africa; the oldest are from eastern Africa. Let's consider some features that this early hominid inherited from its more apelike ancestors: agile arms with full-circle rotation at the shoulder, depth perception through three-dimensional stereoscopic vision (which required a reduced nose), branch holding, dextrous hands with apposable thumbs, corresponding increases in the area of the brain concerned with these functions, and coordination of the eyes and hands through further development of the brain.

How had *Ramapithecus* diverged from its ancestors? Its jaw was relatively massive and contained large, flattened molars that were reinforced with a relatively large amount of enamel. The canine teeth were relatively reduced in size, and this allowed side-to-side chewing movements. These changes in dentition indicate a shift in diet to vegetable matter—roots, seeds, stems, nuts, and fruits—and also insects.

Between 15 and 20 million years ago, ancestors of *Ramapithecus* apparently invaded woodlands and clearings and began to spend more time on the ground. *Ramapithecus* and the modern apes (gibbons, orangutans, gorillas, and chimpanzees) probably diverged from common ancestors during this period. Bipedalism, or habitual upright walking, probably evolved about 8 million years ago. This freed the hands and allowed hominids to carry and manipulate things, hurl missiles, and perhaps even invent hand gestures.

Australopithecus

Fossil evidence dating from over 3.5 to 2.0 million years B.P. (before the present) indicates that three and possibly four hominid species coexisted in East Africa. Two of these species, **Australopithecus robustus** and **A. africanus,** were specialized for a herbivorous existence, with massive jaws, large teeth, and powerful jaw muscles required to eat coarse vegetation. These chewing muscles were attached to a pronounced ridge (sagittal crest) on top of the head (Fig. 42-1). There is also evidence that remnant populations of *Ramapithecus* persisted.

Homo habilis

The fourth hominid species is considered to be our direct ancestor and has been named **Homo habilis** ("able man"). This hominid stood upright, was 1.5 to 2 m tall, and had a brain larger than that of the other hominids.

The fossils of these two genera of hominids (*Australopithecus* and *Homo*) are sufficiently unlike to indicate that they represent an adaptive radiation from *Ramapithecus* stock about 5–6 million years ago. However, they are sufficiently similar to indicate that they must have utilized different food resources. The **competitive exclusion principle** of ecology states that no two species can exploit the same resources in the same place for an indefinite length of time; eventually the better competitor ousts its rival from that habitat. The coexistence of at least three hominid species in East Africa over a period of several million years indicates that they did not compete directly for food resources.

The dentition of *Homo habilis* is not as massive as that of the australopithecines, and there is no sagittal crest on top of the skull. The first **stone tools** date from about 3 million years ago, and their increasing coincidence near animal bones indicate that meat was becoming a more important item in our ancestors' diet. The concentration of stone implements and fossils in certain locations indicates the existence of social campsites.

The brain size of *Homo* fossils also becomes progressively larger in more recent fossils. Paleoanthropologists correlate these trends with the emergence of a **hunting–gathering culture** among our ancestors.

Figure 42-1. Skulls of man and his ancestors. One can see a progression from the apelike skulls with large sagittal crests and jaws to that of modern man with a pronounced forehead, chin, and reduced jaw. Skulls include (A) *Australopithecus robustus.* (B) *A. africanus.* (C) *Homo erectus,* Peking man. (D) *H. sapiens neanderthalensis,* Rhodesia man. (E) *H. sapiens,* intermediate form. (F) *H. sapiens,* modern man (Cro-Magnon).

Using analogies with primitive human cultures that survive today, a **division of labor** is predicted in which males hunted and scavenged animal food while females gathered plant foods and raised the young. Such a culture necessitates cooperative sharing, and the acquisition of communication and social skills necessary to maintain this venture is seen as the principal driving force of subsequent hominid evolution.

Homo erectus

The australopithecines and *Homo habilis* were replaced about 1 million years ago by **Homo erectus,** an evolutionary offshoot of *H. habilis.* There is fossil evidence that *H. erectus* evolved about 1.5 million years ago in Africa and migrated into Europe and Asia about 1.0 million years ago. As was true for their immediate ancestors, *H. erectus* were probably good hunters as they were one of the few diurnal predators and had effective weapons.

H. erectus also acquired the use of fire as a tool. Fire was apparently used by some to frighten away bears from the shelter of caves. Fire could be used as a weapon against the cold as well. Another defense used against the cold by *H. erectus* was animal skins for clothing. Now, previously hostile environments could be exploited. Fire was put to another important use, that of cooking food. This expanded the range of both plant and animal foods that could be digested. Cooked foods in general require less chewing, probably leading to natural selection for reduction of both teeth and jaws.

Homo sapiens

Homo sapiens appeared about 300,000 (or perhaps half a million) years ago. The earliest fossils are found in East Africa and Europe. The *H. sapiens* skull shows

A

B

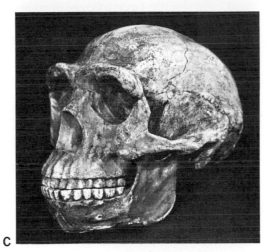

C

D

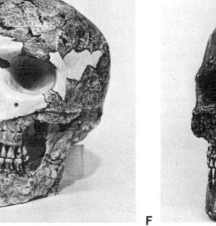

E

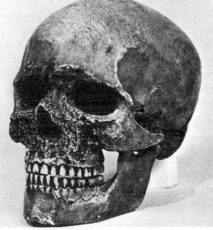

F

further reduction of the jaw and a prominent forehead that housed a larger brain (Fig. 41-1).

For mammals, humans are relatively hairless. A widely accepted theory is that hair loss evolved to facilitate heat loss. Early hominids evolved in a hot tropical climate and were active diurnally (probably to avoid competition and encounters with large crespuscular carnivores such as lions). Sustained activity in such a hot environment requires an efficient mechanism to lose heat from the body, especially because bipedalism requires more energy than does quadrupedalism. This was accomplished by the evaporation of sweat secreted by integumentary sweat glands. Hair would disrupt this process by impeding the flow of air over the skin.

The reduction in body hair exposed the integument to solar radiation. The amount of pigment (melanin) in human skin probably reflects an evolutionary response to selective pressure for vitamin D regulation. The dermis is able to synthesize previtamin D when struck by ultraviolet radiation from the sun. Insufficient vitamin D results in rickets (brittle bones), but excess vitamin D results in kidney stones and hardened joints and blood vessels. High concentrations of melanin probably protected early tropical hominids from too much ultraviolet radiation (hence too much viatmin D production).

As hominids migrated into temperate regions where clouds and the atmosphere obstructed some ultraviolet radiation, selection for a reduced level of melanin occurred, thus permitting sufficient vitamin D production. However, exposure to sustained intense sunlight will to some extent increase melanin concentrations (tanning). Eskimos living on foods rich in vitamin D (fish liver) need no additional vitamin D and have evolutionarily acquired increased melanin levels, which along with clothes prevent an overload of the vitamin.

Because skin color (determined by the concentration of melanin) is an obvious trait, this is a common basis for dividing the human population into **races.** However there is no justification for this assumption, as one might just as well choose blood type or some other less obvious trait and find radically different distributions among semi-isolated populations.

The European population was probably driven south by ice sheets during the Great Ice Age, and at the next interglacial period this first European group was replaced by the puzzling **Neanderthals.** Neanderthal people lived from about 100,000 to 35,000 years B.P. They had stocky builds and an enlarged jaw, a long shallow braincase, and a diminutive chin. Neanderthal fossils are known mostly from northern Europe, but similar populations lived in other parts of the world: **Solo man** in southeastern Asia, **Rodesia man** in Africa, and **Palestinus** in southwestern Asia. Neanderthal people were once considered as a separate species (*Homo neanderthalensis*), but they are now more commonly considered to be populations of *Homo sapiens* that exhibited an array of exaggerated physical features. Their brain size was comparable (and even slightly larger) to our own, and the oldest record of a ritual burial is from a Neanderthal site dating 60,000 years ago. This event, the interment of a man on a bed of flowers, provides one of the first indications of human self-awareness.

By the waning of the last glacial period, modern humans known as **Cro-Magnons** appeared all over the Old World, radiating from the Middle East and northern Africa. Nomadic groups of *Homo sapiens* probably interbred, resulting in several relatively similar populations among which there may have been cultural exchange. About 30,000–50,000 years ago, *H. sapiens* migrated across the Bering Straits and entered North America.

Persistent evidence of painting and other nonutilitarian activities date from about 50,000 years ago. Cro-Magnon people decorated caves with renderings of animals, a concrete indication that culture and communication had progressed to the use of symbols.

Morphological evolution was being superseded by cultural evolution—human survival depended on culture. **Agriculture** and **animal domestication** (beginning with the goat) started about 10,000 years ago in southern Asia and the **Fertile Crescent** in Asia Minor. These cultural developments were apparently independent events, but they soon spread through neighboring populations.

The development of agriculture had profound effects on *H. sapiens.* The freedom to roam as nomadic hunter-gathers was lost, as crops had to be tended and harvested. Larger and more stationary communities resulted—the beginnings of **urbanization.** Agricultural communities could provide more food, allowing population increases. However, this brought new problems: Dense human enclaves provided a favorable habitat for disease organisms; population increases demanded still larger food supplies; the removal of forests and planting of homogeneous crops provided favorable habitats for certain herbivorous insects. **Deforestation,** combined with grazing by domesticated animals, created favorable habitats for weed plants, and stationary dwellings created niches for domiciliated animals such as the rat, mouse, cockroach, and bedbug.

Human Impact on the Environment

Human disregard for the complexities of nature has repeatedly compounded problems with cultural advances that have provided only short-term solutions. Fighting diseases that result from urbanization served to increase overpopulation. Fighting insect pests with pesticides is poisoning the environment (through the food chain) and enhancing the evolution of hardier pests. We prevent floods from occurring after deforestation by building dams that hold back nutrient-rich silt from the lowlands that the dams are designed to protect. Fertilizers, used to remedy this situation, pollute the water supply, while the still water behind the dam provides excellent breeding grounds for mosquitoes and other disease-bearing organisms. There are countless other examples of shortsighted solutions to lingering problems, and some of these are considered in the paragraphs that follow. Our populations are now approaching an important new stage in our cultural development: ecological foresight. To some extent, assessment of environmental impact of human activities is now possible, and so we can make rational decisions about our actions.

Human survival depends on meeting basic biological requirements for life such as food, water, air, and living space. When human populations were small and dispersed, the environment could provide these essentials and absorb human wastes. Increasing population size and urbanization have strained this delicate ecological balance and resulted in pollution of water, air, soil, and food. We will now explore some ways in which we have mishandled and overexploited our natural heritage. Ecological awareness is not the mourn-

ing of a single squashed ant, it is concern for the total environment upon which we all depend. The survival of the human species is by no means assured. In the following paragraphs, several of the more immediate problems are discussed.

Water

Water on earth is confined mostly to the ocean (over 90 percent). The remainder is frozen in ice caps and glaciers (2 percent), as water vapor in the atmosphere (0.001 percent), and as fresh water in soil, lakes, and rivers (0.01 percent). Freshwater sources for human populations come primarily from rain, sleet, hail, and snow. Rainwater ultimately comes from the great ocean reservoir. Water at the ocean surface is evaporated by solar radiation into the air. When these moisture-laden air masses encounter cooler continental air, water vapor condenses into rain and falls to the ground. This water usually flows down gravitational grades into rivers and then into the ocean. This is the major circuit of the hydrological cycle (Chapter 40). Human populations may never exhaust the total renewable supply of fresh water, but local supplies are depleted in many areas (such as in the southwestern United States) to such an extent that man-made systems must transport water over great distances (e.g., the California aqueduct system). We also may pollute it to such an extent that it becomes useless to us and toxic to many other organisms.

Water is absolutely necessary for human life (about 1 liter per day per person). In the United States, about 72 billion liters of drinking water are used per year. Actually we use much more water than this for bathing, preparing food, flushing toilets, watering lawns and gardens, washing clothes, and so on. And such domestic uses represent only a small fraction of the water used for agriculture, power production, and industry (e.g., cooling and waste disposal). Some of the effects of these uses are deoxygenation, heat loading, turbidity, and toxicity.

Organic Pollution and Deoxygenation

Many aquatic bacteria break down organic matter, utilizing oxygen in the process. Their populations are usually limited by the amount of organic substance available. When the amount of organic matter increases, the bacteria proliferate, and the oxygen content of the water may be depleted.

Organic wastes are a major constituent of human sewage, paper mills, petroleum refineries, food processors (canneries and frozen-food packagers), and other industries. These organic wastes may upset the natural ecological balance of the waterways into which they are dumped. If oxygen depletion due to bacterial decomposition proceeds at a faster rate than natural aeration, fish and other animals (and plants) will suffocate as the water becomes anaerobic.

This may also result from a less direct type of pollution. Phosphate and nitrate nutrients that enter the water from agricultural runoff (fertilizers) and from sewage (urine and detergents) support population blooms of algae that form mats at the sunlit water surface (eutrophication). These plants shade the subsurface waters, causing the death of plants below them. The dead plants sink to the bottom and are decomposed by oxygen-depleting bacteria. Decomposition releases nutrients into the water that continue to foster the growth of algae at the surface. Meanwhile, the subsurface water becomes inhospitable for aerobic animals and may become deoxygenated enough to kill fish.

So far, new freshwater supplies produced from the desalinization of seawater are too expensive to be carried out on a large scale. Because we have only a limited freshwater supply, it makes good sense not to ruin it. This involves a combination of conservation and treatment. Much water could be saved by recycling water within the household: for example, retaining bath water in tanks to be used for flushing toilets and other domestic uses.

Treatment of sewage before it is returned to the environment is becoming increasingly necessary in populated areas. Primary treatment facilities remove suspended particles and solids. Secondary treatment facilities remove organic wastes through the use of bacteria and aerations. Tertiary treatment involves the removal of dissolved inorganic nutrients including industrial wastes and generally consists of some method of precipitating the unwanted chemicals and afterward removing the precipitate from the water.

Heat Loading

Warm water cannot hold as much dissolved oxygen and other gases as cold water. Many industries, notably the electric power industry, use water for cooling purposes and then dump the superheated water into the environment. This heat loading has a deoxygenating effect that can contribute to the problems just cited.

A century ago, water falling over a dam provided the power to turn the turbines. This method did not heat the water much, though blocking a river had other ecological effects. Larger demands for electricity resulted in (and were created by) steam-powered turbines. Steam was first produced by heating water with wood or fossil fuels: oil, coal, and gas. Recent technology uses the heat from nuclear reactions. In any event, the steam is recondensed as it passes over pipes containing cooling water. The cooling water that is heated by this countercurrent exchange is returned to the environment. For each unit of electricity generated, nuclear power plants heat about twice as much coolant as fossil-fuel power plants.

Heat, in addition to causing deoxygenation, kills plants and animals that are drawn into coolant pipes. They are also killed by chemicals used to eliminate growths on the surface of the pipes. Many organisms depend on temperature signals for their life cycles and may emerge at an inopportune time of the year. By means of the alteration of food webs, our diet can be affected. The depletion of cool-water lake fish is one example.

Some efforts have been made to curb heat pollution, the most effective and expensive of which is the dry tower. A dry tower simply passes cool air over pipes containing hot water, thereby transferring the heat to the air.

Turbidity

Turbidity is the clouding of water by many small suspended particles such as clay or steel dust (Fig. 42-2). Turbidity reduces the penetration of light necessary for photosynthesis. The reduction in primary production has obvious effects on aquatic food webs. Furthermore, when the particles settle out, they suffocate the living organisms on the bottom.

Figure 42-2. Turbidity, caused by the effluent of this factory, will diminish sunlight available to photosynthetic primary producers that are indispensable to many food webs. When these primary producers are endangered, so too are all the organisms that depend on them for life's ultimate energy source, the sun. Downstream towns also use this water for domestic purposes, drinking among them.

Settling ponds are now used by many and would cure this problem that is perpetuated by many industries and poor agricultural practices.

Toxins, Mutagens, and Carcinogens

Our modern industrial society has reached its present position with the help of a large variety of chemical aids. Modern agricultural productivity is much higher than ever imagined just a few decades ago; this is largely attributable to the widespread use of **pesticides.** The profusion of synthetic products such as plastics, building materials, textiles, artificial preservatives, pesticides, and so on has allowed significant escape from man's previous dependence on natural products. Unfortunately, recent results in medical research have revealed an alarming increase in the occurrence of many cancers. They have indicated that many of the chemicals that we use can cause cancer, mutations that lead to genetic defects in our children and direct toxic effects on the people who use or come in contact with these chemicals where they work. The mutagens and carcinogens are especially difficult to recognize because their effects are not immediate; they jeopardize both the longevity of individuals and the health of future generations. News reports are continually filled with such reports: the carcinogenic effects of certain artificial preservatives and sweeteners (e.g., saccharin), the rise of lung cancer in workers that come in contact with asbestos or other cancers caused by a large variety of chemicals in the plastics and textile industries, the toxic effects of pesticides on the people who harvest fruits and vegetables, and so on.

Although much has been done to regulate the use of such chemicals by the Food and Drug Administration and the Environmental Protection Agency (such as banning the use of the pesticide DDT), there are still many thousands of dangerous pesticides and other chemicals still in use, and many new chemicals are made each year. An encouraging trend is the use of natural insect predators to control insect pests, allowing a reduction in the use of pesticides. The "innocent until proven guilty" philosophy concerning the use of chemicals (except those added to food during processing) poses an impossible task for the EPA, which must prepare an involved and expensive legal indictment for each chemical tested individually. Obviously, new legislation is needed if we are to curb these problems.

The disposal of chemical wastes is another aspect of this problem. A recent U.S. government investigation reported that over 35 billion kg of chemical wastes are produced each year and that there are over 30,000 hazardous waste sites in the United States. Many occurrences of chemical leakage from rusting barrels into the air or water supply have been linked to sicknesses in the surrounding neighborhoods.

Certain metallic wastes such as mercury and lead interfere with vital processes and, hence, are poisonous to organisms that live in water containing enough of these substances. The metals are dumped directly into waters from plating industries, among others. Mercury from discarded batteries, coal smoke, dental wastes, and other sources eventually enters the waters. Although there has always been some mercury in the environment, the levels in the ground, water, air, and organisms are rising. The average human intake of mercury in industrialized countries is about ten times what it was 40 years ago. In the United States a large percentage of the swordfishes and some tuna were seized by the government because they exceeded legal limits of mercury concentration (0.5 parts per million). Like many other toxins, mercury becomes incorporated into the tissues of living organisms and so becomes concentrated in higher levels of the food chain.

The Greenhouse Effect

Human combustion of fossil fuels—coal, oil, and gas—has increased the carbon dioxide content of the atmosphere significantly. Since the 1880s, carbon dioxide has increased from 290 ppm (parts per million) to 330 ppm of air. The carbon dioxide concentration will theoretically rise to 385 ppm by the end of this century. Widespread deforestation is contributing to this trend, as is the rapid industrialization of the Third World.

Carbon dioxide does not affect the short-wave solar radiation that penetrates the earth's atmosphere, but it does absorb the long-wave aradiation that the earth radiates back into space. This absorption heats the atmosphere and may have a great impact on world climate. Since the beginning of the Industrial Revolution, the mean world temperature has increased 0.5°C, and another rise of 0.05°C is expected by the year 2000. Warming of the surface waters of the ocean will decrease the solubility of carbon dioxide, releasing more of this gas into the atmosphere. The rate at which it is occurring and the potential consequences of this trend is under active investigation by the world scientific community. A rise in mean world temperature of only a few degrees might melt the polar ice

caps and flood heavily populated coastal regions throughout the world. Long before the oceans rising, the meteorological balance of the earth would be upset to an extent enough to drastically change climates and the distribution of agricultural areas.

Radioactive Pollution

Nuclear power has been proposed as an alternative energy source in the face of dwindling supplies of fossil fuels and the growing awareness of carbon dioxide pollution. However, nuclear power has severe disadvantages: over a billion liters of highly radioactive wastes are presently stored in underground tanks. These wastes will take thousands of years to decay. The large amount of wastes produced in nuclear plants (breeder reactors included), even if they are condensed into solids, will require vast areas of land for their disposal. The reactors themselves leak small amounts of radiation to the environment, but these are insignificant compared with the amounts of radioactive dust produced during the mining and processing of uranium. The dust finds its way into our water supplies and eventually into crops, livestock, and people. Radiation is a known source of mutation, and its carcinogenic effects are under active investigation. The breeder reactors (not used commercially in the United States) use large amounts of plutonium, a small fraction of which, if brought together through any number of accidents, could form a critical mass (the amount of plutonium necessary to explode). Although there are many safeguards to prevent this from occurring, it can happen accidentally or through terrorist activity. Another problem is that nuclear power plants have to be sealed for thousands of years after their operational life span of only 20 years or so. No matter how well constructed, and how well designed the safety mechanisms of reactors may be, there is always the problem of human error. This was dramatically demonstrated by the accident at Three Mile Island, Pennsylvania, in 1979. Since that time, a large number of individuals have staged public protests against the use of nuclear power.

Clearly, alternative energy sources should be explored. Solar energy, tidal energy (using the differences between low and high tide levels), geothermal energy, together with improved storage batteries will supply more of our energy desires. However, it will take something like the annihilation of matter (fusion reactors) to provide a large-scale high-energy source, and this appears to be well in the future if it can be harnessed safely and economically at all.

The Growth of Human Population

In Chapter 40, the concepts of exponential growth and carrying capacity were introduced.

The carrying capacity for the human population is difficult to determine. Not only has the versatility of the human species allowed exploitation of virtually every terrestrial habitat, but cultural achievements have had dramatic effects on the carrying capacity. Enormous population increases have resulted from improvements in agriculture and medicine. It is estimated that a hunting-and-gathering way of life could support a world population of about 1 billion people. The beginnings of agriculture and animal husbandry about 10,000 years ago allowed a concentration of food resources that has increased the carrying capacity many times. Recent medical advances have reduced infant mortality and greatly extended life expectancy, resulting in an explosive population increase (Fig. 40-16).

Thomas Malthus predicted that such geometric population increases would lead to catastrophe because increased food production would lag population growth. However, he had not foreseen the development of large-scale agriculture and the colonization of the Americas and Australia, which offered a temporary reprieve. Still he was basically correct, for now over two thirds of the world population is hungry. With improvements in crop plants, irrigation of new land for agriculture, and harvesting more marine foods, we might be able to feed today's population, but, by the time these means are developed, the population may have doubled again! Furthermore, the concerted effects of various types of pollution are lowering the carrying capacity of the environment to support our growing population.

According to some estimates, the earth can carry a population of only 1 billion humans comfortably. Population increases are greatest in South America and Asia, although the population also continues to increase in the United States and Europe and other parts of the world. In an industrialized society, the population usually increases at a slower rate than in agricultural cultures, in which children provide farm labor and old-age insurance.

Advising Third World people to practice birth control is not the only solution because the average American consumes many times the resources that an individual in an agricultural country does. Clearly, we must curb our own consumption of resources as well as our population size. Government incentives such as tax breaks for a limited family size could help to slow our population growth. Changes in social attitudes would also help. If women are encouraged to enter careers other than homemaking, we might expect a decline in the birth rate. The technological means for birth control are available: oral contraceptives, diaphragms and spermicidal foams, prophylactics, intrauterine devices, sterilization, and abortion (Fig. 42-3).

Can our traditions change in time to avert Malthus's grim predictions? Perhaps they can, opening the door to a stable and perhaps improved existence.

Eugenics—A control mechanism of human growth?

Eugenics is the study of ways to improve humanity through hereditary manipulations (Chapter 38). This basically means the encouragement of breeding between individuals with "desirable" traits and the deterrence of matings between individuals with "undesirable" traits—such qualities are rarely subject to any general consensus, however.

Early eugenicists familiar with Darwin's theory of evolution sought to introduce social selection through selective mating. Their proposals were colored by cultural and class biases, focused on preserving civilization by controlling reproduction in "inferior classes and races." Racism found in eugenics a "scientific" rationale for genocide and slavery. The former is exemplified by the Nazis, who, in the name of race improvement, murdered millions of "inferior" human beings.

The "pure race" concept was devised without reference to genetics and environmental factors that modify gene expression. Human traits vary and mix from one region to another, and the choice of important characteristics is arbitrary for determining "race." Minor differences between relatively isolated human populations add versatility to the human gene pool

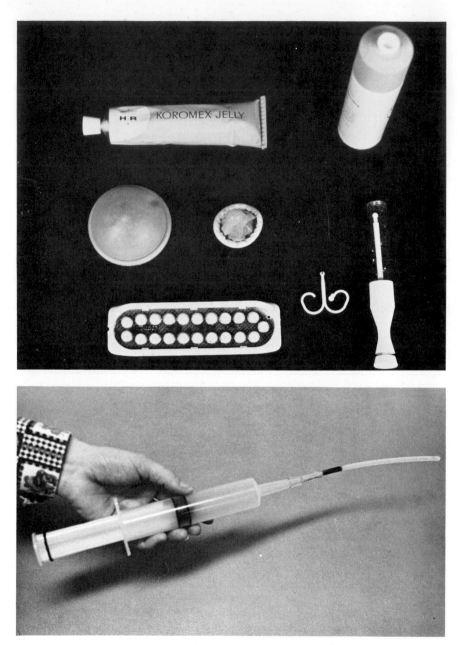

Figure 42-3. Birth control devices include, in the top photograph going clockwise from the middle foreground, the I.U.D., birth control pills (both are very effective), the diaphragm (that fits over the cervix) that is used with the next item, spermacidal jelly. Next is the spermacidal foam and the foam injector. In the center is the prophylactic or condom that fits over the male penis. In addition to these devices, surgical means of birth control include vasectomy in which the male vas deferens is cut and tied to prevent the flow of sperm into the seminal fluid, and tubal ligation, the severance and tying of each of the female Fallopian tubes. Bottom photograph shows the Karmin canula. This instrument is used to extract the fetus gently in early pregnancy. The plastic tube on the syringe is small and passes through the cervix. The syringe provides suction (though a suction machine is usually used for abortions after the fifth week of pregnancy).

through gene flow. The monumental evolutionary importance of genetic versatility becomes apparent when one considers that sexual reproduction results in astronomically more genetic combinations than does asexual reproduction and that sexuality is almost universal in the animal kingdom (and in some protists).

Any systematic attempt to alter the human gene pool would require value judgments as to which traits are desirable. Many eugenic schemes have been concerned with increasing the average intelligence. However, it is not known how many and which genes affect intelligence, nor is there a general consensus on what "intelligence" really is. Definitive ways of testing intelligence without language, cultural, or economic biases have yet to be developed. Differences based on response to I.Q. tests* in various ethnic and socioeconomic groups were once assumed to indicate genetic intelligence hierarchies. However, that controversial idea has lost much support.

For example, numerous studies have shown that environmental parameters such as early nutrition (that affects central nervous system development), health care, cultural surroundings, testing situations (e.g., the color and sex of the tester), familiarity with tests in general, and many other environmental factors have a profound influence on I.Q. test scores. Hence, the results of these tests are useless as a yardstick of intrinsic intelligence.

In spite of these findings some people still advocate sterilization of "inferiors." This diverts energy from meeting the educational, medical, and nutritional needs of the neglected segments of society. Under adverse environmental conditions, full genetic potential cannot be reached. Genetic manipulation is an exercise in futility until we meet the basic needs that ensure each person optimal development. The means to accomplish this are within our reach. Distribution of goods and services to meet developmental needs would certainly fulfill the hopes of the eugenicist by increasing intelligence and health.

The desire to increase average intelligence is based on the belief that this will somehow improve the quality of human existence. Yet, meanwhile, women and many minorities are often restricted in the social roles available to them.

Today, eugenics is more effective in limiting specific genetic traits that cause human suffering, such as hemophilia and sicklecell anemia. This is accomplished through identification of heterozygous carriers and subsequent counseling of potential parents.

Summary

Homo sapiens is a unique animal. No other animal has had such a tremendous impact on the environment. Our evolutionary history has resulted in a highly developed nervous system enabling development of a powerful technology. At breakneck speed we have spread over the earth through advances in cultural evolution.

Our technological power has far outpaced evolutionary restraints so that we now sit precariously perched atop a barricade against nature. We have prolonged life through the control of predators and diseases, but we now face the effects of overpopulation. We produce food in ways that permanently upset ecosystems. We build industrial empires and then squander irreplaceable resources and foul our nests with pollution in the process. However, we cannot ignore nature forever. Unless people drastically change their priorities, an ecological catastrophe of epic proportions is probable.

What is obstructing humanity may be its concept of being apart from, and superior to, nature. If we can overcome this notion of separateness, limit our population size, and act with ecological awareness, we may yet live a life of biological peace.

Review Questions

1. Discuss the theory about human hair loss relative to other mammals. What bearing does this have on skin pigmentation?
2. Discuss the ways in which water can become deoxygenated and how this condition can affect the local fauna.
3. If we assume that the current rate of increase of

---

* I.Q. (intelligence quotient) $\frac{\text{Mental age}}{\text{Chronological age}} \times 100$

The average score for the test population is defined and set at 100.

the human population ($r = 0.02$) continues, how long would it take for us to reach $N = 5 \times 10^{15}$ (a figure at which there would be about ten people per square meter—people would be pressed very tightly together standing up over the entire surface of the earth, oceans included)?

4. Summarize human evolution or pick a specific human adaptation and discuss its relevance to human evolution.
5. Make assumptions about the future and describe the possible evolution of future species of hominids.
6. Design a "utopia" for humans and support your formulations with biological principles and illustrations.

Selected References

Bodmer, W. F. *Genetics, Evolution, and Man.* San Francisco: W. H. Freeman and Company, Publishers, 1976.

Campbell, B. G. *Human Evolution,* 2nd ed. Chicago: Aldine Publishing Co., 1974.

Clark, W. E. Le Gros. *The Antecedents of Man.* Chicago: Quadrangle Books, Inc., 1959.

Darwin, C. G. *The Next Million Years.* New York: Dolphin Books, Doubleday and Co., Inc., 1952.

Detwyler, T. R. *Man's Impact on Environment.* New York: McGraw-Hill Book Company, 1971.

Dobzhansky, Th. *Mankind Evolving.* New Haven, Conn.: Yale University Press, 1962.

Hardin, G. *Population, Evolution, and Birth Control.* San Francisco: W. H. Freeman and Company, Publishers, 1969.

King, J. C. *The Biology of Race.* New York: Harcourt Brace Jovanovich, 1971.

Krenkel, P. A., and F. L. Parker (eds.). *Biological Aspects of Thermal Pollution.* Louisville, Ky.: Vanderbilt University Press, 1969.

Linton, R. M. *Terracide.* Boston-Toronto: Little Brown and Company, 1970.

Meadows, D. H., D. L. Meadows, J. Randers, and W. W. Behrens, III. *The Limits to Growth.* New York: Universe Books, 1972.

Odum, H. T. *Environment, Power, and Society.* New York: Wiley-Interscience, 1971.

Odum, H. T., and Odum, E. C. *Energy Basis for Man and Nature.* New York: McGraw-Hill Book Company, 1976.

Oxnard, C. *Form and Pattern in Human Evolution: Some Mathematical, Physical, and Engineering Approaches.* Chicago: University of Chicago Press, 1973.

Oxnard, C. *Uniqueness and Diversity in Human Evolution: Morphometric Studies of Australopithecines.* Chicago: University of Chicago Press, 1975.

Strickberger, M. W. *Genetics.* New York: Macmillan Publishing Co., Inc., 1976.

Tuttle, R. (ed.). *The Functional and Evolutionary Biology of Primates.* Chicago. Aldine-Atherton, 1972.

Wagner, R. H. *Environment and Man.* New York: W. W. Norton and Company, Inc., 1971.

Wilson, E. O., and W. H. Bossert. *A Primer of Population Biology.* Stanford, Conn.: Sinauer Associates, Inc., Publishers, 1971.

Classification Appendix

This appendix is designed to provide a general taxonomic orientation to zoology. The particular system used here is a synthesis of several others and reflects the current preferences of the author. This classification system covers all animals and animal-like protists. The first part is a brief synopsis of the kingdoms and phyla. This is followed by a detailed phylogeny of the phylum Chordata, showing approximate times of origin and extinction for all groups discussed in the text, and a detailed outline of the taxonomic scheme used in this book.

A Taxomony of Animals and Animal-like Protists

Phylum Sarcomastigophora. Unicellular organisms that may reside in a colony of similar cells but do not differentiate into tissues and are usually microscopic in size. Locomotion by flagella, pseudopodia, or cilia; single type of nucleus; usually no spore formation; sexuality, when it occurs, by syngamy. Three superclasses:

Superclass 1. Mastigophora. These are solitary or colonial and bear one or more flagella in the adult stage. They may be amoeboid in shape but are generally covered with a pellicle. Reproduction is asexual by binary fission. Many are parasitic. Two classes:

Class 1. Phytomastigophorea. Photosynthetic flagellates, some saprophytic or partially holozoic. Chromatophores usually present; often a red eyespot; only one, two, or a few flagella; mostly free living. Nine orders:

Order 1. Chrysomonadida. Small; one to three flagella; some colonial. *Uroglena*. (part of botanical class Chrysophyceae, phylum Chrysophyta).

Order 2. Silicoflagellida. One or no flagellum; siliceous internal skeleton (part of botanical class Chrysophyceae, phylum Chrysophyta).

Order 3. Cocolithophorida. Two flagella; calcareous platelets cover body; marine (part of botanical class Haptophyceae, phylum Chrysophyta).

Order 4. Heterochlorida. Two unequal flagella; amoeboid forms.

Order 5. Cryptomonadida. Two flagella arising from a depression and usually one or two chromatophores. *Chilomonas* (same as botanical phylum Cryptophyta).

Order 6. Dinoflagellida. Two flagella in grooves along the cellulose membrane; mostly marine, with brown or yellow plastids. *Noctiluca* (same as botanical class Dinophyceae, phylum Pyrrhophyta).

Kingdom	Branch	Grade	Level	Superphylum	Phylum
Monera					
Fungi					
Plantae					
Protista	Protozoa				Sarcomastigophora
					Ciliophora
					Sporozoa
					Cnidospora
Animalia	Mesozoa				Mesozoa
	Parazoa				Porifera
	Eumetazoa				
		Radiata			Cnidaria
					Ctenophora
		Bilateria			
			Acoelomata		Platyhelminthes
					Nemertina
			Pseudocoelomata		
					Rotifera
					Gastrotricha
					Kinorhyncha
					Nematoda
					Nematomorpha
					Acanthocephala
					Entoprocta
					Gnathostomulida
			Eucoelomata		
				Lophophorates	Ectoprocta
					Brachiopoda
					Phoronida
				Schizocoela	Mollusca
					Annelida
					Sipunculida
					Echiurida
					Priapulida*
					Onychophora
					Tardigrada
					Pentastomida
					Arthropoda
				Enterocoela	Chaetognatha
					Pogonophora
					Echinodermata
					Hemichordata
					Chordata

*Mechanism of coelom formation is unknown but is close to Schizocoela in other characteristics.

Order 7. Euglenida. Usually one or two flagella emerging from anterior reservoir; a cytostome and cytopharynx; often chromatophores and an eyespot. *Euglena; Phacus* (same as botanical phylum Euglenophyta).

Order 8. Chloromonadida. Two flagella with one trailing, both originating near an apical cleft. *Gonyostomum* (same as botanical phylum Chloromonadophta).

Order 9. Phytomonadida. Two to four flagella; solitary or colonial. *Chlamydomonas, Volvox* (same as botanical order Volvocales, class Chlorophycea, phylum Chlorophyta).

Class 2. Zoomastigophorea. Heterotrophic flagellates. No chromatophores; one to many flagella; mostly parasitic. Nine orders:

Order 1. Choanoflagellida. One anterior flagellum arising from a thin collar; free living.

Order 2. Bicosoecida. Two flagella (one for attachment to shell); free living.

Order 3. Rhizomastigida. Amoeboid with one to four flagella mostly free living. *Mastigamoeba.*

Order 4. Kinetoplastida. One to four flagella; mostly parasitic.

Order 5. Retortamonadida. Two to four flagella (one turned back and associated with cytostome); parasitic.

Order 6. Diplomonadida. Bilateral symmetry; eight flagella; parasitic. *Giardia.*

Order 7. Oxymonadida. Four or more flagella (some turning back and sticking to the body); some sexual reproduction; parasitic.

Order 8. Trichomonadida. Four to six flagella (one associated with an undulating membrane); parasitic. *Trichomonas.*

Order 9. Hypermastigida. Many flagella; often very complex; commensal intestinal inhabitants of termites. *Spirotrichonympha.*

Superclass 2. Opalinata. Many cilialike organelles in oblique rows; two to many similar nuclei; no cytostome; life cycle involves sexual stage with unequal-sized gametes (anisogametes); all parasitic. One order:

Order 1. Opalinida. With the characteristics of the superclass. *Opalina.*

Superclass 3. Sarcodina. Pseudopodia usually present; body naked or with various tests (skeletons); asexual reproduction by fission; mostly free living. Three classes:

Class 1. Rhizopodea. Typically creeping forms with lobose or fine pseudopodia. Five subclasses:

Subclass 1. Lobosia. Pseudopodia lobose; naked or with test. *Amoeba, Arcella.*

Subclass 2. Filosia. Fine pseudopodia that branch.

Subclass 3. Granuloreticulosia. Fine anastomosing pseudopodia; foraminiferans with simple or chambered perforated shells. *Globigerina.*

Subclass 4. Mycetozoia. Slime molds. Complex life cycle involving a sexual stage and multiple fission.

Subclass 5. Labyrinthulid. Groups of these form slime nets in soil and on marine algae.

Class 2. Piroplasmea. Amoeboid, round or rod shaped; locomotion by flexion or gliding; asexual reproduction by binary or multiple fission. Parasites of vertebrate red blood cells; often carried by ticks. *Babesia.*

Class 3. Actinopodea. Typically free floating with pseudopodia radiating from spherical body; naked or with test; both sexual and asexual reproduction; gametes with flagella. Four subclasses:

Subclass 1. Radiolaria. Marine; often spherical; pseudopodia raylike; protoplasm divided into inner and outer parts by a perforated capsule; usually with a siliceous skeleton. *Sphaerozoum.*

Subclass 2. Acantharia. Marine; thin membranous central capsule without pores; skeleton of radial spines composed of strontium sulfate. *Acanthometra.*

Subclass 3. Heliozoia. Freshwater actinopods; pseudopodia thin, radially arranged, and usually supported by axial siliceous spines; no test; spherical. *Actinophrys.*

Subclass 4. Proteomyxidia. No test or skeleton; mostly parasitic; sometimes with flagellated swarmer stages. *Vampyrella.*

Phylum Ciliophora. Protozoans that possess cilia or ciliary organelles at some stage in their life cycle; with a large macronucleus and one or more smaller micronuclei. Reproduction is asexual (binary fission) and sexual (conjugation). Most are free living in fresh water or the ocean, but many are parasites. One class:

Class 1. Ciliatea. With the characteristics of the phylum. Four subclasses:

Subclass 1. Holotrichia. Cilia typically of equal length all over the body; without specialized buccal cilia. Seven orders:

Order 1. Gymnostomatida. No oral cilia; simple arrangement of cilia; cytostome opens directly to outside. *Chilodonella.*

Order 2. Trichostomatida. Highly asymmetrical; vestibule present. *Balantidium.*

Order 3. Chonotrichida. Adults lack cilia and are vase shaped; asexual reproduction by budding; attachment to crustaceans by stalk.

Order 4. Apostomatida. Cilia arranged in spirals; polymorphic life cycles involving marine crustacean hosts.

Order 5. Astomatida. Uniform cilia; no cytostome; parasitic in oligochaetes.

Order 6. Hymenostomatida. Cilia on body uniform; with buccal cavity; usually small. *Paramecium.*

Order 7. Thigmotrichida. Tufts of thigmotactic cilia near anterior of body; parasitic in bivalve mollusks.

Subclass 2. Peritrichia. Cilia in sessile adults confined to buccal area; often attached by contractile stalk; mostly colonial. *Vorticella.*

Subclass 3. Suctoria. Ciliated when young; "tentacles" in sessile, stalked adult stage without cytostome. *Podophyra.*

Subclass 4. Spirotrichia. General reduction of body cilia but well-developed buccal cilia; numerous ciliary membranelles typically wind clockwise around the cytostome. Six orders:

Order 1. Heterotrichida. Large ciliates with uniform body cilia or with the body encased within a lorica. *Stentor.*

Order 2. Oligotrichida. Small ciliates with conspicuous buccal membranelles extending around apical part of body; mostly marine. *Halteria.*

Order 3. Tintinnida. Loricate, with conspicuous extensible oral membranelles; mostly marine. *Tintinnus.*

Order 4. Entodiniomorphida. Commensal within digestive tracts of herbivorous mammals; digest cellulose; with prominent oral membranelles. *Entodinium.*

Order 5. Odontostomatida. Small group of laterally compressed wedge-shaped spirotrichs. *Saprodinium.*

Order 6. Hypotrichida. Dorsoventrally flattened spirotrichs with body cilia reduced to ventral compound organelles called cirri. *Euplotes.*

Phylum Sporozoa. Unicellular parasites that typically form spores (without polar capsules); among the most widely distributed of all parasites; members of almost every large group in the animal kingdom are parasitized by one or more sporozoan species. Highly modified for parasitic existence; no locomotor organelles, mouth, anal pore, or water-expulsion vesicle. Three classes:

Class 1. Telosporea. Spores produced at end of life of trophozoite; sexual and asexual reproduction. Two subclasses:

Subclass 1. Gregarinia. Common parasites of insects; at first intracellular, but adult often free in cavities; large trophozoite. *Monocystis.*

Subclass 2. Coccidia. Parasites of vertebrates and invertebrates; adults usually intracellular and small. *Plasmodium.*

Class 2. Toxoplasmea. Without spores; asexual reproduction; gliding locomotion. *Toxoplasma.*

Class 3. Haplosporea. With spores; asexual reproduction by schizogony; may have pseudopodia. *Coelosporidium.*

Phylum Cnidospora. Parasitic sporozoans forming spores with one to four polar capsules, with a coiled polar filament; probably a polyphyletic grouping. Two classes:

Class 1. Myxosporidea. Many cells form a spore; one or more sporoplasms; with more than one valve. Three orders:

Order 1. Myxosporida. Generally two sporoplasms, two polar capsules, and two valves. Principally parasites of fish. *Myxidium.*

Order 2. Actinomyxida. Usually three polar capsules, three valves, and many sporoplasms. Found in the coelomic cavity or gut of aquatic annelids. *Triactinomyxon.*

Order 3. Helicosporida. Three sporoplasms surrounded by spiral filament; one valve. Parasites in the larvae of flies and mites. *Heliosporidium.*

Class 2. Microsporidea. One cell forms a spore; one sporoplasm and one valve. Spores extremely small; usually with one polar capsule; insects most frequently infected. *Thelohania.*

Phylum Mesozoa. Mesozoans are small wormlike endoparasites composed of about 13 ciliated external cells surrounding a few axial reproductive cells. Complex life cycles, but fate of dispersal larvae is unknown. Probably an early offshoot of metazoans that have adapted to parasitism. Two orders:

Order 1. Dicyemida. Adults live in nephridial cavity of cephalopods.

Order 2. Orthonectida. Parasitize various other marine invertebrates.

Phylum Porifera. Sponges are the most primitive multicellular animals. Adults are all sessile animals that exhibit radial or irregular symmetries; often colonial. Neither tissues nor organs are present, the cells acting mostly independently. Body wall has two layers: outer epidermis and inner layer of choanocytes between which is a gelatinous mesenchyme traversed by wandering amoebocytes. The body is supported by a skeleton of calcareous or silicieous spicules and/or spongin. No mouth; flagellated choanocytes draw water through pores and canals that pierce the body wall. Food particles are phagocytosed by choanocytes. Water is expelled through central spongocoel and osculum. Three grades of body organization: asconoid, syconoid, and leuconoid. The 10,000 living species of sponges are mostly marine, except for about 150 freshwater species. Three classes:

Class 1. Calcarea. Mostly small shallow-water species with skeletons of calcium carbonate spicules. Generally simple, but all three grades of body organization are represented. *Leucosolenia; Scypha.*

Class 2. Hexactinellida. Chiefly tropical deep-sea species. Six-pointed siliceous spicules often fused into a latticelike framework. Syconoid or simple leuconoid. *Euplectella,* Venus's flower basket.

Class 3. Demospongiae. Contains most of modern species; all leuconoid and irregular; often massive and brightly colored. Skeleton composed of siliceous spicules, spongin fibers, both, or neither. Three subclasses:

Subclass 1. Tetractinellida. Without spongin, spicules four rayed (textraxon) or absent. *Halisarca; Chondrosia; Geodia.*

Subclass 2. Monaxonida. With or without spongin, spicules monaxon; most common of all sponges. *Cliona,* boring sponges; *Halichondria,* the common "crumb of bread" sponge; *Microciona; Haliciona,* the "finger" sponge; *Spongilla,* a freshwater sponge.

Subclass 3. Keratosa. No spicules; skeleton of well-developed spongin fibers. *Spongia* and *Hippospongia,* the commercial bath sponges.

Phylum Cnidaria. Contains hydras, jellyfishes, sea anemones, corals, and many others. All have a central gastrovascular cavity surrounded by a body wall with three layers. The epidermis is derived from ectoderm and the gastrodermis from endoderm. The intervening mesoglea is a protein matrix that may contain amebocytes. Many cell types but no true tissues. All have stinging cnidoblasts or cnidocytes called nematocysts. About 10,000 species have been described; most are marine. Polypoid or medusoid body types, often with metagenesis. Many colonial, sometimes with marked polymorphism. Three classes:

Class 1. Hydrozoa. Hydroid polyps and medusae with a velum; the polypoid generation often predominates; mesoglea noncellular; solitary or colonial; mostly marine. *Hydra,* a freshwater polyp; *Obelia,* a branching colonial hydroid; *Gonionemus,* a hydromedusa with no polypoid stage; *Physalia,* a pelagic polymorphic colony.

Class 2. Scyphozoa. True jellyfish; without velum; usually eight notches in umbrellar margin; mesoglea abundant and somewhat cellular; polypoid stage absent or reduced. All jellyfish are marine and often have powerful nematocysts. *Aurelia; Chironex,* the sea wasp; *Cyanea,* the lion's mane.

Class 3. Anthozoa. Corals, sea anemones, sea fans, and many others; 6000 described species. Solitary or colonial; sessile polyps only, with no medusoid stage; distal end, an oral disc; well-developed pharynx leads into a gastrovascular cavity divided into radial compartments by mesenteries; mesoglea cellular; calcareous exoskeleton sometimes secreted. Two subclasses:

Subclass 1. Alcyonaria. Eight pinnate tentacles; eight mesenteries; one siphonoglyph; mostly colonial; skeleton present. *Tubipora,* organ-pipe coral; *Alcyonium,* a soft coral; *Gorgonia,* sea fans; *corallium,* precious red coral.

Subclass 2. Zoantharia. With more than eight tentacles; solitary or colonial; with or without skeleton. *Metridium,* a sea anemone; *Astrangia;* a true stony coral; *Cerianthus,* a burrowing anemone; *Antipathes,* a black or thorny coral.

Phylum Ctenophora. Ctenophores are biradially symmetrical marine carnivores with colloblasts (adhesive cells) and sense organs. Locomotion by eight rows of ciliated comb plates. The digestive system may contain anal pores. Less than 100 known species, but ctenophores are common in marine plankton. *Mnemiopsis*, a comb jelly; *Cestus veneris*, Venus's girdle.

Phylum Platyhelminthes. Platyhelminths are free-living and parasitic flatworms. All are bilaterally symmetrical acoelomate metazoans. The body is flattened dorsoventrally, consisting of a series of duplicated parts (tapeworm). No anus (usually) or skeletal, circulatory, or respiratory systems; an excretory system with flame cells. Cephalization with a head, sense organs, and a central nervous system, which consists of a brain and two longitudinal nerve cords. Most are hermaphroditic; more than 10,000 species. Three classes, two of which are entirely parasitic:

Class 1. Turbellaria. Mostly free-living flatworms; epidermis at least partly ciliated with rod-shaped rhabdites and many mucous glands; no cuticle or suckers. About 3000 species described, mostly marine. Five orders often recognized:

Order 1. Acoela. Small marine flatworms; simple (probably primitive) body organization; no enteron. *Polychoerus; Convoluta*.

Order 2. Rhabdocoela. Small flatworms with simple, unbranched gut. Marine, freshwater, and terrestrial species. *Stenostomum*.

Order 3. Alloeocoela. Moderately sized, predominantly marine flatworms; intestine straight or with short branches; complex reproductive systems. *Prorhynchus*.

Order 4. Tricladida. Moderately large freshwater, marine, and terrestrial flatworms; intestine of three main trunks, each with many lateral branches. *Dugesia*, freshwater planarian; *Bipalium*, a terrestrial triclad; *Bdelloura*, lives on gills of horseshoe crabs.

Order 5. Polycladida. Large and highly flattened marine flatworms; central digestive cavity with many irregular branches. Most littoral, but a few pelagic. *Stylochus; Notoplana*.

Class 2. Trematoda. Parasitic flukes; cuticle with no external cilia; gut with two main branches; unsegmented; with one or more suckers for attachment to host; about 5000 species recognized so far. Two subclasses:

Subclass 1. Monogenea. External parasites of aquatic vertebrates; direct life cycle with no alternation of hosts. *Benedenia*.

Subclass 2. Digenea. Internal parasites of vertebrates, with complex life cycles involving larval hosts, generally mollusks. An asexual generation in life cycle. *Fasciola hepatica*, sheep liver fluke; *Opisthorchis (Clonorchis)*, Chinese liver fluke.

Class 3. Cestoda. Endoparasitic tapeworms; cuticle without cilia; no mouth or gut; body divided into reproductive segments (proglottids) that bud from anterior end; live in host's intestine and may attain lengths of 9 m; over 200 species described. *Taenia solium*, parasitizes pigs and humans.

Phylum Nemertinea (Rhynchocoela). Ribbon worms, with a long eversible proboscis used to capture prey; acoelomate, bilaterally symmetrical, unsegmented, and dorsoventrally flattened. They possess a complete digestive tract, a simple closed circulatory system, and a nervous system with ganglia. Mostly free-living marine worms that live in littoral sands and muds; about 600 recognized species. *Lineus*.

Phylum Rotifera. Common freshwater rotifers; distinguished by an anterior crown of cilia, the corona; mostly microscopic, with diverse body forms; eggs may develop parthenogenically; about 2000 known species. *Ephiphanes.*

Phylum Gastrotricha. Common microscopic marine and freshwater metazoans; body shaped like flattened bottle, with cuticle elaborated with spines and scales; 1500 described species. *Chaetonotus.*

Phylum Kinorhyncha. Minute marine pseudocoelomates that burrow through muddy bottoms of coastal waters; spiny cuticle divided into 13 distinct segments; about 100 known species. *Campylodera.*

Phylum Nematoda. Roundworms. Second largest invertebrate phylum, with over 100,000 described species; marine, freshwater, or terrestrial; body cylindrical and tapered at both ends; no cilia; possess pseudocoel; triangular pharynx; syncytial epidermis, most are less than 1 mm; tough resistant cuticle covers body wall, which has only longitudinal muscles; digestive tract a straight tube with mouth and anus at opposite ends of body; no protonephridia; free-living or parasitic. *Ascaris; Trichinella.*

Phylum Nematomorpha. Horsehair worms. Very long, slender, cylindrical pseudocoelomates; free-living adults do not feed; juveniles parasitize arthropods; 230 described species in freshwater habitats. *Paragordius.*

Phylum Acanthocephala. Spiny-headed worms. Parasitic pseudocoelomates that infest the intestines of vertebrates; anterior probosis covered with recurved spines; no digestive system; arthropods serve as intermediate hosts; about 400 species described. *Macracanthorhynchus.*

Phylum Entoprocta. Sessile, stalked pseudocoelomates that inhabit marine coastal waters; attached stalk bears a tentacle-crowned head; tentacles surround both mouth and anus; free-swimming larva similar to trochophore; about 90 species described. *Loxosoma.*

Phylum Gnathostomulida. Acoelomate marine worms, usually less than 1 mm long; worldwide distribution. Probably over 100 species.

Phylum Ectoprocta (Bryozoa). Major phylum of sessile, mostly marine animals; a ring of ciliated tentacles encircling the mouth but not the anus of the U-shaped digestive tract; most colonial, with exoskeletons. *Bugula; Plumatella.*

Phylum Brachiopoda. Lamp shells. Marine bivalves with asymmetric dorsal and ventral shells; some attached by muscular stalk (peduncle); lophophore of two coiled ridges with ciliated tentacles. Oldest genus known (from 400 million years ago), *Lingula.*

Phylum Phoronida. Small wormlike, sessile, marine lophophorates that live in tubes; with free-swimming trochophorelike larva (actinotroch). *Phoronis.*

Phylum Mollusca. Mollusks are unsegmented protostomes (except the class Monoplacophora). The body consists of a cephalized head, a muscular foot, and a visceral mass. The dorsal body wall extends laterally to form the mantle, which typically secretes a calcareous shell; the mantle cavity contains gills or may be adapted into an air-breathing lung. The coelom is reduced, but the open circulatory system is extensive; the nervous, digestive, and excretory systems are also well developed. Feeding organ a radula, except in pelecypods. Cleavage is typically spiral and determinate; trochophore and sometimes veliger larval stages. One of the most successful and diverse phyla, with marine, freshwater, and terrestrial representatives. Seven classes, four of which contain the vast majority of species:

Class 1. Monoplacophora. Originally described from fossilized shells, but living representatives were discovered in 1952; limpetlike shell, but broad foot attaches to shell by six pairs of muscles; paired auricles and ventricles, and five pairs of gills. It is uncertain as to whether this metamerism is primitive or secondarily derived. *Neopilina*.

Class 2. Amphineura. (Polyplacophora). Chitons. Elliptical bilateral body; head greatly reduced; large flat foot; shell a middorsal row of eight plates, surrounded by a fleshy girdle; radula used to rasp microorganisms from intertidal rocks; trochophore larva; about 600 modern species. *Tonicella; Chiton*.

Class 3. Aplacophora. Wormlike benthic mollusks without a shell; live in mud or as commensals on cnidarians; with radula; about 50 species described. *Neomenia*.

Class 4. Gastropoda. Snails, slugs, whelks, limpets, abalone, and others undergo tortion during early development. Foot flat for creeping; head distinct, with eyes and one or two tentacle pairs; shell if present is in one piece, usually spiral but may be uncoiled, reduced, or absent; radula present in all but a few parasitic species; trochophore and usually veliger larvae. Over 35,000 described living species divided into three subclasses:

Subclass 1. Prosobranchia. Mostly marine snails. Respiration usually by gills that are situated in the mantle cavity anterior to the heart. Largest and most primitive subclass. *Strobus gigas*, giant conch; *Littorina*, periwinkle; *Fissurella*, keyhole limpets; *Haliotus*, abalones; *Crepidula*, slipper or boat shells.

Subclass 2. Opisthobranchia. Strictly marine. Shell reduced or absent; gills, if present are situated posterior to the heart; often with surface gills; hermaphroditic. *Clione*, pteropods or sea butterflies, with foot expanded into two fins used in swimming; *Aplysia*, the sea hare; many colorful sea slugs, nudibranchs.

Subclass 3. Pulmonata. Mostly terrestrial and freshwater snails and slugs. No gills; vascularized mantle cavity serves as a pulmonary chamber (lung); shell usually present, sometimes reduced or absent; one or two pairs of tentacles; mostly vegetarian, a few carnivorous. Hermaphroditic; yolky eggs without larval stages. *Lymnea*, a freshwater snail; *Helix*, European garden snails, introduced into America; *Arion*, slugs with no shell; *Limax*, slug with rudimentary shell in mantle; *Testacella haliotidea*, a slug that lives in greenhouses and preys on earthworms.

Class 5. Scaphopoda. Tooth shells or tusk shells. Marine; body enclosed in conical shell, open at both ends; foot elongated and used for digging; no gills; delicate tentacles; trochophore larva; about 200 modern species. *Dentalium*, from Cape Cod northward.

Class 6. Pelecypoda (Lamellibranchia or Bivalvia). Bivalve mollusks: clams, mussels, oysters, scallops, shipworms, and others. Mostly marine but some common freshwater species; typically with shell consisting of two valves; no head or radula; foot usually wedge shaped and adapted for digging in sand and mud; usually two gills on each side of the mantle cavity; over 26,000 living species arranged into four orders:

Order 1. Protobranchia. Primitive, with gills not used for feeding; mantle margins united ventrally and posteriorly with openings for siphon and foot; usually two equal adductor muscles; large oral palps are feeding organs; entirely marine. *Yoldia; Nucula*.

Order 2. Filibranchia. Gills used for feeding as well as respiration; palps small;

siphons lacking or only slightly developed; anterior adductor muscle small or absent; large posterior adductor muscle; typically with byssal gland for attachment to solid objects. *Modiolus* and *Mytilus*, common mussels; *Pecten*, scallops; *Anomia*, jingle shells.

Order 3. Eulamellibranchia. Gills also used for filter feeding, but more complex than in filibranchs, with adjacent filaments fused; siphons generally well developed; adductor muscles equal, or anterior muscle smaller. Marine, brackish, and freshwater species in this largest pelecypod order. *Mercenaria (Venus)*, quahog or hard-shell clam; *Ensis*, razor clams; *Mya*, soft-shell clam; *Teredo* shipworm; *Ostrea*, oyster; *Cardium*, cockles; *Unio*, a freshwater clam.

Order 4. Septibranchia. Deep-water marine bivalves, with gills modified into muscular membranes that pump water through mantle cavity. *Poromya*; *Cuspidaria*.

Class 7. Cephalopoda. Squids, cuttlefishes, octopods, and nautiluses. All marine; head and foot fused and surrounded by prehensile tentacles; shell generally reduced in modern forms; pelagic or crawling; swimming by forcing jet of water through funnel as mantle contracts; eyes large and often complex; radula present; dioecious, with complex reproductive behavior that involves transfer of spermatophores; the over 600 modern species are divided into two subclasses:

Subclass 1. Nautiloidea (Tetrabranchia). Calcareous shell, closely coiled and divided into internal chambers; numerous tentacles without suckers; eyes without lens; no chromatophores; no ink sac; two pairs of gills and two pairs of nephridia. *Nautilus pompilius*, the chambered nautilus.

Subclass 2. Coleoidea (Dibranchia). Internal shell reduced or absent; one pair of gills; one pair of nephridia; eight or ten tentacles, with suckers or hooks; eyes with lens; chromatophores and ink sac present. *Loligo*, common squid; *Architeuthis*, giant squid; *Sepia*, cuttlefish; *Octopus*.

Phylum Annelida. Segmented worms; the spacious body cavity a true coelom (schizocoel); homonomous metamerism; a dorsal brain and a pair of ventral nerve cords with typically a pair of ganglia in each segment; the digestive tract a straight tube with an anterior mouth and a posterior anus; flexible cuticle covers body wall, which has segmentally arranged outer circular and inner longitudinal muscle layers; chitinous setae usually present; pair of nephridia in each metamere; gametes are derived from mesoderm. Cleavage is generally spiral and determinate; trochophore larva in primitive aquatic forms. About 7000 described species. Three classes:

Class 1. Polychaeta. Marine; parapodia well developed and provided with setae that are variously modified; prostomium and first few segments sometimes highly cephalized; sexes usually separate; larva typically a trochophore. Largest and most diverse annelid group; 4000 species adapted to crawling, swimming, and sessile lives. *Neanthes virens*, the clam worm; *Tomopterus*, a pelagic polychaete; *Arenicola*, the lugworm; *Chaetopterus*, the parchment tube worm. Groups of uncertain affinities, but closest to polychaeta:

Archiannelida. A heterogeneous assemblage of minute worms that live between sand grains in shallow marine waters; primitive, simplified, or neotenic; sometimes classified as an order of polychaetes. *Dinophilus*.

Myzostomida. Small group of minute, aberrant worms that are commensals or parasites on echinoderms, particularly crinoids. *Myzostomum*.

Class 2. Oligochaeta. Earthworms. Most burrow through soil; some freshwater and a few marine species. Body streamlined for burrowing, without parapodia or cephalic tentacles and with few setea. Internal metamerism highly developed; hermaphrodites cross-fertilized; eggs deposited in cocoon, direct development without trochophore larva; 3100 species. *Lumbricus*, an earthworm; *Tubifex*, an oligochaete that lives in polluted fresh waters.

   Class 3. Hirudinea. Parasitic or predatory leeches; mostly freshwater, but some marine and a few terrestrial. Most specialized annelids; without parapodia, cephalic appendages, or setae; metamerism reduced; septa absent, and coelom reduced by encroachment of connective tissue; body with 33 segments plus prostomium; a posterior and often an anterior sucker; hermaphroditic; about 300 species described. *Hirudo*, the medicinal leech.

Phylum Sipunculida. About 300 species of marine "peanut worms" that burrow through intertidal and subtidal bottoms; body divided into introvert and trunk with U-shaped gut; minor protostomes. *Sipunculus*.

Phylum Echiuridea. About 100 species of marine worms; with extensible proboscis and sausage-shaped trunk; live in semipermanent U-shaped or blind burrows. *Echiurus.*

Phylum Priapulida. Cylindrical predacious marine worms with eversible pharynx; they are problematic concerning evolutionary relationships. A coelom is identified, but manner of formation is uncertain. Less than ten species described.

Phylum Onychophora. Terrestrial protostomes with many features that suggest evolutionary ties with annelids and arthropods. *Peripatus*.

Phylum Tardigrada. Water bears. Microscopic; usually live on terrestrial mosses and lichens; four pairs of clawed legs; no respiratory, excretory, or circulatory organs; undergo cryptobiosis. *Macrobiotus*.

Phylum Pentastomida. A wormlike group of parasites (usually on reptiles) that exhibit many annelid and arthropod characteristics: metamerism, a pair of clawed appendages, annelidlike nervous system. Larvae infect fish as intermediate hosts; less than 70 species known.

Phylum Arthropoda. Segmented protostomes with hard exoskeletons and jointed appendages; cephalized nervous system with ventral nerve cord; reduced coelom but extensive hemocoel; open circulatory system with dorsal heart; heteronomous metamerism and tagmosis; diversified musculature; must molt exoskeleton periodically to accommodate growth; striated muscles; no cilia or flagella; yolky eggs often undergo superficial cleavage; development direct or with complex larvae, never a simple trochophore. This is the largest and most diverse of all animal phyla, with more than 1 million species described so far. Arthropods account for 80 percent of all animal species, and 90 percent of arthropod species are insects. Three subphyla; subphylum Trilobitomorpha contains the extinct trilobites; extant subphyla:

Subphylum 1. Chelicerata. Anterior cephalothorax (prosoma) and posterior abdomen; no antennae; most anterior appendages are pincerlike chelicerae, in front of mouth and used for feeding; first postoral appendages are pedipalps, variously modified; no mandibles. Three classes:

   Class 1. Merostomata. Aquatic; five or six pairs of abdominal segments modified for respiration; spikelike posterior telson. Two subclasses:

      Subclass 1. Xiphosura. Closest living relatives of trilobites; marine; horseshoe-shaped cephalothorax with chelicerae; pedipalps, four pairs of walking legs, and chilaria; cephalothorax broadly hinged to abdomen; abdomen with gill books;

telson long and spikelike, only five modern species. *Limulus*, the king, or horseshoe, crab.

Class 2. Arachnida. Predominantly terrestrial chelicerates: spiders, scorpions, mites, and others. Respiration by book lungs and tracheae; prosoma with chelicerae, pedipalps, and four pairs of walking legs; abdominal segments often reduced or fused with prosoma. Many have silk glands and/or poison glands. The 50,000 described species are divided into eight orders:

Order 1. Scorpionida. Scorpions. Elongated body; large chelate pedipalps, long abdomen of 13 segments; poison sting at end of telson. Most primitive arachnids, with serially repeated heart ostia, digestive glands, and ventral nerve ganglia. *Centruroides*, from Arizona.

Order 2. Uropygida. Whip scorpions. First pair walking legs elongated into many-jointed tactile organs; tropical or subtropical; nocturnal. *Mastigoprocteus*, the garoon.

Order 3. Araneae. Spiders. Cephalothorax and abdomen joined by a narrow waist; abdomen reduced to four fused segments; small chelicerae contain a poison duct in the terminal fang; book lungs, some with tracheae; silk glands open through spinnerets on posterior abdomen; chiefly terrestrial. Some common spider families are:

1. Theraphosidae. Tarantulas. *Dugesiella*, of our South and Southwest.
2. Ctenizidae. Trap-door spiders. *Bothriocyrtum*.
3. Dictynidae. Hackled-band weavers. *Dictyna*, of wide distribution.
4. Pholcidae. Long-legged spiders. *Pholcus*, of wide distribution.
5. Theridiidae. Comb-footed spiders. *Theridion*, common house spider that builds cobwebs in the corners of rooms; *Lactrodectus*, the black widow.
6. Linyphiidae. Sheet-web weavers. *Linyphia*, the filmy dome spider.
7. Araneidae. Orb weavers. *Argiope*, the black-and-orange garden spider that builds its web in grass and among bushes exposed to sunlight and in the hub of which it stands head down.
8. Agelenidae. Funnel-web weavers. *Agelenopsis*, which builds its web most commonly in grass but also among shrubbery and stone fences.
9. Lycosidae. Wolf spiders. *Lycosa*, our largest species.
10. Thomisidae. Crab spiders. *Misumena*.
11. Salticidae. Jumping spiders. *Salticus*, common on fences and the outside of buildings.

Order 4. Palpigrada. Palpigrades. Minute, primitive arachnids that live in warm soils; abdomen segmented; long caudal filament with bristles. *Koenenia*.

Order 5. Pseudoscorpionida. Pseudoscorpions. Minute; pedipalps scorpionlike; flattened abdomen. *Chelifer*.

Order 6. Solpugida. Sun spiders. Head and thorax distinct; large chelate chelicerae; respiration by tracheae. *Eremobates*, southern states west of Mississippi.

Order 7. Opiliones. Harvestmen or daddy longlegs. Body short and ovoid; fused cephalothorax broadly attached to segmented abdomen; pedipalps long and leglike; walking legs usually very long and slender. *Leiobunum*.

Order 8. Acarina. Ticks and mites. Small; body short and thick; cephalothorax and unsegmented abdomen fused; free living or parasitic; chelicerae modified for piercing or chelate; worldwide distribution. Some principal families are:

1. Argasidae. Soft ticks. *Ornithodoros hermsi*, vector of relapsing fever in humans.
2. Ixodidae. Hard ticks. *Boöphilus annulatus*, vector of Texas fever in cattle.
3. Demodecidae. Follicle mites. *Demodex folliculorum*, human face mite.
4. Dermanyssidae. Chicken mites and others. *Dermanyssus gallinae*, the chicken mite, a pest of economic importance.
5. Sarcoptidae. Itch mites. *Sarcoptes scabiei*, human seven-year itch mite; causes mange in mammals.
6. Trombidiidae. Harvest mites and chiggers. *Trombicula alfreddugesi*, North American chigger mite or red bug.
7. Tetranychidae. Red spider mites. *Tetrarhynchus telarius*, red spider.
8. Hydrachnellidae. Freshwater mites. *Hydrachna*, parasitic on aquatic insects.

Class 3. Pycnogonida. Sea spiders. Marine; body small; abdomen rudimentary. Chelate chelicerae; proboscis around mouth; sensory pedipalps; ovigerous legs; four pair of very long and thin walking legs. Common and cosmopolitan throughout ocean; most feed on corals, hydroids, and bryozoans. *Pycnogonum*.

Subphylum 2. Mandibulata. Contains the vast majority of living arthropods; the head region bears one or two pairs of preoral antennae, mandibles that flank the mouth, and typically two pairs of accessory mouthparts called maxillae. Includes the predominantly aquatic crustaceans and the predominantly terrestrial insects, centipedes, and millipedes. Four classes:

Class 1. Crustacea. Mandibulate arthropods with two pairs of antennae; the second maxillae do not fuse together to form a lower lip or labium; appendages are biramous (composed of two branches); primarily aquatic. The body is primitively divided into head and trunk regions, but in most groups tagmatization has produced two differentiated trunk regions: an anterior thorax (cephalothorax) and posterior abdomen. The thorax is generally covered by a dorsal carapace, and a terminal telson bears the anus. The number of trunk segments and the types of appendages vary widely, and there is a general trend toward reduction of segments and specialization of appendages. Development is more or less direct, often with nauplius and other complex planktonic larval stages. Eight subclasses, seven of which are often called entomostracans: these have a simple organization, small or minute size, variable trunk segments, no abdominal appendages, and usually a free-swimming nauplius larva. The other subclass, Malacostraca, contains two thirds of the crustacean species, including most of the familiar types.

Subclass 1. Branchiopoda. Primitive crustaceans with leaflike respiratory appendages on thorax; small filter feeders generally restricted to fresh waters; sometimes with laterally compressed carapace enclosing trunk and its appendages. *Artemia*, brine shrimp; *Triops*, tadpole shrimp; *Daphnia*, water flea.

Subclass 2. Ostracoda. Body completely encased within a carapace that forms a bivalved shell; most are minute filter feeders that creep over marine substrata. *Cypricercus*.

Subclass 3. Copepoda. One of the most numerous (numbers of individuals) metazoans; important primary and secondary consumers in aquatic food chains.

Antennules often elongated for swimming; three pairs of maxillipeds and five pairs of swimming thoracic appendages; no compound eyes but median larval eye persists in adult. Free living, commensal, or parasitic. *Calanus*, common in marine plankton; *Cyclops*, fresh water; *Parapandarus*, ectoparasite on fishes.

Subclass 4. Cirripedia. Barnacles. Adults sessile or parasitic. Sessile barnacles typically encased within calcareous shells; no compound eyes or heart; attach by millions to objects in shallow marine water and intertidal zone. *Balanus*, acorn barnacle; *Lepas*, gooseneck barnacle; *Sacculina*, endoparasite of crabs.

Subclass 5. Cephalocarida. Minute crustaceans that live between sand grains; very primitive; only four species recognized. *Hutchinsoniella*, from Long Island Sound.

Subclass 6. Mystacocarida. Minute primitive crustaceans that inhabit intertidal sand; only three described species. *Derocheilocaris*, from Cape Cod.

Subclass 7. Branchiura. Fish lice. Highly specialized ectoparasites on skin and gills of freshwater and marine fishes. *Argulus*.

Subclass 8. Malacostraca. Contains 18,000 of the 26,000 crustacean species. Head composed of six fused somites; thorax has eight variously fused somites; abdomen almost always has six somites plus telson. Head and thorax may be fused into cephalothorax. Female genital openings on sixth thoracic somite; male openings on eighth. Five superorders:

Superorder 1. Leptostraca. With seven abdominal somites. Cosmopolitan in mud and seaweeds along the seashore. *Nebalia*.

Superorder 2. Syncarida. Primitive, aberrant freshwater crustaceans found in the Amazon, European caves, and Australia. *Anaspides*.

Superorder 3. Hoplocarida. Marine stomatopods or mantis shrimp. Second thoracic appendages modified for raptorial feeding. *Squilla*.

Superorder 4. Peracarida. Females form a brood punch under abdomen in which eggs develop directly into miniature adults. Six orders:

Order 1. Mysidacea. Look like miniature shrimp; small, slender, and graceful; with a statocyst on each uropod; mostly marine and important in food chains. *Mysis*.

Order 2. Cumacea. Hooded shrimp; carapace covers half of thorax and extends as hood over head; tiny; burrow in marine sand and mud. *Diastylis*.

Order 3. Thermosbaenacea. Tiny, blind crustaceans known only from certain hot springs and caves; with dorsal brood pouch; only three species recognized.

Order 4. Tanaidacea. Similar to isopods, but with carapace overhanging first two thoracic segments; most inhabit marine littoral zone. *Apseudes*.

Order 5. Isopoda. One of the most successful crustacean groups; marine, freshwater, and terrestrial representatives; body dorsoventrally flattened; no carapace; abdomen reduced; about 4000 species described. *Ligia*, from rocky intertidal zone; *Gnathia*, ectoparasite on marine fishes; *Ligidium*, terrestrial pill bugs, sow bugs, or wood lice.

Order 6. Amphipoda. Body laterally compressed; no carapace; last three pairs of abdominal appendages modified for jumping; predominantly marine, but some in fresh water; over 3600 species described. *Gammarus*, common aquatic amphipods; *Orchestoidea,* beachhopper; *Paracyamus,* whale lice.

Superorder 5. Eucarida. Largest malacostracan group that contains the most familiar crustaceans; all have a solid cephalothorax covering the entire head and thorax; 8600 described species. Two orders:

Order 1. Euphausiacea. Krill. Pelagic marine crustaceans with long shrimp-like bodies; no maxillipeds, as all eight pairs of thoracic appendages are used for swimming, respiration, and filter feeding. *Euphausia,* major food of great blue and humpbacked whales.

Order 2. Decapoda. The first three pairs of thoracic appendages are modified as mouthparts, leaving five pairs of legs that are used for walking, swimming, grasping, and other purposes. Two suborders:

Suborder 1. Natantia. Shrimps and prawns. Body laterally compressed, with well-developed swimming appendages. *Crago* and *Penaeus,* common commercial shrimp; *Crangon,* snapping shrimp.

Suborder 2. Reptantia. Lobsters, burrowing shrimp, sand crabs, true crabs, and others. Body generally dorsoventrally compressed, with appendages modified for crawling; abdomen sometimes reduced. *Panulirus,* spiny lobster; *Homarus,* American lobster; *Cambarus,* freshwater crayfish; *Callianassa,* ghost shrimp; *Emerita,* sand crab; *Pagurus,* hermit crab; *Cancer,* rock crab; *Callinectes,* blue crab.

Class 2. Insecta. Primarily terrestrial mandibulates with body divided into three tagmata: head, thorax, and abdomen. With a single pair of preoral antennae, a shelflike labrum forming an upper lip, a pair of jawlike mandibles, a pair of food-handling maxillae, and a labium derived from the second maxillae forming a lower lip. Thorax bears three pairs of walking legs and typically two pairs of wings—the only invertebrates capable of flight. Abdominal appendages reduced to terminal reproductive structures. Malpighian tubules are excretory organs; uric acid crystals excreted; respiration generally by tracheae. Reproduction by copulation; gradual development or complete metamorphosis. Insects have exploited virtually every terrestrial habitat, and many have secondarily invaded fresh waters. Over 750,000 species recognized but many more remain to be described. About 29 orders are grouped into two subclasses:

Subclass 1. Apterygota. Primitive wingless insects; contains five orders:

Order 1. Protura. Minute insects that live in humid soil litter and humus; lack antennae and eyes; forelegs used as tactile appendages. *Eosentomon.*

Order 2. Aptera. Campodeids and japygids. Small, soft-bodied insects that live in damp leaf mold and soil; long antennae, no eyes. *Campodea.*

Order 3. Microcoryphia. Bristletails. Long Caudal filament; inhabit soil litter. *Machilis.*

Order 4. Collembola. Springtails. Jumping organ generally present on ventral side of third and fifth abdominal segments; live in soil and feed on decayed vegetable matter. *Achorutes,* snow flea.

Order 5. Thysanura. Silverfish. Small insects with elongated bodies, conspicuous antennae, abdominal cerci, and a caudal filament. *Lepisma.*

Subclass 2. Pterygota. Insects with wings or secondarily wingless; often divided into two series:

Series 1. Hemimetabola. Winged insects that undergo gradual development: hatch as nymphs or naiads that are similar in appearance to adults; gradual development of external wing buds and reproductive organs occurs during a series of subsequent molts. Fourteen orders:

Order 1. Ephemerida. Mayfies. Two pairs of membranous, triangular wings; spend most of lives as aquatic naiads, with tracheal gills; adults live only several hours, reproduce, and die. *Ephemera*, an important fish food.

Order 2. Odonata. Dragonflies and damselflies. Carnivorous aquatic naiads with caudal gills; adults flying predators with two pairs of membranous wings, large compound eyes and chewing mouthparts. *Anax*.

Order 3. Dictyoptera. Cockroaches, mantids, walking sticks. Moderately large insects with antennae, eyes, and chewing mouthparts; wingless or with leathery forewings covering membranous hindwings. *Blatta*, a cockroach; *Stagomantis*, a mantid.

Order 4. Isoptera. Termites. Social insects living in colonies organized around a caste system of reproductive individuals, soldiers, and workers; ingest cellulose and fungi. *Reticulitermes*.

Order 5. Orthoptera. Grasshoppers, crickets, and locusts. Large hindwings adapted for jumping; leathery forewings cover and protect membranous hindwings; chewing mouthparts. *Melanoplus*.

Order 6. Dermaptera. Earwigs. Caudal cerci modified to form pincerlike forceps; vestigial forewings and membranous hindwings, or wingless; nocturnal omnivores. *Forcicula*.

Order 7. Grylloblatodea. Small wingless insects with long cerci; inhabit soil litter that is snow covered most of year. *Grylloblatta*.

Order 8. Embioptera. Webspinners. Colonial insects that form silken webs under stones in tropical regions; social aggregations with no true division of labor. *Embia*.

Order 9. Plecoptera. Stoneflies. Carnivorous naiads live in rapid streams; some breath with trachael gills. Adults do not feed or are herbivorous; with two pairs of membranous wings folded over the back; soft bodies; prominent abdominal cerci. *Allocapnia*.

Order 10. Zoraptera. Minute insects that form social aggregations; winged and wingless members; no division of labor; biting mouthparts. *Zorotypus*.

Order 11. Psocoptera. Booklice. Wingless or with two pairs of membranous wings that have few, prominent veins; wings held over body like sides of a roof; chewing mouthparts; ingest vegetable debris. *Liposcelis*.

Order 12. Phthiraptera. True lice. Small to medium sized, dorso-ventrally flattened and wingless insects that live as ectoparasites on birds and mammals. Chewing lice consume feathers, hair, and epidermal scales. Sucking lice feed on blood of mammals; may transmit diseases, including typhus. *Menopon*, common chicken louse; *Pediculus*, body louse.

Order 13. Thysanoptera. Thrips. Small elongated insects with rasping mouthparts; wingless or with two pairs of narrow wings; herbivores or carnivores; peculiar development, with early nymphlike stages and a preadult pupal stage. *Thrips tabaci*, onion thrips.

Order 14. Hemiptera. Bugs. With piercing and sucking mouthparts; labium forms a jointed beak in which the slender piercing maxillae and mandibles move; wingless or with two pairs of similar membranous wings, or fore-wings usually have a leathery base; adapted to a wide variety of habitats and habits, with appendages modified for running, jumping, digging, grasping, or swimming. Some of the approximately 60 families are

Cicadidae. Cicadas.
Cercopidae. Spittle insects.
Membracidae. Treehoppers.
Cicadellidae. Leafhoppers.
Aphididae. Plant lice.
Aleyrodidae. Whiteflies.
Coccidae. Scale insects.
Miridae. Leaf bugs.
Cimicidae. Bedbugs.
Reduviidae. Assassin bugs.
Tingidae. Lace bugs.
Lygaeidae. Chinch bugs.
Coreidae. Squash bugs.
Pentatomidae. Stink bugs.
Corixidae. Water boatmen.
Notonectidae. Backswimmers.
Belostomatidae. Water bugs.
Gerridae. Water striders.

Series 2. Holometabola. Winged insects that undergo complete metamorphosis; characterized by an egg–larval–pupal–adult sequence of developments. Includes ten orders:

Order 1. Hymenoptera. Ants, bees, wasps, and others. Often form large colonies in which there is division of labor among several specialized types of individuals; different castes are induced by a combination of heredity and diet; complex behaviors coordinate colonies; strong exoskeletons; mouthparts modified for chewing, lapping, or sucking; two pairs of stiff membranous wings, or wings reduced or absent; abdomen with narrow waist or pedicel. Females usually with sting, piercer, or saw; some parasitic on other insects. Some of the approximately 110 families are

Tenthredinidae. Sawflies.
Braconidae. Braconid wasps.
Ichneumonidae. Icheumon wasps.
Cynipidae. Gall wasps.
Chalcididae. Chalcid wasps.
Formicidae. Ants.
Vespidae. Vespid wasps.
Sphecidae. Sphecid wasps.
Andrenidae. Mining bees.
Megachilidae. Leaf-cutter bees.
Bombidae. Bumblebees.
Apidae. Honeybees.

Order 2. Coleoptera. Beetles. Largest insect group, with 300,000 described species; tough or hard exoskeleton; chewing mouthparts; usually with two pairs of wings; forewings hard and sheathlike (elytra), and membranous hindwings fold under elytra; distributed in many habitats, especially in warmer regions; virtually no animal or plant materials are exempt from their attack. Some of the approximately 185 families are

Cicindelidae. Tiger beetles.
Carabidae. Ground beetles.
Hydrophilidae. Water scavengers.
Silphidae. Carrion beetles.
Staphylinidae. Rove beetles.
Dermestidae. Skin beetles.
Tenebrionidae. Darkling beetles.
Coccinellidae. Ladybird beetles.
Scarabaeidae. Scarab beetles.
Dytiscidae. Predacious diving beetles.
Gyrinidae. Whirligig beetles.
Lampyridae. Fireflies.
Meloidae. Blister beetles.
Elateridae. Click beetles.
Buprestidae. Metallic wood borers.
Lucanidae. Stag beetles.
Cerambycidae. Long-horned wood-boring beetles.
Chrysomelidae. Leaf beetles.
Curculionidae. Snout beetles.
Scolytidae. Bark or engraver beetles.

Order 3. Rhaphidoidea. Snakeflies. Long bodies with slender necklike prothorax; biting mouthparts and conspicuous compound eyes; two pairs of similar net-veined wings; larvae and adults prey upon insect pests.

Order 4. Neuroptera. Lacewings. Two pairs of large net-veined wings held rooflike over the back; larvae (aphis lions) prey upon aphids. *Chrysopa*.

Order 5. Mecoptera. Scorpionflies. Chewing mouthparts at end of beak; antennae long and slender; wingless or with two pairs of long narrow, membranous wings; some males with bulbous genital capsule at tip of abdomen resembling sting of scorpion; generally carnivorous. *Panorpa*.

Order 6. Megaloptera. Dobsonflies. Aquatic larvae with lateral abdominal gills; adults frequent fresh waters, called "fish flies"; two pairs of many-veined wings held rooflike or flat over body.

Order 7. Trichoptera. Caddisflies. Aquatic naiads construct portable cases from sand grains; naiads mostly herbivorous; vestigial mouthparts in adult; two pairs of membranous wings clothed with long silky hairs. *Phryganea*.

Order 8. Siphonaptera. Ectoparasites of birds and mammals, with piercing and sucking mouthparts; wingless; body laterally compressed; head small; no compound eyes; legs adapted for leaping; feed on blood; sometimes spread diseases. *Ctenocephalides*.

Order 9. Diptera. Flies. Piercing, sucking, or lapping mouthparts forming proboscis; membranous forewings functional in flight; hindwings reduced to knoblike structures called halteres; larvae known as maggots; larval skin sometimes serves as a cocoon and called a puparium. Some of the approximately 140 families are

Tipulidae. Crane flies.

Chironomidae (Tendipediadae). Midges.
Psychodidae. Sand flies.
Culicidae. Mosquitoes.
Cecidomyidae. Gall gnats.
Syrphidae. Flower flies.
Trypetidae. Fruit flies.
Drosophilidae. Small fruit flies.
Oestridae. Bot flies.
Calliphoridae. Blow flies.
Simuliidae. Black flies.
Tabanidae. Deer flies.
Bombyliidae. Bee flies.
Asilidae. Robber flies.
Sarcophagidae. Flesh flies.
Tachinidae. Tachinid flies.
Muscidae. House flies.
Hippoboscidae. Louse flies.
Braulidae. Bee lice.

Order 10. Lepidoptera. Moths and butterflies. Typically with two pairs of membranous wings covered with overlapping scales; butterfly wings held straight up vertically; moth wings generally held flat against back; siphoning proboscis coiled underneath head; herbivorous larvae have chewing mouthparts and are called caterpillars; moths spin silken cocoon; butterfly pupa protected by hard outer skin called a chrysalis. A few of the approximately 120 families are

Tineidae. Clothes moths.
Tortricidae. Leaf rollers.
Sphingidae. Hawk moths.
Geometridae. Measuring worm moths.
Lymantriidae. Tussock moths.
Noctuidae. Owlet moths.
Arctiidae. Tiger moths.
Citheroniidae. Royal moths.
Saturniidae. Giant silkworm moths.
Bombycidae. Silkworm moths.
Hesperiidae. Skippers.
Papilionidae. Swallowtails.
Pieridae. Whites and sulphurs.
Nymphalidae. Four-footed butterflies.
Lycaenidae. Blues, coppers, and hairstreaks.

Class 3. Diplopoda. Millipedes. Terrestrial mandibulates with head and a long trunk with many segments, each (except first three) of which bears two pairs of legs; slow-moving herbivorous scavengers in soil litter; coil up into ball when disturbed. *Spirobolus*.

Class 4. Chilopoda. Centipedes. Terrestrial mandibulates characterized by a head and a long trunk with many segments, each of which bears a single pair of legs; fast-moving, nocturnal predators; first thoracic appendages modified into poison jaws. *Scutigera*.

Phylum Chaetognatha. Arrow worms. Bilaterally symmetrical deuterostomes with three body regions (head, trunk, and tail); lateral and caudal fins present; complete intestinal tract with mouth and bristles; nervous system with eyes and other sensory organs; no circulatory, respiratory, or excretory organs; monoecious; common in marine plankton. *Sagitta.*

Phylum Pogonophora. Beard worms; live in a secreted tube in the ocean bottom; with closed circulatory system and epidermal nervous system; no respiratory or digestive system; crowned by mass of tentacles. *Siboglinum.*

Phylum Echinodermata. Radially symmetrical as adults; bilaterally symmetrical larvae. Body typically divided into five ambulacra; uncephalized; calcareous endoskeleton; complex nervous system; no excretory or respiratory systems; complex coelom includes a unique water vascular system. Deuterostomes; cleavage holoblastic and radial; free-swimming ciliated larva. Five classes:

Class 1. Crinoidea. Sea lilies and feather stars. Adults usually with five branched rays with pinnules; cuplike calyx; tube feet suckerless; about 80 of the 650 modern species possess a stalk for temporary or permanent attachment; many fossil species. *Antedon.*

Class 2. Asteroidea. Starfishes. Adults typically with five rays not sharply marked off from central disc; open ambulacral grooves with suckered tube feet; madreporite aboral; respiration by dermal branchiae. *Asterias; Pisaster.*

Class 3. Echinoidea. Sea urchins, heart urchins, sand dollars. Body hemispherical, egg or disc shaped; no free rays; skeleton of calcareous plates forms a solid test bearing movable spines; usually three-jawed pedicellariae; tube feet with suckers. *Stronglylocentrotus; Echinarachnius.*

Class 4. Ophiuroidea. Brittle stars and basket stars. Typically with five unbranched or branched rays sharply set off from distinct central disc; rays flexible; no ambulacral grooves; no pedicellariae; madreporite oral. *Ophioderma.*

Class 5. Holothurioidea. Sea cucumbers. Adult body long, ovoid, and soft, with muscular wall; retractile tentacles around mouth; body wall usually contains calcareous plates; no rays; no spines or pedicellariae; tube feet usually reduced; cloaca usually with respiratory tree. *Thyone; Stichopus.*

Phylum Hemichordata. Two classes of wormlike deuterostomous animals.

Class 1. Enteropneusta. Acorn worm. Wormlike with many gill slits.

Class 2. Pterobranchia. Very small hemichordates with one pair of gills or none.

Phylum Chordata. The phylum Chordata includes the vertebrate animals such as the mammals, birds, reptiles, amphibians, fishes, elasmobranchs, cyclostomes, and a number of marine forms that are less well known. These animals are characterized by a skeletal axis, the notochord, at some stage in the life cycle; paired gill pouches connecting the pharynx with the exterior at some stage in the life cycle; and a dorsal nerve cord that contains a cavity or system of cavities. There are three subphyla:

Subphylum 1. Urochordata (Tunicata). Tunicates (sea squirts) and a number of other marine forms. Notochord in tail of larva; absent in adult.

Subphylum 2. Cephalochordata. Two families of fishlike animals called lancelets.

Subphylum 3. Vertebrata. Vertebrates Cranium, vertebrae and visceral arches present.

Superclass 1. Agnatha. No jaws or paired appendages.

Class Cyclostomata. Lampreys and hagfishes. Cold-blooded (poikilotherm), fishlike vertebrates without scales, jaws, or lateral fins.

Superclass 2. Gnathostoma. Jaws, modified from first pair of visceral arches; paired appendages; dioecious.

Class Chondrichthyes. The chief characteristics of this class are the presence of a cartilaginous skeleton; persistent notochord; placoid scales; spiral valve in intestine; two-chambered heart; claspers in male; no gill cover, pyloric ceca, or air bladder; mouth, a transverse opening on ventral side of head; tail heterocercal. Approximately 600 species. Two subclasses:

Subclass 1. Elasmobranchii. Sharks. Slender and cylindrical, with gill slits on the side. Two orders:

Order 1. Squaliformes. Carnivorous planktonic feeders; voracious. *Squalus acanthias,* dogfish shark.

Order 2. Rajiiformes. Rays and skates. Flattened dorsoventrally, with gill slits beneath. These are highly specialized sharks, adapted for life on the bottom of the seas. The rays are ovoviviparous; they have a long whiplike tail, usually without a trace of a caudal fin, and near the midlength of the tail is a long sharp-pointed barbed spine, connected with poison glands. The skates are oviparous; they have a short thick "tail" without the poison spine, and the caudal fin is represented by a low dermal fold. *Dasyatis sabina,* sting ray.

Subclass 2. Holocephali. Elephant fishes and chimaeras. The latter are grotesque-looking creatures named after the fire-breathing monster of Greek mythology. One olfactory sac; gills covered by operculum; no spiracles or cloaca; adults nearly scaleless. *Chimaera.*

Class 3. Osteichthyes. The bony fishes. About 40,000 living species are known. Of these, about 168 families and about 3300 species occur in North America. Two subclasses:

Subclass 1. Sarcopterygii. Fleshy-finned fishes. Two orders:

Order 1. Dipnoi. Lungfishes.

Order 2. Crossopterygii. Lobe-finned fishes. *Latimeria,* the coelocanth.

Subclass 2. Actinopterygii. Ray-finned fishes. Four living superorders:

Superorder 1. Polyteri. Primitive ray-finned fish. *Polypterus.*

Superorder 2. Chondrostei. Much of skeleton is cartilaginous; heavy ganoid scales. *Acipenser,* sturgeons.

Superorder 3. Holostei. Skeletons of bone and cartilage; ganoid scales; consists of gars and bowfin.

Superorder 4. Teleostei. True bony fish. Small scales and homocercal tails. Some of the approximately 40 orders are

Order 1. Clupeiformes. Herrings, tarpons, salmons.

Order 2. Cypriniformes. Minnows, catfishes.

Order 3. Anguilliformes. True eels.

Order 4. Beloniformes. Flying fishes.

Order 5. Gadiformes. Cod fishes.

Order 6. Percopsiformes. Pirate and trout perches.

Order 7. Perciformes. True perches.

Order 8. Muligiformes. Barracudas.

Order 9. Pleuronectiformes. Flatfishes.

Order 10. Lophiformes. Angler fishes.

Order 11. Tetraodontiformes. Puffers, ocean sunfishes.

Class 4. Amphibia About 2000 different species of living amphibians are known, a number very much smaller than that of the other principal classes of vertebrates. Approximately 60 belong to the order Apoda (Gymnophiona), about 200 to the Caudata, and approximately 1740 to the Salientia. Three orders:

Order 1. Apoda (Gymnophiona) Caecilians. Wormlike; no limbs or limb girdles; sometimes with small scales embedded in skin; tail short or absent. Family Caeciliidae. *Ichthyophis*, blindworm.

Order 2. Caudata (Urodela). Tailed amphibians. With a tail; without scales; usually two pairs of limbs. Seven families:

Family 1. Cryptobranchidae. Hellbenders. *Cryptobranchus alleganiensis*, American hellbender.

Family 2. Ambystomidae. *Ambystoma tigrinum*, tiger salamander.

Family 3. Salamandridae. Salamanders and newts. *Diemictylus viridescens*, Easter newt.

Family 4. Amphiumidae. Congo eels.

Family 5. Plethodontidae. Lungless salamanders. *Plethodon cinereus*, redbacked salamander.

Family 6. Proteidae. Mud puppies. *Necturus*.

Family 7. Sirenidae. Sirens. Eel-shaped amphibians having small forelimbs but lacking hindlimbs and pelvis and having permanent external gills as well as lungs.

Order 3. Anura. Tailless amphibians. Without tail; without scales; two pairs of limbs; without external gills or gill openings in adult. Four families:

Family 1. Pelobatidae. Spade-foot toad. *Scaphiopus*.

Family 2. Bufonidae. True toads. *Bufo terrestris*.

Family 3. Hylidae. Tree frogs. *Hyla*.

Family 4. Ranidae. True frogs. *Rana pipiens*, leopard frog.

Class 5. Reptilia. Reptiles are cold-blooded vertebrates covered with horny scales or plates; their digits are usually provided with claws; the majority possess functional legs; they breathe by lungs. The 7000 or more species of living reptiles may be grouped into four orders:

Order 1. Chelonia (Testudinata). Turtles, terrapins, and tortoises. Body encased in bony capsule; jaws without teeth. Some of the 12 families are

Family 1. Chelydridae. Snapping turtles. *Chelydra serpentina*, snapping turtle; *Kinosternon*, musk turtles.

Family 2. Testudinidae. Tortoises and most turtles. *Chrysemys picta*, painted turtle.

Family 3. Cheloniidae. Sea turtles. *Chelonia mydas*, green turtle.

Family 4. Dermochelyidae. Leatherback turtle. *Dermochelys coriacea*, leatherback turtle.

Family 5. Trionychidae. Soft-shelled turtles. *Amyda spinifera*, soft-shelled turtle.

Order 2. Rhynchocephalia. One genus of New Zealand lizardlike reptiles. Vertebrae biconcave, often containing remains of notochord; parietal organ present. *Sphenodon*.

Order 3. Squamata. Lizards and snakes. Reptiles usually with horny epidermal scales; quadrate bones movable. Two suborders:

Suborder 1. Sauria. Lizards. Cloacal opening transverse; paired copulatory organs; usually well-developed limbs; rami of lower jaw united. Some of the 20 living families:

Family 1. Gekkonidae. Geckos. *Hemidactylus turcicus*, warty gecko.

Family 2. Iguanidae. New World lizards. *Anolis carolinesis*, American "chameleon" or green anole.

Family 3. Agamidae. Old World lizards. *Draco volans*, "flying" lizard (dragon).

Family 4. Chamaeleonidae. Chameleons. *Chamaeleo chamaeleon*, true Chameleon.

Family 5. Lacertidae. Old World lizards. *Lacerta viridis*, green lizard.

Family 6. Scincidae. Skinks. *Eumeces antracinus*.

Family 7. Amphisbaenidae. Worm lizards. *Rhineura floridana*, Florida worm lizard.

Family 8. Helodermatidae. Beaded lizards. *Heloderma suspectum*, gila monster.

Family 9. Anguidae. Alligator and glass lizards. *Ophisaurus ventralis*, glass snake.

Suborder 2. Serpentes. Snakes. Elongated; no limbs; cloacal opening transverse; copulatory organs paired; without movable eyelids, tympanic cavity, urinary bladder, and pectoral arch; rami of lower jaw connected by ligament. Some of the ten living families are:

Family 1. Leptotyphlopidae. Blind snakes. *Leptotyphlops dulcis*, Texas blind snake.

Family 2. Boidae. Pythons and boas. *Boa constrictor*, boa constrictor.

Family 3. Colubridae. Harmless snakes. *Thamnophis sirtalis*, garter snake.

Family 4. Viperidae. Pit vipers, rattlesnakes. *Crotalus horridus*, timber rattlesnake; *Agkistrodon contortix*, copperhead; and *Agkistrodon piscivorus*, water moccasin.

Order 4. Crocidilia. Crocodiles, alligators, gavials, and caimans. Nostrils paired at end of snout; cloacal opening longitudinal. Three families:

Family 1. Gavialidae. Gavials. *Gavialis gangeticus*, Indian gavial.

Family 2. Alligatoridae. Alligators and caimans. *Alligator mississippiensis*, American alligator.

Family 3. Crocodylidae. Crocodiles. *Crocodylus actus*.

Class 6. Aves. The birds form a more homogeneous class of vertebrates than do the reptiles and cannot be separated into a few well-defined groups. The structural differences that distinguish the orders, families, genera,and species are for the most part so slight as to make it impossible to state them in a brief and clear manner. There are more than 8600 species of birds, and 27 living orders:

Order 1. Struthioniformes. Ostriches. *Struthio camelus*, African ostrich.

Order 2. Rheiformes. Rheas. *Rhea americana*, American ostrichlike bird.

Order 3. Causariiformes. Cassowaries and emus. *Causarius uniappendiculatus*, cassowary.

Order 4. Apterygiformes. Kiwis. *Apteryx australis*, kiwi.

Order 5. Tinaminiformes. Tinamous. *Rhynchotus rufescens*, great tinamou.

Order 6. Sphenisciformes. Penguins. *Spheniscus demersus*, cape penguin.

Order 7. Gaviiformes. Loons. *Gavia immer*, loon.

Order 8. Podicipediformes. Grebes. *Podilymbus podiceps*, pied-billed grebe.
Order 9. Procellariiformes. Albatrosses, petrels, shearwaters, and fulmars. *Diomedea nigripes*, black-footed albatross.
Order 10. Pelecaniformes. Tropic birds, pelicans, cormorants,gannets,darters, boobies. *Pelecanus erythrorhynchos*, white pelican.
Order 11. Ciconiiformes. Herons, bitterns, storks,ibises, spoonbills, flamingos. *Ardea herodias*, great blue heron.
Order 12. Anseriformes. Screamers, swans, geese, and ducks. *Anas platyrhynchos*, mallard duck.
Order 13. Falconiformes. Vultures secretary birds, falcons, kites, eagles, hawks, caracaras, condors, and buzzards. *Haliaeetus leucocephalus*, bald eagle.
Order 14. Galliformes. Pheasants, grouse, ptarmigans, partridges, quails, turkeys, and the hoatzin. *Bonasa umbellus*, ruffed goose.
Order 15. Gruiformes. Limpkins, gallinules, coots, rails, and cranes. *Rallus elegans*, king rail.
Order 16. Charadriiformes. Shorebirds. Great auk, jacanas, oystercatchers, plovers, sandpipers, skuas, gulls, terns, skimmers, auks, murres, and puffins. *Charadrius vociferus*, killdeer.
Order 17. Columbiformes. Pigeons and doves. *Zenaidura macroura*, mourning dove.
Order 18. Psittaciformes. Parrots, parakeets, macaws. *Conuropsis carolinensis*, Carolina paroquet.
Order 19. Cuculiformes. Cuckoos, anis, and roadrunners. *Coccyzus americanus*, yellow-billed cuckoo.
Order 20. Strigiformes. Owls. *Bubo virginianus*, great horned owl.
Order 21. Caprimulgiformes. Goatsuckers, nighthawks, whippoorwills, oil birds. *Chordeiles minor*, nighthawk.
Order 22. Apodiformes. Swifts and hummingbirds. *Chaetura pelagica*, chimney swift.
Order 23. Coliiformes. Colies or mousebirds. *Colius*.
Order 24. Trogoniformes. Trogons and quetzals. *Trogon ambiguus*, coppery-tailed trogon.
Order 25. Coraciiformes. Kingfishers, rollers, hornbills, motmots, and bee eaters. *Megaceryle alcyon*, belted kingfisher.
Order 26. Piciformes. Woodpeckers, puffbirds, jacamars, toucans, barbets. *Colaptes auratus*, southern flicker.
Order 27. Passeriformes. Perching birds. *Tyrannus tyrannus*, kingbird. North American (and Hawaiian) families are as follows:
1. Cotingidae. Cotingas.
2. Tyrannidae. Tyrant flycatchers.
3. Alaudidae. Larks.
4. Hirundinidae. Swallows.
5. Corvidae. Crows, jays, and magpies.
6. Paridae. Titmice and chickadees.
7. Sittidae. Nuthatches.
8. Certhiidae. Creepers.
9. Chamaeidae. Wrentits.
10. Troglodytidae. Wrens.

    11. Mimidae. Trashers, mockingbirds, and catbirds.
    12. Turdidae. Thrushes, robins, and bluebirds.
    13. Sylviidae. Kinglets and gnatcatchers.
    14. Cinclidae. Dippers (water ouzels).
    15. Motacillidae. Wagtails and pipits.
    16. Bombycillidae. Waxwings.
    17. Ptilogonatidae. Silky flycatchers.
    18. Laniidae. Shrikes.
    19. Sturnidae. Starlings.
    20. Vireonidae. Vireos.
    21. Drepanididae. Hawaiian honeycreepers.
    22. Parulidae. Wood warblers and bananaquits.
    23. Ploceidae. Weaver finches.
    24. Icteridae. Blackbirds and grackles.
    25. Thraupidae. Tanagers.
    26. Fringillidae. Sparrows, finches, grosbeaks, and buntings.

Class 7. Mammalia. Mammals are warm blooded; possess hair at some stage of existence; and have mammary glands in the female, which secrete milk to nourish the young. Approximately 5000 species. Two subclasses.

Subclass 1. Prototheria. Egg-laying mammals. They are confined to Australia, Tasmania, and New Guinea. The two oviducts open directly into a cloaca along with the intestine and urethra, as in birds and reptiles; in certain respects the skeleton agrees with that of the reptiles. One order:

    Order Monotremata. Monotremes. *Ornithorhynchus anatinus*, duck-billed platypus; *Tachyglossus*, spiny anteater.

Subclass 2. Theria. Mammals bearing the young alive. Two infraclasses:

    Infraclass 1. Metatheria. The young are born in a very immature condition and are carried in a marsupium or pouch; usually no placenta. One order:

        Order 1. Marsupialia. Marsupials. They possess abdominal pouches or marsupia in which they carry their immature young. *Didelphis virginiana* American opossum; *Phascolarctus*, koala.

    Infraclass 2. Eutheria. Mammals with an efficient placenta attached to wall of uterus. The young are born in an advanced stage. Seventeen orders.

        Order 1. Insectivora. Insectivores. Moles and shrews.

        Order 2. Dermoptera. Flying lemurs." *Galeopithecus volans*, flying lemur.

        Order 3. Chiroptera. Bats. *Myotis lucifugus*, little brown bat; flying mammals.

        Order 4. Primates. An aboreal offshoot of the primitive placental stock. Two suborders:

            Suborder 1. Prosimii. Lemurs. *Lemur varius*.

            Suborder 2. Anthropoidea. Monkeys, apes, and humans. *Homo sapiens*.

        Order 5. Carnivora. Carnivores. The flesh-eating mammals. *Mephitis mephitis, skunk.*

        Order 6. Pinnipedia. Aquatic carnivores. Seals and walruses.

        Order 7. Hyracoidea. Hyraxes of Africa and Asia Minor. Rabbitlike habits, but actually ungulates. *Hyrax*.

        Order 8. Proboscidea. Elephants. *Elephas maximus*, Indian elephant.

        Order 9. Sirenia. Sea cows. *Trichechus manatus*, manatee.

Order 10. Perissodactyla. Odd-toed hooved mammals. *Tapirella bairdii,* Baird's tapir.

Order 11. Artiodactyla. Even-toed hooved mammals. *Antilocapra americana,* pronghorn antelope.

Order 12. Edentata. Sloths and armadillos. *Dasypus novemcinctus,* nine-banded armadillo, southern Texas.

Order 13. Pholidota. Scaly anteaters or pangolins. *Manis pangolin,* from Africa and southeastern Asia.

Order 14. Tubulidentata. Aardvarks. *Orycteropus,* an anteater, but not related to the previous order. These animals are confined to Africa.

Order 15. Cetacea. Whales, dolphins, and porpoises.

Suborder 1. Odontoceti. Toothed whales, dolphins and porpoises. *Physeter catadon,* sperm whale.

Suborder 2. Mysticeti. Whalebone (baleen) whales. *Baleana mysticetus,* right whale.

Order 16. Rodentia. Rodents. *Castor canadensis,* beaver.

Order 17. Lagomorpha. Hares, rabbits, and pikas (conies). *Lepus californicus,* black tailed hare.

A Phylogeny of Phylum Chordata

*Numbers are millions of years before present

ERAS	CENOZOIC		PALEOZOIC							MESOZOIC			

Quaternary Period*

Periods (millions of years): Cambrian 500, Ordovician 440, Silurian 395, Devonian 345, Mississ-ippian 310, Pennsyl-vanian 280, Permian 225, Triassic 195, Jurassic 135, Cretaceous 65, Tertiary 2

ANCESTRIAL CHORDATE

Class Agnatha
- Urochordata (Amphioxus)
- Cephalochordata
- S.C. Cyclostomata (Lampreys & Hagfishes)

Palcodermi
- S.C. Ostracolermi (Jawed Armored Fishes)

Chondrichthyes
- S.C. Holocophali (Ratfish)
- O. Squaliformes (Sharks)
- O. Raiiformes (Rays)
- S.C. Elasmobrainchii

Osteichthyes
- S.C. Actinopterygii
- S.O. Polypteri (Polypterus & Bowfin)
- S.O. Chondrostei (Paddlefish & Sturgeon)
- S.O. Holostei (Gars)
- S.C. Prototheia (Bony Fishes)
- O. Crossopterygii (Coelocanth)
- O. Dipno (Lungfish)

Amphibia
- S.C. Labyrinthodontia
- O. Apoda (Caecilians)
- O. Caudata (Tailed Amphi)
- O. Anura (Frogs, Toads)

Reptilia
- O. Cotylosauria
- S.C. Anapsida
- S.C. Lepidosauria
- O. Sauropterygia
- O. Ichthyopterygia
- S.C. Archosauria
- O. Mesosauria
- O. Pelysosauria
- S.C. Synapsida
- O. Therapsida
- O. Chelonia (Turtles)
- O. Rhynchocephalia (Tuatara)
- O. Squamata (Snakes & Lizards)
- (Ichthyosaurs)
- O. Pterosauria
- O. Crccodilia (Crocodiles)
- O. Ornithiscia
- O. Saurichia
- O. Thecodontia
- (Plesicsaurs)

Aves
- (27 Orders)

Mammalia
- S.C. Theria
- S.O. Paleonisci
- S.O. Teleostei
- O. Monotremata
- Infraclass Metatheria — O. Marsupiala
- Infraclass Eutheria — (17 Orders)

Glossary

This glossary is included as an aid to understanding the meaning of zoological terms; frequent reference to it will prove very helpful. It is strongly recommended that the beginning zoology student make a serious systematic study of this list of words in an effort to gain facility in the use of the vocabulary in this field.

When primary and secondary stress marks are required, primary stress is indicated by a double prime (″) and secondary by a single prime (′).

Definitions of terms not included in the glossary can be found in the text; see the index for page numbers. Also, taxonomic names are omitted as they are covered in the classification appendix and the text.

Ab″do-men. Body region that contains the viscera; in mammals, extending from the diaphragm to the pelvis.

Ab-duc″tor. A muscle that draws a body part away from the midline of the body.

A′bi-o-gen″i-sis. Any form of the idea that living organisms may arise from nonliving matter by spontaneous generation. See **biogenesis.**

A′bi-o′tic. Physiochemical, as opposed to living.

Ab-o′ral. Opposite the mouth.

A-bridged″ learning. Insight.

Ab-sorp″tion. Taking in of fluids or other substances by cells.

A-bys″sal zone. Ocean zone below 2000 m; all aquatic environments below **bathyal** zone.

Ac-cli″ma-tize. To become habituated to a different environment.

Ac-com′mo-da″tion. Ability of the eyes to automatically change from farsighted to nearsighted vision (and vice versa) when distance to observed object changes.

Ac-cre″tion. Process of growth in nonliving matter in which material is added to the outside.

A-cen″tric. Having no center; specifically, a type of mitosis in which a centriole is absent.

A-ce-ta″bu-lum. Cup-shaped socket of the hipbone into which the femur fits.

Ac′e-tyl-cho″line. A neural transmitter substance liberated at nerve axon endings; involved in conduction of nerve impulses across synapses, and associated with many parasympathetic nerves.

A″chro-ma′tic fi″gure. Metaphase plate chromosomal line-up.

A-cic″u-lum. Needlelike chitinous bristle found within parapodium of polychaete.

A″cid. A compound that yields hydrogen ions to a base in a chemical reaction; having a pH below 7.

Ac″i-nus. One of the small terminal sacs in a lung or multicellular gland.

Ac′quired″ char″ac-ter. Modification of the body that occurs during the life of an individual as a result of environmental conditions.

Ac′ro-meg″a-ly. Disease caused by the excessive secretion of the anterior pituitary after the bones have reached full growth; characterized by overgrowth of the mandible, phalangeal cartilages and other areas.

A-cro″mi-on. Pertaining to the prolongation of the spine of the scapula, forming the point of the shoulder.

A′cro-so″mal cap. Covering over the head of a sperm; derived from Golgi apparatus.

Ac″ro-some. Structure at tip of a sperm head that makes contact with the ovum at the time of fertilization. It contains many proteolytic enzymes that are needed for ovum penetration.

Ac-ti-va″tion e″ner-gy. The energy necessary to break or form a bond or bonds in a chemical reaction.

Ac′tive tran″sport. Movement of a substance through a membrane with the help of a carrier molecule; requires energy.

Ad-ap-ta″tion. Fitness for the environment; may result from selection upon morphology, physiology, or behavior of an organism; the process by which the population becomes fitted to its environment.

A-dap″tive ra′di-a″tion. Evolutionary divergence of a single species into several ecological niches, resulting in speciation, sometimes of strikingly different forms, each adapted to a particular habitat.

Ad-duc″tor. A muscle that draws a structure toward the body midline.

Ad″e-nine. A purine nitrogen base; component of the nucleotides in nucleic acids.

A-den″o-sine. An organic compound containing **adenosine** and sugar ribose, that function in intercellular energy transfers when it is phosphorylated as **ADP** (adenosine diphosphate) and **ATP** (adenosine triphosphate). Adenenosine monophosphate (AMP) is a **nucleotide.**

ADH. Antidiuretic hormone, secreted by posterior pituitary; presence increases water reabsorption in the kidney.

A″di-pose tis″sue. Fat tissue, composed of polygonal or rounded cells. Mostly for energy storage.

ADP. (adenosine diphosphate). **ATP** (adenosine triphosphate) with one less phosphate, of lower energy than ATP.

Ad-re″nal gland. Endocrine gland situated on or near the kidney. Composed of the adrenal medulla and adrenal cortex, which secrete different hormones.

Ad-re″nal-ine. See **epinephrine.**

Ad-ren-er″gic. A type of nerve that releases an adrenalinelike substance from the axon terminal during the transmission of impulses across synapses.

Ad-sorp″tion. Adhesion of a very thin layer of liquid, dissolved substance, or gas molecules to a solid surface.

Ae″ri-al. Inhabiting or frequenting the air.

A′er-o″bic. The ability to live and grow only in the presence of free oxygen.

Aes″ti-vate. See **estivate.**

Af″fer-ent. Carrying to or toward a certain region (e.g., the afferent branchial arteries of the shark carry blood to the gills). Opposite of **efferent.**

Ag-glu″tin-a′tion. Clumping or cohesion of suspended molecules, particles, or cells.

Ag-glu″tin-in. See **antibody.**

Ag-glu-tin″o-gen. See **antigen.**

Air blad″der. In fishes, a thin-walled sac, positioned dorsally, that serves as a float; a hydrostatic organ.

Al″bin-ism (L. *albus*, white). A condition in which normal pigment is lacking, such as in the skin, hair, eyes. Albinism in rats and humans is a typical Mendelian recessive characteristic.

Al-bi″no. An individual, such as a white rat, lacking normal pigmentation.

Al-bu″men. White portion of the reptile and bird egg, surrounding the yolk and zygote, and supplying food for the embryo.

Al-i-men″tary. Pertaining to digestion or to the digestive tract.

Al″ka-li. Any base or hydroxide that is soluble in water and can neutralize acids; having a pH above 7.

Al-lan-to″is. An extraembryonic membrane arising as an outgrowth of the cloaca in reptiles, birds, and mammals.

Al-lele″. One of a series of alternate genes having the same locus in homologous chromosomes; different alleles determine the same trait but code for different proteins.

Al″pha (α) mo-tor neu″rons. Neurons leaving the spinal cord and synapsing with skeletal muscle fibers.

Al-ter-na″tion of gen-er-a″tions. See **metagenesis.**

Al-ti-tu″din-al mi′gra″tion. Migration perpendicular to the earth's surface (e.g., along a mountain side).

Al-ve″o-lus. A small cavity, such as the tiny air sacs in the mammalian lung; the secreting portion of an alveolar gland; or the socket of a tooth.

Am-bu-la″crum. A groove lined with tube feet in each foot of some echinoderms.

A-moe″bo-cyte. Amorphous, independent cell that moves by pseudopodia.

A-mi″no ac″id. An organic compound containing an amino group (NH₂) and an acid group (COOH); **proteins** are polymers of amino acids.

A′mi-to″sis. Direct nuclear division in which the nucleus constricts and separates into two portions; while the cell is in the interphase stage, no condensed chromosomes, asters, or spindle fibers are formed.

Am′mo-coe″te. Larva of lamprey.

Am′ni-o-cen-te″sis. A procedure in which the karyotype of a fetus is monitored by removal of some amniotic fluid. The cells in the fluid are stained, and a karyotype made.

Am″ni-on. Innermost membrane that encloses the embryo in reptiles, birds, and mammals.

Am′ni-o″te. Group of vertebrates—reptiles, birds, and mammals—that develop an amnion and an allantois.

A-moe″boid. Pertaining to cell movements, resembling those of the amoeba, by means of pseudopodia.

Am″phi-bla′stu-la. Free-swimming sponge larva.

Am′phi-coe″lous. A structure that is concave at both ends; applied to some vertebrae.

Am-pul″la. A small bladder-shaped enlargement; sacs attached to the tube feet of echinoderms.

Am″y-lase. An enzyme that helps break down polysaccharides into smaller sugar subunits.

Am′y-lop″sin. A pancreatic amylase; an enzyme produced by the pancreas that digests carbohydrates.

An-ab″o-lism. The constructive phase of metabolism in which cells build protoplasm from food materials.

An′ae-ro″bic. Ability to live and grow in the absence of free oxygen.

A-nal″o-gous. Body parts similar in function but not necessarily alike in their embryonic origin.

An-am′ni-o″te. A vertebrate having no amnion; includes cyclostomes, fishes, and amphibians.

An″a-phase. The stage in mitosis when chromosomes move from the equatorial plate to opposite ends of the mitotic spindle.

A-nas′to-mo″sis. A union or joining together, as of two or more blood vessels, nerves, or other structures.

A-nat″o-my. The study of the structure of animals and plants.

A″ni-mal pole. In telolecithal eggs, the nonyolky end where cytoplasm congregates.

An″i-on. A negatively charged ion.

An-i′so-ga″-my. Sexual fertilization in which the male and female gametes are of unequal size; also called heterogamy.

An′ky-lo″sis. A union or knitting together of two or more bones or parts of bones.

An-nel″i-da. Phylum of segmented worms such as polychaetes, earthworms, and leeches.

An″nu-lus. Any ring or ringlike structure.

An-ten″na. Movable sensory appendage on the head of insects, myriapods, and crustaceans; insects and myriapods have one pair, crustaceans have two pairs.

An-ten″nules. The first antennae of crustaceans.

An-te″ri-or. Pertaining to the front or head end of an organism.

An′thro-po-mor″phic. Ascribing human characteristics to other organisms.

An″ti-bo′dy. A protein produced in an animal body that reacts with a specific foreign substance, or antigen.

An′ti-co″don. A sequence of three nucleotides on a tRNA molecule (see **RNA**) that is complementary to the **codon** in an mRNA.

An′ti-di-u-re″tic hor″mone. See **ADH.**

An″ti-gen. A foreign substance, usually a protein or cell, that causes the production of specific antibodies within an animal body.

An″trum (L., cavity). A cavity or chamber, referring especially to one within a bone.

A″nus. Posterior opening of the digestive tract.

A-or″ta. Large artery responsible for distributing blood throughout the body.

A-or″tic arch. One of paired arteries that connects the ventral and dorsal aortas in the pharynx or gill region.

Ap″er-ture. An orifice or opening.

Ap″i-cal. Referring to the end or outermost part.

A″po-en′zyme. Portion of an enzyme composed of protein.

Ap″o-pyle. Pore leading from the flagellated chambers into the central cavity in leuconoid sponges.

Ap-pend″age. Portion of the body that projects and has a free end, such as limbs.

Ap′pen-dic″u-lar skel″e-ton. Bones of the limbs and limb girdles of vertebrates.

Ap-pen″dix. See **vermiform appendix.**

A″pter-i-a. Area of bird skin on which contour feathers do not grow.

Ap″ter-ous. Wingless.

A-quat″ic. Of or pertaining to water; living in water.

A″que-ous hu″mor. Clear jellylike liquid within the eye between the cornea and the lens.

A-rach″noid membrane (Gr. *arachne*, spider; *eidos*, form). The middle of the three meninges covering the vertebrate brain and spinal cord; very fine and delicate.

Ar-bo″re-al. Pertaining to trees or living in trees.

Ar″bor vit″ae. Internal white matter within the cerebellum; resembles a multibranched tree.

Ar-chen″ter-on. Primitive digestive sac of a metazoan embryo, formed during gastrulation.

Arch″e-o-cyte. Amoeboid cell that receives, digests, and transports food; may also differentiate into other types of cells.

A′re-o″la. Distal extent of the iris; closed cell of forewings of some Lepidoptera; hourglass-shaped longitudinal strip of crayfish carapace posterior to cervical groove; pigmented area around nipple.

A″ri-sto′tles′s lan″tern. In echinoderm class Echinoidea, complex pentagonal chewing apparatus that protudes from mouth.

Ar-ter″i-ole. Small artery.

Ar-ter′i-o-scler-o″sis. Reduction of elasticity of small arteries.

Ar″ter-y. A muscular blood vessel that carries blood away from the heart.

Ar″thro-pod. Segmented protostomes with inflexible exoskeletons and jointed appendages; arachnids, insects, crustaceans, etc.

Ar-tic″u-late′. To join, as between two segments or bones; often flexible.

Ar′ti-fi″cial clas′si-fi-ca″tion. Classification based on characters of convenience without relation to phylogenetic relationships.

As-bes-to″sis. Lung disease caused by inhalation of asbestos fibers.

As-con″oid. Simplest type of sponge structure in which ostia open directly into the spongocoel, which is lined with collar cells.

-ase. Suffix added to indicate an enzyme.

A-sex″u-al re-pro-duc″tion. Reproduction not involving haploid gametes.

As-sim′i-la″tion. Changing of digested foods and other materials into protoplasm.

As-sor″ta-tive ma″ting. Type of mating in which mates are chosen, as opposed to random mating (panmixia).

As″ter. A starlike figure formed during mitosis, composed of the centrosome and the microtubules radiating from it.

A-sym″me-try (Gr. *a*, without; *syn*, with; *metron*, measure). Condition in which opposite sides of an animal are not alike; without symmetry.

At″las (Gr., giant). First cervical vertebra, upon which the skull rests and nods.

At″oll. Horseshoe- or ring-shaped belt of coral islands surrounding a central lagoon.

A-to″mic weight. Combined weights of protons and neutrons of a given atom; characterizes elements and isotopes.

ATP. Adenosine triphosphate, an energy-rich molecule that plays a major role in metabolic energy transfers.

A″tri-al cav″i-ty. Cavity in protochordates and some caudate larvae (e.g., tunicates, *Amphioxus*, frog tadpoles) through which water is exhausted after passing through the gill slits.

A″tri-o-pore. The opening from the atrial cavity to the exterior in *Amphioxus*.

A′tri-o-ven-tri″cu-lar node. A group of specialized cardiac muscle cells lying within the interventricular septum, which, when stimulated, send an impulse to the base of the ventricles, initiating contraction of the ventricles.

A′tri-o-ven-tri″cu-lar valves. Heart valves between the atria and the ventricles; includes the tricuspid and bicuspid valves.

A″tri-um. Chamber or cavity; may refer to the atria of the heart or to the specialized cavity of *Amphioxus* that contains the internal organs.

A″tro-phy. Wasting away or withering of the body or any of its parts.

Au″di-to-ry Au″ri-cle. Pertaining to the organ or sense of hearing; the eighth cranial nerve. Ear-shaped body part or lobelike appendage (e.g., lateral flap near planarian eye, mammalian heart atrium).

Au-to″ly-sis′. Process by which a cell is destroyed by its own lysosomes.

Au-to-nom″ic. Independent; self-governing.

Au-to-nom″ic nerv″ous sys″tem. Division of nervous system controlling involuntary visceral functions; composed of sympathetic and parasympathetic subdivisions.

Au′to-some. Any chromosome except a sex chromosome.

Au-tot″o-my. Self-mutilation; the automatic breaking off of a trapped body part.

Au′to-tro″phy. Nutrition in which an organism manufactures its own food from simple inorganic molecules (e.g., photosynthesis and chemosynthesis).

AV node. See **atrioventricular node.**

Ax″i-al skel″e-ton. Part of the vertebrate skeleton that consists of the skull, vertebrae, sternum, and ribs.

Ax″is. Second cervical vertebra upon which the skull turns; an imaginary line on either side of which body parts are symmetrically arranged.

Ax″on. Fiber of a nerve cell that conducts impulses away from the cell body; synapses with dendrites or somas of other neurons or with muscle or gland cells.

Az″y-gos. Unpaired anatomical structure, such as the azygos vein of mammals.

Back″bone. The vertebral column, especially if composed of bone.

Back″cross. The mating of a hybrid to either of its parents.

Bal″anced car″ri-er. An individual whose karyotype appears abnormal but who nevertheless has a normal chromosomal complement; usually the result of a translocation.

Bar″bule. Processes emanating from barbs of a feather and having hooklets.

Barr bo″dy. In the cells of a human female, a dark-staining portion consisting of an inactive X chromosome.

Bar″ri-er. Any type of obstruction—physical, chemical, or biological—that prevents the migration of animals or an extension of a species range.

Bar″ri-er reef. Coral reef separated from shore by a wide, deep channel.

Ba″sal disc (disk). Flattened aboral end of some cnidarian polyps that attaches to a substratum by secretion of a sticky substance.

Base. A substance that gives off hydronium ions in solution. See **Alkali.**

Base″ment mem″brane. Any membrane upon which an epithelium lies.

Ba″so-phil. Type of multinucleate, nonphagocytic leucocyte.

Ba″thy-al. Ocean zone between the **enphotic** and **abyssal** zones; from about 200 m to 2000 m.

Be′hav″ior. Reactions of the whole organism to its environment.

Ben″thos. Organisms living on or in the bottom of the ocean or fresh waters, from shoreline down to the greatest depths.

Be-ri′be-ri″. Deficiency disease caused by lack of vitamin B_1 in the diet.

Bi-cus″pid valve. In mammalian heart, the bilobed valve between the left atrium and ventricle.

Bi-la″ter-al clea″vage. Unequal-sized blastomeres are confined to separate areas in the morula; yields a bilaterally symmetrical morula.

Bi-lat″er-al sym″me-try. Arrangement of the parts of an organism in such a way that the right and left halves of the body are mirror images of each other.

Bile. A fluid that is secreted by the liver in vertebrates; emulsifies fats in small intestine.

Bile duct. Duct that transports bile from the gallbladder to the small intestine.

Bi″na-ry fis″sion. Type of asexual reproduction by means of which an organism divides into two approximately equal parts.

Bi-no″mi-al. Having two names; in taxonomic nomenclature the first name is the genus and the second is the specific name.

Bi′o-gen″e-sis. Theory that living things are produced only from living things; the opposite of abiogenesis.

Bi′o-ge-net″ic law (Gr. *bios*, life; *genesis*, production). Principle that an animal may repeat in its embryonic development some of the corresponding stages of its ancestors; ontogeny recapitulates phylogeny.

Bi-ol″o-gy. Science of life; it includes botany, zoology, and all the fields that study living organisms.

Bi'o-lu′mi-nes″cence. Production of light as the result of chemical reactions in living organisms.

Bi″ome. A habitat zone such as a grassland or tundra, resulting from interaction of climate, biota, and substratum.

Bi-o″ta. Organisms composing the community of a certain region.

Bi″pin-nar″ia. Larva of asteroid echinoderms with ciliated bands resembling wings.

Bi-ra″di-al sym″me-try. Condition in which an animal has radially arranged parts that lie half on one side and half on the other side of a median longitudinal plane (e.g., ctenophores).

Bi-ra″mous. Crustacean appendage with protopodite bearing an endopodite and an exopodite; the two branches are often highly specialized.

Bi-va″lent. In pachynema stage of meiosis, the four chromatids formed from two homologous chromosomes all joined by a common centromere; also called a tetrad.

Blas″to-coel. Cavity present within the blastula.

Blast″o-derm. Blastodisk, or its outgrowth; in superficial cleavage (e.g., in insects), the outer cellular layer of the blastula.

Blast″o-disc. Cytoplasmic cap of telolecithal eggs that overlies yolk material; the animal pole that undergoes cleavage.

Blas″to-mere. A cleavage cell; any one of the cells in an embryo from the first cleavage division to the beginning of gastrulation.

Blas″to-pore. Porelike opening from the archenteron or gastrula cavity to the exterior.

Blast″o-style. Central axis of a cnidarian gonangium.

Blas″tu-la. Early developmental stage produced by cleavage, usually a hollow ball of cells.

Bleph'a-ro-plast″. Basal granule, basal body, or kinetosome of flagellum or cilium.

Blood. Fluid circulating in vascular system of many animals.

Bo″dy ca″vi-ty. Cavity located between body wall and gut of an animal.

Bo″lus. Lump of masticated (chewed) food plus saliva.

Bone. Hard connective tissue composed mostly of collagen and calcium and phosphorus salts.

Book lungs. Paired respiratory structures present in various arachnids.

Bo″re-al for-est. See **coniferous forest.**

Bot″tle-neck ef-fect″. A type of genetic drift that results from a rapid decrease in population size; the gene pool is drastically reduced, thus enhancing the gene frequencies of the remaining alleles.

Bow″man's cap″sule. Cup-shaped end of a kidney tubule that forms around a glomerulus.

Bra″chi-al. Belonging or pertaining to the upper part of a vertebrate forelimb.

Bran″chi-al. Refering to the gills or gill region.

Bran'chi-os″te-gite. Portion of the exoskeleton that covers the gills in higher crustaceans (e.g., crayfish).

Breath″ing. Inhaling and exhaling air; the mechanics forcing air over the surface of the lungs.

Breed. Subspecies; also to selectively mate.

Bron″chi-ole. Bronchial tubule terminating in an alveolar sac.

Bron″chus. Either of the two main branches of the trachea of a lung-breathing vertebrate.

Brow″ni-an move″ment. Movement of small particles or cells in solution or suspension due to collisions with water or other solvent; especially visible in protists.

Buc″cal. Pertaining to the mouth or oral cavity.

Bud. Developing lateral branch of an organism such as the hydra; usually used to denote a young organism produced by budding.

Bud″ding. Production of offspring by outpocketing of body wall.

Buf″fer. Substance that prevents a noticeable change of pH in a solution to which an acid or base is added.

Bul'bo-u-re″th-ral gland. Paired glands whose secretions contribute to seminal fluid; also known as Cowper's gland.

Bul″bus ar-te′ri-o″sus. Enlarged bulblike base of the ventral aorta, occurring chiefly in the bony fishes.

Bys″sus. Strong, proteinaceous threads secreted by mussels for attachment to rock surfaces.

Cae″cum. A cavity open at one end; for example, the pouch that is the beginning of the large intestine.

Cal-car"e-ous Composed of or containing calcium carbonate ($CaCO_3$) or other carbonates.

Cal-cif"er-ous glands. Structures that lie at the sides of the esophagus of earthworms.

Cal'o-rie. The amount of heat required to raise the temperature of 1 gram of water by 1°C; a unit of heat.

Ca-nal"of Schlemm. Passage through which old eye humors exit; blockage of the canal of Schlemm leads to increased intraocular pressure (glaucoma) as new humors are secreted.

Can"cel-lous. Porous structure, especially referring to bone; spongy.

Cap"il-la-ry. One of many minute branches of blood vessels that carry blood to tissues in many animals.

Car"a-pace. Hard dorsal shell of turtles and some crustaceans.

Car'bo-hy"drase. An enzyme promoting synthesis or breakdown of a carbohydrate.

Car'bo-hy"drate. An organic compound, such as sugar or starch, composed of carbon, hydrogen, and oxygen atoms, with the general formula $(CH_2O)_n$.

Car"di-ac. Pertaining to the heart.

Car"ni-vore. An organism that eats other animals.

Ca-rot"id. Principal artery leading to the head; one on each side.

Car"pals. Bones of the wrist.

Car"ri-er. Chemical substance that bonds to a substance to facilitate passage through a membrane; in genetics, an individual that is heterozygous for a certain recessive trait or an individual infected with a give pathogen.

Car"ry-ing ca-pa"city. The population size supportable by a particular environment; determined by environmental factors such as food supplies, living space, predators, competition, etc.

Car"ti-lage'. Connective tissues that contain homogeneous cells in a nonliving matrix that may contain collagen fibers.

Car-til-a"gin-ous joint. Joint without a cavity; semirigid, with either fibrocartilage or hyaline cartilage between the bones.

Caste. Any one of the individual types within a colony of social insects that carries out a specific function.

Cas-tra"tion. Removal of the gonads, especially male.

Ca-tab"o-lism. Breaking down or destructive phase of metabolism; the metabolic processes in which chemical breakdown occurs; for example, respiration.

Ca"ta-lyst'. A substance (e.g., an enzyme) that accelerates a chemical reaction, but undergoes no permanent chemical change itself.

Cat"i-on. Positively charged ion.

Cau"dal. Pertaining to the tail or the posterior part of the body.

Cell. Small unit of protoplasm surrounded by a plasma membrane and containing one or more nuclei.

Cel'lu-lar. Pertaining to or consisting of cells.

Cel'lulase. An enzyme that digests cellulose.

Cel'lu-lose. Carbohydrate polymer that forms cell wall of plants; wood.

Ce"no-zo"ic. Geological era following Mesozoic, from approximately 75 million years ago to the present time. The age of mammals and birds.

Cen'tral nerv"ous sys"tem. The brain and the spinal cord.

Cen"tri-ole. Cytoplasmic organelle that lies near the nucleus and forms the spindle poles during cell division.

Cen'tro-le"ci-thal. Referring to an egg with the yolk confined to the center, leading to superficial cleavage; occurs in insects.

Cen'tro-mere (kinetochore). A constricted region of the chromosome to which the spindle fiber attaches.

Cen"tro-some. A small differentiated area of cytoplasm containing the centriole.

Cen"trum. Ventral body portion of a vertebra that bears the spinous and transverse processes.

Ce-phal"ic. Pertaining to or situated near the head.

Ceph'a-li-za"tion. The tendency toward the centralization of important parts, as the sense organs, in the head region.

Ceph'a-lo-tho"rax. Body division that is formed by the fusion of the head and thorax in some arthropods.

Cere. Patch of swollen skin at base of some avian beaks.

Ce're-bel"lum. The division of the brain that is associated with muscular coordination in the higher vertebrates.

Ce-re″bral. Pertaining to the brain as a whole or more specifically to the cerebral hemispheres and their activities.

Ce-re″bral aq″ueduct. Canal between the third and fourth ventricles of the vertebrate brain.

Ce-re″brum. The large, lobed, anterior part of the brain, which in primates is the seat of thought, memory, and other higher mental functions.

Cer″vi-cal. Pertaining to the neck region.

Cer″vix. Necklike structure, as the uterine cervix of a typical mammal.

Char″ac-ter. Distinguishing structure or function; any trait of an organism.

Chei-lo″sis. Inflammation and soreness at the corner of the mouth; may be caused by vitamin B2 deficiency.

Che-lic″er-a. Anterior pair of appendages in horseshoe crabs, arachnids, and pycnogonids.

Che″li-ped. A crustacean appendage with a distal pincerlike claw, such as the first and second walking legs of a crayfish.

Che″mo-re-cep-′tor. A receptor neuron sensitive to specific chemical changes or gradients.

Che-mo-syn″the-sis. Type of autotrophic nutrition in certain bacteria in which energy is derived from the breakdown of inorganic materials.

Che-mo-taxis. Behavioral response of an organism to chemical stimulation; involves directional movement.

Chi-as″ma. Usually pertaining to the vertebrate optic chiasma (chiasm), where the optic nerves from the retinas of each eye cross on their way to the brain.

Chi-as-ma″ta. Position of crossover on chromosomes during meiosis.

Chi″tin. Complex organic substance occurring in the exoskeleton of arthropods and some other animals.

Chlo′ro-cru-o″rin. Green blood pigment present in the plasma of certain polychaetes.

Chlo″ro-gogue. Excretory cell of certain annelids; release wastes into coelom, stores food.

Chlo″ro-phyll. Green pigments in plants that are essential for photosynthesis.

Chlo″ro-plast. A chromatophore containing chlorophyll.

Cho″a-na. A funnel, as the opening between nasal passages and pharynx.

Cho″a-no-cyte. Flagellated collar cells found in sponges.

Cho′le-cys′to-kin″in. An intestinal hormone that stimulates the gallbladder to empty.

Cho′li-ner″gic. Referring to a nerve fiber that releases acetycholine from its axon into a synaptic cleft.

Cho-lin-es″ter-ase. Enzyme that inactivates acetylcholine.

Chon″drin. A constituent of the matrix that is found in simple cartilage.

Chon″dro-cra″ni-um. A cartilaginous skull, as in sharks.

Chor″dae ten″din-ae. Stringlike cords attached to the myocardial wall and the atrioventricular valves; help keep valves shut when ventricles contract.

Chor′date. Pertaining to the phylum Chordata; animals having a notochord present at some time in their life history, a dorsally located tubular nerve cord, and gill slits present at some time in their life history.

Cho′ri-on. Extraembryonic membrane of a reptile, bird, or mammal; in mammals it contributes to the placenta; also outer shell of insect egg.

Chor″oid coat. Middle layer of eyeball, heavily pigmented and vascularized.

Chro′ma-tid. Either one of the two duplicate copies of a chromosome in cell division.

Chro″ma-tin. The stainable protoplasmic substance in the nucleus of a cell that condenses into conspicuous chromosomes during mitosis.

Chro′ma″to-phore. Specialized pigment-bearing cell type responsible for color markings on many animals.

Chro″mo-some. Deeply staining body visible under the microscope in the cell nucleus; composed of nucleic acids and protein; consist essentially of genes arranged in linear order.

Ci″li-a-ry bo″dy. Anterior portion of choroid coat that suspends the lens.

Ci″li-um. Microscopic hairlike, protoplasmic organelle projecting from the surface of certain cells and capable of movement.

Cir-cad″i-an rhy″thm. Cycle with a period of about 24 hours.

Cir″ri. Small slender projections or appendages appearing almost like tentacles except for their position.

Class. The main subdivision of a phylum.

Clas″si-cal con-di″tion-ing. A learning paradigm under which an unconditioned stimulus is paired with a conditioned stimulus to elicit the same response.

Clav″i-cle. The human collarbone or its homolog in other vertebrates.

Cleav″age. Series of early divisions between zygote and blastula.

Cleav″age fur″row. Line of cytoplasmic division during cytokinesis.

Cli″ma″tic mi′gra″tion. Movement to a preferred climatic zone.

Cli″max com-mu″ni-ty. A stable ecological community that has reached a state of dynamic equilibrium.

Cli-tel″lum. A thickened glandular portion of the body of an earthworm used in the formation of the cocoon.

Cli″tor-us. Female sexual organ of erectile tissue; homologous to the male penis.

Clo-a″ca. The common posterior passageway or cavity into which the intestine, kidneys, and gonads discharge their products; found in fishes, amphibians, reptiles, and birds and may also appear in embryonic stages of mammals; also in some invertebrates, as in many insects.

Clone. Offspring produced by asexual reproduction of a single animal, with same genotype as parent.

Clot. Conglomeration of fibrin and erythrocytes that serves to decrease or stop blood flow from ruptured vessel; an embolism.

Cly″pe-us. Rectangular sclerite ventral to the frons of an insect head.

Cni″do-blast. Type of cell in which a nematocyst develops in cnidarians.

Cni″do-cil. Hairlike trigger projecting from the outer margin of a cnidoblast.

CoA. Coenzyme A; an important carrier molecule in many biochemical reactions.

Co-a′gu-la″tion. Process of unfolding of protein molecules resulting in their association as an insoluble mass, such as in the formation of blood clots.

Coch″le-a. Coiled structure within inner ear of mammals that houses auditory receptors on the organ of Corti.

Co-coon″. A protective case surrounding a mass of eggs, a larva, a pupa, or an adult animal; often silken.

Co′dom″i-nance. In genetics, the case in which two alleles are both equally expressive, with neither recessive to the other

Co′don. A particular sequence of three nucleotides in an mRNA molecule that represents an amino acid (according to the genetic code).

Coe″lom. Body cavity lined with tissue of mesodermal origin.

Coe″no-sarc. Inner, cellular part of certain hydroids as distinguished from the nonliving outer perisarc.

Co-en″zyme. An organic substance associated with and activating an enzyme.

Co-fac″tor. An inorganic substance (usually an ion) associated with and activating an enzyme.

Cold″-blood″ed. See **poikilotherm.**

Col″la-gen. A polymeric protein that forms some of the fibers in tendons and connective tissue, including that which is incorporated into bone.

Col″lo-blast. Type of cell in ctenophore tentacles capable of adhesion.

Col″loid. A state of matter in which particles larger than single molecules are distributed throughout a (dispersion) medium such as a liquid, gas, or solid.

Co″lon. Anterior portion of the large intestine of a vertebrate.

Co″lo-ny. Group of individuals, unicellular or multicellular, of the same species, that has developed from a common parent and remain organically attached or held together; also used for insect societies; opposite of solitary.

Col′u-mel″la. A long supportive or skeletal structure.

Com-men″sal-ism. A nonparasitic association of two different species in which one (commensal) is benefited and the other (host) is neither benefited nor harmed.

Com″mis-sure. A group of connective nerve fibers uniting two similar structures in the two sides of the brain or spinal cord, or nerve tracts connecting nerve centers elsewhere.

Com-mu″ni-ty. A more or less complex group of plants or animals that occupies a particular area.

Com″pound. A combination of atoms in particular ratios held together by chemical bonding. Also, any combination of similar subparts to form a whole, as in compound eye.

Com-pressed″. Flattened laterally or dorsoventrally.

Con-di″tioned re-sponse″ (CR). In classical conditioning, the response that was originally the unconditioned response but, after conditioning, is elicited by the conditioned stimulus.

Con-di″tioned sti″mu-lus (CS). In classical conditioning, the stimulus that is made to replace the unconditioned stimulus—eliciting the conditioned response.

Co-ni″fer-ous for-est. Boreal forest, taiga; forest of evergreen trees.

Con-ju-ga″tion. Method of sexual reproduction in which two ciliates unite, exchange nuclear material, and then divide, as in the paramecium.

Con-junc-ti″va. The mucous membrane over the cornea of the eye, continuous with the inner lining of the eyelid.

Con′san-gui″ni-ty. The mating of genetically related individuals ("blood relatives").

Con-sum″ers. In an ecosystem, those heterotrophs that obtain energy from the tissues of producers or other consumers.

Con-ti-nen″tal drift. The theory that states that large blocks or plates of the earth's crust are moving with respect to each other; the continental (land) portions thus seem to move also.

Con-ti-nen″tal is″land. An island that was once attached to a nearby continent; normally with fauna characteristic of that continent.

Con″tour fea″ther. Feather with stiff shaft and vane; appear on exterior.

Con-trac″tile vac″u-ole. See **water-expulsion vesicle.**

Co″nus ar-te′ri-o″sus. Expanded cone-shaped structure of the right ventricle that empties into the aorta.

Con-ver″gence. Morphological similarity (analogous) in distantly related forms.

Con-ver″gent e′vo-lu″tion. Evolution toward a common appearance but from different ancestry, as in the evolution of similar body form in dolphins (a mammal) and fish.

Con′vo-lu″ted. A coiled or folded condition.

Cop′u-la″tion. Sexual union of two individuals involving the transfer of sperm from the male to the female.

Cor″a-coid. One of the bones of the pectoral girdle of many vertebrates, especially terrestrial vertebrates.

Co″ri-um. Inner dermal portion of the skin.

Cor″ne-a. Anterior transparent window of the eye.

Cor″ne-a-gen. Type of cell that secretes the cornea.

Co-ro″na. A ring or circle of tentacles or cilia.

Cor-po″ra al-la″ta. Endocrine glands behind the brain in insect head that secretes juvenile hormone necessary for larval molt.

Cor″pus cal-los″um. Tract of transverse nerve fibers uniting the cerebral hemispheres in mammals.

Cor″pus lu″te-um. Ruptured follicle in mammalian ovary that secretes progesterone and estrogen.

Cor″pus-cle. Small, round structure such as renal corpuscle or red blood corpuscle.

Cor″tex. Outer portion of a structure.

Cor′ti-sone. Drug that may stabilize membranes; a corticosteroid.

Cos″tal car″ti-lage. Cartilages that attach the ribs to the sternum.

Coun″ter-cur-rent mul″ti-pli′er sys″tem. In the kidney, the osmotic gradient based mostly on sodium ion concentration.

Co-va″lent bond. Bond between two atoms formed by sharing electron(s).

Cow″per's gland. See **bulbourethral gland.**

Cos-op″o-dite. Joint of an arthropod appendage that lies next to the body.

CR. See **conditioned response.**

Cra″ni-al. Pertaining to brain or skull, such as cranial nerves.

Cra″ni-um. Part of the vertebrate skull that encloses the brain; the braincase.

Cre-pus″cu-lar. Referring to those organisms whose activity is limited to periods of dim light at dawn and dusk.

Cre″tin-ism. Human abnormality resulting from hypoactivity of the thyroid gland from birth or youth.

Cri′noid. Class of primitive echinoderms to which crinoids and sea lilies belong.

Cris″tae. Membranes inside the mitochondrion upon which numerous metabolic reactions occur.

Crop. Expanded anterior part of a digestive tract, specialized for storage.

Cross-fer′ti-li-za″tion. Fertilization of a haploid egg

from one individual with a haploid sperm from another individual; produces diploid zygote by recombination of genes.

Cross"ing o"ver. Process in which homologous chromosomes break and exchange corresponding segments during meiosis.

CS. See **conditioned stimulus.**

Cten"oid scale. Fish scale, the posterior toothed edge of which extends out from under the preceding scale.

Cu"pu-la. Gelatinous mass on tips of sensory hairs of lateral line sense organs in fishes.

Cu-ta"ne-ous. Refers to a thin, noncellular, outermost covering of an organism.

Cu-ti"cu-la. Modern term for "cuticle," preferred because "cuticle" implies inactivity.

Cu-tic"u-lin. Substance making up epicuticle (secreted during molting).

Cy-clo"sis. Rotary streaming movement of protoplasm in certain cells.

Cy"dip-pid. Ctenophore larva.

Cy'pho-nau"tes. Ectoproct larva.

Cyst. A dormant organism enclosed in a thickened resistant wall; a sac or bladderlike structure.

Cy-sti"cer-cus. Bladder worm; tapeworm larva containing an invaginated scolex.

Cy"to-chrome. Iron-containing hydrogen carriers that are part of the electron transport system in aerobic respiration.

Cy'to-kin-e"sis. Cleavage of cytoplasm following nuclear division; starting with invagination at equatorial plate.

Cy-tol"o-gy. Science that deals with the structure of cells.

Cy'to-phar"ynx. Pharynx or gullet of a protist such as *Paramecium.*

Cy'to-plasm. Protoplasm of a cell exclusive of the nucleus.

Cy"to-sine. Nitrogen base found in nucleic acids and nucleotides.

Cy"to-stome. Cell mouth of a ciliate and some other protists.

Dac"tyl. Referring to the finger or toe.

Dac"ty-lo-zoo'id. Sensory polyp of polymorphic hydrozoan colony.

Dar"win-ism. Darwin's theory that species have evolved by natural selection.

Daugh"ter cells. Two cells formed by the division of one "parent" cell.

De-cid"u-ous fo-rest. Forest with deciduous trees that lose their leaves annually.

De'com-po"sers. In an ecosystem, those organisms that obtain energy from dead tissues of producers and consumers; usually fungi, bacteria, and certain protists.

De"dif-fer-en-ti-a"tion. In regeneration, the cells proximal to the injury become unspecialized; they will later differentiate into the structure to be formed.

Def-e-ca"tion. Elimination of waste material from the digestive tract of an animal.

De-layed" le"thal gene. A lethal gene that affects the carrier after the attainment of reproductive age; such lethal genes may be transmitted to offspring.

De-le"tion mu-ta"tion. Genetic mutation in which one or more codons are eliminated, resulting in shorter protein.

Deme. A small interbreeding population that inhabits a very restricted area.

De-na"ture. To disrupt the secondary or tertiary structure of a protein molecule, thereby destroying its biological activity.

Den"drite. Fiber of a neuron that conducts impulses toward the cell body.

De-ni"tri-fy. The process of converting nitrates to ammonia and molecular nitrogen.

Den"ta-ry. One of the lower jawbone(s) of a vertebrate; the only jawbone of a mammal.

Den"tine. Calcareous substance; accounts for major portions of teeth and placoid scales.

Den-ti"tion. Complement of teeth, or general masticatory apparatus.

De-ox"y-ri'bose. Sugar component of deoxyribonucleic acid (DNA).

De'plas-mo"ly-sis. A swelling or bursting of a cell due to an increase in intracellular pressure when the cell is placed in a hypotonic solution.

De-pressed". Flattened vertically from above.

Der"mal. Pertaining to the skin, especially the inner connective tissue layers of the vertebrate skin.

Der"mis. The inner living layer of the skin, lying below the epidermis.

Des-ic-ca"tion. Drying, dehydration.

Deu'te-ro-stomes. Animals in which the blastopore forms the anus, cleavage is usually indeterminate and radial, and the mesoderm and coelom form as outpocketings of gut.

Dex"tral. Coiling in clockwise direction, such as in a snail shell.

Di-a-be"tes mel-li"tus. Abnormal condition in humans characterized by excessive sugar in blood and urine due to insufficient insulin production.

Di'a-kin-e"sis. In meiosis, the stage at which all cytoplasmic divisions occur.

Di-al"y-sis. The separation of crystalloids and colloids in solution by means of their unequal diffusion through certain natural or artificial membranes.

Di"a-phragm. Sheet-like muscle forming a partition between the thoracic and abdominal cavities in mammals; or dividing membrane.

Di-a"phy-sis. The long, main body of a bone, as differentiated from the **epiphyses.**

Di-as"to-le. Relaxation phase of atria or ventricles when they fill with blood.

Di-en-ceph"a-lon. Region of the adult vertebrate brain just posterior to the cerebrum; in the embryo, the second of two divisions of the prosencephalon.

Dif'fer-en'ti-a"tion. Process whereby cells and tissues become specialized for specific functions during development.

Dif-fu"sion. Movement of molecules from a region of high concentration to one of lower concentration, brought about as a consequence of their kinetic energy.

Di-ges"tion. Chemical conversion of complex food material into simpler soluble forms that can be absorbed.

Dig"it. A finger or toe.

Di-gi"ti-grade. Walking on the toes.

Di-hy"brid. Offspring of parents that differ in two traits (characters); an individual that is hybrid (heterozygous) at two loci.

Di-mor"phism. Difference in size, structure, form, color, etc., between two types of individuals of the same species.

Di-oe"cious. Having the male and female reproductive organs in separate individuals.

Di'pleu-rul"a. Free-swimming larva of echinoderms and hemichordates; hypothetical ancestor of most deuterostomes.

Di"plo-blas'tic. Tissue derived from two embryonic germ layers, ectoderm and endoderm.

Dip"loid. Referring to the number of chromosomes in somatic cells, which is twice the number of chromosomes in haploid eggs or sperms.

Di-plo-ne"ma stage. In meiosis, the stage at which the chromosomes of each bivalent split; chiasmata may appear.

Di"pole. The partial separation of electrical charge in an atom or molecule.

Di-rect"ion-al se-lec"tion. Selection for traits at one distribution extreme, thus shifting the distribution mean in that direction.

Di-sac"cha-ride'. Twelve-carbon sugars, composed of two monosaccharides.

Dis-coi"dal clea"vage. See **meroblastic egg.**

Dis-rup"tive se-lec"tion. Selection against the distribution mean and for the extremes, resulting in diverging traits.

Dis-sim'i-la"tion. Disintegration of protoplasm, principally by oxidation.

Dis-so'ci-a"tion. Formation of free ions during the dissolving of a salt in water.

Dis"tal. Away from the point of attachment; for example, the hand is the distal part of the arm; opposite of **proximal.**

Di-ur"nal. Pertaining to the time of daylight; opposite of nocturnal.

Di-ver"gent e'vo-lu"tion. Evolution of dissimilar characteristics in descendants from common ancestors.

Di'ver-ti"cu-lum. Saclike projection of a tubular organ.

DNA. Deoxyribonucleic acid; heritable genetic material that codes for protein synthesis.

DNA po-ly"mer-ase. An enzyme promoting DNA synthesis.

Dom"i-nance. Attribute of genes; full effect of gene is expressed regardless of its allele on homologous chromosome.

Dom"i-nant trait. A trait that appears as the result of either a single or a double "dose" of a particular gene; in contrast to the **recessive trait,** which develops only when both members of a pair of allelic genes are alike.

Dor"sal. Pertaining to the back; opposed to *ventral* in a bilaterally symmetrical animal.

Down fea"ther. Shaftless, insulating feather.

Duct. Tube other than a lymphatic or blood vessel through which a liquid or other product of metabolism is carried.

Duct"less gland. Any gland that secretes a hormone

directly into the bloodstream; an endocrine gland.

Duc″tus ar-te-ri-o″sus. Embryonic duct in mammals that conducts blood to aorta from pulmonary artery.

Du′o-de″num. First part of the small intestine next to the stomach; so named because its length is approximately twelve fingerbreadths.

Du″ra ma″ter. Outermost membrane covering the brain and spinal cord.

Dy-na″mic e′qui-li″bri-um. State at which all changes cancel each other out, so that no net change occurs.

Ec-dy″sone. Insect hormone that initiates molting.

Ec-dy″sis. Molting; the shedding of the outer cuticular covering of an arthropod or other organism to accommodate growth.

ECG. See **electrocardiogram.**

E′co-lo″gi-cal suc-ces″sion. A linear sequence of changes of species composition in an ecological community.

E-col″o-gy. Study of the interrelationships of organisms and their environment.

E′co-sys′tem. The biotic and abiotic components of a particular area (e.g., a pond, a forest).

Ec″to-com-men″sal. Pertaining to an organism that lives on the external surface of another organism, the host, without either benefiting or injuring it.

Ec″to-derm. Outer layer of cells in the gastrula; gives rise to the epidermis, sense organs, and nervous system.

Ec′to-par″a-site. A parasite that lives on the body surface or gills of its host.

Ec″to-plasm. Layer of cytoplasm nearest the surface of a cell.

Ec″to-therm. An animal that obtains heat directly from the environment; see **cold-blooded.**

E-de″ma. An abnormal watery swelling, usually in distal body parts.

EEG. See **electroencephalogram.**

Ef-fect″or. Muscles and glands that carry out responses formulated by the central nervous system.

Ef″fer-ent. Conveying outward or away from a structure; opposite of **afferent.**

E-gest″. To discharge unusable food or residues from the digestive tract.

Egg. Ovum; nonmotile female gamete.

E-jac″u-la-to′ry duct. Duct conveying semen to the urethra from the seminal vesicle; also a portion of the male nematode and insect reproductive tracts.

EKG. See **electrocardiogram.**

E-lec″tri-cal po-ten″tial. Separation of charge.

E-lec′tro-car″di-o-gram. Mechanical tracing based on heart contractile movements; also called EKG and/or ECG.

E-lec″tro-en-ceph″a-lo-gram″. Graphic record of variations in the electrical potential of a brain; also called **EEG.**

E-lec″tro-lyte. Substance that dissociates into ions in a water solution and thus allows electric conduction through the solution; also the ions themselves.

E-lec″tron. Negatively charged subatomic particle with atomic weight = 1/1800 that of a proton or neutron.

E-lec″tron mi″cro-graph. Enlarged photograph of an object taken with an electron microscope.

E-lec″tron mi″cro-scope. Instrument in which a beam of electrons focused by means of a magnetic field (magnetic lens) is used to produce an enlarged image of a minute object on a fluorescent screen or photographic plate.

E-lec′tro-nega-ti″vity. The tendency of some atoms to attract electrons; the greater electronegativity, the greater the affinity for electrons.

El″e-ment. Natural or man-made types of matter that compose all materials of the universe, singly or in combination; also the set of all atoms each with the same numbers of protons.

El′e-phan-ti-a″sis. Human disease caused by a nematode (*Wuchereria bancrofti*) that obstructs lymph ducts; eventually results in grotesquely enlarged body parts.

Em″bol-us. Blood clot within a blood vessel; an embolism results if the embolus blocks a vessel.

Em″bol-y. Invaginative gastrulation.

Em″bry-o. Young animal that is passing through developmental stages, usually within the egg membranes or within the maternal uterus.

Em′bryo-og″e-ny. Development of an organism.

Em′bry-on″ic mem″branes. Cellular membranes formed during embryonic development (e.g., amnion, chorion, allantois).

E-mul″sion. A fluid formed by the suspension of an oily or resinous liquid in another liquid.

E-na″mel. Dense white covering of the crown of a tooth; the hardest substance produced in the animal body.

En-cyst″. To become enclosed in a sac (cyst).

En-cyst″ment. Process whereby an animal becomes enclosed in a protective envelope.

En-dem″ic. Ecologically, found only in a particular region.

En′der-gon″ic. Requiring energy, as in a chemical reaction.

En″do-crine. Pertaining to the **ductless glands,** which produce many hormones.

En″do-cu′ticle. During molting, the last exoskeletonous layer to be secreted.

En″do-cy-to″sis. See **phagocytosis.**

En″do-derm. Innermost layer of the early embryo that gives rise to the lining of the digestive tract; sometimes called entoderm.

En-dog″e-nous. Coming from within.

En′do-par″a-site. A parasite living within the body of its host.

En′do-pep″tid-ase. An enzyme secreted by many organisms to digest proteins by cutting within the amino acid chain.

En″do-plasm. Within a cell; the cytoplasm that is surrounded by ectoplasm.

En″do-plas″mic re-ti″cu-lum. Commonly known as ER, a complex system of interconnecting channels within a cell, made up of unit membranes.

En-dop″o-dite. Inner branch of a biramous crustacean appendage.

En′do-skel″e-ton. A supporting structure on the inside of an animal, of cartilage, bone, or other material.

En′do-some. The central mass, consisting largely of chromatin material, in the nucleus of certain protozoans.

En″do-style. Ciliated groove in the ventral surface of the pharynx in *Amphioxus* and some other protochordates; also larval lampreys.

En′do-the″li-um. Cellular membrane that lines the blood vessels, heart, and lymphatic vessels of vertebrates.

En″do-therm. Individual that produces its own body heat metabolically; it is also a **homeotherm** and is frequently called warm-blooded.

En′do-ther″mic re-ac″tion. Chemical reaction requiring heat.

En″er-gy. Exertion or the capacity for any particular kind of work.

En″er-gy py″ra-mid. Model describing energy relationships within community in which energy content per unit time of each trophic level is estimated; moving toward apex, amount of energy decreases.

En-ter″ic. Adjectival form of enteron.

En′te-ro-kin″ase. Vertebrate intestinal enzyme responsible for converting trypsinogen to trypsin.

En″ter-on. Digestive tract, especially in cnidarians (coelenterates).

En-vi″ron-ment. The total of physical, chemical, and biological conditions surrounding an organism.

En″zyme. A proteinaceous substance produced by living cells that catalyzes specific chemical reactions but does not itself undergo significant change; speeds up chemical reactions.

E′o-si″no-phil. A type of leukocyte, nonphagocytic, one or bilobed nucleus.

E-phy″ra. Free-swimming larval stage in scyphozoan coelenterates.

E-pi″bo-ly. Gastrulation in which micromeres grow over and envelop macromeres of the vegetal pole.

Ep′i-cra″ni-um. Largest sclerite of the head in the grasshopper and related forms.

E′pi-cu″ti-cle. During molting, those epidermal cells secreted under the old exoskeleton; made of cuticulin.

Ep-i-der″mis. Outer cellular layer covering the external surface of a metazoan; it secretes the cuticle on some animals but is composed of dead cells in humans.

E′pi-di″dy-mis. Coiled structure containing efferent tubules of mammalian testis and serves in sperm storage.

Ep′i-glot″tis. Flap of cartilaginous tissue in mammals that closes the air passage to the lungs when swallowing.

E″pi-mor′phic re-gen-er-a″tion. Regeneration where only a small portion of the organism is replaced.

Ep′i-neph″rine. Hormone secreted by the medulla of the adrenal glands; also called adrenaline.

E′pi″phy-seal plate. Epiphysis; growing end of a long bone.

Ep′iph″y-sis. Growing end of a long bone; also the pineal body of a vertebrate.

E-pi-po″dite. A long slender structure, fastened to the protopodite of the walking leg of a crustacean.

Epi-the″li-um. Usually a sheet of cells covering either external or internal body surfaces.

—, columnar. Epithelium that is relatively "tall" and is found in stomach and intestines.

—, **cuboidal.** Composed of cells shaped like cubes; found in glands and ducts.

—, **cuboidal ciliated.** Epithelium of the cuboidal variety, with cilia on the free surface; frequently found in ducts.

—, **flagellated.** Epithelium with flagella on free surface; dulipodia; found in digestive cavity of some simple metazoans.

—, **pseudostratified.** Epithelium composed of more than one layer; layers are not even; cells may be of various shapes.

—, **simple.** Unilaminar, usually composed of flat cells.

—, **stratified.** Composed of more than one layer; the layers are usually even; found in skin, urethra, etc.

E'qua-to"ri-al plate. Linear arrangement of chromosomes in the middle plane of the spindle during mitotic cell division.

E'qui-li"bri-um. A state of balance, in which nothing changes, or dynamic, in which changes in opposite directions cancel each other out.

E'qui-li"bri-um con"stant. "K" indicative of relative concentrations of products and reactants at chemical equilibrium. See also **dynamic equilibrium.**

E-ryth"ro-blas-to'sis fe-tal"is. Condition in infants in which, because of mother–father Rh factor incompatibility. Fetus' erythrocytes are destroyed in the presence of Rh⁺ antibodies produced by the mother.

E'ryth"ro-cyte. Red blood cell.

E-so"pha-gus. The gullet; the tube extending from the pharynx to the stomach; also a part of the digestive tract of certain invertebrates, though a pharynx and/or stomach may not be present.

Es"ti-vate. To pass through hot and dry periods in a dormant state; sometimes spelled aestivate.

Es-tra"di-ol. Female hormone that stimulates uterine hypertrophy; secreted by mature Graafian follicles.

Es"tro-gen. A type of related vertebrate female sex hormones; effects the reproductive system and causes secondary sexual characteristics.

Es"trus. Mammalian egg production and fertilizability.

Eth"moid. A small facial bone forming the upper wall of the nasal passageway.

E-tho"lo-gy. The study of animal behavior based on comparisons, the observation of patterns, and the consideration of environmental adaptations.

Eu-car'-y-ote. A major division of cell type; a cell (or organism containing such cells) that contains a membrane-bound nucleus; all animals, plants, and fungi are eucaryotic. See **procaryote.**

Eu-coe-lom-a"ta. Group of metazoans in which true coeloms exist; includes protostomes, deuterostomes, and lophophorates.

Eu-gen"ics. Application of the knowledge of heredity to the improvement of the human species.

Eu-pho"tic zone. Upper 200 m or less of ocean where photosynthesis may occur; above the **bathyal** zone.

Eu-sta"chi-an tube. Tube leading from middle ear to pharynx in higher vertebrates, permitting the equalization of air pressure between the middle ear and environment.

Eu-then"ics. Science of improving the human species by providing the best possible environment.

E-vag'i-na"tion. An outpocketing or cavity formed by an outgrowth.

E-vis"cer-ate. To remove the internal organs.

Ev'o-lu"tion, or-gan'ic. Process by which organisms have changed through time, both structurally and functionally, due to changes in allele frequencies of a gene pool.

Ex-cre"tion. Discharge of metabolic wastes; also the substance discharged.

Ex'er-gon"ic. Yielding energy, as in a chemical reaction.

Ex"o-crine. Type of gland that releases its secretions through ducts.

Ex"o-cu"ti-cle. Structure containing pigment granules secreted under the new epicuticle during molting.

Ex-op"o-dite. External branch of a typical biramous crustacean appendage.

Ex'o-skel"e-ton. A supporting structure on the outside of an animal body.

Ex'o-ther"mic. Giving off heat, as in an exothermic chemical reaction.

Ex-per"i-ment. A group of observations conducted within the limits of the scientific method.

Ex'pi-ra"tion. The expulsion of air from the lungs of a vertebrate; exhalation.

Ex-ten"sor. Any muscle that straightens out or extends a body part; the opposite of **flexor.**

Ex-ter″nal res′pi-ra″tion. Exchange of gases (CO_2 and O_2) between the respiratory structure and the blood (circulatory fluid).

Ex′ter-o-cep′tor. Sense organ capable of receiving stimuli from the external environment.

Ex′tra-cell″u-lar. Outside of the cell, or cells.

Ex″tra-fu′sal fiber. In muscles, the large fibers making up the bulk of the muscle.

Ex-um″brel-la. Convex aboral surface of a medusa.

F_1, F_2, etc. First filial, second filial, etc.; notation referring to successive generations following cross-breeding.

Fac″et. The external window of an individual ommatidium in a compound eye.

Fa-cil″i-ta-ted trans″port. Movement of a substance through a membrane with the help of a carrier that electrically neutralizes the substance.

Fam″i-ly. The principal subdivision of an order.

Fas″ci-a. A band of connective tissue that covers and supports or binds parts together.

Fat tis″sue. See **adipose tissue.**

Fa-tigue″. In the muscular system, the term describes the result of a buildup of waste products such as CO_2; lactic acid impedes proper muscular function.

Fau″na. Referring to animal life of a given period or region.

Fe″ces. The indigestible, unabsorbed residue of digestion.

Fe″mur. The thighbone; adjective form is "femoral."

Fer′men-ta″tion. Fuel combustion without oxygen, as in anaerobic respiration.

Fer′ti-li-za″tion. Union of a haploid ovum and a haploid sperm to form a diploid zygote.

Fe″tus. Advanced stages of a mammalian embryo in the womb.

Fi″ber. Protoplasmic thread or filament produced by cells and extending from them; also a cell, such as a nerve or muscle fiber.

Fi″bril. Thread or filament produced by cells and located within them.

Fi″brin. An insoluble blood protein and major component of vertebrate blood clots.

Fi-brin″o-gen. Soluble protein that, when in contact with thrombin, forms fibrin, which forms the meshwork of the blood clot.

Fib″rous joint. Joint without a cavity; bones are directly apposed.

Fi″bu-la. Narrower of two bones in vertebrate hindlimb between knee and ankle.

Fi″lo-plume. Hairlike feather with relatively few barbs at tip.

Fi′lo-pod″i-um. Filamentous pseudopodium of some sarcodine protozoans.

Fin. Extension of body of aquatic animal, used for locomotion or steering.

Fis″sion. Asexual reproduction by division into two or more parts approximately equal in size.

Fis″sure. Furrow, cleft, or slit.

Fla-gel″lum. Long, motile cytoplasmic projection containing $9 + 2$ arrangement of microtubule pairs.

Flame cell. Terminal invertebrate excretory cell containing a group of beating cilia.

Flex″or. A muscle whose function is to bend or flex a joint, thereby decreasing the angle between the two body parts that it spans.

Float″ing ribs. The two most caudal ribs, which do not attach to the sternum.

Fo-li-a″ceous appendage. Flattened leaflike appendage; may be multilobed.

Fol″li-cle. A cellular sac or covering; the sac of cells in the vertebrate ovary that surrounds each ovum.

Fol″li-cle sti″mu-la-ting hor″mone. FSH; hormone produced by anterior pituitary that stimulates certain of the Graafian follicles to mature.

Food chain. Linear transfer of chemical energy between trophic levels in a community (e.g., plant to herbivore to carnivore to decomposer).

Food vac″u-ole. Intracellular digestive organelle.

Food web. Complex trophic interactions within a community; includes alternatives and reflects a more natural situation than the simple food chain model.

Fo-ra″men. A natural opening in a bone or membrane through which blood vessels and nerves pass.

For-a″men mag″num. Cranial foramen through which spinal cord passes.

Fo-ra″men of Mon-ro″. (after Alexander Monro); A passageway between the lateral ventricles and third ventricle of the brain.

Fos″sa. Pit or depression in a bone.

Fos″sil. Remains or other indications of prehistoric forms of life.

Foun"der ef-fect". The result of genetic drift in which a new population is founded by relatively few individuals; hence, their limited gene pool is the basis for all future generations.

Fo"vea cen-tra"lis. Optic center of mammalian retina where only cone cells are present.

Frag-men-ta"tion. Type of asexual reproduction in which parts of an organism break off and form new individuals.

Frame"shift mu-ta"tion. Genetic mutation in which one, two, (three plus one)$_n$, or (three plus two)$_n$ base pairs are deleted. Result: the code is disturbed, and different amino acids may be coded for; usually results in premature termination of protein synthesis.

Free liv'ing. Not parasitic or attached; compare with **parasite** and **sessile.**

Frin"ging reef. Coral ridge with no navigable channel between reef and shoreline.

Frons. Anterior portion of an insect's head.

Fron"tal. The forward or leading portion of an organism; a pair of bones in the mammalian skull.

Fron'to-pa-ri"e-tal. Wall of long flat bones forming the roof of the cranium, as in the frog.

FSH. See **follicle-stimulating hormone.**

Func"tion. The action of any part of a plant or animal, usually given in terms of the adaptive value of the part(s) to the organism.

Fun"ctio-nal group. Group of atoms giving a characteristic property to a compound.

Fun-ic"u-lus. A cord, such as the umbilical cord, or tracts of nerve fibers in the spinal cord.

G$_1$. Portion of the cell cycle after telophase of mitosis and before chromosomal replication.

G$_2$. Phase of the cell cycle after chromosomal replication and before prophase of mitosis.

Ga"mete. A haploid reproductive cell, generally an ovum (egg) or sperm.

Ga-met"o-cyte. A potential gamete of some merozoites that undergoes no further development as long as it remains in host; harmless to humans, but if sucked into a mosquito, becomes active; also, a developing gamete.

Gam'e-to-gen"e-sis. The process of development of gametes.

Gan"gli-on. Group or mass of nerve cell bodies, located outside the central nervous system in vertebrates; in invertebrates, ganglia occur within the central nervous system.

Gan"oid. Referring to rhomboid- or diamond-shaped fish scales, enamel covered.

Gas'tric. Pertaining to the stomach.

Gas"trin. Mammalian hormone produced by stomach wall when in the presence of proteinaceous food.

Gas"troc-ne"mi-us. Large muscle on the posterior side of the lower leg of a vertebrate; the calf.

Gas"tro-coel. Primitive digestive cavity of metazoan embryo formed by gastrulation.

Gas"tro-der"mis. Lining of coelenterate digestive cavity.

Gas"trovas"cu-lar. Serving the function of both digestion and circulation.

Gas"tru-la. Stage in development in which the embryo usually consists of two germs layers (ectoderm and endoderm) with a cavity surrounded by endoderm.

Gas"tru-la"tion. Process by which gastrula is formed, often by invagination of the blastula.

Gel. Jellylike substance formed by a colloidal solution while in the solid phase.

Gem'mule. Multicellular vegetative bud of certain freshwater sponges.

Ge"na. Lateral portion of an insect's head; feathered portion of avian mandible.

Gene. Unit of heredity that is transmitted in the chromosome and that codes for the synthesis of a specific protein; controls the development of a trait (character).

Gene flow. The flow into a gene pool of alleles from another population through migration and interbreeding; changes the original allele frequencies.

Ge-net"ic code. A set of relationships between particular triplet sequence of nucleotides of nucleic acids and particular amino acids that allows the genetic information in genes to be translated into proteins; see **codon, anticodon.**

Ge-net"ic drift. The changes in allele frequencies that result from chance and random samples; the smaller the deme, the higher the probability of genetic drift.

Gen-e"tics. Study of the transmission, from parents to offspring, of developmental potentialities (genes) and how they come to expression.

Gen″i-tal. Pertaining to the reproductive organs of either sex.

Ge″nome. The complete set of genes found in an individual organism, usually referring to the material nature of genetic information.

Gen″o-type. The entire informational content of the genetic complement of an individual; the set of alleles in every one of its cells; compare with **phenotype.**

Ge″no-ty-pic ra″ti-o. The ratio of genotypes resulting from a particular cross.

Ge′nus. The taxonomic subdivision of a family; usually composed of several similar species; genus names are latinized, capitalized, and italicized or underlined.

Ge-o-tro″pism. Behavioral response of an organism to gravity.

Germ cell. A gamete.

Germ layer. One of the primary cell layers in an embryo: ectoderm, endoderm, or mesoderm; each tends to produce distinct body parts.

Ger′min-al disc. See **blastodisc.**

Ges-ta″tion. Period between fertilization and birth of a mammal.

Gill. Respiratory organ in many aquatic animals, consisting of an elaborated, thin-walled, vascularized set of tissues.

Gill ar″ches. Walls adjacent to the gill slits, which bears the gills.

Gill fil″a-ment. A fingerlike subdivision of a gill; various gills may include rows of gill filaments.

Gill ra″kers. Spiny ossifications attached to the four gill arches that serve to filter large particles including food from the incurrent water.

Gill slit. (pharyngeal cleft). Series of paired openings in the wall of the pharynx of chordates.

Giz″zard. Muscular part of the digestive tract, as in earthworms, insects, and birds, used for grinding ingested food.

Gland. One or many associated cells that secrete or excrete one or more special substances.

Glau-co″ma. Increased intraocular pressure caused by blockage of the canal of Schlemm.

Glob″u-lin. A class of proteins present in vertebrate blood plasma; included in this class are the immunoglobulins, which function in the **immune system** and include **antibodies.**

Glo-chi″di-a. Bivalve larvae of freshwater clams.

Glo-mer″u-lus. Small coiled mass of capillaries contained in each Bowman's capsule of the kidney.

Glot″tis. Opening from the pharynx into the larynx of a terrestrial vertebrate.

Glu″cose. A six-carbon sugar; the main form in which carbohydrates are transported to the cells.

Glu″tin-ant, oval. A type of cnidarian nematocyst used in food capture, anchorage, and locomotion.

Gly″cer-ine. A three-carbon organic compound prepared by the hydrolysis of fats and oils.

Gly″co-gen. Polysaccharide stored in the liver, muscles, and some other tissues; "animal starch."

Gly-co″ly-sis. Anaerobic carbohydrate respiration; breakdown of glucose to pyruvic acid.

Gob″let cell. Modified epithelial cell that secretes mucus; found in the respiratory, digestive, and other organ systems of vertebrates.

Goi″ter. Enlargement of thyroid gland that results from deficiency of iodine in diet.

Gol″gi bo″dy. Cytoplasmic organelle contributing to the production of certain secretions.

Go″nad. Reproductive organ in which gametes are produced, either ovary, testis, or ovotestis.

Go-nan″gi-um. Reproductive individual of a polymorphic hydroid colony.

Gon-dwa″na-land. Immense land mass thought to have existed in early Cenozoic times; composed of South America, Africa, Australia, New Zealand, and Antartica. It is the southern portion of **Pangea** after the latter broke apart by continental drift. See **Laurasia.**

Gon′o-the″ca. Firm external covering of a gonangium.

Go″no-zoo′id. A reproductive polyp of a hydroid colony.

Graa″fi-an fol″li-cle. Ovarian follicles containing developing ova.

Gray cre″scent. Portion of the gastrula that induces mesoderm, endoderm, and ectoderm; becomes dorsal lip of blastopore during gastrulation; the primary organizer.

Gray mat″ter. Nonmyelinated neurons in the central nervous system.

Green gland. The excretory organ, containing nephridia, of some crustaceans, such as crayfishes, shrimps, and crabs.

Gre-gar″i-ous. Living in company, as in flocks and herds.

Gua″nine. A nitrogenous base used in the nucleotides of nucleic acids.

Gul″let. Synonym for **esophagus.**

H⁺. Hydronium ion; usually hydrated in aqueous solution: H_3O^+.

Hab″i-tat. Particular environment in which an animal lives.

Ha-bit″u-a′tion. Decreased response to a frequently encountered stimulus.

Hair. A slender, horny, nonliving, filamentous covering over the skin of most mammals, with insulating and tactile functions; any filamentous projection serving as a tactile organ in invertebrates.

Hap″loid. The number of chromosomes typically found in a mature gamete; half the diploid number.

Ha″rem. Group of females, normally kept for copulatory purposes.

Hal′lux. First digit of an avian foot, directed backward; the first digit of the hindlimb of mammals, known as the big toe in humans.

Hav-er″sian ca-nal. After Havers, an English physician; L. *canalis*, water pipe. One of the concentric canals in bone that houses blood vessels and nerves.

Hec′to-cot″y-lus. Modified arm of male cephalopod mollusk employed in sperm transfer to female.

He′li-co-tre″ma. At the apical end of the cochlea, area in which the basilar membrane ends.

He-li-o-tro″pism. See **phototropism.**

He″lix. Anything of spiral shape (e.g., shell of a snail).

He″mal. Referring to the blood or blood vascular system.

Heme. Iron-containing red pigment, found in many proteins, such as hemoglobin.

He″mo-coel. Spacious blood sinus functioning as a body cavity as in mollusks and arthropods.

He′mo-cy″a-nin. Bluish respiratory pigment containing copper, found in various invertebrates.

He″mo-glo-bin. An iron-containing pigment in red blood cells of vertebrates capable of carrying oxygen and some carbon dioxide; in some invertebrates, it is dissolved in the plasma.

He′mo-phil″i-a. Abnormal inherited condition in humans in which there is a delayed clotting of the blood.

Hen″son's node. The anterior end of the primitive streak in vertebrate embryology.

He″pa-rin. Protein-containing molecule capable of preventing blood clotting.

He-pat″ic. Pertaining to the liver.

He-pat″ic por″tal sys″tem. System of veins leading from the digestive tract and entering the vertebrate liver.

Her-biv″ore. An animal that feeds only on plants.

Her-ed″i-ty. The transmission of all morphological, biochemical, and behavioral characteristics from the parent; the information for inherited traits is contained in the genes of the organism.

Her-maph″ro-dite. An individual producing both male and female gametes.

Het′er-o-cer″cal. Pertaining to an asymmetric tail in which the vertebral column extends into the dorsal portion, as in sharks.

Het″er-o-crine gland. Gland with a combination of exocrine and endocrine activity.

He″te-ro-dont. Having differing types of teeth.

Het′er-o-mor″pho-sis. Regeneration of a new body part that differs from the structure that it replaces.

Het′er-on″o-mous. Condition in which the segments of an animal are not similar; with metameres specialized for different functions in various parts of the body.

Het″er-o-tro′phy. Nutrition involving the ingestion or absorption of organic material synthesized by other organisms. See **autotrophy.**

Het′er-o-zy″gote. An individual having two different alleles at a particular locus; expression may be dominant or codominant; two types of gametes produced; compare with **homozygote.**

Hi′ber-na″tion. Passing through the winter in a dormant state.

Hind″gut. In arthropods, posterior portion of digestive tract, lined with cuticle.

His″ta-mine. Powerful dilator of the capillaries; it is found in all animal tissues.

His′to-gen″e-sis. Referring to the origin, development, and differentiation of the tissues of an organism.

His-tol″o-gy. Branch of anatomy that deals with the microscopic structure of tissues of an organism.

Hol-lan″dric in-her″i-tance. Sex-linked inheritance in which the Y chromosome contains the gene in question; hairy ears in males is the only trait thus far identified as probably hollandrically transmitted.

Ho′lo-blas″tic egg. An egg that divides completely into blastomeres during cleavage.

Ho′lo-phy″tic. Autotrophic nutrition found in green plants and some mastigophorans; involves photosynthesis.

Hol′o-zo″ic. Type of nutrition, found in most animals, that involves ingestion and digestion of organic matter; a subtype (like saprozoic) of **heterotrophic.**

Ho′me-o-sta″sis. All processes whereby a cell or organism seeks to maintain a steady state; involves positive and negative feedback mechanisms.

Ho″me-o-therm. Individual that maintains a constant body temperature due to endogenous heat production; usually, but not necessarily, also an **endotherm.**

Ho″min-id′. Extant or extinct human or humanlike primate.

Ho′mo-cer″cal. Pertaining to the type of tail that is externally and internally symmetrical.

Ho″mo-dont. Having uniformly similar teeth.

Ho′mol″o-gous. Referring to body parts derived from a common origin. See **homology** and compare with **analogous.**

Ho′mol″o-gous chro″mo-somes. Pairs of chromosomes carrying genes affecting the same traits; one is paternal and the other maternal; they come together in synapsis.

Ho-mol″o-gy. Basic similarity; structural likeness of an organ or part of one kind of animal with the comparable unit in another, resulting from descent from a common ancestry; these organs may or may not have the same function.

Ho′mo-zy″gote. An individual in which both alleles are the same at a particular locus on homologous chromosomes; compare with **heterozygote.**

Hook″let. Hooked structures that hold together the feather barbs.

Hor″mone. A chemical regulator or coordinator secreted by an endocrine gland and having a specific action on some target organ or organs at a distance from the gland; carried by the blood or other body fluids.

Host. Organism that provides food, shelter, or other benefits to another organism.

Host cells. In cnidarians, the epitheliomuscular cells that house nematocysts.

Hu″mer-us. Bone of the vertebrate upper arm.

Hu″mor-al. Pertaining to body fluids, such as blood or lymph.

Hu″mus. Organic part of soil.

Hy″a-line. Glassy or semitransparent; applied to the clear substance of protoplasm and to a type of cartilage.

Hy′brid. Individual resulting from union of sperm and egg that differs in one or more genes, a heterozygote; also the (usually sterile) offspring of two diverging species.

Hy″dranth. Expanded end of branch of a hydroid colony specialized for vegetative function.

Hy″dra-tion. Surrounded by H_2O molecules; usually occurs because of polarity.

Hy′dro-caul″us. Main stalklike stem of a hydroid colony.

Hy″dro-gen bond. A weak attraction between a covalently bound hydrogen atom and an **electronegative** atom, usually oxygen or nitrogen; the bond is important in maintaining the structure of nucleic acids and proteins.

Hy′droid. Used as an adjective, pertains to the Hydrozoa; used as a noun it refers to the polypoid form of a hydrozoan, as distinguished from the medusoid form.

Hy-drol″y-sis. Process by which a complex compound is digested into one or more simpler compounds through a reaction with water molecules.

Hy′dro″ni-um i″on. See H^+.

Hy′dro-rhi″za. Basal portion of a hydroid colony, often branched and rootlike; used for attachment to the substratum.

Hy″dro-sta′tic or″gan. See **air bladder.**

Hy′dro-the″ca. Transparent membrane that extends from the perisarc and surrounds the main part of a hydranth.

Hy′drox″ide i″on. See OH^-.

Hy″oid. Referring to bones and cartilages at or near the base of the tongue.

Hy″per-cy′cle. A cycle whose components are also cycles; used in reference to complex cyclic chemical reactions that occurred on the early earth and were necessary for the origin of life from nonliving chemical reactions.

Hy-per-o″pia. Farsightedness; a condition in which

the eyeball is relatively long, so that to the point of focus lies in front of the retina.

Hy″per-par″a-sit-ism. Condition in which parasites are parasitized by other parasites; may be secondary, tertiary, etc.

Hy′per-ton″ic. In terms of a living cell, the concentration of water molecules is greater inside the cell than outside; therefore more water molecules diffuse out of the cell membrane than into it, resulting in the shrinking of the cell.

Hy-per″tro-phy. Abnormal increase in the size of an organ or other body part.

Hyp″no-tox′in. A poison contained in nematocysts.

Hy″po-blast. Germinal layer that spreads down and under the trophoblast, forming the yolk sac.

Hy′po-cer″cal. Referring to the caudal fin of fishes, in which the notochord or vertebral column extends into the ventral portion of the fin, which is larger than the dorsal portion.

Hy-po″phy-sis. Pituitary gland. See **pituitary body** and **infundibulum.**

Hy″po-stome. Region surrounding the mouth in cnidarians.

Hy′po-thal″a-mus. Region of the forebrain that contains various autonomic nervous system centers.

Hy-poth″e-sis. A tentative solution to a problem that can be tested by experimentation.

Hy′po-ton″ic. In terms of a living cell, the concentration of water molecules is greater on the outside of the cell than inside; therefore more water molecules diffuse into than out of the cell, resulting in the swelling or bursting of the cell.

-idae. Suffix indicating family.

I-den″ti-cal twins. Two individuals developed from a single fertilized egg and therefore having identical sets of genes; also called monozygotic twins.

Il″e-um. Posterior and longest part of the small intestine of a mammal.

I′li-o-ce″cal valve. Valve between the ileum and the colon.

I-ma″go. Adult sexually mature insect.

Im′mune sys-tem. A subdivision of the circulatory and lymphatic systems; the set of leukocytes involved in the production of antibodies and the phagocytosis of foreign matter.

Im′mu-no-glo′bu-lin. A class of **globulins** used in the **immune** system that is important in the recognition

of foreign substances; soluble immunoglobulins are called **antibodies.**

In″breed-ing. Crossing or mating of closely related individuals, such as first cousins or brother and sister.

In″com-plete″ dom″i-nance. Partial dominance; in genetics, the case in which the phenotypes of heterozygotes are distinguishable from those of homozygotes.

In″cus. Second ear bone, lying between malleus and stapes.

In-duc″tion. Embryonic process in which one body tissue causes another tissue to differentiate; also, causal relationship between (usually) chemical substance and portion of embryo differentiating into a specific tissue; also, synthetic reasoning from observation to probable conclusion.

In-ert″ gas. Elements with filled electron shells, therefore highly unreactive: He, Ne, Ar, Kr, Xe, and Rn.

In′flam-ma″tion. Condition in which blood vessels are dilated, leukocytes congregate, and edema occurs; normally a response to injured tissue.

In-fun-dib″u-lum. Stalklike down-pushing of the diencephalon of the brain, which, along with the embryonic hypophysis, will give rise to the pituitary gland of the adult.

In-gest″. To take food and water into the digestive system of an animal.

In-her″i-tance. The sum of all the genes and characters transmitted from parents to offspring.

In′hi-bi″tion. In neurophysiology, the decrement or prevention of neural transmission across a synapse.

In-sec″ti-vore. A type of **carnivore,** an organism that feeds on insects.

In-ser″tion. Place of attachment of a muscle to a movable part, in contrast to the origin.

In-ser″tion mu-ta″tion. Genetic mutation in which an extra amino acid is coded because an additional base triplet is included in the chromosome.

In′spi-ra″tion. Intake of air to the lungs or other internal breathing organs.

In″star. Stage between consecutive insect molts.

In″stinct. Inherited response, often complex, invoked by a particular stimulus and leading to a particular result.

In′stru-men″tal con-di″tion-ing. See **operant conditioning.**

In″su-lin. Hormone secreted by the islets of Langerhans of the pancreas; promotes uptake of glucose from blood by body cells.

In-te″gu-ment. Outer covering, especially the skin, of a vertebrate.

In-ter-car-ti-la″gin-ous bone. Bone formed from an intermediate cartilage model, which was originally formed of mesenchyme.

In′ter-cel″u-lar. Between cells.

In-ter-me″din. In some vertebrates, a hormone secreted by intermediate lobe of pituitary gland; major activity is chromatophore control.

In-ter″nal res′pi-ra″tion. Exchange of gases between the blood and body cells.

In-ter″nal se-cre″tion. A hormone, the product of an endocrine gland.

In′ter-neur″on. A nerve cell that connects two other neurons, often a sensory cell to a motor neuron or a cell of the **central nervous system.**

In″ter-phase. State of cell that is not undergoing mitotic division.

In′ter-ven-tri″cu-lar sep″tum. Partition between the ventricles in the heart of higher vertebrates.

In-tes″tine. Part of the digestive tract posterior to the stomach; in animals without a stomach, the intestine is usually the digestive tract posterior to the ingestive region.

In′tra-cel″lu-lar. Within cells.

In″tra-fu′sal fiber. In muscles, the thin fibers also known as spindle fibers, on to which stretch receptors are attached.

In″tro-vert. Narrow anterior part of a sipunculid that may be muscularly withdrawn inward; also, distal end of a bryozoan bearing lophophore, which also may be withdrawn.

In-vag″i-nate. To infold; folding or inpushing of a layer of cells into a cavity, such as when a blastula begins gastrulation.

In-ver″tase. Sucrase; enzyme hydrolyzing sucrose to glucose and fructose.

In-ver″te-brate. An animal without vertebrae (a backbone).

In′vo-lu″tion. The process of rolling or turning in of cells over a rim, as during gastrulation.

I″on. Charged atom because of loss or gain of electrons.

I-o″nic bond. Bond between two or more atoms that involves a transfer of electrons.

I′on-i-za″tion. The addition or removal of electrons from atoms.

Ir′ri-ta-bil″ity. Ability to respond to stimuli, one of the fundamental characteristics of protoplasm.

I-so″ga-my. Sexual fusion in which gametes are structurally alike, not differentiated into male and female; uncommon.

I′so-lec″i-thal. Pertaining to eggs that have yolk distributed evenly throughout the cytoplasm.

I″so-mers. Different forms of the same molecule; the numbers of constituent atoms are the same, but the bonding patterns are different; isomers usually have different chemical properties, such as glucose and fructose.

I′so-ton″ic. In terms of a living cell, concentration of water molecules is the same inside and outside the cell; therefore, the water molecules pass into and out of the cell in equal numbers, results in no change in cell size.

I″so-tope. Any of two or more forms of an element differing from one another in atomic weight but not in chemical properties; variation is in neutron number; may be radioactive.

Jaw. Oral structure normally used for chewing or holding food.

Je-ju″num. Part of the intestine extending from the duodenum to the ileum in a vertebrate.

Joint. Junction of two separate bones or other hard structures; the point of articulation.

Jug″u-lar. Pertaining to the throat, such as the jugular vein.

Ju″ve-nile hor″mone. Hormone produced by the corpora allata in insects; prevents an imaginal molt.

Ka″ry-o-plasm. The nucleus, including the nuclear membrane and all that it encloses.

Ka″ry-o-type. Photographic representation of the chromosomal complement of an individual; usually the chromosomes are arranged into identifiable groups with common morphological characteristics.

Keel. Distinct median ridge of avian sternum onto which flight muscles attach.

Ker″a-tin. Protein formed by certain epidermal tissues, such as vertebrate skin.

Kid″ney. Chief organ for the excretion of nitrogenous wastes in vertebrates; also loosely applied to analogous organs in other animals.

Kin″dred. Pedigree; chart showing mating relationships within a given family over several generations.

Kin-e″sis. An undirected locomotor response.

Kin-es-the″tic. Referring to the sense of awareness of the position and movements of the body.

Ki-ne″to-some. An organelle similar to centriole; located at base of flagellum or cilium; also called **basal body.**

Klein″fel-ter's syn″drome. In humans, a chromosomal abnormality in humans in which the individual has two X and one Y chromosome (XXY); phenotypically a male, the individual is sterile and may evidence secondary female characteristics.

La″bi-al. Pertaining to the lips.

La″bi-um. A lip, specifically the fused lower lip of a insect.

La″brum. Upper lip of the insect's mouth.

Lac″ri-mal. Pertaining to tears, the tear (lacrimal) gland, or associated structures.

Lac″te-al. Pertains to milk; often refers to the lymph vessels of the villi of vertebrate small intestine.

Lac′to-gen″ic. Milk-producing; lactogenic hormone (prolactin) of vertebrate pituitary gland.

Lac″tose. Milk sugar.

La-cu″na. Small cavity or space, particularly in cartilage or bone, that in life contains a cartilage or bone cell.

La-mel″la (L., small plate). A thin sheetlike layer.

Lar″va. An immature stage in the life cycle of various animals, that reach the adult form by undergoing a metamorphosis.

Lar″ynx. Organ situated between the trachea and the base of the tongue, into which the glottis opens and containing the vocal cords; larynx is typically found in all lunged vertebrates except birds.

Lat″er-al. The side of the body; at each side of the median line.

La″ter-al line. Sense organ in fishes and some amphibians extending bilaterally and longitudinally; also, in nematodes, hypodermal longitudinal thickening.

La′ti-tu″din-al mi′gra″tion. Migration in a north–south direction.

Lau-ra″sia. The northern portion of **Pangea** after it broke up by continental drift; it included North America and Eurasia (minus India). See **Gondwanaland.**

Le-ci-tha″li-ty. Pertaining to oocyte yolk concentration.

Lep-to-ne″ma. In meiosis, the stage resembling mitotic prophase, that is, when chromosomes become visible.

Le″thal gene. Gene that is capable of bringing about death; may be either dominant or recessive and kill the organism at any stage of its development.

Leu-ke″mi-a. Blood cancer characterized by excessive leukocyte production.

Leu″ko-cyte. White blood cell or corpuscle, especially important in the lymphatic and immune systems.

Le″va-tor. A muscle (in general) that raises a body part.

LH. See **lutenizing hormone.**

Lig″a-ment. Tough, fibrous band of tissue connecting bones at a joint.

Li″my. Containing calcium salts, particularly calcium carbonate.

Lin″gu-al. Pertaining to the tongue (e.g., the lingual artery).

Link″age. The tendency for certain traits (characters) to remain together in heredity because their gene loci are located on the same homologous chromosomes.

Li″pase. A fat-splitting enzyme.

Li″pid. A class of water insoluble compounds including most notably the fats and sterols.

Lith″o-sphere. The solid, rocky component of the earth; the earth's crust.

Lit″to-ral. Ocean floor from the shore to the edge of the continental shelf; also, the region of shoreline between high-land low-tide levels.

Lo′co-mo″tion. Movement involving the organism as a whole.

Loop of Hen″le. Nephron tubule in medulla of kidney; most water resorption occurs here.

Lo″pho-phore. Anterior tentacle-bearing crown of certain coelomates that functions in food capture.

Lo″ri-ca. Protective covering secreted by certain organisms, such as some ciliate protozoans.

Lu-cif″er-ase. Enzyme that contributes to bioluminescent light production in certain animals.

Lum″bar. Pertaining to the "small of the back," just posterior to the ribs.

Lu″men. Internal cavity within a body structure, such as the lumen of the intestine, a gland, or a blood vessel.

Lu'mi-nes"cence. Production of light as a result of chemical reactions in cells; bioluminescence.

Lung. An internal organ for respiration.

Lu"ten-i-zing hor"mone (LH). Hormone secreted by the anterior pituitary that stimulates formation of corpus luteum in ♀; interstitial cells in ♂, allowing the ovum to escape.

Lym-pha"tic sys"tem. System of vessels and nodes in vertebrates that lead from tissue spaces to large veins entering heart; a part of the circulatory system that recovers interstitial fluid and acts in body defense.

Lymph node. Organs consisting lymphatic tissue, that are filtration sites of body fluids and poroduce antibodies.

Lymph"o-cyte. Vertrbrate white blood cell with round or kidney-shaped nucleus; produces antibodies.

Lyse. To break up, usually chemically.

Ly"so-some. Cytoplasmic organelle that contains digestive enzymes.

Mac"ro-mere. The relatively large, yolk-laden cells that are produced along with micromeres during the cleavage of certain animals.

Mac'ro-nu"cleus. Large nucleus found in ciliates that controls all activities except reproduction.

Ma-dre"po-rite. Strainerlike cover to the opening of the water vascular system in echinoderms.

Mag"got. Wormlike legless larva of a fly (Diptera).

Ma-la"ri-a. Any of several fevers, each produced by a specific sporozoan parasite that invades the red blood corpuscles of various mammals and birds.

Mal"leus. First ear bone, attached to tympanum.

Mal'pigh"ian tubule. Excretory glands that open into anterior hindgut of terrestrial arthropods.

Mal"tose. Twelve-carbon disaccharide formed by two glucose molecules.

Man"di-ble. The lower jaw in vertebrates; also, lateral jawlike mouthpart of an arthropod.

Man"tle. Fold of the body wall that encloses the soft structures of an animal such as a mollusk and secretes the shell.

Ma-nu"bri-um. A structure projecting from the middle of the subumbrellar surface of the medusa and bearing the mouth at its free end.

Ma-rine". Of or pertaining to the ocean.

Mar-su"pi-al. Mammal that possesses a pouch.

Mass py'ra-mid. Model describing communities by plotting total biomass (standing crop) in each trophic level at any one time.

Ma"tax. Toothed chewing apparatus found in rotifer pharynx.

Mas'ti-ca"tion. Chewing food with the teeth or other mouthparts.

Ma-ter"nal. Pertaining to a mother.

Ma"trix. In animal histology, an intercellular substance; the noncellular substance of connective tissue as in bone and cartilage.

Max-il"la. One of several postoral mouthparts of arthropods; in the vertebrates, the large bones of the upper jaw.

Max-il"li-ped. One of the first three pairs of thoracic appendages in crustaceans.

Me-a"tus. Small passage or canal, as in the external auditory meatus, the canal leading to the eardrum.

Me"di-an. Refers to the midline or near the middle of the body.

Me-dul"la. Inner region of a gland or other structure.

Me-dul"la ob-lon-ga"ta. Most posterior division of the vertebrate brain, communicates with spinal cord.

Med"ul-lary plate, groove, and tube. Synonymous with neural plate, groove, or tube; three successive stages in the embryonic development of the vertebrate central nervous system.

Me-dus"a. Free-swimming cnidarian with bell-shaped body ringed by tentacles; a jellyfish able to produce either eggs or spermatozoa.

Mei-o"sis. Division of the diploid germ cell to produce haploid gametes; characterized by two divisions in which the number of chromosomes is reduced by half.

Mei-o"tic drive. Changes in allelic frequencies that result from genes that inhibit their alleles during meiosis.

Meis"sner's cor"pus-cle. A dermal touch receptor.

Mel"an-in. Dark-brown pigment found in the cytoplasmic granules of chromatophore cells and in hair.

Membrane. Structure surrounding an organelle, cell, or organ; most are semipermeable.

—, fluid mosaic model of. The lipid molecules within a bylayer are not motionless but may move with respect to each other (i.e., they are fluid); membrane proteins are suspending within the bylayer in a pattern (mosaic) that is also flexible (fluid).

—, **mucous.** Membrane made of a layer of simple or stratified epithelium; anchored to basal connective tissue.

—, **serous.** Derived from mesoderm and supported by thin connective tissue; lines coelomic cavities of vertebrates.

—, **unit model of.** A cell membrane is thought of as a "sandwich," with a lipid layer enclosed by two protein layers.

Mem″bran-ous la″by-rinth. That portion of the inner ear that constitutes the sacculus and utriculus and houses the equilibrium receptors.

Me-nin″ges. The three membranes (dura mater, arachnoid, and pia mater) that covers the brain and spinal cord.

Men″o-pause. Time when menstrual cycle ceases in human females and reproduction is no longer possible.

Men′stru-a″tion. Discharge of blood and uterine tissue in humans and apes at the end of a menstrual cycle in which fertilization did not take place.

Mer′o-blas″tic. Egg cleavage in which yolk is left undivided as only part of the cytoplasm divides.

Mer-o″zo-ite. Mature trophozoite after mitotic division.

Mes′en-ceph″a-lon. Midbrain of vertebrates.

Mes″en-chyme. Loose embryonic connective tissue derived chiefly from mesoderm.

Mes′en-ter-y. A thin double-walled sheet of peritoneum that supports the organs in the abdominal cavity of vertebrates; also one of the partitions in the gastrovascular cavity of anthozoan cnidarians.

Mes″o-blast. Blastula cell(s) from which mesoderm develops.

Mes″o-derm. The middle layer of embryonic cells, between the ectoderm and endoderm.

Mes′o-gle″a. Jellylike substance lying between the epidermis and gastrodermis in coelenterates; also other jellylike, generally noncellular layers, as in sponges.

Mes′o-neph″ric duct. Duct leading from the mesonephric type of kidney to the cloaca; also known as the archinephric or wolfian duct.

Mes′o-neph″ros. Type of vertebrate kidney; the functional kidney in adult cyclostomes, fishes, and amphibians; present in embryos of reptiles, birds, and mammals.

Me′so-tho″rax. Middle segment of the insect thorax.

Me-ta″bo-lism. The sum total of the chemical reactions that occur within the cells of an organism.

Me-ta″bo-lite. Any chemical participating in the metabolic process.

Me″ta-car″pals. Proximal bones of the hand; forming the palm.

Me″ta-chron″ous. Pertaining to a wavelike, successive beating of a ciliated surface.

Me′ta-gen″e-sis. Alternation of a sexual with an asexual stage in the life cycle of a cnidarian such as *Obelia.*

Me″ta-mere. One of a series of body segments; a somite.

Me-tam″er-ism. Condition in which the body of an animal is made up of a succession of homologous parts (metameres).

Met′a-mor″pho-sis. Marked structural change or transformation during development, such as from larva to adult.

Me′ta-ne-phrid″i-um. Tubular organ of excretion, such as in the earthworm, leading from the ciliated funnel found in the coelom to the exterior.

Me″ta-phase. Stage of mitosis during which the chromosomes are lined up in the equatorial plane of the spindle.

Me″ta-pleu″ral folds. Two folds on the ventral surface of *Amphioxus,* extending from the mouth to the atriopore.

Me′ta-tar″sals. Proximal bones of the foot, between ankle and toes.

Me′ta-tho″rax. Posterior segment of insect thorax.

Me′ta-zo″a. All multicellular animals in which there is a differentiation of the somatic (body cells) and reproductive cells; as opposed to unicellular animal-like protists.

Me′ten-ceph″a-lon. Anterior part of hindbrain in vertebrates; includes the **pons** and the **medulla.**

Mi-celle″. Globule of lipid that forms in water because of the insolubility of lipid in the water.

Mi″cron. One-thousandth part of a millimeter.

Mi″cro-nu″cle-us. In ciliates, the smaller nucleus used in reproductive functions.

Mi″cro-tu″bule. Literally, very small tube; intracellular fibers.

Mid″gut. In arthropods, the middle portion of the digestive tract, not lined with cuticle; also in vertebrate embryo, rudiment of the intestine.

Mi-gra"tion. Movement of a population (usually in groups) from one region to another.

Mi"li-li-ter. One-thousandth part of a liter; equivalent to a cubic centimeter; abbreviated ml.

Mi"mi-cry. Adaptive resemblance of certain animals to others or surroundings resulting in protection for the mimic from predators.

Mi"ne-ral. A solid inorganic substance or compound, usually composed of one type of element or compound.

Mir'a-cid"i-um. Larval stage of a fluke; develops from an egg and gives rise to sporocyst larva.

Mis"sense mu-ta"tion. Gene mutation in which one base pair is altered so that an amino acid different from that originally coded for is placed in the protein being formed.

Mit'o-chon"dri-a. Small spherical or rodlike cytoplasmic organelles associated with important metabolic reactions in a cell.

Mi-to"sis. A type of cell division during which the chromosomes replicate and each daughter cell receives the same genotype as the parent cell.

Mi"tral valve. See **biscuspid valve.**

Mo-da"li-ty. Qualitative aspect of physical sensations, such as the modalities of taste and smell.

Mo"lar. Permanent mammalian posterior teeth.

Mol"e-cule. Compound in which chemical bonds hold two or more atoms together.

Molt. To cast off the exoskeleton; to shed portions of the skin, feathers, or hair.

Mon"o-cyte. Type of leukocyte; large, phagocytic, uninucleate.

Mo-noc"cious. See **hermaphrodite.**

Mon'o-hy"brid. A heterozygote with respect to one particular pair of allelic genes, such as Aa.

Mo"no-hy"brid cross. In genetics, a cross in which only one trait is varied.

Mon'o-phy-let"ic. Evolved from a single ancestral type.

Mon'-sac"cha-ride. A simple sugar.

Mor'phal-lac"tic re-gen-er-a"tion. Regeneration in which a large portion of the organism is replaced (e.g., in *Hydra*, where several cells may regenerate an entire individual).

Mor'pho-gen"e-sis. The process of developing form and size.

Mor-phol"o-gy. Study of the form and structure of organisms.

Mor"u-la. Solid ball of cells resulting from cleavage of the zygote.

Mo-til"i-ty. Exhibiting or having the capabilities of movement.

Mo"tive. Internal stimulus; may be psychological, physiological, or both.

Mo"tor neu"ron. A nerve cell that synapses with muscle fibers; nervous stimulation of a motor neuron elicits muscle contraction.

Mu"cin. A glycoprotein forming mucus in solution.

Mu-co"sa. A membrane that secretes mucus.

Mu"cous. Pertaining to mucus.

Mu"cous mem"brane. Consists of an epithelium with some subepithelial connective tissue that lines cavities that communicate with the exterior; secretes lubricating mucus.

Mu"cron. In sporozoans, an anterior structure in some individuals that is a modification of a portion of the ectoplasm.

Mu"cus. Viscous secretion containing mucin; a product of mucous glands.

Mul"ti-ple fis"sion. Type of asexual reproduction in which the nucleus performs several mitotic divisions before cytokinesis occurs.

Mus"cu-la'ris. Thin layer of smooth muscle between the tunica propria and the submucosa in the wall of the esophagus, stomach, and intestine.

Mu-ta"tion. A change in a gene or chromosome that alters its expressed protein; broadly used to include all kinds of hereditary variations resulting from gross chromosome changes; mutations may also occur in somatic cells, but these are not hereditable unless the individual reproduces asexually.

Mu-ta"tion pres"sure. The change in allele frequencies that results from mutations; these are usually small changes, but they provide the variations upon which selection and other factors operate.

Mu"tu-al-ism. An association between individuals of two species that is beneficial to both.

My"e-lin. The insulating fatty substance surrounding the axon of some neurons.

My"o-car"di-um. The main bulk of the heart, composed of cardiac muscle.

My"o-cyte. Contractile cell found in poriferans; surrounding pores and oscula, they close these openings when stimulated.

My'o-fi"bril. One of the numerous longitudinal fibrils contained within the protoplasm of the muscle cell or fiber; composed of actin or myosin.

My″o-mere. Muscle somite or segment.

My″o-neme. Type of contractile fibril found in certain protists.

My-o″pi-a. Nearsightedness; a condition in which the eyeball is relatively short, so that the point of focus lies behind the retina.

My″o-sin. A protein of muscle that constitutes the thick filaments and contributes to contraction.

Myx′e-de″ma. Thyroid deficiency disease characterized by swelling within and beneath the skin.

Na″cre. Mother-of-pearl; the iridescent layer of a mollusk shell nearest the mantle and secreted by it.

NAD. Nicotinamide adenine dinucleotide, a hydrogen carrier in cellular respiration and other biochemical reactions.

NADP. Nicotinamide adenine dinucleotide phosphate, a hydrogen carrier used primarily in biosynthetic reactions.

Nai″ad. An aquatic nymph.

Na″res. Openings of the air passages, both external and internal, in the head of a vertebrate.

Na″sal. Pertaining to the nose or nostrils.

Na″u-ral par′the-no-gen″e-sis. Development of an egg without fertilization during the normal life cycle of an animal, as in aphids.

Nat″u-ral se-lec″tion. Process whereby the environment acting on the phenotype indirectly determines which genotypes will be represented in the next generation's gene pool; survival of the fittest.

Nau″pli-us. First larval stage of many crustaceans.

Nek″ton. A collective term for all animals that actively swim in the ocean waters.

Ne-ma″to-cyst. One of the stinging capsules found in cnidarians; each is produced by a single cell, a cnidoblast.

Ne-o″te-ny. Sexually mature state in an organism that retains many larval characteristics.

Ne-phri-″di-o-pore. External opening of an excretory tubule or nephridium.

Ne-phri″di-um. Tubular osmoregulatory and excretory structure characteristic of many invertebrates.

Neph-ro-mix″i-um. Reproductive–excretory structure of certain polychaetes.

Ne″phron. The functional unit of the vertebrate kidney.

Neph″ro-stome. The funnel-shaped ciliated opening at the coelomic end of a nephridium through which fluid enters the tubule.

Ne-ri″tic. Pertaining to the open waters above the continental shelf.

Nerve. Bundle of nerve fibers located outside the central nervous system.

Nerve cord. Solid strand of neurons forming part of the central nervous system; usually with ganglia.

Nerve net. Epidermal motor and sensory nerve plexus (network) in cnidarians and some platyhelminths.

Neu″ral. Pertaining to the nervous system.

Neu″ral can-al″. Canal through the vertebrae; the canal formed by the neural arches.

Neu″ral plate, groove, and **tube.** Three successive stages in the development of the central nervous system in vertebrate embryos.

Neu″ral spine. The dorsal projection of a vertebra; also known as the spinous process.

Neu″ri-lem″ma (neurolemma). Outermost sheath of a nerve fiber; important in nerve regeneration.

Neu″ro-coele. Cavity in a chordate nerve cord.

Neu″ron. A nerve cell, including the cell body (soma), dendrites, and axon.

Neu″tro-phil. Type of leukocyte; phagocytic, with a multilobed nuclei.

Neu″ro-mast. Aggregation of sensory cells, sensory hairs, and cupulae in the lateral line organ of fishes and some amphibians.

Neu″tron. Uncharged subatomic particle in the nucleus of an atom with atomic weight $\cong 1$.

Nex″al jun″ction. The point between some cells in which the intercellular gap is extremely narrow.

Ni″a-cin. A vitamin of the B complex.

Niche (ecological). The constellation of environmental factors to which a species is adapted or that is required by a species.

Nic″ti-ta′ting mem″brane. Transparent, thin membrane possessed by many vertebrates that opens and closes laterally across the cornea; third eyelid.

Ni-da-men″tal. Pertaining to gland in female cephalopods that secretes a protective egg capsule.

Nit″ri-fy. To convert ammonia and nitrites to nitrates, as is done by nitrifying bacteria.

No′ble gas. See **inert gas.**

Noc-tur″nal. Referring to activity at night.

Node of Ran″vi-er. Unmyelinated gaps along a myelinated axon.

Non′as-sor″ta-tive ma″ting. Mating where mate selection is random.

Non′dis-junc″tion. During meiosis, the failure of equal partitioning of all chromosomes, resulting in an extra chromosome in one gamete and a missing chromosome in another; the cause of certain genetic diseases if the mutant gamete is used.

Non″sense mu-ta″tion. A type of missense mutation in which a terminating codon results; this results in premature termination of protein synthesis.

No″to-chord. Characteristic cylindrical supporting rod in chordates, dorsal to digestive tract and ventral to nerve cord; either surrounded or supplanted by the vertebrae in most vertebrates.

Nu-cle″ic ac″ids. Polymeric molecules made up of nucleotide sequences; DNA and RNA; they contain genetic information by means of the genetic code.

Nu-cle″ol-us. A spherical well-defined body found within the nucleus of many cells; site of rRNA synthesis.

Nu′cle-o-pro″te-in. A complex made up of nucleic acid and protein, (e.g., a chromosome).

Nu″cle-o-tide. A molecule that consists of a phosphate group, a five-carbon sugar, and a purine or a pyrimidine nitrogen base.

Nu″cle-us. A membrane-bound body within the eucaryotic cell that contains the chromosomes.

Nurse cell. In female poriferans, amoebocyte or choanocyte that transfers a sperm cell to site of ovum and then releases it; also, a cell that nourishes a developing ovum.

Nu″tri-ent. Any inorganic or organic substance used in metabolism.

Nu-tri″tion. Sum of the processes concerned in the growth, maintance, and repair of the living body as a whole or of its constituent parts.

Nyc-ta-lo″pi-a. Night blindness.

Ob-jec″tive. The lens or lenses of a microscope nearest the object under observation.

Ob′ser-va″tion. A perception, either quantitative or qualitative, of a set of phenomena (the set may have only one member).

Oc-cip″i-tal. Pertaining to the base of the vertebrate skull.

O′ce-a″nic is″land. Island of volcanic origin; its faunal and floral complement depends on the mobility of the colonizing animals.

O′ce-a″nic re″gion. Deep waters beyond the neritic zone; past the continent shelf.

O′cel″lus. A simple type of eye in many invertebrates, especially insects.

Oc″u-lar. Eyepiece of the microscope; pertaining to the eye.

OH⁻. Hydroxide ion.

Ol-fac″to-ry. Pertaining to the sense of smell.

O′li-go-le″ci-thal. Egg with yolk in low concentration and rather evenly dispersed.

Om-ma-tid″i-um. One of the elongated rodlike units of an arthropod compound eye.

Om-ni″vore. An animal that eats both plants and animals.

On-to″ge-ny. Entire developmental history of an individual organism.

O″o-cyst. Swelling in mosquito stomach wall caused by implantation of an ookinete; produces sporozoites.

O′o-cyte. Egg mother cell from which are produced, by the first meiotic division, the secondary oocyte and the first polar body.

O′o-ge″ne-sis. Process of formation of haploid ova.

O″o-go″ni-um. In animals, the primordial egg cell from which primary oocytes are produced by mitosis.

O″o-ki′nete. Movable sporozoan zygote found within a mosquito or suitable host.

O″pen sys″tem. A system that is not self-sufficient; most biological or zoological systems are open as they require energy from the environment.

O′per-ant con-di″tion-ing. A learning paradigm under which a response is reinforced, leading to its more frequent occurrence.

O-per″cu-lum. Structure covering the gills of fishes and tadpoles; also the plate serving to cover the opening of some snail shells.

Oph-thal″mic. Pertaining to the eye.

Op″sin. Part of the visual pigment rhodopsin; the protein portion that combines with retinal, the light-absorbing component.

Op″tic. Pertaining to the eye or to sight.

Op″tic chi″asm. (Gr. *chiasma*, two crossed lines). Part of the optic tract in which the neurons from the medial retinas intersect on their way to the brain.

Op"tic lobes. Thickenings on the dorsal surface of the midbrain (mesencephalon).

Op"tic nerves. Nerves that travel from the eye to the brain.

O"ral. Pertaining to the mouth.

O"ral groove. Groove having ciliated ridges that starts at the anterior end and runs posteriorly to the cell mouth (cytostome) in some protists, such as *Paramecium.*

Or"bit. Socket in the vertebrate skull surrounding the eye.

Or"gan. A group of tissues associated in the body to perform one or more complex functions.

Or"gan of Cor"ti. Auditory organ within the cochlea.

Or-gan-elle". A membrane-bound structure within protoplasm of a cell, specialized to perform a certain function.

Or-gan"ic com"pound. A molecule containing carbon atoms, usually bonded to hydrogen, oxygen, nitrogen, and other atoms.

Or"gan-ism. Any living individual, plant or animal.

Or'ga-nog"e-ny. Process of the formation of specialized tissues and organ systems during embryonic development.

Or"i-gin. End of a muscle that remains relatively fixed during contraction; also, the ancestry or derivation of an organism.

Or'tho-gen"e-sis. The observed tendency to evolve consistently in the same direction; unsubstantiated hypothesis that evolution proceeds toward a definite predetermined goal.

Os"cu-lum. Relatively large external opening of the central cavity (spongocoel) through which water leaves a sponge.

Os-mo"sis. The diffusion of water through a semipermeable cell membrane.

Os-mo"tic pres"sure. The aquatic force exerted because of differential solute concentrations in water solvent.

Os-phra"di-um. A chemoreceptor area in incurrent siphon of some mollusks; epithelial, frequently yellow.

Os"se-ous. Pertaining to bone.

Os"si-cle. A small bone or bonelike supporting structure.

Os"te-o-blast. Mesenchyme cell that has differentiated into a protein matrix-secreting cell.

Os'te-ol"o-gy. Study of bones.

Os'te-o-scler-o"sis. Pathological condition in which two or all three auditory ossicles cease to be movable because joints have become sclerotized.

Os"ti-um. A small mouthlike opening such as the anterior end of the oviduct in vertebrates; in general, an opening in both invertebrates and vertebrates, usually guarded by a valve or circular muscle.

O"to-lith. A limy particle in the inner ear of vertebrates or in the equilibrium organ of some invertebrates; indicates orientation by displacing sensory hairs.

O"va-ry. Female gonad in which the ova develop.

O"vi-duct. A tube that conveys ova from the ovary to the uterus or to the exterior.

O-vi"pa-rous. Referring to those organisms whose eggs hatch outside the body of the mother; egg-laying animals.

O'vi-po"si-tor. Organ of female insects that aids in depositing eggs.

O'vo-vi-vi"pa-rous. Referring to those organisms that have their eggs hatch within the parent's body but are *not* nourished by the mother's bloodstream through a placenta.

O-vu-la"tion. Release of mature ova from the ovary.

O"vum (L., egg). A nonmotile female gamete; an egg.

Ox'i-da"tion. Chemical change in which a molecule loses one or more electrons; sometimes involves combining with oxygen; usually an exothermic reaction.

O"xy-he'mo-glo-bin. The molecule formed when hemoglobin combines with oxygen in the lungs.

P. Referring to the parental generation.

Pa-chy-ne"ma. In meiosis, the stage where the homologous chromosomes twist around each other, followed by their splitting into chromatids, forming tetrad.

Pa-cin"i-an cor"pus-cle. A type of pressure receptor in the dermis.

Pal"a-tine. Bone that serves as an anterior brace of the upper jaw in the skull of the frog and other vertebrates.

Pa'le-on-to"lo-gy. The science that deals with ancient life of the earth as revealed by fossils, impressions, and other remains found in sedementary rocks.

Pa'le o zo"ic. Geological era occurring between the Pre-Cambrian and the Mesozoic and dating from approximately 500 to 200 million years ago.

Palp. Projecting process in some invertebrates, sensory in function, often near the mouth.

Pal"pus. Palp that is a process of an appendage, as in insects.

Pam"pas. South American grassland or prairie.

Pan"cre-as. An exocrine gland that discharges digestive enzymes into the intestine; also contains endocrine cells that secrete insulin and glucagon.

Pan-ge"a. The single land mass during the Permian that included most of the area of the present continents. See **Laurasia** and **Gondwanaland.**

Pan'to-then"ic acid. B-complex vitamin occurring in coenzyme A.

Pa-pil"la. Any small nipple-shaped elevation.

Pa-ra-dox"i-cal sleep. Sleep state in which rapid eye movements occur; the EEG tracing looks as if the individual were awake.

Pa-ra-my"lum. Starchlike substance; reserve food material produced by photosynthesis in phytoflagellates such as *Euglena.*

Par'a-po"di-um. Flattened, moveable, paired appendages on the body segments of many polychaete annelids that are used for locomotion and respiration.

Par"a-site. Organism that lives during the whole or a phase of its life upon or within another organism (host) from which it derives nourishment.

Par'a-sphe"noid. Bone forming the floor of the vertebrate cranium.

Par'a-sym"pa-thet"ic. Subdivision of the vertebrate autonomic nervous system that is chiefly concerned with maintaining internal homeostasis.

Pa'ra-thor"mone. A secretion of the parathyroid glands; stimulates osteoclasts to release calcium and phosphorus into blood.

Par'a-thy"roid. One of several (usually four) small endocrine glands, closely associated with the thyroid gland of vertebrates.

Pa-ren-chy"ma. Type of loose, spongy, connective tissue found in some invertebrates; in the vertebrates, the specific tissue component of an organ, such as liver hepatic cells.

Pa-ri"e-tal. Pertaining to the inner mesodermal lining of the coelom; also, a bone (one of a pair) of the vertebrate cranium just posterior to the fronta.

Pa ri"e tal bone. A pair of bones located just posterior to the frontal bone.

Par'the-no-gen"e-sis. The production of offspring from unfertilized eggs; see **natural parthenogenesis.**

Par'the-no-go-nid"i-um. Cell that produces a miniature colony through asexual methods of multiplication in such forms as *Volvox.*

Par"tial dom"i-nance. See **incomplete dominance.**

Pa-tel"la. Kneecap.

Pa-ter"nal. Pertaining to the father.

Path'o-gen"ic. Disease producing.

Pathology. Study of abnormal (diseased) structures and abnormal functioning of life processes.

Pec"ten. In insects, a comblike arrangement of setae; also, comblike structure in eyes of most birds and reptiles.

Pec"tin. A carbohydrate that frequently coats plant cells.

Pec'to-ral. Pertaining to the upper thoracic area.

Pec"to-ral gir"dle. Group of bones and cartilage connecting the forelimbs to the axial skeleton in vertebrates.

Ped"al. Pertaining to the feet or aboral base of an organism.

Ped'i-cel-la"ri-a. Small pincer or scissorlike processes on the surface of certain echinoderms, such as sea fishes and sea urchins.

Ped"i-cle. Narrow waist between cephalothorax and abdomen in spiders and wasps.

Ped"i-gree. See **kindred.**

Ped"i-palp. Either of the second pair of appendages in arachnids; often sensory and sometimes used in seizing prey, as in scorpions.

Pe'do-gen"e-sis. Reproduction by larvae.

Pe"dun-cle. A stalk or stem.

Pe-lag"ic. Inhabiting the open water away from shore, as in the ocean.

Pel"li-cle. Protective layer on the surface of some protozoans, for example, *Paramecium.* Comparable to plasma membrane.

Pel"vic. Group of bones connecting the bones of the hindlimbs to the axial skeleton in the vertebrates.

Pen. Feather-shaped (usually) internal skeleton of a squid.

Pen'e-trance. In genetics, the extent to which a particular gene is expressed phenotypically.

Pe″nis. Male copulatory organ for conveying sperm to the female genital tract.

Pen″e-trant. Largest type of cnidarian nematocyst, containing a coiled tube and spines; used in prey capture.

Pen′ta-dac″tyl. Having five fingers, toes, or digits.

Pep″sin. An enzyme concerned with protein digestion in animals.

Pep″tide. Bond formed when two amino acids join end to end resulting in a dipeptide; a polypeptide results from joining of many amino acid units and is the primary component of a protein molecule.

Per-ei″o-pods. Walking legs of a crustacean.

Per″i-blast. In fishes, the layer of cells under the blastocoel.

Per′i-car″di-um. Membranous sac surrounding the heart; the part in contact with the heart is the visceral pericardium, the other is the parietal pericardium.

Per′i-os″te-um. Connective tissue sheath that covers the surface of a bone.

Per′i-o″stra-cum. Outer proteinaceous horny layer of mollusk shells; protective in function. Reduces acid erosion.

Pe-riph″er-al ner″vous sys″tem. Nervous tissue located outside the central nervous system (brain and spinal cord).

Per″i-sarc. Outer transparent membrane that encloses the living coenosarc of a hydroid.

Per′i-stal″sis. Type of smooth muscle contraction in which a wave of contraction follows a wave of relaxation, thus pushing material through a hollow organ, especially the digestive tract.

Per′i-to-ne″um. Thin mesodermal membrane that lines the coelom.

pH. Notation for indicating the relative concentration of hydrogen ions in a solution; values range from 0 to 14; the lower the value, the greater the number of hydrogen ions; it is equal to the negative logarithm of the hydrogen ion concentration.

Pha″go-cyte. Type of white blood cell that engulfs and digests bacteria and other foreign materials.

Pha′go-cy-to″sis. Engulfment of particles by the cell membrane.

Pha″go-some. Result of phagocytosis; a food vacuole.

Pha-lan″ges. Bones of the digits; singular, phalanx.

Pha-ryn″ge-al cleft. Slit in the wall of the pharynx; same as gill slit, gill cleft, or branchial slit.

Phar″ynx. Anterior portion of the digestive tract between the mouth cavity and the esophagus, often muscular; also the gill region of many aquatic vertebrates; sometimes with teeth in invertebrates.

Phe″no-type. Expressed traits of an individual, as contrasted to genotype; the sum total of the structural and physiological characteristics of an individual.

Phe″no-ty-pic ra″tio. The ratio of phenotypes resulting from a particular cross.

Pher″o-mone. In insects, an emitted chemical that acts as a chemical message to others of its species for communication and location of mates.

Phos″pha-gens. Compounds that store high-energy phosphates for use by the muscular system.

Phos′pho-li″pid. A form of lipid with a phosphate group replacing one fatty acid of a normal lipid.

Pho′to-per″i-od. The length of daylight; varies with the season and latitude.

Pho″to-re-cept′or. Light-sensitive cell or organ.

Pho′to-syn″the-sis. Formation of glucose from carbon dioxide and water by chlorophyll in the presence of light.

Pho-to-tro″pism. Behavioral response of an animal to light stimuli.

Phre″nic. Pertaining to the diaphragm.

Phy-lo″ge-ny. Ancestral or evolutionary history of an organism or group of organisms.

Phy″lum. Any of the main taxonomic divisions into which the animal kingdom is divided; each characterized by a distinct body plan.

Phy-si-o″lo-gy. Science dealing with living processes in organisms.

Pi″a ma′ter. Thin innermost membrane covering the brain and spinal cord.

Pig″ment. Colored substances of some cells and tissues

Pin″e-al. A structure located on the roof of the brain of vertebrate animals; phylogenetically associated with a median eye.

Pin″na. External ear.

Pin″ule. A featherlike structure, as in crinoid arm pinnules.

Pi′no-cy-to″sis. Cellular intake of fluid by small imaginations in the cell membrane that pinch off within the cytoplasm.

Pi-tu″i-tary bo″dy. An endocrine gland located on the ventral surface of the brain; composed of anterior, posterior, and sometimes intermediate lobes; also known as the hypophysis.

Pla-cen"ta. Organ that attaches the fetus of mammals to the uterine wall; serves in fetal nourishment, respiration, and excretion; made of maternal and fetal tissues. Produces chorionic gonadotropin which perpetuates corpora lutea during pregnancy.

Pla"coid scale. Toothlike scale found in cartilaginous fishes, with basal dentine and rearward-pointing spine.

Plains. Western North American grasslands; short grass, as opposed to tall grasses of the prairie.

Plank"ton. Floating or drifting aquatic organisms, mostly microscopic.

Plan"ti-grade. Walking with the whole sole of the foot bearing on the ground, as in humans.

Plan"u-la. Ciliated free-swimming larval stage of most coelenterates.

Plas"ma. Liquid part of the blood.

Plas"ma mem"brane. External membrane covering the cytoplasm of a cell; see **membrane.**

Plas"ma-gel'. Relatively rigid cytoplasm.

Plas-mo"ly-sis. Decrease in intracellular pressure; occurs if the cell is placed in hypertonic solution; results in the collapse of the cell.

Plas"ma-sol'. Relatively liquid cytoplasm.

Plas"tid. A cytoplasmic body found in plant cells; contains pigments or food reserves.

Plas"tron. Ventral plate (e.g., a turtle shell).

Plate"let. In blood, the thrombocytes that, when ruptured, begin the clotting sequence.

Plei-o-tro"phic. When a gene affects more than one phenotypic trait in phenotype.

Pleur"a. Membrane that covers lungs and lines inner wall of thorax.

Pleur"al. Pertaining to the cavity, that portion of the coelom that contains the lungs; the membrane covering the lung and lining the pleural cavity.

Pleur"on. Lateral portion of a typical segment of some arthropods such as the crayfish.

Plex"us. Network, chiefly of nerves or blood vessels.

Plu"te-us. The ciliated, planktonic echinoid and ophiuroid larva.

Pneu-ma"to-phore. Air-filled float of siphonophoran hydroid colonies.

PNS. See **peripheral nervous system.**

Poi"ki-lo-ther'm. A "cold-blooded" animal that lacks internal temperature controls; also, an **ectotherm.**

Po"lar bo"dy. Small nonfunctional cell formed during meiosis of the oocyte.

Po-la"ri-ty. *Chem.:* Separation of positive and negative charge in a molecule. *Zool.:* Presence of distinctly different ends diametrically opposed to each other.

Po"ly-em"bry-o'ny. One egg producing many embryos.

Po"ly-mer. A long-chained molecule formed by the covalent linkage of smaller subunits, called monomers.

Po"ly-mor"phism. Occurrence of more than one form in a single species; when only two such forms occur, dimorphism is the term usually applied; polymorphism implies several different forms.

Po"lyp. The form of a cnidarian having the shape of an elongated cylinder fastened at the aboral end, with mouth and tentacles at the free oral end.

Po"ly-pep"tide. A polymer of amino acid monomers; the basic polymer of **proteins.**

Po"ly-phy-le"tic. Derived from more than one ancestor; with distinctly different evolutionary histories.

Po"ly-sac"cha-ride. A carbohydrate composed of numerous monosaccharide subunits, such as glycogen, starch, or cellulose.

Pons. Part of brainstem; a routing station for nerve impulses passing between medulla and higher brain centers.

Po-pu-la"tion. Group of individuals of the same species living within the same geographical area.

Pore. A hole.

Po"ro-cyte. Water intake cell of certain sponges, characterized by canal passing through it.

Por'phy-rop"sin. In freshwater fishes, a retinal photopigment based on vitamin A_2.

Por"tal vein. A vein that branches into capillaries before reaching the heart, as in the hepatic portal or renal portal vein.

Pos-te"ri-or (L., latter). The tail; toward the hind or rear end; opposite of anterior.

P sec"tion. One of the basic units of an electrocardiogram, representing atrial depolarization.

Prai"rie. North American grasslands with long grass; as opposed to plains, with short grass.

Pre-co"cial. Referring to those organisms that attain early maturity, particularly certain birds and mammals.

Pre-da"cious. Capturing living animals for food.

Pre″da-tor. Animal that preys upon other animals for its food.

Pre-hen″sile. Referring to an appendage adapted for holding or grasping.

Pre-max-il″lae. A pair of anterior bones of the upper vertebrate jaw.

Pri′ma-ry or″gan-i-zer. See **gray crescent.**

Pri″mi-type. Early stage or type.

Pri″mi-tive streak. Structure developing on the dorsal edge of the blastodisc due to the migration of mesodermal cells along this area.

Pri-mor″di-al. First in order of time; the primitive form.

Pri-mor″di-um. Earliest stage in the formation of a body organ or part.

Prin″ci-ple. A scientific theory, fact, or law of wide application.

Pro′blem. The realization of ignorance about a given subject.

Pro-bos″cis. Tubular extension of the nose, lips, or pharynx. The extended beaklike mouthparts of some insects.

Pro-car′y-ote. A major division of cell types; a cell that contains no membrane-bound organelles (especially no nucleus); bacteria and blue-green algae constitute the procaryotes. See **encaryote.**

Pro″cess. A protrusion or projecting part of an organism or structure.

Proc-to″de-um. Terminal part of digestive tract, near anus, lined with ectoderm.

Pro-du″cers. The autotrophs (photosynthetic and chemosynthetic organisms) in a community.

Pro-ges″ter-one. Vertebrate hormone secreted by the corpus luteum. See **Placenta.**

Pro-glot″tid. Tapeworm reproductive segment.

Pro-neph″ros. The first kidney embryologically formed in a vertebrate.

Pro-no″tum. Dorsal surface, including sclerites, of an insect prothorax.

Pro-nu″cle-us. In a fertilized egg, one of two haploid nuclear bodies, the male pronucleus and female pronucleus, the fusion of which results in the diploid nucleus of the zygote.

Pro″phase. The first stages of mitosis or meiosis during which the chromosomes become distinctly visible.

Pro′pri-o-cep″tor. A sense receptor responsive to internal stimuli.

Pros″o-pyle. One of the surface pores opening into the flagellated chamber of a poriferan.

Pros″tate gland. Gland surrounding the neck of the bladder and urethra in the male mammal, that secretes a large portion of the seminal fluid.

Pro-sto″mi-um. Anterior portion of the first segment of annelids, such as the earth worm, that overhangs the mouth.

Pro-tan″dry. Production of sperm and then ova by the same gonad during a later reproductive period.

Pro″te-in. An organic compound always containing nitrogen, carbon, oxygen, and hydrogen and often other elements; made up of a polymer amino acids (**polypeptide**) that contains the structural or enzymatic properties in all living systems.

Pro″te-in-ase. Enzyme that promotes the synthesis or hydrolysis of proteins.

Pro″tho-ra″cic gland. In insects, a gland that secretes ecdysone during molting.

Pro″thor″ax. Anterior segment of the insect thorax.

Pro″to-cer″cal fin. A type caudal fin in which the vertebrae or notochord continues to the tip of the fin.

Pro″ton. Positively charged subatomic particle in nucleus, with atomic weight ∼ 1.

Pro″to-ne-phri″di-um. Excretory organ of certain invertebrates, with blind inner end branched or with one terminal cell.

Pro″to-plasm. See **cytoplasm.**

Pro-to″po-dite. The basal portion, usually composed of two segments of a biramous appendage, of a crustacean.

Pro′ven-tri″cu-lus. Anterior portion of avian stomach; in insects, the gizzard; in earthworms, the crop preceding the gizzard.

Prox″i-mal. Nearer the point of attachment of the body structure; opposite of distal.

Pseu″do-coel. A body cavity not completely lined with a membrane derived from mesoderm.

Pseu′do-po″di-a. Blunt temporary protoplasmic projections that are pushed out from an amoeba or amoebalike cell in feeding and locomotor activities.

Psy′cho-gen″ic. Evolution caused by internal forces (unsubstantiated) as opposed to natural selection; also, a condition produced in the mind, as a disease originating from a mental state.

Pter″y-la. Area of bird skin on which contour feathers grow; feather tract.

Pty″a-lin. An enzyme in the saliva in some vertebrates that acts on starch, also called salivary amylase.

Pu″bis. Anteroventral area of pelvic girdle.

Pul″mo-nar-y. Pertaining to the lung.

Pul″vil-us. Fleshy pad or lobe on the terminal segment of some insect legs.

Pun-nett″ square. A matrix used in determining expected genotypes from a particular genetic cross.

Pu″pa. Developmental stage between the larva and imago in certain insects, usually encapsulated or in cocoon.

Pu″pil. Hole in the eye through which light passes; appears black because of pigment cells that line the eyeball.

Pur-kin″ge fi″bers. Myocardial conducting tissue that forms a network in the ventricular walls.

Py″go-style. Fused "tail" vertebrae of birds.

Py-lo″ric. Pertaining to the pylorus.

Py′lor″ic cae″ca. A portion of the starfish digestive tract that branches into each ray and is responsible for the secretion of digestive enzymes.

Py-lor″ic valve. Valve at the posterior end of the stomach.

Py-lo″rus. Opening to the intestine from the stomach.

Py-re″noid. In some chloroplasts, a center for the formation of a starchlike substance called paramylum.

QRS com″plex. One of the basic units of an electrocardiogram tracing representing the combined ventricular depolarization and atrial repolarization.

Quad-ra′to-ju″gal. A bone located in the posterior portion of the upper jaw in some vertebrates.

Qua″dru-ped. Four-footed animal.

Queen. Reproductive female of social insects, as bees and ants.

Ra″di-al ca-nal″. A canal radiating from ring canal of echinoderms, supplying podia; also certain gastrovascular canals in cnidarian medusae.

Ra″di-al clea″vage. Cleavage in which successive tiers of equal-sized blastomeres lie directly above and below one another.

Ra″di-al sym″metry. The condition in which similar parts are arranged about a common center, like the spokes of a wheel.

Ra′di-o-ge-net″ics. The study of the effects of radiation upon heredity.

Ra′di-o-ul′na. Fused radius and ulna bone, as in frogs.

Ra″di-us. The bone of the lower arm located on the thumb side of vertebrates.

Ra″du-la. Rasping organ in alimentary tract of many mollusks.

Ra″mus. A branch or branching part.

Re′ca-pit′u-la″tion the″ory (L. *re*, again; *caput*, head; Gr. *theoria*, a beholding). Theory that the individual in its development passes through the ancestral history of the species; that is, ontogeny repeats phylogeny.

Re-cep″tor. A sensory neuron specialized to respond to a particular stimulus.

Re-ces″sive trait. Trait that is expressed only when both members of an allelic pair of genes are alike, as in the homozygous condition; compare with **dominant trait;** the term recessive is also applied to alleles.

Rec″tum. Terminal portion of the vertebrate large intestine; posterior intestinal region in some higher invertebrates.

Red blood cell. Cell that functions as a carrier of hemoglobin; **erythrocyte.**

Re″di-a. One of several types of larvae in the life cycle of most trematodes.

Re″dox re-ac″tion. Combined oxidation and reduction reactions.

Re-duc″tion. Gain of an electron; usually an energy-consuming reaction.

Re-duc″tion di-vi″sion. During meiosis, each of the two cell divisions that halve the chromosome number.

Re″flex act. A relatively simple, automatic response to a stimulus that is independent of the higher nerve centers of the brain.

Re-gen-er-a″tion. Replacement by growth of a body part that has been lost.

Re-gen-er-a″tion blas-te″ma. Bulge in which the proliferating cells of the injured part that accumulate during regeneration.

REM sleep (rapid eye movement). Sleep state in which eyes move rapidly.

Re″nal. Pertaining to the kidney.

Ren-ette″. In some nematodes, an excretory cell with short tube attached.

Ren"nin. A stomach enzyme that acts on casein, a protein of milk.

Re-pli-ca"tion. In DNA synthesis, the formation of a new, identical DNA strand; the resulting two strands are each made up of half old and half new genetic material.

Re'pro-duc"tion. Production by an organism of others of its kind.

Re'pro-duc"tive mi-gra"tion. Migration to an area for breeding purposes; also called gametic migration.

Res'pi-ra"tion. Use of oxygen by a cell; this usually termed cellular or internal respiration; compare with **external respiration.**

Re-sponse". Reaction to a stimulus, either internal or external.

Re-tic"u-lar for-ma"tion. Sensory filter in the midbrain responsible for general activation.

Re-ti"cu-lum. Network of filaments, fibrils, or fibers within cytoplasm.

Re"ti-na. Light-sensitive layer of an eye.

Re-tin-al". Part of rhodopsin, a visual pigment; an aldehyde of vitamin A_1 that combines with the protein opsin.

Re-tin"u-la. Group of elongated pigment cells in basal portion of arthropod compound eye.

Rhab"dites. Rodlike bodies in the epidermis of certain flatworms.

Rhab"dom. A rodlike structure formed from the inner surfaces of adjacent sensory cells in the ommatidium of a compound arthropod eye; runs down center of retinula.

Rhab"do-mere. Receptive area of a retina cell, one of the component parts of the rhabdom.

Rh fac"tor. Type(s) of antigens found in some red blood corpuscles; first found in red blood cells of the rhesus monkey, hence the name Rh positive and Rh negative, denoting presence or absence of this factor.

Rho-dop"sin. Retinal pigment present in rods; sensitive to low levels of illumination.

Ri"bo-flavin. Constituent of the vitamin B complex, important in cell metabolism.

Ri"bo-some. RNA-containing multiprotein complex; the cytoplasmic organelle that is the site of protein synthesis.

Rick-ett"si-a. Parasitic microorganisms of insects and ticks.

RNA. Ribonucleic acid; may be transfer RNA (tRNA), messenger RNA (mRNA), or ribosomal RNA (rRNA); all participate in protein synthesis.

Ros"trum. An anteriorly projecting beak, as in the crayfish.

Ro"ta-tor. Type of voluntary muscle that brings about the rotary motion of one body part on another.

Ru-di-men"ta-ry. Pertaining to a body part not completely developed.

Ru"gae. Ridges or folds, as in the lining of the vertebrate stomach.

Ru"mi-nant. Cud-chewing mammal, as the cow, with complex stomachs.

SA node. See **sinoatrial node.**

Sac"cu-lus. Portion of the membranous labyrinth of the inner ear; contains gravity receptors.

Sa"crum. Posterior portion of vertebral column that is attached to the pelvic girdle.

Sa"git-tal. In a bilaterally symmetrical animal, pertaining to the median anteroposterior plane.

Sa'li-va-ry. Pertaining to the saliva-secreting glands of the mouth.

Salt. The result of an acid–base reaction.

Sa"pro-troph. An organism that obtains organic matter in solution from dead plant and animal tissues.

Sar'co-lem"ma. Plasma membrane surrounding a striated muscle fiber.

Sar"co-mere. Basic unit of muscle contraction.

Sar"co-plasm. Cytoplasm surrounding myofilaments in striated muscle fibers.

Sa-van"na (sa-van"nah). A broad belt of South American, African, and Australian grassland interspersed with trees.

Sca"pu-la. Shoulder blade in a vertebrate.

Sca"ven-ger. Animal that feeds on dead organisms that it has not killed.

Schi-zo"go-ny. A type of protistan multiple asexual fission.

Sci'en-tif"ic me"thod. The scientific method involves, primarily, the formulation of hypotheses on the basis of a relatively small amount of knowledge, testing the correctness of the hypothesis by securing more facts by observation, arranging the facts in some orderly manner to determine relationships, and then drawing logical conclusions.

Sci'en-tif"ic name. Binomial designation of an animal, plant, or protist species; the generic plus the species terms together.

Scle″rite. Hardened plate of the body wall bounded by sutures or joints, as in arthropods.

Scler″o-blast. Amoeboid cells in sponges that secrete spicules.

Scler-o″tic coat. The "white" of the eye; the outside layer of the eyeball.

Sco″lex. Small anterior portion of a tapeworm, for attachment to vertebrate intestine.

Scro″tum. External skin pouch found in most mammals, that contains the testes.

Scute. Horny outer plates on turtle shell.

Se-ba″ceous glands. Skin glands, usually found in connection with the hair follicles of mammals that produce an oily secretion that lubricates the skin and hair.

Se″bum. The fatty secretion of the sebaceous glands.

Sec″ond-ary sex″u-al char″ac-ters. Features other than the gonads and related organs in which males and females of a species differ.

Se″con-da″ry sper-mat″o-cytes. In meiosis, the cells formed after the first meiotic division.

Se-cre″tion. Production by protoplasm of a substance that is released from the cell for use by the organism; also the substance produced.

Scd″en-ta-ry. Staying in one place.

Seg″ment. One somite of a metameric animal.

Seg-men-ta″tion, hom-o″no-mous. All metameres (segments) are similar.

—, **heteronomous.** Metameres are not similar.

Self-fer-tiliza″tion. A type of fertilization in which an ovum is fertilized by a sperm from the same hermaphroditic individual.

Se″men. Fluid that carries the sperm from the males of most animals.

Sem′i-cir″cu-lar can-als″. Canals in the vertebrate inner ear, contain sense organs of equilibrium.

Se″mi-nal. Pertaining to spermatozoa.

—, **re-cep″ta-cle.** Saclike organ in several types of animals that receive and store sperm after copulation; considered a part of the female genital system.

—, **ves″i-cles.** Saclike organs in several types of animals in which sperm are stored before copulation; considered a part of the male genital system.

Se′mi-ni″fer-ous tu″bule. A lengthy, coiled tube found in the testes in which spermatogenesis occurs.

Sense or″gan. An organ sensitive to a particular type of stimulus.

Sen″so-ry cell. Cell that is very sensitive to stimuli; a receptor; usually a specialized type of neuron in higher organisms.

Sep″tum. Partition that separates two cavities or two structures.

Ser″i-al ho-mo″lo-gy. Occurrence of homologous structures in different segments of the same individual.

Se-ro″lo-gy. Study of antibody–antigen interactions.

Se-ro″sa. Outer extraembryonic membrane of reptiles and birds.

Se″rous. Pertaining to, producing, or resembling serum.

Se″rous coat. The peritoneal covering of the visceral organs.

Ser-to″li cells. Found in the testes of vertebrates, they are accessory cells that seem to provide nourishment for developing sperm.

Se″rum. The part of the blood plasma that separates from a clot and is without cells or fibrin, a plasma protein.

Ses″sile. Attached; not free moving; sedentary.

Se″tae. Bristles, such as those on the body wall of the earthworm.

Scx chro″mo-somes. The X and Y chromosomes that are especially concerned with the determination of sex.

Sex-li″mi-ted cha″rac-ter. Trait belonging to only one sex; commonly a secondary sex characteristic.

Sex-linked cha″rac-ter. Character with its gene located in the sex chromosome.

Sex″u-al di′mor″phism. Phenomenon of two sexes of a given species differing in secondary sex characteristics.

Sex″u-al re′pro-duc″tion. Reproduction involving random union of haploid gametes.

Sex″u-al u″nion. Temporary union of male and female during copulation.

Shaft. Slender, middle portion of a long bone or feather.

Sig″moid curve. S-shaped curve.

Si-li″ceous. Referring to a structure that contains silicon dioxide; also known as silica.

Sin″i-stral. Coiling in a counterclockwise direction, as in a gastropod shell.

Si′no-a″tri-al node. In the heart, the pacemaker cells

near the tricuspid valve; it initiates a wave of contraction across the atria.

Si″nus. A thin-walled cavity.

Si″nus gland. In some arthropods, a gland at the base of the eyestalk that controls chromatophore dispersion; under the influence of the X gland.

Si″nus ven-o″sus. In fishes, amphibians, and reptiles, a large chamber receiving blood from the veins and transmitting it to the right atrium.

Si″nu-soid. Large, irregular, tortuous blood spaces in liver and bone marrow; comparable to capillaries but their lumens are often wider.

Si″phon. Tube that draws in or expels fluids.

Si-pho″no-glyph. Flagellated groove in the sea anemone that draws a water current into the gastrovascular cavity.

Si′phon-o-zoo″id. Specialized polymorphic individual that produces water currents to aid internal circulation in certain hydrozoan colonies.

Ske″le-ton. External or internal hardened framework of an animal body, giving support and protection to the soft parts.

Slime gland. In gastropods, a gland at the anterior end of the foot that deposits a mucoid secretion over substratum.

Sol. A colloidal dispersion in a liquid.

So″li-ta-ry. Living alone; not a member of a colony or group.

Sol″ute. A substance dissolved in a solvent, of lesser concentration than the solvent.

Sol″vent. The substance of greatest concentration in a solution.

So″ma. The entire body, exclusive of the reproductive cells.

So-ma″tic ner″vous sys″tem. The portion of the nervous system that controls voluntary acts.

So″mite. One of the serial segments or metameres of a metameric animal; an organ or organ rudiment.

Spat. Microscopic spherical ciliated larval stage of an oyster.

Spe-cial-i-za″tion. In cells, any changes that lead to less generalized functions; in organisms, adapting to a particular niche.

Spe′ci-a″tion. Evolutionary change in population characteristics, usually highly correlated with environmental pressures, that results in the formation of a new species.

Spe″cies. A group or groups of actually or potentially interbreeding natural populations that resemble one another closely and are reproductively isolated from other such groups.

Sperm. A haploid male gamete; also known as a spermatozoon; usually motile.

Sper″ma-ry. The male reproductive gland of hydras.

Sperm′a-the″ca. Seminal receptacle in some female insects.

Sper″ma-tid. Male haploid germ cell immediately before assuming the form of a sperm.

Sper′ma-to-gen″e-sis. The process dealing with the formation of spermatozoa.

Sper-mat-o-go″n-ium. Male sex cell before it begins differentiating into a sperm; immature male gamete.

Sper-ma″to-phore. Packet of sperm transferred by the male to the female of several invertebrate groups and a few vertebrates, primarily salamanders.

Sper′ma-to-zo″a. Mature male gametes or sperm.

Sphinc″ter. Ring of smooth muscle surrounding a tube or opening, which, by its contraction, closes the lumen.

Spi″cule. One of many mineral structures that compose the skeletal framework of a sponge.

Spi″cules, pen″i-al. In male *Ascaris*, two chitinous rods extending from the cloacal opening, used in copulation.

Spi″nal col″umn. A series of vertebrae that enclose the spinal cord in the vertebrates.

Spi″nal cord. Part of the central nervous system extending the length of the spinal column and continuing anteriorly into the brain of vertebrates.

Spin″dle. Fingerlike structures on the abdomen of spiders, having tiny tubes at their bases, from which a fluid issues that hardens as it comes in contact with air to form silk threads.

Spin″dle fi″bers. The collection of proteinaceous fibers that attaches to the chromosomes and draw them to the opposite poles of a cell during meiosis and mitosis; composed of a polymeric association of individual spindle protein subunits.

Spi″ra-cle. In insects, an external opening of the tracheal respiratory system; in the cartilaginous fishes, the modified first gill slit; also in tadpoles, the atriopore.

Spi″ral valve. In some amphibians, a flattened twisted structure in the truncus arteriosus; also, in sharks and certain fishes, the spiral intestinal partition.

Splanch″nic. Pertaining to the viscera.

Spleen. A large organ, characteristic of most vertebrates, lying near the stomach; functioning in red blood cell storage, the destruction of old red blood cells, and is also important in body defense.

Spon″gin. Proteinaceous fibers of some poriferan skeletons, such as the bath sponges.

Spon″go-blast. Flask-shaped poriferan cells that secrete spongin.

Spon″go-coel. Central cavity of a sponge.

Spore. Special reproductive body of some protists, usually protected by a resistant covering and capable of developing independently into a new individual.

Spo″ro-cyst. Larval stage in the fluke life cycle, developing from a miracidium larva and giving rise to numerous rediae.

Spo-ro″zo-ite. Young infective sporozoan.

Spor′u-la″tion. Process of forming spores, whereby reproduction is by a form of multiple fission.

Squa-mo″sal. T-shaped bone of the vertebrate skull, bracing posterior end of the jaws.

Squa″mous cells. Relatively flat epithelial cells.

Sta″bi-li-zing se-le″ct-ion. Selection against the distribution extremes, thus strengthening the distribution mean.

Sta″pes. Third ear bone, attached to the incus and the oval window.

Sta″to-blast. Chitinous shell secreted by, for example, certain freshwater bryozoans.

Sta″to-cyst. Organ of equilibrium in some crustaceans, flatworms, etc.

Sta″to-lith. Solid body within a statocyst.

Ste-ap″sin. Enzyme of the pancreas that acts on fats.

Steppe. Russian grasslands.

Ster-co″ral poc-ket. In arachnids, a caecum (pocket) connected to the rectum; stores feces.

Ster″num. Bone or bones of the ventral side of vertebrate thorax; the breastbone.

Ster″ol, ster″oid. Organic compound with four carbon rings; includes vitamin D, cholesterol, adrenocorticotropic hormones, and sex hormones.

Stig″ma. A sensitive pigment or eyespot in certain protozoans.

Stim″u-lus. Change in the external or internal environment of an animal that brings about a response.

Stol″on. A rootlike structure that fixes certain colonial animals, such as hydroids, to a substratum.

Sto-mo″de-um. Portion of mouth cavity lined with ectoderm.

Stra″ti-fied. Arranged in layers, one above the other.

Stra″tum com-pac″tum. Inner layer of the corium or dermis of the skin.

Stra″tum cor″ne-um. Outer layers of cells in the epidermis of the vertebrate skin.

Stra″tum ger′mi-na′ti″vum. Mitotic layer of cells in the epidermis of the skin of the vertebrate.

Stra″tum spon-gi-o″sum. Outer layer of connective tissue in the dermic (corium) of a vertebrate.

Stri″a-ted. Cross-striped, as in skeletal muscle fibers.

Stro-bi-la″tion. The process of budding in segments of the sessile scyphozoan larvae resulting in free-swimming ephrya larvae.

Style. Elongated body part, often pointed at one end.

Sub-cla″vi-an. Under the collar bone (clavicle); refers to a particular muscle, blood vessel, or other structure.

Sub-cu-ta″ne-ous. Just beneath the skin, as subcutaneous connective tissue.

Sub″le-thal gene. A lethal gene that affects the carrier between birth and the attainment of reproductive age.

Sub″strate. A substance acted upon by an enzyme.

Sub-um″brel-la. Concave, oral surface of cnidarian medusae.

Sul″cus. A fissure or groove.

Sum-ma″tion. The collective effect of several synaptic events on the same dendrite or muscle fiber; may be spatial or temporal.

Su″ture. Special fibrous joint occurring exclusively in the skull; tightest type of joint.

Swim″mer-et′. An abdominal appendage that functions as a swimming organ and aids in respiration and carrying eggs in adults.

Sym″bi-ont. Organism living with another in a symbiotic relationship.

Sym′bi-o″sis. Living together of two different species; an inclusive term to cover mutualism, commensalism, and parasitism.

Sym″me-try. Characteristic of geometrial equivalence of two or more parts of an organism.

Sy-nap″sis. In meiosis, during zygonema stage of

prophase, the side-by-side pairing of homologous chromosomes.

Sy-nap″tic cleft (or **sy″napse**). The space, normally more than 200 Å, across which transmitting chemicals diffuse during neural transmission.

Sy-nap″tic knob. Structure on the tip of a axon that contains synaptic vesicles.

Sy-nap″tic ves″i-cle. Structures within the synaptic knobs that contain transmitter chemicals that, when released, either depolarize or hyperpolarize dendrites.

Syn-cy″ti-um. Undivided mass of cytoplasm containing many nuclei; a product of nuclear division without cell division.

Syn″ga-my. Gametes united in pairs, forming a zygote.

Sy-no″vi-al joint. Diathroid joint; freely moveable in certain directions; synovial fluid between the bones, which are bound by ligaments.

Syn″the-sis. Bonding of at least two molecules that results in one larger molecule.

Sy″rinx. Vocal organ of birds.

Sys″tem. Group of organs concerned with the same general function, such as circulation or digestion.

Sys-te″mic. That part of the circulatory system not involved directly in external respiration but consists of the circulation through the rest of the body.

Sys-tem″ic arch. Any large artery of a vertebrate carrying blood from the heart to the dorsal aorta.

Sys″to-le. Contraction phase of the atria or ventricles of the heart.

T sec″tion. One of the basic units of an electrocardiogram, representing ventricular repolarization.

Tac″tile. Pertaining to the sense of touch.

Tag″ma. Body region, such as the thorax or abdomen, composed of several specialized segments.

Tag-mo″sis. Union of body segments into functional groups.

Ta-i″ga. See **coniferous forest.**

Ta″pe-tum lu″ci-dum. In organisms such as the cat, a reflective layer behind the retina that enhances vision in dim illumination.

Tar″sals. Bones of the ankle.

Tar″sus. The distal end of the leg of an insect, consisting of one or more segments; an ankle bone of vertebrates.

Ta″xis. An orientational movement; for example, a

taxis may be toward a desirable or away from an undesirable condition.

Ta″xon. Category of animal classification, such as species, genus, phylum, etc.

Ta-xon″o-my. Science that deals with the classification of organisms.

Teat. A nipple.

Tec-ton″ic plates. Crustal blocks upon which the continents drift due to convection currents in the earth's mantle.

Te′len-ceph″a-lon. Forebrain of vertebrates.

Te′le-o″lo-gy. Unscientific method of thinking; the use of purpose in which a result is goal directed and serves as a cause in the explanation of natural phenomena.

Te′lo-le″ci-thal. Pertaining to those eggs that have a large amount of yolk in the vegetal portion of the cell.

Te″lo-phase. Any of the final phases of mitosis in which the cell divides and the daughter nuclei are formed.

Tel″son. Caudal extension of the last abdominal appendage of a crustacean.

Tem″plate. Mold or pattern for formation of a duplicate, as in gene replication.

Tem″por-al lobe. Region of the vertebrate cerebrum containing speech and hearing centers.

Ten″don. Tough, cordlike connective tissue that attaches a muscle to a bone.

Ten″ta-cle. A flexible extension from the body of many animals such as the hydra; used in grasping and movement.

Ten-ta″lo-cyst. Sense organs of some cnidarians.

Ter″gum. Dorsal portion of the exoskeleton of any body segment in the arthropods, such as the grasshopper.

Ter″min-a-ting co″don. A DNA base triplet that causes the cessation of protein synthesis; the terminating codon does not code for any amino acid.

Ter-res″tri-al. Living on land.

Ter′rit-tor-i-al″i-ty. The behavior of maintaining a personal living space, usually for breeding purposes, by keeping out other members of the same species.

Test. A shell or other durable external covering; also an experiment used to substantiate or disprove a hypothesis.

Tes″tis. Male gonad in which sperms are formed.

Tes·to"ste·rone. Male hormone secreted by the testes that influences sexual behavior, development of male accessory genitalia, and secondary sex characteristics, such as body hair and sex urge.

Te"ta·nus. A condition in which the muscle is in complete, unceasing contraction.

Te"trad. See **bivalent.**

Te"tra·pod. Four-limbed vertebrate. **TFN.** (trifluoronitrophenyl). Chemical poison affecting especially lampreys.

Tha"la·mus. Portion of vertebrate forebrain concerned with sensory coordination.

The"ory. Possible explanation of natural phenomena for which there is some evidence but not enough for proof.

Ther'mo·tax"is. Behavioral response of an organism to heat.

Thigmo·tro"pism. Behavioral response of an organism to contact.

Tho·rac"ic. Pertaining to the thorax or chest.

Tho"rax. Major division of an animal just posterior to the neck or posterior to the head if no neck is present.

Throm"bin. Enzyme formed from prothrombin in the presence of thrombokinase during vertebrate blood clotting; converts fibrogen to fibrin.

Throm'bo·kin"ase. Thromboplastin; released from platelets in clotting sequence; with prothrombin and calcium, forms thrombin.

Throm"bus. Blood clot within circulatory system.

Thy"mine. A **nucleotide,** found only in DNA.

Thy"mus. Two-lobed organ located just below the thyroid of mammals that is involved with the production of the T cells needed by the immune system.

Thy"roid. Endocrine gland in the neck of vertebrates; it secretes thyroxine that influences growth and metabolism.

Thy·rox"ine. Hormone secreted by the thyroid gland.

Ti"bi·a. The larger medial bone of the vertebrate lower leg; the part between the femur and tarsus in insects, between patella and tarsus in arachnids, etc.

Tie"de·mann bo"dies. In echinoderms, the spherical pouches on inner wall of ring canal in echinoderms; may be involved in amoebocyte production.

Tis"sue. Group of cells of similar structure that perform a specialized function.

Tox"in. Poison of animal or plant origin.

Tra·be"cu·la. Inward process or septum, usually formed of connective tissue; also, rod or bundle of fibers.

Tra"che·a. Windpipe of vertebrates; an air tube in terrestrial arthropods.

Tra"che·oles. Fine tracheae in intimate contact with the tissue that they service; found in many terrestrial arthropods.

Trait. Term loosely used by geneticists as a synonym for a phenotypic character.

Tran·scrip"tion. The synthesis of RNA molecules from DNA molecules.

Trans·duc"tion. Transfer of genetic material by means of a virus from one bacterium to another.

Trans"fer re·ac"tion. Chemical reaction involving the removal of one portion of a molecule by a "carrier".

Trans·la"tion. Referring to the expression of genetic information (string of nucleotides) as functional proteins (strings of amino acids) according to the genetic code; equivalent to protein synthesis.

Trans·lo·ca"tion. A type of chromosomal abnormality in which part or all of one chromosome is attached to another chromosome.

Trans·verse" proc"ess. One of the lateral projections of a vertebra.

Tri"cho·cyst. One of the organelles lying in the ectoplasm and producing hairlike fibers in ciliates such as *Paramecium.*

Tri·cus"pid valve. In mammalian heart, the trilobed valve between the right atrium and right ventricle.

Tri·ge"mi·nal. Fifth cranial nerve.

Trip'lo·blas"tic. Referring to the tissues of a body being derived from three primary germ layers: ectoderm, mesoderm, and endoderm.

Tri"so·my. A chromosomal mutation leading to an individual possessing three copies instead of the normal two copies (diploidy) of a given chromosome.

Tro·chan"ter. Second segment of the insect leg, located between the coxa and femur.

Tro"cho·phore. Free-swimming ciliated larva of some protostomes.

Tro·pho·blast. Germinal layer from which chorion develops.

Tro-pho"zo-ite. Sporozoite in the phase of its life cycle in which its principal activity is nutrition and growth.

Tro"pism. Antiquated term for orientational and locomotory responses. See **kinesis** and **taxis.**

Trun"cus ar-ter"i-o'sus. In lower vertebrates, base of aorta.

Tryp"sin. Enzyme that promotes digestion of protein.

Tube feet. Tubular organs of locomotion found in the ambulacral grooves of sea stars and other echinoderms; podia.

Tu"ber-cule. Knoblike elevation or process.

Tun"dra. Treeless biome with frozen ground and sparse fauna for most of the year.

Tur"gor. Cell distension owing to its fluid content.

Tur"ner's syn"drome. In humans, a chromosomal abnormality where the individual has one X chromosome and no Y chromosome (XO).

Tym'pan"ic mem"brane. Eardrum or tympanum.

Typh"lo-sole. A median dorsal internal fold in the intestinal wall of several types of animals, including earthworms.

UCR. See **unconditioned response.**

UCS. See **unconditioned stimulus.**

Ul"na. The bone of the little finger side of the forearm in tetrapod vertebrates.

Um-bi'li-cal cord. The cordlike connection or umbilicus, between the embryo or fetus of a mammal and the placenta, composed mainly of blood vessels and connective tissue; umbilicus.

Um"bo. An elevated portion near the hinge of a bivalve shell.

Un'con-di"tioned re-sponse" (UCR). In classical conditioning, the response that naturally follows the unconditioned stimulus; it becomes the conditioned response at the completion of conditioning.

Un'con-di"tioned sti"mu-lus (UCS). In classical conditioning, the stimulus that naturally elicits the response in question (UCR).

Un"gu-late. Mammals with hooves.

Un-gu"li-grade. A type of locomotion in which the hooved mammal is supported only through the nails (hooves) of the digits.

U'ni-ra"mous (appendage). Arthropod appendage with only one branch.

U'ra-cil. An nitrogen base in RNA that is analogous to DNA's thymine.

U-re"a. Principal nitrogenous metabolic waste of most mammals.

U"re-ter. Tube that drains urine from the kidney into the urinary bladder or cloaca.

U-re"thra. Tube that carries urine from the bladder to the outside in mammals; also carries seminal fluid in males.

U"rine. Liquid waste excreted by the kidneys.

U'ro-gen"ni-tal. Pertaining to the organs of both the urinary and the reproductive systems taken collectively.

U'ro-pod. Flattened lateral abdominal appendages of the crayfish and other crustaceans.

U"ter-us (L., womb). Enlarged portion of an oviduct in which at least part of the development of an animal takes place; technically, the term uterus is applicable only to animals in which the embryo becomes attached to the wall of the organ.

U-tri"cu-lus. Portion of the membranous labyrinth of the inner ear; contains the gravity receptors.

U'ro-style. Terminal rodlike bone of the frog spinal column.

Va"cu-ole. Small membrane-bound cavity in the cytoplasm filled with a liquid and/or other materials.

Va-gi"na. Distal end of the female reproductive tract; it often receives the copulatory organ of the male during the mating act.

Va"gus. Tenth cranial nerve in vertebrates; innervates heart.

Valve. Any structure that limits or closes an opening; also, one of the two shells of a typical bivalve mollusk or the invertebrate group.

Vane. Flat, distal portion of a feather.

Va'ri-a"tion. Difference in structure or function exhibited by individuals of the same species.

Va-ri"e-ty. In taxonomy, a division of a species; a group of individuals within an interbreeding population that differs in some minor respect from the rest of the species.

Vas. Small tubular duct.

Va"sa ef'fer-en"ti-a. Small ducts carrying sperms from the testes to the kidney in the frog or to the ductus epididymis or similar tubules in higher vertebrates.

Vas de″fer-ens. Duct that carries sperms away from the testis.

Vas″cu-lar. Pertaining to vessels that carry blood or lymph.

Va′so-mo″tion. The constriction and/or dilation of blood vessels.

Vas′o-mo″tor cen″ter. A center in the medulla of the brain; it is responsible for arterial tonus.

Va′so-mo″tor nerves. Nerves that control the contraction and expansion of blood vessels.

Ve″ge-tal he″mi-sphere or **pole.** Part of an egg or zygote containing abundant yolk material.

Vein″. Blood vessel that carries blood toward the heart.

Ve″lar ten″ta-cles. In cephalochordates, tentacles projecting from the mouth region; act as strainers.

Veld. South African grassland.

Ve″li-ger. Larval stage after trochophore in many mollusks.

Ve″lum. Membranous band of tissue on the subumbrella surface of hydrozoan jellyfishes.

Ve″nous. Pertaining to veins.

Ven″tral. Pertaining to the belly; opposite of dorsal.

Ven″tri-cle. Any of the small chambers in the anatomy of animals; specifically in the heart, a chamber from which blood is distributed; in the brain, any of the several larger subdivisions of the central space.

Ven-tri″cu-lus. In insects, the stomach.

Ven″ule. Small vein.

Ver″mi-form ap-pen″dix. A slender pouch projecting from the caecum of the large intestine of some mammals.

Ver″te-bra. One osseous unit of the vertebrate spinal column.

Ver″te-bral col″umn. Series of vertebrae in a vertebrate animal; backbone.

Ver″te-brate. Referring to those chordates that have a vertebral column.

Ver″tex. Dorsal region of an insect's head.

Ves″sel. The general term for a tubelike structure conveying fluid, such as blood or lymph.

Ves″ti-bule. An outer cavity with an entrance to a (usually) larger, deeper cavity.

Ves-ti″gi-al. Degenerate structure that was more developed or functional in ancestral groups.

Vi-bris″sa. Stiff hair in or close to the nostril; a whisker.

Vil′lus. A minute, fingerlike projection; especially those on the intestinal lining of vertebrates that increase the absorptive surface.

Vi″rus. Noncellular, parasitic, submicroscopic structure, composed of a protein shell and core of nucleic acid.

Vis″cer-a. The collective term for the organs within the body cavities, especially the abdomen.

Vis″cer-al ske″le-ton. Supporting framework of vertebrate jaws and gill arches.

Vi″su-al. Pertaining to the sense of sight.

Vi″tal ca-pa″ci-ty. Total quantity of air that can be expelled from the lungs after they have been fully inflated.

Vi″ta-min. Any of a number of organic substances that are essential in minute quantities for normal growth and function.

Vi″tre-ous. Glassy in appearance, such as the aqueous and vitreous humors of the eye.

Vi″tre-ous hu″mor. Jellylike liquid within the eye between the lens and the retina.

Vi″tel-line. Pertaining to yolk; also, a noncellular membrane just inside the shell of an egg.

Vi-vi″pa-rous. Referring to those organisms that give birth to living young that develop from eggs within the body of the mother and are nourished from her bloodstream, such as is found in most mammals.

Vol″vent. Small type of cnidarian nematocyst, containing a thread armed with spines; used in prey capture.

Vul″va. External portions of the female reproductive organs.

Warm-blood″ed. See **endotherm, homeotherm.**

Wa″ter-ex-pul″sion ves″i-cle. An osmoregulatory organelle that secretes water from the cytoplasm of many protists and simple metazoans; also known as a contractile vacuole.

Wax. A form of lipid useful in organism protection.

White blood cell. Colorless blood cell (**leukocyte**).

Wolff″fi-an ducts (after the German anatomist Wolff). Mesonephric ducts.

X, Y chro″mo-somes. Chromosomes concerned especially with the determination of sex; in some animals the females have two X chromosomes and the males one X and one Y chromosome; certain others, moths for example, have XX males and XY females; a few animals are known without a Y chromosome, with XO males and XX females.

X gland. In some arthropods, an endocrine gland that passes secretions (hormones) along axons to the sinus gland at the base of the eyestalk.

Y chro"mo-some. See **X, Y chromosomes.**

Yolk. Stored nutritive materials in the egg and utilized for nourishment of the embryo.

Yolk plug. Near the end of gastrulation, the circular spot of yolk left visible, around which the blastopore lies.

Yolk sac. The sac that encloses the yolk or, for example, in humans, the enclosed cytoplasm (there is no yolk).

Zo'o-ge-o"gra-phy. Branch of zoology dealing with the geographic distribution of animals.

Zo"oid. One of the members of a hydroid colony; often in a restricted sense, a particular kind of individual, as hydranth or gonangium; also, subordinate individual formed by transverse fission in such forms as the planarian.

Zo-o"lo-gy. The study of animal life.

Zy-go-ne"ma stage. In meiosis, stage at which synapsis occurs.

Zy"gote. Diploid cell that results from the fusion of two haploid gametes.

Index